Asymptotic Modeling of Atmospheric Flows

R. Zeytounian

Asymptotic Modeling of Atmospheric Flows

With 6 Figures

Springer-Verlag
Berlin Heidelberg New York
London Paris Tokyo Hong Kong

Professor Dr. Radyadour Zeytounian
Université de Lille I, Laboratoire de Mécanique, de Lille,
F-59655 Villeneuve d'Ascq Cédex, France

Translated by

Lesly Bry
2 ave. des Prunus, F-77770 Fontaine-le-Port, France

ISBN 3-540-19404-5 Springer-Verlag Berlin Heidelberg New York
ISBN 0-387-19404-5 Springer-Verlag New York Berlin Heidelberg

Library of Congress Cataloging-in-Publication Data. Zeytounian, R. Kh. (Radyadour Kh.), 1928– [Modèles asymptotiques de la mécanique des fluides. English] Asymptotic modeling of atmospheric flows / R. Zeytounian ; [translated by Lesly Bry]. p. cm. Translation of: Les modèles asymptotiques de la mécanique des fluides. Includes bibliographical references.
ISBN 0-387-19404-5 (U. S. : alk. paper)
1. Fluid mechanics. 2. Asymptotic expansions. 3. Atmospheric physics. 4. Dynamic meteorology. I. Title. II. Title: Atmospheric flows. QC145.2.Z4813 1990 532–dc20

Printing: Weihert-Druck GmbH, D-6100 Darmstadt
Binding: J. Schäffer GmbH & Co. KG., 6718 Grünstadt
2155/3150-543210 – Printed on acid-free paper

Preface

The present work is not exactly a "course", but rather is presented as a monograph in which the author has set forth what are, for the most part, his own results; this is particularly true of Chaps. 7–13.

Many of the problems dealt with herein have, since the school year 1975–76, been the subject of a series of graduate lectures at the "Université des Sciences et Techniques de Lille I" for students preparing for the "Diplôme d'Etudes Approfondies de Mécanique (option fluides)".

The writing of this book was thus strongly influenced by the author's own conception of meteorology as a fluid mechanics discipline which is in a privileged area for the application of singular perturbation techniques. It goes without saying that the modeling of atmospheric flows is a vast and complex problem which is presently the focal point of many research projects. The enormity of the topic explains why many important questions have not been taken up in this work, even among those which are closely related to the subject treated herein. Nonetheless, the author thought it worthwhile for the development of future research on the modeling of atmospheric flows (from the viewpoint of theoretical fluid mechanics) to bring forth a book specifying the problems which have already been resolved in this field and those which are, as yet, unsolved.

It is hoped that the reader, whether he/she be a theoretician or a numerical analysis specialist, will profit from this work and will not be disappointed with it. In any case, we will be thoroughly satisfied if the reader can be brought to appreciate a certain method of tackling the problems posed by the modeling of atmospheric phenomena.

The atmosphere is essentially a gaseous cover enveloping the earth and is held in place by its own weight. By virtue of its compressibility, this cover becomes constantly thinner as the altitude increases, until it is barely discernible from the interplanetary environment.

For the fluid mechanics specialist, the atmosphere is a continuous medium in motion. It is acknowledged that the atmospheric air (assumed dry) is a perfect gas with constant specific heats. Hence, the equations governing atmospheric "flows" are the Navier–Stokes equations written for a heavy, compressible, baroclinic fluid in rotation and conducting heat, while taking into account heat input.

The aim of the present work is the analysis of the asymptotic structure of atmospheric flows with a view to obtaining (asymptotic) models for these (atmospheric) flows. The governing idea is that these models must be obtained as significant degeneracies of the Navier–Stokes equations. Generally speaking, it just so happens that the various dimensionless parameters which come into play in the Navier–Stokes equations are in fact small or large singular perturbation parameters.

Throughout the present work, we have limited our attention to atmospheric flows for which the characteristic horizontal scale is much smaller than the mean radius of the earth. In this way, these flows can be defined in a system of Cartesian coordinates associated with the plane normal to the gravitational acceleration (the so-called tangent plane approximation). Therefore, in the energy equation of the Navier–Stokes system, the source term (heat density received per unit volume from outside sources, e.g., by radiation) can be assumed known.

It is interesting to note that mathematical studies are generally carried out on "typical problems" which are relatively unrealistic. In contrast, studies which are more physical and realistic in nature have not as yet been defined formally in the sense of singular perturbation techniques. It is precisely via a formal method of "applied mathematics" that we have attempted to analyze herein these atmospheric flows while using as a starting basis the Navier–Stokes equations.

In order to obtain approximate models for short-range weather forecasting, a kind of filtering must be carried out in the Navier–Stokes equations. This filtering consists in the elimination of local interference phenomena which handicap the numerical codes without significantly influencing the forecast itself. The first models were created nearly fifty years ago [see, for example, the book by Monin (1972) or the articles by Phillips (1970) and Zeytounian (1976)]. Among others can be cited the so-called primitive equation model and the quasi-geostrophic model. These two models were based on the existence of a small dimensionless parameter: in the case of the former, a low ratio of vertical/horizontal scales; for the latter, the so-called Kibel number is low [in reference to the article by Kibel (1940)]. Present-day numerical codes and data processing techniques would allow the treatment of models which are somewhat simplified in comparison to the primitive equation model by taking advantage of the low value of the Mach number (which is a small, natural, dimensionless parameter for atmospheric flows). More elaborate models than the quasi-geostrophic model could be dealt with by constructing a correction for the latter in the same way that in the 1960's a correction for the classical Prandtl boundary layer was constructed [e.g., see Van Dyke (1962)]. Unfortunately, the very existence of these powerful numerical codes and data processing techniques has for some twenty years channeled research in the direction of numerical experiments while ignoring asymptotic modeling. For several years now, we have been focusing on the latter [Zeytounian (1970, 1974, 1976, 1977, 1983, 1985); Guiraud and Zeytounian (1979, 1980, 1982)]. It is as yet too early to predict the future of this particular research orientation but it is the author's opinion that it deserves a systematic exploration. The present work is, in fact, a first attempt at a systematic review of the recent results obtained in the field of asymptotic modeling.

Our purpose here is to initiate a process which does not seem to have sufficiently attracted the attention of scientists. [Although Pedlosky (1979), in a certain sense, constitutes an exception, our aim is distinct from his, despite the fact that our methods are quite similar.] This process involves the use of methods of formal asymptotic analysis for carrying out asymptotic modeling, i.e., for building approximate models based on various physical situations. We do not, of course, affirm that this is the only method, or even the most efficient one, for deriving such models. We do, however, feel that when such a procedure is feasible it should be undertaken. As a matter of fact, the application of this method implies that the approximate model is associated with an asymptotic expansion procedure which, in principle, makes it possible to improve the approximation obtained with the model used by advancement in the hierarchy of approximations.

It must be kept in mind that, at the present time, modeling, i.e., the translation into correctly expressed mathematical terms of a complex physical situation, has become very important in the realm of scientific research in many fields. This is particularly true of fluid mechanics and thus also of meteorology, which is considered a discipline of fluid mechanics taken as a whole.

I am well aware that this book is very personal, one might even say "impassioned". I hope that the reader will be indulgent with me, for it is said in French that "passion is the motor of all great things". I have tried throughout this book to put "ideas" into concrete form and also sometimes to give key-words. Very often, these naturally become both very general and obvious once they have been expressed concretely. Much of knowledge, however, is based on simple truths which are exceedingly difficult to put into words.

Finally, it is obvious that when a certain number of ideas have been initiated in a certain domain, assuming that these ideas are valid, the author of such ideas would like to see others share his/her enthusiasm and begin a research movement. In our particular field of fluid mechanics, so many problems remain unresolved, and yet their applications are vital to the knowledge and protection of our environment.

My first acknowledgment is to Prof. Dr. W. Beiglböck who accepted my manuscript (written in French) for publication in English by Springer-Verlag. I am also indebted to my colleague and friend, Professor J.P. Guiraud of the Université de Paris VI. Chapters 11, 12 and 13 were largely inspired by the notes written by him for the course: "Models for Atmospheric Flows", given at the I.C.M.S. at Udine (Italy) in October 1983.

Lille
January 1990 *R. Zeytounian*

Contents

1. Introduction

Application of the Newtonian fluid mechanics equations to the theoretical prediction of atmospheric flows was first accomplished resolutely by Kibel (1940) who proposed a simplification of these equations (which were, in fact, none other than the Euler equations for an inviscid, non-heat-conducting fluid) based on the asymptotic quasi-geostrophic expansion. Such a procedure would allow the filtering out of both the internal acoustic and gravity waves within the approximation of these equations.

It must be understood that the fluid mechanics equations are relatively general which means that their solutions can describe a whole class of atmospheric wave motions. This is precisely what explains the failure in the 1920's of the Englishman Richardson (1922) in his attempt at a "numerical" prediction. Quite naturally, certain waves – e.g., short internal acoustic waves – are meaningless from the meteorological viewpoint which justifies, in particular, quasi-static filtering. On the other hand, waves having dimensions like those of Rossby (1939) with a frequency directly proportional to the variability of the local Coriolis parameter $l = 2\Omega_0 \sin\varphi$ relative to φ (where $\Omega_0 = 7.292 \times 10^{-5}$ radians per second and φ is the algebraic latitude) are long and relatively slow waves. The latter constitute the basic driving force of the atmospheric machine and appear as fundamental elements in weather forecasting models. It should thus be clear why it is, in general, highly desirable to eliminate the solutions corresponding to short, fast waves while at the same time, preserving the long, slow wave solutions.

It should be pointed out that Kibel's asymptotic quasi-geostrophic expansion is based on the existence of a small parameter Ki – the so-called *Kibel number* – which for "short-range" weather prediction (corresponding to a time duration t_0 such that $\mathrm{Ki} = (t_0 l_0)^{-1} \ll 1$, with l_0, a constant value of the *Coriolis* parameter), makes it possible to work towards the solution of fluid mechanics equations by using asymptotic expansions based on this small parameter Ki.

Concerning quasi-static filtering, it can be said to consist in replacing the equation for the vertical movement (in the direction of the altitude) by the hydrostatic balance between the vertical pressure gradient and the force of gravity. We shall see later that the quasi-geostrophic and quasi-static approximations are essential for describing atmospheric flows on so-called synoptic scales which are characterized by a horizontal length scale of the order of 10^6 m. It is these atmospheric flows which are related to short-range weather prediction.

The Navier-Stokes equations describe the evolutions for the following fields: the velocity $\boldsymbol{u}$, the temperature T, and the density ϱ. It is thus necessary to know in sufficient detail these independent fields at an initial given time $t = t^0$ throughout the entire atmosphere. This means that we must know the initial situation of the atmosphere so as to have a well-posed Cauchy problem for the further determination of the space-time evolution of the atmosphere.

At the present time, it is practically impossible to solve this Cauchy problem which determines the "real" weather. This impossibility is, first and foremost, related to approximations that would have to be made in order to deal numerically with such a problem, given the fact that today's powerful computer is as yet only capable of treating approximations of the basic equations. Like in all fluid mechanics problems, we are thus led to select the atmospheric flows to be analyzed according to their dimensions in time and space. Then, appropriate equations are used to describe these selected flows (model equations): this process of selection is generally referred to as "filtering". The scales of time and space combined with the parameters describing the properties of the atmosphere lead to the defining of dimensionless numbers: these include those of Strouhal, Reynolds, Mach, Rossby, Froude, Prandtl, Boussinesq, Kibel, Ekman and others. It turns out then that filtering is closely linked to the concept of limiting flows when these dimensionless numbers are assumed small or large compared to unity. Moreover, these numbers represent singular perturbation parameters and as such, imply the application of singular perturbation techniques (see Chap. 5). The latter involve mostly matched asymptotic expansions [see the book by Van Dyke (1964)] and, to some extent, the technique of multiple scales [discussed in Cole (1968) and Nafeh (1973)]. Any topics related to the physics of the atmosphere can be followed up in the book by Houghton (1977). In the books by Scorer (1978) and Eskinazi (1975), an approach to the phenomena of the atmosphere from the fluid mechanics point of view can be found.

In what follows, we will restrict our discussion to atmospheric flows for which the horizontal length scale L is "much smaller" than the mean radius a_0 of the earth ($a_0 \cong 6367\,\mathrm{km}$). Based on this hypothesis ($L/a_0 \ll 1$), one can describe to a very good approximation the atmospheric flows within a system of Cartesian coordinates associated with the plane normal to the gravitational acceleration $\boldsymbol{g}$ (of magnitude g), resulting from the Newtonian gravity force and the centrifugal force due to the earth's rotation. As the vector of rotation of the earth $\boldsymbol{\Omega}$ is directed from south to north according to the axis of the poles, it can be expressed as follows:

$$\boldsymbol{\Omega} = \Omega_0 \boldsymbol{e} \equiv |\boldsymbol{\Omega}|(\sin \varphi\, \boldsymbol{k} + \cos \varphi\, \boldsymbol{j}) \quad , \tag{1.1}$$

where φ is the algebraic latitude of the point P_0 of observation on the earth's surface around which the atmospheric flow is analyzed ($\varphi > 0$ in the northern hemisphere). The unit vectors directed to the east, north, and zenith in the opposite direction from $\boldsymbol{g}$ are denoted $\boldsymbol{i}$, $\boldsymbol{j}$ and $\boldsymbol{k}$ respectively (see Fig. 1.1).

If the atmospheric air (assumed dry) is considered as a perfect gas with constant specific heats c_p and c_v, then the following state equation can be written for the pressure p:

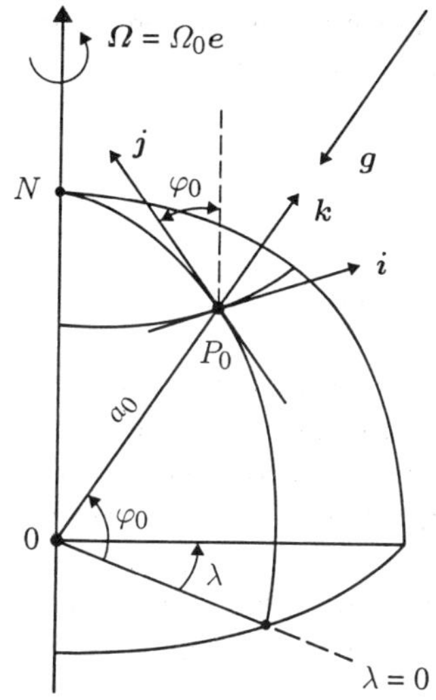

Fig. 1.1

$$p = R\varrho T \quad , \tag{1.2}$$

where

$$R = c_p - c_v \equiv c_p \frac{\gamma - 1}{\gamma} \cong 287 \frac{\mathrm{m}^2}{\mathrm{s}^2\,\mathrm{degree}}$$

and

$$\gamma = \frac{c_p}{c_v} \cong 1.4 \quad .$$

It can be seen that except near the earth's surface and in other very limited regions which separate air masses of different (thermal) natures, the effects of turbulence and heat transfer appear negligible – that is, at least in the analysis of short-range phenomena (a few hours to a few days in duration). Therefore, in many situations it is possible to use the Euler equations which govern these adiabatic phenomena of the free atomosphere.

The above explains why, for the sake of simplicity, the laws of vertical exchange by turbulence will be considered as identical to those of exchange by molecular agitation. Of course, in this case, values 10^4 to 10^6 times larger must be used for the transfer coefficients. Concerning heat transfer, we will focus our attention on 1) turbulent heat conduction (in the form of a classical Fourier law), 2) the dissipation function ϕ, which is a quadratic form of velocity and, 3) the radiation $\mathcal{R}$, (assumed known) which is the heat received per unit volume from the exterior environment via radiation.

For the following analysis, it is very useful to postulate the existence of a so-called standard atmosphere which is assumed to exist on a day to day basis in the form of a thermodynamic reference situation. This will be solely a function of the altitude with the actual weather having only a slight effect. If z_∞ is the altitude of this standard atmosphere, p_∞ its pressure, ϱ_∞ its density and T_∞ its (absolute) temperature, then we can write:

$$p_\infty = R\varrho_\infty T_\infty \quad ;$$
$$\frac{dp_\infty}{dz_\infty} + g\varrho_\infty = 0 \quad ; \tag{1.3}$$
$$k(T_\infty)\frac{dT_\infty}{dz_\infty} + \mathcal{R}_\infty(T_\infty) = 0 \quad ,$$

where k is the turbulent thermal conduction coefficient and $\mathcal{R}_\infty(T_\infty)$ can be identified with the (assumed) known radiation. If we suppose that the standard temperature $T_\infty(z_\infty)$ remains bounded when $z_\infty \to +\infty$, then we can determine from (1.3) the characteristics of the standard atmosphere once $\mathcal{R}_\infty$ is known as a function of T_∞. Further on in this text, we will assume that this is indeed the case [see Sect. 1.4 in the book by Kibel (1963) for more on this subject].

It should be kept in mind that the expression:

$$N^2_\infty(z_\infty) = \frac{g}{T_\infty(z_\infty)}\left\{\frac{\gamma-1}{\gamma}\frac{g}{R} + \frac{dT_\infty}{dz_\infty}\right\} \quad , \tag{1.4}$$

is the square of the Väisälä internal frequency and that the standard atmosphere is statically stable if $N_\infty(z_\infty)$ is real, i.e., if:

$$-\frac{dT_\infty}{dz_\infty} < \frac{\gamma-1}{\gamma}\frac{g}{R} \quad . \tag{1.5}$$

Let us emphasize once again that we are considering here atmospheric flows for which L (the horizontal characteristic scale) complies with the relation:

$$\frac{RT_\infty(0)}{g} \leq L \ll a_0 \cong 6367\,\mathrm{km} \quad ; \tag{1.6}$$

where $RT_\infty(0)/g \equiv H_\infty$ is the altitude of the so-called homogeneous atmosphere which is equivalent to approximately 8 km. If we remain within the troposphere, then H_∞ can be the characteristic vertical scale.

The value $L \cong 10^6$ m characterizes synoptic scale flows which are associated with short-range weather prediction. $L \cong 10^5$ m represents medium-scale flows (regional mesoscale phenomenon), and, finally, $L \cong 10^4\,\mathrm{m} \sim H_\infty$ typifies local scale flows ("lee wave" type phenomena, in particular).

2. The Equations

In all that follows, the Euler independent variables will be used, i.e., the time t and the position vector $\boldsymbol{r}$, whose origin is at the center of the earth and which is directed towards the exterior following the radii of the globe. Let $\boldsymbol{u}(t, \boldsymbol{r})$ be the velocity vector of a fluid particle in a moving frame which is generally characterized by the vector $\boldsymbol{v}_0$ of its origin and the vector $\boldsymbol{\Omega}$ of angular velocity of rotation.

If p, ϱ and $\hat{\boldsymbol{T}}$ are the pressure, density and the stress tensor respectively, then the vector equation for the relative motion of a heavy, compressible, viscous fluid can be written in the form [Zeytounian (1974); see Sect. 1.10]:

$$\varrho\left\{\frac{D\boldsymbol{u}}{Dt} + 2\boldsymbol{\Omega} \times \boldsymbol{u} + \frac{d\boldsymbol{v}_0}{dt} + \frac{d\boldsymbol{\Omega}}{dt} \times \boldsymbol{r}\right\}$$
$$= -\nabla p + \varrho\left\{\boldsymbol{f} + |\boldsymbol{\Omega}|^2 \boldsymbol{r}_\perp\right\} + \nabla \cdot \hat{\boldsymbol{T}} \quad , \tag{2.1}$$

where the normal component to $\boldsymbol{\Omega}$ has been designated by the subscript $\perp$, and the "true" gravitational force (the Newtonian attraction force) by $\boldsymbol{f}$. We note that:

$$\frac{D}{Dt} = \frac{\partial}{\partial t} + \boldsymbol{u} \cdot \nabla \quad ; \tag{2.2}$$

in particular, for any orthogonal coordinate system , we have:

$$\boldsymbol{u} \cdot \nabla \boldsymbol{u} = (\nabla \times \boldsymbol{u}) \times \boldsymbol{u} + \tfrac{1}{2}\nabla(\boldsymbol{u} \cdot \boldsymbol{u}) \quad . \tag{2.3}$$

If $\boldsymbol{g}$ denotes the gravitational acceleration (the so-called force of gravity), then:

$$\boldsymbol{f} + |\boldsymbol{\Omega}|^2 \boldsymbol{r}_\perp = \boldsymbol{g} \quad . \tag{2.4}$$

For our purposes, $\boldsymbol{v}_0$ and $\boldsymbol{\Omega}$ are fixed, time-independent vectors and thus, instead of (2.1) it follows that:

$$\varrho\left(\frac{D\boldsymbol{u}}{Dt} + 2\boldsymbol{\Omega} \times \boldsymbol{u}\right) + \nabla p - \varrho \boldsymbol{g} = \nabla \cdot \hat{\boldsymbol{T}} \quad . \tag{2.5}$$

By definition (continuous media with an infinitely short memory), the deformation state for a fluid is characterized by the tensor of deformation rate $\hat{\boldsymbol{D}}$ whose components are (in a Cartesian orthonormal reference frame):

$$d_{ij} = \frac{1}{2}\left(\frac{\partial u_i}{\partial x_j} + \frac{\partial u_j}{\partial x_i}\right) \quad , \quad i, j = 1, 2, 3 \quad ; \tag{2.6}$$

the u_i and the x_i being the components of $\boldsymbol{u}$ and $\boldsymbol{r}$ respectively in the Cartesian (orthonormal) system. In all that follows, we have in mind Newtonian fluids for which the components of tensor $\hat{T}$ are:

$$\tau_{ij} = \lambda d_{kk}\delta_{ij} + 2\mu d_{ij} \quad , \tag{2.7}$$

where δ_{ij} is the Kronecker delta symbol ($\delta_{ij} \equiv 0$, if $i \not\equiv j$ and, $\delta_{ij} \equiv 1$ if $i \equiv j$) and also λ and μ are the Lamé viscosity coefficients. In the so-called "Navier-Stokes" fluids, λ and μ of (2.7) are functions of the (absolute) temperature T and satisfy the inequalities (thermodynamically admissible processes):

$$\lambda + \tfrac{2}{3}\mu \geq 0 \quad , \quad \mu \geq 0 \quad . \tag{2.8}$$

If we suppose λ and μ to be constant (in this case, we will use a "0" subscript) and to satisfy the so-called Stokes hypothesis, $(3\lambda_0 + 2\mu_0 = 0)$, then:

$$\nabla \cdot \hat{T} = \mu_0 \left\{ \nabla^2 \boldsymbol{u} + \tfrac{1}{3}\nabla(\nabla \cdot \boldsymbol{u}) \right\} \quad . \tag{2.9}$$

It has already been acknowledged that the pressure p can be determined from the equation of state of a perfect gas with constant specific heats:

$$p = R\varrho T \quad . \tag{2.10}$$

The density ϱ satisfies the continuity equation:

$$\frac{D \log \varrho}{Dt} + \nabla \cdot \boldsymbol{u} = 0 \quad . \tag{2.11}$$

From the equation of state (2.10), atmospheric flows can be seen to be baroclinic. Thus we must write an equation for the temperature T which is, in fact, the energy equation. The latter will be written as follows:

$$c_p \varrho \frac{DT}{Dt} - \frac{Dp}{Dt} = -\nabla \cdot \boldsymbol{q} + \hat{T} : \hat{D} + Q \quad , \tag{2.12}$$

where: (1) $\boldsymbol{q}$ designates the heat flux vector corresponding to the turbulent transfer of internal energy and, (2) $\phi \equiv \hat{T} : \hat{D}$ is the dissipation function, i.e., the rate of dissipation due to friction, and (3) Q represents a volumic rate of energy supply coming from sources outside of the fluid – mainly from radiation (denoted $\mathcal{R}$). Let us recall that $\hat{T}$ and $\hat{D}$ are tensors of second order and that the product $\hat{T} : \hat{D} = \tau_{ij} d_{ij} = \phi$, which is the inner product of $\hat{T}$ and $\hat{D}$, leads to the scalar ϕ which is a quadratic form of the velocity $\boldsymbol{u}$ since:

$$\phi = 9Kd^2 + 2\mu D_{ij} D_{ij} \quad , \tag{2.13}$$

where $3K = 3\lambda + 2\mu$, $d_{ij} = d\delta_{ij} + D_{ij}$, $D_{kk} \equiv 0$.

Usually, $\boldsymbol{q}$ is represented in the form of a Fourier law:

$$\boldsymbol{q} = -k\nabla T \quad , \quad k \geq 0 \quad , \tag{2.14}$$

where k is the turbulent thermal conduction coefficient (heat conductivity). If we assume Q to be given, then by taking into account (2.6, 7, 9, 14), the equations (2.5, 10–12) form a closed system for determining the functions $\boldsymbol{u}$, p, ϱ and T.

2.1 The Euler Equations

In the adiabatic case, the Euler equations can be employed when flows are analyzed in the free atmosphere:

$$\begin{aligned}
&\frac{D\boldsymbol{u}}{Dt} + 2\boldsymbol{\Omega} \times \boldsymbol{u} + \frac{1}{\varrho}\nabla p = \boldsymbol{g} \quad ; \\
&p = R\varrho T \quad ; \\
&\frac{D \log \varrho}{Dt} + \nabla \cdot \boldsymbol{u} = 0 \quad ; \\
&c_p \varrho \frac{DT}{Dt} = \frac{Dp}{Dt} \quad .
\end{aligned} \tag{2.15}$$

A detailed analysis of these equations can be found in our Notes [Zeytounian (1974)]. It will suffice here to mention just a few meaningful results relative to system (2.15).

Let the (relative) vorticity be:

$$\boldsymbol{\omega} = \nabla \times \boldsymbol{u} \quad ; \tag{2.16}$$

since the force of gravity $\boldsymbol{g}$ is derived from a function of forces U_0 ($\boldsymbol{g} \equiv -\nabla U_0$), the following equation of the vorticity can be obtained from (2.15):

$$\frac{\partial}{\partial t}(\boldsymbol{\omega} + 2\boldsymbol{\Omega}) + \nabla \times [(\boldsymbol{\omega} + 2\boldsymbol{\Omega}) \times \boldsymbol{u}] = \frac{1}{\varrho^2}(\nabla \varrho \times \nabla p) \tag{2.17}$$

and the relation $\nabla \varrho \times \nabla p \not\equiv 0$ indicates that our flow is in baroclinic evolution [as opposed to barotropic when $p = p(\varrho)$].

If we now introduce the specific entropy $s = s(p, \varrho)$, our perfect gas with constant specific heats can be stated in the form:

$$s = c_v \log (p/\varrho^\gamma) \quad , \tag{2.18}$$

in which case, the last equation of (2.15) becomes:

$$\frac{Ds}{Dt} = 0 \quad . \tag{2.19}$$

The above demonstrates that the specific entropy remains constant along the trajectories and so the flow is said to be in adiabatic evolution. It is also said that s is a lagrangian invariant for flows governed by the Euler equations [for more on this topic, see Chap. 2 in Zeytounian (1974)].

Another consequence of the Euler equations is the following energy equation:

$$\varrho \frac{DH}{Dt} = \frac{\partial p}{\partial t} \quad , \tag{2.20}$$

with H, which is called total specific enthalpy, being defined by

$$H = \frac{|\boldsymbol{u}|^2}{2} + c_p T + U_0 \quad . \tag{2.21}$$

Thus in a steady flow, H is conserved along the streamlines [such a steady flow is sometimes called isoenergetic; see Chap. 8 in Zeytounian (1974)]. Equation (2.17) can also be written as follows:

$$\frac{D}{Dt}\left(\frac{\omega+2\Omega}{\varrho}\right)-\left(\frac{\omega+2\Omega}{\varrho}\cdot\nabla\right)\boldsymbol{u}=\frac{1}{\varrho}\left(\nabla p\times\nabla\frac{1}{\varrho}\right) \quad .$$

Let μ be a scalar which is an arbitrary function of the space and time variables. By multiplying the above equation by the vector $\nabla\mu$, and after some mathematical rearrangement, the so-called Ertel relation (1942) is obtained:

$$\frac{D}{Dt}\left(\frac{\omega+2\Omega}{\varrho}\cdot\nabla\mu\right)-\frac{\omega+2\Omega}{\varrho}\cdot\nabla\left(\frac{D\mu}{Dt}\right)$$
$$=\frac{1}{\varrho}\left(\nabla p\times\nabla\frac{1}{\varrho}\right)\cdot\nabla\mu \quad . \tag{2.22}$$

In particular, if we take $\mu\equiv s$, then by taking into account (2.18, 19), we get:

$$\frac{D}{Dt}\left(\frac{\omega+2\Omega}{\varrho}\cdot\nabla s\right)=0 \quad . \tag{2.23}$$

Thus it can be seen that the expression $\varrho^{-1}(\omega+2\Omega)\cdot\nabla s$, called Ertel's potential vorticity, is a new lagrangian invariant for non-viscous flows in adiabatic evolution which are governed by the Euler equations (2.15). It will be noted that besides the two lagrangian invariants (2.19) and (2.23), three new independent lagrangian invariants can be constructed. These five lagrangian invariants, each of which is independent of the others, correspond to the five Euler equations (2.15). Relation (2.23) plays a major role in the analysis of flows linked to atmospheric circulation dominated by the effect of the earth's rotation, the latter being the main contributor to the formation of vorticity in such flows. Nonetheless, when dealing with atmospheric flows, it is often more worthwhile to work with the following potential temperature rather than the specific entropy s:

$$\theta=T\left(\frac{p_\infty(0)}{p}\right)^{R/c_p}\equiv\frac{1}{R}[p_\infty(0)]^{(\gamma-1)/\gamma}\exp\left(\frac{s}{c_p}\right)$$

and then

$$\frac{D}{Dt}\left(\frac{\omega+2\Omega}{\varrho}\cdot\nabla\theta\right)=0 \quad . \tag{2.24}$$

It will be remarked that in the free atmosphere in adiabatic evolution, the creation and variation of vorticity is caused by several factors: (1) first of all, a variation of vorticity is caused by the stretching effect of the vortex lines by the deformation due to the velocity field; (2) there is also the effect of the term of the right-hand side of (2.17) which corresponds to the creation of vorticity (via the baroclinic effect) which is due to the non-collinear character of the vectors ∇p and $\nabla\varrho$; (3) and finally, in the relative flow, we observe the creation of vorticity proportional to the contraction velocity of the area $\sum_p$ enclosed by the projection L_p of the fluid curve L in the plane of the equator (the projection being carried out parallel to the south-north axis bearing the angular velocity vector of rotation $\boldsymbol{\Omega}$).

2.1.1 Steady Flows

For a steady flow in adiabatic evolution of an inviscid fluid which is assumed continuous, we have the so-called Vaszonyi equation:

$$\boldsymbol{u} \times (\boldsymbol{\omega} + 2\boldsymbol{\Omega}) = \nabla H - T\nabla s \quad , \tag{2.25}$$

where functions H and s comply with:

$$\boldsymbol{u} \cdot \nabla H = 0 \quad , \quad \boldsymbol{u} \cdot \nabla s = 0 \quad . \tag{2.26}$$

This implies that the surfaces H = const and s = const are stream surfaces. More generally speaking, let ψ = const and χ = const be two families of stream surfaces of our continuous steady flow. Because $\boldsymbol{u} \cdot \nabla\psi = \boldsymbol{u} \cdot \nabla\chi = 0$, we have:

$$\varrho\boldsymbol{u} = \nabla\psi \times \nabla\chi \quad , \tag{2.27}$$

once consideration is also given to the continuity equation [Giese (1951), Yih (1957), Zeytounian (1966)].

From (2.27), it follows that instead of (2.26), we can write:

$$H = \frac{|\boldsymbol{u}|^2}{2} + c_pT + U_0 = A(\psi, \chi) \quad ; \tag{2.28a}$$

$$s = B(\psi, \chi) \quad , \tag{2.28b}$$

with A and B conserved along the stream lines. Equation (2.25) can then be written:

$$\boldsymbol{u} \times (\boldsymbol{\omega} + 2\boldsymbol{\Omega}) = \left(\frac{\partial A}{\partial\psi} - T\frac{\partial B}{\partial\psi}\right)\nabla\psi + \left(\frac{\partial A}{\partial\chi} - T\frac{\partial B}{\partial\chi}\right)\nabla\chi \quad ,$$

and by taking into account (2.27), we find:

$$\begin{aligned} (\boldsymbol{\omega} + 2\boldsymbol{\Omega}) \cdot \nabla\psi &= \varrho\left(\frac{\partial A}{\partial\chi} - T\frac{\partial B}{\partial\chi}\right) \quad ; \\ (\boldsymbol{\omega} + 2\boldsymbol{\Omega}) \cdot \nabla\chi &= -\varrho\left(\frac{\partial A}{\partial\psi} - T\frac{\partial B}{\partial\psi}\right) \quad . \end{aligned} \tag{2.29}$$

Equations (2.28, 29) yield four first integrals of the flow governed by the steady Euler equations.

2.2 The Tangent Plane Approximation

It is generally acknowledged that when the horizontal scale L is of the order of 10^6 m or smaller, atmospheric flows can be located in a system of Cartesian coordinates linked to the plane normal to $\boldsymbol{g}$. This approximation is based on the fact that:

$$\delta = \frac{L}{a_0} \ll 1 \quad , \tag{2.30}$$

and is called the tangent plane or even the "flat earth" approximation. We will come back to the justification of this approximation (which is, in fact, related to the β-plane approximation) in Sect. 2.3. We only wish to show here how the Navier-Stokes equations can be written with this approximation. The unit vectors $\boldsymbol{i}$ and $\boldsymbol{j}$ introduced in Fig. 1.1 are, respectively, tangent to the parallel of latitude (directed towards the east), and to the meridian (directed towards the north); the unit vector $\boldsymbol{k}$ is oriented in the opposite direction from $\boldsymbol{g}$. These three unit vectors are attached at the point P_0 in the vicinity of which the atmospheric flow is analyzed. We denote by x, y and z the Cartesian coordinates of a point of the flow relative to the axes P_0x, P_0y, P_0z whose directions coincide with $\boldsymbol{i}$, $\boldsymbol{j}$, $\boldsymbol{k}$. Finally, u, v and w (with dimensions) are the components of the (relative) velocity vector $\boldsymbol{u}$ along the axes P_0x, P_0y and P_0z of the local system with the point P_0 as the origin. In this case, if we assume the turbulent coefficients to be constant (we then denote them by μ_0, λ_0 and k_0), and also if we acknowledge the Stokes hypothesis ($3\lambda_0+2\mu_0 = 0$), then we can write the following Navier-Stokes equations:

$$\begin{aligned}
&\frac{Du}{Dt} - 2\Omega_0 \sin\varphi\, v + 2\Omega_0 \cos\varphi\, w + \frac{1}{\varrho}\frac{\partial p}{\partial x} = \frac{\mu_0}{\varrho}\left\{\Delta u + \frac{1}{3}\frac{\partial}{\partial x}(\operatorname{div}\boldsymbol{u})\right\} \quad ; \\
&\frac{Dv}{Dt} + 2\Omega_0 \sin\varphi\, u + \frac{1}{\varrho}\frac{\partial p}{\partial y} = \frac{\mu_0}{\varrho}\left\{\Delta v + \frac{1}{3}\frac{\partial}{\partial y}(\operatorname{div}\boldsymbol{u})\right\} \quad ; \\
&\frac{Dw}{Dt} - 2\Omega_0 \cos\varphi\, u + \frac{1}{\varrho}\frac{\partial p}{\partial z} + g = \frac{\mu_0}{\varrho}\left\{\Delta w + \frac{1}{3}\frac{\partial}{\partial z}(\operatorname{div}\boldsymbol{u})\right\} \quad ; \\
&\frac{D\log\varrho}{Dt} + \operatorname{div}\boldsymbol{u} = 0 \quad ; \\
&c_p\varrho\frac{DT}{Dt} - \frac{Dp}{Dt} = k_0\Delta T + \phi + \frac{\partial \mathcal{R}}{\partial z} \quad ; \\
&p = R\varrho T \quad ,
\end{aligned} \tag{2.31}$$

where

$$\frac{D}{Dt} = \frac{\partial}{\partial t} + u\frac{\partial}{\partial x} + v\frac{\partial}{\partial y} + w\frac{\partial}{\partial z} \tag{2.32}$$

is the convective derivative in the local moving frame (P_0; x, y, z) connected with the earth and

$$\Delta = \frac{\partial^2}{\partial x^2} + \frac{\partial^2}{\partial y^2} + \frac{\partial^2}{\partial z^2} \quad , \qquad \operatorname{div}\boldsymbol{u} = \frac{\partial u}{\partial x} + \frac{\partial v}{\partial y} + \frac{\partial w}{\partial z} \quad .$$

It should be pointed out that in the right-hand side of the energy equation of system (2.31), there appears on the one hand, the dissipation function:

$$\phi = -\tfrac{2}{3}\mu_0(\operatorname{div}\boldsymbol{u})^2 + 2\mu_0\left\{\left(\frac{\partial u}{\partial x}\right)^2 + \left(\frac{\partial v}{\partial y}\right)^2 + \left(\frac{\partial w}{\partial z}\right)^2 + \tfrac{1}{2}\left[\left(\frac{\partial u}{\partial y} + \frac{\partial v}{\partial x}\right)^2 + \left(\frac{\partial u}{\partial z} + \frac{\partial w}{\partial x}\right)^2 + \left(\frac{\partial v}{\partial z} + \frac{\partial w}{\partial y}\right)^2\right]\right\} \tag{2.33}$$

and, on the other hand, the term $\partial\mathcal{R}/\partial z$ related to the radiation that can be expressed in this form once it is acknowledged that the earth's surface is simulated by the equation $z = 0$. When such is the case, the boundary conditions can be written:

$$u = v = w = 0 \quad , \quad -k_0\frac{\partial T}{\partial z} = \mathcal{R} \quad , \quad \text{on} \quad z = 0 \quad . \tag{2.34}$$

Knowledge of the functions u^0, v^0, w^0, ϱ^0 and T^0, at an initial time $t = 0$ (initial data of the Cauchy evolution problem) makes it possible to determine the future evolution of the atmosphere at relatively close times since (2.31) explicitly gives the derivatives of these functions in relation to the time t:

$$u = u^0, \quad v = v^0 \quad , \quad w = w^0 \quad , \quad \varrho = \varrho^0 \quad , \quad T = T^0 \quad , \quad \text{for } t = 0 \quad . \tag{2.35}$$

For the time being, we will put aside the very subtle task of formulating the other boundary conditions. Equations (2.31), together with conditions (2.34, 35) constitute a sufficiently realistic physical basis for analyzing and modeling atmospheric flows.

2.3 The So-called β-Plane Approximation

Let us now return to the Euler equations (2.15). First of all, we want to write them explicitly in a system of spherical coordinates r, λ, φ, where r is the distance from a point at the center 0 of the earth, φ is the latitude and λ, the longitude (see Fig. 2.1). The components of the (relative) velocity $\boldsymbol{u}$ with respect to the directions $\boldsymbol{i}$, $\boldsymbol{j}$ and $\boldsymbol{k}$ (parallel, meridian and zenith) are u, v and w.

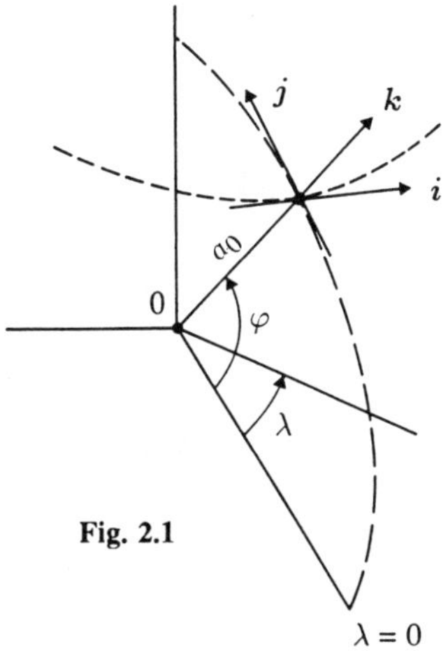

Fig. 2.1

The following equations then result:

$$\frac{du}{dt}+\frac{uw}{r}-\frac{uv}{r}\operatorname{tg}\varphi-2\Omega_0\sin\varphi\, v+2\Omega_0\cos\varphi\, w + \frac{1}{\varrho r\cos\varphi}\frac{\partial p}{\partial\lambda}=0 \quad ; \tag{2.36}$$

$$\frac{dv}{dt}+\frac{wv}{r}+\frac{u^2}{r}\operatorname{tg}\varphi+2\Omega_0\sin\varphi\, u+\frac{1}{\varrho r}\frac{\partial p}{\partial\varphi}=0 \quad ; \tag{2.37}$$

$$\frac{dw}{dt}-\frac{u^2+v^2}{r}-2\Omega_0\cos\varphi\, u+\frac{1}{\varrho}\frac{\partial p}{\partial r}+g=0 \quad ; \tag{2.38}$$

$$\frac{d\varrho}{dt}+\varrho\left\{\frac{\partial w}{\partial r}+2\frac{w}{r}+\frac{1}{r\cos\varphi}\frac{\partial(v\cos\varphi)}{\partial\varphi}+\frac{1}{r\cos\varphi}\frac{\partial u}{\partial\lambda}\right\}=0 \quad ; \tag{2.39}$$

$$c_p\varrho\frac{dT}{dt}-\frac{dp}{dt}=0 \quad ; \quad p=R\varrho T \quad , \tag{2.40}$$

where

$$\frac{d}{dt}\equiv\frac{\partial}{\partial t}+\frac{u}{r\cos\varphi}\frac{\partial}{\partial\lambda}+\frac{v}{r}\frac{\partial}{\partial\varphi}+w\frac{\partial}{\partial r} \quad . \tag{2.41}$$

The second step consists in converting to curvilinear coordinates:

$$x=(a_0\cos\varphi_0)\lambda \quad , \quad y=a_0(\varphi-\varphi_0) \quad , \quad z=r-a_0 \quad , \tag{2.42}$$

and hence:

$$\frac{\partial}{\partial\lambda}=a_0\cos\varphi_0\frac{\partial}{\partial x} \quad ; \quad \frac{\partial}{\partial\varphi}=a_0\frac{\partial}{\partial y} \quad , \quad \frac{\partial}{\partial r}\equiv\frac{\partial}{\partial z} \quad , \tag{2.43}$$

since $\varphi_0=\text{const}$ is a reference latitude. The origin of the system x, y, z is in P_0 (see Fig. 1.1) and corresponds to $\lambda=0$, $\varphi\equiv\varphi_0$ and $r\equiv a_0$. The velocities u, v and w thus correlate to the direction east (x), north (y) and ascending vertical (z).

The dimensionless quantities (with bars) must now be introduced:

$$\begin{gathered}\bar{x}=\frac{x}{L} \quad , \quad \bar{y}=\frac{y}{L} \quad , \quad \bar{z}=\frac{z}{H} \quad , \quad \bar{t}=\frac{t}{t_0} \quad ; \\ \bar{u}=\frac{u}{U_0} \quad , \quad \bar{v}=\frac{v}{U_0} \quad , \quad \bar{w}=\frac{w}{W_0} \quad , \quad \bar{p}=\frac{p}{p_\infty(0)} \quad , \\ \bar{\varrho}=\frac{\varrho}{\varrho_\infty(0)} \quad , \quad \bar{T}=\frac{T}{T_\infty(0)} \quad ; \quad p_\infty(0)=R\varrho_\infty(0)T_\infty(0) \quad .\end{gathered} \tag{2.44}$$

Moreover, as $\varphi=y/a_0+\varphi_0=(L/a_0)\bar{y}+\varphi_0\equiv\delta\bar{y}+\varphi_0$, we must develop $\sin\varphi$, $\cos\varphi$ and $\operatorname{tg}\varphi$ in the vicinity of φ_0 since in all that follows it is assumed that:

$$\delta\equiv\frac{L}{a_0}\ll 1.$$

It thus follows that:

$$\sin\varphi = \sin\varphi_0\left\{1 + \delta\overline{y}\,\mathrm{cotg}\,\varphi_0 - \delta^2\frac{\overline{y}^2}{2} + \dots\right\} \quad ;$$

$$\cos\varphi = \cos\varphi_0\left\{1 - \delta\overline{y}\,\mathrm{tg}\,\varphi_0 - \delta^2\frac{\overline{y}^2}{2} + \dots\right\} \quad ; \tag{2.45}$$

$$\mathrm{tg}\,\varphi = \mathrm{tg}\,\varphi_0\left\{1 + \delta\overline{y}\frac{1}{\cos\varphi_0\,\sin\varphi_0} + \delta^2\overline{y}^2\frac{1}{\cos^2\varphi_0} + \dots\right\} \quad .$$

Finally, as $r = a_0 + z = a_0[1 + (H/a_0)\overline{z}] = a_0(1 + \varepsilon\delta\overline{z})$, we have

$$r = a_0\overline{r} \quad , \quad \overline{r} = 1 + \varepsilon\delta\overline{z} \quad , \tag{2.46}$$

where

$$\varepsilon = \frac{H}{L} \quad . \tag{2.47}$$

By making use of (2.42–46), we first find that:

$$\frac{d}{dt} = \frac{U_0}{L}\left\{\mathrm{S}\frac{\partial}{\partial\overline{t}} + \frac{1 + \delta\overline{y}\,\mathrm{tg}\,\varphi_0 + \dots}{1 + \varepsilon\delta\overline{z}}\overline{u}\frac{\partial}{\partial\overline{x}} + \frac{1}{1 + \varepsilon\delta\overline{z}}\overline{v}\frac{\partial}{\partial\overline{y}} + \frac{\eta}{\varepsilon}\overline{w}\frac{\partial}{\partial\overline{z}}\right\}, \tag{2.48}$$

where

$$\mathrm{S} = \frac{L}{t_0U_0} \quad , \quad \eta = \frac{W_0}{U_0} \quad . \tag{2.49}$$

The Strouhal number S characterizes the unsteady effects.

Therefore, for any value of ε, be it much smaller than or equal to unity (it is generally acknowledged that when $\varepsilon \ll 1$, $\varepsilon \equiv \eta$), it follows from (2.48) that *when* $\delta \to 0$:

$$\begin{aligned}\frac{d}{dt} &= \frac{U_0}{L}\left\{\mathrm{S}\frac{\partial}{\partial\overline{t}} + \overline{u}\frac{\partial}{\partial\overline{x}} + \overline{v}\frac{\partial}{\partial\overline{y}} + \frac{\eta}{\varepsilon}\overline{w}\frac{\partial}{\partial\overline{z}}\right\}\\ &\equiv \frac{U_0}{L}\mathrm{S}\frac{D}{D\overline{t}} \quad ,\end{aligned} \tag{2.50}$$

$$\frac{D}{D\overline{t}} = \frac{\partial}{\partial\overline{t}} + \frac{1}{\mathrm{S}}\left(\overline{u}\frac{\partial}{\partial\overline{x}} + \overline{v}\frac{\partial}{\partial\overline{y}} + \frac{\eta}{\varepsilon}\overline{w}\frac{\partial}{\partial\overline{z}}\right) \quad , \tag{2.51}$$

and once again we find the expression (in non-dimensional form) of the convective derivative in the local moving reference frame which was brought up during the tangent plane approximation discussion of Sect. 2.2.

Next, we can state (2.36) in the following form:

$$\begin{aligned}&\mathrm{S}\frac{D\overline{u}}{D\overline{t}} + \frac{\delta}{1 + \varepsilon\delta\overline{z}}\eta\overline{u}\,\overline{w} - \frac{1 + \delta\overline{y}(\cos\varphi_0\,\sin\varphi_0)^{-1} + \dots}{1 + \varepsilon\delta\overline{z}}\delta\,\mathrm{tg}\,\varphi_0\overline{u}\,\overline{v}\\ &\quad - \frac{1}{\mathrm{Ro}}\left(1 + \frac{\delta}{\mathrm{tg}\,\varphi_0}\overline{y}\right)\overline{v} + \frac{1}{\mathrm{Ro}}\eta\left(\frac{1}{\mathrm{tg}\,\varphi_0} - \delta\overline{y}\right)\overline{w}\\ &\quad + \dots + \frac{1}{\overline{\varrho}}\frac{1}{\gamma\mathrm{M}_\infty^2}\frac{1 + \delta\overline{y}\,\mathrm{tg}\,\varphi_0 + \dots}{1 + \varepsilon\delta\overline{z}}\frac{\partial\overline{p}}{\partial\overline{x}} = 0 \quad ,\end{aligned} \tag{2.52}$$

where

$$\mathrm{Ro} = \frac{U_0}{L l_0} \quad , \quad l_0 \equiv 2\Omega_0 \sin \varphi_0 \quad ,$$
$$\mathrm{M}_\infty^2 = \frac{U_0^2}{\gamma p_\infty(0)/\varrho_\infty(0)} \quad . \tag{2.53}$$

The Rossby number Ro characterizes the effect of the earth's rotation, while the Mach number M_∞ characterizes the compressibility effect.

It will be remarked that if the local Coriolis parameter $l = 2\Omega_0 \sin \varphi$ is introduced, then we can define the parameter:

$$\beta_0 = \left(\frac{1}{a_0} \frac{dl}{d\varphi} \right)_{\varphi=\varphi_0} \equiv \frac{2\Omega_0 \cos \varphi_0}{a_0} \quad , \tag{2.54}$$

and in this case

$$\frac{\delta}{\mathrm{tg}\, \varphi_0} \equiv \frac{\beta_0 L}{l_0} \quad , \quad \frac{\beta_0 L / l_0}{\mathrm{Ro}} \equiv \frac{\beta_0 L^2}{U_0} = \beta \quad . \tag{2.55}$$

For the synoptic scale phenomena ($U_0 \cong 10\,\mathrm{m/s}$ and $L \cong 10^6\,\mathrm{m}$) we have the following:

$$\beta = \mathrm{O}(1). \tag{2.56}$$

However, it must be pointed out that for these same phenomena, the following also holds:

$$\varepsilon \ll 1 \quad , \quad \eta = \varepsilon \quad . \tag{2.57}$$

Therefore, when $\delta \to 0$, the following non-dimensional equation must be written:

$$\mathrm{S} \frac{D\overline{u}}{D\overline{t}} - \frac{1}{\mathrm{Ro}} \overline{v} - \beta \overline{y}\, \overline{v} + \frac{1}{\mathrm{Ro}} \frac{\eta}{\mathrm{tg}\, \varphi_0} \overline{w} + \frac{1}{\overline{\varrho}} \frac{1}{\gamma \mathrm{M}_\infty^2} \frac{\partial \overline{p}}{\partial \overline{x}} = 0 \quad . \tag{2.58}$$

Strictly speaking, it must be acknowledged that $\delta \mathrm{tg}\, \varphi_0 \ll 1$. Nonetheless, when $\mathrm{Ro} \ll 1$, we must come back to (2.52) seeing that in this case, the following can be said to hold true:

$$\frac{\delta}{\mathrm{Ro}} = \mathrm{O}(1) \Rightarrow L \cong \sqrt{\frac{U_0 a_0}{l_0}} \quad . \tag{2.59}$$

The term $-\beta \overline{y}\, \overline{v}$ of (2.58) is a sequel of the earth's sphericity when $\delta \to 0$. The latter appears in the β plane approximation as the variation of the Coriolis parameter l with the latitude φ. Given the fact that in first approximation, the parameter δ does not affect any derivative, it can be expected not to produce any singular perturbation phenomena. It must be understood that various approaches exist for further dealing with the problem: for example, the usual procedure is to carry out $\delta \to 0$ from the start and then perform diverse limiting processes relative to ε, M_∞, Ro etc., during the modeling phase. Another possibility is to consider simultaneously δ, ε, M_∞ or Ro as small parameters: we will adopt this procedure when considering the quasi-geostrophy associated with $\text{Ro} \to 0$.

Let us now backtrack to (2.37) which can be expressed in the following form:

$$S\frac{D\overline{v}}{D\overline{t}} + \frac{\delta}{1+\varepsilon\delta\overline{z}}\eta\overline{v}\,\overline{w} + \frac{1+\delta\overline{y}(\cos\varphi_0 \sin\varphi_0)^{-1}+\ldots}{1+\varepsilon\delta\overline{z}}\delta\,\text{tg}\,\varphi_0\,\overline{u}^2 + \ldots$$
$$+\frac{1}{\text{Ro}}\left(1+\frac{\delta}{\text{tg}\,\varphi_0}\overline{y}\right)\overline{u} + \frac{1}{\overline{\varrho}}\frac{1}{\gamma M_\infty^2}\frac{1}{1+\varepsilon\delta\overline{z}}\frac{\partial\overline{p}}{\partial\overline{y}} = 0 \quad , \tag{2.60}$$

and when $\delta \to 0$, it naturally follows that:

$$S\frac{D\overline{v}}{D\overline{t}} + \frac{1}{\text{Ro}}\overline{u} + \beta\overline{y}\,\overline{u} + \frac{1}{\overline{\varrho}}\frac{1}{\gamma M_\infty^2}\frac{\partial\overline{p}}{\partial\overline{y}} = 0 \quad . \tag{2.61}$$

Equation (2.38) can be rewritten as follows:

$$\varepsilon\eta S\frac{D\overline{w}}{D\overline{t}} - \varepsilon\delta\frac{\overline{u}^2+\overline{v}^2}{1+\varepsilon\delta\overline{z}}$$
$$-\frac{1}{\text{Ro}}\frac{\varepsilon}{\text{tg}\,\varphi_0}\overline{u} + \frac{\varepsilon\delta}{\text{Ro}}\overline{y}\,\overline{u} + \frac{1}{\overline{\varrho}}\frac{1}{\gamma M_\infty^2}\frac{\partial\overline{p}}{\partial\overline{z}} + \frac{\varepsilon}{\text{Fr}_L^2} = 0 \quad , \tag{2.62}$$

where

$$\text{Fr}_L = U_0/\sqrt{gL} \tag{2.63}$$

is the Froude number based on L.

In (2.62), carrying out the limiting process $\delta \to 0$ is not as straightforward since in many situations, ε, η and Ro are also small parameters. When $\varepsilon \ll 1$, and more precisely if:

$$\varepsilon \ll \text{Ro} \approx \delta \ll 1 \quad ,$$

then instead of (2.62), we obtain hydrostatic balance:

$$\frac{\partial\overline{p}}{\partial\overline{z}} + \text{Bo}\,\overline{\varrho} = 0 \quad , \tag{2.64}$$

where

$$\text{Bo} = \varepsilon\frac{\gamma M_\infty^2}{\text{Fr}_L^2} = \frac{H}{RT_\infty(0)/g} \tag{2.65}$$

is the Boussinesq number. On the other hand, if

$$\varepsilon = \eta \cong 1 \quad , \quad \delta \ll 1 \quad , \quad \text{Ro} \gg 1 \quad ,$$

then we have:

$$S\frac{D\overline{w}}{D\overline{t}} + \frac{1}{\overline{\varrho}}\frac{1}{\gamma M_\infty^2}\frac{\partial\overline{p}}{\partial\overline{z}} + \frac{\text{Bo}}{\gamma M_\infty^2} = 0 \quad . \tag{2.66}$$

The continuity equation (2.39) becomes:

$$S\frac{D\overline{\varrho}}{D\overline{t}} + \overline{\varrho}\left\{\frac{1+\delta\overline{y}\,\text{tg}\,\varphi_0+\ldots}{1+\varepsilon\delta\overline{z}}\frac{\partial\overline{u}}{\partial\overline{x}} + \frac{\partial\overline{v}}{\partial\overline{y}} + \frac{\eta}{\varepsilon}\frac{\partial\overline{w}}{\partial\overline{z}} + 2\delta\eta\frac{\overline{w}}{1+\varepsilon\delta\overline{z}}\right.$$
$$\left.-\delta\,\text{tg}\,\varphi_0\,\overline{y}\frac{\partial\overline{v}}{\partial\overline{y}} - \delta\,\text{tg}\,\varphi_0\,\overline{v} + \ldots\right\} = 0 \tag{2.67}$$

and when $\delta \to 0$, we recover the local form of the continuity equation.

Finally, instead of (2.40), it follows that:

$$\overline{\varrho}\mathrm{S}\frac{D\overline{T}}{D\overline{t}} - \frac{\gamma-1}{\gamma}\mathrm{S}\frac{D\overline{p}}{D\overline{t}} = 0 \quad , \quad \overline{p} = \overline{\varrho}\overline{T} \quad . \tag{2.68}$$

We will now finish with this rather detailed justification of the β-plane approximation. Our main aim has been to demonstrate the difficulties encountered when several small parameters are simultaneously taken into account in the problem.

Let us recall that in the non-dimensional equations (2.58, 61, 66), S is the Strouhal number, ε is the quasi-static parameter, Ro is the Rossby number and M_∞ is the Mach number.

A detailed analysis of the β-plane approximation initially due to Rossby (1939) can be found in the books of Leblond and Mysak (1978) (see pp. 17–22), and Pedlosky (1979) (see Sect. 6.2). On the same subject, the two articles by Veronis (1963) may likewise prove worthwhile.

To conclude this section, let it be said that in what is to follow we have considered (except in particular questions mainly associated with the limiting case $\mathrm{Ro} \to 0$) the β-plane approximation as consisting in taking the local Coriolis parameter $l = 2\Omega_0 \sin\varphi$ as being approximately equal to $l_0 + \beta_0 y$ (with dimensions) where $l_0 \equiv 2\Omega_0 \sin\varphi_0$ and $\beta_0 = 2\Omega_0 \cos\varphi_0/a_0$. The dimensional analysis implies that $|\beta_0 y| \ll |l_0|$, when $L \gg U_0/|l_0|$, but $L \ll a_0 \cong 6367\,\mathrm{km}$.

2.4 Different Forms of the Euler Equations

Given the results of Sect. 2.3, we can thus retain the following form for the Euler equations on the hypothesis that $\varepsilon \equiv \eta$ and with the understanding that all quantities are dimensionless:

$$\mathrm{S}\frac{D\overline{u}}{D\overline{t}} - \left(\frac{1}{\mathrm{Ro}} + \beta\overline{y}\right)\overline{v} + \frac{1}{\mathrm{Ro}}\frac{\varepsilon}{\mathrm{tg}\,\varphi_0}\overline{w} + \frac{1}{\overline{\varrho}}\frac{1}{\gamma\mathrm{M}_\infty^2}\frac{\partial\overline{p}}{\partial\overline{x}} = 0 \quad ; \tag{2.69a}$$

$$\mathrm{S}\frac{D\overline{v}}{D\overline{t}} + \left(\frac{1}{\mathrm{Ro}} + \beta\overline{y}\right)\overline{u} + \frac{1}{\overline{\varrho}}\frac{1}{\gamma\mathrm{M}_\infty^2}\frac{\partial\overline{p}}{\partial\overline{y}} = 0 \quad ; \tag{2.69b}$$

$$\varepsilon^2\mathrm{S}\frac{D\overline{w}}{D\overline{t}} - \frac{1}{\mathrm{Ro}}\frac{\varepsilon}{\mathrm{tg}\,\varphi_0}\overline{u} + \frac{1}{\overline{\varrho}}\frac{1}{\gamma\mathrm{M}_\infty^2}\frac{\partial\overline{p}}{\partial\overline{z}} + \frac{\mathrm{Bo}}{\gamma\mathrm{M}_\infty^2} = 0 \quad ; \tag{2.69c}$$

$$\mathrm{S}\frac{D\overline{\varrho}}{D\overline{t}} + \overline{\varrho}\left(\frac{\partial\overline{u}}{\partial\overline{x}} + \frac{\partial\overline{v}}{\partial\overline{y}} + \frac{\partial\overline{w}}{\partial\overline{z}}\right) = 0 \quad ; \tag{2.69d}$$

$$\overline{\varrho}\mathrm{S}\frac{D\overline{T}}{D\overline{t}} - \frac{\gamma-1}{\gamma}\mathrm{S}\frac{D\overline{p}}{D\overline{t}} = 0 \quad ; \tag{2.69e}$$

$$\overline{p} = \overline{\varrho}\overline{T} \quad . \tag{2.69f}$$

The above non-dimensional Euler equations form a closed system for $\overline{u}$, $\overline{v}$, $\overline{w}$, $\overline{\varrho}$, $\overline{T}$ and $\overline{p}$. Within this system come into play the dimensionless parameters S, Ro, β, γ, M_∞, ε, Bo and also Ro tg φ_0. It will be noted that:

$$\mathrm{S\,Ro} \equiv \mathrm{Ki} = 1/l_0 t_0 \quad .$$

2.4.1 The Euler Equations for $\overline{u}$, $\overline{v}$, $\overline{w}$, π, ω, and ϑ

In the atmosphere, the meaningful functions from the thermodynamics point of view are those which take into account the "deviations" relative to the standard atmosphere discussed at the end of Chap. 1. It thus makes sense to introduce the following thermodynamic perturbations into (2.69):

$$\pi = \frac{\overline{p} - \overline{p}_\infty}{\overline{p}_\infty} \quad , \quad \omega = \frac{\overline{\varrho} - \overline{\varrho}_\infty}{\overline{\varrho}_\infty} \quad , \quad \vartheta = \frac{\overline{T} - \overline{T}_\infty}{\overline{T}_\infty} \quad , \tag{2.70}$$

where $\overline{p}_\infty$, $\overline{\varrho}_\infty$ and $\overline{T}_\infty$ are non-dimensional with respect to $p_\infty(0)$, $\varrho_\infty(0)$ and $T_\infty(0)$ respectively. The latter are functions of:

$$\overline{z}_\infty = \frac{z_\infty}{RT_\infty(0)/g} = \mathrm{Bo}\frac{z}{H} = \mathrm{Bo}\,\overline{z} \quad , \tag{2.71}$$

where $\mathrm{Bo} = Hg/RT_\infty(0)$.

Hence, for the functions $\overline{u}$, $\overline{v}$, $\overline{w}$, π, ω and ϑ, the following dimensionless Euler equations result:

$$(1+\omega)\left\{\mathrm{S}\frac{D\overline{u}}{D\overline{t}} - \left(\frac{1}{\mathrm{Ro}} + \beta\overline{y}\right)\overline{v} + \frac{1}{\mathrm{Ro}}\frac{\varepsilon}{\mathrm{tg}\,\varphi_0}\overline{w}\right\} + \frac{\overline{T}_\infty(\overline{z}_\infty)}{\gamma M_\infty^2}\frac{\partial \pi}{\partial \overline{x}} = 0; \tag{2.72a}$$

$$(1+\omega)\left\{\mathrm{S}\frac{D\overline{v}}{D\overline{t}} + \left(\frac{1}{\mathrm{Ro}} + \beta\overline{y}\right)\overline{u}\right\} + \frac{\overline{T}_\infty(\overline{z}_\infty)}{\gamma M_\infty^2}\frac{\partial \pi}{\partial \overline{y}} = 0 \quad ; \tag{2.72b}$$

$$(1+\omega)\left\{\varepsilon^2\mathrm{S}\frac{D\overline{w}}{D\overline{t}} - \frac{1}{\mathrm{Ro}}\frac{\varepsilon}{\mathrm{tg}\,\varphi_0}\overline{u}\right\} + \frac{\overline{T}_\infty(\overline{z}_\infty)}{\gamma M_\infty^2}\frac{\partial \pi}{\partial \overline{z}} - (1+\omega)\frac{\mathrm{Bo}}{\gamma M_\infty^2}\vartheta = 0 \quad ; \tag{2.72c}$$

$$\mathrm{S}\frac{D\omega}{D\overline{t}} + (1+\omega)\left(\frac{\partial \overline{u}}{\partial \overline{x}} + \frac{\partial \overline{v}}{\partial \overline{y}} + \frac{\partial \overline{w}}{\partial \overline{z}}\right) - (1+\omega)\mathrm{Bo}\left(1 + \frac{d\overline{T}_\infty(\overline{z}_\infty)}{d\overline{z}_\infty}\right)\frac{\overline{w}}{\overline{T}_\infty(\overline{z}_\infty)} = 0 \quad ; \tag{2.72d}$$

$$(1+\omega)\mathrm{S}\frac{D\vartheta}{D\overline{t}} - \frac{\gamma-1}{\gamma}\mathrm{S}\frac{D\pi}{D\overline{t}} + (1+\pi)\mathrm{Bo}\left(\frac{\gamma-1}{\gamma} + \frac{d\overline{T}_\infty(\overline{z}_\infty)}{d\overline{z}_\infty}\right)\frac{\overline{w}}{\overline{T}_\infty(\overline{z}_\infty)} = 0 \quad ; \tag{2.72e}$$

$$\pi = \omega + \vartheta + \omega\vartheta \quad . \tag{2.72f}$$

The expression (with dimensional quantities)

$$N_\infty^2(z_\infty) = \frac{g}{T_\infty(z_\infty)}\left\{\frac{\gamma-1}{\gamma}\frac{g}{R} + \frac{dT_\infty(z_\infty)}{dz_\infty}\right\} \tag{2.73}$$

gives us the square of the Väisälä internal frequency.

Let $N_\infty^2(0)$ be a characteristic value of $N_\infty^2(z_\infty)$. In this case, we can introduce:

$$\overline{N}_\infty^2(\overline{z}_\infty) \equiv \frac{N_\infty^2[(RT_\infty(0)/g)\overline{z}_\infty]}{N_\infty^2(0)} \quad \text{and} \quad \alpha_\infty^0 \equiv \frac{H}{g}N_\infty^2(0) \quad , \tag{2.74}$$

which enables us to write as a substitute for (2.72e), the following equation for ϑ:

$$(1+\omega)\mathrm{S}\frac{D\vartheta}{D\overline{t}} - \frac{\gamma-1}{\gamma}\mathrm{S}\frac{D\pi}{D\overline{t}} + (1+\pi)\alpha_\infty^0\overline{N}_\infty^2(\overline{z}_\infty)\overline{w} = 0 \quad . \tag{2.75}$$

The dimensionless parameter α_∞^0 can be interpreted as being a measure of static stability. When $\overline{N}_\infty(\overline{z}_\infty)$ is real, the standard atmosphere is in a state of static stability, i.e.:

$$-\frac{d\overline{T}_\infty(\overline{z}_\infty)}{d\overline{z}_\infty} < \frac{\gamma-1}{\gamma} \quad . \tag{2.76}$$

It will be noted that for atmospheric flows, the values π, ω and ϑ are small compared to unity. Moreover, in (2.72), two vertical scales come into play and their ratio is equal to Bo. Therefore, when Bo $\ll 1$, we sense that a double-scale phenomenon relative to the altitude will appear (see Sect. 6.2).

We will see in Chap. 8 that a priveleged situation exists corresponding to:

$$\frac{\mathrm{Bo}}{\mathrm{M}_\infty} \equiv \frac{\gamma\varepsilon\mathrm{M}_\infty}{\mathrm{Fr}_L^2} = \frac{gH}{U_0}\sqrt{\frac{\gamma}{RT_\infty(0)}} \equiv \hat{B} = \mathrm{O}(1) \quad , \tag{2.77}$$

when Bo and M_∞ tend towards zero, both simultaneously and at the same velocity. This situation leads to a Boussinesq approximation according to Zeytounian (1974) if we postulate the asymptotic representation:

$$\begin{aligned} &\overline{u} = \overline{u}_0 + \ldots \quad , \quad \overline{v} = \overline{v}_0 + \ldots \quad , \quad \overline{w} = \overline{w}_0 + \ldots \quad , \\ &\pi = \mathrm{M}_\infty^2\pi_2 + \ldots \quad , \quad \omega = \mathrm{M}_\infty\omega_1 + \ldots \quad , \quad \vartheta = \mathrm{M}_\infty\vartheta_1 + \ldots \quad . \end{aligned} \tag{2.78}$$

2.4.2 The Euler Equations for $\overline{u}$, $\overline{v}$, $\overline{w}$, $\overline{\Pi}$, and $\overline{\theta}$

Before formulating (2.24) in Sect. 2.1, we introduced the potential temperature:

$$\theta = T\left(\frac{p_\infty(0)}{p}\right)^{R/c_p} \equiv T_\infty(0)\overline{T}/\overline{p}^{(\gamma-1)/\gamma} \quad . \tag{2.79}$$

Let us introduce in the following, the non-dimensional functions:

$$\overline{\theta} = \frac{\theta}{T_\infty(0)} = \overline{T}\,\overline{\Pi} \quad , \quad \overline{\Pi} = \overline{p}^{-(\gamma-1)/\gamma} \quad , \tag{2.80}$$

in such a way that

$$\overline{T} = \overline{\theta}/\overline{\Pi} \quad , \quad \overline{p} = \overline{\Pi}^{\gamma/(\gamma-1)} \quad , \quad \overline{\varrho} = \frac{\overline{\Pi}^{1/(\gamma-1)}}{\overline{\theta}} \quad . \tag{2.81}$$

By making use of (2.81), instead of the system of Euler equations (2.69), we obtain the following equations for $\overline{u}$, $\overline{v}$, $\overline{w}$, $\overline{\Pi}$ and $\overline{\theta}$:

$$\mathrm{S}\frac{D\overline{u}}{D\overline{t}} - \left(\frac{1}{\mathrm{Ro}} + \beta\overline{y}\right)\overline{v} + \frac{1}{\mathrm{Ro}}\frac{\varepsilon}{\mathrm{tg}\,\varphi_0}\overline{w} + \frac{1}{(\gamma-1)\mathrm{M}_\infty^2}\overline{\theta}\frac{\partial\overline{\Pi}}{\partial\overline{x}} = 0 \quad ; \tag{2.82a}$$

$$\mathrm{S}\frac{D\overline{v}}{D\overline{t}} + \left(\frac{1}{\mathrm{Ro}} + \beta\overline{y}\right)\overline{u} + \frac{1}{(\gamma-1)\mathrm{M}_\infty^2}\overline{\theta}\frac{\partial\overline{\Pi}}{\partial\overline{y}} = 0 \quad ; \tag{2.82b}$$

$$\mathrm{S}\varepsilon^2\frac{D\overline{w}}{D\overline{t}} - \frac{1}{\mathrm{Ro}}\frac{\varepsilon}{\mathrm{tg}\,\varphi_0}\overline{u} + \frac{1}{(\gamma-1)\mathrm{M}_\infty^2}\overline{\theta}\frac{\partial\overline{\Pi}}{\partial\overline{z}} + \frac{\mathrm{Bo}}{\gamma\mathrm{M}_\infty^2} = 0 \quad ; \tag{2.82c}$$

$$\mathrm{S}\frac{D\overline{\theta}}{D\overline{t}} = 0 \quad ; \tag{2.82d}$$

$$(\gamma-1)\left(\frac{\partial\overline{u}}{\partial\overline{x}} + \frac{\partial\overline{v}}{\partial\overline{y}} + \frac{\partial\overline{w}}{\partial\overline{z}}\right) + \mathrm{S}\frac{D\log\overline{\Pi}}{D\overline{t}} = 0 \quad . \tag{2.28e}$$

In particular, in the situation:

$$\mathrm{S} \equiv 1, \quad \varepsilon \equiv 1 \quad , \quad \mathrm{Ro} \equiv \infty \quad ,$$

we can rewrite (2.82) in the form

$$\begin{aligned}
&(\gamma-1)\mathrm{M}_\infty^2\frac{D\overline{u}}{D\overline{t}} + \overline{\theta}\frac{\partial\overline{\Pi}}{\partial\overline{x}} = 0 \quad ; \quad (\gamma-1)\mathrm{M}_\infty^2\frac{D\overline{v}}{D\overline{t}} + \overline{\theta}\frac{\partial\overline{\Pi}}{\partial\overline{y}} = 0 \quad ; \\
&(\gamma-1)\mathrm{M}_\infty^2\frac{D\overline{w}}{D\overline{t}} + \overline{\theta}\frac{\partial\overline{\Pi}}{\partial\overline{z}} + \frac{\gamma-1}{\gamma}\mathrm{Bo} = 0 \quad ; \quad \frac{D\overline{\theta}}{D\overline{t}} = 0 \quad ; \\
&(\gamma-1)\left(\frac{\partial\overline{u}}{\partial\overline{x}} + \frac{\partial\overline{v}}{\partial\overline{y}} + \frac{\partial\overline{w}}{\partial\overline{z}}\right) + \frac{D\log\overline{\Pi}}{D\overline{t}} = 0 \quad .
\end{aligned} \tag{2.83}$$

We thus rediscover the equations used by Ogura and Phillips (1962).

2.4.3 The So-called Primitive Equations

Let us return to the Euler equations (2.69). In (2.69c), we notice that an unusually favorable situation exists which corresponds to:

$$\varepsilon \to 0 \quad , \tag{2.84}$$

as long as we consider atmospheric flows outside of an equatorial zone (i.e., for φ_0 sufficiently different from the value $\varphi_0 = 0$, so that the term $\varepsilon/\mathrm{tg}\,\varphi_0$ disappears with $\varepsilon \to 0$).

Indeed, when $\varepsilon \to 0$, (2.69c) degenerates strongly and becomes

$$\frac{\partial \overline{p}}{\partial \overline{z}} + \mathrm{Bo}\,\overline{\varrho} = 0 \quad . \tag{2.85}$$

We thus recover the equation describing the hydrostatic balance.

By taking advantage of (2.85), like Eliassen (1949), we can introduce a new system of Cartesian coordinates directly related to the isobaric surfaces $\overline{p} = \text{const}$. In this system of coordinates:

$$\tau \equiv \overline{t} \quad , \quad \xi \equiv \overline{x} \quad , \quad \eta \equiv \overline{y} \quad , \quad \zeta \equiv \overline{p} \quad , \tag{2.86}$$

the reduced (non-dimensional) pressure ζ becomes one of the independent variables. As for

$$\overline{z} = \mathcal{H}(\tau, \xi, \eta, \zeta) \quad , \tag{2.87}$$

the reduced local (non-dimensional) altitude of the absolute topography of the isobaric surfaces ζ = const, it becomes one of the unknown functions just like the following:

$$\begin{aligned} &U \equiv \overline{u} \quad , \quad V \equiv \overline{v} \quad , \\ &W \equiv \frac{D\overline{p}}{D\overline{t}} = \frac{\mathrm{Bo}}{\mathrm{S}}\overline{\varrho}\left(\mathrm{S}\frac{\partial \mathcal{H}}{\partial \tau} + U\frac{\partial \mathcal{H}}{\partial \xi} + V\frac{\partial \mathcal{H}}{\partial \eta} - \overline{w}\right) \quad . \end{aligned} \tag{2.88}$$

It comes to light that:

$$\begin{aligned} &\frac{\partial}{\partial \overline{x}} = \frac{\partial}{\partial \xi} + \mathrm{Bo}\overline{\varrho}\frac{\partial \mathcal{H}}{\partial \xi}\frac{\partial}{\partial \zeta} \quad ; \quad \frac{\partial}{\partial \overline{y}} = \frac{\partial}{\partial \eta} + \mathrm{Bo}\overline{\varrho}\frac{\partial \mathcal{H}}{\partial \eta}\frac{\partial}{\partial \zeta} \quad ; \\ &\frac{\partial}{\partial \overline{t}} = \frac{\partial}{\partial \tau} + \mathrm{Bo}\overline{\varrho}\frac{\partial \mathcal{H}}{\partial \tau}\frac{\partial}{\partial \zeta} \quad ; \quad \mathrm{Bo}\overline{\varrho}\frac{\partial \mathcal{H}}{\partial \zeta} = -1 \quad . \end{aligned} \tag{2.89}$$

Hence, it follows that:

$$\mathrm{S}\frac{D}{D\overline{t}} = \mathrm{S}\frac{\partial}{\partial \tau} + U\frac{\partial}{\partial \xi} + V\frac{\partial}{\partial \eta} + \mathrm{S}W\frac{\partial}{\partial \zeta} \quad . \tag{2.90}$$

This being the case, the continuity equation (2.69d) takes on an "incompressible" form:

$$\frac{\partial U}{\partial \xi} + \frac{\partial V}{\partial \eta} + \mathrm{S}\frac{\partial W}{\partial \zeta} = 0 \quad . \tag{2.91}$$

In addition, (2.69c) is replaced by (2.85) which becomes a relation that makes it possible to determine the temperature:

$$\overline{T} = -\mathrm{Bo}\,\zeta\frac{\partial \mathcal{H}}{\partial \zeta} \tag{2.92}$$

and likewise, (2.69f) is transformed in such a way that the density can be determined:

$$\overline{\varrho} = \zeta/\overline{T} \quad . \tag{2.93}$$

Finally, by taking into account (2.90, 92), we obtain the following for U, V and $\mathcal{H}$ in place of (2.69a–c):

$$\mathrm{S}\frac{\partial U}{\partial \tau} + U\frac{\partial U}{\partial \xi} + V\frac{\partial U}{\partial \eta} + \mathrm{S}W\frac{\partial U}{\partial \zeta} - \left(\frac{1}{\mathrm{Ro}} + \beta\eta\right)V + \frac{\mathrm{Bo}}{\gamma \mathrm{M}_\infty^2}\frac{\partial \mathcal{H}}{\partial \xi} = 0 \quad ;$$

$$\mathrm{S}\frac{\partial V}{\partial \tau} + U\frac{\partial V}{\partial \xi} + V\frac{\partial V}{\partial \eta} + \mathrm{S}W\frac{\partial V}{\partial \zeta} + \left(\frac{1}{\mathrm{Ro}} + \beta\eta\right)U + \frac{\mathrm{Bo}}{\gamma \mathrm{M}_\infty^2}\frac{\partial \mathcal{H}}{\partial \eta} = 0\,; \quad (2.94)$$

$$\left(\mathrm{S}\frac{\partial}{\partial \tau} + U\frac{\partial}{\partial \xi} + V\frac{\partial}{\partial \eta}\right)\frac{\partial \mathcal{H}}{\partial \zeta} + \mathrm{S}\frac{W}{\zeta}\left\{\frac{\partial}{\partial \zeta}\left(\zeta\frac{\partial \mathcal{H}}{\partial \zeta}\right) - \frac{\gamma - 1}{\gamma}\frac{\partial \mathcal{H}}{\partial \zeta}\right\} = 0 \quad .$$

Equations (2.91, 94) form the so-called primitive equations for U, V, W and $\mathcal{H}$.

A slip condition must be superimposed on these equations:

$$\mathrm{S}\frac{\partial \mathcal{H}}{\partial \tau} + U\frac{\partial \mathcal{H}}{\partial \xi} + V\frac{\partial \mathcal{H}}{\partial \eta} + \mathrm{S}W\frac{\partial \mathcal{H}}{\partial \zeta} = 0 \quad , \quad \text{on} \quad \mathcal{H}(\tau, \xi, \eta, \zeta) = 0 \quad . \quad (2.95)$$

If it is supposed that

$$\mathcal{H} = \mathcal{H}_0(\zeta) + \mathcal{H}'(\tau, \xi, \eta, \zeta) \quad (2.96)$$

one can then write the following equation for $\mathcal{H}'$ which naturally issues from the last equation of (2.94):

$$\left(\mathrm{S}\frac{\partial}{\partial \tau} + U\frac{\partial}{\partial \xi} + V\frac{\partial}{\partial \eta}\right)\frac{\partial \mathcal{H}'}{\partial \zeta} + \mathrm{S}\left\{\frac{\partial}{\partial \zeta}\left(\zeta\frac{\partial \mathcal{H}'}{\partial \zeta}\right) - \frac{\gamma - 1}{\gamma}\frac{\partial \mathcal{H}'}{\partial \zeta}\right\}\frac{W}{\zeta}$$
$$+\mathrm{S}K_0(\zeta)\frac{W}{\zeta^2} = 0 \quad , \quad (2.97)$$

where

$$K_0(\zeta) = \zeta\left\{\frac{d}{d\zeta}\left(\zeta\frac{d\mathcal{H}_0}{d\zeta}\right) - \frac{\gamma - 1}{\gamma}\frac{d\mathcal{H}_0}{d\zeta}\right\} \quad . \quad (2.98)$$

Slip condition (2.95) becomes:

$$\mathrm{S}W\frac{d\mathcal{H}_0}{d\zeta} = -\left\{\mathrm{S}\frac{\partial}{\partial \tau} + U\frac{\partial}{\partial \xi} + V\frac{\partial}{\partial \eta} + \mathrm{S}W\frac{\partial}{\partial \zeta}\right\}\mathcal{H}' \quad ,$$

on $\quad \mathcal{H}_0(\zeta) + \mathcal{H}'(\tau, \xi, \eta, \zeta) = 0 \quad .$ (2.99)

It will be observed that the first two equations (2.94) imply in fact that:

$$\mathcal{H}' = \mathrm{M}_\infty^2 \mathcal{H}_2 \quad (2.100)$$

and thus, when $\mathrm{M}_\infty \to 0$, the slip condition (2.99) can be linearized. We will return to this matter in Chap. 5 where we will see, in particular, under what conditions the linearization is valid within the framework of an asymptotic theory.

2.5 The Non-dimensional Non-adiabatic Equations

First of all, let us backtrack to (2.31). According to the results of Sect. 2.3 and on the assumption that $\varepsilon \equiv \eta$ [see (2.47, 49)], we can write the following non-dimensional, non-adiabatic equations based on (2.31):

$$\overline{\varrho}\left\{\mathrm{S}\frac{D\overline{u}}{D\overline{t}} - \left(\frac{1}{\mathrm{Ro}} + \beta\overline{y}\right)\overline{v} + \frac{1}{\mathrm{Ro}}\frac{\varepsilon}{\mathrm{tg}\,\varphi_0}\overline{w}\right\} + \frac{1}{\gamma \mathrm{M}_\infty^2}\frac{\partial \overline{p}}{\partial \overline{x}}$$
$$= \frac{1}{\mathrm{Re}}\left\{\overline{\Delta}_2\overline{u} + \frac{1}{\varepsilon^2}\frac{\partial^2\overline{u}}{\partial\overline{z}^2} + \frac{1}{3}\frac{\partial}{\partial\overline{x}}(\overline{\mathcal{D}}_3)\right\} \quad ; \tag{2.101a}$$

$$\overline{\varrho}\left\{\mathrm{S}\frac{D\overline{v}}{D\overline{t}} + \left(\frac{1}{\mathrm{Ro}} + \beta\overline{y}\right)\overline{u}\right\} + \frac{1}{\gamma \mathrm{M}_\infty^2}\frac{\partial \overline{p}}{\partial \overline{y}}$$
$$= \frac{1}{\mathrm{Re}}\left\{\overline{\Delta}_2\overline{v} + \frac{1}{\varepsilon^2}\frac{\partial^2\overline{v}}{\partial\overline{z}^2} + \frac{1}{3}\frac{\partial}{\partial\overline{y}}(\overline{\mathcal{D}}_3)\right\} \quad ; \tag{2.101b}$$

$$\overline{\varrho}\left\{\varepsilon^2\mathrm{S}\frac{D\overline{w}}{D\overline{t}} - \frac{1}{\mathrm{Ro}}\frac{\varepsilon}{\mathrm{tg}\,\varphi_0}\overline{u}\right\} + \frac{1}{\gamma \mathrm{M}_\infty^2}\frac{\partial \overline{p}}{\partial \overline{z}} + \frac{\mathrm{Bo}}{\gamma \mathrm{M}_\infty^2}\overline{\varrho}$$
$$= \frac{1}{\mathrm{Re}}\left\{\varepsilon^2\overline{\Delta}_2\overline{w} + \frac{\partial^2\overline{w}}{\partial\overline{z}^2} + \frac{1}{3}\frac{\partial}{\partial\overline{z}}(\overline{\mathcal{D}}_3)\right\} \quad ; \tag{2.101c}$$

$$\mathrm{S}\frac{D\overline{\varrho}}{D\overline{t}} + \overline{\varrho}\overline{\mathcal{D}}_3 = 0 \quad ; \tag{2.101d}$$

$$\overline{\varrho}\mathrm{S}\frac{D\overline{T}}{D\overline{t}} - \frac{\gamma-1}{\gamma}\mathrm{S}\frac{D\overline{p}}{D\overline{t}} = \frac{1}{\mathrm{Pr}}\frac{1}{\mathrm{Re}}\left\{\overline{\Delta}_2\overline{T} + \frac{1}{\varepsilon^2}\frac{\partial^2\overline{T}}{\partial\overline{z}^2}\right\}$$
$$+\frac{\gamma-1}{\varepsilon^2\mathrm{Re}}\mathrm{M}_\infty^2\overline{\phi} + \frac{1}{\mathrm{Pr}}\frac{1}{\mathrm{Re}}\frac{\mathrm{Bo}\sigma_{00}}{\varepsilon^2}\frac{\partial\overline{\mathcal{R}}}{\partial\overline{z}} \quad ; \tag{2.101e}$$

$$\overline{p} = \overline{\varrho}\overline{T} \quad , \tag{2.101f}$$

where

$$\overline{\Delta}_2 = \frac{\partial^2}{\partial\overline{x}^2} + \frac{\partial^2}{\partial\overline{y}^2} \quad ; \quad \overline{\mathcal{D}}_3 = \frac{\partial\overline{u}}{\partial\overline{x}} + \frac{\partial\overline{v}}{\partial\overline{y}} + \frac{\partial\overline{w}}{\partial\overline{z}} \quad ,$$

$$\overline{\phi} = \left(\frac{\partial\overline{u}}{\partial\overline{z}} + \varepsilon^2\frac{\partial\overline{w}}{\partial\overline{x}}\right)^2 + \left(\frac{\partial\overline{v}}{\partial\overline{z}} + \varepsilon^2\frac{\partial\overline{w}}{\partial\overline{y}}\right)^2 + \varepsilon^2\left\{\left(\frac{\partial\overline{u}}{\partial\overline{y}} + \frac{\partial\overline{v}}{\partial\overline{x}}\right)^2\right.$$
$$\left. + 2\left[\left(\frac{\partial\overline{u}}{\partial\overline{x}}\right)^2 + \left(\frac{\partial\overline{v}}{\partial\overline{y}}\right)^2 + \left(\frac{\partial\overline{w}}{\partial\overline{z}}\right)^2\right] - \frac{2}{3}\overline{\mathcal{D}}_3^2\right\} . \tag{2.102}$$

In (2.101), the characteristic Reynolds number appears:

$$\mathrm{Re} = \frac{\varrho_\infty(0)U_0L}{\mu_0} \tag{2.103}$$

along with the radiation parameter:

$$\sigma_{00} = \frac{R}{gk_0}\mathcal{R}_\infty(T_\infty(0)) \quad . \tag{2.104}$$

In dimensionless form, the thermal balance equation [the second of the boundary conditions (2.34)] becomes:

$$\frac{\partial \overline{T}}{\partial \overline{z}} + \mathrm{Bo}\sigma_{00}\overline{\mathcal{R}} = 0 \quad , \quad \text{on} \quad \overline{z} = 0 \quad . \tag{2.105}$$

It will be remarked that the ratio

$$\frac{\mathrm{Ro}}{\mathrm{Re}} = \mathrm{Ek} \tag{2.106}$$

is the Ekman number. Lastly, $\mathrm{Pr} = \mu_0 c_p/k_0$ is the Prandtl number.

Naturally, the dimensionless parameters introduced throughout this chapter are not the only ones used in the analysis of atmospheric flows considered hereafter. Nevertheless, it did not seem reasonable here to give the complete inventory of dimensionless parameters which will appear in the coming chapters. It should be pointed out however, that a relevant discussion on the similarity and dimensionless numbers can be found in the monograph by Birkhoff (1960).

A final remark concerning the non-dimensional form of the initial conditions corresponding to (2.35) is called for. We have:

$$\begin{aligned} &\overline{u} = \overline{u}^0 \quad , \quad \overline{v} = \overline{v}^0 \quad , \quad \overline{w} = \frac{1}{\varepsilon}\overline{w}^0 \quad , \\ &\overline{\varrho} = \overline{\varrho}^0 \quad , \quad \overline{T} = \overline{T}^0 \quad , \quad \text{when } \overline{t} = 0 \quad , \end{aligned} \tag{2.107}$$

since, by hypothesis, the three components of the initial velocity are reduced in relation to U_0. In view of (2.107), the limiting case $\varepsilon \to 0$ discussed in Subsect. 2.4.3 can be expected to lead to a singular perturbation problem. We will come back to this important question in Chap. 7.

Analogously to the procedure followed in Subsect. 2.4.1, for (2.101), we can introduce the thermodynamic perturbation π, ω and ϑ defined by (2.70). In this case, it is sufficient to add the following terms to the right-hand sides of (2.72a–c) respectively:

$$\begin{aligned} &\frac{1}{\overline{\varrho}_\infty}\frac{1}{\mathrm{Re}}\left\{\overline{\Delta}_2\overline{u} + \frac{1}{\varepsilon^2}\frac{\partial^2\overline{u}}{\partial\overline{z}^2} + \frac{1}{3}\frac{\partial}{\partial\overline{x}}(\overline{\mathcal{D}}_3)\right\} \quad , \\ &\frac{1}{\overline{\varrho}_\infty}\frac{1}{\mathrm{Re}}\left\{\overline{\Delta}_2\overline{v} + \frac{1}{\varepsilon^2}\frac{\partial^2\overline{v}}{\partial\overline{z}^2} + \frac{1}{3}\frac{\partial}{\partial\overline{y}}(\overline{\mathcal{D}}_3)\right\} \quad , \\ &\frac{1}{\overline{\varrho}_\infty}\frac{\varepsilon^2}{\mathrm{Re}}\left\{\overline{\Delta}_2\overline{w} + \frac{1}{\varepsilon^2}\frac{\partial^2\overline{w}}{\partial\overline{z}^2} + \frac{1}{3}\frac{1}{\varepsilon^2}\frac{\partial}{\partial\overline{z}}(\overline{\mathcal{D}}_3)\right\} \quad , \end{aligned} \tag{2.108}$$

in order to obtain the non-dimensional, non-adiabatic equations of motion for $\overline{u}$, $\overline{v}$, $\overline{w}$, π, ω and ϑ.

As regards the energy equation, the following term must be added to the right-hand side of the adiabatic equation (2.72e):

$$\frac{1}{\overline{\varrho}_\infty}\frac{1}{\mathrm{Pr}}\frac{1}{\mathrm{Re}}\left\{\overline{\Delta}_2\vartheta+\frac{1}{\varepsilon^2}\left(\frac{\partial^2\vartheta}{\partial\overline{z}^2}+\frac{2\mathrm{Bo}}{\overline{T}_\infty(\overline{z}_\infty)}\frac{d\overline{T}_\infty(\overline{z}_\infty)}{d\overline{z}_\infty}\frac{\partial\vartheta}{\partial\overline{z}}\right.\right.$$
$$\left.\left.+\frac{\mathrm{Bo}^2}{\overline{T}_\infty(\overline{z}_\infty)}\frac{d^2\overline{T}_\infty(\overline{z}_\infty)}{d\overline{z}_\infty^2}\vartheta\right)\right\}+\frac{\gamma-1}{\overline{\varrho}_\infty\overline{T}_\infty}\frac{\mathrm{M}_\infty^2}{\varepsilon^2\mathrm{Re}}\overline{\phi}\quad, \tag{2.109}$$

once it is supposed that in (2.101e), $\overline{\mathcal{R}}$ identifies with $\overline{\mathcal{R}}_\infty(\overline{T}_\infty)$ in such a way that:

$$\frac{d^2\overline{T}_\infty}{d\overline{z}_\infty^2}+\sigma_{00}\frac{d\overline{\mathcal{R}}_\infty}{d\overline{z}_\infty}=0\quad. \tag{2.110}$$

Finally, instead of condition (2.105), we obtain:

$$\frac{\partial\vartheta}{\partial\overline{z}}+\mathrm{Bo}\frac{d\overline{T}_\infty}{d\overline{z}_\infty}\vartheta=0\quad,\quad\text{on}\quad\overline{z}=0\quad, \tag{2.111}$$

since $\overline{T}_\infty(0)\equiv 1$. It will be remarked that (2.110) is indeed equivalent to the third relation in (1.3) since we have admitted the following in order to obtain (2.111):

$$\frac{d\overline{T}_\infty}{d\overline{z}_\infty}+\sigma_{00}\overline{\mathcal{R}}_\infty=0\quad,\quad\text{on}\quad\overline{z}_\infty\equiv\mathrm{Bo}\,\overline{z}=0\quad.$$

Naturally, the hypothesis concerning the turbulent exchange coefficients (μ_0 and k_0 constant) is not very acceptable physically speaking. We can hence attempt a less restrictive hypothesis which consists in supposing that μ and k are only functions of $T_\infty(z_\infty)$: $\mu_\infty=\mu(T_\infty)$ and $k_\infty=k(T_\infty)$.

To close this chapter, one final remark on the subject of $\varepsilon\to 0$ is called for. Supposing that simultaneously with $\varepsilon\to 0$, we also have $\mathrm{Re}\to\infty$ so that $\varepsilon^2\mathrm{Re}\equiv\mathrm{Re}_\perp=\mathrm{O}(1)$, then without making the β plane approximation (and even less, the tangent plane approximation), we can obtain the so-called "quasi-static" equations (see Chap. 7) which are valid for the entire globe and describe large-scale atmospheric motions ($\delta\cong 1$).

If we assume that $\eta\equiv\varepsilon$, then these equations take on the following form:

$$\overline{\varrho}\left\{\mathrm{S}\frac{\partial\overline{u}}{\partial\overline{t}}+\frac{\delta}{\cos\varphi}\overline{u}\frac{\partial\overline{u}}{\partial\lambda}+\delta\overline{v}\frac{\partial\overline{u}}{\partial\varphi}+\overline{w}\frac{\partial\overline{u}}{\partial\overline{z}}-\delta\,\mathrm{tg}\,\varphi\,\overline{u}\,\overline{v}-\frac{1}{Ro}\sin\varphi\,\overline{v}\right\}$$
$$+\frac{1}{\gamma\mathrm{M}_\infty^2}\frac{\delta}{\cos\varphi}\frac{\partial\overline{p}}{\partial\lambda}=\frac{1}{\mathrm{Re}_\perp}\frac{\partial}{\partial\overline{z}}\left(\overline{\mu}\frac{\partial\overline{u}}{\partial\overline{z}}\right)\quad; \tag{2.112a}$$

$$\overline{\varrho}\left\{\mathrm{S}\frac{\partial\overline{v}}{\partial\overline{t}}+\frac{\delta}{\cos\varphi}\overline{u}\frac{\partial\overline{v}}{\partial\lambda}+\delta\overline{v}\frac{\partial\overline{v}}{\partial\varphi}+\overline{w}\frac{\partial\overline{v}}{\partial\overline{z}}+\delta\,\mathrm{tg}\,\varphi\,\overline{u}^2+\frac{1}{Ro}\sin\varphi\,\overline{u}\right\}$$
$$+\frac{\delta}{\gamma\mathrm{M}_\infty^2}\frac{\partial\overline{p}}{\partial\varphi}=\frac{1}{\mathrm{Re}_\perp}\frac{\partial}{\partial\overline{z}}\left(\overline{\mu}\frac{\partial\overline{v}}{\partial\overline{z}}\right)\quad; \tag{2.112b}$$

$$\frac{\partial\overline{p}}{\partial\overline{z}}+\mathrm{Bo}\overline{\varrho}=0\quad,\quad\overline{p}=\overline{\varrho}\overline{T}\quad; \tag{2.112c}$$

$$\mathrm{S}\frac{\partial\overline{\varrho}}{\partial\overline{t}} + \frac{\delta}{\cos\varphi}\overline{u}\frac{\partial\overline{\varrho}}{\partial\lambda} + \delta\overline{v}\frac{\partial\overline{\varrho}}{\partial\varphi} + \overline{w}\frac{\partial\overline{\varrho}}{\partial\overline{z}}$$

$$+\overline{\varrho}\left\{\frac{\partial\overline{w}}{\partial\overline{z}} + \frac{\delta}{\cos\varphi}\frac{\partial\overline{u}}{\partial\lambda} + \frac{\delta}{\cos\varphi}\frac{\partial}{\partial\varphi}(\overline{v}\cos\varphi)\right\} = 0 \quad ; \tag{2.112d}$$

$$\overline{\varrho}\left\{\mathrm{S}\frac{\partial\overline{T}}{\partial\overline{t}} + \frac{\delta}{\cos\varphi}\overline{u}\frac{\partial\overline{T}}{\partial\lambda} + \delta\overline{v}\frac{\partial\overline{T}}{\partial\varphi} + \overline{w}\frac{\partial\overline{T}}{\partial\overline{z}}\right\}$$

$$-\frac{\gamma-1}{\gamma}\left\{\mathrm{S}\frac{\partial\overline{p}}{\partial\overline{t}} + \frac{\delta}{\cos\varphi}\overline{u}\frac{\partial\overline{p}}{\partial\lambda} + \delta\overline{v}\frac{\partial\overline{p}}{\partial\varphi} + \overline{w}\frac{\partial\overline{p}}{\partial\overline{z}}\right\}$$

$$= \frac{1}{\mathrm{Pr}}\frac{1}{\mathrm{Re}_\perp}\frac{\partial}{\partial\overline{z}}\left(\overline{k}\frac{\partial\overline{T}}{\partial\overline{z}}\right) + \frac{\gamma-1}{\mathrm{Re}_\perp}\mathrm{M}_\infty^2\overline{\mu}\left[\left(\frac{\partial\overline{u}}{\partial\overline{z}}\right)^2 + \left(\frac{\partial\overline{v}}{\partial\overline{z}}\right)^2\right]$$

$$+\frac{1}{\mathrm{Pr}}\frac{\mathrm{Bo}\sigma_{00}}{\mathrm{Re}_\perp}\frac{\partial\overline{\mathcal{R}}}{\partial\overline{z}} \quad , \tag{2.112e}$$

where

$$Ro \equiv U_0/2\Omega_0 L \tag{2.113}$$

is related to the Rossby number [defined in (2.53)] by the relation

$$\mathrm{Ro}\sin\varphi_0 = Ro \quad .$$

3. Internal Waves and Filtering

For problems concerning the modeling of atmospheric flows, it is essential to fully understand the structure of the acoustic and gravity internal waves in relation to the limiting values of the dimensionless parameters ε, M_∞, Bo, Ro and α^0_∞. To this end, it is sufficient to consider the adiabatic flows described by (2.72).

Given the fact that for both $\varepsilon \ll 1$ (in this case, $\varepsilon \ll \mathrm{Ro}$), and for $\varepsilon \cong 1$ (then $\mathrm{Ro} \cong \infty$), the terms coupled with $\varepsilon/\mathrm{Ro}\,\mathrm{tg}\,\varphi_0$ are negligible in (2.72a,c) (at least for all φ_0 very different from $\varphi_0 \equiv 0$), they will not be retained in the following simplified analysis. [For the taking into account of these terms, the article by Kurgansky (1980) can be consulted.] We will also consider the case $\beta \equiv 0$ in (2.72a,b).

Upon these assumptions, we consider "small" motions relative to a reference situation where the velocity is zero. After linearization, the following linear equations are obtained for the dimensionless perturbations u', v', w', π' and ϑ':

$$\mathrm{S}\frac{\partial u'}{\partial \bar{t}} - \frac{1}{\mathrm{Ro}}v' + \frac{\overline{T}_\infty}{\gamma \mathrm{M}^2_\infty}\frac{\partial \pi'}{\partial \bar{x}} = 0 \quad ;$$

$$\mathrm{S}\frac{\partial v'}{\partial \bar{t}} + \frac{1}{\mathrm{Ro}}u' + \frac{\overline{T}_\infty}{\gamma \mathrm{M}^2_\infty}\frac{\partial \pi'}{\partial \bar{y}} = 0 \quad ;$$

$$\varepsilon^2 \mathrm{S}\frac{\partial w'}{\partial \bar{t}} + \frac{\overline{T}_\infty}{\gamma \mathrm{M}^2_\infty}\frac{\partial \pi'}{\partial \bar{z}} = \frac{\mathrm{Bo}}{\gamma \mathrm{M}^2_\infty}\vartheta' \quad ; \tag{3.1}$$

$$\mathrm{S}\left(\frac{\partial \pi'}{\partial \bar{t}} - \frac{\partial \vartheta'}{\partial \bar{t}}\right) + \frac{\partial u'}{\partial \bar{x}} + \frac{\partial v'}{\partial \bar{y}} + \frac{\partial w'}{\partial \bar{z}} = \frac{\mathrm{Bo}}{\overline{T}_\infty}\left(1 + \frac{d\overline{T}_\infty}{d\bar{z}_\infty}\right)w' \quad ;$$

$$\mathrm{S}\frac{\partial}{\partial \bar{t}}\left(\vartheta' - \frac{\gamma - 1}{\gamma}\pi'\right) + \alpha^0_\infty \overline{N}^2_\infty(\bar{z}_\infty)w' = 0 \quad ,$$

$$\omega' \equiv \pi' - \vartheta' \quad ,$$

where

$$\overline{T}_\infty \equiv \overline{T}_\infty(\bar{z}_\infty) \quad .$$

The last equation of the above system may, of course, be replaced by:

$$\mathrm{S}\frac{\partial}{\partial \bar{t}}\left(\frac{\gamma - 1}{\gamma}\pi' - \vartheta'\right) = \frac{\mathrm{Bo}}{\overline{T}_\infty}\left(\frac{\gamma - 1}{\gamma} + \frac{d\overline{T}_\infty}{d\bar{z}_\infty}\right)w' \quad . \tag{3.2}$$

System (3.1) being of fifth order in $\bar{t}$, five initial conditions must be prescribed for $\bar{t} = 0$. Moreover, the following slip condition must also be imposed:

$$w' = 0 \quad , \quad \text{on} \quad \bar{z} = 0 \quad . \tag{3.3}$$

Lastly, it will be noted that $\bar{z}_\infty = \mathrm{Bo}\,\bar{z}$. The other boundary conditions will be discussed a little further on.

3.1 The Case of $d\overline{T}_\infty/d\bar{z}_\infty \equiv 0$. The Wave Equation

When $d\overline{T}_\infty(\bar{z}_\infty)/d\bar{z}_\infty \equiv 0$, it is simple to eliminate all the unknown functions from system (3.1) – all but one, of course. For instance, if the latter is π', then the following wave equation results:

$$\left\{ \mathrm{S}^2 \frac{\partial^2}{\partial \bar{t}^2} \left[\varepsilon^2 \mathrm{M}_\infty^2 \mathrm{S}^2 \frac{\partial^2}{\partial \bar{t}^2} - \mathrm{ID} \right] - \mathrm{IB} \right\} \mathrm{S} \frac{\partial \pi'}{\partial \bar{t}} = 0 \quad , \tag{3.4}$$

where

$$\begin{aligned} \mathrm{ID} &= \varepsilon^2 \overline{\Delta}_2 + \frac{\partial^2}{\partial \bar{z}^2} - \mathrm{Bo}\frac{\partial}{\partial \bar{z}} - \frac{\varepsilon^2 \mathrm{M}_\infty^2}{\mathrm{Ro}^2} \quad , \\ \mathrm{IB} &= \frac{\mathrm{Bo}}{\gamma \mathrm{M}_\infty^2} \alpha_\infty^0 \overline{\Delta}_2 + \frac{1}{\mathrm{Ro}^2}\left(\frac{\partial^2}{\partial \bar{z}^2} - \mathrm{Bo}\frac{\partial}{\partial \bar{z}} \right) \quad , \end{aligned} \tag{3.5}$$

since $\overline{N}_\infty^2(0) \equiv 1$ and, in fact, $\alpha_\infty^0 \equiv \mathrm{Bo}(\gamma - 1)/\gamma$ in the case considered here. Thus, we again find that

$$\overline{N}_\infty^2(0) \equiv \frac{g^2}{RT_\infty(0)} \frac{\gamma - 1}{\gamma} \quad ,$$

when $dT_\infty/dz_\infty \equiv 0$ and in this case $\overline{T}_\infty(\bar{z}_\infty) \equiv 1$.

For the case of heavy ($\mathrm{Bo} \neq 0$), stratified ($\alpha_\infty^0 \not\equiv 0$), rotating ($\mathrm{Ro} \not\equiv \infty$) fluids, the wave equation (3.4) generalizes the classical aerodynamics acoustic wave equation.

First of all, initial conditions must be assigned to this wave equation (3.4). We suppose that

$$\text{for } \bar{t} = 0, \text{ the values of } \pi', \ \frac{\partial \pi'}{\partial \bar{t}}, \ \frac{\partial^2 \pi'}{\partial \bar{t}^2}, \ \frac{\partial^3 \pi'}{\partial \bar{t}^3}, \ \frac{\partial^4 \pi'}{\partial \bar{t}^4} \tag{3.6}$$

are given.

By elimination the perturbation ϑ' from the third and last equations of (3.1), the following relation is acquired:

$$\gamma \mathrm{M}_\infty^2 \varepsilon^2 \mathrm{S}^2 \frac{\partial^2 w'}{\partial \bar{t}^2} + \mathrm{Bo}\alpha_\infty^0 \overline{N}_\infty^2(\bar{z}_\infty) w' + \left(\overline{T}_\infty \frac{\partial}{\partial \bar{z}} - \frac{\gamma - 1}{\gamma}\mathrm{Bo} \right) \mathrm{S} \frac{\partial \pi'}{\partial \bar{t}} = 0$$

which explains why, to (3.4), we should add the condition:

$$\left(\frac{\partial}{\partial \bar{z}} - \frac{\gamma - 1}{\gamma}\mathrm{Bo} \right) \mathrm{S} \frac{\partial \pi'}{\partial \bar{t}} = 0 \quad , \quad \text{on} \quad \bar{z} = 0 \quad . \tag{3.7}$$

Boundary conditions in $\overline{x}$ and $\overline{y}$ also need to be written. For the simplified analysis being made here, it is sufficient to postulate the existence of a solution for (3.4) which can be "represented" as follows:

$$\frac{\partial \pi'}{\partial \overline{t}} = \int\int_{-\infty}^{+\infty} \Pi_{nm}(\overline{t}, \overline{z}) \exp[\mathrm{i}(n\overline{x} + m\overline{y})]\, dn\, dm \quad , \tag{3.8}$$

where $K^2 = n^2 + m^2$ is the square of the horizontal wave number related to the motion in the plane $(\overline{x}, \overline{y})$.

The vertical structure in $\overline{z}$ of the solution of (3.4) is thus supposed to be of the form:

$$\Pi_{nm}(\overline{t}, \overline{z}) = \exp\left(\frac{\mathrm{Bo}}{2}\overline{z}\right) \int_0^{\infty} \chi_{nm}\left(\frac{\overline{t}}{\mathrm{S}}, \lambda\right) \mathcal{L}_\lambda(\overline{z}) d\lambda \quad , \tag{3.9}$$

where λ and $\mathcal{L}_\lambda(\overline{z}) \equiv \mathcal{L}(\overline{z}; \lambda)$ are respectively the eigenvalues and eigenfunctions of the equation:

$$\mathcal{L}''_\lambda + \left(\lambda - \frac{\mathrm{Bo}^2}{4}\right)\mathcal{L}_\lambda = 0 \quad . \tag{3.10}$$

Furthermore, these eigenfunctions $\mathcal{L}_\lambda(\overline{z})$ must satisfy the condition

$$\mathcal{L}'_\lambda + \frac{2-\gamma}{2\gamma}\mathrm{Bo}\mathcal{L}_\lambda = 0 \quad , \quad \text{on} \quad \overline{z} = 0 \quad , \tag{3.11}$$

which results from (3.7) once (3.8, 9) have been taken into account.

The "Sturm-Liouville-type" equation (3.10) must be solved on the reference interval $[0, \infty)$, the point $\overline{z} = 0$ being regular whereas the ∞ is a limiting point. Firstly, we must be convinced of the necessity of imposing a condition at ∞, and if indeed such a condition is called for, we must be able to formulate one correctly! For the time being, we will set aside this subtle question to which we will have the occasion to return in Chap. 7. It is a most crucial question since it is what demonstrates that the set of eigenfunctions $\{\mathcal{L}_\lambda(\overline{z})\}$ indeed forms a total system. In what follows, we will assume such to be the case.

Moreover, we can readily be convinced that the set of eigenvalues λ of (3.10) which satisfies (3.11) covers the interval $(\mathrm{Bo}^2/4, \infty)$ across a continuous spectrum:

$$\mathrm{Bo}^2/4 < \lambda < \infty \quad , \tag{3.12}$$

to which the following isolated eigenvalue must be added:

$$\lambda_0 = \frac{\gamma - 1}{\gamma^2}\mathrm{Bo}^2 \quad . \tag{3.13}$$

Finally, let us also take note of the fact that to this eigenvalue λ_0 corresponds the eigenfunction:

$$\mathcal{L}_{\lambda_0}(\overline{z}) = \exp\left(-\frac{2-\gamma}{2\gamma}\mathrm{Bo}\overline{z}\right) \quad , \tag{3.14}$$

which is a solution of (3.10, 11). The continuous spectrum (3.12) however, corresponds to the following solution of the same equation (3.10) with (3.11):

$$\cos(\mathrm{Bo}\mu\bar{z}) - \frac{2-\gamma}{2\gamma}\sin(\mathrm{Bo}\,\mu\bar{z}) \quad , \tag{3.15}$$

with

$$\mu^2 = \frac{\lambda_\mu}{\mathrm{Bo}^2} - \frac{1}{4} \quad ,$$

i.e.:[1]

$$\mathrm{Bo}\mu = \left(\lambda_\mu - \frac{\mathrm{Bo}^2}{4}\right)^{1/2} \quad . \tag{3.16}$$

Since we have assumed that the eigenfunctions $\mathcal{L}(\bar{z};\,\lambda_\mu)$ form a total system, we can represent the functions intervening in the initial conditions (3.6) for $\bar{t}=0$ in a form analogous to (3.8), and then (3.9). In this case, the following Cauchy problem follows for the function $\chi_{nm}(\tau;\,\lambda_\mu)$:

$$\left\{\varepsilon^2\mathrm{M}_\infty^2\frac{\partial^4}{\partial\tau^4} + \left[\left(\frac{\varepsilon\mathrm{M}_\infty}{\mathrm{Ro}}\right)^2 + \varepsilon^2K^2 + \lambda_\mu\right]\frac{\partial^2}{\partial\tau^2} + \left[\frac{\lambda_\mu}{\mathrm{Ro}^2} + K^2\frac{\mathrm{Bo}}{\gamma\mathrm{M}_\infty^2}\alpha_\infty^0\right]\right\}\chi_{nm} = 0, \tag{3.17}$$

$$\tau \equiv \frac{\bar{t}}{\mathrm{S}} = 0 \; : \text{ the values of } \chi_{nm},\ \frac{\partial\chi_{nm}}{\partial\tau},\ \frac{\partial^2\chi_{nm}}{\partial\tau^2} \text{ and } \frac{\partial^3\chi_{nm}}{\partial\tau^3} \text{ are given} \quad . \tag{3.18}$$

The solution of (3.17), valid for all values of τ is:

$$\chi_{nm}(\tau;\,\lambda_\mu) = \sum_{s=1}^{4} C_{nm}^s(\lambda_\mu)\exp[r_{nm}^s(\lambda_\mu)\tau] \quad , \tag{3.19}$$

where the arbitrary constants C_{nm}^s [which are functions of the eigenvalues λ_μ, such that (3.16) leads to real values of μ] are determined in the usual manner from the initial data (3.18). It is thus easy to see that:

$$r_{nm}^1 \equiv -r_{nm}^2 = \mathrm{i}\sigma_\mathrm{a} \quad ; \quad r_{nm}^3 \equiv -r_{nm}^4 = \mathrm{i}\sigma_\mathrm{g} \quad , \tag{3.20}$$

where $\mathrm{i}=\sqrt{-1}$.

The values of σ_a and σ_g are given by the relation:

$$\sigma_{\mathrm{a,g}}^2 = \mathcal{A}_{K,\lambda_\mu}\{1 \pm (1-\delta)^{1/2}\} \quad , \tag{3.21}$$

where

[1] It will be remarked that for every $\lambda_\mu - \mathrm{Bo}^2/4 > 0$ which is a double eigenvalue, the solution (3.15) is an eigenfunction bounded at infinity for $\bar{z} \to +\infty$.

$$\mathcal{A}_{K,\lambda_\mu} = \frac{1}{2\mathrm{M}_\infty^2}\left\{\left(\frac{\mathrm{M}_\infty}{\mathrm{Ro}}\right)^2 + K^2 + \frac{\lambda_\mu}{\varepsilon^2}\right\} \quad , \tag{3.22}$$

$$\delta = \frac{1}{\mathcal{A}_{K,\lambda_\mu}^2 \mathrm{M}_\infty^2}\left\{\frac{\lambda_\mu}{(\varepsilon \mathrm{Ro})^2} + K^2 \frac{\alpha_\infty^0}{\varepsilon^2}\frac{\mathrm{Bo}}{\gamma \mathrm{M}_\infty^2}\right\} \quad . \tag{3.23}$$

In (3.21), the sign "+" corresponds to the frequency σ_a of the internal acoustic waves, whereas the sign "−" corresponds to the frequency σ_g of the internal gravity waves.

If we take into consideration more specifically the case of horizontal waves for which $w' \equiv 0$ in (3.1), then instead of (3.4) we obtain:

$$\left\{\mathrm{S}^2\frac{\partial^2}{\partial \bar{t}^2} + \frac{1}{\mathrm{Ro}^2} - \frac{1}{\mathrm{M}_\infty^2}\overline{\Delta}_2\right\}\mathrm{S}\frac{\partial \pi'}{\partial \bar{t}} = 0 \tag{3.24}$$

and therefore the frequency of these plane horizontal waves is:

$$\sigma_\mathrm{p}^2 = \frac{1}{\mathrm{Ro}^2} + \frac{K^2}{\mathrm{M}_\infty^2} \quad . \tag{3.25}$$

It should be pointed out that these plane horizontal waves are purely acoustic in nature. In fact, if the following limiting process is carried out in (3.21–23),

$$\lambda_\mu \to 0 \quad , \quad \mathrm{Bo} \to 0 \quad , \tag{3.26}$$

(which characterizes a barotropic atmosphere), we then recover:

$$\sigma_\mathrm{a}^2 \to \sigma_\mathrm{p}^2 \quad , \quad \text{but} \quad \sigma_\mathrm{g}^2 \to 0 \quad .$$

It will be seen that due to the hypothesis $\beta \equiv 0$, one of the eigenfrequencies (which corresponds to the so-called Rossby waves as we will see in Chap. 4) degenerates and is equal to zero.

It is generally recognized that, in fact, the eigenvalue $\lambda_0 = (\gamma - 1)\mathrm{Bo}^2/\gamma^2$ and the eigenfunction $\mathcal{L}_{\lambda_0}(\bar{z}) = \exp[-(2-\gamma)\mathrm{Bo}\bar{z}/2\gamma]$ describe the barotropic component of the motion. The frequency σ_p which satisfies (3.25) is associated with the internal acoustic waves in a barotropic, quasi-static atmosphere.

3.2 The Vertical Structure of the Internal Waves

In order to analyze the vertical structure of the internal waves, it is, of course, necessary to suppose that:

$$\frac{d\overline{T}_\infty(\bar{z}_\infty)}{d\bar{z}_\infty} \not\equiv 0 \quad . \tag{3.27}$$

On the other hand, it is possible for us to assume that:

$$\mathrm{Ro} \equiv \infty \tag{3.28}$$

and to consider only the two-dimensional case in the plane of $(\overline{x}, \overline{z})$, i.e., to admit that in (3.1):

$$v' \equiv 0 \quad , \quad \frac{\partial \pi'}{\partial \overline{y}} \equiv 0 \quad , \quad \frac{\partial v'}{\partial \overline{y}} \equiv 0 \quad . \tag{3.29}$$

Thus, after elimination ω', (3.1) is replaced by the following equations for u', w', π' and ϑ':

$$\begin{aligned}
&\mathrm{S}\frac{\partial u'}{\partial \overline{t}} = -\frac{\overline{T}_\infty}{\gamma \mathrm{M}_\infty^2}\frac{\partial \pi'}{\partial \overline{x}} \quad ; \qquad \varepsilon^2\mathrm{S}\frac{\partial w'}{\partial \overline{t}} = -\frac{\overline{T}_\infty}{\gamma \mathrm{M}_\infty^2}\frac{\partial \pi'}{\partial \overline{z}} + \frac{\mathrm{Bo}}{\gamma \mathrm{M}_\infty^2}\vartheta' \quad ;\\
&\mathrm{S}\frac{\partial \pi'}{\partial \overline{t}} = -\gamma\left(\frac{\partial u'}{\partial \overline{x}} + \frac{\partial w'}{\partial \overline{z}}\right) + \frac{\mathrm{Bo}}{\overline{T}_\infty}w'; \\
&\mathrm{S}\frac{\partial \vartheta'}{\partial \overline{t}} = -(\gamma - 1)\left(\frac{\partial u'}{\partial \overline{x}} + \frac{\partial w'}{\partial \overline{z}}\right) - \frac{1}{\overline{T}_\infty}\frac{d\overline{T}_\infty}{d\overline{z}}w' \quad ,
\end{aligned} \tag{3.30}$$

since

$$\mathrm{Bo}\frac{d\overline{T}_\infty(\overline{z}_\infty)}{d\overline{z}_\infty} \equiv \frac{d\overline{T}_\infty(\mathrm{Bo}\overline{z})}{d\overline{z}} \quad ,$$

it being understood that in (3.30):

$$\overline{T}_\infty \equiv \overline{T}_\infty(\mathrm{Bo}\overline{z}) \quad .$$

We can now introduce the divergence:

$$\mathcal{D}_2' = \frac{\partial u'}{\partial \overline{x}} + \frac{\partial w'}{\partial \overline{z}} \quad . \tag{3.31}$$

In this case, by differentiation relative to $\overline{t}$ and by combining the first and third equations of (3.30), we find the following:

$$\left(\mathrm{S}^2\frac{\partial^2}{\partial \overline{t}^2} - \frac{\overline{T}_\infty}{\mathrm{M}_\infty^2}\frac{\partial^2}{\partial \overline{x}^2}\right)\mathcal{D}_2' = \left[\frac{\partial}{\partial \overline{z}}\left(\mathrm{S}^2\frac{\partial^2}{\partial \overline{t}^2}\right) - \frac{\mathrm{Bo}}{\gamma \mathrm{M}_\infty^2}\frac{\partial^2}{\partial \overline{x}^2}\right]w' \quad . \tag{3.32}$$

In order to obtain a second equation linking $\mathcal{D}_2'$ and w', it is sufficient to eliminate π' and ϑ' from the last three equations of (3.30). The result is:

$$\left(\varepsilon^2\mathrm{S}^2\frac{\partial^2}{\partial \overline{t}^2} + \frac{\mathrm{Bo}}{\gamma \mathrm{M}_\infty^2}\frac{\partial}{\partial \overline{z}}\right)w' = \left(\frac{\overline{T}_\infty}{\mathrm{M}_\infty^2}\frac{\partial}{\partial \overline{z}} - \frac{\gamma - 1}{\gamma}\frac{\mathrm{Bo}}{\mathrm{M}_\infty^2}\right)\mathcal{D}_2' \quad . \tag{3.33}$$

It is a simple matter to check that the operator in the right-hand side of (3.32) commutes with the one in the left-hand side of (3.33). Hence, we can eliminate w' from (3.32, 33) and obtain the following for $\mathcal{D}_2'$:

$$\begin{aligned}
&\left\{\varepsilon^2\mathrm{M}_\infty^2\mathrm{S}^4\frac{\partial^4}{\partial \overline{t}^4} - \overline{T}_\infty(\mathrm{Bo}\overline{z})\left[\varepsilon^2\frac{\partial^2}{\partial \overline{x}^2} + \frac{\partial^2}{\partial \overline{z}^2}\right]\mathrm{S}^2\frac{\partial^2}{\partial \overline{t}^2}\right.\\
&+\left[\mathrm{Bo} - \frac{d\overline{T}_\infty(\mathrm{Bo}\overline{z})}{d\overline{z}}\right]\frac{\partial}{\partial \overline{z}}\left(\mathrm{S}^2\frac{\partial^2}{\partial \overline{t}^2}\right) - \frac{\mathrm{Bo}}{\gamma \mathrm{M}_\infty^2}\left[\frac{\gamma - 1}{\gamma}\mathrm{Bo}\right.\\
&\left.\left.+\frac{d\overline{T}_\infty(\mathrm{Bo}\overline{z})}{d\overline{z}}\right]\frac{\partial^2}{\partial \overline{x}^2}\right\}\mathcal{D}_2' = 0 \quad .
\end{aligned} \tag{3.34}$$

It is also observed that from (3.32, 33), the following relation can be obtained between w' and $\mathcal{D}_2'$,

$$\varepsilon^2 S^4 \frac{\partial^4 w'}{\partial \bar{t}^4} + \left(\frac{\mathrm{Bo}}{\gamma M_\infty^2}\right)^2 \frac{\partial^2 w'}{\partial \bar{x}^2}$$
$$= \left\{ \frac{\overline{T}_\infty}{M_\infty^2} \frac{\partial}{\partial \bar{z}} \left(S^2 \frac{\partial^2}{\partial \bar{t}^2} \right) - \frac{\mathrm{Bo}}{M_\infty^2} S^2 \frac{\partial^2}{\partial \bar{t}^2} + \frac{\mathrm{Bo}}{\gamma M_\infty^2} \frac{\overline{T}_\infty}{M_\infty^2} \frac{\partial^2}{\partial \bar{x}^2} \right\} \mathcal{D}_2' \quad , \tag{3.35}$$

which does not call for any derivative in $\bar{z}$ of w'. As the coefficients of (3.34) are functions of $\bar{z}$, it is pertinent to study the solution of this equation in the form:

$$\mathcal{D}_2'(\bar{t}, \bar{x}, \bar{z}) = \tilde{\mathcal{D}}_2'(\bar{z}) \exp\left\{ \mathrm{i}\left(k\bar{x} - \omega \frac{\bar{t}}{S} \right) \right\} \quad ; \tag{3.36}$$

In like manner, we might have, for example:

$$w'(\bar{t}, \bar{x}, \bar{z}) = \tilde{w}'(\bar{z}) \exp\left\{ \mathrm{i}\left(k\bar{x} - \omega \frac{\bar{t}}{S} \right) \right\} \quad .$$

It can be seen that in the expression of exp{ }, $\mathrm{i} = \sqrt{-1}$. So for $\tilde{\mathcal{D}}_2'(\bar{z})$ the following ordinary second order differential equation then results:

$$\overline{T}_\infty(\mathrm{Bo}\bar{z}) \frac{d^2 \tilde{\mathcal{D}}_2'}{d\bar{z}^2} - \left[\mathrm{Bo} - \frac{d\overline{T}_\infty(\mathrm{Bo}\bar{z})}{d\bar{z}} \right] \frac{d\tilde{\mathcal{D}}_2'}{d\bar{z}} + \left\{ \varepsilon^2 M_\infty^2 \omega^2 - \overline{T}_\infty(\mathrm{Bo}\bar{z}) \varepsilon^2 k^2 \right.$$
$$\left. + \frac{\mathrm{Bo}}{\gamma M_\infty^2} \frac{k^2}{\omega^2} \left[\frac{\gamma - 1}{\gamma} \mathrm{Bo} + \frac{d\overline{T}_\infty(\mathrm{Bo}\bar{z})}{d\bar{z}} \right] \right\} \tilde{\mathcal{D}}_2' = 0 \quad . \tag{3.37}$$

The boundary conditions in $\bar{z}$ for (3.37) must now be formulated.

First of all, between $\tilde{\mathcal{D}}_2'(\bar{z})$ and $\tilde{w}'(\bar{z})$, the following relation ensues from (3.35):

$$\left\{ \varepsilon^2 \omega^4 - \left(\frac{\mathrm{Bo}}{\gamma M_\infty^2} \right)^2 k^2 \right\} \tilde{w}'(\bar{z})$$
$$= -\frac{\overline{T}_\infty}{M_\infty^2} \omega^2 \frac{d\tilde{\mathcal{D}}_2'}{d\bar{z}} + \frac{\mathrm{Bo}}{M_\infty^2} \left(\omega^2 - \frac{\overline{T}_\infty}{\gamma M_\infty^2} k^2 \right) \tilde{\mathcal{D}}_2' \quad . \tag{3.38}$$

Thus, when:

$$\omega^4 \neq \left(\frac{\mathrm{Bo}}{\gamma M_\infty^2} \right)^2 \frac{k^2}{\varepsilon^2} \quad , \tag{3.39}$$

the ground condition (assumed flat, without relief) can be written as follows:

$$\left(\frac{d\tilde{\mathcal{D}}_2'}{d\bar{z}} \right)_{\bar{z}=0} = \mathrm{Bo}\left(1 - \frac{1}{\gamma M_\infty^2} \frac{k^2}{\omega^2} \right) \tilde{\mathcal{D}}_2'(0) \quad . \tag{3.40}$$

The above coincides with the slip condition $\tilde{w}' = 0$ on $\bar{z} = 0$. Lastly, it is noteworthy that $\overline{T}_\infty(0) \equiv 1$.

As regards the second condition in $\overline{z}$ for (3.37), the following can serve as a guide: the solutions of the wave equation (3.37) must form a total system such that via these solutions, we can expand for the Cauchy problem, an arbitrary initial condition to be imposed on (3.34), which is of fourth order in $\overline{t}$. Otherwise stated, the system of eigenfunctions of (3.37) corresponding to the various eigenvalues ω^2 must, for fixed k, be a total system of functions of $\overline{z}$. So as to be able to formulate this constraint as a boundary condition, (3.37) must be written in self-adjoint form in which the first derivative does not intervene. In this case, when using the new variables, the conditions at the upper boundary constitute a condition 1) which imposes bounds on the solution, and 2) which makes it possible to retain only one of the two linearly independent solutions of the self-adjoint equation.

In order for (3.37) to be brought to a self-adjoint form, the following transformations in variable and function suffice:

$$\zeta = \mathrm{Bo} \int_0^{\overline{z}} \frac{d\overline{z}}{\overline{T}_\infty(\mathrm{Bo}\overline{z})} \equiv \varphi(\overline{z}) \quad , \tag{3.41}$$

$$Z(\zeta) = \exp\{-\zeta/2\}\tilde{\mathcal{D}}_2'(\varphi^{-1}(\zeta)) \quad . \tag{3.42}$$

In particular, for a standard isothermal atmosphere, it follows that:

$$\zeta \equiv \overline{z}_\infty \quad , \quad \varphi^{-1}(\zeta) \equiv \frac{\zeta}{\mathrm{Bo}} = \overline{z} \quad .$$

The change to the new variable ζ makes it possible to eliminate, first, the coefficient of derivatives and then, thanks to (3.42), the term with the first derivative in ζ. Finally, the following equation for the new function $Z(\zeta)$ replaces (3.37):

$$Z'' + \left\{-\frac{1}{4} - \frac{\varepsilon^2 k^2}{\mathrm{Bo}^2}\overline{T}_\infty(\overline{z}_\infty) + \mathcal{G}(\omega^2; \overline{z}_\infty)\right\} Z = 0 \quad , \tag{3.43}$$

where

$$\mathcal{G}(\omega^2; \overline{z}_\infty) = \frac{\varepsilon^2 \mathrm{M}_\infty^2}{\mathrm{Bo}^2}\omega^2 \overline{T}_\infty(\overline{z}_\infty) + \frac{k^2}{\gamma \mathrm{M}_\infty^2}\frac{1}{\omega^2}\overline{T}_\infty(\overline{z}_\infty)\overline{\Gamma}_\infty(\overline{z}_\infty) \quad , \tag{3.44}$$

with

$$\overline{\Gamma}_\infty(\overline{z}_\infty) = \frac{\gamma - 1}{\gamma} + \frac{d\overline{T}_\infty(\overline{z}_\infty)}{d\overline{z}_\infty} \quad . \tag{3.45}$$

If Bo = 0(1), then of course:

$$\overline{z}_\infty = \mathrm{Bo}\overline{z} = \mathrm{Bo}\varphi^{-1}(\zeta)$$

and in this case:

$$\overline{T}_\infty(\overline{z}_\infty) \equiv \tau_\infty(\zeta) \quad , \quad \overline{\Gamma}_\infty(\overline{z}_\infty) \equiv \chi_\infty(\zeta) \quad .$$

However, if $\mathrm{Bo} \ll 1$, then ζ and $\overline{z}_\infty$ are two distinct vertical scales and in this case, a vertical structure in double scale appears when $\mathrm{Bo} \to 0$.

In accordance with (3.40), the following boundary condition must be prescribed for (3.43):

$$Z'_{\zeta=0} = \left\{\frac{1}{2} - \frac{1}{\gamma \mathrm{M}_\infty^2}\frac{k^2}{\omega^2}\right\} Z(0) \quad . \tag{3.46}$$

To this same equation, we can now add the second boundary condition in ζ which follows:

$$|Z(\infty)| < \infty \quad . \tag{3.47}$$

This last condition enables us to retain only one of the two linearly independent solutions of (3.43). For $\tilde{\mathcal{D}}'_2(\overline{z})$, condition (3.47) can be interpreted in relation to (3.37) as a constraint on the growth of $\tilde{\mathcal{D}}'_2(\overline{z})$ af infinity:

$$|\tilde{\mathcal{D}}'_2(\overline{z})| < \text{const. } \exp\left\{\frac{\mathrm{Bo}}{2}\int_0^{\overline{z}} \frac{d\overline{z}}{\overline{T}_\infty(\mathrm{Bo}\overline{z})}\right\} \quad ,$$
$$\text{when} \quad \overline{z} \to +\infty \quad . \tag{3.48}$$

For every fixed k, the eigenfrequency ω^2 must be determined in such a way that the boundary problem (3.43, 46, 47) [or (3.37, 40, 48)] has a unique solution different from zero. Solving this problem becomes rather complicated due to the fact that the eigenfrequency ω^2 also comes into play in the boundary condition on the ground [either (3.46) or (3.40)]. For applications to the case $\overline{T}_\infty(\overline{z}_\infty) =$ const., the book by Dikij (1969) may be consulted. To our knowledge, it does not appear that the above-cited problem has ever been thoroughly analyzed. The solution to this problem is closely related to the value of the dimensionless parameters ε^2, M_∞^2, Bo, γ which characterize different types of atmospheric flows.

The case $\varepsilon \to 0$ (quasi-static approximation), in particular, greatly simplifies the boundary problem which can then be written in the form:

$$-Z_0'' + \tfrac{1}{4} Z_0 = \lambda^2 \psi(\zeta) Z_0 \quad ; \tag{3.49a}$$

$$Z'_0|_{\zeta=0} = [\tfrac{1}{2} - \lambda^2] Z_0(0) \quad ,$$
$$|Z_0(\infty)| < \infty \quad , \tag{3.49b}$$

where

$$\lim_{\varepsilon\to 0} Z(\zeta;\, \varepsilon, \mathrm{M}_\infty, \mathrm{Bo}, \gamma) \equiv Z_0(\zeta;\, \lambda) \quad , \quad \text{and}$$

$$\lambda^2 = \frac{k^2}{\gamma \mathrm{M}_\infty^2}\frac{1}{\omega^2} \quad , \quad \psi(\zeta) \equiv \tau_\infty(\zeta)\chi_\infty(\zeta) \quad . \tag{3.50}$$

The limiting case $\mathrm{M}_\infty \to 0$, with all the other parameters remaining fixed at the order unity, is very delicate and demands a special analysis (see Chap. 12). In this case $\mathcal{G}(\omega^2;\, \mathrm{Bo}\varphi^{-1}(\zeta))$, given by (3.44) tends to infinity along with $\mathrm{M}_\infty^2 \to 0$.

In contrast, the following case can be considered

$$\mathrm{M}_\infty \to 0 \quad , \quad k \to 0 \quad ; \quad \frac{k}{\mathrm{M}_\infty} = \mathrm{O}(1) \quad , \tag{3.51}$$

which corresponds to the limiting case of very long horizontal waves in a weakly compressible atmosphere.

The case of very short horizontal waves

$$k \to \infty \tag{3.52}$$

leads to a singular perturbation problem:

$$\nu^2 Z'' + \left\{ \mu^2 \psi(\zeta) - \frac{\varepsilon^2}{\mathrm{Bo}^2} \tau_\infty(\zeta) + \nu^2 \left[\frac{\varepsilon^2}{\gamma \mathrm{Bo}^2} \frac{1}{\mu^2} \tau_\infty(\zeta) - \tfrac{1}{4} \right] \right\} Z = 0 \quad ; \tag{3.53a}$$

$$\nu^2 Z'|_{\zeta=0} + \left(\mu^2 - \frac{\nu^2}{2} \right) Z(0) = 0 \quad ,$$

$$|Z(\infty)| < \infty \quad , \tag{3.53b}$$

when $\nu^2 \equiv 1/k^2 \to 0$. Here

$$\left(\mu^2 = \frac{1}{\gamma \mathrm{M}_\infty^2} \frac{1}{\omega^2} \right) \quad .$$

Another interesting limiting case is:

$$\mathrm{M}_\infty \to 0 \quad , \quad \gamma \to \infty \quad ; \quad \gamma \mathrm{M}_\infty^2 \equiv \hat{M}^2 = \mathrm{O}(1) \quad , \tag{3.54}$$

which is related to the isochoric approximation. In this case:

$$\mathcal{G}\left(\omega^2; \mathrm{Bo}\varphi^{-1}(\zeta)\right) \equiv \hat{\lambda}^2 \psi(\zeta) \quad ,$$

$$\hat{\lambda}^2 = \frac{k^2}{\hat{M}^2} \frac{1}{\omega^2} \quad .$$

In Chap. 8, we will return to the limiting case:

$$\mathrm{Bo} \to 0 \quad , \quad \mathrm{M}_\infty \to 0 \quad ; \quad \frac{\mathrm{Bo}}{\mathrm{M}_\infty} = \hat{B} = \mathrm{O}(1) \quad , \tag{3.55}$$

which leads to the so-called *Boussinesq* equations. In particular, the method used to study (linearized) wave propagation by applying the multiple scale technique [see the book by Nayfeh (1973)] makes it possible to take into account the variation of the Väisälä frequency with altitude. In this case, the motion of the atmosphere is the result of the modulation of a local wave by the slow variation with altitude of the parameters of the standard atmosphere. Just such an application as inspired by Bois (1979), can be found in Sect. 6.2.

3.3 Filtering

When remaining within the bounds of the tangent plane approximation, we notice that three types of waves exist in an adiabatic atmosphere: internal acoustic waves with a frequency σ_a, internal gravity waves with a frequency σ_g and plane acoustic waves with a frequency σ_p for which vertical motions are absent.

Filtering, in fact, consists in eliminating the solutions which correspond to short, fast, internal waves so as to preserve only the solutions corresponding to long, slow waves of the type called *Rossby* waves (see Chap. 4) whose frequency σ_R is directly proportional to the parameter β [see (2.55)]. These *Rossby* waves constitute the basic driving force behind the atmospheric machine and appear as fundamental elements for the supporting weather models. Filtering is closely linked to the concept of limiting flows which can be obtained when the dimensionless parameters characterizing the atmosphere in motion tend towards their limiting values (i.e., zero or infinity). Based on the Navier-Stokes equations for the atmosphere, filtering thus leads to the selectioning of model equations – i.e., to the constructing of models. Therefore, filtering an equation means constructing a model. Chapters 11 to 13 are devoted to the topic of modeling. However, before constructing any models, the nature of the approximations which filter the equations must be thoroughly understood.

3.3.1 Quasi-static Filtering

When $\varepsilon \to 0$, the wave equation (3.4) degenerates strongly and, as a result, instead of (3.21–23), we have:

$$\lim_{\varepsilon \to 0} \sigma_a^2 = \infty \quad ; \quad \lim_{\varepsilon \to 0} \sigma_g^2 \equiv \sigma_{gq-s}^2 = \frac{1}{\mathrm{Ro}^2} + \frac{K^2}{\lambda_\mu} \frac{\gamma - 1}{\gamma^2} \left(\frac{\mathrm{Bo}}{\mathrm{M}_\infty} \right)^2 \quad . \tag{3.56}$$

Therefore, all the internal acoustic waves are filtered out and the internal gravity waves are greatly altered. Among the latter, only the internal gravity waves corresponding to $K \ll 1$, or $\lambda_\mu \gg 1$ remain intact. When $\varepsilon \to 0$, the loss of the fourth derivative with respect to time $\bar{t}$ in (3.4) shows that the double limiting process:

$$\varepsilon \to 0 \quad , \quad \bar{t} \to 0$$

is singular. In the vicinity of $\bar{t} = 0$, an initial layer must be introduced and the problem of adjustment to hydrostatic balance must be formulated. This makes it possible to state the "right" initial conditions for the *primitive* equations which were dealt with in Subsect. 2.4.3. The work by Guiraud and Zeytounian (1982) can be cited as a relevant reference – we will come back to this matter in Chap. 7.

It is essential to understand that because the short, horizontal, internal gravity waves are filtered out, the limiting process $\varepsilon \to 0$ leads to a singular perturbation problem even in the steady case ($\mathrm{S} \equiv 0$). In particular, steady lee waves cannot be described within the framework of quasi-static filtering. In order to understand how short, horizontal, internal gravity waves are filtered out when $\varepsilon \to 0$, it appears necessary, once again, to apply a multiple scale technique. For the time being, when $\varepsilon \to 0$, it is not quite understood how the transition from a process with short internal gravity waves to one with long internal gravity waves operates for the steady processes.

It will also be noticed that quasi-static filtering does not alter the frequency σ_{p} of the plane waves.

3.3.2 Filtering of Waves with Frequency $\sigma_{\mathrm{g\,q-s}}$

One of the following two limiting processes can be applied to the primitive equations:

$$\mathrm{S\,Ro} = \mathrm{Ki} \to 0 \quad , \quad M_\infty \to 0 : \quad \frac{\mathrm{Bo}}{\gamma \mathrm{S}} \left(\frac{\mathrm{Ki}}{M_\infty} \right)^2 \equiv \lambda_0 = O(1) \quad , \tag{3.57}$$

or

$$\mathrm{Ki} = O(1) \quad , \quad M_\infty \to 0 \quad . \tag{3.58}$$

According to (3.56), both cases lead to:

$$\sigma^2_{\mathrm{g\,q-s}} = \infty \quad , \tag{3.59}$$

and no internal waves remain.

Two new adjustment problems thus arise: the first, in conjuction with (3.57), concerns adjustment to geostrophy; the second, related to (3.58), deals with adjustment to the so-called balance equation.

Although the geostrophy adjustment problem has now been thoroughly analyzed, [see Chap. 4 in the book by Kibel (1963) and also the article by Guiraud and Zeytounian (1980)] the "balance equation" adjustment problem of the quasi-solenoidal model has not. We will return to this question in Chap. 12.

Solving this last adjustment problem would, in particular, permit us to shed light on the coherence of the quasi-solenoidal model with respect to time – wherein lies a fundamental problem for the understanding of the limiting process (3.58). It is essential to comprehend that all other constraints aside, the limiting process $M_\infty \to 0$ applied to the Navier-Stokes equations for the atmosphere leads to a strong degeneracy of these equations and poses a number of other problems still unsolved. Most specifically, the influence of the vanishing internal acoustic waves must be properly taken into account.

In classical aerodynamics, when the Coriolis and gravity forces are not taken into accout, thanks to the works by Viviand (1970) and Crow (1970), as well as Zeytounian and Guiraud (1980) [see also the review by Zeytounian (1983)] the above-mentioned influence can be accounted for by using singular perturbation techniques. The latter now only need to be generalized for atmospheric flows.

It must also be stressed that when applied to the complete Euler equations, both limiting processes (3.57, 58) lead to elliptical limit equations! Therefore, the hyperbolic nature of the initial Euler equations is lost – this change in character constitutes one of the main causes of the difficulties associated with (3.57,58).

This loss of hyperbolicity also occurs when proceeding from the Euler equations to the primitive equations. We will come back to this point in Chap. 7.

3.3.3 "Boussinesq" Filtering

First of all, let us note that if in (3.56), which yields the frequency of the quasi-static (internal) gravity waves, we carry out the limiting process (3.55) in the *Boussinesq* manner, then we find:

$$\lim_{\substack{\mathrm{Bo}=\hat{B}\mathrm{M}_\infty \\ \mathrm{M}_\infty\to 0;\ \hat{B}=\mathrm{O}(1)}} \sigma^2_{\mathrm{g\,q-s}} = \frac{1}{\mathrm{Ro}^2} + \frac{K^2}{\lambda_\mu}\frac{\gamma-1}{\gamma^2}\hat{B}^2 \equiv \sigma^2_{\mathrm{g\,q-s,B}} \quad , \tag{3.60}$$

i.e., an expression analogous to (3.56).

The following Boussinesq filtering:

$$\mathrm{Bo} = \hat{B}\mathrm{M}_\infty \quad , \quad \hat{B} = \mathrm{O}(1) \quad , \quad \mathrm{M}_\infty \to 0 \quad , \tag{3.61}$$

once again, leads to a strongly degenerated wave equation which results in:

$$\lim_{\substack{\mathrm{Bo}=\hat{B}\mathrm{M}_\infty \\ \mathrm{M}_\infty\to 0;\ \hat{B}=\mathrm{O}(1)}} \sigma^2_{\mathrm{a}} = \infty \quad ;$$

$$\lim_{\substack{\mathrm{Bo}=\hat{B}\mathrm{M}_\infty \\ \mathrm{M}_\infty\to 0;\ \hat{B}=\mathrm{O}(1)}} \sigma^2_{\mathrm{g}} \equiv \sigma^2_{\mathrm{g\,B}} = \frac{\lambda_\mu/\mathrm{Ro}^2 + \hat{B}^2(\gamma-1)/\gamma^2}{K^2\varepsilon^2 + \lambda_\mu} \quad . \tag{3.62}$$

Hence, all the internal acoustic waves are again filtered out, while the internal gravity waves are greatly modified. In addition, it can be seen that:

$$\lim_{\varepsilon\to 0} \sigma^2_{\mathrm{g\,B}} \equiv \sigma^2_{\mathrm{g\,B,q-s}} = \sigma^2_{\mathrm{g\,q-s,B}} \quad . \tag{3.63}$$

Due to the loss of the fourth derivative with respect to time $\bar{t}$ in the wave equation (3.4) when the limiting process (3.61) is executed, it is seen that the vicinity of $\bar{t} = 0$ is singular. As the double limiting process

$$\mathrm{Bo} = \hat{B}\mathrm{M}_\infty \quad , \quad \hat{B} = \mathrm{O}(1) \quad , \quad \mathrm{M}_\infty \to 0 \quad , \quad \bar{t} \to 0 \quad ,$$

is not regular, the adjustment problem to the *Boussinesq* state must be formulated in the vicinity of $\bar{t} = 0$. This should make it possible to obtain the correct initial conditions for the Boussinesq unsteady equations which are valid when $\bar{t} = \mathrm{O}(1)$.

Morever, the contraint $\hat{B} = \mathrm{O}(1)$ implies that

$$\frac{RT_\infty(0)}{g} \gg H = \hat{B}\frac{U_0}{g}\sqrt{\frac{RT_\infty(0)}{\gamma}} \equiv H_{\mathrm{B}} \quad , \tag{3.64}$$

which means that the Boussinesq limit equations remain rigorously valid only in an atmospheric layer whose thickness H_B is of the order of a few hundred meters.

If, in particular, in these *Boussinesq* equations (we will work them out for the general case in Chap. 8), we assume that $\varepsilon = O(1)$, then $L \cong H = H_B$ which means that they can only describe atmospheric flows locally: thus arises the problem of how the *Boussinesq* solution behaves at infinity! In the steady adiabatic case [see *Guirauds*'s (1979) note], it turns out that a radiation condition in the manner of *Sommerfeld* must be imposed on the *Boussinesq* solution at infinity. However, it is not yet known (even in the adiabatic case) what outer equations can be associated with the *Boussinesq* equations which would match the latter from the point of view of the matched asymptotic expansion-theory. For the steady two-dimensional adiabatic case, as well as for the problem of lee waves, the above was resolved by Guiraud and Zeytounian (1979). It has been confirmed that lee waves generated near an obstacle (which is small compared to the thickness of the troposphere) persist at quite a distance due to the radiation condition. Consequently, only a double-scale outer representation makes it possible to describe the steady evolution of these lee waves.

Once again, the transition of Euler equations into Boussinesq equations via (3.61) changes the character of the limit equations which are no longer hyperbolic.

As a matter of fact, the Boussinesq limiting process (3.61) eliminates any influence of the standard atmosphere (i.e., of stratification) on the atmospheric flow under consideration. This is particularly clear in (2.72e) which becomes:

$$S\frac{D}{D\bar{t}}(\vartheta + \hat{B}\bar{\Gamma}_\infty(0)\bar{z}) = 0 \quad , \tag{3.65}$$

with $\bar{\Gamma}_\infty(0) \equiv (\gamma - 1)/\gamma + (d\bar{T}_\infty/d\bar{z}_\infty)_{\bar{z}_\infty} \equiv \mathrm{Bo}_{\bar{z}=0}$, being a constant once (3.61) and (2.78) have been taken into account.

3.3.4 Isochoric Filtering (Quasi-Incompressible)

The Boussinesq limiting process (3.61) leads us, in fact, to the consideration of a "quasi-incompressible" atmosphere in that 1) the atmospheric flow is incompressible, and 2) the oscillations in density are ignored except when they are coupled with the Archimedes force.

Nevertheless, it is possible to "attain" a quasi-incompressible atmosphere by considering the following isochoric limiting process [see (3.54)]:

$$\gamma \to \infty \quad , \quad M_\infty \to 0 \quad ; \quad \gamma M_\infty^2 \equiv \hat{M}^2 = O(1) \quad , \tag{3.66}$$

where $\gamma \to \infty$ means that:

$$c_p = O(1) \quad , \quad c_v \to 0 \quad , \quad \gamma = c_p/c_v \quad .$$

It will be noted that $\hat{M}^2 \equiv U_0^2 \varrho_\infty(0)/p_\infty(0) = 0(1)$ implies that in the case of the isochoric limit (3.66), the following value for $p_\infty(0)$ can be chosen:

$$p_\infty(0) = U_0^2 \varrho_\infty(0) \quad . \tag{3.67}$$

The above relation is what generally characterizes incompressible processes. Naturally, all internal acoustic waves disappear when isochoric filtering takes place. As regards internal gravity waves, if we assume that $\varepsilon \equiv 1$ and $\mathrm{Ro} \equiv \infty$, then:

$$\sigma_{\mathrm{g}} \to \sigma_{\mathrm{g\,is}} = \pm \frac{1}{\mathrm{Fr}^2} \frac{K}{(K^2 + \lambda_\mu)^{1/2}} \quad . \tag{3.68}$$

Hence, (when the Coriolis force is ignored) isochoric internal gravity waves are directly related to the force of gravity – more precisely, to its magnitude – and these waves must satisfy the following condition:

$$\lambda_\mu \cong K^2 \quad .$$

In an adiabatic isochoric atmosphere the density is a lagrangian invariant, i.e., it is conserved along the isochoric trajectories, and therefore comes into play in the force of inertia. In addition, while carrying out the isochoric limit (3.66), we can see that:

$$\overline{N}^2_\infty(\overline{z}_\infty) \Rightarrow \; - \frac{1}{N^2_\infty(0)} \frac{g^2}{U_0^2} \frac{d \log \overline{\varrho}_\infty(\overline{z}_\infty)}{d\overline{z}_\infty} \equiv \overline{N}^2_{\infty,\mathrm{is}}(\overline{z}_\infty) \quad , \tag{3.69}$$

once (3.68) has been selected. In this case, the following equation is obtained for the isochoric perturbation ω_{is} of the density in an adiabatic atmosphere:

$$\mathrm{S}\frac{D}{D\bar{t}}\left\{\log(1 + \omega_{\mathrm{is}}) - \alpha^0_\infty \int \overline{N}^2_{\infty,\mathrm{is}}(\mathrm{Bo}\overline{z})d\overline{z}\right\} = 0 \quad . \tag{3.70}$$

We will return to (3.70) in Chap. 9 which deals with the isochoric approximation.

When the Boussinesq approximation is being expounded, the isochoric approximation is often carried out implicitly at the same time – this is particularly true within the framework of problems related to the dynamics of the oceans [see, for example, Sect. 2.4 in Phillips' book (1977)]. We will see in Chap. 8 that when the Boussinesq approximation is correctly carried out at the very start, then it is superfluous to execute the isochoric limit (3.66) since the former will have already sufficiently simplified the basic equations. On the other hand, after (3.66), for an even greater simplification, we can carry out (3.61) – which is equivalent to effecting the Boussinesq approximation but with unity mandatorily replacing $(\gamma - 1)/\gamma$.

A general review of isochoric flows can be found in the paper by Truesdell (1954). Finally, the book by Yih (1980) gives an in-depth exposition on stratified flows based mainly on the hypothesis of the isochoricity of flows.

3.3.5 Deep Convection Filtering ("Anelastic")

It ensues from all of the above that each time $\mathrm{M}_\infty \to 0$, another parameter takes on its limiting value (either $\mathrm{Bo} \to 0$, or $\mathrm{Ro} \to 0$, or even $\gamma \to \infty$). It is precisely this complementary constraint which enables us to avoid falling into the dead-end of a limit system, free of internal acoustic waves and too strongly degenerate just like the one in the classical quasi-solenoidal model which will be analyzed in Chap. 10.

However, it turns out that another case exists which makes it possible to obtain a "significant" limit system when $M_\infty \to 0$: it is the case associated with the so-called deep convection filtering (also called "anelastic" filtering in the English-speaking world). Let us suppose that Bo = O(1) and let us consider the following limiting process:

$$M_\infty \to 0 \quad , \quad \alpha_\infty^0 \to 0 \quad ; \quad \frac{\alpha_\infty^0}{\gamma M_\infty^2} \equiv \hat{\alpha} = O(1) \quad . \tag{3.71}$$

The necessity of the similarity condition which comes into play in (3.71) is related to the fact that when Bo = O(1), then instead of (2.78), we must consider for π, ω and ϑ the expansions:

$$\pi = M_\infty^2 \pi_2 + \dots \quad , \quad \omega = M_\infty^2 \omega_2 + \dots \quad , \quad \vartheta = M_\infty^2 \vartheta_2 + \dots \quad . \tag{3.72}$$

In this case, (2.75) is the least degenerate if:

$$\alpha_\infty^0 / M_\infty^2 = O(1) \quad .$$

Another approach can be taken from (3.43), more precisely from the last term of $\mathcal{G}(\omega^2; \overline{z}_\infty)$. In fact, we have:

$$\overline{\Gamma}_\infty(\overline{z}_\infty) \equiv \frac{\overline{T}_\infty(\overline{z}_\infty)}{\mathrm{Bo}} \alpha_\infty^0 \overline{N}_\infty^2(\overline{z}_\infty) \quad , \tag{3.73}$$

and consequently, the ratio $\alpha_\infty^0/\gamma M_\infty^2$ appears in the expression (3.44). Finally, this ratio reappears in (3.23) for δ which enters into (3.21) for the frequencies $\sigma_{a,g}$ of the internal waves. When Bo $\cong$ 1, the constraint $\hat{\alpha}$ = O(1) requires the characteristic frequency $N_\infty(0)$ to satisfy the relation:

$$N_\infty^2(0) \cong \frac{U_0^2}{H^2} \ll \gamma \frac{g}{H} \quad . \tag{3.74}$$

This means that the (adiabatic) deep convection equations which are obtained from the general equations (2.72) subject to the conditions (3.71, 72) will contain internal wave solutions for frequencies smaller than U_0/H. Once again, the high frequency internal acoustic waves, in particular, will be filtered out. It must also be pointed out that when $\varepsilon \equiv 1$ and Ro $\equiv \infty$, then the similarity condition $\hat{\alpha}$ = O(1) leads us to choose the following as the characteristic time:

$$t_0 = N_\infty^{-1}(0) \cong H/U_0 \quad . \tag{3.75}$$

It is likewise noteworthy that applying the limiting process Bo $\to$ 0 to the deep convection limit equations (see Chap. 10) with Bo = O(1), leads to the classical Boussinesq equations of Chap. 6. It can thus be said that the Boussinesq are "shallow" convection equations [the Soviets call them the equations of free convection – in particular, see Sect. 56 in Landau and Lifshitz (1959)].

Finally, it can be pointed out that a similar constraint to the one written in (3.74) is to be found in the work of Ogura and Phillips (1962) who employ (2.83). The limiting process (3.71) was introduced by Zeytounian (1974).

3.4 Conclusions and Bibliographical References

Quite obviously, only the very basic elements of the theory of internal waves in the atmosphere have been presented in this chapter. We will, of course, have the opportunity to return to other aspects of this theory in the coming chapters.

At the present time, there are a number of books on waves. To start with, let us cite the book by Beer (1974) where a thorough exposition on the different types of atmospheric waves can be found. Of course, there is the classic book by Lighthill (1978) on the waves in fluids (in particular, Chaps. 1–4). Many references are given in Lighthill's book on the theory of acoustic waves and the theory of waves in stratified fluids. Concerning the latter, Yih's book (1980) contains a report on small amplitude waves (Chap. 2). In Chap. 5 of Phillips' book (1977), is a very relevant discussion on internal waves. Whitham (1974) gives a very thorough overview of wave phenomena in general. Among the Russian-language books, we can first cite the book by Dikij (1969) which deals with the theory of oscillations in the earth's atmosphere, in a very general manner. Next is the book by Karpman [English edition by Pergamon Press (1975)], where we can find a theory on non-linear waves in various dispersive media which is based on the ideas developed, in particular, by Whitham. On this same subject, we can also consult the reports of the discussion organized by Lighthill (1967). One application of this theory quite naturally concerns atmospheric waves; in Chap. 6, we will see just such an application which was inspired by the relatively recent analysis of Bois (1979). The articles by Grimshaw (1972, 1975) deal more particularly with internal gravity waves. We must not overlook Eckart's book (1960) which contains a survey of a classical linear theory of wave motions in the atmosphere (and in the oceans). Finally, a lagrangian approach is to be found in Tolstoy's book (1973). Regarding this approach, it should be noted that based on (3.30), the following two equations can be written for the velocity components u' and w':

$$\bar{\varrho}_\infty S^2 \frac{\partial^2 u'}{\partial \bar{t}^2} = -\frac{\mathrm{Bo}}{\gamma M_\infty^2}\bar{\varrho}_\infty \frac{\partial w'}{\partial \bar{x}} + \frac{\partial}{\partial \bar{x}}\left\{\frac{\bar{\varrho}_\infty \bar{T}_\infty}{M_\infty^2}\left(\frac{\partial u'}{\partial \bar{x}} + \frac{\partial w'}{\partial \bar{z}}\right)\right\} \quad ;$$

$$\bar{\varrho}_\infty S^2 \varepsilon^2 \frac{\partial^2 w'}{\partial \bar{t}^2} = \frac{\mathrm{Bo}}{\gamma M_\infty^2}\bar{\varrho}_\infty \frac{\partial u'}{\partial \bar{x}} + \frac{\partial}{\partial \bar{z}}\left\{\frac{\bar{\varrho}_\infty \bar{T}_\infty}{M_\infty^2}\left(\frac{\partial u'}{\partial \bar{x}} + \frac{\partial w'}{\partial \bar{z}}\right)\right\} \quad .$$

If each of these equations is now integrated once with respect to $\bar{t}$, and if we take into account that:

$$u' = \frac{\partial \xi}{\partial \bar{t}} \quad , \quad w' = \frac{\partial \zeta}{\partial \bar{t}} \quad ,$$

where ξ and ζ are the displacement components of a fluid element (starting from a state of equilibrium) with respect to $\bar{x}$ and $\bar{z}$, respectively, we find:

$$\bar{\varrho}_\infty S^2 \frac{\partial^2 \xi}{\partial \bar{t}^2} - \frac{1}{M_\infty^2}\left\{\frac{\partial}{\partial \bar{x}}(\bar{\varrho}_\infty \bar{T}_\infty \bar{\Delta}) - \frac{\mathrm{Bo}}{\gamma}\bar{\varrho}_\infty \frac{\partial \zeta}{\partial \bar{x}}\right\} = 0 \quad ;$$

$$\bar{\varrho}_\infty S^2 \varepsilon^2 \frac{\partial^2 \zeta}{\partial \bar{t}^2} - \frac{1}{M_\infty^2}\left\{\frac{\partial}{\partial \bar{z}}(\bar{\varrho}_\infty \bar{T}_\infty \bar{\Delta}) + \frac{\mathrm{Bo}}{\gamma}\bar{\varrho}_\infty \frac{\partial \xi}{\partial \bar{x}}\right\} = 0 \quad ,$$

where

$$\overline{\Delta} = \frac{\partial \xi}{\partial \overline{x}} + \frac{\partial \zeta}{\partial \overline{z}} \quad , \quad \frac{\partial \overline{\Delta}}{\partial \overline{t}} \equiv \frac{\partial u'}{\partial \overline{x}} + \frac{\partial w'}{\partial \overline{z}} \quad .$$

It is interesting to note at this point that both of the above equations, for ξ and ζ, can be obtained from the classical Euler-Langrangian equations [see, in particular, Chap. 5 of the Zeytounian notes (1974)] once the expression of the lagrangian density (written in non-dimensional form) has been taken into account:

$$\mathcal{L} = \frac{1}{2}\overline{\varrho}_\infty \mathrm{S}^2 \left[\left(\frac{\partial \xi}{\partial \overline{t}} \right)^2 + \left(\frac{\partial \zeta}{\partial \overline{t}} \right)^2 \right] - \frac{1}{2} \frac{1}{\mathrm{M}_\infty^2} \overline{\varrho}_\infty \overline{T}_\infty \overline{\Delta}^2$$
$$+ \frac{1}{2} \frac{\mathrm{Bo}}{\gamma \mathrm{M}_\infty^2} \zeta^2 \frac{d\overline{\varrho}_\infty}{d\overline{z}_\infty} + \overline{\varrho}_\infty \frac{\mathrm{Bo}}{\gamma \mathrm{M}_\infty^2} \zeta \overline{\Delta} \quad .$$

With the above expression of $\mathcal{L}$ as a starting point, a Whitham-type (linear) theory can be developed for internal atmospheric waves.

4. Rossby Waves

As has already been mentioned, one of the basic features of atmospheric motions is the relative "smallness" of the speed of the wind compared to the speed of sound. Due to this difference, we must consider, on the one hand, slow moving waves having speeds of the same order as the wind, and on the other hand, fast internal waves with speeds similar to the speed of sound.

The fundamental properties of slow, real synoptic processes and, in particular, their evolution in time are, in fact, related to these slow waves.

In the tangent plane approximation (see Sect. 2.2), these slow waves can be taken into account if we suppose that the Coriolis parameter $l = 2\Omega_0 \sin\varphi$ is a function of the y coordinate along the meridian. More precisely (see Sect. 2.3), after derivation of the corresponding equations, it must be assumed that in the evolution equations, $l = l_0$ and $\beta = dl/dy \equiv \beta_0$ with l_0 and β_0 being constant values [see, in particular, (2.54)] – which means making a transition to the β plane.

In the model equations, such a procedure brings out the long, slow waves which are the so-called *Rossby* waves (1939). It was, in fact, Rossby who discovered the latter in the particular case of a quasi-incompressible, barotropic atmosphere. It is precisely these slow Rossby waves which are important for medium- and long-range weather forecasting; for very short-range forecasting, they are of less importance.

4.1 An Evolution Equation for Rossby Waves

Our starting point here will be the primitive equations of Subsect. 2.4.3 [see (2.91–94)]. We introduce the Kibel number:

$$\mathrm{Ki} \equiv \mathrm{S\,Ro} = \frac{1}{t_0 l_0} \quad , \tag{4.1}$$

and the parameter:

$$\lambda_0 = \frac{\mathrm{Bo}}{\gamma \mathrm{S}} \left(\frac{\mathrm{Ki}}{\mathrm{M}_\infty} \right)^2 \quad . \tag{4.2}$$

Equations (2.91–94) can thus be rewritten in the following form:

$$\mathrm{Ki}\left\{\frac{\partial U}{\partial \tau}+\frac{1}{\mathrm{S}}\left(U\frac{\partial U}{\partial \xi}+V\frac{\partial U}{\partial \eta}\right)+W\frac{\partial U}{\partial \zeta}\right\}-\left(1+\frac{\beta}{\mathrm{S}}\mathrm{Ki}\eta\right)V+\frac{\lambda_0}{\mathrm{Ki}}\frac{\partial \mathcal{H}}{\partial \xi}=0 \quad ;$$

$$\mathrm{Ki}\left\{\frac{\partial V}{\partial \tau}+\frac{1}{\mathrm{S}}\left(U\frac{\partial V}{\partial \xi}+V\frac{\partial V}{\partial \eta}\right)+W\frac{\partial V}{\partial \zeta}\right\}+\left(1+\frac{\beta}{\mathrm{S}}\mathrm{Ki}\eta\right)U+\frac{\lambda_0}{\mathrm{Ki}}\frac{\partial \mathcal{H}}{\partial \eta}=0 \quad ;$$

$$\mathrm{Bo}\zeta\frac{\partial \mathcal{H}}{\partial \zeta}+\overline{T}=0 \quad , \quad \overline{\varrho}=\frac{\zeta}{\overline{T}} \quad ;$$

$$\frac{\partial U}{\partial \xi}+\frac{\partial V}{\partial \eta}+\mathrm{S}\frac{\partial W}{\partial \zeta}=0 \quad ; \tag{4.3}$$

$$\mathrm{Ki}\left\{\frac{\partial \overline{T}}{\partial \tau}+\frac{1}{\mathrm{S}}\left(U\frac{\partial \overline{T}}{\partial \xi}+V\frac{\partial \overline{T}}{\partial \eta}\right)+W\left(\frac{\partial \overline{T}}{\partial \zeta}-\frac{\gamma-1}{\gamma}\frac{\overline{T}}{\zeta}\right)\right\}=0 \quad .$$

For the remainder of this section, the following situation holds:

$$\mathrm{Ki} \ll 1 \quad , \quad \mathrm{M}_\infty \ll 1 \quad ; \quad \lambda_0 = \mathrm{O}(1) \quad ; \tag{4.4}$$

and we will assume that $\mathrm{S} = \mathrm{O}(1)$ and $\mathrm{Bo} = \mathrm{O}(1)$.

The variables τ, ξ, η and ζ being fixed when $\mathrm{Ki} \to 0$, let us develop the solution of the reduced equations (4.3) for U, V, W, $\mathcal{H}$ and $\overline{\varrho}$, $\overline{T}$ in the form of the following (outer) asymptotic expansions:

$$\begin{aligned}
U &= u_0 + \mathrm{Ki}u_1 + \dots \quad ;\\
V &= v_0 + \mathrm{Ki}v_1 + \dots \quad ;\\
W &= w_0 + \mathrm{Ki}w_1 + \dots \quad ;\\
\mathcal{H} &= \mathcal{H}_0 + \mathrm{Ki}\mathcal{H}_1 + \mathrm{Ki}^2\mathcal{H}_2 + \dots \quad ;\\
\overline{T} &= T_0 + \mathrm{Ki}T_1 + \mathrm{Ki}^2T_2 + \dots \quad ;\\
\overline{\varrho} &= \varrho_0 + \mathrm{Ki}\varrho_1 + \mathrm{Ki}^2\varrho_2 + \dots \quad .
\end{aligned} \tag{4.5}$$

First of all, it is noticed that:

$$\frac{\partial \mathcal{H}_0}{\partial \xi}=\frac{\partial \mathcal{H}_0}{\partial \eta}=0 \quad , \quad T_0=-\mathrm{Bo}\zeta\frac{d\mathcal{H}_0}{d\zeta}$$

and therefore, it would appear reasonable to take:

$$T_0 \equiv T_0(\zeta) \quad , \quad \mathcal{H}_0 \equiv \mathcal{H}_0(\zeta) \Rightarrow \varrho_0=\frac{\zeta}{T_0(\zeta)} \quad . \tag{4.6}$$

If (4.5, 6) are taken into account, then it follows from the last equation in (4.3) that:

$$w_0\left(\frac{dT_0}{d\zeta}-\frac{\gamma-1}{\gamma}\frac{T_0}{\zeta}\right)=0 \quad . \tag{4.7}$$

In what follows, we will assume that:

$$\frac{dT_0}{d\zeta} \not\equiv \frac{\gamma-1}{\gamma}\frac{T_0}{\zeta} \equiv \frac{1}{\varrho_0(\zeta)}\frac{\gamma-1}{\gamma} \quad , \tag{4.8}$$

and in this case, we have:

$$w_0 \equiv 0 \quad . \tag{4.9}$$

Hence, when $\mathrm{Ki} \to 0$, it follows that:

$$\frac{\partial u_0}{\partial \xi} + \frac{\partial v_0}{\partial \eta} = 0 \quad , \tag{4.10}$$

which is in perfect agreement with the so-called geostrophic relations:

$$v_0 = \lambda_0 \frac{\partial \mathcal{H}_1}{\partial \xi} \quad , \quad u_0 = -\lambda_0 \frac{\partial \mathcal{H}_1}{\partial \eta} \quad , \tag{4.11}$$

since $\lambda_0 = \text{const}$. We thus recover, at order zero, the quasi-geostrophic balance which is, in fact, a strong degeneracy of system (4.3) in the sense that the function $\mathcal{H}_1(\tau, \xi\, \eta, \zeta)$ remains undetermined. In order to obtain the equation governing the evolution of $\mathcal{H}_1$ as a function of τ, ξ, η and ζ, we must resort to the first order equations in Ki. The latter can be easily obtained from (4.3) and can be written in the following form [it will be remarked that $\beta = O(1)$; see (2.55)]:

$$\begin{aligned}
&\frac{\partial u_0}{\partial \tau} + \frac{1}{\mathrm{S}}\left(u_0 \frac{\partial u_0}{\partial \xi} + v_0 \frac{\partial u_0}{\partial \eta}\right) - \frac{\beta}{\mathrm{S}} \eta v_0 = v_1 - \lambda_0 \frac{\partial \mathcal{H}_2}{\partial \xi} \quad ; \\
&\frac{\partial v_0}{\partial \tau} + \frac{1}{\mathrm{S}}\left(u_0 \frac{\partial v_0}{\partial \xi} + v_0 \frac{\partial v_0}{\partial \eta}\right) + \frac{\beta}{\mathrm{S}} \eta u_0 = -u_1 - \lambda_0 \frac{\partial \mathcal{H}_2}{\partial \eta} \quad ; \\
&\frac{\partial u_1}{\partial \xi} + \frac{\partial v_1}{\partial \eta} + \mathrm{S}\frac{\partial w_1}{\partial \zeta} = 0 \quad ; \quad T_1 = -\mathrm{Bo}\, \zeta \frac{\partial \mathcal{H}_1}{\partial \zeta} \quad ; \\
&\frac{\partial T_1}{\partial \tau} + \frac{1}{\mathrm{S}}\left(u_0 \frac{\partial T_1}{\partial \xi} + v_0 \frac{\partial T_1}{\partial \eta}\right) - \mathrm{Bo}\frac{K_0(\zeta)}{\zeta} w_1 = 0 \quad ,
\end{aligned} \tag{4.12}$$

where

$$\begin{aligned}
K_0(\zeta) &= -\frac{\zeta}{\mathrm{Bo}}\left(\frac{dT_0}{d\zeta} - \frac{\gamma - 1}{\gamma}\frac{T_0}{\zeta}\right) \\
&= \zeta\left\{\frac{d}{d\zeta}\left(\zeta \frac{d\mathcal{H}_0}{d\zeta}\right) - \frac{\gamma - 1}{\gamma}\frac{d\mathcal{H}_0}{d\zeta}\right\}
\end{aligned}$$

is assumed positive [see (2.98)].

By taking advantage of (4.11), from (4.12) we can express the first order velocity components in the following form:

$$\begin{aligned}
u_1 &= -\lambda_0 \left\{\frac{\partial \mathcal{H}_2}{\partial \eta} + \frac{D_0}{D\tau}\left(\frac{\partial \mathcal{H}_1}{\partial \xi}\right) - \frac{\beta}{\mathrm{S}} \eta \frac{\partial \mathcal{H}_1}{\partial \eta}\right\} \quad ; \\
v_1 &= -\lambda_0 \left\{-\frac{\partial \mathcal{H}_2}{\partial \xi} + \frac{D_0}{D\tau}\left(\frac{\partial \mathcal{H}_1}{\partial \eta}\right) + \frac{\beta}{\mathrm{S}} \eta \frac{\partial \mathcal{H}_1}{\partial \xi}\right\} \quad ; \\
w_1 &= -\frac{\zeta^2}{K_0(\zeta)} \frac{D_0}{D\tau}\left(\frac{\partial \mathcal{H}_1}{\partial \zeta}\right) \quad ,
\end{aligned} \tag{4.13}$$

after having introduced the convective derivative in the "quasi-geostrophic" flow:

$$\frac{D_0}{D\tau} = \frac{\partial}{\partial \tau} + \frac{1}{S}\left(u_0 \frac{\partial}{\partial \xi} + v_0 \frac{\partial}{\partial \eta}\right)$$

$$= \frac{\partial}{\partial \tau} + \frac{\lambda_0}{\mathrm{S}}\left(\frac{\partial \mathcal{H}_1}{\partial \xi}\frac{\partial}{\partial \eta} - \frac{\partial \mathcal{H}_1}{\partial \eta}\frac{\partial}{\partial \xi}\right)$$

$$= \frac{\partial}{\partial \tau} + \frac{\lambda_0}{\mathrm{S}}\frac{D(\mathcal{H}_1;\, \cdot)}{D(\xi, \eta)} \quad . \tag{4.14}$$

If we now introduce the expressions for u_1, v_1 and w_1 from (4.13) into the continuity equation of the first order system (4.12), the following equation in $\mathcal{H}_1$ is obtained:

$$\boxed{\frac{D_0}{D\tau}(\wedge\, \mathcal{H}_1) = -\frac{\lambda_0 \beta}{\mathrm{S}}\frac{\partial \mathcal{H}_1}{\partial \xi}} \quad , \tag{4.15}$$

once the following identity has been taken into account:

$$\frac{\partial}{\partial \xi}\left[\frac{D_0}{D\tau}\left(\frac{\partial \mathcal{H}_1}{\partial \xi}\right)\right] + \frac{\partial}{\partial \eta}\left[\frac{D_0}{D\tau}\left(\frac{\partial \mathcal{H}_1}{\partial \eta}\right)\right] \equiv \frac{D_0}{D\tau}\left(\frac{\partial^2 \mathcal{H}_1}{\partial \xi^2} + \frac{\partial^2 \mathcal{H}_1}{\partial \eta^2}\right)$$

and the "quasi-geostrophic Laplacian" has been introduced:

$$\wedge = \lambda_0\left(\frac{\partial^2}{\partial \xi^2} + \frac{\partial^2}{\partial \eta^2}\right) + \mathrm{S}\frac{\partial}{\partial \zeta}\left(\frac{\zeta^2}{K_0(\zeta)}\frac{\partial}{\partial \zeta}\right) \quad . \tag{4.16}$$

We thus return to the classical equation which governs quasi-geostrophic flows. This equation (4.15) shows that the expression $\wedge\, \mathcal{H}_1 + (\beta/\lambda_0)\eta$ is preserved along the quasi-geostrophic trajectories: it is thus a lagrangian invariant for quasi-geostrophic flows. This property is not merely coincidental, for it can be demonstrated, in fact, that (4.15) can be directly obtained from the Ertel invariant (2.24).

Equation (4.15) is first order in τ and therefore, the quasi-geostrophic approximation filters out all the internal gravity waves which were still present in the primitive equations (4.3). Thus the analysis carried out at the beginning of Subsect. 3.3.2 is confirmed. In order to determine the first order perturbations of the function $\mathcal{H}$, $\mathcal{H}_1$, we need only know the initial value of $\mathcal{H}_1$ for $\tau = 0$. This value $\mathcal{H}_1^0\ (\xi, \eta, \zeta)$ can be obtained from the problem describing the adjustment to geostrophy which will be analyzed in Sect. 6.1 and in Chap. 11.

Furthermore, the slip condition for (4.15) must be formulated. According to (2.95), the exact slip condition is the following:

$$\mathrm{S}\frac{\partial \mathcal{H}}{\partial \tau} + U\frac{\partial \mathcal{H}}{\partial \xi} + V\frac{\partial \mathcal{H}}{\partial \eta} + \mathrm{S}W\frac{\partial \mathcal{H}}{\partial \zeta} = 0 \quad , \quad \text{on} \quad \mathcal{H}(\tau, \xi, \eta, \zeta) = 0 \quad .$$

But according to (4.5), $\mathcal{H} = 0$ means that:

$$\mathcal{H}_0(\zeta) + \mathrm{Ki}\mathcal{H}_1 + \mathrm{Ki}^2\mathcal{H}_2 + \ldots = 0$$

or else

$$\zeta = \zeta_s(\tau, \xi, \eta; \mathrm{Ki}) = \zeta_{s0} + \mathrm{Ki}\zeta_{s1} + \ldots \quad . \tag{4.17}$$

The first term ζ_{s0} = const of the expansion (4.17) of $\zeta_s(\tau, \xi, \eta; \mathrm{Ki})$ can be taken as equal to unity if it is recognized that in the reduced basic equations, the variable ζ was made dimensionless via ζ_{s0}. In this case, $\zeta = 1$ is a solution of $\mathcal{H}_0(\zeta) = 0$ and

$$\zeta_{s1}(\tau, \xi, \eta) = -\left(\frac{\mathcal{H}_1}{d\mathcal{H}_0/d\zeta}\right)_{\zeta=1} . \tag{4.18}$$

Hence, the slip condition can be written as follows:[1]

$$\mathrm{S}\frac{\partial\mathcal{H}_1}{\partial\tau} + u_0\frac{\partial\mathcal{H}_1}{\partial\xi} + v_0\frac{\partial\mathcal{H}_1}{\partial\eta} + \mathrm{S}w_1\frac{d\mathcal{H}_0}{d\zeta} = 0 \quad , \quad \text{on} \quad \zeta = 1 \quad .$$

However, thanks to (4.11), $u_0(\partial\mathcal{H}_1/\partial\xi) + v_0(\partial\mathcal{H}_1/\partial\eta) \equiv 0$, and if the value of w_1 as a function of $\mathcal{H}_1$ is taken into account, according to (4.13) we obtain the following boundary condition (on the ground) for $\mathcal{H}_1$ which must be imposed on (4.15):

$$\frac{\partial\mathcal{H}_1}{\partial\tau} + \frac{1}{\mathrm{Bo}}\frac{T_0(1)}{K_0(1)}\frac{D_0}{D\tau}\left(\frac{\partial\mathcal{H}_1}{\partial\zeta}\right) = 0 \quad , \quad \text{on} \quad \zeta = 1 \quad . \tag{4.19}$$

Of course, a second boundary condition in ζ for $\zeta \to 0$ must be added to this same (4.15) – we will come back to this problem in Chap. 7. For the time being, it should be noted that in accordance with the type of operator $\wedge$, from (4.16), the expression:

$$\frac{\zeta^2}{K_0(\zeta)}\left(\frac{\partial\mathcal{H}_1}{\partial\zeta}\right)^2 + \frac{\lambda_0}{\mathrm{S}}\left[\left(\frac{\partial\mathcal{H}_1}{\partial\xi}\right)^2 + \left(\frac{\partial\mathcal{H}_1}{\partial\eta}\right)^2\right] \quad ,$$

can be required to decrease sufficiently rapidly when $\zeta \to 0$ and $\xi^2 + \eta^2 \to +\infty$.

4.2 Rossby Waves in Linear Theory

Let us consider, once again, small motions of the adiabatic atmosphere with respect to a reference situation where the velocity is zero (the state of rest). After linearization, we obtain the following equation derived from (4.15) for the perturbation $\mathcal{H}_1'$:

$$\frac{\mathrm{S}}{\lambda_0}\frac{\partial}{\partial\tau}\left\{\lambda_0\left(\frac{\partial^2\mathcal{H}_1'}{\partial\xi^2} + \frac{\partial^2\mathcal{H}_1'}{\partial\eta^2}\right) + \mathrm{S}\frac{\partial}{\partial\zeta}\left(\frac{\zeta^2}{K_0(\zeta)}\frac{\partial\mathcal{H}_1'}{\partial\zeta}\right)\right\} + \beta\frac{\partial\mathcal{H}_1'}{\partial\xi} = 0 \, . \tag{4.20}$$

The linearized boundary condition on the ground associated with the above linear equation (4.20), is obtained from (4.19) and can take the form:

[1] It will be seen that at the zeroth order, we have: $w_0 = 0$, on $\zeta = 1$ which is in good agreement with (4.9). We want to specify that the slip condition written on $\zeta = 1$ for $\mathcal{H}_1$ results from a double linearization with respect to Ki which takes into account (4.5) and the fact that $\zeta = 1 + \mathrm{Ki}\zeta_{s1} + \ldots$.

$$\left(1+\frac{\mathrm{Bo}T_0(1)}{K_0(1)}\frac{\partial}{\partial\zeta}\right)\frac{\partial\mathcal{H}_1'}{\partial\tau}=0 \quad , \quad \text{on} \quad \zeta=1 \quad . \tag{4.21}$$

The solution to (4.21) can be sought in the form:

$$\mathcal{H}_1'(\tau,\xi,\eta,\zeta)=\Re\left\{\phi_1(\zeta)\exp\left[\mathrm{i}\left(m\xi+n\eta-\frac{\sigma}{\mathrm{S}}\tau\right)\right]\right\} \quad . \tag{4.22}$$

In this case, if μ is an eigenvalue associated with the vertical structure such that:

$$\frac{d}{d\zeta}\left(\frac{\zeta^2}{K_0(\zeta)}\frac{d\phi_1}{d\zeta}\right)+\mu\phi_1=0 \quad , \tag{4.23}$$

then the frequency of the Rossby waves σ_{R}, satisfies the classical relation:

$$\sigma_{\mathrm{R}}=\frac{-\beta m}{\lambda_0(m^2+n^2)+\mathrm{S}\mu}. \tag{4.24}$$

The group velocity for these Rossby waves in the direction of ξ (i.e., from west to east), is therefore:

$$C_{\mathrm{g}\xi}^{\mathrm{R}}=\frac{\beta[\lambda_0 m^2-(\lambda_0 n^2+\mathrm{S}\mu)]}{[\lambda_0(m^2+n^2)+\mathrm{S}\mu]^2} \quad . \tag{4.25}$$

In order to determine the function $\phi_1(\zeta)$, as well as μ, two boundary conditions must be imposed on (4.23): one in $\zeta=1$ and one for $\zeta\to 0$. This means that a Sturm-Liouville eigenvalue problem must be solved. In Chap. 7, we will have the opportunity of analyzing this eigenvalue problem in detail and we will specify, in particular, the behavior condition to be imposed for $\zeta\to 0$.

To be sure, the solutions of the form (4.22) are not very realistic as they do not take into consideration the variability in amplitude of the Rossby waves with τ, ξ and η. As a general rule, this amplitude varies slowly with τ, ξ and η and thus we can successfully "apply" the so-called multiple-scale method. To this end, let us return to the linear equation (4.20) and study its solution in the more general form:

$$\mathcal{H}_1'(\tau,\xi,\eta,\zeta)=\phi_1(\zeta)\mathcal{P}(\tau,\xi,\eta). \tag{4.26}$$

By taking into account (4.23), the following wave equation is obtained for determining the function $\mathcal{P}(\tau,\xi,\eta)$:

$$\mathrm{S}\frac{\partial}{\partial\tau}\left[\frac{\partial^2\mathcal{P}}{\partial\xi^2}+\frac{\partial^2\mathcal{P}}{\partial\eta^2}-\frac{\mu\mathrm{S}}{\lambda_0}\mathcal{P}\right]+\frac{\beta}{\lambda_0}\frac{\partial\mathcal{P}}{\partial\xi}=0 \quad . \tag{4.27}$$

Let now $\Delta\ll 1$ be a small parameter which measures the "smallness" of the time and space variations (in the ξ,η plane) of the Rossby waves.

In the multiple scale method, two systems of variables are considered: one for the so-called fast variables

$$\tilde{\tau}\equiv\tau \quad , \quad \tilde{\xi}\equiv\xi \quad , \quad \tilde{\eta}\equiv\eta \quad , \tag{4.28}$$

and the so-called slow variables

$$T = \Delta\tau \quad , \quad X = \Delta\xi \quad , \quad Y = \Delta\eta \quad . \tag{4.29}$$

Then, it is supposed that the unknown function $\mathcal{P}$ is explicitly a function of both systems of variables,

$$\mathcal{P} = \overline{\mathcal{P}}(\tilde{\tau}, \tilde{\xi}, \tilde{\eta}; X, Y, T; \Delta) \quad , \tag{4.30}$$

and this naturally yields:

$$\frac{\partial \mathcal{P}}{\partial \tau} = \frac{\partial \overline{\mathcal{P}}}{\partial \tilde{\tau}} + \Delta \frac{\partial \overline{\mathcal{P}}}{\partial T} \quad ; \quad \frac{\partial \mathcal{P}}{\partial \xi} = \frac{\partial \overline{\mathcal{P}}}{\partial \tilde{\xi}} + \Delta \frac{\partial \overline{\mathcal{P}}}{\partial X} \quad ; \tag{4.31a}$$

$$\frac{\partial \mathcal{P}}{\partial \eta} = \frac{\partial \overline{\mathcal{P}}}{\partial \tilde{\eta}} + \Delta \frac{\partial \overline{\mathcal{P}}}{\partial Y} \quad . \tag{4.31b}$$

Therefore, the following equation replaces (4.27) for the function $\overline{\mathcal{P}}$:

$$\begin{aligned}
&\mathrm{S}\frac{\partial}{\partial \tilde{\tau}}\left[\frac{\partial^2 \overline{\mathcal{P}}}{\partial \tilde{\xi}^2} + \frac{\partial^2 \overline{\mathcal{P}}}{\partial \tilde{\eta}^2} - \frac{\mu \mathrm{S}}{\lambda_0}\overline{\mathcal{P}}\right] + \frac{\beta}{\lambda_0}\frac{\partial \overline{\mathcal{P}}}{\partial \tilde{\xi}} \\
&= -\Delta\left\{\mathrm{S}\frac{\partial}{\partial T}\left[\frac{\partial^2 \overline{\mathcal{P}}}{\partial \tilde{\xi}^2} + \frac{\partial^2 \overline{\mathcal{P}}}{\partial \tilde{\eta}^2} - \frac{\mu \mathrm{S}}{\lambda_0}\overline{\mathcal{P}}\right]\right. \\
&\quad \left. + 2\mathrm{S}\left[\frac{\partial}{\partial X}\left(\frac{\partial^2 \overline{\mathcal{P}}}{\partial \tilde{\xi}\partial \tilde{\tau}}\right) + \frac{\partial}{\partial Y}\left(\frac{\partial^2 \overline{\mathcal{P}}}{\partial \tilde{\eta}\partial \tilde{\tau}}\right)\right] + \frac{\beta}{\lambda_0}\frac{\partial \overline{\mathcal{P}}}{\partial X}\right\} \\
&\quad - \mathrm{S}\Delta^2\left\{\frac{\partial}{\partial T}\left(2\frac{\partial^2 \overline{\mathcal{P}}}{\partial X \partial \tilde{\xi}} + 2\frac{\partial^2 \overline{\mathcal{P}}}{\partial Y \partial \tilde{\eta}}\right) + \frac{\partial}{\partial \tilde{\tau}}\left(\frac{\partial^2 \overline{\mathcal{P}}}{\partial X^2} + \frac{\partial^2 \overline{\mathcal{P}}}{\partial Y^2}\right)\right\} \\
&\quad - \mathrm{S}\Delta^3\left\{\frac{\partial}{\partial T}\left(\frac{\partial^2 \overline{\mathcal{P}}}{\partial X^2} + \frac{\partial^2 \overline{\mathcal{P}}}{\partial Y^2}\right)\right\} \quad .
\end{aligned} \tag{4.32}$$

Since $\Delta \ll 1$, the solution of (4.32) is sought in the form of a uniformly valid expansion:

$$\overline{\mathcal{P}} = \overline{\mathcal{P}}_0 + \Delta \overline{\mathcal{P}}_1 + \Delta^2 \overline{\mathcal{P}}_2 + \ldots \quad . \tag{4.33}$$

At the order zero, we have

$$\mathrm{S}\frac{\partial}{\partial \tilde{\tau}}\left[\frac{\partial^2 \overline{\mathcal{P}}_0}{\partial \tilde{\xi}^2} + \frac{\partial^2 \overline{\mathcal{P}}_0}{\partial \tilde{\eta}^2} - \frac{\mu \mathrm{S}}{\lambda_0}\overline{\mathcal{P}}_0\right] + \frac{\beta}{\lambda_0}\frac{\partial \overline{\mathcal{P}}_0}{\partial \tilde{\xi}} = 0 \quad ,$$

and it is obvious that by way of analogy with (4.22), a solution can be postulated in the form:

$$\overline{\mathcal{P}}_0 = \Re\left\{\mathcal{A}_0(X, Y, T)\exp\left[\mathrm{i}\left(k\tilde{\xi} + l\tilde{\eta} - \frac{\sigma}{\mathrm{S}}\tilde{\tau}\right)\right]\right\} \quad , \tag{4.34}$$

where the amplitude $\mathcal{A}_0$ is an unknown among the slow variables. Of course:

$$\sigma(k^2 + l^2) + \frac{\beta}{\lambda_0}k + \frac{\mu \mathrm{S}}{\lambda_0}\sigma = 0 \quad ; \quad \sigma \equiv \sigma_{\mathrm{R}} \quad . \tag{4.35}$$

After taking advantage of (4.34), the following equation results to the order Δ:

$$\mathrm{S}\frac{\partial}{\partial\tilde{\tau}}\left[\frac{\partial^2\overline{\mathcal{P}}_1}{\partial\tilde{\xi}^2}+\frac{\partial^2\overline{\mathcal{P}}_1}{\partial\tilde{\eta}^2}-\frac{\mu\mathrm{S}}{\lambda_0}\overline{\mathcal{P}}_1\right]+\frac{\beta}{\lambda_0}\frac{\partial\overline{\mathcal{P}}_1}{\partial\tilde{\xi}}$$

$$=-\left\{\mathrm{S}\frac{\partial}{\partial T}\left(\frac{\partial^2\overline{\mathcal{P}}_0}{\partial\tilde{\xi}^2}+\frac{\partial^2\overline{\mathcal{P}}_0}{\partial\tilde{\eta}^2}-\frac{\mu\mathrm{S}}{\lambda_0}\overline{\mathcal{P}}_0\right)\right.$$

$$\left.+2\mathrm{S}\frac{\partial}{\partial X}\left(\frac{\partial^2\overline{\mathcal{P}}_0}{\partial\tilde{\xi}\partial\tilde{\tau}}\right)+2\mathrm{S}\frac{\partial}{\partial Y}\left(\frac{\partial^2\overline{\mathcal{P}}_0}{\partial\tilde{\eta}\partial\tilde{\tau}}\right)+\frac{\beta}{\lambda_0}\frac{\partial\overline{\mathcal{P}}_0}{\partial X}\right\}$$

$$=\left(k^2+l^2+\frac{\mu\mathrm{S}}{\lambda_0}\right)\left\{\mathrm{S}\frac{\partial\mathcal{A}_0}{\partial T}-\frac{2\sigma k+\beta/\lambda_0}{k^2+l^2+\mu\mathrm{S}/\lambda_0}\frac{\partial\mathcal{A}_0}{\partial X}\right.$$

$$\left.-\frac{2\sigma l+\beta/\lambda_0}{k^2+l^2+\mu\mathrm{S}/\lambda_0}\frac{\partial\mathcal{A}_0}{\partial Y}\right\}\exp\left[\mathrm{i}\left(k\tilde{\xi}+l\tilde{\eta}-\frac{\sigma}{\mathrm{S}}\tilde{\tau}\right)\right] \quad . \tag{4.36}$$

We now must reason as follows which is conceptually the basis of the multiple scale method: as we want to obtain an asymptotic approximation which is uniformly valid for all the values of τ, ξ and η, i.e., of $\tilde{\tau}$, $\tilde{\xi}$ and $\tilde{\eta}$, and, in particular for the large values of these fast variables, we must cancel the term in the right-hand side of (4.36) proportional to $\exp[\mathrm{i}(k\tilde{\xi}+l\tilde{\eta}-\sigma\tilde{\tau}/\mathrm{S})]$, which is referred to as a "secular" term (as in celestial mechanics). We are thus led to the following:

$$\mathrm{S}\frac{\partial\mathcal{A}_0}{\partial T}-\frac{2\sigma k+\beta/\lambda_0}{k^2+l^2+\mu\mathrm{S}/\lambda_0}\frac{\partial\mathcal{A}_0}{\partial X}-\frac{2\sigma l+\beta/\lambda_0}{k^2+l^2+\mu\mathrm{S}/\lambda_0}\frac{\partial\mathcal{A}_0}{\partial Y}=0 \quad . \tag{4.37}$$

It should be emphasized that a common characteristic of the various applications of the multiple scale method is that at each level in the hierarchy of approximations, indeterminates are encountered which can be removed by internal coherence conditions when the approximation is extended. We feel that the above constitutes a sufficient basis for understanding the forthcoming applications. For further details, the reader can, however, consult Nayfeh's book (1973), as well as the Germain Conferences (given in 1973 at the Houches Theoretical Physics Summer School; these conference notes were edited by Balian and Peube in 1977).

Let us return to (4.37) for the "slow variable" amplitude $\mathcal{A}_0(X, Y, T)$ and introduce the group velocity vector $\boldsymbol{C}_{\mathrm{g}}^{\mathrm{R}}$ in the horizontal plane. The component of $\boldsymbol{C}_{\mathrm{g}}^{\mathrm{R}}$ can be written respectively when taking into account (4.35).

$$\begin{aligned} C_{\mathrm{gX}}^{\mathrm{R}} &= -\frac{2k\sigma\lambda_0+\beta}{(k^2+\lambda^2)\lambda_0+\mu\mathrm{S}}=\beta\frac{k^2-l^2-\mu\mathrm{S}/\lambda_0}{(k^2+l^2+\mu\mathrm{S}/\lambda_0)^2} \quad ; \\ C_{\mathrm{gY}}^{\mathrm{R}} &= -\frac{2l\sigma\lambda_0+\beta}{(k^2+l^2)\lambda_0+\mu\mathrm{S}}=2\frac{\beta kl}{(k^2+l^2+\mu\mathrm{S}/\lambda_0)^2} \quad . \end{aligned} \tag{4.38}$$

Equation (4.37) can thus be replaced by:

$$\mathrm{S}\frac{\partial\mathcal{A}_0}{\partial T}+\boldsymbol{C}_{\mathrm{g}}^{\mathrm{R}}\cdot\boldsymbol{D}\mathcal{A}_0=0 \quad , \tag{4.39}$$

where

$$\boldsymbol{D}=\frac{\partial}{\partial X}\boldsymbol{i}+\frac{\partial}{\partial Y}\boldsymbol{j} \quad ; \quad \boldsymbol{C}_g^R=C_{\mathrm{gX}}^{\mathrm{R}}\boldsymbol{i}+C_{\mathrm{gY}}^{\mathrm{R}}\boldsymbol{j} \quad .$$

Therefore,

$$\mathcal{A}_0 = \mathcal{A}_0\left(\boldsymbol{r} - \frac{T}{\mathrm{S}}\boldsymbol{C}_{\mathrm{g}}^{\mathrm{R}}\right) \quad , \tag{4.40}$$

where $\boldsymbol{r}$ is the position vector in the X, Y horizontal plane. It will be noticed that from (4.35) and (4.38), it follows that:

$$C_{\mathrm{gX}}^{\mathrm{R}} = \frac{\partial\sigma}{\partial k} \quad , \quad C_{\mathrm{gY}}^{\mathrm{R}} = \frac{\partial\sigma}{\partial l} \quad ; \quad \boldsymbol{C}_{\mathrm{g}}^{\mathrm{R}} = \frac{\partial\sigma_{\mathrm{R}}}{\partial k}\boldsymbol{i} + \frac{\partial\sigma_{\mathrm{R}}}{\partial l}\boldsymbol{j} \quad . \tag{4.41}$$

Moreover, let us define:

$$\boldsymbol{K} = k\boldsymbol{i} + l\boldsymbol{j} \quad , \quad |\boldsymbol{K}| = \sqrt{k^2 + l^2} \equiv K \quad ; \tag{4.42}$$

$\boldsymbol{K}$ is the wave vector and, at a given moment, is perpendicular to the lines of constant wave phase. If Ξ designates the phase:

$$\Xi = k\xi + l\eta - \frac{\sigma}{\mathrm{S}}\tau \quad ; \tag{4.43}$$

then we have:

$$\sigma = -\mathrm{S}\frac{\partial\Xi}{\partial\tau} \quad .$$

In addition, the so-called phase velocity, i.e., the propagation velocity of the phase Ξ in the direction of the wave vector $\boldsymbol{K}$, is then:

$$\begin{aligned} C_{\mathrm{ph}}^{\mathrm{R}} &= -\mathrm{S}\frac{\partial\Xi}{\partial\tau} \Big/ \left|\frac{\partial\Xi}{\partial\xi}\boldsymbol{i} + \frac{\partial\Xi}{\partial\eta}\boldsymbol{j}\right| \\ &= \sigma_{\mathrm{R}}/|\boldsymbol{K}| \\ &= \frac{\sigma_{\mathrm{R}}}{(k^2 + l^2)^{1/2}} \quad . \end{aligned} \tag{4.44}$$

Hence:

$$\sigma_{\mathrm{R}} = C_{\mathrm{ph}}^{\mathrm{R}} K$$

and consequently:

$$\begin{aligned} \boldsymbol{C}_{g}^{\mathrm{R}} &= \frac{\boldsymbol{K}}{K}C_{\mathrm{ph}}^{\mathrm{R}} + K\left(\frac{\partial C_{\mathrm{ph}}^{\mathrm{R}}}{\partial k}\boldsymbol{i} + \frac{\partial C_{\mathrm{ph}}^{\mathrm{R}}}{\partial l}\boldsymbol{j}\right) \\ &= \boldsymbol{C}_{\mathrm{ph}}^{\mathrm{R}} + K\left(\boldsymbol{D}_K C_{\mathrm{ph}}^{\mathrm{R}}\right) \quad , \end{aligned} \tag{4.45}$$

where

$$\boldsymbol{D}_K \equiv \frac{\partial}{\partial k}\boldsymbol{i} + \frac{\partial}{\partial l}\boldsymbol{j} \quad .$$

Therefore:

$$\boldsymbol{C}_{\mathrm{g}}^{\mathrm{R}} \not\equiv \boldsymbol{C}_{\mathrm{ph}}^{\mathrm{R}} \quad , \tag{4.46}$$

which means that the Rossby waves are *dispersive*.

We again specify that the hypothesis $\lambda_0 = O(1)$ brings about the constraint:

$$\frac{gH}{U_0 L t_0 l_0^2} \cong 1 \quad .$$

However, it is also necessary that: $S = L/t_0 U_0 \cong 1$ and thus we obtain:

$$L \cong \frac{(gH)^{1/2}}{l_0} \approx 2 \cdot 10^6\,\mathrm{m} \quad . \tag{4.47}$$

4.3 Rossby Waves in a So-called Barotropic Atmosphere

Let us backtrack to the fundamental equation (4.15) obtained in Sect. 4.1. It can be rewritten in the following form:

$$\frac{D_0}{D\tau}\left[\lambda_0\left(\frac{\partial^2 \mathcal{H}_1}{\partial \xi^2} + \frac{\partial^2 \mathcal{H}_1}{\partial \eta^2}\right) + \frac{\beta}{\lambda_0}\eta\right] + S\frac{\partial}{\partial \zeta}\left[\frac{D_0}{D\tau}\left(\frac{\zeta^2}{K_0(\zeta)}\frac{\partial \mathcal{H}_1}{\partial \zeta}\right)\right] = 0\,, \tag{4.48}$$

since the differential operators:

$$\frac{\partial}{\partial \tau} \quad \text{and} \quad \frac{D_0}{D\tau} = \frac{\partial}{\partial \tau} + \frac{\lambda_0}{S}\left(\frac{\partial \mathcal{H}_1}{\partial \xi}\frac{\partial}{\partial \eta} - \frac{\partial \mathcal{H}_1}{\partial \eta}\frac{\partial}{\partial \xi}\right)$$

can be inverted. If (4.48) is integrated through the entire thickness of the atmosphere from $\zeta = 1$ to $\zeta = 0$, then it follows that:

$$\int_0^1 \frac{D_0}{D\tau}\left(\lambda_0 \Delta_{\xi\eta}\mathcal{H}_1 + \frac{\beta}{\lambda_0}\eta\right)d\zeta + S\frac{D_0}{D\tau}\left(\frac{\zeta^2}{K_0(\zeta)}\frac{\partial \mathcal{H}_1}{\partial \zeta}\right)\bigg|_0^1 = 0 \quad ,$$

where $\Delta_{\xi\eta} = (\partial^2/\partial\xi^2) + (\partial^2/\partial\eta^2)$. When $\zeta \to 0$, we impose the following at the upper limit of the atmosphere:

$$\lim_{\zeta \to 0} \frac{D_0}{D\tau}\left(\frac{\zeta^2}{K_0(\zeta)}\frac{\partial \mathcal{H}_1}{\partial \zeta}\right) = 0 \quad , \tag{4.49}$$

which also gives us a plausible condition for (4.15) when $\zeta \to 0$. Although such a condition is certainly necessary, it may be insufficient in order for the boundary value problem associated with (4.15) to be well-posed.

In this case, if (4.19) is put to good use, we can write:

$$\int_0^1 \frac{D_0}{D\tau}\left(\lambda_0 \Delta_{\xi\eta}\mathcal{H}_1 + \frac{\beta}{\lambda_0}\eta\right)d\zeta = \frac{S\mathrm{Bo}}{T_0(1)}\frac{\partial \mathcal{H}_1}{\partial \tau}\bigg|_{\zeta=1} \quad . \tag{4.50}$$

In a barotropic atmosphere, the equation corresponding to (4.50) can be obtained by supposing that $\mathcal{H}_1 \equiv \mathcal{H}_1^*(\tau, \xi, \eta)$. Thus:

$$\frac{D_0^*}{D\tau}\left(\lambda_0 \Delta_{\xi\eta}\mathcal{H}_1^* + \frac{\beta}{\lambda_0}\eta\right) - \frac{S\mathrm{Bo}}{T_0(1)}\frac{\partial \mathcal{H}_1^*}{\partial \tau} = 0$$

or even:

$$\frac{D_0^*}{D\tau}\left\{\lambda_0^2\Delta_{\xi\eta}\mathcal{H}_1^* + \beta\eta - \frac{\mathrm{SBo}}{T_0(1)}\lambda_0\mathcal{H}_1^*\right\} = 0 \quad , \tag{4.51}$$

since

$$\frac{D_0^*}{D\tau}\mathcal{H}_1^* \equiv \frac{\partial\mathcal{H}_1^*}{\partial\tau} \quad , \quad \text{with} \quad \frac{D_0^*}{D\tau} = \frac{\partial}{\partial\tau} + \frac{\lambda_0}{\mathrm{S}}\frac{D(\mathcal{H}_1^*;\,\cdot)}{D(\xi,\eta)} \quad .$$

Therefore, the *Rossby* waves do not have a vertical structure in a barotropic atmosphere but rather, are horizontal, plane, unsteady waves. In linear theory, we can always return to the barotropic case by carrying out a decomposition according to the eigenfunctions of (4.23).

A solution to (4.51) can be sought in the form:

$$\mathcal{H}_1^*(\tau,\,\xi,\,\eta) = -U_{00}^*\eta + h^*(\tau,\,\xi,\,\eta) \quad , \tag{4.52}$$

where U_{00}^* = const is dimensionless. If $U_{00}^* \equiv +1$, then there is a basic current directed westwards; if $U_{00}^* \equiv -1$, then this current is directed eastwards.

For $h^*(\tau,\,\xi,\,\eta)$, from (4.51,52), it follows that:

$$\left(\frac{\partial}{\partial\tau} + \frac{\lambda_0}{S}U_{00}^*\frac{\partial}{\partial\xi}\right)\left[\lambda_0\Delta_{\xi\eta}h^* - \frac{\mathrm{SBo}}{T_0(1)}h^*\right] + \frac{\lambda_0}{\mathrm{S}}\left[\frac{\beta}{\lambda_0} + \frac{\mathrm{SBo}}{T_0(1)}U_{00}^*\right]\frac{\partial h^*}{\partial\xi}$$
$$+\frac{\lambda_0}{\mathrm{S}}\left(\frac{\partial h^*}{\partial\xi}\frac{\partial}{\partial\eta} - \frac{\partial h^*}{\partial\eta}\frac{\partial}{\partial\xi}\right)\left[\lambda_0\Delta_{\xi\eta}h^* - \frac{\mathrm{SBo}}{T_0(1)}h^*\right] = 0 \quad . \tag{4.53}$$

This quasi-linear equation in $h^*(\tau,\,\xi,\,\eta)$ has one exact plane wave solution in the form:

$$h^* = A_{00}\cos(k\xi + l\eta - \sigma\tau) \tag{4.54}$$

given the fact that:

$$\lambda_0\Delta_{\xi\eta}h^* - \frac{\mathrm{SBo}}{T_0(1)}h^* \equiv -\left\{\lambda_0(k^2+l^2) + \frac{\mathrm{SBo}}{T_0(1)}\right\}h^*$$

which causes the quasi-linear terms in (4.53) to disappear identically. Thus, when:

$$\sigma = \sigma_R^* \equiv \frac{(\lambda_0/\mathrm{S})k[\lambda_0U_{00}^*(k^2+l^2) - \beta/\lambda_0]}{\lambda_0(k^2+l^2) + \mathrm{SBo}/T_0(1)} \quad , \tag{4.55}$$

the solution to (4.53) is indeed (4.54) with A_{00} an arbitrary constant amplitude.

Let us return to (4.51) which will now be rewritten in the following form:

$$\frac{\partial}{\partial\tau}\left(\Delta_{\xi\eta}\mathcal{H}_1^* - \frac{\mathrm{SBo}}{\lambda_0T_0(1)}\mathcal{H}_1^*\right) + \frac{\lambda_0}{\mathrm{S}}\left\{\frac{\partial\mathcal{H}_1^*}{\partial\xi}\frac{\partial}{\partial\eta}(\Delta_{\xi\eta}\mathcal{H}_1^*)\right.$$
$$\left.-\frac{\partial\mathcal{H}_1^*}{\partial\eta}\frac{\partial}{\partial\xi}(\Delta_{\xi\eta}\mathcal{H}_1^*)\right\} + \frac{\beta}{\mathrm{S}\lambda_0}\frac{\partial\mathcal{H}_1^*}{\partial\xi} = 0 \quad . \tag{4.56}$$

We multiply this equation by $\mathcal{H}_1^*$ and assume that the motion disappears at infinity. In this case, after integration over the entire ξ,η plane, we find:

$$\frac{\partial}{\partial\tau}\iint\left\{\frac{(\nabla_{\xi\eta}\mathcal{H}_1^*)^2}{2} + \frac{\mathrm{SBo}}{\lambda_0T_0(1)}\frac{\mathcal{H}_1^{*2}}{2}\right\}d\xi\,d\eta = 0 \quad , \tag{4.57}$$

which is a conservation of energy equation.

Furthermore, if this same (4.56) is multiplied by the expression $\Delta_{\xi\eta}\mathcal{H}_1^* - \mathrm{SBo}\mathcal{H}_1^*/\lambda_0 T_0(1)$, we obtain:

$$\frac{\partial}{\partial\tau}\frac{1}{2}\left\{\Delta_{\xi\eta}\mathcal{H}_1^* - \frac{\mathrm{SBo}}{\lambda_0 T_0(1)}\mathcal{H}_1^*\right\}^2$$

$$= -\frac{\lambda_0}{2\mathrm{S}}\frac{\partial}{\partial\eta}\left[\frac{\partial\mathcal{H}_1^*}{\partial\xi}\left(\Delta_{\xi\eta}\mathcal{H}_1^* - \frac{\mathrm{SBo}}{\lambda_0 T_0(1)}\mathcal{H}_1^*\right)^2\right]$$

$$+\frac{\lambda_0}{2\mathrm{S}}\frac{\partial}{\partial\xi}\left[\frac{\partial\mathcal{H}_1^*}{\partial\xi}\left(\Delta_{\xi\eta}\mathcal{H}_1^* - \frac{\mathrm{SBo}}{\lambda_0 T_0(1)}\mathcal{H}_1^*\right)^2\right]$$

$$-\frac{\beta}{2\mathrm{S}\lambda_0}\frac{\partial}{\partial\xi}\left[\left(\frac{\partial\mathcal{H}_1^*}{\partial\xi}\right)^2 - \left(\frac{\partial\mathcal{H}_1^*}{\partial\eta}\right)^2 + \frac{\mathrm{SBo}}{\lambda_0 T_0(1)}\mathcal{H}_1^{*2}\right]$$

$$-\frac{\beta}{\mathrm{S}\lambda_0}\frac{\partial}{\partial\eta}\left(\frac{\partial\mathcal{H}_1^*}{\partial\eta}\frac{\partial\mathcal{H}_1^*}{\partial\xi}\right) \quad . \tag{4.58}$$

Consequently, if the motion disappears at infinity, the integral over the entire ξ, η plane yields:

$$\frac{\partial}{\partial\tau}\iint\left\{\Delta_{\xi\eta}\mathcal{H}_1^* - \frac{\mathrm{SBo}}{\lambda_0 T_0(1)}\mathcal{H}_1^*\right\}^2 d\xi\, d\eta = 0 \quad , \tag{4.59}$$

which demonstrates that the potential "enstrophy" also remains constant. It will be noted that as a general rule, the result (4.59) is not valid for a bounded domain due to the fact that, in this case, the integrated terms in the right-hand side of (4.58) do not necessarily disappear. This right-hand side of (4.58) appears as the divergence of the vector:

$$\boldsymbol{Q} = \left\{\frac{\lambda_0}{2\mathrm{S}}\left[\frac{\partial\mathcal{H}_1^*}{\partial\eta}\left(\Delta_{\xi\eta}\mathcal{H}_1^* - \frac{\mathrm{SBo}}{\lambda_0 T_0(1)}\mathcal{H}_1^*\right)^2\right.\right.$$

$$\left.\left. -\frac{\beta}{2\mathrm{S}\lambda_0}\left[\left(\frac{\partial\mathcal{H}_1^*}{\partial\xi}\right)^2 - \left(\frac{\partial\mathcal{H}_1^*}{\partial\eta}\right)^2 + \frac{\mathrm{SBo}}{\lambda_0 T_0(1)}\mathcal{H}_1^{*2}\right]\right\}\boldsymbol{i}$$

$$+\left\{-\frac{\lambda_0}{2\mathrm{S}}\frac{\partial\mathcal{H}_1^*}{\partial\xi}\left(\Delta_{\xi\eta}\mathcal{H}_1^* - \frac{\mathrm{SBo}}{\lambda_0 T_0(1)}\mathcal{H}_1^*\right)^2 - \frac{\beta}{\mathrm{S}\lambda_0}\frac{\partial\mathcal{H}_1^*}{\partial\eta}\frac{\partial\mathcal{H}_1^*}{\partial\xi}\right\}\boldsymbol{j} \quad .$$

Although $\boldsymbol{Q}$ can be zero at infinity for motion in a non-bounded plane, its normal component is not necessarily cancelled on every finite boundary.

On the other hand, when a bounded area is considered in the ξ, η plane, it can be proved that:

$$\frac{D_0^*}{D\tau}\overline{E} \equiv \frac{D_0^*}{D\tau}\left\{\frac{1}{2}\iint\left[(\boldsymbol{\nabla}_{\xi\eta}\mathcal{H}_1^*)^2 + \frac{\mathrm{SBo}}{\lambda_0 T_0(1)}\mathcal{H}_1^{*2}\right]d\xi d\eta\right\} = 0 \quad . \tag{4.60}$$

In fact,

$$\frac{D_0^*\overline{E}}{D\tau} = \iint \left[\nabla_{\xi\eta}\mathcal{H}_1^* \cdot \nabla_{\xi\eta}\frac{\partial \mathcal{H}_1^*}{\partial\tau} + \frac{\mathrm{SBo}}{\lambda_0 T_0(1)}\mathcal{H}_1^*\frac{\partial\mathcal{H}_1^*}{\partial\tau}\right] d\xi\, d\eta$$
$$= \iint \frac{\partial}{\partial\tau}\left[-\Delta_{\xi\eta}\mathcal{H}_1^* + \frac{\mathrm{SBo}}{\lambda_0 T_0(1)}\mathcal{H}_1^*\right]\mathcal{H}_1^* d\xi\, d\eta$$
$$+ \oint \mathcal{H}_1^*\nabla_{\xi\eta}\frac{\partial\mathcal{H}_1^*}{\partial\tau}\cdot \boldsymbol{\nu}\, d\sigma \quad , \tag{4.61}$$

where $\boldsymbol{\nu}$ is the unit vector of the normal at the boundary of the considered bounded area, and $d\sigma$ is the arc element along this boundary. From (4.11), we know that $\lambda_0\mathcal{H}_1^*$ plays the role of a stream function for a two-dimensional barotropic incompressible flow. Consequently, it is only natural to adopt the slip condition $\mathcal{H}_1^* = \text{const}$ as a condition at the boundary and also to require the circulation to be constant along this boundary:

$$\oint \nabla\mathcal{H}_1^* \cdot \boldsymbol{\nu}\, d\sigma = \text{const} \quad . \tag{4.62}$$

Thus, in (4.61) the contour integral disappears and there remains only

$$\frac{D_0^*\overline{E}}{D\tau} = -\iint \frac{\partial\Omega_1^*}{\partial\tau}\mathcal{H}_1^*\, d\xi\, d\eta \quad ,$$

since the following quasi-geostrophic eddy potential has been introduced:

$$\Omega_1^* \equiv \Delta_{\xi\eta}\mathcal{H}_1^* + \frac{\beta}{\lambda_0^2}\eta - \frac{\mathrm{SBo}}{\lambda_0 T_0(1)}\mathcal{H}_1^* \quad , \tag{4.63}$$

where β, S, λ_0 are constant.

Therefore, once (4.51) has been taken advantage of, we have:

$$\frac{D_0^*}{D\tau}\overline{E} = \frac{\lambda_0}{\mathrm{S}}\iint\left(\frac{\partial\mathcal{H}_1^*}{\partial\xi}\frac{\partial\Omega_1^*}{\partial\eta} - \frac{\partial\mathcal{H}_1^*}{\partial\eta}\frac{\partial\Omega_1^*}{\partial\xi}\right)\mathcal{H}_1^*\, d\xi\, d\eta$$
$$= \frac{\lambda_0}{2\mathrm{S}}\iint\left(-\frac{\partial\mathcal{H}_1^{*2}}{\partial\eta}\frac{\partial\Omega_1^*}{\partial\xi} + \frac{\partial\mathcal{H}_1^{*2}}{\partial\xi}\frac{\partial\Omega_1^*}{\partial\eta}\right) d\xi d\eta$$
$$= -\frac{\lambda_0}{2\mathrm{S}}\iint\left[\frac{\partial}{\partial\eta}\left(\mathcal{H}_1^{*2}\frac{\partial\Omega_1^*}{\partial\xi}\right) - \frac{\partial}{\partial\xi}\left(\mathcal{H}_1^{*2}\frac{\partial\Omega_1^*}{\partial\eta}\right)\right] d\xi\, d\eta$$
$$= \frac{\lambda_0}{2\mathrm{S}}\oint \mathcal{H}_1^{*2}\left(\frac{\partial\Omega_1^*}{\partial\xi}d\xi + \frac{\partial\Omega_1^*}{\partial\eta}d\eta\right)$$
$$= \frac{\lambda_0}{2\mathrm{S}}\oint \mathcal{H}_1^{*2}d\Omega_1^* = 0 \quad ,$$

since $\mathcal{H}_1^* = \text{const}$ at the frontier of the bounded area.

4.4 On the Problem of Hydrodynamic Instability

Let us consider a basic flow having a purely zonal velocity (i.e., directed along the circles of latitude) and which is expressed from a geostrophic stream function:

$$\overline{u}_0(\eta, \zeta) = -\lambda_0 \frac{\partial \overline{\mathcal{H}}_1}{\partial \eta} \quad , \tag{4.64}$$

$\overline{\mathcal{H}}_1 = \overline{\mathcal{H}}_1(\eta, \zeta)$. This basic current is naturally assumed to be a solution to (4.15). We consider the evolution of a perturbation $h_1(\tau, \xi, \eta, \zeta)$ of this basic flow, i.e.:

$$\mathcal{H}_1(\tau, \xi, \eta, \zeta) = \overline{\mathcal{H}}_1(\eta, \zeta) + h_1(\tau, \xi, \eta, \zeta) \quad . \tag{4.65}$$

If (4.65) is inserted into (4.15), the following equation is obtained for h_1:

$$\left(\frac{\partial}{\partial \tau} + \frac{1}{\mathrm{S}}\overline{u}_0 \frac{\partial}{\partial \xi}\right) q_1 + \frac{\lambda_0}{\mathrm{S}} \frac{\partial h_1}{\partial \xi} \frac{\partial \Pi_0}{\partial \eta}$$
$$+ \frac{\lambda_0}{\mathrm{S}} \left(\frac{\partial h_1}{\partial \xi}\frac{\partial q_1}{\partial \eta} - \frac{\partial h_1}{\partial \eta}\frac{\partial q_1}{\partial \xi}\right) = 0 \quad , \tag{4.66}$$

where $q_1(\tau, \xi, \eta, \zeta)$ is the eddy potential (quasi-geostrophic and baroclinic) of perturbation defined by:

$$q_1 = \lambda_0 \left(\frac{\partial^2 h_1}{\partial \xi^2} + \frac{\partial^2 h_1}{\partial \eta^2}\right) + \mathrm{S}\frac{\partial}{\partial \zeta}\left(\frac{\zeta^2}{K_0(\zeta)}\frac{\partial h_1}{\partial \zeta}\right) \quad , \tag{4.67}$$

whereas $\partial \Pi_0/\partial \eta$ is the gradient along the meridian of the eddy potential of the basic flow:

$$\Pi_0 = \frac{\beta}{\lambda_0 \mathrm{S}}\eta + \lambda_0 \frac{\partial^2 \overline{\mathcal{H}}_1}{\partial \eta^2} + \mathrm{S}\frac{\partial}{\partial \zeta}\left(\frac{\zeta^2}{K_0(\zeta)}\frac{\partial \overline{\mathcal{H}}_1}{\partial \zeta}\right) \quad . \tag{4.68}$$

One of the fundamental questions to be clarified is how the (given) structure of $\overline{u}_0$ determines the evolution of the perturbation field h_1. We note that:

$$\frac{\partial \Pi_0}{\partial \eta} \equiv \frac{\beta}{\lambda_0 \mathrm{S}} - \frac{\partial^2 \overline{u}_0}{\partial \eta^2} - \frac{\mathrm{S}}{\lambda_0}\frac{\partial}{\partial \zeta}\left(\frac{\zeta^2}{K_0(\zeta)}\frac{\partial \overline{u}_0}{\partial \zeta}\right) \quad .$$

Stated more precisely, this means that given the basic flow $\overline{u}_0(\eta, \zeta)$, the behavior of h_1 stemming from (4.65) must be elucidated in order to determine whether it increases or decreases. If it *increases*, then the *instability* of $\overline{u}_0$ with respect to h_1 is ascertained. $\overline{u}_0$ can be said to be "truly" *stable* only when it is stable with respect to *all* h_1. On the contrary, instability takes place if $\overline{u}_0$ is unstable for even one h_1.

The equation dealing with h_1 – (4.66) – is quasi-linear. Generally speaking, it is quite difficult to study the behavior of its solution under adequate boundary conditions! Thus, the *linear* case is often adopted and it is assumed that $|h_1| \ll 1$. In this way, the following equation which governs the linear stability problem can replace (4.66):

$$\left(\frac{\partial}{\partial \tau} + \frac{1}{\mathrm{S}}\overline{u}_0 \frac{\partial}{\partial \xi}\right) q_1' + \frac{\lambda_0}{S}\frac{\partial \Pi_0}{\partial \eta}\frac{\partial h_1'}{\partial \xi} = 0 \quad . \tag{4.69}$$

In order to solve (4.69) under adequate conditions in η and ζ, we can set the following:

$$h_1'(\tau, \xi, \eta, \zeta) = \Re\left\{\tilde{h}_1(\eta, \zeta)\, e^{ik(\xi - c\tau)}\right\} \quad , \tag{4.70}$$

where the zonal wave number k must be real since h_1' must remain finite for all $\xi \to \pm\infty$: it can be assumed that $k > 0$. On the other hand, the phase velocity c can be written in the form:

$$c = c_r + ic_i \quad ; \quad i \equiv \sqrt{-1} \quad , \tag{4.71}$$

and therefore, the following replaces (4.70):

$$h_1' = \Re\left\{\tilde{h}_1(\eta, \zeta)\, e^{ik(\xi - c_r\tau)}\, e^{kc_i\tau}\right\} \quad . \tag{4.72}$$

From (4.69–70), the following equation results for $\tilde{h}_1(\eta, \zeta)$:

$$(\overline{u}_0 - Sc)\left\{S\frac{\partial}{\partial\zeta}\left(\frac{\zeta^2}{K_0(\zeta)}\frac{\partial\tilde{h}_1}{\partial\zeta}\right) + \lambda_0\left(\frac{\partial^2\tilde{h}_1}{\partial\eta^2} - k^2\tilde{h}_1\right)\right\}$$
$$+\lambda_0\frac{\partial\Pi_0}{\partial\eta}\tilde{h}_1 = 0 \quad . \tag{4.73}$$

As a general rule, boundary conditions in η and ζ must be superimposed on (4.73). It turns out that these conditions are homogenous and hence, the corresponding linear stability problem usually has only the trivial solution which is identically zero. One exception is when k and c are linked by a relation depending on the profile of $\overline{u}_0\ (\eta, \zeta)$ which can be called the dispersion relation of the problem. For a fixed profile $\overline{u}_0$, the dispersion relation allows a sequence of complex roots in c if k is fixed. If $c_i < 0$ for all the roots (we remark that k is real and positive), then the perturbations (called normal modes) *attenuate* exponentially as a function of time, and the Rossby waves are stable for the type of perturbations considered. On the other hand, if $c_i > 0$ for *at least* one normal mode, then the Rossby waves are unstable for the perturbations of wave number k. In the barotropic case when $\overline{u}_0 \equiv \overline{u}_0^*(\eta)$, the instability process is related essentially to the existence of the term $d^2\overline{u}_0^*/d\eta^2$: the situation is then referred to as *barotropic instability*. However, when $\overline{u}_0 = \overline{u}_0\ (\eta, \zeta)$ – baroclinic case – the vertical shearing is an important cause of instability. The corresponding process gives us the *baroclinic instability*.

The Eady model (1949), with

$$\overline{u}_0 \equiv \zeta \quad , \quad \beta \equiv 0 \quad , \tag{4.74}$$

is a simple and very good example of baroclinic instability. In this case, we have

$$\frac{\partial\Pi_0}{\partial\eta} \equiv 0 \quad . \tag{4.75}$$

Details concerning the Eady model can be found on pages 456–64, Sect. 7.7 of Pedlosky's book (1979).

Furthermore, Kuo (1949) gives us a necessary condition for barotropic instability which is expressed in the form:

$$c_i \int_{-1}^{+1} \frac{|A|^2}{|\overline{u}_0^* - Sc|^2} \left[\frac{\beta}{S} - \lambda_0 \frac{d^2 \overline{u}_0^*}{d\eta^2} \right] d\eta = 0 \quad , \tag{4.76}$$

once the solution of (4.73) has been sought out [where $\overline{u}_0 \equiv \overline{u}_0^*(\eta)$ and $\lambda_0(\partial \Pi_0 / \partial \eta) \equiv \beta / S - \lambda_0 (d^2 \overline{u}_0^* / d\eta^2)$] in the form:

$$\tilde{h}_1(\eta, \zeta) = A(\eta) \phi_1(\zeta) \quad .$$

The following lateral boundary conditions must also be imposed:

$$A(\eta) = 0 \quad , \quad \text{for} \quad \eta = \pm 1 \quad .$$

In Sect. 4.3, it was seen that for the barotropic case, we have the conservation of energy law (4.60). This is, of course, insufficient for elucidating the problem of barotropic instability. It is true that it follows from (4.60) that the solution $\mathcal{H}_1^*$ cannot increase too strongly with time, but it can, nonetheless, depart from the basic steady state. However, it can be demonstrated that with the notation (4.63), the integral

$$J = \iint F(\Omega_1^*) d\xi \, d\eta \quad , \tag{4.77}$$

(where F is an arbitrary function) is also conserved due to (4.51) which shows that Ω_1^* is conserved along the quasi-geostrophic trajectories. Thus, $\overline{E}$ and J are conserved and we can form:

$$\mathcal{L} = \overline{E} + J \quad .$$

The function F must now be selected in such a way that $\mathcal{L}$ admits of a relative minimum for the basic steady flow, relative to which barotropic stability is sought. To this end, it is sufficient that the variation δF be zero and the second variation $\delta^2 F$ be positive: this will guarantee stability. This method is accredited to Fjortoft (1950) and was generalized by Arnold (1965).

In particular, it can be demonstrated that if

$$\overline{\Omega_1^*} = \frac{d^2 \overline{\mathcal{H}_1^*}}{d\eta^2} + \frac{\beta}{\lambda_0^2} \eta - \frac{\text{SBo}}{\lambda_0 T_0(1)} \overline{\mathcal{H}_1^*}(\eta)$$

varies monotonically from one pole to the other – if it decreases from the north towards the south – then the basic zonal steady flow $\overline{\mathcal{H}_1^*}(\eta)$ is stable [see Kuo (1949)]. It will be noticed that this last condition is, in fact, realized as a limiting case. In reality, meteorological situations often occur for which this condition is not satisfied: in this case, there is barotropic instability. This is due to the fact that this stability condition is very general.

4.5 Conclusions and Bibliographical References

To be sure, there was never any question of our undertaking a complete accounting of the Rossby waves within the relatively limited number of pages of the present chapter. Let us point out from the start that just such an accounting is to be found in Chaps. 3, 6 and 7 of Pedlosky's book (1979). Concerning the hydrodynamic stability theory, the short book by Dikij (1976) is recommended. The important article by Kuo (1973) is also to be stressed for its original report on the dynamics of quasi-geostrophic flows and the theory of instability. As regards quasi-geostrophic flow, we will return to this matter in Chap. 11 where a complete theory will be expounded which includes the problem of adjustment to geostrophy and the problem of the Ekman layer, as well as the ageostrophic model.

It can be pointed out that the rather colorful name of planetary waves is given to the oscillations of a thin liquid layer (the atmosphere) spread over a rotating sphere (the earth) and limited at the top by a free surface (the tropopause). The thin liquid layer simulates, in fact, the troposphere. By taking advantage of the thinness of the liquid layer, the oscillations in question can be shown to reduce to the classical Rossby waves if the tangent plane and β plane approximations are carried out, and also the duration of these oscillations is assumed to be on the order of several days. The Rossby waves are said to be, in first approximation, non-divergent planetary waves in a plane tangent to the earth.

The recent article by Ahmed and Eltayeb (1980) can be consulted on matters concerning the propagation, reflection, transmission and stability of atmospheric Rossby waves while taking into account gravity waves in a β plane in the presence of a zonal flow sheared in the direction of the latitude.

Various indications on the dynamics of Rossby waves can be found in the books by Pedlosky (1979), Friedlander (1980), Robert and Soward (1978) and Beer (1974).

Let us emphasize here that the barotropic model attempts to simulate the behavior of the entire atmosphere by analyzing motions at only one level. By doing so, the otherwise very real effects of the vertical motions are not taken into account. The fundamental differences between the barotropic and baroclinic models stem from the fact that in the latter, the winds vary from one plane to another and the temperature distribution does not depend solely on the pressure distribution. The baroclinic model thus represents the atmosphere in a more precise manner than the barotropic model.

The Rossby waves are actually large atmospheric currents which meander around the globe. They are big, long westerly waves which are essential to "long-range" weather forecasting. The latter requires that the entire globe be studied while taking into account its sphericity and also non-adiabatic phenomena of the atmosphere. As already stated, such questions go beyond the scope of this book and thus, only a somewhat "academic" account of the Rossby waves is given in the present chapter. In first approximation, non-adiabatic phenomena can, of course, be ignored since they are significant especially in the planetary boundary-layer whose thickness does not exceed a few hundred meters. According to (2.112), this thickness is, in fact, proportional to $(\mathrm{Ek}_\perp)^{1/2}$ with ($\mathrm{Ro} = Ro/\sin\varphi_0$)

$$\mathrm{Ek}_\perp \equiv \mathrm{Ro}/\mathrm{Re}_\perp \quad .$$

Let us then return to (2.112) with the hypothesis $\mathrm{Re}_\perp \equiv \infty$. We will also assume that:

$$Ro \ll 1 \quad , \quad \mathrm{S} \gg 1 \quad , \quad \mathrm{M}_\infty \ll 1,$$

in such a way that:

$$Ki \equiv \mathrm{S}Ro = \mathrm{O}(1) \quad , \quad \Lambda_0 = \frac{\delta}{\gamma} Ro^2/\mathrm{M}_\infty^2 = \mathrm{O}(1) \quad ;$$

i.e., $t_0 \cong (2\Omega_0)^{-1}$, which implies the consideration of motions that vary on a daily basis and:

$$L \cong RT_\infty(0)/4a_0\Omega_0^2 \quad .$$

The Boussinesq number Bo remaining of the order unity ($H \cong RT_\infty(0)/g$), we seek solution in the form:

$$\overline{u} = u + \dots \quad ; \quad \overline{v} = v + \dots \quad ; \quad \overline{w} = w + \dots \quad ;$$

$$\overline{p} = p_0(\overline{z}) + Ro\pi + \dots \quad ;$$

$$\overline{\varrho} = \varrho_0(\overline{z}) + Ro\omega + \dots \quad ;$$

$$\overline{T} = T_0(\overline{z}) + Ro\theta + \dots \quad ,$$

with

$$\frac{dp_0}{d\overline{z}} + \mathrm{Bo}p_0 = 0 \quad , \quad p_0 = \varrho_0 T_0 \quad .$$

The equations describing the planetary waves in the free atmosphere outside the boundary layers then follow from system (2.112) after applying the limiting process $Ro \to 0$. These equations with respect to u, v, w, π and ω are written in the following form:

$$\varrho_0\left(Ki\frac{\partial u}{\partial \overline{t}} - v\sin\varphi\right) + \Lambda_0\frac{1}{\cos\varphi}\frac{\partial\pi}{\partial\lambda} = 0 \quad ;$$

$$\varrho_0\left(Ki\frac{\partial v}{\partial \overline{t}} + u\sin\varphi\right) + \Lambda_0\frac{\partial\pi}{\partial\varphi} = 0 \quad ;$$

$$\frac{\partial\pi}{\partial\overline{z}} + \mathrm{Bo}\omega = 0 \quad ;$$

$$Ki\frac{\partial\omega}{\partial\overline{t}} + \frac{d\varrho_0}{d\overline{z}}w = -\varrho_0\chi \quad ;$$

$$Ki\frac{\partial\pi}{\partial\overline{t}} + \frac{dp_0}{d\overline{z}}w = -\gamma p_0\chi \quad ,$$

where

$$\chi \equiv \frac{\delta}{\cos\varphi}\frac{\partial u}{\partial\lambda} + \frac{\delta}{\cos\varphi}\frac{\partial(v\cos\varphi)}{\partial\varphi} + \frac{\partial w}{\partial\overline{z}} \quad .$$

It will be remarked that in the above equations, we have ignored the force terms linked to the dynamics and thermodynamics of the tides (see Chap. 5 in Beer's book 1974). The solution to these equations with respect to $\overline{t}$ can be sought in the form of periodic oscillation modes in $\exp(\mathrm{i}\omega\overline{t})$; the variables with respect to $\overline{z}$ and (λ, φ) will then be separated. This leads to an equation for the vertical structure and a horizontal equation in λ and φ – that of Laplace[2] which describes the atmospheric tides, i.e., the planetary waves. The reader interested in the theory of atmospheric tides is first and foremost, referred to the book by Chapman and Lindzen (1970), as well as the articles by the following authors: Dikij (1961), Miles (1974), Pekeris (1975), Longuet-Higgins (1968) and Longuet-Higgins and Pond (1970).

In conclusion, it is also pointed out that (4.15) which governs baroclinic quasi-geostrophic flows can have "solitary wave" type solutions which are related to non-linear effects. More precisely, the existence of these solitary Rossby waves is partially linked to the β effect and the shear effects of the zonal basic flow, and possibly even to the relief. On this subject, the articles by the following authors can be of interest: Long (1964), Redekopp (1977), Berestov and Monin (1980) and Redekopp and Weidman (1978). We will only mention here that it is the dispersive character of the Rossby waves [see (4.46)] interacting with the non-linear effects which enables the formation of solitary Rossby waves in the majority of cases. A review of the works devoted to solitary waves can be found in Miles (1980).

[2] According to the original paper of 1775. Another relevant source is Eckart (1960; Sects. 38, 93–100).

5. A Presentation of Asymptotic Methods

At this point, we can take some time out before tackling the *approximations* (Chaps. 7–10), and the *models* (Chaps. 11–13).

As a matter of fact, the first four chapters of the present book constitute a sort of initiation to the asymptotic analysis of atmospheric flows. This asymptotic analysis which will be systematically used throughout Chaps. 6–13, is principally based on the *method of matched asymptotic expansions* (MMAE) and on the *multiple scale method* (MSM).

From the equations obtained in Chap. 2 – which are none other than the Navier-Stokes or Euler equations – the MMAE and the MSM make it possible to 1) give an asymptotic formulation of various approximations usually employed in the study of atmospheric flows (mainly the following approximations: the quasi-static, Boussinesq, isochoric, and deep convection), and 2) to obtain, in a rational manner, a whole series of asymptotic models which simulate different types of atmospheric motions (e.g., quasi-geostrophic, ageostrophic, quasi-solenoidal, meso and local models). One of the biggest problems in asymptotic analysis is the reestablishing asymptotically of a well-posed boundary and initial value problem. From this viewpoint, the models obtained in Chap. 11 are exemplary. We will see that, unfortunately, such is not always the case (e.g., the quasi-solenoidal model of Chap. 12). It must be fully understood that the systematic study of all approximations and models of interest to the meteorologist is a monumental task and we cannot hope to be exhaustive in the coming chapters. Certain approximations and models are merely outlined in this book; others should be studied in depth in the future.

Therefore, what follows appears as a first attempt to "asymptotize" dynamical meteorology. We feel that parallel to a "practical" meteorology whose goal is mainly to (numerically) predict the weather, we should develop a dynamic meteorology of fluids which would be considered as one of the branches of theoretical fluid dynamics. In our opinion, this return of meteorology to the family of fluid mechanics should be of value to both meteorologists and fluid mechanics specialists. Nevertheless, as defined by singular perturbation methods, the obtaining of rational and consistent models is not always obvious. For example, modeling low Mach number atmospheric flows in a rational manner is currently an unsolved problem! Although this can be done for classical aerodynamics problems – as was stated in Sect. 3.3 – such modeling is closely related to the degeneracy of the hyperbolic Euler equations into elliptical-type equations which entails the local formulation of adjustment problems with respect to time. It is important to understand that in the majority of cases, the establishment of models is an intuitive, heuristic matter and so it is not clear how to insert the model under consideration into a hierarchy of rational approximations which in turn result from the general equations chosen at the beginning (either the Navier-Stokes or Euler equations). From this point of view, the approach presented in Chaps. 7–13 enables the rationalization of certain models apparently considered until now as mere "ad hoc" approximations.

It seems obvious that an improvement in weather forecasting depends largely on the obtaining of more "efficient" models and not only on the development of numerical techniques of analysis and calculation as is thought by certain specialists in the field of numerical weather forecasting. We feel, in particular, that the development of accurate asymptotic models, no longer of first order but of second order [e.g., of the type corresponding to the ageostrophic model recently obtained by Guiraud and Zeytounian (1980)], as well as the taking into account more systematically of uniformly valid asymptotic representations should, in the course of the next decade, permit us to rethink the problem of short-range weather prediction and also to initiate regional and local forecasting.

The science of meteorology and, more particularly, numerical weather prediction seem to be suffering today from an excess of "experimentation". Thus, the realistic modeling of atmospheric phenomena is lagging behind.[1] We are of the opinion, however, that only conceptually coherent theoretical modeling can bring to light the time problems to be solved in order to achieve a significant improvement in the reliability of predictions. Of course, it must not be forgotten that such modeling must be a mathematical expression of real atmospheric phenomena permitting their interpretation. Thus is it necessary from the start to choose sufficiently realistic equations and conditions which reflect the essential characteristics of atmospheric phenomena such as gravity, compressibility, stratification, rotation and baroclinity.

The theoretician of fluid mechanics now has at his disposal conceptual tools which permit the modeling of atmospheric phenomena – above all, we naturally have in mind the asymptotic techniques which have proven so decisive in fluid mechanics [(on this subject, see the course given by Germain (1977) at the 1973 "Ecole d'Eté de Physique Théorique des Houches")]. We believe that these asymptotic techniques should find new applications in the special field of meteorology – a meaningful illustration of this tendency can be found in the thesis by Bois (1979).

The idea of presenting a somewhat systematic survey on the asymptotic modeling of atmospheric flows goes back to 1974 when upon invitation by Prof. W. Fiszdon of the Polish Academy of Sciences, we drafted our general conference[2] for the XIIth Symposium on Advanced Problems and Methods in Fluid Mechanics (Bialowieza, Poland, Sept. 8–13, 1975). The exposé in Chaps. 6–13 is not of the "geophysical fluid dynamics" type which is currently in fashion.[3]

---

[1] A very convincing document is the "Exposés sur la Météorologie" published in the "Comptes Rendus des Séances du l'Académie des Sciences", Tome 291, supplement to the "Vie Académique", July, 1980.

[2] The title is "La Météorologie du point de vue du mécanicien des fluides" published in Warsaw in 1976, in "Fluid Dynamics Transactions", vol. 8, pages 289–352.

[3] As examples, the following can be cited: Pedlosky (1979), Roberts and Soward (1978), Reid (1971), Morel (1973) and Friedlander (1980).

Our aim is to analyze the approximations and models from the viewpoint of their internal coherence and with respect to the basic equations so as to clarify the mathematical problems posed by the application of these models to weather forecasting. A typical example is the case of the primitive equations which are currently being used for short-range numerical weather prediction. Until recently, it was not known which initial conditions should be assigned to these equations. With the work of Guiraud and Zeytounian (1982), this question was given a theoretical solution which remains to be applied to forecasting in an operative manner.

Generally speaking, we feel that the time has come to re-examine, at least from the theoretician's point of view, the presentation of dynamical meteorology, by keeping in mind the objectives of asymptotic representation. The present book is an attempt at a first step in this direction. We hope that the reader – whether he/she be a theoretician or numerical analyst – will want to read the developments which are presented and will not be disappointed with them. In any case, we will have fully reached our goal if the reader can be brought to appreciate a certain way of tackling the problems posed by the dynamics of the atmosphere. If even some of our professor-researcher colleagues should rethink the presentation of their courses/research in terms of asymptotic modeling as a result of the time enthusiastically devoted to the drafting of this book, then we will be perfectly satisfied.

Before getting down to the heart of the matter, a concise survey of the MMAE and the MSM is presented briefly with a view to expanding the curriculum of the student preparing a degree in theoretical mechanics.

5.1 The Matched Asymptotic Expansions Method

In many cases, a "crude" asymptotic expansion with respect to a small parameter $\alpha \ll 1$, does not constitute a uniformly valid approximation in our space-time area of interest, or else it is only valid in intervals of the space-time coordinates which are too limited to produce really worthwhile results. The problem is then said to be one of *singular perturbations*. Two methods for studying the latter are concisely presented in what follows: they are the MMAE and the MSE. Concerning singular perturbation methods, quite a thorough report from the point of view of mathematical analysis can be found in the book by Eckhaus (1979). Regarding the applications of these methods, we can cite the French article by Germain (1977), as well as the recent English book by Kevorkian and Cole (1981). Morever, Chap. 6 of the present book consists in an illustration via from brief examples of the application of the MMAE and the MSE to meteorological problems.

First of all, we write our basic equations (see Chap. 2) in the symbolic form:

$$\mathcal{E}^{(i)}(D, \mathcal{U}; \lambda, \alpha) = 0 \quad , \quad i = 1, 2, \ldots, I \quad . \tag{5.1}$$

In this system, D is any derivative with respect to one of the four independent variables t, x, y and z; $\mathcal{U}(t, x, y, z; \alpha)$ designates the set of functions characterizing the solution of (5.1) and λ designates a set of dimensionless parameters other than the small principal dimensionless parameter α.

To be sure, initial conditions (for $t = 0$) and boundary conditions (on $z = 0$, when the relief is not taken into account and also at infinity) must be joined to system (5.1). Hence, the solution to (5.1) is sought in a space-time domain $\mathcal{D}$ defined by:

$$\mathcal{D}: \{t \geq 0 \quad , \quad z \geq 0 \quad , \quad -\infty \leq (x, y) \leq +\infty\} \quad . \tag{5.2}$$

In order for the relief to be taken into account, it must be "rectified" by an appropriate change in variables in such a way that the earth's surface is simulated by a plane wall.

An asymptotic expansion of order m in a closed subdomain $\mathcal{D}_1$ of $\mathcal{D}(\mathcal{D}_1 \subset \mathcal{D})$ of the solution $\mathcal{U}$ is a series of the form:

$$E_m\mathcal{U} = \sum_{p=0}^{m-1} \delta_p(\alpha)\mathcal{U}_p(t, x, y, z; \alpha) \quad , \tag{5.3}$$

where the $\delta_p(\alpha)$ are functions of α, $\alpha \in [0, \alpha_0]$ such that[4]

$$\delta_{p+1} = \mathrm{o}(\delta_p) \quad . \tag{5.4}$$

The set $\{\delta_p\}$ is said to be an asymptotic sequence of gauge functions. In fact, when we write (5.3), it is implicitly recognized that:

$$|\mathcal{U}_p| = \mathrm{O}(1) \quad . \tag{5.5}$$

More precisely, $E_m\mathcal{U}$ is an asymptotic approximation of $\mathcal{U}$ of order m in $\mathcal{D}_1$ if:

$$\mathcal{U} - E_m\mathcal{U} = \mathrm{O}(\delta_m) \quad . \tag{5.6}$$

In this case, $\mathcal{D}_1 \subset \mathcal{D}$ is said to be the domain of validity of the asymptotic expansion $E_m\mathcal{U}$ of $\mathcal{U}$.

An asymptotic (expansion) approximation $E_m\mathcal{U}$ of $\mathcal{U}$ in $\mathcal{D}_1 \subset \mathcal{D}$ is called regular if the $\mathcal{U}_p$ do not depend on α. Then the $\mathcal{U}_p(t, x, y, z)$ can be defined by successive limiting processes:

$$\mathcal{U}_p = \lim_{\alpha \to 0} \left\{ \frac{\mathcal{U} - \sum_{n=0}^{p-1} \delta_n(\alpha)\mathcal{U}_n}{\delta_p} \right\} \quad , \quad p = 0, 1, \ldots, m-1 \quad . \tag{5.7}$$

The functions $\mathcal{U}_p$ are generally obtained as solutions to problems (*not* "stiff") which are easier to solve than the one associated with the basic equations (5.1) wherein $\mathcal{U}$ is initially defined (which is often a "stiff" problem).[5] It will be

---

[4] Let us specify the meaning of the so-called *Landau* notations O and o which have already been employed in Chaps. 1–4:

$$f = \mathrm{o}(g) \Leftrightarrow \lim_{\alpha \to 0} f/g = 0 \quad ,$$
$$f = \mathrm{O}(g) \Leftrightarrow \lim_{\alpha \to 0} f/g < +\infty \quad .$$

[5] On this topic, the recent book by Miranker (1981) can be cited.

remarked that the terms $\mathcal{U}_0, \mathcal{U}_1, \mathcal{U}_2, \ldots$ are respectively terms of order zero, one, two, ... It is also said that $E_m\mathcal{U}$ is the expansion to the order $O(\delta_{m-1})$.

We will not attempt here to prove the validity of (5.6) which, in any case, is only possible in very particular situations [see the book by Eckhaus (1979)]. However, if in $\mathcal{D}_1$, the validity of (5.5) can be confirmed, then it will be tempting to believe that (5.6) does indeed occur in $\mathcal{D}_1$. This would, of course, be merely an assumption.

In actual practice, regular asymptotic expansions – which we will also call *main* expansions – are obtained by the repeated application of the limiting process:

$$\lim_P = \lim \{\alpha \to 0,\ t,\ x,\ y,\ z \text{ remaining fixed of order } O(1)\} \tag{5.8}$$

which is termed the main limit. It must be clearly understood that it is exceptional for the main (regular) expansion,

$$E_m^P\mathcal{U} = \sum_{p=0}^{m-1} \delta_p^P(\alpha)\mathcal{U}_p^P(t,\ x,\ y,\ z) \quad , \tag{5.9}$$

to have a domain of validity $\mathcal{D}_1 \equiv \mathcal{D}$. When, however, such is the case, then $E_m\mathcal{U}$ is said to be an asymptotic expansion (or approximation) of $\mathcal{U}$ of order m *uniformly valid* in $\mathcal{D}$.

Let us now return to (5.1) and suppose that $\lim\limits_P$ and D permute as many times as is necessary. This means that if we call D^k any partial derivative of order k, then:

$$E_m^P D^k\mathcal{U} \equiv D^k E_m^P\mathcal{U} \quad . \tag{5.10}$$

If the following limits exist:

$$\mathcal{U}_0^P = \lim_P \frac{\mathcal{U}}{\delta_0^P} \quad , \quad \mathcal{U}_1^P = \lim_P \frac{\mathcal{U} - \delta_0^P\mathcal{U}_0^P}{\delta_1^P} \quad , \ldots \quad , \tag{5.11}$$

then we can find gauges $\gamma_n^{(i)}(\alpha)$ such that:

$$\lim_P \frac{1}{\gamma_0^{(i)}} \mathcal{E}^{(i)}(D, \mathcal{U};\ \lambda,\ \alpha) = \mathcal{E}_0^{(i)}(D, \mathcal{U}_0^P;\ \lambda) \quad ; \tag{5.12a}$$

$$\lim_P \frac{1}{\gamma_1^{(i)}} \left[\mathcal{E}^{(i)}(D, \mathcal{U};\ \lambda,\ \alpha) - \gamma_0^{(i)}\mathcal{E}_0^{(i)}(D, \mathcal{U}_0^P;\ \lambda)\right]$$

$$= \mathcal{E}_1^{(i)}(D, \mathcal{U}_0^P, \mathcal{U}_1^P;\ \lambda) \quad , \tag{5.12b}$$

– –

This is tantamount to a formal development of the left-hand side of (5.1) in the form:

$$\mathcal{E}^{(i)}(D, \mathcal{U}; \lambda, \alpha) = \sum_{r=0}^{R-1} \gamma_r^{(i)}(\alpha)\mathcal{E}_r^{(i)}((D, \mathcal{U}_0^P, \mathcal{U}_1^P, \ldots, \mathcal{U}_r^P; \lambda) + \mathrm{O}\big(\gamma_R^{(i)}(\alpha)\big). \quad (5.13)$$

Thus, if (5.10, 11) occur, then we necessarily have the hierarchy of systems of main equations:

$$\mathcal{E}_0^{(i)}(D, \mathcal{U}_0^P; \lambda) = 0 \quad ; \quad \mathcal{E}_1^{(i)}(D, \mathcal{U}_0^P, \mathcal{U}_1^P; \lambda) = 0 \quad , \quad \ldots \quad (5.14)$$

It should be noted that by writing (5.13), it is implicitly postulated that in $\mathcal{E}_r^{(i)}$ intervenes only $\mathcal{U}_0^P, \mathcal{U}_1^P, \ldots, \mathcal{U}_r^P$ and their derivatives. This property will be obvious in all the applications treated in Chaps. 6–13. Thus it can be seen at this stage that the first equation (5.14) (which is, in fact a system of equations of order zero) considerably limits the choice of the term $\mathcal{U}_0^P$ of order zero from the main expansion (5.9). This choice can be further restricted by making use (during the main limiting process (5.8)) of both the initial and boundary conditions deduced from the conditions posed initially. However, it must be immediately pointed out that in most of the cases considered in Chaps. 6–13, this main limiting process (5.8), where t, x, y and z remain fixed of the order unity when $\alpha \to 0$, leads to the filtering out of the short internal acoustic and gravity waves (as was seen in Sect. 3.3). The unfortunate consequence of this filtering out is that in the system of limiting equations of order zero:

$$\mathcal{E}_0^{(i)}(D, \mathcal{U}_0^P; \lambda) = 0 \quad , \quad (5.15)$$

a part of the partial derivatives with respect to time t disappear. This means that the initial conditions which were imposed on (5.1) cannot be automatically applied to (5.15). We will find a typical example of this situation in Chaps. 7 and 11. Therefore, in this case, we are led to favor a so-called initial layer in the vicinity of the initial time $t = 0$, which in turn brings us to introduce a local limiting process:

$$\lim_{l} = \lim \, \{\alpha \to 0, \tilde{t}, x, y, z \text{ remaining fixed of order O(1)}\} \quad , \quad (5.16)$$

where the short time is defined by:

$$\tilde{t} = \frac{t}{\nu(\alpha)} \quad , \quad \nu(\alpha) \to 0 \quad , \quad \text{with} \quad \alpha \to 0 \quad . \quad (5.17)$$

Here, in fact, it is always considered that:

$$\nu(\alpha) \equiv \alpha^a \quad , \quad a > 0 \quad \text{a real number} \quad . \quad (5.18)$$

The short time $\tilde{t} = t/\alpha^a$ is adequate for describing the adjustment phenomenon of the initial conditions which makes it possible to formulate the "right" initial conditions for the system of (main) limiting equations of order zero (5.15).

The introduction of the short time $\tilde{t}$ into (5.1) leads to the system of equations:

$$\tilde{\mathcal{E}}^{(i)}\left(\frac{\partial}{\partial \tilde{t}}, D_e, \tilde{\mathcal{U}}; \lambda, \alpha\right) = 0 \quad , \quad i = 1, 2, \ldots, I \quad , \tag{5.19}$$

where D_e designates a partial derivative with respect to one of the space coordinates x, y, z. The derivative $\partial/\partial\tilde{t}$ will be treated separately since it plays a privileged role in the hierarchy of systems of local equations [see (5.24)]. Finally:

$$\tilde{\mathcal{U}}(\tilde{t}, x, y, z; \alpha) \equiv \mathcal{U}(\nu(\alpha)\tilde{t}, x, y, z; \alpha) \quad . \tag{5.20}$$

By the repeated application of the local limiting process (5.16), a local expansion of order n is obtained:

$$E_n^l \mathcal{U} = \sum_{p=0}^{n-1} \tilde{\delta}_p(\alpha)\tilde{\mathcal{U}}_p(\tilde{t}, x, y, z) \tag{5.21}$$

which like (5.9) is regular and thus unique:

$$\tilde{\mathcal{U}}_0 = \lim_l \frac{\mathcal{U}}{\tilde{\delta}_0} \quad , \quad \tilde{\mathcal{U}}_1 = \lim_l \frac{\mathcal{U} - \tilde{\delta}_0\tilde{\mathcal{U}}_0}{\tilde{\delta}_1} \quad , \ldots \quad . \tag{5.22}$$

If the series of operations carried out previously on (5.1, 9, 11) is now applied to (5.19, 21, 22), while at the same time taking into account the permutation relations analogous to (5.10), then the following is obtained:

$$\tilde{\mathcal{E}}^{(i)} = \sum_{s=0}^{S-1} \tilde{\gamma}_s^{(i)}(\alpha)\tilde{\mathcal{E}}_s^{(i)}\left(\frac{\partial}{\partial \tilde{t}}, D_e, \tilde{\mathcal{U}}_0, \tilde{\mathcal{U}}_1, \ldots, \tilde{\mathcal{U}}_s; \lambda\right) + \mathrm{O}(\tilde{\gamma}_S^{(i)}(\alpha)) \quad . \tag{5.23}$$

It can thus be concluded that the successive terms of the local expansion (5.21) verify a hierarchy of systems of local equations:

$$\tilde{\mathcal{E}}_0^{(i)}\left(\frac{\partial}{\partial \tilde{t}}, D_e, \tilde{\mathcal{U}}_0; \lambda)\right) = 0 \quad ; \quad \tilde{\mathcal{E}}_1^{(i)}\left(\frac{\partial}{\partial \tilde{t}}, D_e, \tilde{\mathcal{U}}_0, \tilde{\mathcal{U}}_1; \lambda\right) = 0 \quad , \tag{5.24}$$

- -

To the following system of local limiting equations of order zero:

$$\tilde{\mathcal{E}}_0^{(i)}\left(\frac{\partial}{\partial \tilde{t}}, D_e, \tilde{\mathcal{U}}_0; \lambda\right) = 0 \quad , \tag{5.25}$$

we can apply the initial conditions at $\tilde{t} = 0$ which are directly deduced from those of the initial problem related to (5.1).

Having reached this stage, two fundamental questions arise: 1) the first concerns the choice of the gauge $\nu(\alpha)$, i.e., for our purposes, the exponent a in (5.18); 2) the second concerns the pursuit of conditions which would complete those already acquired so that all of the successive terms in the main expansion could be determined and possibly even those of the local expansion in the vicinity of $t = 0$. The choice of gauges in each expansion will come about quite naturally in the cases being considered once we have the rules with which to answer the second above-cited question. Of course, it is sufficient here to state only the steps necessary for the implementation of the construction procedure of the expansions corresponding to the specific cases with which we will have to deal.

Let us begin by studying the complementary conditions in t. These are conditions for the main expansion, for $t = 0$, and for the local expansion, for $\tilde{t} = +\infty$ of an initial, temporal layer. These complementary conditions must not be sought among the initial conditions of the starting initial value problem associated with (5.1), the latter having already been used for the problem related to (5.25). Rather, these conditions should be looked for among what are known as matching conditions which are conditions of compatibility or internal coherence between the two expansions. It is precisely to the important role played by these matching conditions that the MMAE owes its name. For the moment, only the limited matching rule due to Van Dyke (1964) will be given. Let $E_m^P \mathcal{U}$ and $E_n^l \mathcal{U}$ be two approximations which match together. E_m^P and E_n^l can be considered as two operators applied to $\mathcal{U}$ (assumed known, for the time being). The main expansion of $\mathcal{U}$, (5.9), is said to match with the local expansion $\mathcal{U}$, (5.21) via the restricted matching rule if:

$$E_m^P E_n^l \mathcal{U} = E_n^l E_m^P \mathcal{U} \quad . \tag{5.26}$$

Otherwise stated: *the main expansion with m terms of the local expansion with n terms is equivalent to the local expansion with n terms of the main expansion with m terms.*

It will be remarked that in the cases considered in Chaps. 6–13, the asymptotic sequences will always be of the form $\{\alpha^m\}$, $m = 0, 1, 2, \ldots$ The following matching condition of order zero is then a particular case of (5.26):

$$\lim_{\tilde{t}\to\infty} \left(\lim_{l} \tilde{\mathcal{U}} \right) = \lim_{t\to 0} \left(\lim_{P} \mathcal{U} \right) \quad .$$

or even:

$$\tilde{\mathcal{U}}_0(\infty, x, y, z) = \mathcal{U}_0^P(0, x, y, z) \quad . \tag{5.27}$$

If we assume that $E_m^P \mathcal{U}$ and $E_n^l \mathcal{U}$ satisfy the rule (5.26) and we form the expression:

$$C_{m,n}\mathcal{U} = E_m^P \mathcal{U} + E_n^l \mathcal{U} - E_m^P E_n^l \mathcal{U} \quad , \tag{5.28}$$

we will obtain:

$$E_m^P C_{m,n}\mathcal{U} \equiv E_m^P \mathcal{U} \quad , \quad E_n^l C_{m,n}\mathcal{U} \equiv E_n^l \mathcal{U} \quad . \tag{5.29}$$

We say then that $C_{m,n}\mathcal{U}$ is a *composite expansion* whose domain of validity encompasses all the other domains of validity of the expansions of which it is composed.

Let us again point out that the notion of matching $E_m^P \mathcal{U}$ with $E_n^l \mathcal{U}$ is, in fact, based on the *Kaplun theorem of extension* [see, Kaplun (1967) edited by Lagerstrom]. The latter states that: *each of the expansions $E_m^P \mathcal{U}$ or $E_n^l \mathcal{U}$ is valid in an extended domain, but may be of a less valid order than either* δ_{m-1}^P

or $\tilde{\delta}_{n-1}$. A priori, nothing is known about these domains of validity, but it is acknowledged that two such extended domains can be found which *overlap* and then $E_m^P\mathcal{U}$ and $E_n^l\mathcal{U}$ can be said to match.

We can now turn our attention to the choice of the gauge $\nu(\alpha) \equiv \alpha^a$ which determines the short time $\tilde{t} = t/\alpha^a$. If the gauge $\nu(\alpha)$ is changed to $\nu'(\alpha)$, then the local expansion (5.21) naturally becomes $E_n^{l'}\mathcal{U}$. If it happens that:

$$E_n^{l'}E_n^l\mathcal{U} \equiv E_n^{l'}\mathcal{U} \quad , \tag{5.30}$$

then we say that the approximation defined by $E_n^{l'}\mathcal{U}$ is contained in the one defined by $E_n^l\mathcal{U}$. An approximation is said to be *significant* only if it is not contained in any other. Thus, the problem is to determine which gauge $\nu(\alpha)$ leads to a significant approximation. The answer is linked to the concept of significant degeneracy. To resolve this question, let us backtrack a bit and write (5.25) in the form:

$$\tilde{\mathcal{E}}_{\nu,0}^{(i)}\left(\frac{\partial}{\partial \tilde{t}}, D_e, \tilde{\mathcal{U}}_{\nu,0}^j; \lambda\right) = 0 \quad . \tag{5.25a}$$

We will say that $\tilde{\mathcal{E}}_{\nu,0}^{(i)}$ is the local degeneracy associated with the local time variable $\tilde{t} = t/\nu(\alpha)$ and with the choice of the first gauge $\tilde{\delta}_0^j(\alpha)$. The superscript j in $\mathcal{U}^j$ is intended to bring out the fact that $\mathcal{U}$ is a matrix variable. The degeneracy is an operation which starting from an expression involving $\mathcal{U}^j$ and its partial derivatives in t, x, y, z, joins to this expression another one which involves a value $\tilde{\mathcal{U}}_{\nu,0}^j$ via the introduction of gauges such as:

$$\mathcal{U}^j = \tilde{\delta}_0^j(\alpha)\tilde{\mathcal{U}}_{\nu,0}^j \quad , \tag{5.31}$$

as well as its partial derivatives in the system of variables $\tilde{t}$, x, y, z.

Generally speaking, two short times can be considered: $\tilde{t}_1 = t/\nu_1(\alpha)$ and $\tilde{t}_2 = t/\nu_2(\alpha)$, as well as two gauges $\tilde{\delta}_{0,1}^j(\alpha)$ and $\tilde{\delta}_{0,2}^j(\alpha)$ such that:

$$\mathcal{U}^j = \tilde{\delta}_{0,1}^j(\alpha)\tilde{\mathcal{U}}_{\nu_1,0}^j \quad \text{and} \quad \mathcal{U}^j = \tilde{\delta}_{0,2}^j\tilde{\mathcal{U}}_{\nu_2,0}^j \quad .$$

It is obvious that we can then find a $\nu(\alpha) \equiv \nu_1(\alpha)/\nu_2(\alpha)$ such that:

$$\tilde{t}_2 = \nu(\alpha)\tilde{t}_1 \quad ,$$

and a gauge $\tilde{\delta}_{0,0}^j(\alpha)$ such that:

$$\tilde{\delta}_{0,2}^j = \tilde{\delta}_{0,0}^j\tilde{\delta}_{0,1}^j \quad .$$

Let $\tilde{\mathcal{E}}_{\nu_1,0}^{(i)}$ and $\tilde{\mathcal{E}}_{\nu_2,0}^{(i)}$ be the corresponding degeneracies. The change defined by $\nu(\alpha)$ and $\tilde{\delta}_{0,0}^j$ can then be applied to the degeneracy $\tilde{\mathcal{E}}_{\nu_1,0}^{(i)}$ and a third degeneracy is obtained. If the latter happens to coincide with $\tilde{\mathcal{E}}_{\nu_2,0}^i$, then the degeneracy $\tilde{\mathcal{E}}_{\nu_2,0}^{(i)}$ will be said to be contained in the degeneracy $\tilde{\mathcal{E}}_{\nu_1,0}^{(i)}$. A degeneracy will be considered significant in a certain category of degeneracies (related to the exponent a, for example) only if it is not contained in any other.

We can now formulate the following rule (which has been confirmed for the cases considered here): the function $\mathcal{U}^j$ being defined for a certain problem, let $E_n^l \mathcal{U}^j$ be a local approximation of $\mathcal{U}^j$ which is valid in a certain domain of validity. Contained in the latter will be the domain of validity of a significant degeneracy such that the terms composing $E_n^l \mathcal{U}^j$ will be solutions of systems of equations of the hierarchy associated with the degeneracy. Therefore, the significant approximation can be found by considering only the significant degeneracy which plays an essential role in the construction of the local approximation by greatly limiting the uncertainty in the choice of the gauge $\nu(\alpha)$.

It must also be pointed out that several local expansions can exist in the same problem simultaneously. In the present context, it is basically a question of three types of local expansions, viz:

a) in the vicinity of $t = 0$,

b) in the vicinity of $z = 0$,

c) in vicinity of infinity when $x^2 + y^2 \rightarrow \infty$, $z \rightarrow \infty$.

In Chap. 11, we will see, for example, that an initial layer in the vicinity of $t = 0$ and also a boundary layer in the vicinity of $z = 0$ must both be considered at the same time. It is appropriate to assign a local expansion to each of these layers. There is, however, only one main expansion during the actual construction stage. On the other hand, in Chap. 13, we will be confronted with an asymptotic situation necessitating the introduciton of an outer layer rejected to infinity: it is a critical study of the so-called Long model (1953) for lee waves [see Guiraud and Zeytounian (1979)]. It turns out that this Long model yields only one main representation as defined by the MMAE. Furthermore, it remains valid only when 1) the characteristic Mach number associated with the basic flow at upstream infinity is small (which is not, of course, a constraint for atmospheric flows), and 2) when the size of the obstacle (which engenders the lee waves) is small in comparison to the thickness of the troposphere. The lee waves initiated near the obstacle persist at a considerable distance and with J.P. Guiraud, we have demonstrated how an outer representation, which is necessarily in double scale, makes it possible to describe the steady evolution of these two-dimensional lee waves in an adiabatic atmosphere. Thus are we confronted with a cumulative or secular-type singular perturbation problem. As the behavior for away affects the outer representation, an MSM is called for.

5.2 The Multiple-Scale Method

The fundamental characteristics of the MSM can be summarized as follows: When the data of a problem show that the small parameter α is the ratio of two scales (of time or space), the MSM consists in first introducing two variables constructed with these scales (one of them possibly being "distorted"). Next, the formal expansion of the solution in α is considered, each coefficient of the expansion being a function of the two variables introduced (e.g., of time) which are considered as independent during the entire calculation. In order to completely determine a coefficient of this expansion, it is not enough to solve the equation where it appears for the first time. The indeterminants which necessarily remain are chosen by making sure that the equation in which the next term appears will lead to a solution which does not destroy, but on the contrary, best guarantees the validity of the approximation which is sought.

From a certain a priori knowledge of the solution, we generally assume that $\mathcal{U}$ depends on the variables t and $x \equiv (x, y, z)$ so that a rapid variation having a repetitive character analogous to an oscillatory phenomenon is made to appear. This variation is itself modulated from one moment to another and from one point to another. Such a structure is described by:

$$\mathcal{U}(t, x) = \mathcal{U}^*\left(\chi(t, x); t, x\right) \quad , \tag{5.32}$$

where χ is a so-called fast intermediary variable because the function χ varies rapidly as a function of t and x. If only dimensionless variables are used, then this property can be characterized by writing:

$$\frac{\partial \chi}{\partial t} = -\frac{\omega}{\alpha} \quad , \quad \nabla \chi = \frac{\boldsymbol{\lambda}}{\alpha} \quad , \tag{5.33}$$

where ω and $|\boldsymbol{\lambda}|$ are of the order unity. From (5.33), we deduce:

$$\frac{\partial \mathcal{U}}{\partial t} = -\frac{\omega}{\alpha}\frac{\partial \mathcal{U}^*}{\partial \chi} + \frac{\partial \mathcal{U}^*}{\partial t} , \tag{5.34a}$$

$$\nabla \mathcal{U} = -\frac{1}{\alpha}\frac{\partial \mathcal{U}^*}{\partial \chi}\boldsymbol{\lambda} + \nabla \mathcal{U}^* \quad ; \tag{5.34b}$$

with analogous formulas for the higher order derivatives. By substituting these expressions for the derivatives into the basic equations, the small parameter α is introduced into the latter (even if it did not appear initially). We are thus led to such an approximation of $\mathcal{U}^*$ via an expansion, for example:

$$\begin{aligned} \mathcal{U}^* &= \mathcal{U}_0^* + \alpha \mathcal{U}_1^* + \dots \quad ; \\ \chi &= \chi_0 + \alpha \chi_1 + \dots \quad . \end{aligned} \tag{5.35}$$

By substitution within the equations (where expressions of the type (5.34) have previously been introduced) and by setting at zero the terms proportional to the successive powers of α, a hierarchy of systems of equations is obtained for $\mathcal{U}_0^*, \mathcal{U}_1^*, \dots$ The first system in this hierarchy determines at best $\mathcal{U}_0^*$ in its dependence with respect to χ_0, but not with respect to t and x. It is usually while seeking to determine $\mathcal{U}_1^*$ – or even other higher order terms – that the dependence of $\mathcal{U}_0^*$ with respect to t and x is prescribed by the cancelling of the secular terms, i.e., of terms which in $\mathcal{U}_1^*$ do not remain bounded when χ_0 increases indefinitely. Indeed, if we want (5.35) to cover an interval O(1) in variations of t and x, then because of (5.33), this corresponds to a variation of χ_0 which is $O(1/\alpha)$.

We feel that the preceding should give the reader a sufficient idea of the MSM for understanding the applications which will be presented in the coming chapters. For complementary information, the reader might consult, for example, Nayfeh (1973), or the recent book by Kevorkian and Cole (1981). For our part, we will conclude this short report on the MSM with some brief remarks. It is obvious that a privileged domain of applicability of this method is the study of the propagation of slowly modulated waves [see the third part of Germain's article, (1977)]. As was pointed out in Sect. 4.2, $\mathcal{U}_0^*(\chi_0)$ appears, in this case, as a plane wave when the variation in t and x is ignored, whereas the latter gives the modulation of the amplitude and possibly even of the phase of the wave in question. This, however, is not the only example of application.

For the sake of simplicity, the case of several small parameters has not been included in the above although it will come up many times in what follows. In any case, we can always return to the case considered by taking α as the main small parameter and considering the others as functions thereof via the similarity relations. This type of case, with both a small and large parameter, will be encountered in Chap. 7. The reader might consult Darrozes (1972) for a more in-depth look into this problem.

6. Some Applications of the MMAE and MSM

In the following pages, four applications of the MMAE and MSM to specific meteorologial problems are presented.

The first application aims at showing how the MMAE enables us to formulate the inital condition for $\tau = 0$ associated with (4.15) which describes a quasi-geostrophic flow. This question is taken up again in a general context in Chap. 11 which deals with quasi-geostrophic and ageostrophic models.

The second application is related to the Boussinesq approximation. It illustrates in a convincing manner the possibilities of the MSM for describing wave phenomena. The whole of Chap. 8 deals with the Boussinesq approximation.

The third application concerns the hydrostatic limiting process when the space-time variables take on different limiting values. This application will be of value in Chap. 13.

Finally, via a "triple deck" model, the fourth application will make it possible to clarify the perturbing influence of a thermally nonhomogeneous local site on an Ekman boundary layer. We will see in Chap. 11 how the Ekman boundary layer naturally comes into question in an asymptotic theory where Ro and Ek are small compared to unity.

6.1 Application of the MMAE to Adiabatic Flows with Small Kibel Numbers

Let us go back to the primitive equations (4.3). We have seen that when the following main limiting process is carried out:

$$\lim_{\mathrm{Ki}\to 0}^{P} = \left\{\mathrm{Ki}\to 0 \quad , \quad \mathrm{M}_\infty \to 0 \quad ; \quad \lambda_0 = \mathrm{O}(1) \quad ; \quad \tau, \xi, \eta, \zeta \ \text{fixed}\right\} \quad , \tag{6.1}$$

then:

$$\begin{aligned}
U &= -\lambda_0 \frac{\partial \mathcal{H}_1}{\partial \eta} + \mathrm{O}(\mathrm{Ki}) \quad , \\
V &= +\lambda_0 \frac{\partial \mathcal{H}_1}{\partial \xi} + \mathrm{O}(\mathrm{Ki}) \quad , \\
W &= -(\mathrm{Ki}\zeta^2/K_0(\zeta)) \frac{D_0}{D\tau}\left(\frac{\partial \mathcal{H}_1}{\partial \zeta}\right) + \mathrm{O}(\mathrm{Ki}^2) \quad , \\
\overline{T} &= -\mathrm{Bo}\zeta \frac{d\mathcal{H}_0}{d\zeta} - \mathrm{Ki}\,\mathrm{Bo}\zeta \frac{\partial \mathcal{H}_1}{\partial \zeta} + \mathrm{O}(\mathrm{Ki}^2) \quad , \\
\mathcal{H} &= \mathcal{H}_0(\zeta) + \mathrm{Ki}\mathcal{H}_1 + \mathrm{O}(\mathrm{Ki}^2) \quad ,
\end{aligned} \tag{6.2}$$

and $\mathcal{H}_1(\tau, \xi, \eta, \zeta)$ satisfies (4.15):

$$\frac{\mathrm{S}}{\lambda_0} \frac{D_0}{D\tau}(\wedge \mathcal{H}_1) + \beta \frac{\partial \mathcal{H}_1}{\partial \xi} = 0 \quad . \tag{6.3}$$

In order to solve (6.3), an initial condition in $\tau = 0$ is necessary. It is emphasized right away that the system of primitive equations (4.3) is third order in τ and thus the main limiting process (6.1) eliminates two orders in τ. This leads to (6.3) where only *one* derivative in τ remains. Hence it follows that:

$$\overset{P}{\lim_{\mathrm{Ki} \to 0}} \left(\lim_{\tau \to 0} \right) \neq \lim_{\tau \to 0} \left(\overset{P}{\lim_{\mathrm{Ki} \to 0}} \right) \quad , \tag{6.4}$$

which shows the vicinity of $\tau = 0$ to be singular during the limiting process (6.1).

So that the limiting flow in the vicinity of $\tau = 0$ can be studied when $\mathrm{Ki} \to 0$ and $\lambda = 0(1)$, the following short time must be introduced:

$$\tilde{\tau} = \frac{\tau}{\mathrm{Ki}} = l_0 t \quad , \quad l_0 = 2\Omega_0 \sin \varphi_0 \quad . \tag{6.5}$$

Thus, the characteristic short time $\tilde{t} \equiv 1/l_0$ is on the order of 3 to 4 hours at a mean latitude ($|l_0| \approx 10^{-4}\,\mathrm{sec}^{-1}$). It is precisely during this interval that the real wind (in a quasi-static atmosphere) adjusts[1] to the geostrophic wind and all the internal gravity waves still present in the primitive equations (4.3) are filtered out. First of all, we have to rewrite (4.3) with the new variables $\tilde{\tau}, \xi, \eta, \zeta$ and then carry out the local limiting process:

$$\overset{l}{\lim_{\mathrm{Ki} \to 0}} = \left\{ \mathrm{Ki} \to 0, \, \mathrm{M}_\infty \to 0; \;\; \lambda_0 = \mathrm{O}(1); \;\; \tilde{\tau}, \xi, \eta, \zeta \text{ fixed} \right\} \quad . \tag{6.6}$$

On (6.6), the following local asymptotic representation must be superimposed:

$$\begin{aligned} U &= \tilde{U}_0 + \mathrm{O}(\mathrm{Ki}) \quad , \\ V &= \tilde{V}_0 + \mathrm{O}(\mathrm{Ki}) \quad , \\ W &= \tilde{W}_0 + \mathrm{O}(\mathrm{Ki}) \quad , \\ \mathcal{H} &= \mathcal{H}_0(\zeta) + \mathrm{Ki}\tilde{\mathcal{H}}_1 + \mathrm{O}(\mathrm{Ki}^2) \quad , \\ \overline{T} &= -\mathrm{Bo}\zeta \frac{d\mathcal{H}_0}{d\zeta} - \mathrm{Ki}\,\mathrm{Bo}\zeta \frac{\partial \tilde{\mathcal{H}}_1}{\partial \zeta} + \mathrm{O}(\mathrm{Ki}^2) \quad , \end{aligned} \tag{6.7}$$

where $\tilde{U}_0$, $\tilde{V}_0$, $\tilde{W}_0$ and $\tilde{\mathcal{H}}_1$ satisfy the following local limiting system:

$$\begin{aligned} &\frac{\partial \tilde{U}_0}{\partial \tilde{\tau}} - \tilde{U}_0 = -\lambda_0 \frac{\partial \tilde{\mathcal{H}}_1}{\partial \xi} \quad , \qquad && \frac{\partial \tilde{V}_0}{\partial \tilde{\tau}} + \tilde{V}_0 = -\lambda_0 \frac{\partial \tilde{\mathcal{H}}_1}{\partial \eta} \quad , \\ &\frac{\partial \tilde{U}_0}{\partial \xi} + \frac{\partial \tilde{V}_0}{\partial \eta} + \mathrm{S}\frac{\partial \tilde{W}_0}{\partial \xi} = 0 \quad , && \frac{\partial}{\partial \tilde{\tau}} \left(\frac{\partial \tilde{\mathcal{H}}_1}{\partial \zeta} \right) + \frac{K_0(\zeta)}{\zeta^2} \tilde{W}_0 = 0 \quad . \end{aligned} \tag{6.8}$$

---

[1] A detailed analysis of the adjustment to geostrophy problem can be found in the work by W. Blumen: *Reviews of Geophysics and Space Physics*, Vol. 10, no. 2, pp. 485–528, 1972.

For the MMAE to be applicable, it is necessary that:

$$\lim_{\tilde{\tau}\to\infty}\begin{pmatrix}\tilde{U}_0\\ \tilde{V}_0\\ \tilde{W}_0\\ \tilde{\mathcal{H}}_1\end{pmatrix} = \lim_{\tau\to 0}\begin{pmatrix}-\lambda_0\frac{\partial\mathcal{H}_1}{\partial\eta}\\ +\lambda_0\frac{\partial\mathcal{H}_1}{\partial\xi}\\ 0\\ \mathcal{H}_1\end{pmatrix} \quad . \tag{6.9}$$

The above is possible only if we assume that:

$$\frac{\partial\tilde{U}_0}{\partial\tilde{\tau}}\to 0 \quad , \quad \frac{\partial\tilde{V}_0}{\partial\tilde{\tau}}\to 0 \quad , \quad \frac{\partial\tilde{\mathcal{H}}_1}{\partial\tilde{\tau}}\to 0 \quad , \quad \text{when } \tilde{\tau}\to\infty \quad . \tag{6.10}$$

From (6.8), the following equation results for $\tilde{\mathcal{H}}_1$

$$\mathrm{S}\frac{\partial^3}{\partial\tilde{\tau}^2\partial\zeta}\left(\frac{\zeta^2}{K_0(\zeta)}\frac{\partial\tilde{\mathcal{H}}_1}{\partial\zeta}\right) + \wedge\tilde{\mathcal{H}}_1$$

$$= \frac{\partial V^0}{\partial\xi} - \frac{\partial U^0}{\partial\eta} + \mathrm{S}\frac{\partial}{\partial\zeta}\left(\frac{\zeta^2}{K_0(\zeta)}\frac{\partial H^0}{\partial\zeta}\right) \quad , \tag{6.11}$$

once these initial conditions have been imposed on the limiting system (6.8):

$$\tilde{\tau} = 0: \qquad \tilde{U}_0 = U^0 \quad , \quad \tilde{V}_0 = V^0 \quad , \quad \mathcal{H}_1 = H^0 \quad . \tag{6.12}$$

A slip condition at $\zeta = 1$ must be added to (6.11) since the adjustment to geostrophy takes place essentially outside of the boundary layers (which are not taken into account in the present formulation). An argument similar to the one which led to (4.19) then yields:

$$\frac{\partial}{\partial\tilde{\tau}}\left\{\tilde{\mathcal{H}}_1 + \frac{1}{\mathrm{Bo}}\frac{T_0(1)}{K_0(1)}\frac{\partial\tilde{\mathcal{H}}_1}{\partial\zeta}\right\} = 0 \quad , \quad \text{on } \zeta = 1 \quad . \tag{6.13}$$

A second condition in ζ must bound the solution of (6.11) at infinity when $\zeta\to 0$. Finally, the following two initial conditions must be added to (6.11):

$$\tilde{\tau} = 0 : \tilde{\mathcal{H}}_1 = H^0 \quad ;$$

$$\mathrm{S}\frac{\partial}{\partial\zeta}\left[\frac{\zeta^2}{K_0(\zeta)}\frac{\partial}{\partial\zeta}\left(\frac{\partial\tilde{\mathcal{H}}_1}{\partial\tilde{\tau}}\right)\right] = \frac{\partial U^0}{\partial\xi} + \frac{\partial V^0}{\partial\eta} \quad . \tag{6.14}$$

The problem thus lies in accepting that the solution to problem (6.11–14) indeed leads to a solution $\mathcal{H}_1$ such that $(\partial\tilde{\mathcal{H}}_1/\partial\tilde{\tau})\to 0$, with $\tilde{\tau}\to\infty$. If such is the case, according to the matching condition (6.9), we can assign to (6.3) the initial condition:

$$\wedge\mathcal{H}_1\Big|_{\tau=0} = \frac{\partial V^0}{\partial\xi} - \frac{\partial U^0}{\partial\eta} + \mathrm{S}\frac{\partial}{\partial\zeta}\left(\frac{\zeta^2}{K_0(\zeta)}\frac{\partial H^0}{\partial\zeta}\right) \quad . \tag{6.15}$$

To be assured that $\partial\tilde{U}_0/\partial\tilde{\tau}$ and $\partial\tilde{V}_0/\partial\tilde{\tau}$ do indeed tend towards zero when $\tilde{\tau}\to\infty$, we can take advantage of the equations:

$$\begin{aligned}\frac{\partial^2\tilde{U}_0}{\partial\tilde{\tau}^2}+\tilde{U}_0 &= -\lambda_0\left(\frac{\partial^2\tilde{\mathcal{H}}_1}{\partial\tilde{\tau}\partial\xi}+\frac{\partial\tilde{\mathcal{H}}_1}{\partial\eta}\right) \quad ; \\ \frac{\partial^2\tilde{V}_0}{\partial\tilde{\tau}^2}+\tilde{V}_0 &= -\lambda_0\left(\frac{\partial^2\tilde{\mathcal{H}}_1}{\partial\tilde{\tau}\partial\eta}-\frac{\partial\tilde{\mathcal{H}}_1}{\partial\xi}\right) \quad ,\end{aligned} \tag{6.16}$$

to which the following initial conditions must be added:

$$\tilde{\tau}=0: \begin{cases} \tilde{U}_0=U^0 \quad , & \dfrac{\partial\tilde{U}_0}{\partial\tilde{\tau}}=V^0-\lambda_0\dfrac{\partial H^0}{\partial\xi} \quad , \\ \tilde{V}_0=V^0 \quad , & \dfrac{\partial\tilde{V}_0}{\partial\tilde{\tau}}=-U^0-\lambda_0\dfrac{\partial H^0}{\partial\eta} \quad . \end{cases} \tag{6.17}$$

6.2 Double-Scale Structure of the Boussinesq Waves: Linear Theory [2]

In the context of an adiabatic atmosphere, the basic equations of Subsect. 2.4.1 can be adopted. In Subsect. 3.3.3, it was indicated that according to Zeytounian (1974), the Boussinesq approximation is associated with the limiting process [see (3.61)]:

$$\mathrm{Bo}=\hat{B}\mathrm{M}_\infty \quad , \quad \hat{B}=\mathrm{O}(1) \quad \text{and} \quad \mathrm{M}_\infty\to 0 \quad , \tag{6.18}$$

all the while assuming that:

$$\pi=\mathrm{M}_\infty^2\hat{\pi} \quad , \quad \omega=\mathrm{M}_\infty\hat{\omega} \quad , \quad \vartheta=\mathrm{M}_\infty\hat{\vartheta} \quad . \tag{6.19}$$

By taking into account (6.18, 19) and by introducing the "slow" vertical coordinate:

$$\zeta=\frac{\overline{z}_\infty}{\hat{B}}=\mathrm{M}_\infty\overline{z} \quad , \tag{6.20}$$

the following generalized Boussinesq-type equations can replace system (2.72):

$$\mathrm{S}\frac{D\overline{u}}{D\overline{t}}+\frac{\overline{T}_\infty(\hat{B}\zeta)}{\gamma}\frac{\partial\hat{\pi}}{\partial\overline{x}}=-\mathrm{M}_\infty\hat{\omega}\mathrm{S}\frac{D\overline{u}}{D\overline{t}} \quad ; \tag{6.21a}$$

$$\mathrm{S}\frac{D\overline{v}}{D\overline{t}}+\frac{\overline{T}_\infty(\hat{B}\zeta)}{\gamma}\frac{\partial\hat{\pi}}{\partial\overline{y}}=-\mathrm{M}_\infty\hat{\omega}\mathrm{S}\frac{D\overline{v}}{D\overline{t}} \quad ; \tag{6.21b}$$

$$\varepsilon^2\mathrm{S}\frac{D\overline{w}}{D\overline{t}}+\frac{\overline{T}_\infty(\hat{B}\zeta)}{\gamma}\frac{\partial\hat{\pi}}{\partial\overline{z}}-\frac{\hat{B}}{\gamma}\hat{\vartheta}=-\mathrm{M}_\infty\hat{\omega}\left[\varepsilon^2\mathrm{S}\frac{D\overline{w}}{D\overline{t}}-\frac{\hat{B}}{\gamma}\hat{\vartheta}\right] \quad ; \tag{6.21c}$$

[2] The following was strongly influenced by the thesis of Bois (1979). The reader can consult Chaps. 3 and 4 of this work (Université P. et M. Curie, Paris 6, Jussieu).

$$\frac{\partial\overline{u}}{\partial\overline{x}}+\frac{\partial\overline{v}}{\partial\overline{y}}+\frac{\partial\overline{w}}{\partial\overline{z}} = -\mathrm{M}_\infty\left[\mathrm{S}\frac{D\hat{\omega}}{D\overline{t}}+\hat{\omega}\left(\frac{\partial\overline{u}}{\partial\overline{x}}+\frac{\partial\overline{v}}{\partial\overline{y}}+\frac{\partial\overline{w}}{\partial\overline{z}}\right)\right.$$
$$\left.-\left(\hat{B}+\frac{d\overline{T}_\infty(\hat{B}\zeta)}{d\zeta}\right)\frac{\overline{w}}{\overline{T}_\infty(\hat{B}\zeta)}\right]+\mathrm{M}_\infty^2\hat{\omega}\left(\hat{B}+\frac{d\overline{T}_\infty(\hat{B}\zeta)}{d\zeta}\right)\frac{\overline{w}}{\overline{T}_\infty(\hat{B}\zeta)} \quad ; \tag{6.21d}$$

$$\mathrm{S}\frac{D\hat{\vartheta}}{D\overline{t}}+\left[\hat{B}\frac{\gamma-1}{\gamma}+\frac{d\overline{T}_\infty(\hat{B}\zeta)}{d\zeta}\right]\frac{\overline{w}}{\overline{T}_\infty(\hat{B}\zeta)}$$
$$= -\mathrm{M}_\infty\left[\hat{\omega}\mathrm{S}\frac{D\hat{\vartheta}}{D\overline{t}}-\frac{\gamma-1}{\gamma}\mathrm{S}\frac{D\hat{\pi}}{D\overline{t}}\right]$$
$$-\mathrm{M}_\infty^2\hat{\pi}\left[\hat{B}\frac{\gamma-1}{\gamma}+\frac{d\overline{T}_\infty(\hat{B}\zeta)}{d\zeta}\right]\frac{\overline{w}}{\overline{T}_\infty(\hat{B}\zeta)} \quad ; \tag{6.21e}$$

$$\hat{\vartheta}+\hat{\omega} = \mathrm{M}_\infty(\hat{\pi}-\hat{\omega}\hat{\vartheta}) \quad . \tag{6.21f}$$

Equations (6.21) "generalizes" the Boussinesq equations in so much as they are obtained according to Zeytounian (1974) when:

$$\mathrm{M}_\infty \to 0 \quad , \quad \overline{t},\ \overline{x},\ \overline{y},\ \overline{z} \ \text{ fixed} \tag{6.22}$$

provided (6.20) has been taken into account. They are thus written:

$$\mathrm{S}\frac{D\overline{u}_\mathrm{B}}{D\overline{t}}+\frac{1}{\gamma}\frac{\partial\hat{\pi}_\mathrm{B}}{\partial\overline{x}} = 0 \quad ; \qquad \mathrm{S}\frac{D\overline{v}_\mathrm{B}}{D\overline{t}}+\frac{1}{\gamma}\frac{\partial\hat{\pi}_\mathrm{B}}{\partial\overline{y}} = 0 \quad ;$$
$$\varepsilon^2\mathrm{S}\frac{D\overline{w}_\mathrm{B}}{D\overline{t}}+\frac{1}{\gamma}\frac{\partial\hat{\pi}_B}{\partial\overline{z}}-\frac{\hat{B}}{\gamma}\hat{\vartheta}_\mathrm{B} = 0 \quad ;$$
$$\frac{\partial\overline{u}_\mathrm{B}}{\partial\overline{x}}+\frac{\partial\overline{v}_\mathrm{B}}{\partial\overline{y}}+\frac{\partial\overline{w}_\mathrm{B}}{\partial\overline{z}} = 0 \quad ; \tag{6.23}$$
$$\mathrm{S}\frac{D\hat{\vartheta}_\mathrm{B}}{D\overline{t}}+\hat{B}\left[\frac{\gamma-1}{\gamma}+\left(\frac{d\overline{T}_\infty}{d\overline{z}_\infty}\right)_{\overline{z}_\infty=0}\right]\overline{w}_\mathrm{B} = 0 \quad ;$$
$$\hat{\vartheta}_\mathrm{B} = -\hat{\omega}_\mathrm{B} \quad ,$$

once it has been understood that $\overline{T}_\infty(0)\equiv 1$ since we are dealing with dimensionless functions. In (6.21), it was assumed that $\beta\equiv 0$ and $\mathrm{Ro}\equiv\infty$. Finally, in (6.23), we have:

$$\mathrm{S}\frac{D}{D\overline{t}} = \mathrm{S}\frac{\partial}{\partial\overline{t}}+\overline{u}_\mathrm{B}\frac{\partial}{\partial\overline{x}}+\overline{v}_\mathrm{B}\frac{\partial}{\partial\overline{y}}+\overline{w}_\mathrm{B}\frac{\partial}{\partial\overline{z}} \quad .$$

As:

$$\frac{\zeta}{\overline{z}} = \mathrm{M}_\infty \ll 1 \quad ,$$

a multiple-scale method is called for. However, experience has shown that $\overline{z}$ cannot be taken as a fast variable since the latter should be a non-linear function of the slow variable. Hence:

$$\zeta = \mathrm{M}_\infty \overline{z} \quad , \quad \xi = \frac{\varphi(\zeta)}{\mathrm{M}_\infty} \quad , \tag{6.24}$$

where the function $\varphi(\zeta)$ should be determined during the research into the double scale solution.

Let us call u^*, v^*, w^*, π^*, ω^* and ϑ^* the unknowns of our problem which are supposed to be functions of t, x, y, ζ and ξ, and also of M_∞ in such a way that[3]:

$$\frac{\partial}{\partial z} = \frac{d\varphi}{d\zeta}\frac{\partial}{\partial \xi} + \mathrm{M}_\infty \frac{\partial}{\partial \zeta} \quad . \tag{6.25}$$

Since we are only dealing with a linear theory here, the expansions, supposed uniformly valid, are written as follows:

$$\begin{pmatrix} u^* \\ v^* \\ w^* \\ \pi^* \\ \omega^* \\ \vartheta^* \end{pmatrix} \equiv \mathcal{U}^* = \eta\left\{\mathcal{U}^*_{00} + \mathrm{M}_\infty \mathcal{U}^*_{01} + \mathrm{O}(\eta)\right\} \quad , \tag{6.26}$$

where $\mathcal{U}^*_{00}$ and $\mathcal{U}^*_{01}$ are functions of t, x, y, ζ and ξ. In the expansion (6.26), η designates the second small parameter which characterizes the amplitude – assumed small in linear theory – of the Boussinesq waves.

If now (6.26) is substituted into (6.21) while taking into account (6.25), and if the terms proportional to η, then ηM_∞, are identified, two limiting systems of equations are obtained with respect to $\mathcal{U}^*_{00}$ and $\mathcal{U}^*_{01}$. It should be noted that for the forthcoming linear study, it is not necessary to write the terms in (6.26) whose order in M_∞ is greater than one; the same holds for the terms of order η^2 and $\eta^2\mathrm{M}_\infty$.

To the order $(0,0)$, the limiting equations for u^*_{00}, v^*_{00}, w^*_{00}, π^*_{00}, ω^*_{00} and ϑ^*_{00} have the form:

$$\begin{aligned}
&\mathrm{S}\frac{\partial u^*_{00}}{\partial t} + \frac{T_\infty(\hat{B}\zeta)}{\gamma}\frac{\partial \pi^*_{00}}{\partial x} = 0 \quad ; \\
&\mathrm{S}\frac{\partial v^*_{00}}{\partial t} + \frac{T_\infty(\hat{B}\zeta)}{\gamma}\frac{\partial \pi^*_{00}}{\partial y} = 0 \quad ; \\
&\varepsilon^2\mathrm{S}\frac{\partial w^*_{00}}{\partial t} + \frac{T_\infty(\hat{B}\zeta)}{\gamma}\frac{\partial \pi^*_{00}}{\partial \xi}\frac{d\varphi}{d\zeta} - \frac{\hat{B}}{\gamma}\vartheta^*_{00} = 0 \quad ; \\
&\frac{\partial u^*_{00}}{\partial x} + \frac{\partial v^*_{00}}{\partial y} + \frac{d\varphi}{d\zeta}\frac{\partial w^*_{00}}{\partial \xi} = 0 \quad ; \\
&\mathrm{S}\frac{\partial \vartheta^*_{00}}{\partial t} + \left[\hat{B}\frac{\gamma-1}{\gamma} + \frac{dT_\infty(\hat{B}\zeta)}{d\zeta}\right]\frac{w^*_{00}}{T_\infty(\hat{B}\zeta)} = 0 \quad ; \\
&\omega^*_{00} = -\vartheta^*_{00} \quad .
\end{aligned} \tag{6.27}$$

[3] All bars and hats are eliminated from the dimensionless values.

It is pointed out right away that a two-dimensional problem in x, ζ, ξ can, in fact, be considered since the Coriolis force has been ignored (this is the procedure in what follows). The solution to (6.27) can be sought in the form of a progressive wave:

$$\mathcal{U}^*_{00} = \tilde{\mathcal{U}}^*_{00}(\xi, \zeta)\exp(\mathrm{i}\theta) \quad , \tag{6.28}$$

with

$$\theta = \sigma\frac{t}{\mathrm{S}} - kx \quad , \tag{6.29}$$

in the two-dimensional case.

Thus, for $\tilde{u}^*_{00}$, $\tilde{w}^*_{00}$, $\tilde{\pi}^*_{00}$ and $\tilde{\vartheta}^*_{00}$ which depend on ξ and ζ, the following equations replace (6.27):

$$\begin{aligned}
\tilde{u}^*_{00} &= \frac{k}{\sigma}\frac{1}{\gamma}T_\infty(\hat{B}\zeta)\tilde{\pi}^*_{00} \quad ; \\
\mathrm{i}\tilde{w}^*_{00} &= -\frac{1}{\sigma}\frac{1}{\gamma}T_\infty(\hat{B}\zeta)\frac{d\varphi}{d\zeta}\frac{\partial\tilde{\pi}^*_{00}}{\partial\xi} + \frac{\hat{B}}{\sigma\gamma}\tilde{\vartheta}^*_{00} \quad ; \\
\tilde{u}^*_{00} &= -\mathrm{i}\frac{1}{k}\frac{d\varphi}{d\zeta}\frac{\partial\tilde{w}^*_{00}}{\partial\xi} \quad ; \\
\tilde{\vartheta}^*_{00} &= \frac{\mathrm{i}}{\sigma}\left[\hat{B}\frac{\gamma-1}{\gamma} + \frac{dT_\infty(\hat{B}\zeta)}{d\zeta}\right]\frac{\tilde{w}^*_{00}}{T_\infty(\hat{B}\zeta)} \quad ,
\end{aligned} \tag{6.30}$$

where it has been assumed that $\varepsilon \equiv 1$. From (6.30), a single equation is deduced for $\tilde{w}^*_{00}$:

$$\sigma^2\left(\frac{d\varphi}{d\zeta}\right)^2\frac{\partial^2\tilde{w}^*_{00}}{\partial\xi^2} + k^2\left[\Gamma_\infty(\zeta) - \sigma^2\right]\tilde{w}^*_{00} = 0 \quad , \quad \text{where} \tag{6.31}$$

$$\Gamma_\infty(\zeta) = \frac{\hat{B}}{\gamma}\left[\hat{B}\frac{\gamma-1}{\gamma} + \frac{dT_\infty(\hat{B}\zeta)}{d\zeta}\right]\frac{1}{T_\infty(\hat{B}\zeta)} \quad . \tag{6.32}$$

In (6.31), we can introduce the wave number ω_B which is defined by:

$$\omega^2_\mathrm{B} = \frac{k^2}{(d\varphi/d\zeta)^2}\left[\frac{\Gamma_\infty(\zeta)}{\sigma^2} - 1\right] \quad . \tag{6.33}$$

The fact that ω_B can arbitrarily be chosen constant [otherwise it would suffice to replace ξ by $\tilde{\xi} \equiv \omega_\mathrm{B}\xi$ in (6.24)] demonstrates that ω_B can be taken equal to unity. Thus:

$$\frac{d\varphi}{d\zeta} = \frac{k}{\sigma}\sqrt{\Gamma_\infty(\zeta) - \sigma^2} \quad . \tag{6.34}$$

This formula indicates in what way the variable $\xi = \varphi(\zeta)/\mathrm{M}_\infty$ depends on the slow variable $\zeta = \mathrm{M}_\infty z$.

The solution to the equation:

$$\frac{\partial^2\tilde{w}^*_{00}}{\partial\xi^2} + \tilde{w}^*_{00} = 0 \quad . \tag{6.35}$$

is oscillatory

$$\tilde{w}^*_{00+} = \mathcal{A}_{00}(\zeta)\,\mathrm{e}^{\mathrm{i}\xi} + \mathcal{B}_{00}(\zeta)\,\mathrm{e}^{-\mathrm{i}\xi} \quad . \tag{6.36}$$

Thus, in this case: $\Gamma_\infty(\zeta) > \sigma^2$.

However, the solution to:

$$\frac{\partial^2 \tilde{w}^*_{00}}{\partial \xi^2} - \tilde{w}^*_{00} = 0 \tag{6.37}$$

is non-periodic:

$$\tilde{w}^*_{00-} = \mathcal{C}_{00}(\zeta)\,\mathrm{e}^{\xi} + \mathcal{D}_{00}(\zeta)\,\mathrm{e}^{-\xi} \quad . \tag{6.38}$$

In this case: $\Gamma_\infty(\zeta) < \sigma^2$.

When

$$\Gamma_\infty(\zeta_0) \equiv \sigma^2 \tag{6.39}$$

for $\zeta = \zeta_0$, there is a cut-off phenomenon at the point $\zeta = \zeta_0$ which leads to a turning point problem (see, for example, Nayfeh 1973, p. 335–360). The problem thus arises of knowing how the transition can take place between the two types of solutions (6.36) and (6.38) valid on either side of the point $\zeta = \zeta_0$ in such a way as to verify (6.39).

The functions $\mathcal{A}_{00}(\zeta)$, $\mathcal{B}_{00}(\zeta)$, $\mathcal{C}_{00}(\zeta)$ and $\mathcal{D}_{00}(\zeta)$ are undetermined at this stage [to order (0,0)] and the manner in which they depend on ζ is obtained by cancelling in the limiting equations to order (0,1), the terms which give rise to secular terms.

To order (0,1), the limiting equations for u^*_{01}, w^{**}_{01}, π^*_{01}, ω^*_{01} and ϑ^*_{01} are (again in the two-dimensional case):

$$\begin{aligned}
&\mathrm{S}\frac{\partial u^*_{01}}{\partial t} + \frac{T_\infty(\hat{B}\zeta)}{\gamma}\frac{\partial \pi^*_{01}}{\partial x} = 0 \quad ; \\
&\varepsilon^2\mathrm{S}\frac{\partial w^*_{01}}{\partial t} + \frac{T_\infty(\hat{B}\zeta)}{\gamma}\frac{\partial \pi^*_{01}}{\partial \xi}\frac{d\varphi}{d\zeta} - \frac{\hat{B}}{\gamma}\vartheta^*_{01} = -\frac{T_\infty(\hat{B}\zeta)}{\gamma}\frac{\partial \pi^*_{00}}{\partial \zeta} \quad ; \\
&\frac{\partial u^*_{01}}{\partial x} + \frac{d\varphi}{d\zeta}\frac{\partial w^*_{01}}{\partial \xi} = \left[\hat{B} + \frac{dT_\infty(\hat{B}\zeta)}{d\zeta}\right]\frac{w^*_{00}}{T_\infty(\hat{B}\zeta)} \\
&\qquad - \left(\mathrm{S}\frac{\partial \omega^*_{00}}{\partial t} + \frac{\partial w^*_{00}}{\partial \zeta}\right) \quad ; \\
&\omega^*_{01} = \pi^*_{01} - \vartheta^*_{01} \quad ; \\
&\mathrm{S}\frac{\partial \vartheta^*_{01}}{\partial t} + \left[\hat{B}\frac{\gamma-1}{\gamma} + \frac{dT_\infty(\hat{B}\zeta)}{d\zeta}\right]\frac{w^*_{01}}{T_\infty(\hat{B}\zeta)} = \frac{\gamma-1}{\gamma}\mathrm{S}\frac{\partial \pi^*_{00}}{\partial t} \quad .
\end{aligned} \tag{6.40}$$

Once again, it is supposed that:

$$\begin{aligned}
&\mathcal{U}^*_{01} = \tilde{\mathcal{U}}^*_{01}(\xi, \zeta)\exp(\mathrm{i}\theta) \quad , \\
&\theta = \sigma\frac{t}{\mathrm{S}} - kx \quad .
\end{aligned} \tag{6.41}$$

Then (when $\varepsilon \equiv 1$), we find just one equation for $\tilde{w}^*_{01}$:

$$\sigma^2 \left(\frac{d\varphi}{d\zeta}\right)^2 \frac{\partial^2 \tilde{w}^*_{01}}{\partial \xi^2} + k^2 \left[\Gamma_\infty(\zeta) - \sigma^2\right] \tilde{w}^*_{01}$$
$$= \sigma^2 \frac{d\varphi}{d\zeta} \left\{ -2 \frac{\partial^2 \tilde{w}^*_{00}}{\partial \zeta \partial \xi} + \left[\hat{B} + \frac{dT_\infty(\hat{B}\zeta)}{d\zeta}\right] \frac{1}{T_\infty(\hat{B}\zeta)} \frac{\partial \tilde{w}^*_{00}}{\partial \xi} - \frac{d^2\varphi/d\zeta^2}{d\varphi/d\zeta} \frac{\partial \tilde{w}^*_{00}}{\partial \xi} \right\} . \tag{6.42}$$

The writing of the second term of (6.42) takes into account (6.30) which makes it possible to express $\tilde{\pi}^*_{00}$ and $\tilde{\omega}^*_{00} \equiv -\tilde{\vartheta}^*_{00}$ as functions of $\tilde{w}^*_{00}$.

If we consider the case: $\Gamma_\infty(\zeta) > \sigma^2$ while taking into account (6.34) as well as the corresponding oscillatory solution (6.36), then instead of (6.42), the following equation for $\tilde{w}^*_{01}$ is found:

$$\frac{\partial^2 \tilde{w}^*_{01}}{\partial \xi^2} + \tilde{w}^*_{01} = -\mathrm{i} \left(\frac{d\varphi}{d\zeta}\right)^{-1} \left[\left\{ 2 \frac{d\mathcal{A}_{00}}{d\zeta} - \left[\left(\hat{B} + \frac{dT_\infty(\hat{B}\zeta)}{d\zeta}\right) \frac{1}{T_\infty(\hat{B}\zeta)} - \frac{d^2\varphi/d\zeta^2}{d\varphi/d\zeta} \right] \mathcal{A}_{00} \right\} \mathrm{e}^{\mathrm{i}\xi} - \left\{ 2 \frac{d\mathcal{B}_{00}}{d\zeta} - \left[\left(\hat{B} + \frac{dT_\infty(\hat{B}\zeta)}{d\zeta}\right) \frac{1}{T_\infty(\hat{B}\zeta)} - \frac{d^2\varphi/d\zeta^2}{d\varphi/d\zeta} \right] \mathcal{B}_{00} \right\} \mathrm{e}^{-\mathrm{i}\xi} \right] , \tag{6.43}$$

whose solution is:

$$\tilde{w}^*_{01} = \mathcal{A}_{01}(\zeta)\,\mathrm{e}^{\mathrm{i}\xi} + \mathcal{B}_{01}(\zeta)\,\mathrm{e}^{-\mathrm{i}\xi} - \frac{1}{2}\sigma^2 \frac{d\varphi}{d\zeta} \left[\xi \{ \quad \} \mathrm{e}^{\mathrm{i}\xi} + \xi \{ \quad \} \mathrm{e}^{-\mathrm{i}\xi} \right] .$$

As the expansion (6.26) must be uniformly valid regardless of the values of the independent variables, it is necessary to cancel the secular terms proportional to $\xi\,\mathrm{e}^{\mathrm{i}\xi}$ and $\xi\,\mathrm{e}^{-\mathrm{i}\xi}$. Thus result two relations for determining the functions $\mathcal{A}_{00}(\zeta)$ and $\mathcal{B}_{00}(\zeta)$ which were unknown in the solution (6.36):

$$2 \frac{d\mathcal{A}_{00}}{d\zeta} = -\frac{d}{d\zeta} \left\{ \log \left[\varrho_\infty(\hat{B}\zeta) \frac{d\varphi}{d\zeta} \right] \right\} \mathcal{A}_{00} \quad ;$$
$$2 \frac{d\mathcal{B}_{00}}{d\zeta} = -\frac{d}{d\zeta} \left\{ \log \left[\varrho_\infty(\hat{B}\zeta) \frac{d\varphi}{d\zeta} \right] \right\} \mathcal{B}_{00} \quad , \tag{6.44}$$

as long as the following has been taken into account:

$$\left(\hat{B} + \frac{dT_\infty(\hat{B}\zeta)}{d\zeta}\right) \frac{1}{T_\infty(\hat{B}\zeta)} \equiv -\frac{d \log \varrho_\infty(\hat{B}\zeta)}{d\zeta} . \tag{6.45}$$

Finally, we obtain the expressions:

$$\mathcal{A}_{00}(\zeta) = \frac{A^0}{[\varrho_\infty(\hat{B}\zeta) d\varphi/d\zeta]^{1/2}} \quad ; \quad \mathcal{B}_{00}(\zeta) = \frac{B^0}{[\varrho_\infty(\hat{B}\zeta) d\varphi/d\zeta]^{1/2}} , \tag{6.46}$$

where A^0 and B^0 are two integration constants.

When the non-periodic case corresponding to $\Gamma_\infty(\zeta) < \sigma^2$ is considered, it is seen that the determining of $\mathcal{C}_{00}(\zeta)$ and $\mathcal{D}_{00}(\zeta)$ by cancelling out the secular terms leads to a system analogous to (6.44) and the following is obtained:

$$\mathcal{C}_{00}(\zeta) = \frac{C^0}{[\varrho_\infty(\hat{B}\zeta)d\varphi/d\zeta]^{1/2}} \quad ; \quad \mathcal{D}_{00}(\zeta) = \frac{D^0}{[\varrho_\infty(\hat{B}\zeta)d\varphi/d\zeta]^{1/2}} \quad . \tag{6.47}$$

The two preceding solutions of $\tilde{w}^*_{00}$: $\tilde{w}^*_{00+}$ and $\tilde{w}^*_{00-}$ are valid on either side of the point ζ_0 such that $\zeta = \zeta_0$. Beyond the vicinity of the point $\zeta = \zeta_0$, the equation studied is:

$$\frac{\partial^2 \tilde{w}^*_{00}}{\partial \xi^2} + \frac{k^2[\Gamma_\infty(\zeta) - \sigma^2]}{\sigma^2(d\varphi/d\zeta)^2}\tilde{w}^*_{00} = 0 \quad . \tag{6.48}$$

It is then said that we are dealing with a "turning point" problem which must be solved by the MMAE via an asymptotic study in the vicinity of the turning point $\zeta = \zeta_0$. We must write:

$$\Gamma_\infty(\zeta) = \sigma^2 + (\zeta - \zeta_0)\left.\frac{d\Gamma_\infty}{d\zeta}\right|_{\zeta=\zeta_0} \quad , \tag{6.49}$$

and assume that $(d\Gamma_\infty/d\zeta)_{\zeta=\zeta_0} < 0$.

In the vicinity of $\zeta = \zeta_0$, the solution can be written in the form:

$$\mathcal{U} = \hat{\mathcal{U}}(z; \mathrm{M}_\infty)\exp(\mathrm{i}\theta) \quad , \tag{6.50}$$

with $\theta = \sigma t/\mathrm{S} - kx$. Generally speaking, we have to start with the linearized equations (6.21) [since according to (6.26), we are dealing with a linear theory with $\eta \ll \mathrm{M}_\infty$]. For the functions $\hat{u}_0$, $\hat{w}_0$, $\hat{\pi}_0$, $\hat{\omega}_0$ and $\hat{\vartheta}_0$, the following linear system then results:

$$\mathrm{i}\sigma\hat{u}_0 - \frac{\mathrm{i}k}{\gamma}T_\infty(\hat{B}\zeta)\hat{\pi}_0 = \mathrm{o}(M_\infty) \quad ;$$

$$\varepsilon^2\mathrm{i}\sigma\hat{w}_0 + \frac{1}{\gamma}T_\infty(\hat{B}\zeta)\frac{\partial\hat{\pi}_0}{\partial z} - \frac{\hat{B}}{\gamma}\hat{\vartheta}_0 = \mathrm{o}(\mathrm{M}_\infty) \quad ;$$

$$\begin{aligned}-\mathrm{i}k\hat{u}_0 + \frac{\partial\hat{w}_0}{\partial z} &= -\mathrm{M}_\infty\mathrm{i}\sigma\hat{\omega}_0 \\ &\quad - \mathrm{M}_\infty\left(\hat{B} + \frac{dT_\infty(\hat{B}\zeta)}{d\zeta}\right)\frac{\hat{w}_0}{T_\infty(\hat{B}\zeta)} + \mathrm{o}(\mathrm{M}_\infty) \quad ;\end{aligned} \tag{6.51}$$

$$\mathrm{i}\sigma\hat{\vartheta}_0 + \frac{\gamma}{\hat{B}}\Gamma_\infty(\zeta)\hat{w}_0 = \mathrm{i}\sigma\mathrm{M}_\infty\frac{\gamma-1}{\gamma}\hat{\pi}_0 + \mathrm{o}(\mathrm{M}_\infty) \quad ;$$

$$\hat{\vartheta}_0 + \hat{\omega}_0 = \mathrm{M}_\infty\hat{\pi}_0 + \mathrm{o}(\mathrm{M}_\infty) \quad .$$

From (6.51), all functions except $\hat{w}_0$ may be eliminated and a single equation for the latter can be obtained. By taking into consideration (6.45), this equation can be written in the following form:

$$\frac{\partial^2 \hat{w}_0}{\partial z^2} + k^2\left[\frac{\Gamma_\infty(\zeta)}{\sigma^2} - 1\right]\hat{w}_0$$

$$= -\mathrm{M}_\infty \frac{d}{d\zeta}\log\bigl(\varrho_\infty(\hat{B}\zeta)\bigr)\frac{\partial \hat{w}_0}{\partial z} + \mathrm{o}(\mathrm{M}_\infty) \quad , \tag{6.52}$$

where we must set $\zeta \equiv \mathrm{M}_\infty z$ and expand $\Gamma_\infty(\zeta)$ in the vicinity of ζ_0 with the help of (6.49). In this way, $\hat{w}_0(z;\mathrm{M}_\infty)$ satisfies the equation:

$$\frac{d^2\hat{w}_0}{dz^2} + \mathrm{M}_\infty \frac{k^2}{\sigma^2}\chi_\infty(\zeta_0)(z - z_0)\hat{w}_0 + \mathrm{M}_\infty \psi_\infty(\zeta_0)\frac{\partial \hat{w}_0}{\partial z} = \mathrm{o}(\mathrm{M}_\infty) \quad ; \tag{6.53}$$

$$\chi_\infty(\zeta_0) = \left.\frac{d\Gamma_\infty(\zeta)}{d\zeta}\right|_{\zeta=\zeta_0} \quad , \quad \psi_\infty(\zeta_0) = \left.\frac{d\log\left(\varrho_\infty(\hat{B}\zeta)\right)}{d\zeta}\right|_{\zeta=\zeta_0} \quad ,$$

in the vicinity of ζ_0.

We must now characterize the thickness of the region where the turning point problem is significant to the order being considered. Let us then introduce into (6.53) the inner variable:

$$\hat{z} = \mathrm{M}_\infty^\alpha (z - z_0) \tag{6.54}$$

and expand $\hat{w}_0$ in the form:

$$\hat{w}_0 = \hat{w}_{00} + \mathrm{M}_{00}^\beta \hat{w}_{01} + \ldots \quad . \tag{6.55}$$

To determine α and β, the "principle of the least degeneracy" is applied which consists in conserving the maximum number of terms in the equation resulting from (6.53–55):

$$\frac{d^2\hat{w}_{00}}{d\hat{z}^2} + \mathrm{M}_\infty^{1-3\alpha}\frac{k^2}{\sigma^2}\chi_\infty(\zeta_0)\hat{z}\hat{w}_{00} + \mathrm{M}_\infty^{1-\alpha}\psi_\infty(\zeta_0)\frac{d\hat{w}_{00}}{d\hat{z}} + \mathrm{M}_\infty^\beta \frac{d^2\hat{w}_{01}}{d\hat{z}^2}$$
$$+ \ldots = 0 \quad .$$

It is observed that the following must be imposed:

$$1 - 3\alpha = 0 \quad , \quad 1 - \alpha = \beta$$

and therefore

$$\alpha = \tfrac{1}{3} \quad , \quad \beta = \tfrac{2}{3} \quad . \tag{6.56}$$

Thus, for $\hat{w}_{00}(\hat{z})$, the so-called Airy type equation results:

$$\frac{d^2\hat{w}_{00}}{d\hat{z}^2} + \frac{k^2}{\sigma^2}\chi_\infty(\zeta_0)\hat{z}\hat{w}_{00} = 0 \quad , \tag{6.57}$$

whose solution can be written in the form[4], $\bigl((\chi_\infty(\zeta_0) < 0\bigr)$,

[4] It is recalled that the classical Airy equation is written in the form $y'' - xy = 0$ and the solution is $y(x) = a\mathrm{Ai}(x) + b\mathrm{Bi}(x)$ where Ai and Bi are the Airy functions of the first and second kind. When $x \to -\infty$, these functions have an oscillatory behavior; whereas when $x \to +\infty$, their behavior is non-periodic, i.e., exponential.

$$W_{00}(\mathrm{z}) = a\mathrm{Ai}(\mathrm{z}) + b\mathrm{Bi}(\mathrm{z}) \quad , \tag{6.58}$$

where

$$\mathrm{z} = \left(\frac{k}{\sigma}\right)^{2/3} |\chi_\infty(\zeta_0)|^{1/3} \hat{z} \quad ;$$

$$W_{00}(\mathrm{z}) \equiv \hat{w}_{00}\left(\frac{\mathrm{z}}{(k/\sigma)^{2/3}|\chi_\infty(\zeta_0)|^{1/3}}\right) \quad . \tag{6.59}$$

The solution $\tilde{w}^*_{00+}$ for $\zeta < \zeta_0$ must now be matched with the solution $\tilde{w}^*_{00-}$ for $\zeta > \zeta_0$ via the inner solution (6.58) which is valid in the vicinity of $\zeta = \zeta_0$. We have:

$$\tilde{w}^*_{00+} = \left[\varrho_\infty(\hat{B}\zeta)\frac{d\varphi}{d\zeta}\right]^{-1/2} \left\{A^0\,\mathrm{e}^{\mathrm{i}\xi} + B^0\,\mathrm{e}^{-\mathrm{i}\xi}\right\} \quad , \text{ for } \zeta < \zeta_0 \quad ;$$

$$\tilde{w}^*_{00-} = \left[\varrho_\infty(\hat{B}\zeta)\frac{d\varphi}{d\zeta}\right]^{-1/2} \left\{C^0\,\mathrm{e}^{\mathrm{i}\xi} + D^0\,\mathrm{e}^{-\mathrm{i}\xi}\right\} \quad , \text{ for } \zeta > \zeta_0 \quad ,$$

with

$$\frac{d\varphi}{d\zeta} = k\left|1 - \frac{\Gamma_\infty(\zeta)}{\sigma^2}\right|^{1/2} \quad .$$

Let us first have a look at how the matching of $\tilde{w}^*_{00+}$ with the inner solution (6.58) takes place. It is necessary that[5]:

$$\lim_{\mathrm{z}\to-\infty} W_{00}(\mathrm{z}) = \lim_{\substack{\zeta\to\zeta_0 \\ (\zeta<\zeta_0)}} \tilde{w}^*_{00+} \quad . \tag{6.60}$$

When $\mathrm{z} \to -\infty$, the asymptotic expansion of the Airy functions is (see, for example, *The Airy Integral*, by J.C.P. Miller, Cambridge, The University Press, 1946):

$$\mathrm{Ai}(\mathrm{z}) \sim \pi^{-1/2}(-\mathrm{z})^{-1/4} \sin(\psi + \pi/4) \quad ; \tag{6.61a}$$

$$\mathrm{Bi}(\mathrm{z}) \sim \pi^{-1/2}(-\mathrm{z})^{-1/4} \cos(\psi + \pi/4) \quad , \tag{6.61b}$$

with

$$\psi = \tfrac{2}{3}(-\mathrm{z})^{3/2} \quad \text{i.e.}$$

$$\begin{aligned}\psi &= \frac{2}{3}\frac{k}{\sigma}|\chi_\infty(\zeta_0)|^{1/2}(-\hat{z})^{3/2} \\ &= \frac{2}{3}\frac{k}{\sigma}|\chi_\infty(\zeta_0)|^{1/2}\frac{|\zeta - \zeta_0|}{\mathrm{M}_\infty}, \quad \text{because } \zeta < \zeta_0 \quad .\end{aligned}$$

---

[5] We must take into account the fact that according to (6.24, 54, 56), we have

$$\hat{z} = \mathrm{M}^\alpha_\infty(z - z_0) = \mathrm{M}^{1/3}_\infty\frac{\zeta - \zeta_0}{\mathrm{M}_\infty} = \frac{\zeta - \zeta_0}{\mathrm{M}^{2/3}_\infty}$$

and thus

$$\mathrm{z} = \left(\frac{k}{\sigma}\right)^{2/3}|\chi_\infty(\zeta)|^{1/3}\frac{\zeta - \zeta_0}{\mathrm{M}^{2/3}_\infty}$$

is indeed an inner variable with respect to the outer variables $\zeta_+(\zeta_+ < \zeta_0)$ and $\zeta_-(\zeta_- > \zeta_0)$.

Therefore, for the region $\zeta < \zeta_0$, the outer expansion is:

$$W_{00}(z) \sim \frac{\mathcal{L}_0 M_\infty^{1/6}}{\pi^{1/2}} |\zeta - \zeta_0|^{-1/4} \frac{\sqrt{2}}{2} \{(a-b) \sin \psi + (a+b) \cos \psi\} \quad , (6.62)$$

where the following has been set:

$$\mathcal{L}_0 \equiv \left\{ k^{2/3} \sigma^{-2/3} |\chi_\infty(\zeta_0)|^{1/3} \right\}^{-1/4} \quad . \qquad (6.63)$$

The inner expansion of $\tilde{w}^*_{00+}$ must now be specified when $\zeta \to \zeta_0$ and $\zeta < \zeta_0$. In the vicinity of $\zeta = \zeta_0$, we have (for $\zeta < \zeta_0$):

$$\xi = \frac{\varphi(\zeta)}{M_\infty} = \frac{1}{M_\infty} \int_0^\zeta \frac{d\varphi(u)}{du} du \quad , \quad \varphi(0) \equiv 0 \quad ,$$

or in a different form:

$$\xi = \frac{1}{M_\infty} \int_0^{\zeta_0} \frac{d\varphi(u)}{du} du + \frac{1}{M_\infty} \int_{\zeta_0}^{\zeta} \frac{d\varphi(u)}{du} du = \frac{\varphi_{00}}{M_\infty} + \frac{1}{M_\infty} \int_{\zeta_0}^{\zeta} \frac{d\varphi(u)}{du} du \quad ,$$

$$\varphi_{00} \equiv \int_0^{\zeta_0} \frac{d\varphi(u)}{du} du \quad .$$

However,

$$\frac{d\varphi(u)}{du} = k|1 - \Gamma_\infty(u)\sigma^{-2}|^{1/2} \quad , \qquad \text{where}$$

$$\Gamma_\infty(u) = \sigma^2 + (u - \zeta_0) \left. \frac{d\Gamma_\infty}{du} \right|_{u=\zeta_0} \quad ,$$

and thus, after integration:

$$\xi = \frac{\varphi_{00}}{M_\infty} + \frac{k}{\sigma M_\infty} |\chi_\infty(\zeta_0)|^{1/2} \int_{\zeta_0}^{\zeta} |u - \zeta_0|^{1/2} du$$

$$= \frac{\varphi_{00}}{M_\infty} - \frac{k}{\sigma M_\infty} |\chi_\infty(\zeta_0)|^{1/2} \frac{2}{3} |\zeta - \zeta_0|^{3/2} \quad , \quad \zeta < \zeta_0 \quad .$$

Let us now define:

$$\xi = \frac{\varphi_{00}}{M_\infty} - \psi \quad , \quad \varphi_{00} = k \int_0^{\zeta_0} |1 - \Gamma_\infty(u)\sigma^{-2}|^{1/2} du \quad . \qquad (6.64)$$

The inner expansion of the oscillatory solution $\tilde{w}^*_{00+}$ is then deduced:

$$\tilde{w}^*_{00+} \sim \left[\varrho_\infty(\hat{B}\zeta) \frac{d\varphi}{d\zeta} \right]^{-1/2} \left\{ \left[(A^0 + B^0) \cos \frac{\varphi_{00}}{M_\infty} \right. \right.$$

$$\left. + i(A^0 - B^0) \sin \frac{\varphi_{00}}{M_\infty} \right] \cos \psi + \left[(A^0 - B^0) \sin \frac{\varphi_{00}}{M_\infty} \right.$$

$$\left. \left. - i(A^0 - B^0) \cos \frac{\varphi_{00}}{M_\infty} \right] \sin \psi \right\} \quad ,$$

or even, since:

$$\left[\varrho_\infty(\hat{B}\zeta)\frac{d\varphi}{d\zeta}\right]^{1/2} \equiv \left[\varrho_\infty(\hat{B}\zeta_0)\right]^{1/2}\left(\frac{\sigma}{k}\right)^{-1/2}|\chi_\infty(\zeta_0)|^{1/4}|\zeta-\zeta_0|^{1/4} \quad ,$$

$$\begin{aligned}\tilde{w}^*_{00+} \sim \sigma^{1/2}\left[k\varrho_\infty(\hat{B}\zeta_0)\right]^{-1/2}|\chi_\infty(\zeta_0)|^{-1/4}|\zeta-\zeta_0|^{-1/4}\Bigg\{\Bigg[\Big(A^0 \\ + B^0\Big)\cos\frac{\varphi_{00}}{\mathrm{M}_\infty} + \mathrm{i}\big(A^0 - B^0\big)\sin\frac{\varphi_{00}}{\mathrm{M}_\infty}\Bigg]\cos\psi \\ + \left[\big(A^0 - B^0\big)\sin\frac{\varphi_{00}}{\mathrm{M}_\infty} - \mathrm{i}\big(A^0 - B^0\big)\cos\frac{\varphi_{00}}{\mathrm{M}_\infty}\right]\sin\psi\Bigg\} \quad . \end{aligned} \tag{6.65}$$

The identification of (6.62, 65) yields the following two relations for a and b:

$$\begin{aligned} a+b &= \frac{\mathrm{M}_\infty^{-1/6}(2\pi\sigma)^{1/2}}{\mathcal{L}_0\left[k\varrho_\infty(\hat{B}\zeta_0)\right]^{1/2}}|\chi_\infty(\zeta_0)|^{-1/4}\Bigg\{\Big(A^0 \\ &\quad + B^0\Big)\cos\frac{\varphi_{00}}{\mathrm{M}_\infty} + \mathrm{i}\big(A^0 - B^0\big)\sin\frac{\varphi_{00}}{\mathrm{M}_\infty}\Bigg\} \quad ; \\ a-b &= \frac{\mathrm{M}_\infty^{-1/6}(2\pi\sigma)^{1/2}}{\mathcal{L}_0\left[k\varrho_\infty(\hat{B}\zeta_0)\right]^{1/2}}|\chi_\infty(\zeta_0)|^{-1/4}\Bigg\{\Big(A^0 \\ &\quad + B^0\Big)\sin\frac{\varphi_{00}}{\mathrm{M}_\infty} - \mathrm{i}\big(A^0 - B^0\big)\cos\frac{\varphi_{00}}{\mathrm{M}_\infty}\Bigg\} \quad . \end{aligned} \tag{6.66}$$

Let us now turn our attention to the region $\zeta > \zeta_0$. The asymptotic expansion of the Airy functions is in this region:

$$\begin{aligned} \mathrm{Ai(z)} &\sim 2^{-1}\pi^{-1/2}\mathrm{z}^{-1/4}\exp(-\psi) \quad , \\ \mathrm{Bi(z)} &\sim \pi^{-1/2}\mathrm{z}^{-1/4}\exp(+\psi) \quad , \end{aligned} \tag{6.67}$$

where

$$\psi = \frac{2}{3}\mathrm{z}^{3/2} \equiv \frac{2}{3}\frac{k}{\sigma\mathrm{M}_\infty}|\chi_\infty(\zeta_0)|^{1/2}|\zeta-\zeta_0|^{3/2} \quad ,$$

since $\zeta > \zeta_0$.

In this case, the following inner expansion of the non-periodic solution $\tilde{w}^*_{00-}$ results when $\zeta \to \zeta_0$ $(\zeta > \zeta_0)$:

$$\begin{aligned}\tilde{w}^*_{00-} \sim \sigma^{1/2}\left[k\varrho_\infty(\hat{B}\zeta_0)\right]^{-1/2}|\zeta-\zeta_0|^{-1/4}|\chi_\infty(\zeta_0)|^{-1/4}\Big\{C^0\,\mathrm{e}^{\varphi_{00}/\mathrm{M}_\infty}\,\mathrm{e}^{\psi} \\ + D^0\,\mathrm{e}^{-\varphi_{00}/\mathrm{M}_\infty}\,\mathrm{e}^{-\psi}\Big\}\end{aligned}$$

and the identification yields:

$$\begin{aligned} a &= \frac{2\mathrm{M}_\infty^{-1/6}(\pi\sigma)^{1/2}}{\mathcal{L}_0\left[k\varrho_\infty(\hat{B}\zeta_0)\right]^{1/2}}|\chi_\infty(\zeta_0)|^{-1/4}D^0\,\mathrm{e}^{-\varphi_{00}/\mathrm{M}_\infty} \quad ; \\ b &= \frac{2\mathrm{M}_\infty^{-1/6}(\pi\sigma)^{1/2}}{\mathcal{L}_0\left[k\varrho_\infty(\hat{B}\zeta_0)\right]^{1/2}}|\chi_\infty(\zeta_0)|^{-1/4}C^0\,\mathrm{e}^{\varphi_{00}/\mathrm{M}_\infty} \quad . \end{aligned} \tag{6.68}$$

The elimination of a and b between (6.66, 68) provides two relations between the four integration constants A^0, B^0, C^0 and D^0:

$$\begin{aligned} &2D^0\,\mathrm{e}^{-\varphi_{00}/\mathrm{M}_\infty} + C^0\,\mathrm{e}^{\varphi_{00}/\mathrm{M}_\infty} \\ &\quad = 2^{1/2}\left\{\left(A^0 - B^0\right)\cos\frac{\varphi_{00}}{\mathrm{M}_\infty} + \mathrm{i}\left(A^0 - B^0\right)\sin\frac{\varphi_{00}}{\mathrm{M}_\infty}\right\} \quad ; \\ &2D^0\,\mathrm{e}^{-\varphi_{00}/\mathrm{M}_\infty} - C^0\,\mathrm{e}^{\varphi_{00}/\mathrm{M}_\infty} \\ &\quad = 2^{1/2}\left\{\left(A^0 - B^0\right)\sin\frac{\varphi_{00}}{\mathrm{M}_\infty} - \mathrm{i}\left(A^0 - B^0\right)\cos\frac{\varphi_{00}}{\mathrm{M}_\infty}\right\} \quad . \end{aligned} \tag{6.69}$$

The above formulas (6.69) make it possible to connect the amplitude coefficients of an oscillatory regime to an adjacent non-periodic regime. An interesting case is one where the function $\Gamma_\infty(\zeta) - \sigma^2$ is zero at the point ζ_0, but non-zero if $\zeta > \zeta_0$. In the case where $(d\Gamma_\infty(\zeta)/d\zeta)_{\zeta=\zeta_0} < 0$, (the case studied above), $\tilde{w}^*_{00-}$ can satisfy a damping condition at infinity (according to the altitude), the coefficient C^0 being then zero. The wave associated with coefficient D^0 is a wave which is rapidly damped, such that qualitatively speaking, we have a wave which is oscillatory for $\zeta < \zeta_0$ and disappears if $\zeta > \zeta_0$. The coefficient D^0 can be eliminated between the two relations of (6.69) (where $C^0 \equiv 0$) which yields a relation between A^0 and B^0. A second relation between A^0 and B^0 can be obtained by writing a slip condition on a lower limiting wall for the considered flow (this last condition should be linearized with respect to the small parameter η which can then be interpreted as a small relative thickness of the wall in comparison with the position $z = 0$).

6.3 Various Hydrostatic Limiting Processes

Once again, we will be considering the context of an adiabatic atmosphere. Hence, everything concerned with the local forecast will be relative only to the dynamic action of the relief of the local site under consideration. The basic equations are thus the Euler equations – e.g., those of Sect. 2.1.

Let it be assumed that the local site is simulated by the specification of its relief in the form[6]:

$$z = h_0 h\left(\frac{x - x_0}{l_0}, \frac{y - y_0}{m_0}\right) \quad , \tag{6.70}$$

where the point $x = x_0$, $y = y_0$ of the plane $z = 0$ serves as the origin for the local site and (l_0, m_0) encloses a finite area on the plane $z = 0$. Finally, $h_0 \equiv \max|h|$, when $x, y \in (l_0, m_0)$.

[6] In fact, the tangent plane approximation is carried out and the β effect ignored. That is to say that the basic equations are those written in Sect.2.2 [Eq. (2.31)], where $\mu_0 \equiv 0$, $k_0 \equiv 0.$, $\phi \equiv 0$ and $\partial\mathcal{R}/\partial z \equiv 0$.

The following slip condition must be prescribed for the Euler equations:

$$w = h_0\left(u\frac{\partial h}{\partial x} + v\frac{\partial h}{\partial y}\right) \quad , \quad \text{on} \quad z = h_0 h \quad . \tag{6.71}$$

To these same equations, the initial conditions for $t = 0$ must be added:

$$u = U^0\,,\ \ v = V^0\,,\ \ w = W^0\,,\ \ T = T^0\,,\ \ \varrho = R^0\,,\quad \text{for} \quad t = 0 \quad , \tag{6.72}$$

where the functions U^0, V^0, W^0, T^0 and R^0 are assumed known and depend on the variables $(x/L,\ y/L,\ z/H)$. We know (see Chap. 5) that the classial, so-called short-range synoptic forecast carried out from the so-called primitive equations of Subsect. 2.4.3 is associated with the characteristic horizontal scale L assumed to be of the order 10^6 m. Since H is the characteristic vertical scale which is of the order of the thickness of the troposphere, we have

$$\varepsilon = \frac{H}{L} \ll 1 \quad .$$

Moreover, initial conditions directly influenced by the local site can be added to these same basic Euler equations and in this case, the following "local" initial conditions can replace (6.72):

$$u = \hat{U}^0\,,\ \ v = \hat{V}^0\,,\ \ w = \hat{W}^0\,,\ \ T = \hat{T}^0\,,\ \ \varrho = \hat{R}^0\,,\ \ \text{for} \ \ t = t^0 \quad , \tag{6.73}$$

where t^0 is an initial time from which the local forecast is calculated. The functions $\hat{U}^0$, $\hat{V}^0$, $\hat{W}^0$, $\hat{T}^0$ and $\hat{R}^0$ are then supposed known functions of

$$\frac{x - x_0}{l_0}\,,\quad \frac{y - y_0}{m_0}\,,\quad \frac{z}{h_0} \quad .$$

To be sure, a more general case is the one where the initial data at $t = 0$ are simultaneously functions of $\big(x/L,\ y/L,\ (x-x_0)/l_0,\ (y-y_0)/m_0,\ z/H \ \text{and} \ z/h_0\big)$. In this situation, the application of the MSM must be considered when $\varepsilon \to 0$. However, we will not here go into this case which demands special attention.

A first, so-called main, non-dimensionalization is now carried out which is intimately related to the primitive equations for the short-range synoptic forecast. We define:

$$\bar{t} = \frac{t}{L/U_0}\,,\quad \bar{x} = \frac{x}{L}\,,\quad \bar{y} = \frac{y}{L}\,,\quad \bar{z} = \frac{z}{H} \quad , \tag{6.74}$$

and

$$\bar{u} = \frac{u}{U_0}\,,\quad \bar{v} = \frac{v}{U_0}\,,\quad \bar{w} = \frac{w}{\varepsilon U_0}\,,\quad \bar{p} = \frac{p}{P_0}\,,\quad \bar{\varrho} = \frac{\varrho}{R_0} \quad , \quad \bar{T} = T/T_0 \quad , \tag{6.75}$$

where the characteristic constant values U_0, R_0, T_0 and $P_0 = RR_0T_0$ are those associated with the initial data (6.72). For $\bar{\boldsymbol{v}} = \bar{u}\boldsymbol{i} + \bar{v}\boldsymbol{j}$, $\bar{w}$, $\bar{p}$, $\bar{\varrho}$ and $\bar{T}$, functions of $\bar{t}$, $\bar{x}$, $\bar{y}$ and $\bar{z}$, the following dimensionless Euler equations result:

$$\bar{\varrho}\left\{\frac{\partial \bar{\boldsymbol{v}}}{\partial \bar{t}} + (\bar{\boldsymbol{v}} \cdot \bar{\boldsymbol{D}})\bar{\boldsymbol{v}} + \bar{w}\frac{\partial \bar{\boldsymbol{v}}}{\partial \bar{z}} + \frac{1}{\text{Ro}}\frac{\sin\varphi}{\sin\varphi_0}(\boldsymbol{k} \times \bar{\boldsymbol{v}}) \right.$$
$$\left. + \frac{\varepsilon}{\text{Ro}}\frac{\cos\varphi}{\sin\varphi_0}\bar{w}\boldsymbol{i}\right\} + \frac{1}{\gamma \text{M}_\infty^2}\bar{\boldsymbol{D}}\bar{p} = 0 \quad ; \tag{6.76a}$$

$$\bar{\varrho}\left\{\varepsilon^2\left[\frac{\partial\bar{w}}{\partial\bar{t}}+\bar{\boldsymbol{v}}\cdot\bar{\boldsymbol{D}}\bar{w}+\bar{w}\frac{\partial\bar{w}}{\partial\bar{z}}\right]-\frac{\varepsilon}{\text{Ro}}\frac{\cos\varphi}{\sin\varphi_0}\bar{\boldsymbol{v}}\cdot\boldsymbol{i}\right\}$$

$$+\frac{1}{\gamma\text{M}_\infty^2}\left(\frac{\partial\bar{p}}{\partial\bar{z}}+\text{Bo}\bar{\varrho}\right)=0 \quad ; \tag{6.76b}$$

$$\frac{\partial\bar{\varrho}}{\partial\bar{t}}+\bar{\boldsymbol{D}}\cdot(\bar{\varrho}\bar{\boldsymbol{v}})+\frac{\partial\bar{\varrho}\bar{w}}{\partial\bar{z}}=0 \quad ; \tag{6.76c}$$

$$\bar{\varrho}\left(\frac{\partial\bar{T}}{\partial\bar{t}}+\bar{\boldsymbol{v}}\cdot\bar{\boldsymbol{D}}\bar{T}\right)-\frac{\gamma-1}{\gamma}\left(\frac{\partial\bar{p}}{\partial\bar{t}}+\bar{\boldsymbol{v}}\cdot\bar{\boldsymbol{D}}\bar{p}\right)$$

$$+\bar{w}\left(\bar{\varrho}\frac{\partial\bar{T}}{\partial\bar{z}}-\frac{\gamma-1}{\gamma}\frac{\partial\bar{p}}{\partial\bar{z}}\right)=0 \quad ; \tag{6.76d}$$

$$\bar{p}=\bar{\varrho}\bar{T} \quad , \tag{6.76e}$$

where $\bar{\boldsymbol{D}}=(\partial/\partial\bar{x})\boldsymbol{i}+(\partial/\partial\bar{y})\boldsymbol{j}$.

The ground condition (6.71) becomes:

$$\bar{w}=\sigma\left\{\bar{u}\frac{\partial h\big(\alpha(\bar{x}-\bar{x}_0)/\varepsilon,\ \beta(\bar{y}-\bar{y}_0)/\varepsilon\big)}{\partial\bar{x}}+\bar{v}\frac{\partial h\big(\alpha(\bar{x}-\bar{x}_0)/\varepsilon,\ \beta(\bar{y}-\bar{y}_0)/\varepsilon\big)}{\partial\bar{y}}\right\}$$

on

$$\bar{z}=\sigma h\left(\alpha\frac{\bar{x}-\bar{x}_0}{\varepsilon},\ \beta\frac{\bar{y}-\bar{y}_0}{\varepsilon}\right) \quad , \tag{6.77}$$

where

$$\sigma=\frac{h_0}{H} \quad , \quad \alpha=\frac{H}{l_0} \quad , \quad \beta=\frac{H}{m_0} \quad , \tag{6.78}$$

are assumed of the order unity.

Since the parameter $\varepsilon=H/L\ll 1$ is the main small parameter of the asymptotic theory being worked out here, we will first consider the following main limiting process:

$$\lim_{\varepsilon\to 0}^{P}\ \equiv\{\varepsilon\to 0 \quad , \quad \text{at } \bar{t},\bar{x},\bar{y},\bar{z} \text{ fixed}\} \quad . \tag{6.79}$$

Generally speaking, if it is supposed that the parameters σ, α, β, Ro, Bo, M_∞ and γ remain fixed of the order unity when (6.79) is carried out, then from (6.76), the primitive equations used by meteorologists for short-range synoptic weather forecasting can be obtained. These equations are as follows:

$$\bar{\varrho}_0\left\{\frac{\partial\bar{\boldsymbol{v}}_0}{\partial\bar{t}}+(\bar{\boldsymbol{v}}_0\cdot\bar{\boldsymbol{D}})\bar{\boldsymbol{v}}_0+\bar{w}_0\frac{\partial\bar{\boldsymbol{v}}_0}{\partial\bar{z}}+\frac{1}{\text{Ro}}\frac{\sin\varphi}{\sin\varphi_0}(\boldsymbol{k}\times\bar{\boldsymbol{v}}_0)\right\}$$

$$+\frac{1}{\gamma\text{M}_\infty^2}\bar{\boldsymbol{D}}\bar{p}_0=0 \quad ; \tag{6.80a}$$

$$\frac{\partial\bar{p}_0}{\partial\bar{z}}+\text{Bo}\bar{\varrho}_0=0 \quad , \quad \bar{p}_0=\bar{\varrho}_0\bar{T}_0 \quad ;$$

$$\frac{\partial \overline{\varrho}_0}{\partial \overline{t}} + \overline{\boldsymbol{D}} \cdot (\overline{\varrho}_0 \overline{\boldsymbol{v}}_0) + \frac{\partial}{\partial \overline{z}} (\overline{\varrho}_0 \overline{w}_0) = 0 \quad ;$$

$$\overline{\varrho}_0 \left(\frac{\partial \overline{T}_0}{\partial \overline{t}} + \overline{\boldsymbol{v}}_0 \cdot \overline{\boldsymbol{D}} \, \overline{T}_0 \right) - \frac{\gamma - 1}{\gamma} \left(\frac{\partial \overline{p}_0}{\partial \overline{t}} + \overline{\boldsymbol{v}}_0 \cdot \overline{\boldsymbol{D}} \overline{p}_0 \right) \tag{6.80b}$$

$$+ \overline{w}_0 \left(\overline{\varrho}_0 \frac{\partial \overline{T}_0}{\partial \overline{z}} - \frac{\gamma - 1}{\gamma} \frac{\partial \overline{p}_0}{\partial \overline{z}} \right) = 0 \quad ,$$

where

$$\begin{pmatrix} \overline{\boldsymbol{v}}_0 \\ \overline{w}_0 \\ \overline{p}_0 \\ \overline{\varrho}_0 \\ \overline{T}_0 \end{pmatrix} = \overset{P}{\lim_{\varepsilon \to 0}} \begin{pmatrix} \overline{\boldsymbol{v}} \\ \overline{w} \\ \overline{p} \\ \overline{\varrho} \\ \overline{T} \end{pmatrix} \quad . \tag{6.81}$$

Only the following ground condition can be superimposed on the primitive equations (6.80):

$$\overline{w}_0 = 0 \quad , \quad \text{for} \quad \overline{z} = 0 \quad , \tag{6.82}$$

since $h(\infty, \infty) \equiv 0$.

As regards the initial conditions (6.72) for the Euler equations (6.76), it is first pointed out that according to (6.74, 76), they can be written in the following dimensionless form:

$$\overline{t} = 0 : \overline{u} = \overline{U}^0 , \quad \overline{v} = \overline{V}^0 , \quad \varepsilon \overline{w} = \overline{W}^0 , \quad \overline{\varrho} = \overline{R}^0 , \quad \overline{T} = \overline{T}^0 \quad . \tag{6.83}$$

However, *after the main limiting process (6.79), only three initial conditions [instead of the five initial conditions of (6.83)] can be added to the primitive equations (6.80):*

$$\overline{t} = 0 : \overline{\boldsymbol{v}}_0 = \overline{\boldsymbol{v}}^0 \quad , \quad \overline{\varrho}_0 = \overline{\varrho}^0 \quad . \tag{6.84}$$

A priori, $\overline{\boldsymbol{v}}^0$ and $\overline{\varrho}^0$ remain undetermined.

These unknowns are determined by the solution to the problem of adjustment to hydrostatic balance recently analyzed by Guiraud and Zeytounian (1982) and which also was the subject of a numerical integration carried out by Outrebon (1981). This analysis brings to light the link between $\overline{U}^0$, $\overline{V}^0$, $\overline{W}^0$, $\overline{R}^0$, $\overline{T}^0$ on the one hand, and $\overline{\boldsymbol{v}}^0$, $\overline{\varrho}^0$ on the other hand. To this end, it must be understood that the vicinity of $\overline{t} = 0$ is singular during the main limiting process (6.79):

$$\lim_{\overline{t} \to 0} \left(\overset{P}{\lim_{\varepsilon \to 0}} \right) \neq \overset{P}{\lim_{\varepsilon \to 0}} \left(\lim_{\overline{t} \to 0} \right) \quad ,$$

and therefore, a new "local" limiting process must be introduced:

$$\overset{l^t}{\lim_{\varepsilon \to 0}} \equiv \left\{ \varepsilon \to 0 \quad , \text{ at } \tilde{t} = \frac{\overline{t}}{\varepsilon} \quad , \quad \overline{x}, \overline{y}, \overline{z} \text{ fixed} \right\} \quad . \tag{6.85}$$

This problem of the adjustment to hydrostatic balance

$$\frac{\partial \overline{p}_0}{\partial \overline{z}} + \mathrm{Bo}\overline{\varrho}_0 = 0 \quad ,$$

is given an in-depth treatment in Sect. 7.5.

For the moment, we will only point out that if:

$$\begin{pmatrix} \tilde{v}_0 \\ \tilde{w}_0 \\ \tilde{p}_0 \\ \tilde{\varrho}_0 \\ \tilde{T}_0 \end{pmatrix} = \lim_{\varepsilon \to 0}{}^{l^t} \begin{pmatrix} \overline{v} \\ \varepsilon\overline{w} \\ \overline{p} \\ \overline{\varrho} \\ \overline{T} \end{pmatrix} \quad , \tag{6.86}$$

then the equations for the problem of the adjustment to hydrostatic balance are as follows:

$$\frac{\partial \tilde{v}_0}{\partial \tilde{t}} + \tilde{w}_0 \frac{\partial \tilde{v}_0}{\partial \overline{z}} = 0 \quad ; \tag{6.87}$$

and

$$\begin{aligned} &\gamma \mathrm{M}_\infty^2 \tilde{\varrho}_0 \left\{ \frac{\partial \tilde{w}_0}{\partial \tilde{t}} + \tilde{w}_0 \frac{\partial \tilde{w}_0}{\partial \overline{z}} \right\} + \frac{\partial \tilde{p}_0}{\partial \overline{z}} + \mathrm{Bo}\tilde{\varrho}_0 = 0 \quad ; \\ &\frac{\partial \tilde{\varrho}_0}{\partial \tilde{t}} + \frac{\partial \tilde{\varrho}_0 \tilde{w}_0}{\partial \overline{z}} = 0 \\ &\tilde{p}_0 = \tilde{\varrho}_0 \tilde{T}_0 \quad ; \\ &\tilde{\varrho}_0 \frac{\partial \tilde{T}_0}{\partial \tilde{t}} - \frac{\gamma - 1}{\gamma} \frac{\partial \tilde{p}_0}{\partial \tilde{t}} + \tilde{w}_0 \left(\tilde{\varrho}_0 \frac{\partial \tilde{T}_0}{\partial \overline{z}} - \frac{\gamma - 1}{\gamma} \frac{\partial \tilde{p}_0}{\partial \overline{z}} \right) = 0 \quad . \end{aligned} \tag{6.88}$$

The system (6.88) is the same as the one which governs the one-dimensional unsteady vertical motions of a heavy gas (perfect, with constant specific heats). It is a hyperbolic quasi-linear system[7] which is characterized by the presence of sound waves and a transfer of the specific entropy.

The five initial conditions of (6.83) can be added to (6.87, 88). The following matching condition (it is assumed that the MMAE is, in fact, applicable which is confirmed by both the numerical results and the linear analysis):

$$\lim_{\tilde{t} \to +\infty} = \lim_{\overline{t} \to 0} \quad , \tag{6.89}$$

enables the determining of the "right" initial conditions for the problem associated with the primitive equations (6.80)[8].

Let us now backtrack to condition (6.82) which shows that *in the primitive equations (6.80) affiliated with the main limiting process (6.79), it is no longer possible to take into account the relief of the local site under consideration.*

Therefore, to reestablish this possibility, a new (second) local limiting process must be devised:

---

[7] It will be remarked that the primitive equations (6.80) *are not* of the hyperbolic type whereas the Euler equations naturally are. It is this change in type (due to the filtering out of the acoustic waves) which makes it necessary to introduce the local limiting process (6.85).

[8] In particular, we must be persuaded that: $\lim_{\tilde{t} \to +\infty} \tilde{w}_0 = 0$.

$$\overset{l_{esp}}{\lim_{\varepsilon\to 0}} \equiv \left\{\varepsilon\to 0 \quad , \text{ at } \bar{t}, \ \hat{x} \equiv \frac{\bar{x}-\bar{x}_0}{\varepsilon}, \ \hat{y} \equiv \frac{\bar{y}-\bar{y}_0}{\varepsilon}, \ \bar{z} \text{ fixed}\right\} \quad . \tag{6.90}$$

If we introduce:

$$\begin{pmatrix} \hat{\boldsymbol{v}}_0 \\ \hat{w}_0 \\ \hat{p}_0 \\ \hat{\varrho}_0 \\ \hat{T}_0 \end{pmatrix} = \overset{l_{esp}}{\lim_{\varepsilon\to 0}} \begin{pmatrix} \overline{\boldsymbol{v}} \\ \varepsilon\overline{w} \\ \overline{p} \\ \overline{\varrho} \\ \overline{T} \end{pmatrix} \quad , \tag{6.91}$$

the following *local steady* equations replace the complete Euler equations (6.76):

$$\left(\hat{\boldsymbol{v}}_0 \cdot \hat{\boldsymbol{D}}\right)\hat{\boldsymbol{v}}_0 + \hat{w}_0\frac{\partial \hat{\boldsymbol{v}}_0}{\partial \bar{z}} + \frac{1}{\hat{\varrho}_0}\frac{1}{\gamma \mathrm{M}_\infty^2}\hat{\boldsymbol{D}}\hat{p}_0 = 0 \quad ; \tag{6.92a}$$

$$\hat{\boldsymbol{v}}_0 \cdot \hat{\boldsymbol{D}}\hat{w}_0 + \hat{w}_0\frac{\partial \hat{w}_0}{\partial \bar{z}} + \frac{1}{\gamma \mathrm{M}_\infty^2}\left(\frac{1}{\hat{\varrho}_0}\frac{\partial \hat{p}_0}{\partial \bar{z}} + \mathrm{Bo}\right) = 0 \quad ; \tag{6.92b}$$

$$\hat{\boldsymbol{D}} \cdot \left(\hat{\varrho}_0\hat{\boldsymbol{v}}_0\right) + \frac{\partial \hat{\varrho}_0\hat{w}_0}{\partial \bar{z}} = 0 \quad ; \tag{6.92c}$$

$$\hat{p}_0 = \hat{\varrho}_0\hat{T}_0 \quad ; \tag{6.92d}$$

$$\hat{\varrho}_0\hat{\boldsymbol{v}}_0 \cdot \hat{\boldsymbol{D}}\hat{T}_0 - \frac{\gamma-1}{\gamma}\hat{\boldsymbol{v}}_0 \cdot \hat{\boldsymbol{D}}\hat{p}_0 + \hat{w}_0\left(\hat{\varrho}_0\frac{\partial \hat{T}_0}{\partial \bar{z}} - \frac{\gamma-1}{\gamma}\frac{\partial \hat{p}_0}{\partial \bar{z}}\right) = 0 \quad , \tag{6.92e}$$

where $\hat{\boldsymbol{D}} \equiv (\partial/\partial\hat{x})\boldsymbol{i} + (\partial/\partial\hat{y})\boldsymbol{j} = \varepsilon\overline{\boldsymbol{D}}$.

To the system of local equations (6.92), we can assign the "right" ground condition:

$$\hat{w}_0 = \sigma\hat{\boldsymbol{v}}_0 \cdot \hat{\boldsymbol{D}}h(\alpha\hat{x}, \beta\hat{y}) \quad , \text{ on } \bar{z} = \sigma h(\alpha\hat{x}, \beta\hat{y}) \quad . \tag{6.93}$$

Since (6.73) has been ignored, there are no initial conditions to be prescribed. On the contrary, with the primitive equations model, there are matching conditions to be imposed with respect to the horizontal variables:

$$\lim_{\substack{|\hat{x}|\to\infty \\ |\hat{y}|\to\infty}} \begin{pmatrix} \hat{\boldsymbol{v}}_0 \\ \hat{w}_0 \\ \hat{p}_0 \\ \hat{\varrho}_0 \\ \hat{T}_0 \end{pmatrix} = \begin{pmatrix} \overline{\boldsymbol{v}}_0(\bar{t}, \bar{x}_0, \bar{y}_0, \bar{z}) \\ \varepsilon\overline{w}_0(\bar{t}, \bar{x}_0, \bar{y}_0, \bar{z}) \\ \overline{p}_0(\bar{t}, \bar{x}_0, \bar{y}_0, \bar{z}) \\ \overline{\varrho}_0(\bar{t}, \bar{x}_0, \bar{y}_0, \bar{z}) \\ \overline{T}_0(\bar{t}, \bar{x}_0, \bar{y}_0, \bar{z}) \end{pmatrix} \quad , \tag{6.94}$$

where $\bar{t}$ plays the role of a parameter which enters into the local problem (6.92, 93) via the "outer field" associated with short-range forecasting from the primitive equations (6.80) at the time $\bar{t}$ (arbitrary but of the order unity) and at the point $(\bar{x}_0, \bar{y}_0)$ of the plane $\bar{z} = 0$. However, with respect to the vertical velocities, (6.91) suggests that[9]:

$$\lim_{\substack{|\hat{x}| \to \infty, \\ |\hat{y}| \to \infty}} \hat{w}_0 = 0 \quad . \tag{6.95}$$

The validity of this last condition (6.95) should be enough to convince us definitively that the MMAE is well-adapted for the present asymptotic description. Of course, it is also necessary that:

$$(\partial \overline{p}_0/\partial \overline{z} + \mathrm{Bo}\overline{\varrho}_0)_{\overline{x}_0, \overline{y}_0} = 0 \quad , \quad \overline{p}_0|_{\overline{x}_0, \overline{y}_0} = (\overline{\varrho}_0 \overline{T}_0)_{\overline{x}_0, \overline{y}_0} \quad . \tag{6.96}$$

Therefore, it can be stated that *in order to make a local (quasi-steady) forecast with respect to a site of given relief (local dynamic forecast), with the origin* (x_0, y_0), *it is first necessary to display, at this point* (x_0, y_0) *and at the chosen time* $t = t_0$, *the short-range synoptic forecast obtained with the primitive equations. This short-range forecast then plays the role of the condition at infinity in all the planes* $\overline{z}$ = const. *for the local problem (6.92, 93).* These conditions at infinity consist in giving $\overline{v}_0(\overline{t}_0, \overline{x}_0, \overline{y}_0, \overline{z})$ and $\overline{T}(\overline{t}_0, \overline{x}_0, \overline{y}_0, \overline{z})$ which are then known from the short-range primitive equations forecast at the point $(\overline{x}_0, \overline{y}_0)$ and at the chosen time $\overline{t} = \overline{t}_0$.

One last remark can be made concerning the link between the limiting processes $\overset{l^t}{\lim}_{\varepsilon \to 0}$ and $\overset{l_{\mathrm{esp}}}{\lim}_{\varepsilon \to 0}$. The following compatibility condition must be satisfied:

$$\lim_{\substack{\overline{t} \to 0 \\ |\hat{x}| \to \infty, |\hat{y}| \to \infty}} \left(\overset{l_{\mathrm{esp}}}{\lim_{\varepsilon \to 0}} \right) = \lim_{\substack{\hat{t} \to \infty \\ \overline{x} \to \overline{x}_0, \overline{y} \to \overline{y}_0}} \left(\overset{l^t}{\lim_{\varepsilon \to 0}} \right) \quad .$$

This means that the initial values in $\overline{t} = 0$ for the primitive equations at the local point of origin $(\overline{x}_0, \overline{y}_0)$ must necessarily coincide with the limiting values at infinity in the planes $\overline{z}$ = const of the local quasi-steady problem at the initial time $\overline{t} = 0$. Otherwise stated, if $\overline{t}_0 = 0$ is chosen in the local quasi-steady problem (6.92, 93), then the "right" initial conditions obtained for the primitive equations from the hydrostatic balance adjustment problem at the point $(\overline{x}_0, \overline{y}_0)$ (which characterizes the local origin of the site considered) must be specified on each plane $\overline{z}$ = const as lateral conditions at infinity for this problem (6.92, 93). Concerning the "right" initial conditions, the reader is referred to Sect. 7.5 and the publication by Guiraud and Zeytounian (1982).

Furthermore, it is observed that, strictly speaking, the asymptotic behavior of the solutions of the local quasi-steady equations (6.92) can naturally be sought when $\hat{r}^2 = \hat{x}^2 + \hat{y}^2 \to \infty$, the predominant term of the behavior being given by the matching with the solution of the primitive equations for $(\overline{x}_0, \overline{y}_0)$. The preceding

[9] From a physics point of view, the validity of (6.95) is related to the effects of skirting the relief in planes $\overline{z}$ = const. as well as to the quasi-linear interactions which by interference, lead to the destruction of the lee waves far downstream of the (bounded) relief of the site being considered. For the plane problem (x, z), the absence of skirting in the planes z = const. leads to a system of lee waves which does not disappear far downstream due to the absence of interference.

is equivalent to linearizing equations (6.92) in the vicinity of the quasi-static solution (a function only of $\overline{z}$):

$$\left\{\overline{v}_0(\overline{t}_0, \overline{x}_0, \overline{y}_0; \overline{z}) \quad , \; 0 \quad , \; \overline{p}_0(\overline{t}_0, \overline{x}_0, \overline{y}_0; \overline{z}) \quad , \; \overline{\varrho}_0(\overline{t}_0, \overline{x}_0, \overline{y}_0; \overline{z})\right\} \quad ,$$

with

$$\left(\frac{d\overline{p}_0}{d\overline{z}} + \mathrm{Bo}\,\overline{\varrho}_0\right)_{\overline{t}_0, \overline{x}_0, \overline{y}_0} = 0 \quad , \quad \overline{T}_0(\overline{t}_0, \overline{x}_0, \overline{y}_0; \overline{z}) = \frac{\overline{p}_0(\overline{t}_0, \overline{x}_0, \overline{y}_0; \overline{z})}{\overline{\varrho}_0(\overline{t}_0, \overline{x}_0, \overline{y}_0; \overline{z})} \quad .$$

For the local, quasi-steady, vertical velocity perturbation, the above leads to an eigenvalue problem in $\overline{z}$ (for the vertical structure). The condition for $\overline{z} \to +\infty$ must then be adequately chosen in order for the corresponding Sturm-Liouville-type problem to be well-posed! We will return to this matter in Chap. 13.

If the initial "local" conditions (6.73) for $\overline{t} = \overline{t}^0$ are to be taken into acccount, then we need a third local limiting process:

$$\lim_{\varepsilon \to 0}^{l^t_{\mathrm{esp}}} \equiv \left\{\varepsilon \to 0 \quad , \text{ at } \hat{t} = \frac{\overline{t} - \overline{t}^0}{\varepsilon} \quad , \hat{x}, \hat{y}, \overline{z} \text{ fixed}\right\} \quad . \tag{6.97}$$

We then define:

$$\begin{pmatrix} \hat{\hat{v}}_0 \\ \hat{\hat{w}}_0 \\ \hat{\hat{p}}_0 \\ \hat{\hat{\varrho}}_0 \\ \hat{\hat{T}}_0 \end{pmatrix} = \lim_{\varepsilon \to 0}^{l^t_{\mathrm{esp}}} \begin{pmatrix} \overline{v} \\ \varepsilon \overline{w} \\ \overline{p} \\ \overline{\varrho} \\ \overline{T} \end{pmatrix} \quad . \tag{6.98}$$

For the variables $(\hat{\hat{\ }}_0)$, the local unsteady equations (of evolution) then result. The latter are obtained from the local quasi-steady system (6.92) if to the first terms of (6.92a–c, e) respectively, the following terms are added:

$$\frac{\partial \hat{\hat{v}}_0}{\partial \hat{t}} \quad , \quad \frac{\partial \hat{\hat{w}}_0}{\partial \hat{t}} \quad , \quad \frac{\partial \hat{\hat{\varrho}}_0}{\partial \hat{t}} \quad , \quad \hat{\hat{\varrho}}_0 \frac{\partial \hat{\hat{T}}_0}{\partial \hat{t}} - \frac{\gamma - 1}{\gamma} \frac{\partial \hat{\hat{p}}_0}{\partial \hat{t}} \quad .$$

A double hat $\hat{\hat{\ }}$ must be imagined instead of a single hat $\hat{\ }$ on all the functions. Initial conditions must be superimposed on these local unsteady equations:

$$\hat{t} = 0 : \; \hat{\hat{v}}_0 = \hat{\overline{V}}^0 , \quad \hat{\hat{w}}_0 = \hat{\overline{W}}^0 , \quad \hat{\hat{T}}_0 = \hat{\overline{T}}^0 , \quad \hat{\hat{\varrho}}_0 = \hat{\overline{R}}^0 \quad ,$$

where the given quantities $\hat{\overline{V}}^0$, $\hat{\overline{W}}^0$, $\hat{\overline{T}}^0$, $\hat{\overline{R}}^0$ are functions of $\hat{x}$, $\hat{y}$ and $\overline{z}$.

When $\hat{t} \to \infty$, we naturally come across the same local quasi-steady problem as before at $\overline{t} = \overline{t}^0$. To a certain extent, the local unsteady problem can be interpreted like the three-dimensional unsteady adjustment problem. The analysis of this unsteady problem makes it possible to establish a local quasi-steady situation as a parametrical function with respect to time, of the short-range synoptic forecast at the local origin $(\overline{x}_0, \overline{y}_0)$ and at the time $\overline{t}^0$. To be sure, if there is not any local structure in the initial data, then the local adjustment is made instantaneously.

In conclusion, it is observed that if for $\bar{t} = \bar{t}^0$, we have initial data relative to our local site which is characterized by the space variables $\hat{x}$, $\hat{y}$ and $\hat{z}$, then a local *unsteady* adjustment problem must be considered in the vicinity of

$$\hat{t} = \frac{\bar{t} - \bar{t}^0}{\varepsilon} = 0 \quad .$$

The latter makes it possible to understand how transition takes place to a local *quasi-steady* state which is valid at time $\bar{t} = \bar{t}^0$ and depends only parametrically on time via the short-range synoptic forecast at the time $\bar{t} = \bar{t}^0$ and the local origin $(\bar{x}_0, \bar{y}_0)$.

6.4 A Triple-Deck Structure Related Local Model[10]

It is common knowledge (see, for example, Kibel, 1963, Sect. 11.3) that in a classical quasi-geostrophic forecast [e.g., via (6.3)], the non-adiabatic action of the surface of the ground – generally assumed flat – is usually simulated by using the so-called Ackerblom model (see Chap. 11). This simulation enables us to take into account the dynamic action of the steady Ekman boundary layer. The latter, also referred to simply as the Ekman layer, exists in the immediate vicinity of the earth's surface and has a dimensional thickness of the order of a kilometer. It is assoiated with $(\nu_0/\Omega_0 \sin\varphi_0)^{1/2}$, where $\nu_0 = \mu_0/\varrho_\infty(0)$. It is noted that this (steady) Ekman layer is directly matched with the classical geostrophic wind which is valid to the order zero (in the definition of the MMAE) outside of the Ekman layer.

If we want to "regionalize" the synoptic forecast, i.e., be able to make a local forecast, then it must be explained how a site of horizontal scale $L_c \ll L$ perturbs the Ekman layer. It turns out that the consistent mathematical formulation of this problem of interaction between the classical Ekman layer and the one induced by the non-homogenous thermal site is closely linked to the order of magnitude of the horizontal scale L_c of the site which can vary between 10^3 m and 10^5 m.

Let then z be the altitude (with dimensions). It is easily confirmed that the variation in velocity within the Ekman layer is intrinsically related to the vertical variable:

$$(z/L_c)/\kappa = \bar{z} \quad , \tag{6.99}$$

where

$$\kappa = \left(\frac{L_c^2 \Omega_0 \sin\varphi_0}{\nu_0}\right)^{-1/2} \equiv \left(\frac{Re}{2Ro}\right)^{-1/2} \quad . \tag{6.100}$$

$Re = L_c U_0/\nu_0$ and $Ro = U_0/2\Omega_0 L_c \sin\varphi_0$ designate respectively the local Reynolds and Rossby numbers based on L_c. It is remarked that the local Reynolds number Re is constituted from the (turbulent) characteristic kinematic synoptic

---

10 Obtained by Zeytounian in 1981, the results in the following paragraph have never been published before.

viscosity ν_0. Given the fact that in the present work, the effects of turbulence are kept to the bare essentials, we did not think it appropriate to take into account a kinematic (turbulent) viscosity related to the nature of the local site.

If the space variables (with dimensions) x, y, z are reduced with respect to L_c, then there appears a local Boussinesq number $Bo = L_c g/RT_\infty(0)$ in the "local" equations which describe the flow induced by the thermally non-homogenous site. As the latter is located in a bounded domain $\mathcal{D}$ of the length scale L_c in the plane $z = 0$, it is assumed that the temperature field T is given in the form:

$$T = T_\infty(0) + \Delta T_0 \Xi\left(\frac{t}{t_c}, \frac{x}{L_c}, \frac{y}{L_c}\right) \quad , \tag{6.101}$$

on $z = 0$ and for $x, y \in \mathcal{D}$.

We have introduced into the condition (6.101), the temperature fluctuation ΔT_0 associated with the function Ξ which is supposed given and which governs the temperature distribution at the surface of the local site (assumed flat); the eventual consideration of a relief is discussed further on in the text. The characteristic local time t_c and the variations of Ξ with time are inherently related. The condition (6.101) brings out the rate of temperature fluctuation $\tau = \Delta T_0/T_\infty(0)$ and the local Strouhal number $S = L_c/U_0 t_c$.

When $L_c \cong 10^5$ m, the following can be adopted:

$$L_c = U_0/\Omega_0 \sin\varphi_0 \Rightarrow 2Ro = 1 \tag{6.102}$$

which leads to:

$$\kappa = Re^{-1/2} \quad . \tag{6.103}$$

Therefore, in this case, the terms related to the Coriolis force must mandatorily be retained in the local equations which describe the flow induced by the thermally non-homogenous site: thus the flow – which can be called mesometeorological – is necessarily three-dimensional. In addition, since $RT_\infty(0)/g \cong 8$ km, it is observed that $Bo \gg 1$ and of the same order as the ratio L_0/L_c. On the contrary, the rate τ and the Mach number $M_\infty = U_0/(\gamma RT_\infty(0))^{1/2}$ always remain small compared to unity since under "normal" meteorological conditions:

$$\Delta T_0 \ll T_\infty(0) \quad , \quad U_0^2 \ll \gamma RT_\infty(0) \quad .$$

Lastly, it is important to note that in the non-adiabatic terms in the right-hand sides of the mesometeorological equations, the inverse of the local Reynolds number $Re^{-1} \equiv \kappa^2$ appears. It is thus clear that when $L_c \cong 10^5$ m, *the Ekman velocity profile is compatible with the thickness of the mesometeorological boundary layer which develops in the vicinity of the thermally non-homogenous site. The thickness of this boundary layer is of the order* $\kappa L_c = L_c Re^{-1/2}$.

On the other hand, if L_c is much smaller than 10^5 m, i.e., of the order of 10^4 m to 10^3 m, then:

$$2Ro \gg 1$$

and it varies between 10 and 150. Therefore, in this case, the terms related to the

Coriolis force in the local equations – at least, to order zero – can be ignored: the flow induced by the site is then only "slightly" three-dimensional. In addition, in this situation, $Re \gg 2Ro$, and hence, in the expression of κ, (6.100), we have:

$$2Ro = Re^{-1/a} \Rightarrow \kappa = Re^{-1/m} \quad , \quad m = \frac{2a}{a-1} > 2 \quad . \tag{6.104}$$

For $U_0 \cong 10\,\mathrm{m/s}$, $\nu_0 \cong 5\,\mathrm{m^2/s}$ and $l_0 = 2\Omega_0 \sin\varphi_0 \cong 1{,}2 \cdot 10^{-4}\,\mathrm{s}^{-1}$, a characteristic value of m, $m_1 = 5$ corresponds to $L_c \cong 10^3\,\mathrm{m}$, whereas for $L_c \cong 10^4\,\mathrm{m}$, this value of m is $m_2 = 3$.

Regarding the local Boussinesq number Bo, for $m = m_1 = 5$, it is observed that $Bo \ll 1$; for $m = m_2 = 3$, we have $Bo \cong 1$. Therefore, when $m = m_1 = 5$, the Boussinesq approximation can be carried out, whereas when $m = m_2 = 3$, this approximation is no longer valid. In any case, *when* L_c *is of the order of* $10^3\,\mathrm{m}$ *to* $10^4\,\mathrm{m}$, *the Ekman velocity profile can no longer be compatible with the thickness of the local boundary layer induced by the thermally non-homogenous site.*

When $L_c \cong 10^3\,\mathrm{m}$, the following can be selected:

$$L_c \cong \frac{U_0}{g}\left(\frac{RT_\infty(0)}{\gamma}\right)^{1/2} \Rightarrow \hat{B} = \frac{Bo}{M_\infty} \cong 1 \quad , \tag{6.105}$$

whereas when $L_c \cong 10^4\,\mathrm{m}$, the following is preferred:

$$L_c \cong \frac{RT_\infty(0)}{g} \Rightarrow Bo \cong 1. \tag{6.106}$$

It is interesting to note that the value $m = m_1 = 5$ is, in fact, the one used by Smith (1973) in his study of laminar flow over a small "hump" on a flat plate. For this study along with the Smith notations, it is enough to introduce $l = \varepsilon^3 L$, $\varepsilon^8 \mathrm{Re} = 1$, where $\mathrm{Re} = U^*_\infty L/\nu^*_\infty$, and a local Reynolds number $U^*_\infty l/\nu^*_\infty = \varepsilon^3 \mathrm{Re} \equiv \varepsilon^{-5}$. In this case, the hump has an expanse equal to l and a maximum height equal to $\varepsilon^2 l$. The Smith study was generalized to the three-dimensional case by Smith, Sykes and Brighton (1973) with also $m_1 = 5$ was also employed in the latter. Sykes (1978), in fact, exploited the previously obtained results by analyzing the effects of stratificaion for a Boussinesq fluid in a boundary layer flow over a small hill, 1.5 km in area and with a maximum height of 60 metres.

The taking into account of the relief of the local site which is supposedly simulated by the equation:

$$z = h_0 h\left(\frac{x}{L_c}, \frac{y}{L_c}\right) \quad ,$$

with $h_0 = \max|h|$ when $x, y \in \mathcal{D}$, brings about the introduction of the parameter:

$$\mu = \frac{h_0}{L_c} \ll 1 \Rightarrow \mu = \mathrm{Re}^{-b} \quad , \quad 0 < b < 1 \quad .$$

Thus appears the small parameter:

$$\mu = \kappa^{bm} \quad . \tag{6.107}$$

In the above-cited works, the authors have chosen $b = 2/5$ which for $m = m_1 = 5$, leads to $\mu = \kappa^2$. In this case, the Sykes data (1978) concerning the importance of the small hill are, in fact, recovered. Moreover, when $m = m_1 = 5$, $\mu = \kappa^2$ is the only option which is consistent with the classical triple deck self-inducing model given by Neiland (1969) and also Stewartson and Williams (1969).[11]

In the forthcoming, a steady two-dimensional problem is considered which is justified when L_c is of the order of a few kilometers. The relief of the site is not taken into account since the main goal here is to clarify just how the Ekman basic profile is perturbed by the presence of a thermal spot situated on the x axis between $x = 0$ and $x = L_c$.

Let $\overline{T} = T/T_\infty(0)$. On $\overline{T}$, the following boundary condition is then imposed:

$$\overline{T} = 1 + \tau\Xi(\overline{x}) \quad , \quad \text{on } \overline{z} = 0 \text{ and } 0 \leq \overline{x} \leq 1 \quad , \tag{6.108}$$

where $\overline{z} = z/L_c$ and $\overline{x} = x/L_c$.

Far upstream of the thermal spot which is simulated by Eq. (6.108), when $x \to -\infty$, the Ekman longitudinal velocity profile will again be encountered (along the x axis). However, since $L_c/L \ll 1$ of the order of κ^{m-2} (in this instance of self-induced coupling, $L_c \cong L\mathrm{Re}^{-3/8}$ since $m = m_1 = 5$), the Ekman longitudinal velocity profile to the order considered here can be written in the following dimensionless form:

$$\overline{V}_\infty\left(\frac{\overline{z}}{\kappa}\right) = 1 - \exp\left(-\frac{\overline{z}}{\kappa}\right)\cos\left(\frac{\overline{z}}{\kappa}\right) \quad ; \quad \frac{\overline{z}}{\kappa} = \zeta, \tag{6.109}$$

and $\lim_{\zeta \to \infty} \overline{V}_\infty(\zeta) = 1$. The characteristic longitudinal velocity is that of the associated geostrophic flow at point $\overline{x} = 0$ and at the level of $\overline{z} = 0$.

According to Zeytounian (1974), the Boussinesq approximation consists in carrying out the triple limiting process:

$$Bo \to 0, \ \mathrm{M}_\infty \to 0, \ \tau \to 0; \ \frac{Bo}{\mathrm{M}_\infty} = \hat{B}, \ \frac{\tau}{\mathrm{M}_\infty} = \lambda \quad , \tag{6.110}$$

with $\hat{B}$ and λ of the order unity, and in seeking the dimensionless functions $\overline{u}$, $\overline{w}$ (velocities), π, ω and ϑ (thermodynamic perturbations relative to the standard state which is assumed as not having an Ekman boundary layer structure) in the form:

$$\overline{u} = \overline{u}_0 + \dots \quad , \quad \overline{w} = \overline{w}_0 + \dots \quad , \quad \pi = \mathrm{M}_\infty^2 \pi_2 + \dots$$
$$\omega = \mathrm{M}_\infty \omega_1 + \dots \quad , \quad \vartheta = \mathrm{M}_\infty \vartheta_1 + \dots \quad . \tag{6.111}$$

If the bars on the dimensionless quantities are eliminated and (6.104, 108, 109) are taken into account, the following reduced local interaction problem results for u_0, w_0, π_2, ω_1 and ϑ_1 as a consequence of (6.110, 111):

[11] The scheme called "triple deck" which via an intermediate layer leads to a self-induced "strong" coupling between the boundary layer and the perfect fluid was invented simultaneously and independently by Neiland and Stewartson and Williams.

$$
\begin{aligned}
&u_0\frac{\partial u_0}{\partial x}+w_0\frac{\partial u_0}{\partial z}+\frac{1}{\gamma}\frac{\partial \pi_2}{\partial x}=\kappa^m\left(\frac{\partial^2 u_0}{\partial x^2}+\frac{\partial^2 u_0}{\partial z^2}\right) \quad ;\\
&u_0\frac{\partial w_0}{\partial x}+w_0\frac{\partial w_0}{\partial z}+\frac{1}{\gamma}\frac{\partial \pi_2}{\partial z}-\frac{\hat{B}}{\gamma}\vartheta_1=\kappa^m\left(\frac{\partial^2 w_0}{\partial x^2}+\frac{\partial^2 w_0}{\partial z^2}\right) \quad ;\\
&\frac{\partial u_0}{\partial x}+\frac{\partial w_0}{\partial z}=0 \quad ;\\
&u_0\frac{\partial \vartheta_1}{\partial x}+w_0\frac{\partial \vartheta_1}{\partial z}+\hat{B}\left[\frac{\gamma-1}{\gamma}+\left.\frac{dT_\infty}{dz}\right|_{z=0}\right]w_0=\frac{\kappa^m}{\mathrm{Pr}}\left(\frac{\partial^2 \vartheta_1}{\partial x^2}+\frac{\partial^2 \vartheta_1}{\partial z^2}\right) \quad ;\\
&\omega_1=-\vartheta_1 \quad ,
\end{aligned}
\tag{6.112}
$$

$$
\begin{aligned}
&z=0:\ u_0=w_0=0 \quad , \quad \vartheta_1=\lambda\Xi(x) \quad , \quad 0\leq x\leq 1 \quad ;\\
&x\to-\infty:\ u_0\to V_\infty(\zeta) \quad , \quad w_0=\pi_2=\vartheta_1=\omega_1\to 0 \quad ,
\end{aligned}
\tag{6.113}
$$

with $V_\infty(\zeta) = 1 - \exp(-\zeta)\cos(\zeta)$. In the equation written for ϑ_1 appears the Prandtl number $\mathrm{Pr} = \varrho_\infty(0)\nu_0 c_p/k_0$, defined from the characteristic turbulent coefficient of thermal conduction k_0. When $\kappa\to 0$, we are naturally led to the consideration in the local problem (6.112, 113), of three vertical scales – at least in the case of self-induced coupling corresponding to $m = m_1 = 5$.

The first scale z is the one introduced to describe the flow of a perfect fluid which is associated with the geostrophic "part" of the Ekman profile:

$$
\lim_{\zeta\to\infty} V_\infty(\zeta)=1 \quad . \tag{6.114}
$$

This scale z leads to the consideration of a so-called *upper* layer.

The second scale $\zeta = z/\kappa$, which is an intermediate scale, makes it possible to take into account practically the entire variation of $V_\infty(\zeta)$, except in the immediate vicinity of $z = 0$. This scale brings us to the consideration of a so-called *intermediate* layer.

Finally, a third scale:

$$
\tilde{z}=z/\kappa^n=\zeta/\kappa^{n-1} \quad , \quad n>1 \quad , \tag{6.115}
$$

(with n to be determined) is associated with the *viscous "sub-layer"* which enables us to take into account the boundary conditions on $z = 0$.

It is emphasized that in the viscous sub-layer, it is necessary that:

$$
u_0\sim\kappa^{n-1}\tilde{u} \quad , \quad \text{since} \quad V_\infty(\zeta)\sim\zeta \quad ,
$$

when $\zeta \equiv \kappa^{n-1}\tilde{z}\to 0$ according to (6.115). Therefore:

$$
\lim_{x\to-\infty}\tilde{u}=\tilde{z} \quad . \tag{6.116}
$$

It is thus observed that the adjustment process of the local flow to the Ekman profile $V_\infty(\zeta)$ is what leads to the setting up of an asymptotic triple-deck model.

When $\lambda = O(1)$ the viscous sub-layer is characterized by:

$$\vartheta_1 \sim \tilde{\theta}(x, \tilde{z}) \quad , \quad u_0 \sim \kappa^{n-1}\tilde{u}(x, \tilde{z}) \tag{6.117}$$

with $\tilde{z} = z/\kappa^n \equiv \zeta/\kappa^{n-1}$, where $\zeta = z/\kappa$ and $n > 1$. Let us consider in this inner layer the following asymptotic representation:

$$\begin{aligned} u_0 &= \kappa^{n-1}\tilde{u} + \dots \; ; \qquad & \pi_2 &= \kappa^c\tilde{\pi} + \dots \; ; \\ w_0 &= \kappa^a\tilde{w} + \dots \; ; \qquad & \vartheta_1 &= \tilde{\theta} + \dots \; . \end{aligned} \tag{6.118}$$

It is then obvious that the following must be selected:

$$2n - 1 = a = c + 1 \quad \text{and} \quad m + 1 = 3n \quad .$$

Moreover, if $c = n$, then instead of the second equation of (6.112), we find:

$$\frac{\partial \tilde{\pi}}{\partial \tilde{z}} = \hat{B}\tilde{\theta} \quad . \tag{6.119}$$

In this case, $n = 2$ and it is necessary that $m = m_1 = 5$. The limiting equation (6.119) is obtained when:

$$c = n = 2 \Rightarrow a = 3 \quad , \quad m = 5 \quad . \tag{6.120}$$

On the other hand, if $c < n$, then

$$\frac{\partial \tilde{\pi}}{\partial \tilde{z}} = 0 \quad . \tag{6.121}$$

In this case, $n < 2$ and $m + 1 < 6$. Hence when the limiting equation (6.121) occurs, it is necessary that:

$$m = 4 \Rightarrow n = \tfrac{5}{3} \quad , \quad c = \tfrac{4}{3} \quad , \quad a = \tfrac{7}{3} \quad . \tag{6.122}$$

Lastly, if $c > n$, then[12]:

$$\tilde{\theta} \equiv 0 \quad ; \tag{6.123}$$

here, $n > 2$ and $m + 1 > 6$. Thus, the relation (6.123) implies that:

$$m = 6 \Rightarrow n = \tfrac{7}{3} \quad , \quad c = \tfrac{8}{3} \quad , \quad a = \tfrac{11}{3} \quad . \tag{6.124}$$

In conclusion, the most significant degeneracy corresponds to the values of (6.120) and leads to the following boundary layer equations which are valid in the lower viscous sub-layer:

$$\begin{aligned} &\tilde{u}\frac{\partial \tilde{u}}{\partial x} + \tilde{w}\frac{\partial \tilde{u}}{\partial \tilde{z}} + \frac{1}{\gamma}\frac{\partial \tilde{\pi}}{\partial x} = \frac{\partial^2 \tilde{u}}{\partial \tilde{z}^2} \quad ; \\ &\frac{\partial \tilde{\pi}}{\partial \tilde{z}} = \hat{B}\tilde{\theta} \quad ; \\ &\frac{\partial \tilde{u}}{\partial x} + \frac{\partial \tilde{w}}{\partial \tilde{z}} = 0 \quad ; \end{aligned} \tag{6.125a}$$

---

[12] This case is strongly degenerated because we can no longer apply the boundary condition for ϑ_1 on $z = 0$ when $\lambda = O(1)$.

$$\tilde{u}\frac{\partial\tilde{\theta}}{\partial x}+\tilde{w}\frac{\partial\tilde{\theta}}{\partial\tilde{z}}=\frac{1}{\mathrm{Pr}}\frac{\partial^2\tilde{\theta}}{\partial\tilde{z}^2} \quad , \tag{6.125b}$$

as long as it has been assumed that $(\gamma-1)/\gamma+(dT_\infty/dz)_0$, as well as the similarity parameters $\hat{B}$ and λ remain of the order unity. The following conditions must now be imposed on (6.125):

$$\begin{aligned}&\text{on} \quad \tilde{z}=0: \ \tilde{u}=\tilde{w}=0 \quad , \ \tilde{\theta}=\lambda\Xi(x) \quad , \ 0\le x\le 1 \quad ,\\ &\text{when} \quad x\to-\infty: \ \tilde{u}\to\tilde{z} \quad , \ \tilde{w}=\tilde{\pi}=\tilde{\theta}\to 0 \quad .\end{aligned} \tag{6.126}$$

The behavior conditions when $\tilde{z}\to+\infty$ still have to be formulated. To this end, the intermediate layer must be considered.

Since $n=2$ and thus, $m=5$, it is obvious – on nearly so – that the following asymptotic representation must be postulated in the intermediate layer:

$$\begin{aligned}u_0 &= V_\infty(\zeta)+\kappa^\varphi\hat{u}+\dots \quad ;\\ w_0 &= \kappa^\psi\hat{w}+\dots \quad ;\\ \pi_2 &= \kappa^2\hat{\pi}+\dots \quad ;\\ \vartheta_1 &= \kappa^\sigma\hat{\theta}+\dots \quad .\end{aligned} \tag{6.127}$$

If we want the longitudinal velocities u_0 to be matched when, respectively, $\tilde{z}\to+\infty$ and $\zeta\to 0$, then we must adopt:

$$\varphi=1 \quad , \quad \psi=1+\varphi=2 \quad . \tag{6.128}$$

However, the following hydrostatic form is once again retained:

$$\frac{\partial\hat{\pi}}{\partial\zeta}=\hat{B}\hat{\theta} \quad , \tag{6.129}$$

in this intermediate layer instead of the second equation of (6.112) if:

$$\sigma\equiv 1 \quad . \tag{6.130}$$

In this case, the equation for ϑ_1 from system (6.112) implies:

$$\frac{\partial\hat{\theta}}{\partial x}=0\Rightarrow\hat{\theta}=\chi(\zeta) \quad , \tag{6.131}$$

with $\chi(\zeta)$ to be determined!

Concerning $\hat{u}(x,\zeta)$ and $\hat{w}(x,\zeta)$, they can be said to satisfy the classical system:

$$V_\infty(\zeta)\frac{\partial\hat{u}}{\partial x}+\frac{dV_\infty}{d\zeta}\hat{w}=0 \quad , \quad \frac{\partial\hat{u}}{\partial x}+\frac{\partial\hat{w}}{\partial\zeta}=0 \quad ,$$

whose solution is:

$$\hat{u}=A(x)\frac{dV_\infty}{d\zeta} \quad , \quad \hat{w}=-\frac{dA}{dx}V_\infty(\zeta) \quad , \tag{6.132}$$

where the function $\kappa A(x)$ must be interpreted as a displacement thickness which, in fact, generates the pressure perturbation. The solution (6.132) does indeed lead to the following behavior conditions for $\tilde{u}$ and $\tilde{w}$ which satisfy equations (6.125) in the lower viscous sub-layer:

$$\tilde{u} \to \tilde{z} + A(x) \quad , \quad \tilde{w} \to -\tilde{z}\frac{dA}{dx} \quad , \quad \text{when} \quad \tilde{z} \to +\infty \quad , \tag{6.133}$$

since $\lim_{\zeta \to 0} dV_\infty/d\zeta = 1$.

Furthermore, (6.129) leads to:

$$\frac{\partial^2 \hat{\pi}}{\partial x \partial \zeta} = 0 \Rightarrow \hat{\pi} = P_1(x) + P_2(\zeta) \tag{6.134}$$

and hence:

$$\hat{\theta} = \frac{1}{\hat{B}}\frac{dP_2}{d\zeta} \equiv \chi(\zeta) \quad . \tag{6.135}$$

However, it is necessary that $\lim_{x \to -\infty} \hat{\pi} = 0$ and thus:

$$P_1(-\infty) = 0 \quad ; \quad P_2(\zeta) \equiv 0 \Rightarrow \chi(\zeta) \equiv 0 \quad .$$

Consequently, $\tilde{\theta}$ and $\tilde{\pi}$ satisfy the behavior:

$$\tilde{\theta} \to 0 \quad , \quad \tilde{\pi} \to \hat{\pi} \equiv P_1(x) \quad , \quad \text{when} \quad \tilde{z} \to +\infty \quad . \tag{6.136}$$

It is also pointed out that the perturbation of the pressure $\tilde{\pi}$ in the viscous lower sub-layer influences the distribution of $A(x)$ whence the singular triple-deck (self-inducted) coupling:

$$\tilde{u}\frac{\partial \tilde{u}}{\partial x} + \tilde{w}\frac{\partial \tilde{u}}{\partial \tilde{z}} + \frac{\hat{B}}{\gamma}\int_\infty^{\tilde{z}} \frac{\partial \tilde{\theta}}{\partial x} d\tilde{z} + \frac{1}{\gamma}\frac{dP_1(x)}{dx} = \frac{\partial^2 \tilde{u}}{\partial \tilde{z}^2} \quad ; \qquad \frac{\partial \tilde{u}}{\partial x} + \frac{\partial \tilde{w}}{\partial \tilde{z}} = 0 \quad ;$$

$$\tilde{\pi} = \hat{B}\int_\infty^{\tilde{z}} \tilde{\theta} d\tilde{z} + P_1(x) \quad ; \tag{6.137a}$$

$$\tilde{u}\frac{\partial \tilde{\theta}}{\partial x} + \tilde{w}\frac{\partial \tilde{\theta}}{\partial \tilde{z}} = \frac{1}{\mathrm{Pr}}\frac{\partial^2 \tilde{\theta}}{\partial \tilde{z}^2} \quad ,$$

$$\begin{aligned}
&\tilde{z} = 0: \ \tilde{u} = 0 \quad , \quad \tilde{w} = 0 \quad , \quad \tilde{\theta} = \lambda \Xi(x) \quad , \quad 0 \le x \le 1 \quad ; \\
&x \to -\infty: \ \tilde{u} \to \tilde{z} \quad , \quad \tilde{w} \to 0 \quad , \quad P_1(x) \to 0 \quad , \quad \tilde{\theta} \to 0 \quad ; \\
&A(-\infty) = \frac{dA(-\infty)}{dx} \equiv 0 \quad ; \\
&\tilde{z} \to +\infty: \ \tilde{u} \to \tilde{z} + A(x) \quad , \quad \tilde{w} \to -\tilde{z}\frac{dA}{dx} \quad , \quad \tilde{\theta} \to 0 \quad .
\end{aligned} \tag{6.137b}$$

The (strong) singular self-induced coupling appears because the problem (6.137) to be solved in the lower viscous layer *does not accept* $P_1(x)$ as data known prior to the resolution (as is the case in classical boundary layer problems). On the contrary, this pressure perturbation $P_1(x)$ must be calculated at the same time as the velocity components $\tilde{u}$ and $\tilde{w}$, as well as the temperature perturbation $\tilde{\theta}$.

Nonetheless, it must be emphasized that this pressure perturbation $P_1(x)$ is not completely arbitrary and that it is connected to the function $A(x)$ through a relation. This relation is obtained via the analysis of the perfect fluid flow in the upper layer.

In this upper layer, it is fairly certain that the following asymptotic representation must be postulated – at least in the case being considered where $n = 2$, i.e., $m = m_1 = 5$:

$$\begin{aligned} u_0 &= 1 + \kappa^2 \overline{u} + \dots \quad ; \\ w_0 &= \kappa^2 \overline{w} + \dots \quad ; \\ \pi_2 &= \kappa^2 \overline{\pi} + \dots \quad ; \\ \vartheta_1 &= \kappa^2 \overline{\theta} + \dots \quad . \end{aligned} \tag{6.138}$$

For $\overline{u}$, $\overline{w}$, $\overline{\pi}$ and $\overline{\theta}$, the following system of linear equations then results:

$$\begin{aligned} &\frac{\partial \overline{u}}{\partial x} + \frac{1}{\gamma}\frac{\partial \overline{\pi}}{\partial x} = 0 \quad ; \\ &\frac{\partial \overline{w}}{\partial x} + \frac{1}{\gamma}\frac{\partial \overline{\pi}}{\partial z} = \frac{\hat{B}}{\gamma}\overline{\theta}; \\ &\frac{\partial \overline{u}}{\partial x} + \frac{\partial \overline{w}}{\partial z} = 0 \quad ; \\ &\frac{\partial \overline{\theta}}{\partial x} + \hat{B}\left[\frac{\gamma - 1}{\gamma} + \left(\frac{dT_\infty}{dz}\right)_0\right]\overline{w} = 0 \quad . \end{aligned} \tag{6.139}$$

By eliminating from (6.139) all the functions except $\overline{\pi}(x, z)$, the classical Helmholtz equation is obtained:

$$\left(\frac{\partial^2}{\partial x^2} + \frac{\partial^2}{\partial z^2} + K_0^2\right)\frac{\partial \overline{\pi}}{\partial x} = 0 \quad , \tag{6.140}$$

where

$$K_0^2 \equiv \frac{\hat{B}^2}{\gamma}\left[\frac{\gamma - 1}{\gamma} + \left(\frac{dT_\infty}{dz}\right)_{z=0}\right] \quad .$$

In the forthcoming, we will assume that $K_0^2 > 0$. At $z = 0$, the following condition must be prescribed for the solution of (6.140):

$$\overline{\pi}(x, 0) = P_1(x) \quad . \tag{6.141}$$

Moreover, the matching of the vertical velocities in the intermediate and upper layers yields:

$$\overline{w}(x, 0) = -\frac{dA}{dx} \quad , \tag{6.142}$$

which leads to:

$$\frac{\partial}{\partial x}\left[\left(\frac{\partial \overline{\pi}}{\partial z}\right)_{z=0}\right] = \gamma\left[K_0^2\frac{dA}{dx} + \frac{d^3 A}{dx^3}\right] \quad . \tag{6.143}$$

Now, $F^*(k)$ is designated as the Fourier transform of the function $F(x)$. In particular, we find:

$$\overline{\pi}^*(k, z) = \int_{-\infty}^{+\infty} \overline{\pi}(x, z)\, \mathrm{e}^{-\mathrm{i}kx}\, dx \quad . \tag{6.144}$$

In this case, by taking advantage of (6.140, 141 and 143), the following is obtained:

$$A^*(k) = \frac{\mathrm{i}\mathcal{N}_0}{\gamma} \frac{P_1^*(k)}{K_0^2 - k^2} \quad , \tag{6.145}$$

where $\mathcal{N}_0$ is determined by:

$$\mathcal{N}_0 = \begin{cases} \mathrm{i}(k^2 - K_0^2)^{1/2} \quad , & |k| > K_0 \quad ; \\ (K_0^2 - k^2)^{1/2} \quad , & |k| < K_0 \end{cases} \quad . \tag{6.146}$$

In (6.146), the decreasing modes are chosen for the first case, whereas in the second case, the modes are preferred for which the energy transport is effected in an upward direction.

To conclude, it is remarked that *the viscous sub-layer problem (6.137) must be solved in conjunction with (6.145), which interlinks P_1 and A in the plane of the Fourier transform.*

If no hypothesis is made concerning the similarity parameter λ which intervenes in the boundary condition on $\tilde{z} = 0$ for $\tilde{\theta}$, then problem (6.137) along with the constraint (6.145) remain quasi-linear and can only be solved numerically. To this end, we most often use an iterative solution based on an adequate initialization of the function $A(x)$[13], with the convergence possibly being ensured by an underrelaxation[14].

Finally, it is pointed out that the case of $m = 6$ and $m = 4$ also pertain to (6.112, 113), but require a separate analysis. The case $m = m_2 = 3$ requires a separate analysis as well, but based on equations other than those of Boussinesq (6.112). Although the recent results obtained by F.T. Smith, P.W.M. Brighton, P.S. Jackson and J.C.R. Hunt (1981) augur well for the future of such analyses, the question remains open for the present

---

[13] This initialization of $A(x)$ can, for example, be inferred from the linear theory corresponding to $\lambda \ll 1$. The latter was recently carried out by R. Kharab within the framework of a Thesis in fluid mechanics at the University of Lille I. This study enabled R. Kharab to obtain a numerical solution of (6.137) for various values of λ.

[14] In the work cited by Sykes (1978) (see pages 233 to 236), we can find the description of such an iterative calculation for the case $\tilde{\theta} \equiv 0$ which comes down to assuming that $\lambda \equiv 0$. When $\lambda \equiv 0$, we are faced with the following: Bo and M_∞ fixed; $\tau \to 0$, then $Bo \to 0$ *and* $\mathrm{M}_\infty \to 0$ such that $Bo/\mathrm{M}_\infty = \hat{B} = O(1)$. Of course, this situation is no longer significant since it is covered in the one corresponding to (6.110). When $\lambda \equiv 0$, we must at least introduce a relief and write the no-slip conditions on $\tilde{z} = h(x)$ while taking into account that $\mu = \kappa^2$ [see (6.107)].

7. The Quasi-static Approximation

As has already been pointed out several times, when atmospheric phenomena are analyzed on what is called a synoptic scale (for which L is of the order of 10^6 m), the anisotropic parameter of the atmosphere $\varepsilon = H/L$ can be estimated to be at least 10^{-2}. The reason for this is that these phenomena take place essentially throughout the entire troposphere. The latter can be characterized by the vertical scale $RT_\infty(0)/g$ which is of the order of 10^4. In such a case, the Boussinesq number Bo remains, of course, of the order unity. Hence, we want to analyze in this chapter the consequences of the limiting process $\varepsilon \to 0$ from the point of view of asymptotic methods. When $\varepsilon \to 0$, the behavior of other dimensionless parameters must be defined as well: most particularly, the Reynolds number, $\mathrm{Re} = U_0 L/\nu_0$, and the Mach number, $\mathrm{M}_\infty = U_0/(\gamma R T_\infty(0))^{1/2}$. Furthermore, when $\varepsilon \to 0$, it is essential to specify whether or not the independent variables $\bar{t}$, $\bar{x}$, $\bar{y}$ and $\bar{z}$ are fixed. And finally, the domain of validity of the limiting process $\varepsilon \to 0$ (main limit) must be delimited in the four-dimensional time-space reference frame. It is understood – the following will apty demonstrate it – that all the problems brought out by the limiting process $\varepsilon \to 0$ have not as yet been resolved. Nonetheless, we think it worthwhile to review what is already known on this topic and this matter will thus form the content of the seven sections of this chapter.

In Sect. 7.1, we obtain the exact quasi-static equations which can be interpreted as boundary layer equations (for synoptic scale phenomena). These equations describe the asymptotic situation: $\varepsilon \to 0$, $\mathrm{Re} \to \infty$, with $\varepsilon^2 \mathrm{Re} \equiv \mathrm{Re}_\perp = O(1)$. Section 7.2 has been devoted to the analysis of the primitive equations which can be deduced from the quasi-static equations of Sect. 7.1 when $\mathrm{Re}_\perp \to \infty$, at fixed values of all the independent variables. It can be noted that these primitive equations do not constitute a significant degeneracy of the general equations when $\varepsilon \to 0$ and $\mathrm{Re} \to \infty$. In Sect. 7.3, we formulate the boundary layer problem in the vicinity of the ground associated with the primitive equations of Sect. 7.2; these boundary layer equations are in fact a degeneracy of the quasi-static equations found in Sect. 7.1. After noting that a small natural parameter – the Mach number M_∞ – remains in the primitive equations of Sect. 7.2, we formulate the simplified primitive equations in Sect. 7.4. A linearized expression of the latter enables us to set up a Sturm-Liouville type problem for the vertical structure. Section 7.5 deals with the important question of which initial conditions must be prescribed for the primitive equations; this in turn leads to the problem of adjustment to hydrostatic balance. Finally, various complementary themes as well as some bibliographical references can be found in Sects. 7.6, 7.

The basic equations in this chapter are the Navier-Stokes equations written in dimensionless form in the local mobile reference frame which was brought up in Sect. 2.2. In this general form, they are rewritten *non-dimensionally* as follows[1]:

$$\varrho\left\{\mathrm{S}\frac{D\boldsymbol{u}}{Dt}+\frac{1}{\mathrm{Ro}}(\boldsymbol{e}\times\boldsymbol{u})\right\}+\frac{1}{\gamma\mathrm{M}_\infty^2}\left(\nabla p+\frac{\mathrm{Bo}}{\varepsilon}\varrho\boldsymbol{k}\right)$$
$$=\frac{1}{\mathrm{Re}}\left\{\frac{1}{\varepsilon^2}\frac{\partial}{\partial z}\left(\mu\frac{\partial\boldsymbol{u}}{\partial z}\right)+\boldsymbol{D}^2\boldsymbol{u}+\frac{1}{3}\nabla(\nabla\cdot\boldsymbol{u})\right\}\quad; \tag{7.1}$$

$$\mathrm{S}\frac{D\varrho}{Dt}+\varrho(\nabla\cdot\boldsymbol{u})=0\quad; \tag{7.2}$$

$$p=\varrho T\quad; \tag{7.3}$$

$$\varrho\mathrm{S}\frac{DT}{Dt}-\frac{\gamma-1}{\gamma}\mathrm{S}\frac{Dp}{Dt}=\frac{1}{\mathrm{Pr}}\frac{1}{\mathrm{Re}}\left\{\frac{1}{\varepsilon^2}\frac{\partial}{\partial z}\left(k\frac{\partial T}{\partial z}\right)+\boldsymbol{D}^2T\right\}$$
$$+\frac{\gamma-1}{\varepsilon^2\mathrm{Re}}\mathrm{M}_\infty^2\mu\phi+\frac{1}{\mathrm{Pr}}\frac{1}{\mathrm{Re}}\frac{\mathrm{Bo}\sigma_{00}}{\varepsilon^2}\frac{\partial\mathcal{R}}{\partial z}\quad. \tag{7.4}$$

In (7.1–4), we have the following

$$\boldsymbol{e}=\frac{\sin\varphi\boldsymbol{k}+\cos\varphi\boldsymbol{j}}{\sin\varphi_0}\quad; \tag{7.5}$$

then

$$\boldsymbol{u}=\boldsymbol{v}+\varepsilon w\boldsymbol{k}\quad,\qquad \boldsymbol{v}=u\boldsymbol{j}+v\boldsymbol{j}\quad;$$
$$\nabla=\boldsymbol{D}+\frac{1}{\varepsilon}\frac{\partial}{\partial z}\boldsymbol{k}\quad,\quad \boldsymbol{D}=\frac{\partial}{\partial x}\boldsymbol{i}+\frac{\partial}{\partial y}\boldsymbol{j}\quad; \tag{7.6}$$
$$\mathrm{S}\frac{D}{Dt}\equiv\mathrm{S}\frac{\partial}{\partial t}+\boldsymbol{v}\cdot\boldsymbol{D}+w\frac{\partial}{\partial z}\quad.$$

Five initial conditions must be joined to this system of equations (7.1–4) of the 5th order in t:

$$t=0:\quad \boldsymbol{v}=\boldsymbol{v}^0\quad,\quad \varepsilon w=w^0\quad,\quad p=p^0\quad,\quad \varrho=\varrho^0\quad. \tag{7.7}$$

For this same system (7.1–4), the following ground conditions are written (the ground is assumed to be flat):

$$z=0:\quad \boldsymbol{v}=0\quad,\quad w=0\quad;\quad \frac{\partial T}{\partial z}+\mathrm{Bo}\sigma_{00}\mathcal{R}=0\quad. \tag{7.8}$$

For the time being, we will ignore the question of boundary conditions relative to the horizontal variables x and y, as well as the behavior condition for $z\to+\infty$. The formulation of these conditions is a very delicate matter: at the present time, few rigorous mathematical results exist [see the recent article by

[1] We have omitted the bars on the dimensionless values.

Gustafsson and Sundstrom (1978)]. Nevertheless, we can say that "intuitively" speaking, the formulation of these conditions must promote the "correct solution" of the mathematical problem which is to be solved and which characterizes the asymptotic situation under consideration.

Finally, it should be noted that if we take advantage of (2.45, 55), then (7.1) coincides with (2.101a–c) with an error $O(\delta)$ ($\delta = L/a_0 \ll 1$) provided that the turbulent dynamic viscosity coefficient μ is constant. The expression of the term ϕ in (7.4) is given by the formula (2.102) in Sect. 2.5.

7.1 The Exact Quasi-static Equations

All the while retaining $\varepsilon \equiv H/L \ll 1$ as the main small parameter, we are naturally led in the general case, to take into consideration the fact that the Reynolds number $\mathrm{Re} = U_0 L/\nu_0 \gg 1$. As a consequence, we must seek out the degeneracy of (7.1–4) with (7.7, 8) when simultaneously

$$\varepsilon \to 0 \quad \text{and} \quad \mathrm{Re} \to \infty \quad . \tag{7.9}$$

However, as we want to obtain a significant degeneracy (as defined in Chap. 5), it turns out that the following similarity relation must be imposed on the double limiting process (7.9)

$$\varepsilon^2 \mathrm{Re} = \mathrm{Re}_\perp \tag{7.10}$$

with

$$\mathrm{Re}_\perp \equiv \frac{H W_0}{\nu_0} = O(1) \quad , \tag{7.11}$$

when $\varepsilon \to 0$. We note that $W_0 \equiv \varepsilon U_0$ designates the characteristic reference velocity for the vertical component w of the relative velocity $\boldsymbol{u}$. Naturally when (7.10) is verified, it is supposed that, together with $\mathrm{Re}_\perp$, all the other dimensionless parameters and reduced undependent variables remain fixed and of the order unity.

We write

$$\lim_{q-s}{}^{P} \equiv \lim \begin{pmatrix} \varepsilon \to 0 \ , \ \mathrm{Re} \to \infty \ ; \ \varepsilon^2 \mathrm{Re} \equiv \mathrm{Re}_\perp \ , \ \mathrm{S} \ , \\ \gamma \, , \mathrm{M}_\infty \, , \mathrm{Pr} \, , \mathrm{Bo} \, , \sigma_{00} \, , t \, , x \, , y \, , z \quad \text{fixed} \end{pmatrix} \quad . \tag{7.12}$$

Let us introduce:

$$(\boldsymbol{v}_0 \, , w_0 \, , p_0 \, , \varrho_0 \, , T_0 \, , \mathcal{R}_0) = \lim_{q-s}{}^{P} (\boldsymbol{v} \, , w \, , p \, , \varrho \, , T \, , \mathcal{R}) \quad . \tag{7.13}$$

There results from the quasi-static development linked to the limiting process (7.12) the following non-dimensional limiting equations for the main terms $\boldsymbol{v}_0$, w_0, p_0, ϱ_0 and T_0:

$$\varrho_0 \left\{ \mathrm{S} \frac{\partial \boldsymbol{v}_0}{\partial t} + (\boldsymbol{v}_0 \cdot \boldsymbol{D}) \boldsymbol{v}_0 + w_0 \frac{\partial \boldsymbol{v}_0}{\partial z} + \frac{1}{\mathrm{Ro}} \frac{\sin \varphi}{\sin \varphi_0} (\boldsymbol{k} \times \boldsymbol{v}_0) \right\} + \frac{1}{\gamma \mathrm{M}_\infty^2} \boldsymbol{D} p_0$$
$$= \frac{1}{\mathrm{Re}_\perp} \frac{\partial}{\partial z} \left(\mu_0 \frac{\partial \boldsymbol{v}_0}{\partial z} \right) \quad ; \tag{7.14}$$

$$\frac{\partial p_0}{\partial z} + \mathrm{Bo}\varrho_0 = 0 \quad ; \tag{7.15}$$

$$\mathrm{S}\frac{\partial \varrho_0}{\partial t} + \boldsymbol{D}\cdot(\varrho_0 \boldsymbol{v}_0) + \frac{\partial \varrho_0 w_0}{\partial z} = 0 \quad ; \tag{7.16}$$

$$p_0 = \varrho_0 T_0 \quad ; \tag{7.17}$$

$$\begin{aligned} &\varrho_0\left(\mathrm{S}\frac{\partial T_0}{\partial t} + \boldsymbol{v}_0\cdot \boldsymbol{D}T_0\right) - \frac{\gamma-1}{\gamma}\left(\mathrm{S}\frac{\partial p_0}{\partial t} + \boldsymbol{v}_0\cdot\boldsymbol{D}p_0\right) \\ &+w_0\left(\varrho_0\frac{\partial T_0}{\partial z} - \frac{\gamma-1}{\gamma}\frac{\partial p_0}{\partial z}\right) = \frac{1}{\mathrm{Pr}}\frac{1}{\mathrm{Re}_\perp}\frac{\partial}{\partial z}\left(k_0\frac{\partial T_0}{\partial z}\right) \\ &+\frac{\gamma-1}{\mathrm{Re}_\perp}\mathrm{M}_0^2\mu_0\left|\frac{\partial \boldsymbol{v}_0}{\partial z}\right|^2 + \frac{1}{\mathrm{Pr}}\frac{\mathrm{Bo}\sigma_{00}}{\mathrm{Re}_\perp}\frac{\partial \mathcal{R}_0}{\partial z} \quad . \end{aligned} \tag{7.18}$$

In (7.14, 18), we assume the coefficients μ_0 and k_0 to be known functions of T_0. The limiting equations (7.14–18) are the quasi-static equations according to Zeytounian (1976).

To these quasi-static equations, we can join ground conditions directly derived from (7.8):

$$z = 0: \quad \boldsymbol{v}_0 = 0 \quad , \quad w_0 = 0 \quad ; \quad \frac{\partial T_0}{\partial z} + \mathrm{Bo}\sigma_{00}\mathcal{R}_0 = 0 \quad . \tag{7.19}$$

The initial conditions cannot be deduced from (7.7) for these quasi-static equations. This is obvious since the degeneracy of the vertical momentum equation leads us to (7.15) which does not contain the term $\partial w_0/\partial t$. Thus we can guess that a singular perturbation phenomenon might exist in the vicinity of $t = 0$ (see Sect. 7.7):

$$\lim_{t\to 0}\lim_{q-s}{}^{P} \neq \lim_{q-s}{}^{P}\lim_{t\to 0} \quad ;$$

this phenomenon being linked to the adjustment to hydrostatic balance expressed by (7.15). The latter will be analyzed in detail in Sect. 7.5 for the adiabatic case ($\mathrm{Re}_\perp \equiv \infty$) which corresponds to the primitive equations. We note that given (7.11), the quasi-static equations (7.14–18) are valid for synoptic atmospheric motions having a characteristic vertical reference velocity

$$W_0 \cong \mathrm{Re}_\perp \nu_0/H \quad . \tag{7.20}$$

Of course, for usual meteorological conditions, $\mathrm{Re}_\perp \gg 1$ and therefore instead of $\lim_{q-s}^{P}$, the following limiting process should be considered for the basic equations (7.1–4):

$$\varepsilon \text{ fixed} \quad , \quad \mathrm{Re}\to\infty \quad , \quad \text{then } \varepsilon\to 0 \quad , \tag{7.21}$$

since

$$\varepsilon^2\,\mathrm{Re} \gg 1 \quad . \tag{7.22}$$

For the time being, let us set aside (7.21) and return to the quasi-static equations (7.14–18). In form, they are typical of boundary layer equations. This can be seen clearly if for (7.14, 18), z is transformed into ζ via

$$\zeta = \sqrt{\mathrm{Re}_\perp} z \tag{7.23}$$

and w_0 into ω_1 via

$$\omega_1 = \sqrt{\mathrm{Re}_\perp} w_0 \quad , \tag{7.24}$$

when $\mathrm{Re}_\perp = \mathrm{O}(1)$, i.e., when

$$H \cong \sqrt{\mathrm{Re}_\perp} \sqrt{\frac{L\nu_0}{U_0}} \cong 10^3\,\mathrm{m} \quad , \tag{7.25}$$

since (under "normal" synoptic conditions)

$$L \cong 10^6\,\mathrm{m} \quad , \quad \nu_0 \cong 5\,\mathrm{m^2/s} \ \text{and} \ U_0 \cong 10\,\mathrm{m/s} \quad .$$

Relation (7.25) demonstrates that our exact quasi-static equations (7.14–18) describe atmospheric motions within what is approximately a one-kilometer thick boundary layer from the ground. Let us go back to the introduction of ζ and ω_1 via (7.23, 24).

We have in fact[2] (in non-dimensional form):

$$z = \frac{z^*}{H} = \frac{z^*/L}{1/\sqrt{\mathrm{Re}}} \frac{1}{\sqrt{\mathrm{Re}_\perp}} = \frac{1}{\sqrt{\mathrm{Re}_\perp}} \frac{Z}{1/\sqrt{\mathrm{Re}}} \equiv \frac{1}{\sqrt{\mathrm{Re}_\perp}} \zeta \quad , \tag{7.26}$$

and also

$$w = \frac{w^*}{W_0} = \frac{w^*/U_0}{1/\sqrt{\mathrm{Re}}} \frac{1}{\sqrt{\mathrm{Re}_\perp}} = \frac{1}{\sqrt{\mathrm{Re}_\perp}} \frac{W}{1/\sqrt{\mathrm{Re}}} \equiv \frac{1}{\sqrt{\mathrm{Re}_\perp}} \omega \quad , \tag{7.27}$$

then

$$\lim_{q-s}{}^{P} w = w_0 \equiv \omega_1/\sqrt{\mathrm{Re}_\perp} \quad , \quad \text{since}$$

$$H \equiv \frac{L}{\sqrt{\mathrm{Re}/\mathrm{Re}_\perp}} \quad . \tag{7.28}$$

Thus, when $\mathrm{Re}_\perp = \mathrm{O}(1)$, the quasi-static equations (7.14–18) can be rewritten in the following form:

$$\varrho_0 \left\{ \mathrm{S} \frac{\partial \boldsymbol{v}_0}{\partial t} + (\boldsymbol{v}_0 \cdot \boldsymbol{D}) \boldsymbol{v}_0 + \omega_1 \frac{\partial \boldsymbol{v}_0}{\partial \zeta} + \frac{1}{\mathrm{Ro}} \frac{\sin\varphi}{\sin\varphi_0} (\boldsymbol{k} \times \boldsymbol{v}_0) \right\}$$
$$+ \frac{1}{\gamma \mathrm{M}_\infty^2} \boldsymbol{D} p_0 = \frac{\partial}{\partial \zeta} \left(\mu_0 \frac{\partial \boldsymbol{v}_0}{\partial \zeta} \right) \quad ; \tag{7.29a}$$

---

[2] It must be kept in mind that all of our values are dimensionless (all bars on these non-dimensional values having been eliminated). Thus in what follows, we will place an asterisk above all values *with* dimensions to avoid any confusion.

$$\frac{\partial p_0}{\partial \zeta} + \frac{\mathrm{Bo}}{\sqrt{\mathrm{Re}_\perp}} \varrho_0 = 0 \quad ;$$

$$\mathrm{S}\frac{\partial \varrho_0}{\partial t} + \boldsymbol{D}\cdot(\varrho_0 \boldsymbol{v}_0) + \frac{\partial \varrho_0 \omega_1}{\partial \zeta} = 0 \quad ;$$

$$p_0 = \varrho_0 T_0 \quad ;$$

$$\varrho_0\left(\mathrm{S}\frac{\partial T_0}{\partial t} + \boldsymbol{v}_0\cdot\boldsymbol{D}T_0\right) - \frac{\gamma-1}{\gamma}\left(\mathrm{S}\frac{\partial p_0}{\partial t} + \boldsymbol{v}_0\cdot\boldsymbol{D}p_0\right)$$
$$+\omega_1\left(\varrho_0\frac{\partial T_0}{\partial \zeta} - \frac{\gamma-1}{\gamma}\frac{\partial p_0}{\partial \zeta}\right) = \frac{1}{\mathrm{Pr}}\frac{\partial}{\partial \zeta}\left(k_0\frac{\partial T_0}{\partial \zeta}\right)$$
$$+(\gamma-1)\mathrm{M}_\infty^2 \mu_0 \left|\frac{\partial \boldsymbol{v}_0}{\partial \zeta}\right|^2 + \frac{1}{\mathrm{Pr}}\frac{\mathrm{Bo}}{\sqrt{\mathrm{Re}_\perp}}\sigma_{00}\frac{\partial \mathcal{R}_0}{\partial \zeta} \quad . \tag{7.29b}$$

The (quasi-static) equations (7.29) can be interpreted as "synoptic boundary layer" equations. The following boundary conditions must be imposed on (7.29) at ground level:

$$\zeta = 0: \quad \boldsymbol{v}_0 = 0 \quad , \quad \omega_1 = 0 \quad ; \quad \frac{\partial T_0}{\partial \zeta} + \frac{\mathrm{Bo}\sigma_{00}}{\sqrt{\mathrm{Re}_\perp}}\mathcal{R}_0 = 0 \quad . \tag{7.30}$$

Equations (7.29) which are valid within the synoptic boundary layer of a thickness

$$H_\nu \equiv \sqrt{\mathrm{Re}_\perp}\sqrt{L\nu_0/U_0}$$

must be associated with outer equations (which are not the primitive equations !!) which are valid in the free adiabatic atmosphere. In order to obtain such equations, the following limiting process must be taken into consideration [as suggested moreover by (7.26,27)] for (7.1–4):

$$\varepsilon \equiv \mathrm{O}(1) \quad , \quad \mathrm{Re} \to \infty \quad , \quad \text{with}$$
$$t,\ x,\ y,\ Z = \varepsilon z \equiv \zeta/\sqrt{\mathrm{Re}} \ \text{ fixed} \quad . \tag{7.31}$$

The following (main) asymptotic representation must be added to the above (7.31):

$$\boldsymbol{v} = \tilde{\boldsymbol{v}}_0 + \ldots \quad , \quad w = \frac{1}{\varepsilon}\tilde{w}_0 + \ldots \quad ,$$
$$p = \tilde{p}_0 + \ldots \quad , \quad \varrho = \tilde{\varrho}_0 + \ldots \quad , \quad T = \tilde{T}_0 + \ldots \quad , \tag{7.32}$$

where $\tilde{\boldsymbol{v}}_0$, $\tilde{w}_0$, $\tilde{p}_0$, $\tilde{\varrho}_0$ and $\tilde{T}_0$ are functions of the independent variables t, x, y and Z and satisfy the Euler equations for the adiabatic atmosphere:

$$\tilde{\varrho}_0\left\{\mathrm{S}\frac{\partial \tilde{\boldsymbol{v}}_0}{\partial t} + (\tilde{\boldsymbol{v}}_0\cdot\boldsymbol{D})\tilde{\boldsymbol{v}}_0 + \tilde{w}_0\frac{\partial \tilde{\boldsymbol{v}}_0}{\partial Z} + \frac{1}{\mathrm{Ro}}\frac{\sin\varphi}{\sin\varphi_0}(\boldsymbol{k}\times\tilde{\boldsymbol{v}}_0) + \frac{1}{\mathrm{Ro}}\frac{\cos\varphi}{\sin\varphi_0}\tilde{w}_0\boldsymbol{i}\right\}$$
$$+\frac{1}{\gamma \mathrm{M}_\infty^2}\boldsymbol{D}\tilde{p}_0 = 0 \quad ; \tag{7.33a}$$

$$\tilde{\varrho}_0\left\{\mathrm{S}\frac{\partial\tilde{w}_0}{\partial t}+(\tilde{\boldsymbol{v}}_0\cdot\boldsymbol{D})\tilde{w}_0+\tilde{w}_0\frac{\partial\tilde{w}_0}{\partial Z}-\frac{1}{\mathrm{Ro}}\frac{\cos\varphi}{\sin\varphi_0}\tilde{\boldsymbol{v}}_0\cdot\boldsymbol{i}\right\}$$

$$+\frac{1}{\gamma\mathrm{M}_\infty^2}\left(\frac{\partial\tilde{p}_0}{\partial Z}+\frac{\mathrm{Bo}}{\varepsilon}\tilde{\varrho}_0\right)=0\quad; \tag{7.33b}$$

$$\mathrm{S}\frac{\partial\tilde{\varrho}_0}{\partial t}+\boldsymbol{D}\cdot(\tilde{\varrho}_0\tilde{\boldsymbol{v}}_0)+\frac{\partial\tilde{\varrho}_0\tilde{w}_0}{\partial Z}=0\quad; \tag{7.33c}$$

$$\tilde{p}_0=\tilde{\varrho}_0\tilde{T}_0\quad; \tag{7.33d}$$

$$\tilde{\varrho}_0\left(\mathrm{S}\frac{\partial\tilde{T}_0}{\partial t}+\tilde{\boldsymbol{v}}_0\cdot\boldsymbol{D}\tilde{T}_0\right)-\frac{\gamma-1}{\gamma}\left(\mathrm{S}\frac{\partial\tilde{p}_0}{\partial t}+\tilde{\boldsymbol{v}}_0\cdot\boldsymbol{D}\tilde{p}_0\right)$$

$$+\tilde{w}_0\left(\tilde{\varrho}_0\frac{\partial\tilde{T}_0}{\partial Z}-\frac{\gamma-1}{\gamma}\frac{\partial\tilde{p}_0}{\partial Z}\right)=0\quad. \tag{7.33e}$$

To the Euler equations (7.33) can be added the initial starting conditions (7.7):

$$t=0:\quad \tilde{\boldsymbol{v}}_0=\boldsymbol{v}^0\quad,\quad \tilde{w}_0=w^0\quad,\quad \tilde{p}_0=p^0\quad,\quad \tilde{\varrho}_0=\varrho^0\quad. \tag{7.34}$$

However, only a slip condition can be prescribed on the flat ground:

$$Z=0:\quad \tilde{w}_0=0\quad. \tag{7.35}$$

Finally, we must write the matching conditions somewhere between the solution of (7.29) when $\zeta\to+\infty$ and the solution of (7.33) when $Z\to 0$. This means that the following behavior conditions can be associated with (7.29):

$$\lim_{\zeta\to\infty}\begin{pmatrix}\boldsymbol{v}_0\\ p_0\\ \varrho_0\\ T_0\end{pmatrix}=\begin{pmatrix}\tilde{\boldsymbol{v}}_{0,0}\equiv\tilde{\boldsymbol{v}}_0(t,x,y,0)\\ \tilde{p}_{0,0}\equiv\tilde{p}_0(t,x,y,0)\\ \tilde{\varrho}_{0,0}\equiv\tilde{\varrho}_0(t,x,y,0)\\ \tilde{T}_{0,0}\equiv\tilde{T}_0(t,x,y,0)\end{pmatrix}\quad. \tag{7.36}$$

Given (7.35), we note that the value of

$$\lim_{\zeta\to+\infty}\omega_1\equiv\omega_{1,\infty}(t,x,y) \tag{7.37}$$

is a result of the calculation of the solution of (7.29), satisfying both (7.30, 36) as well as the initial conditions. Concerning the latter, it is first necessary to obtain the significant degeneracy of (7.1–4) in the vicinity of the ground and at the initial time, by using the limiting process (7.9) together with the similarity relation (7.10), while at the same time taking into account that t/ε^a, $a>0$ (a real number to be determined) retains a fixed value together with x, y and z [see Zeytounian (1980) which deals with the question of initialization of unsteady boundary layer equations such as those written in (7.29)]. It turns out that the initialization must be carried out in such a way that the behavior of the solution of the corresponding unsteady boundary layer equations (for $t\to 0$ and $\zeta\to 0$) be compatile with the behavior of the solution of the related Rayleigh-type equations[3] when:

$$t/\varepsilon^a \longrightarrow \infty\,, \quad \zeta/\varepsilon^a \longrightarrow \infty \quad .$$

Let us backtrack to (7.29). The second of these equations yields:

$$p_0 = -\frac{\mathrm{Bo}}{\sqrt{\mathrm{Re}_\perp}} \int_\infty^\zeta \varrho_0 d\zeta + \tilde{p}_{0,0}(t,\, x,\, y) \quad . \tag{7.38}$$

However, the first of the system of Euler equations (7.33) yields:

$$\frac{1}{\gamma \mathrm{M}_\infty^2} \boldsymbol{D}\tilde{p}_{0,0} = -\,\tilde{\varrho}_{0,0}\left\{ \mathrm{S}\frac{\partial \tilde{\boldsymbol{v}}_{0,0}}{\partial t} + (\tilde{\boldsymbol{v}}_{0,0}\cdot \boldsymbol{D})\tilde{\boldsymbol{v}}_{0,0} + \frac{1}{\mathrm{Ro}}\frac{\sin\varphi}{\sin\varphi_0}(\boldsymbol{k}\times\tilde{\boldsymbol{v}}_{0,0}) \right\} \quad . \tag{7.39}$$

As a consequence of (7.38, 39), we can now rewrite the first equation of (7.29) – describing the synoptic boundary layer – in the form:

$$\begin{aligned} &\varrho_0\left\{ \mathrm{S}\frac{\partial \boldsymbol{v}_0}{\partial t} + (\boldsymbol{v}_0\cdot\boldsymbol{D})\boldsymbol{v}_0 + \omega_1\frac{\partial \boldsymbol{v}_0}{\partial \zeta}\right\} \\ &+\frac{1}{\mathrm{Ro}}\frac{\sin\varphi}{\sin\varphi_0}\left[\boldsymbol{k}\times(\varrho_0\boldsymbol{v}_0 - \tilde{\varrho}_{0,0}\tilde{\boldsymbol{v}}_{0,0})\right] \\ &-\frac{\mathrm{Bo}}{\gamma\mathrm{M}_\infty^2\sqrt{\mathrm{Re}_\perp}}\int_\infty^\zeta \boldsymbol{D}\varrho_0 d\zeta = \frac{\partial}{\partial\zeta}\left(\mu_0\frac{\partial \boldsymbol{v}_0}{\partial\zeta}\right) \\ &+\tilde{\varrho}_{0,0}\left\{\mathrm{S}\frac{\partial \tilde{\boldsymbol{v}}_{0,0}}{\partial t} + (\tilde{\boldsymbol{v}}_{0,0}\cdot\boldsymbol{D})\tilde{\boldsymbol{v}}_{0,0}\right\} \quad . \end{aligned} \tag{7.40}$$

Moreover the last equation of (7.33) yields:

$$\frac{\gamma-1}{\gamma}\mathrm{S}\frac{\partial \tilde{p}_{0,0}}{\partial t} = \tilde{\varrho}_{0,0}\left(\mathrm{S}\frac{\partial \tilde{T}_{0,0}}{\partial t} + \tilde{\boldsymbol{v}}_{0,0}\cdot\boldsymbol{D}\tilde{T}_{0,0}\right) - \frac{\gamma-1}{\gamma}\tilde{\boldsymbol{v}}_{0,0}\cdot\boldsymbol{D}\tilde{p}_{0,0}$$

which makes it possible to express the term

$$-\frac{\gamma-1}{\gamma}\mathrm{S}\frac{\partial p_0}{\partial t} = \frac{\gamma-1}{\gamma}\frac{\mathrm{SBo}}{\sqrt{\mathrm{Re}_\perp}}\int_\infty^\zeta \frac{\partial\varrho_0}{\partial t}d\zeta - \tilde{\varrho}_{0,0}\left(\mathrm{S}\frac{\partial \tilde{T}_{0,0}}{\partial t} + \tilde{\boldsymbol{v}}_{0,0}\cdot\boldsymbol{D}\tilde{T}_{0,0}\right) + \frac{\gamma-1}{\gamma}\tilde{\boldsymbol{v}}_{0,0}\cdot\boldsymbol{D}\tilde{p}_{0,0} \quad , \tag{7.41}$$

[where $\boldsymbol{D}\tilde{p}_{0,0}$ is determined by relation (7.39)] which enters into the final equation of system (7.29).

Thus the synoptic boundary layer equations allow the determining of $\boldsymbol{v}_0$, ϱ_0, ω_1 and T_0 once the solution of the Euler equations (7.33) is known for

[3] In *Howarth*'s paper (1951) can be found all useful information concerning the so-called Rayleigh equations as well as the formulation of the corresponding problem for a compressible, viscous, heat-conducting fluid. Van Dyke (1952) and Zeytounian (1970) can also be consulted. In Sect. 7.7 we will return to this question of initialization of the quasi-static equations (7.29).

$Z = 0$. However, it is obvious that in the vicinity of $Z = 0$, the Euler equations (7.33) degenerate and can be identified with the so-called primitive equations. As a consequence, for all medium-range forecasts deduced from the primitive equations, we can add a "non-adiabatic" medium-range forecast which takes into consideration both radiation and the "local" characteristics of the ground's surface. Such a possibility does not appear to have been brought to light in the present form from the point of view of asymptotic modelling. Quite naturally, it is possible within the context of the present modelling to consider non-adiabatic phenomena (radiation, humidity, thermal balance at the ground) in a more realistic fashion. To accomplish this, one simply has to make sure that the new terms adopted are compatible with those already implicated in the modelling carried out so far in this work.

To complete this section, one last remark concerning the various limiting processes is called for. Limiting process (7.21),

$$\varepsilon \text{ fixed} \quad , \quad \mathrm{Re} \to \infty \quad , \quad \text{then } \varepsilon \to 0 \quad ,$$

for the basic equations (7.1–4) is in fact equivalent to the limiting process

$$\mathrm{Re}_{\perp} \to \infty \quad ,$$

for the exact quasi-static equations (7.14–18). This remark implies that the so-called primitive equations do not represent a significant degeneracy of (7.1–4). In addition, since we are logically led to carry out the limiting process $\mathrm{M}_{\infty} \to 0$ (the Mach number M_{∞} is a natural small parameter for atmospheric flows) for the primitive equations (see Sect. 7.2), the question can thus be brought up whether or not a significant degeneracy can be obtained for the basic equations (7.1–4) when simultaneously:

$$\varepsilon \to 0 \quad , \quad \mathrm{Re} \to \infty \quad \text{and} \quad \mathrm{M}_{\infty} \to 0 \quad ,$$

under the similarity conditions

$$\varepsilon^{\kappa}\mathrm{Re} = \mathrm{O}(1) \quad \text{and} \quad \frac{\varepsilon}{\mathrm{M}_{\infty}^{\sigma}} = \mathrm{O}(1) \quad ,$$

where κ and σ are real positive numbers to be determined ($\sigma \cong 2$ for synoptic-scale phenomena). This question has not as yet been resolved.

7.2 Asymptotic Analysis of the Primitive Equations

Coming back to the exact quasi-static equations (7.14–18), we carry out the limiting process:

$$\mathrm{Re}_{\perp} \to \infty \quad , \quad \text{with } t, x, y, z \quad \text{fixed} \quad . \tag{7.42}$$

In this case, if all other dimensionless parameters remain fixed, we have for the functions:

$$\left(\overline{\boldsymbol{v}}_0, \overline{w}_0, \overline{p}_0, \overline{\varrho}_0, \overline{T}_0\right) = \lim_{\substack{\mathrm{Re}_{\perp} \to \infty \\ \text{with } t,x,y,z \text{ fixed}}} (\boldsymbol{v}_0, w_0, p_0, \varrho_0, T_0) \quad , \tag{7.43}$$

the so-called primitive equations:

$$\mathrm{S}\frac{\partial \overline{\boldsymbol{v}}_0}{\partial t} + (\overline{\boldsymbol{v}}_0 \cdot \boldsymbol{D})\overline{\boldsymbol{v}}_0 + \overline{w}_0 \frac{\partial \overline{\boldsymbol{v}}_0}{\partial z} + \frac{1}{\mathrm{Ro}}\frac{\sin\varphi}{\sin\varphi_0}(\boldsymbol{k} \wedge \overline{\boldsymbol{v}}_0) + \frac{1}{\gamma \mathrm{M}_\infty^2}\frac{1}{\overline{\varrho}_0}\boldsymbol{D}\overline{p}_0 = 0 \quad ;$$

$$\frac{\partial \overline{p}_0}{\partial z} + \mathrm{Bo}\overline{\varrho}_0 = 0 \quad , \quad \overline{p}_0 = \overline{\varrho}_0 \overline{T}_0 \quad ;$$

$$\mathrm{S}\frac{\partial \overline{\varrho}_0}{\partial t} + \boldsymbol{D} \cdot (\overline{\varrho}_0 \overline{\boldsymbol{v}}_0) + \frac{\partial \overline{\varrho}_0 \overline{w}_0}{\partial z} = 0 \quad ; \tag{7.44}$$

$$\overline{\varrho}_0\left(\mathrm{S}\frac{\partial \overline{T}_0}{\partial t} + \overline{\boldsymbol{v}}_0 \cdot \boldsymbol{D}\overline{T}_0\right) - \frac{\gamma - 1}{\gamma}\left(\mathrm{S}\frac{\partial \overline{p}_0}{\partial t} + \overline{\boldsymbol{v}}_0 \cdot \boldsymbol{D}\overline{p}_0\right)$$
$$+\overline{w}_0\left(\overline{\varrho}_0 \frac{\partial \overline{T}_0}{\partial z} - \frac{\gamma - 1}{\gamma}\frac{\partial \overline{p}_0}{\partial z}\right) = 0 \quad .$$

The following condition must be imposed on the ground for these primitive equations (7.44):

$$\overline{w}_0 = 0 \quad \text{at} \quad z = 0 \quad . \tag{7.45}$$

We note that

$$\lim_{\substack{\mathrm{Re}_\perp \to \infty \\ t,x,y,z \text{ fixed}}} \left(\overset{P}{\lim_{q-s}}\right) \quad \text{is equivalent to}$$

$$\lim_{\substack{\varepsilon \to 0 \\ t,x,y,z=Z/\varepsilon \text{ fixed}}} \left\{\lim(\mathrm{Re} \to \infty\,,\ \varepsilon = 0(1)\,;\ t, x, y, Z = \varepsilon z \text{ fixed})\right\} \quad .$$

We can now see that the primitive equations (7.44) do not represent a significant degeneracy of the basic equations (7.1–4) when $\varepsilon \to 0$ and $\mathrm{Re} \to \infty$. They are, rather, mere intermediate between the exact quasi-static equations and the complete Euler equations for the adiabatic atmosphere [which, in fact, correspond to (7.1–4) when $\mathrm{Re} \equiv \infty$, rewritten with $Z = \varepsilon z$ and $\tilde{w}_0 = \varepsilon w$].

The relation

$$Z = \varepsilon z$$

shows that the behavior at infinity of the primitive equations corresponds to the behavior for $Z \to 0$ in the complete Euler equations. Thus we arrive at the important conclusion that the behavior of the complete Euler equations for $Z \to 0$ must allow the correct statement of the behavior at infinity for $z \to \infty$ in the solution of the primitive equations (7.44). Furthermore, it must be kept in mind that when $\varepsilon \to 0$, we have:

$$\tilde{w}_0 = \varepsilon \overline{w}_0 + \ldots$$

but since $\tilde{w}_0 = 0$ for $Z = 0$, we notice that there is no apparent reason to impose a boundary condition on $\overline{w}_0$ when $z \to +\infty$!

When $z \to +\infty$, the behavior of $\overline{w}_0$ must be a natural consequence of the behavior of the primitive equations (of the solution of the equations) when

$z \to +\infty$. Concerning the initial conditions to be imposed on the primitive equations (7.44), we will see in Sect. 7.5 that they can be obtained only from the behavior – for $\hat{t} = t/\varepsilon \to \infty$ – of the adjustment to the hydrostatic balance problem in an adiabatic atmosphere. It must be understood that the combined effects of gravity and the compressibility of the (atmospheric) air tend in a general manner to establish a state of hydrostatic equilibrium on a synoptic scale throughout the entire troposphere. Nonetheless, this state of equilibrium which is the underlying characteristic of the primitive equations is, in fact, only realized "on the average" in time. Hence the question is to know if there exists an adjustment phenomenon to this hydrostatic state such that when $\hat{t} \to \infty$, we again find the quasi-statism for $t = 0$ – we will see in Sect. 7.5 that such is the case.

The singular problem brought out by the formulation of the initial conditions for the primitive equations is linked to the fact that, in contrast to the Euler equations, these primitive equations are not of the hyperbolic type. With Oliger and Sundstrom (1978), we can reestablish the hyperbolic character of these primitive equations by considering a linear variant of these equations and by giving advantage to the vertical structure in the z-direction. We will return to this question in Sects. 7.4 and 7.6.

7.3 The Boundary Layer Phenomenon and the Primitive Equations

We know that the solution of the primitive equations (7.44) with the slip condition (7.45) cannot be valid in the immediate vicinity of the ground (when a no-slip condition and the thermal balance condition must be prescribed). At this level, the following limiting process must be considered instead of (7.42):

$$\mathrm{Re}_\perp \to \infty \quad , \quad \text{with} \quad t,\ x,\ y,\ \hat{z} = \frac{z}{1/\sqrt{\mathrm{Re}_\perp}} \quad \text{fixed} \quad . \tag{7.46}$$

In this case, if once again all the other dimensionless parameters remain fixed, we have for the functions:

$$(\hat{\boldsymbol{v}}_0, \hat{w}_0, \hat{\varrho}_0, \hat{T}_0) = \lim_{\substack{\mathrm{Re}_\perp \to \infty \\ \text{with } t,\, x,\, y,\, \hat{z} \text{ fixed}}} \left(\boldsymbol{v}_0, \frac{w_0}{1/\sqrt{\mathrm{Re}_\perp}}, \varrho_0, T_0 \right) \tag{7.47}$$

the following boundary layer equations (with the exact quasi-static equations (7.14–18) again used as the starting point):

$$\hat{\varrho}_0 \left\{ \mathrm{S} \frac{\partial \hat{\boldsymbol{v}}_0}{\partial t} + (\hat{\boldsymbol{v}}_0 \cdot \boldsymbol{D})\hat{\boldsymbol{v}}_0 + \hat{w}_0 \frac{\partial \hat{\boldsymbol{v}}_0}{\partial \hat{z}} + \frac{1}{\mathrm{Ro}} \frac{\sin\varphi}{\sin\varphi_0} (\boldsymbol{k} \times \hat{\boldsymbol{v}}_0) \right\}$$
$$= \frac{\partial}{\partial \hat{z}} \left(\hat{\mu}_0 \frac{\partial \hat{\boldsymbol{v}}_0}{\partial \hat{z}} \right) - \frac{1}{\gamma \mathrm{M}_\infty^2} \boldsymbol{D} \overline{p}_{0,0} \quad ; \tag{7.48a}$$

$$S\frac{\partial \hat{\varrho}_0}{\partial t} + \boldsymbol{D}\cdot(\hat{\varrho}_0\hat{\boldsymbol{v}}_0) + \frac{\partial \hat{\varrho}_0\hat{w}_0}{\partial \hat{z}} = 0 \quad ;$$

$$\hat{\varrho}_0 = \frac{\overline{p}_{0,0}}{\hat{T}_0} \quad ;$$

$$\hat{\varrho}_0\left(S\frac{\partial \hat{T}_0}{\partial t} + \hat{\boldsymbol{v}}_0\cdot\boldsymbol{D}\hat{T}_0\right) + \hat{\varrho}_0\hat{w}_0\frac{\partial \hat{T}_0}{\partial \hat{z}} \tag{7.48b}$$

$$= \frac{1}{\Pr}\frac{\partial}{\partial \hat{z}}\left(\hat{k}_0\frac{\partial \hat{T}_0}{\partial \hat{z}}\right) + (\gamma-1)M_\infty^2\hat{\mu}_0\left|\frac{\partial \hat{\boldsymbol{v}}_0}{\partial \hat{z}}\right|^2$$

$$+ \frac{\gamma-1}{\gamma}\left(S\frac{\partial \overline{p}_{0,0}}{\partial t} + \hat{\boldsymbol{v}}_0\cdot\boldsymbol{D}\overline{p}_{0,0}\right) \quad .$$

For the boundary layer equations (7.48) associated with the primitive equations (7.44), it must be kept in mind that the terms

$$\boldsymbol{D}\overline{p}_{0,0} = -\gamma M_\infty^2\overline{\varrho}_{0,0}\left\{S\frac{\partial \overline{\boldsymbol{v}}_{0,0}}{\partial t} + (\overline{\boldsymbol{v}}_{0,0}\cdot\boldsymbol{D})\overline{\boldsymbol{v}}_{0,0} + \frac{1}{\mathrm{Ro}}\frac{\sin\varphi}{\sin\varphi_0}(\boldsymbol{k}\times\overline{\boldsymbol{v}}_{0,0})\right\} \tag{7.49a}$$

and

$$S\frac{\partial \overline{p}_{0,0}}{\partial t} = \frac{\gamma}{\gamma-1}\overline{\varrho}_{0,0}\left(S\frac{\partial \overline{T}_{0,0}}{\partial t} + \overline{\boldsymbol{v}}_{0,0}\cdot\boldsymbol{D}\overline{T}_{0,0}\right) - \overline{\boldsymbol{v}}_{0,0}\cdot\boldsymbol{D}\overline{p}_{0,0} \tag{7.49b}$$

are known once the primitive equations (7.44) have been solved since

$$(\overline{\boldsymbol{v}}_{0,0}, \overline{\varrho}_{0,0}, \overline{T}_{0,0}) = \lim_{z\to 0}(\overline{\boldsymbol{v}}_0, \varrho_0, \overline{T}_0) \quad . \tag{7.50}$$

The following boundary conditions must be added to (7.48):

$$\hat{\boldsymbol{v}}_0 = 0 \;,\; \hat{w}_0 = 0 \;,\; \frac{\partial \hat{T}_0}{\partial \hat{z}} = 0 \;,\quad \text{on} \quad \hat{z} = 0 \;, \tag{7.51a}$$

$$\hat{\boldsymbol{v}}_0 \to \overline{\boldsymbol{v}}_{0,0},\; \hat{\varrho}_0 \to \overline{\varrho}_{0,0},\; \overline{T}_0 \to \overline{T}_{0,0} \;,\quad \text{when} \quad \hat{z} \to +\infty \quad . \tag{7.51b}$$

Therefore, when no additional assumption is made in the limiting process (7.46) (concerning, in particular, the radiation parameter σ_{00} which is assumed of the order unity), it turns out that necessarily, the primitive equations (7.44) must be joined to the boundary layer equations (7.48), both systems of equations being significant degeneracies [when $t = O(1)$] of the exact quasi-static equations (7.14–18). Hence, any medium-range numerical forecast obtained from a primitive equations model can be associated with a "non-adiabatic" boundary layer forecast near the earth's surface which is assumed to be an athermanous wall.

It is important to understand that the limiting processes which lead a) to the synoptic boundary layer equations (7.29), and b) to the athermanous wall boundary layer equations (7.48) differ conceptually: as a matter of fact, the boundary layer equations (7.48) are a (significant) degeneracy of the synoptic boundary layer equations (7.29) when

$$\mathrm{Re}_\perp \to \infty \quad , \quad \text{with} \quad t,\, x,\, y \quad \text{and} \quad \zeta \quad \text{fixed}; \quad \hat{z} \equiv \zeta \quad .$$

Likewise, associating (7.29) with the primitive equations which are considered as a limit in the vicinity of $Z = 0$ of the Euler equations (7.33):

$$\varepsilon \to 0 \quad , \quad \text{with} \quad t,\, x,\, y,\, z = \frac{Z}{\varepsilon} \quad \text{fixed} \quad ;$$

$$(\overline{\boldsymbol{v}}_0,\, \overline{w}_0,\, \overline{\varrho}_0,\, \overline{T}_0) = \lim_{\substack{\varepsilon \to 0 \\ t,x,y,z \text{ fixed}}} \left(\tilde{\boldsymbol{v}}_0,\, \frac{\tilde{w}_0}{\varepsilon},\, \tilde{p}_0,\, \tilde{\varrho}_0,\, \tilde{T}_0 \right)$$

is different than joining (7.48) to these same primitive equations (7.44) but which were obtained by applying the limiting process (7.42) to the exact quasi-static equations (7.14–18).

Thus, the primitive equations can be obtained in two different ways: if we start with the general equations (7.1–4), the primitive equations must be associated with the synoptic boundary layer equations (7.29); on the other hand, if the exact quasi-static equations (7.14–18) mark the starting point, then the boundary layer equations (7.48) must be chosen to complement these same primitive equations. Of course, only (7.29, 33) represent significant degeneracies of the basic equations (7.1–4).

7.4 Simplified Primitive Equations

Let us return to the primitive equations (7.44) and try to simplify them by making use of the fact that the Mach number M_∞ is always very small compared to unity in atmospheric flows. But first, let us write, like in Subsect. 2.4.3, our primitive equations in the system of coordinates t, x, y, $\overline{p}_0$ for the unknown functions:

$$\begin{gathered} \overline{\boldsymbol{v}}_0,\; z = \overline{\mathcal{H}}_0(t,\, x,\, y,\, \overline{p}_0) \quad , \\ \overline{\mathcal{W}}_0 = \frac{\mathrm{Bo}}{\mathrm{S}} \overline{\varrho}_0 \left\{ \mathrm{S} \frac{\partial \overline{\mathcal{H}}_0}{\partial t} + \overline{\boldsymbol{v}}_0 \cdot \boldsymbol{D} \overline{\mathcal{H}}_0 - \overline{w}_0 \right\} \quad . \end{gathered} \tag{7.52}$$

Then the solution to the (primitive) equations thus obtained can be sought [see (2.91, 94), for example] in the form of the asymptotic expansions:

$$\overline{\boldsymbol{v}}_0 = \boldsymbol{V}_0 + \dots \quad , \quad \overline{\mathcal{W}}_0 = \omega_0 + \dots \quad , \overline{\mathcal{H}}_0 = \phi_0(\overline{p}_0) + \mathrm{M}_\infty^2 \phi_2 + \dots \quad . \tag{7.53}$$

Let us suppose now that the constant characteristic scale $K_{00}(1)$ of the stratification parameter:

$$K_0(\overline{p}_0) = -\frac{\overline{p}_0}{\mathrm{Bo}} \left(\frac{d\theta_0}{d\overline{p}_0} - \frac{\gamma - 1}{\gamma} \frac{\theta_0}{\overline{p}_0} \right) \equiv K_{00}(1) \tilde{K}_0(\overline{p}_0) \quad , \tag{7.54}$$

where

$$\theta_0(\overline{p}_0) = -\mathrm{Bo}\, \overline{p}_0 \frac{d\phi_0}{d\overline{p}_0} \quad , \tag{7.55}$$

satisfies the following similarity relation:

$$\frac{K_{00}(1)}{\mathrm{M}_\infty^2} \equiv \Gamma_{00} = \mathrm{O}(1) \quad . \tag{7.56}$$

Following these hypotheses, we obtain at the limit $\mathrm{M}_\infty \to 0$, the following system of "simplified" primitive equations for the functions $\boldsymbol{V}_0$, ω_0 and ϕ_2:

$$\mathrm{S}\frac{\partial \boldsymbol{V}_0}{\partial t} + (\boldsymbol{V}_0 \cdot \boldsymbol{D})\boldsymbol{V}_0 + \mathrm{S}\omega_0 \frac{\partial \boldsymbol{V}_0}{\partial \overline{p}_0} + \frac{1}{\mathrm{Ro}}\frac{\sin\varphi}{\sin\varphi_0}(\boldsymbol{k} \times \boldsymbol{V}_0) + \frac{\mathrm{Bo}}{\gamma}\boldsymbol{D}\phi_2 = 0 \quad ; \tag{7.57a}$$

$$\boldsymbol{D}\cdot\boldsymbol{V}_0 + \mathrm{S}\frac{\partial \omega_0}{\partial \overline{p}_0} = 0 \quad ; \tag{7.57b}$$

$$\left(\mathrm{S}\frac{\partial}{\partial t} + \boldsymbol{V}_0\cdot\boldsymbol{D}\right)\frac{\partial \phi_2}{\partial \overline{p}_0} + \frac{\mathrm{S}}{\overline{p}_0}\left\{\frac{\partial}{\partial \overline{p}_0}\left(\overline{p}_0\frac{\partial \phi_2}{\partial \overline{p}_0}\right) - \frac{\gamma-1}{\gamma}\frac{\partial \phi_2}{\partial \overline{p}_0}\right\}\omega_0$$
$$+\mathrm{S}\Gamma_{00}\frac{\tilde{K}_0(\overline{p}_0)}{\overline{p}_0^2}\omega_0 = 0 \quad . \tag{7.57c}$$

The above equations must be solved with the following ground condition:

$$\omega_0 = 0 \Rightarrow \left(\mathrm{S}\frac{\partial}{\partial t} + \boldsymbol{V}_0\cdot\boldsymbol{D}\right)\frac{\partial \phi_2}{\partial \overline{p}_0} = 0 \quad , \quad \text{on} \quad \overline{p}_0 = \overline{p}_{00} \quad , \tag{7.58}$$

where $\overline{p}_{00} = \mathrm{const}$[4] is the solution of the equation $\phi_0(\overline{p}_0) = 0$.

A behavior condition when $\overline{p}_0 \to 0$ and "lateral" conditions relative to the horizontal variables x and y are still needed. However, since Oliger and Sundström's work (1978), we know that the quasi-linear system (7.57) is not of the hyperbolic kind which means that the Cauchy problem is poorly posed for this system of equations. The unfortunate result of this is that the boundary conditions with respect to x and y become very complicated to formulate. As pointed out by Oliger and Sundström in their 1978 article, one way of clarifying the character of the conditions relative to $\overline{p}_0 \to 0$ and x and y is by the linearization of equations (7.57) and the bringing to light of the vertical structure with respect to $\overline{p}_0$.

In the vicinity of a reference situation which is both at rest and thermodynamically stratified in a stable manner ($\tilde{K}_0(\overline{p}_0) > 0$), the simplified linearized primitive equations can, from (7.57), be written in the following form after simple linearization:

$$\mathrm{S}\frac{\partial \boldsymbol{V}_0'}{\partial t} + \frac{\mathrm{Bo}}{\gamma}\boldsymbol{D}\phi_2' + \frac{1}{\mathrm{Ro}}\frac{\sin\varphi}{\sin\varphi_0}(\boldsymbol{k} \times \boldsymbol{V}_0') = 0 \quad ,$$
$$\boldsymbol{D}\cdot\boldsymbol{V}_0' + \mathrm{S}\frac{\partial \omega_0'}{\partial \overline{p}_0} = 0 \quad , \tag{7.59}$$
$$\mathrm{S}\frac{\partial}{\partial t}\left(\frac{\partial \phi_2'}{\partial \overline{p}_0}\right) + \mathrm{S}\Gamma_{00}\frac{\tilde{K}_0(\overline{p}_0)}{\overline{p}_0^2}\omega_0' = 0 \quad .$$

---

[4] In fact, $\overline{p}_{00} = 1$ can be taken since we are dealing with dimensionless values. This will be the procedure in what follows.

The above linear system can again be rewritten as follows after the function ω'_0 has been eliminated from the last two equations of this system (7.59):

$$\mathrm{S}\frac{\partial \boldsymbol{V}'_0}{\partial t} + \frac{1}{\mathrm{Ro}}\frac{\sin\varphi}{\sin\varphi_0}(\boldsymbol{k}\times\boldsymbol{V}'_0) = -\frac{\mathrm{Bo}}{\gamma}\boldsymbol{D}\phi'_2 \quad ; \tag{7.60a}$$

$$\frac{\mathrm{S}}{\Gamma_{00}}\frac{\partial}{\partial t}\left\{\frac{\partial}{\partial \overline{p}_0}\left(\frac{\overline{p}_0^2}{\tilde{K}_0(\overline{p}_0)}\frac{\partial\phi'_2}{\partial\overline{p}_0}\right)\right\} = \boldsymbol{D}\cdot\boldsymbol{V}'_0 \quad . \tag{7.60b}$$

The following linearized slip condition must be assigned to (7.60):

$$\frac{\partial}{\partial t}\left(\frac{\partial\phi'_2}{\partial\overline{p}_0}\right) = 0 \quad , \quad \text{on} \quad \overline{p}_0 = 1 \quad , \tag{7.61}$$

which results trivially from (7.58) once $\overline{p}_{00} \equiv 1$ is adopted.

A second condition in $\overline{p}_0$[5] can be added to (7.60). In linear theory it is written in the following form thanks to the last equation of (7.59):

$$\lim_{\overline{p}\to 0}\left\{\frac{\partial}{\partial t}\left(\frac{\overline{p}_0^2}{\tilde{K}_0(\overline{p}_0)}\frac{\partial\phi'_2}{\partial\overline{p}_0}\right)\right\} = 0 \quad . \tag{7.62}$$

Let us now suppose the linear system (7.60) to be separable. In this case, the solution to system (7.60) can be sought in the classical form of the following Fourier-type expansions:

$$\boldsymbol{V}'_0 = \sum_k \boldsymbol{V}'_{0,k}(t, x, y)\mathcal{L}_k(\overline{p}_0) \quad ; \quad \phi'_2 = \sum_k \phi'_{2,k}(t, x, y)\mathcal{L}_k(\overline{p}_0) \quad . \tag{7.63}$$

For every k, there results a system of hyperbolic equations for $\boldsymbol{V}'_{0,k}$ and $\phi'_{2,k}$ which is written:

$$\begin{aligned} &\mathrm{S}\frac{\partial \boldsymbol{V}'_{0,k}}{\partial t} + \frac{1}{\mathrm{Ro}}\frac{\sin\varphi}{\sin\varphi_0}(\boldsymbol{k}\times\boldsymbol{V}'_{0,k}) + \frac{\mathrm{Bo}}{\gamma}\boldsymbol{D}\phi'_{2,k} = 0 \quad ; \\ &\frac{\mathrm{S}}{\Gamma_{00}}\frac{\partial}{\partial t}(\lambda_k\phi'_{2,k}) + \boldsymbol{D}\cdot\boldsymbol{V}'_{0,k} = 0 \quad , \end{aligned} \tag{7.64}$$

where $\mathcal{L}_k(\overline{p}_0)$ and λ_k are eigenfunctions and eigenvalues (real, distinct and positive) of the following Sturm-Liouville-type eigenvalue problem ($\tilde{K}_0(\overline{p}_0) > 0$):

$$\begin{aligned} &\frac{d}{d\overline{p}_0}\left(\frac{\overline{p}_0^2}{\tilde{K}_0(\overline{p}_0)}\frac{d\mathcal{L}_k}{d\overline{p}_0}\right) + \lambda_k\mathcal{L}_k = 0 \quad ; \quad \left.\frac{d\mathcal{L}_k}{d\overline{p}_0}\right|_{\overline{p}_0=1} = 0 \quad , \\ &\lim_{\overline{p}_0\to 0}\left(\frac{\overline{p}_0^2}{\tilde{K}_0(\overline{p}_0)}\frac{d\mathcal{L}_k}{d\overline{p}_0}\right) = 0 \quad . \end{aligned} \tag{7.65}$$

---

[5] A heuristic argument which leads to the behavior condition (7.62) can be summarized as follows: in adiabatic flow outside of the boundary layers, the energy equation can be written in the dimensional form:

$$\frac{Dp}{Dt} + \gamma p \operatorname{div}\boldsymbol{u} = 0 \quad ,$$

which implies that $Dp/Dt \to 0$ with $p \to 0$ and thus [see (2.88)], according to (7.52), $\overline{W}_0$ also tends to zero when $\overline{p}_0 \to 0$; it is likewise for ω'_0 which indeed leads to (7.62).

As a matter of fact, we know[6] that system (7.60) is separable if the the operator:

$$\mathcal{A} = -\frac{d}{d\overline{p}_0}\left(\frac{\overline{p}_0^2}{\tilde{K}_0(\overline{p}_0)}\frac{d}{d\overline{p}_0}\right) \tag{7.66}$$

is hermitian (i.e., if it coincides with its own adjoint) and is positive defined.

First of all, the above defined-operator is indeed hermitian:

$$\langle \mathcal{L}_k^*, \mathcal{A}\mathcal{L}_k^{**}\rangle = \langle \mathcal{A}\mathcal{L}_k^*, \mathcal{L}_k^{**}\rangle \quad ,$$

regardless of the eigenfunctions $\mathcal{L}_k^*$ and $\mathcal{L}_k^{**}$ of problem (7.65). In fact,

$$\begin{aligned}\langle \mathcal{L}_k^*, \mathcal{A}\mathcal{L}_k^{**}\rangle &\equiv \int_0^1 \mathcal{L}_k^*\mathcal{A}\mathcal{L}_k^{**}d\overline{p}_0 \\ &= -\int_0^1 \mathcal{L}_k^*\frac{d}{d\overline{p}_0}\left(\frac{\overline{p}_0^2}{\tilde{K}_0(\overline{p}_0)}\frac{d\mathcal{L}_k^{**}}{d\overline{p}_0}\right)d\overline{p}_0 \\ &= \int_0^1 \frac{\overline{p}_0^2}{\tilde{K}_0(\overline{p}_0)}\frac{d\mathcal{L}_k^{**}}{d\overline{p}_0}\frac{d\mathcal{L}_k^*}{d\overline{p}_0}d\overline{p}_0 - \left[\frac{\overline{p}_0^2}{\tilde{K}_0(\overline{p}_0)}\mathcal{L}_k^*\frac{d\mathcal{L}_k^{**}}{d\overline{p}_0}\right]_0^1 \quad .\end{aligned}$$

Therefore, thanks to the conditions in $\overline{p}_0 = 1$ and $\overline{p}_0 \to 0$, for the Sturm-Liouville problem (7.65), it can be seen that:[7]

$$\begin{aligned}\langle \mathcal{L}_k^*, \mathcal{A}\mathcal{L}_k^{**}\rangle &= -\int_0^1 \mathcal{L}_k^{**}\frac{d}{d\overline{p}_0}\left(\frac{\overline{p}_0^2}{\tilde{K}_0(\overline{p}_0)}\frac{d\mathcal{L}_k^*}{d\overline{p}_0}\right)d\overline{p}_0 \\ &= \langle \mathcal{A}\mathcal{L}_k^*, \mathcal{L}_k^{**}\rangle \quad .\end{aligned}$$

Moreover, the operator (7.66) is indeed positive defined because for every non-zero eigenfunction $\mathcal{L}_k$:

$$\langle \mathcal{L}_k, \mathcal{A}\mathcal{L}_k\rangle \equiv \int_0^1 \frac{p_0^2}{\tilde{K}_0(\overline{p}_0)}\left(\frac{d\mathcal{L}_k}{d\overline{p}_0}\right)^2 d\overline{p}_0 > 0 \quad ,$$

since, by hypothesis $\tilde{K}_0(\overline{p}_0) > 0$. Thus the eigenvalues λ_k of (7.65) are only real and positive since

$$\langle \mathcal{L}_k, \mathcal{A}\mathcal{L}_k\rangle = \langle \mathcal{L}_k, \lambda_k\mathcal{L}_k\rangle = \lambda_k\langle \mathcal{L}_k, \mathcal{L}_k\rangle$$

---

[6] The reader might consult a classic on physical mathematics. See, for example, the book by Lattes (1967).

[7] In Sect. 7.6 we will in fact see that in $\overline{p}_0 = 0$, the following condition must be imposed:

$$\left[\frac{\overline{p}_0^2}{\tilde{K}_0(\overline{p}_0)}\right]\mathcal{L}_k\frac{d\mathcal{L}_k}{d\overline{p}_0} = 0 \quad ,$$

since the value of $\mathcal{L}_k$ in $\overline{p}_0 = 0$ can be undefined.

is positive. Finally, if $\mathcal{L}_m$ and $\mathcal{L}_n$ are two eigenfunctions of (7.65) associated with the distinct eigenvalues λ_m and λ_n, then they are orthogonal:

$$\int_0^1 \mathcal{L}_m(\overline{p}_0)\mathcal{L}_n(\overline{p}_0)d\overline{p}_0 = 0 \quad ,$$

for $m \neq n$.

Consequently, there exists an (infinite) spectral base whose elements are the eigenfunctions $\mathcal{L}_k$ of the Sturm-Liouville problem (7.65), and this base is orthogonal – it can be normalized ($\langle\mathcal{L}_k, \mathcal{L}_k\rangle$ defines a norm). It can then be demonstrated that the eigenfunction system is complete (the Parseval theorem), i.e., the expansions (7.63) are valid.

In Oliger and Sundström's work (1978), an analysis of a hyperbolic system of the type (7.64) can be found. A discussion concerning the lateral conditions in x and y as a function of the index k is also to be found in this same work as well as that of Kadechnikov (1980), and Miller and Thorpe (1981). The above analysis will not be carried any further, however, in the present book.

7.5 The Hydrostatic Balance Adjustment Problem (in an Adiabatic Atmosphere)

The Euler equations (7.33) are evolution equations for which initial conditions for the set of dependent variables $\tilde{v}_0$, $\tilde{w}_0$, $\tilde{p}_0$ and $\tilde{\varrho}_0$ are called for [see (7.34)]. The limiting process (7.21), or even (7.42), leads to the primitive equations (7.44) for which initial conditions are necessary only with respect to $\overline{v}_0$ and $\overline{\varrho}_0$. Thus, when transferring from the complete Euler equations (7.33) to the primitive equations (7.44), it is no longer possible to prescribe initial conditions with respect to $\overline{p}_0$ and $\overline{w}_0$! It is then perfectly normal to wonder, as did Zeytounian (1976), how the arbitrary initial data (7.34) adjusts to hydrostatic balance. We can easily be persuaded that this adjustment takes place over a duration O(ε) and we will want to obtain the adjustment equations describing this process. We are dealing here with a singular perturbation problem with respect to the small parameter ε which intervenes in both the Euler equations (7.33) *and* the initial conditions (7.34)[8]. This problem falls into the general category of so-called "adjustment" problems which aim at clarifying just how a set of initial data associated with a determined model can be related to another set of initial data associated with a simpler model which is considered as a degeneracy of the model considered at the start.

---

[8] Let us recall that for every fixed t of the order unity, the primitive equations are, in fact, the main limiting equations which are obtained from (7.33) via the main limiting process:

$$\varepsilon \to 0,\ x, y, z = \frac{Z}{\varepsilon} \text{ fixed},\ \text{and}\ (\overline{v}_0, \overline{w}_0, \overline{p}_0, \overline{\varrho}_0, \overline{T}_0) = \lim_{\substack{\varepsilon \to 0 \\ t,x,y,z \text{ fixed}}} \left(\tilde{v}_0, \frac{\tilde{w}_0}{\varepsilon}, \tilde{p}_0, \tilde{\varrho}_0, \tilde{T}_0\right) \ .$$

In order to solve such a problem, Guiraud and Zeytounian (1982) introduced an initial layer in the vicinity of $t = 0$ by distorting the time scale and the unknowns which were initially undefined.

For this hydrostatic balance adjustment problem, the initial layer problem is formulated by defining the change in variables and functions:

$$\begin{aligned} &\hat{t} = \frac{t}{\varepsilon} \quad ; \quad \hat{w} = \varepsilon w \equiv \tilde{w}_0 \quad , \quad \hat{\boldsymbol{v}} = \boldsymbol{v} \equiv \tilde{\boldsymbol{v}}_0 \quad , \\ &\hat{p} = p \equiv \tilde{p}_0 \quad , \quad \hat{\varrho} = \varrho \equiv \tilde{\varrho}_0 \quad , \quad \hat{T} = T \equiv \tilde{T}_0 \quad , \end{aligned} \tag{7.67}$$

where the unknowns $\hat{w}$, $\hat{\boldsymbol{v}}$, $\hat{p}$, $\hat{\varrho}$ and $\hat{T}$ are expressed as functions of the variables $\hat{t}$, x, y and $z = Z/\varepsilon$. By carrying out the change in independent variables:

$$t = \varepsilon \hat{t} \quad , \quad Z = \varepsilon z \quad , \tag{7.68}$$

in the Euler equations (7.33), and the limiting process $\varepsilon \to 0$ with $\hat{t}$, x, y, z and S, Bo, Ro, M_∞, γ, $\sin \varphi_0$ fixed, we obtain a system for $\hat{p}_0$, $\hat{\varrho}_0$, $\hat{T}_0$ and $\hat{w}_0$, and also a transport equation for $\hat{\boldsymbol{v}}_0$. The subscript "0" indicates that the following local limiting process has been executed:

$$\varepsilon \to 0 \quad , \quad \text{with } \hat{t} = \frac{t}{\varepsilon}, x, y, z = \frac{Z}{\varepsilon} \text{ fixed} \tag{7.69}$$

in the Euler equations (7.33).

The above-mentioned transport equation is written:

$$\mathrm{S}\frac{\partial \hat{\boldsymbol{v}}_0}{\partial \hat{t}} + \hat{w}_0 \frac{\partial \hat{\boldsymbol{v}}_0}{\partial z} = 0 \quad . \tag{7.70}$$

For $\hat{p}_0$, $\hat{\varrho}_0$, $\hat{T}_0$ and $\hat{w}_0$ the following system results:

$$\gamma \mathrm{M}_\infty^2 \hat{\varrho}_0 \left\{ \mathrm{S}\frac{\partial \hat{w}_0}{\partial \hat{t}} + \hat{w}_0 \frac{\partial \hat{w}_0}{\partial z} \right\} + \frac{\partial \hat{p}_0}{\partial z} + \mathrm{Bo}\hat{\varrho}_0 = 0 \quad ; \tag{7.71a}$$

$$\mathrm{S}\frac{\partial \hat{\varrho}_0}{\partial \hat{t}} + \frac{\partial \hat{\varrho}_0 \hat{w}_0}{\partial z} = 0 \quad , \tag{7.71b}$$

$$\hat{p}_0 = \hat{\varrho}_0 \hat{T}_0 \quad ; \tag{7.71c}$$

$$\mathrm{S}\hat{\varrho}_0 \frac{\partial \hat{T}_0}{\partial \hat{t}} - \frac{\gamma - 1}{\gamma}\mathrm{S}\frac{\partial \hat{p}_0}{\partial \hat{t}} + \hat{w}_0 \left(\hat{\varrho}_0 \frac{\partial \hat{T}_0}{\partial z} - \frac{\gamma - 1}{\gamma}\frac{\partial \hat{p}_0}{\partial z} \right) = 0 \quad . \tag{7.71d}$$

The above system (7.71) is none other than the one governing the one-dimensional unsteady (vertical) motions of a heavy (perfect) gas (with constant specific heats). It is a quasi-linear hyperbolic system which is characterized by the presence of sound waves and a transport of the specific entropy,

$$\hat{\Sigma}_0 = \log \hat{p}_0 - \gamma \log \hat{\varrho}_0 \quad , \tag{7.72}$$

with the velocity $\hat{w}_0$. The equation:

$$\mathrm{S}\frac{\partial \hat{\Sigma}_0}{\partial \hat{t}} + \hat{w}_0 \frac{\partial \hat{\Sigma}_0}{\partial z} = 0 \quad , \tag{7.73}$$

replaces the last two equations of system (7.71). We can now add to system (7.71), four initial conditions at the instant $\hat{t}=0$ which are defined from those prescribed for the Euler equations (7.33), i.e., according to (7.34):

$$\hat{t}=0:\ \hat{p}_0=p^0 \quad , \quad \hat{\varrho}_0=\varrho^0 \quad , \quad \hat{T}_0=T^0 \quad , \quad \hat{w}_0=w^0 \quad , \qquad \text{with} \tag{7.74}$$

$$T^0 \equiv \frac{p^0}{\varrho^0} \quad .$$

A condition at $\hat{t}=0$ must also be imposed on the transport equation (7.70):

$$\hat{\boldsymbol{v}}_0=\boldsymbol{v}^0 \quad , \qquad \text{for} \quad \hat{t}=0 \quad . \tag{7.75}$$

Finally, a boundary condition at $z=0$ – which is a slip condition – is prescribed for (7.71)

$$\hat{w}_0=0 \quad , \qquad \text{on} \quad z=0 \quad . \tag{7.76}$$

There thus results an initial value problem which by imposing a matching condition:

$$\lim_{\hat{t}\to\infty} \equiv \lim_{t\to 0} \quad ,$$

(which is verified afterwards!) makes it possible to determine the true initial conditions of the problem associated with the primitive equations (7.44). The time of adjustment to hydrostatic balance is:

$$\hat{t}_0=\frac{H}{L}t_0 \quad .$$

However, the quasi-linear character of system (7.71) necessitates a numerical study of this hyperbolic system so that the behavior of the solution can be determined when $\hat{t}$ tends towards "infinity". This behavior was studied (numerically) in the second part of the thesis by Outrebon (1981).

Without going into detail, we will only indicate here that Outrebon's numerical results brought to light three phases in the adjustment process. In the first phase, the unbalance caused by the superposing of perturbing winds on a quasi-static equilibrium state at the initial instant very rapidly results in an inversion of the vertical wind $\hat{w}_0$. This inversion can be explained via the first equation of system (7.71) written for $\hat{t}=0$ as $\hat{w}>0$ and $\partial\hat{w}_0/\partial z>0$. Thus, to the left of the maximum, there results $(\partial\hat{w}_0/\partial\hat{t})_{\hat{t}=0}<0$ which induces the decrease of the vertical wind $\hat{w}_0$. Moreover, this inversion is accompanied by waves which perturb the temperature and pressure for $\hat{t}=0.058$.

In the second phase, the temperature and pressure exhibit regular profiles whereas the vertical wind $\hat{w}_0$ undergoes several additional inversions by decreasing in amplitude when, for example, $\hat{t}=0.180$ and $\hat{t}=0.248$.

The above phase finishes at $\hat{t}=0.385$ for 1200 iterations[9], a state where the

[9] In the thesis by Outrebon (1981) can be found the presentation of the schemes used (directly inspired by those introduced by Lerat and Peyret (1973) and "corrected" by a Fromm-style (1968, 1969) numerical dispersion correction method). These schemes ensure high accuracy despite the great number of iterations necessary for the carrying out of the numerical resolution of the adjustment problem being considered.

pressure and temperature profiles resemble those of the initial state. The third convergence phase corresponds to $\hat{t} = 0.521$ and 1600 iterations, and $\hat{t} = 0.998$ and 3000 iterations. The latter was reached at convergence in a stop test, with a maximum error of the order of 10^{-4}, whereas at $\hat{t} = 0.521$, the maximum in question is only 6×10^{-3}. Hence the adjustment time of the corresponding phenomenon at $\hat{t} \cong 1$ is 864 s.

Ituitively speaking, it is easy to understand that if there is adjustment (which indeed there is judging from Outrebon's numerical results), then a linear variant from (7.71) must enable us to confirm it. As a matter of fact, when $\hat{t} \to \infty$, it can be assumed that the sound waves were radiated towards the upper layers of the atmosphere and that, at a fixed altitude, $\hat{p}_0$, $\hat{\varrho}_0$, $\hat{w}_0$ and $\hat{T}_0$ tend towards $\hat{p}_0^\infty$, $\hat{\varrho}_0^\infty$, 0 and $\hat{T}_0^\infty$, with[10]

$$\frac{d\hat{p}_0^\infty}{dz} + \mathrm{Bo}\,\hat{\varrho}_0^\infty = 0 \quad , \quad \hat{p}_0^\infty = \hat{\varrho}_0^\infty \hat{T}_0^\infty \quad . \tag{7.77}$$

So as to study just how the hydrostatic limiting state is reached, let us define:

$$\begin{aligned} \hat{p}_0 &= \hat{p}_0^\infty(z)(1+\hat{\pi}_0') \quad , \quad \hat{\varrho}_0 = \hat{\varrho}_0^\infty(z)(1+\hat{\omega}_0') \quad , \\ \hat{T}_0 &= \hat{T}_0^\infty(z)(1+\hat{\vartheta}_0') \quad , \quad \hat{w}_0 \equiv \hat{w}_0' \quad , \end{aligned} \tag{7.78}$$

and assume that $\hat{\pi}_0'$, $\hat{\omega}_0'$, $\hat{\vartheta}_0'$ and $\hat{w}_0'$ are small perturbations such that (7.71) can be linearized.

The linearization of system (7.71) yields:

$$\begin{aligned} &\gamma \mathrm{M}_\infty^2 \mathrm{S}\frac{\partial \hat{w}_0'}{\partial \hat{t}} + \hat{T}_0^\infty(z)\frac{\partial \hat{\pi}_0'}{\partial z} - \mathrm{Bo}\,\hat{\vartheta}_0' = 0 \quad ; \\ &\mathrm{S}\frac{\partial \hat{\omega}_0'}{\partial \hat{t}} + \frac{\partial \hat{w}_0'}{\partial z} + \frac{\hat{\Gamma}_0^\infty(z) - \mathrm{Bo}}{\hat{T}_0^\infty(z)}\hat{w}_0' = 0 \quad ; \\ &\mathrm{S}\frac{\partial}{\partial \hat{t}}\left(\frac{\gamma-1}{\gamma}\hat{\pi}_0' - \hat{\vartheta}_0'\right) + \frac{\hat{\Gamma}_0^\infty(z) - \mathrm{Bo}(\gamma-1)/\gamma}{\hat{T}_0^\infty(z)}\hat{w}_0' = 0 \quad ; \\ &\hat{\pi}_0' = \hat{\omega}_0' + \hat{\vartheta}_0' \quad , \end{aligned} \tag{7.79}$$

where we set the following:

$$\hat{\Gamma}_0^\infty(z) \equiv -\frac{d\hat{T}_0^\infty(z)}{dz} \quad . \tag{7.80}$$

The last three equations of system (7.79) make it possible to obtain the following:

$$\mathrm{S}\frac{\partial \hat{\vartheta}_0'}{\partial \hat{t}} + (\gamma-1)\frac{\partial \hat{w}_0'}{\partial z} - \frac{\hat{\Gamma}_0^\infty(z)}{\hat{T}_0^\infty(z)}\hat{w}_0' = 0 \quad , \quad \mathrm{S}\frac{\partial \hat{\pi}_0'}{\partial \hat{t}} + \gamma\frac{\partial \hat{w}_0'}{\partial z} - \frac{\mathrm{Bo}}{\hat{T}_0^\infty(z)}\hat{w}_0' = 0 \quad . \tag{7.81}$$

[10] In all that follows we will not explicitly display the dependence on x, y of the "^" variables considered since it does not come into play in the following expansions. Moreover, it will be noted that this dependence only appears parametrically via the initial values p^0, ϱ^0, w^0, T^0.

It suffices then to apply $S(\partial/\partial\hat{t})$ to the first equation of system (7.79) and to deduct $\partial\hat{\vartheta}_0'/\partial\hat{t}$ and $\partial\hat{\pi}_0'/\partial\hat{t}$ from (7.81) in order to obtain the following single equation for $\hat{w}_0'$:

$$\mathrm{M}_\infty^2 \mathrm{S}^2 \frac{\partial^2 \hat{w}_0'}{\partial \hat{t}^2} - \hat{T}_0^\infty(z) \frac{\partial^2 \hat{w}_0'}{\partial z^2} + \mathrm{Bo} \frac{\partial \hat{w}_0'}{\partial z} = 0 \quad . \tag{7.82}$$

By setting

$$\hat{w}_0' = \phi_0 \exp\left\{ \frac{\mathrm{Bo}}{2} \int_0^z \frac{dz}{\hat{T}_0^\infty(z)} \right\} \quad , \tag{7.83}$$

(7.82) is reduced to the following form:

$$\mathrm{M}_\infty^2 \mathrm{S}^2 \frac{\partial^2 \phi_0}{\partial \hat{t}^2} - \hat{T}_0^\infty(z) \frac{\partial^2 \phi_0}{\partial z^2} + \frac{\mathrm{Bo}}{2\hat{T}_0^\infty(z)} \left(\frac{\mathrm{Bo}}{2} - \hat{\Gamma}_0^\infty(z) \right) \phi_0 = 0 \quad , \tag{7.84}$$

whose (analytical) solution is not obvious since $\hat{T}_0^\infty$ is a function of z!

Therefore, in order to obtain an explicit solution which would enable us to know the behavior of ϕ_0 when $\hat{t} \to \infty$, it is supposed that $\hat{T}_0^\infty(z)$ is constant. This means that:

$$\hat{T}_0^\infty(z) \equiv 1 \quad , \quad \hat{\Gamma}_0^\infty(z) \equiv 0 \quad . \tag{7.85}$$

In this case, by setting:

$$\tau = \frac{\mathrm{Bo}}{2SM_\infty} \hat{t} \quad , \quad \zeta = \frac{\mathrm{Bo}}{2} z \quad , \tag{7.86}$$

the equation reduces to the so-called equation of telegraphy:

$$\frac{\partial^2 \phi_0}{\partial \tau^2} - \frac{\partial^2 \phi_0}{\partial \zeta^2} + \phi_0 = 0 \quad . \tag{7.87}$$

For the latter equation, the solution to the Cauchy initial valaue problem must be obtained:

$$\tau = 0: \quad \phi_0 = \phi^0 \quad , \quad \frac{\partial \phi_0}{\partial \tau} = \phi^1 \quad , \tag{7.88}$$

with ϕ^0 and ϕ^1 given functions of ζ [11] with the ground condition:

$$\phi_0 = 0 \quad , \quad \text{in} \quad \zeta = 0 \quad , \tag{7.89}$$

and the condition that ϕ_0 remains bounded at infinity.

We seek ϕ_0 in the form:

$$\phi_0 = \int_0^\infty \varphi_0(\tau, \nu) \sin \nu\zeta \, d\nu \quad . \tag{7.90}$$

The condition in $\zeta = 0$ is automatically satisfied and:

[11] The functions ϕ^0 and ϕ^1 are easily expressed from the initial values (for $\tau = 0$) which intervene in the conditions (7.74).

$$\frac{d^2\varphi_0}{d\tau^2} + (\nu^2+1)\varphi_0 = 0 \quad ; \quad \tau = 0 : \varphi_0 = \varphi^0 \quad , \quad \frac{d\varphi_0}{d\tau} = \varphi^1 \quad ,$$

and thus,

$$\phi_0 = \int_0^\infty \left\{ \Gamma(\tau, \zeta, \zeta_1)\phi^1(\zeta_1) + \frac{\partial \Gamma(\tau, \zeta, \zeta_1)}{\partial \tau}\phi^0(\zeta_1) \right\} d\zeta_1 \quad , \tag{7.91}$$

where

$$\Gamma(\tau, \zeta, \zeta_1) = \frac{2}{\pi}\int_0^\infty \frac{\sin(\tau\sqrt{\nu^2+1})}{\sqrt{\nu^2+1}} \sin \nu\zeta \sin \nu\zeta_1 \, d\nu \quad , \tag{7.92}$$

is the corresponding Green function which can also be written in the following form:

$$\Gamma(\tau, \zeta, \zeta_1) = K(\tau, \zeta - \zeta_1) - K(\tau, \zeta + \zeta_1) \quad , \tag{7.93}$$

with the help of the basic solution $K(\tau, \zeta)$ of the following problem [see, e.g., Favard (1963)]:

$$\frac{\partial^2 K}{\partial \tau^2} - \frac{\partial^2 K}{\partial \zeta^2} + K = 0 \quad ;$$

$$K(0, \zeta) = 0 \quad , \quad \frac{\partial K(0, \zeta)}{\partial \tau} = \delta(\zeta) \quad , \tag{7.94}$$

where $\delta(\zeta)$ is the Dirac distribution.

It follows then that:

$$K(\tau, \zeta) = \begin{cases} \frac{1}{2} J_0\left(\sqrt{\tau^2 - \zeta^2}\right) \quad , & \tau > |\zeta| \\ 0 & \tau < |\zeta| \end{cases} \tag{7.95}$$

where J_0 is the zeroth order Bessel function.

The behavior of the above-found solution remains to be determined when $\tau \to \infty$. By using the asymptotic behavior of the function J_0, we find for large values of τ:

$$\Gamma(\tau, \zeta, \zeta_1) \cong \frac{1}{\sqrt{2\pi\tau}} \frac{\zeta\zeta_1}{\tau^2} \cos\left(\tau - \frac{\pi}{4}\right) \quad , \tag{7.96}$$

and hence:

$$\phi_0 = O(\hat{t}^{-3/2}), \quad \text{when} \quad \hat{t} \to +\infty \quad . \tag{7.97}$$

It is thus seen in what way $\hat{w}_0'$ tends to zero when $\hat{t} \to \infty$. As a consequence of the behavior in $\hat{t}^{-3/2}$ of $\hat{w}_0'(\hat{t}, z)$ when $\hat{t} \to \infty$, we can define the function $\mathcal{H}'(\hat{t}, z^0)$ which is a solution of the equation of the trajectories:

$$\frac{\partial \mathcal{H}'}{\partial \hat{t}} = \hat{w}_0'\left(\hat{t}, \mathcal{H}'(\hat{t}, z^0)\right) \quad , \tag{7.98}$$

with the initial condition: $\mathcal{H}'(0, z^0) \equiv z^0$. Of course, the following limit exists

$$\mathcal{H}^{\infty}(z^0) = \lim_{\hat{t}\to\infty} \mathcal{H}'(\hat{t}, z^0) \tag{7.99}$$

and so the inverse function $Z^0(z)$ can be defined such that:

$$\mathcal{H}^{\infty}(z^0) \equiv z^{\infty} \Rightarrow z^0 = G^{\infty}(z^{\infty}) \equiv Z^0(z^{\infty}) \quad , \tag{7.100}$$

which yields a correspondance between the initial and final altitudes of the particles during the adjustment phase.

Equation (7.73) shows then that we have:

$$\hat{\Sigma}_0^{\infty}(z) = \Sigma^0(Z^0(z)) \quad , \tag{7.101}$$

where $\Sigma^0 \equiv \log p^0 - \gamma \log \varrho^0$ is the initial (Eulerian) value of the specific entropy. In like manner, (7.70) gives us:

$$\hat{\boldsymbol{v}}_0^{\infty}(z) = \boldsymbol{v}^0(Z^0(z)) \quad . \tag{7.102}$$

Finally, we also have the limiting relations (7.77). We can therefore obtain for the initial values:

$$\hat{p}_0^{\infty}(z) \equiv \overline{p}^0 \quad , \quad \hat{\varrho}_0^{\infty}(z) \equiv \overline{\varrho}^0 \quad , \tag{7.103}$$

the following expressions to be imposed on the primitive equations (7.44):

$$\begin{aligned} \overline{p}^0 &= \left\{1 - \frac{\gamma-1}{\gamma}\int_0^z \exp\left[-\frac{1}{\gamma}\Sigma^0\left(Z^0(z_1)\right)\right] dz_1\right\}^{\gamma/(\gamma-1)} \quad ; \\ \overline{\varrho}^0 &= (\overline{p}^0)^{1/\gamma} \exp\left[-\frac{1}{\gamma}\Sigma^0\left(Z^0(z)\right)\right] \quad . \end{aligned} \tag{7.104}$$

At last, we reach the conclusion that the following three initial conditions must be imposed on the primitive equations (7.44):

$$t = 0: \quad \overline{\boldsymbol{v}}_0 = \boldsymbol{v}^0(Z^0(z) \quad , \quad \overline{\varrho}_0 = \overline{\varrho}^0 \quad , \quad \overline{T}_0 = \overline{p}^0/\overline{\varrho}^0 \quad , \tag{7.105}$$

where $\overline{p}^0$ and $\overline{\varrho}^0$ satisfy relations (7.104) and are directly related to the initial value $\Sigma^0(Z^0(z))$ of the specific entropy imposed at the start in the complete Euler equations.

The adjustment process which has just been qualitatively described is rather special: it demonstrates how an initial situation which is partially incompatible with the hydrostatic balance adjusts. This incompatibility stems from the fact that 1) the initial data of pressure and density [p^0 and ϱ^0, according to (7.74)] do not satisfy the hydrostatic balance relation, and 2) the vertical velocity $\hat{w}_0$ is not O(ε). This incompatibility is, however, only partial in that the horizontal scale of the initial data remains very large compared to the vertical scale which is also the case in the quasi-static approximation.

Another important point must be considered concerning the intervening in the initial data (7.105) for the primitive equations (7.44) of a drift related to the function $\mathcal{H}'(\hat{t}, z^0)$ which satisfies (7.98) as well as the initial condition $\mathcal{H}'(0, z^0) \equiv z^0$. This drift was evaluated numerically by Outrebon in the work on his Thesis in 1981. It turns out that $\mathcal{H}^{\infty}(z^0) \equiv z^{\infty}$ remains of the order of 50 m[12] when z^0 remains between 0 and 3 km, whereas it increases from 200 to 1600 m beyond $z^0 = 3$ km. In terms of pressure, this corresponds to differences varying between 20 and 200 mbars which is rather appreciable considering the uncertainty of meteorological measurements. The importance of the hydrostatic balance adjustment phenomenon for medium-range forecasting based on the numerical solution of the primitive equations (7.44) is thus demonstrated.

7.6 Complementary Remarks 1

During the whole of the present chapter, it has been assumed that the earth's surface – taken to be flat – could be simulated in the local system of coordinates (see Fig. 1.1) by the plane $z = 0$. If, however, we want to take the relief into account, then the first step is to rewrite the basic equations (7.1–4) in a system of local curvilinear coordinates related to the surface which simulates the relief: only then should the quasi-static limiting process be carried out. This procedure is standard practice in aerodynamics when a correct formulation (as defined by the MMAE) is attempted of the boundary layer equations as a local (significant) degeneracy of the Navier-Stokes equations for a low viscosity fluid.

1. Let us return to the primitive equations of Sect. 5.2 and, as in Sect. 5.4, we will write them in the coordinate system t, x, y, $\overline{p}_0$ for the unknown functions $\overline{\boldsymbol{v}}$, $\overline{\mathcal{H}}_0$ and $\overline{\mathcal{W}}_0$ defined in (7.52). We now linearize the equations thus obtained in the vicinity of a reference situation which is at rest and thermodynamically stratified in a stable manner ($K_0(\overline{p}_0) > 0$) by setting:

$$\overline{\boldsymbol{v}}_0 = \boldsymbol{V}_0' \quad , \quad \overline{\mathcal{W}}_0 = \omega_0' \quad , \quad \overline{\mathcal{H}}_0 = \phi_0(\overline{p}_0) + \mathrm{M}_\infty^2 \phi_2' \quad ,$$

where M_∞ is assumed of the order unity. In this case, the following system of linearized primitive equations replaces the system (7.59):

$$\begin{aligned}
&\mathrm{S}\frac{\partial \boldsymbol{V}_0'}{\partial t} + \frac{\mathrm{Bo}}{\gamma}\boldsymbol{D}\phi_2' + \frac{1}{\mathrm{Ro}}\frac{\sin\varphi}{\sin\varphi_0}(\boldsymbol{k} \times \boldsymbol{V}_0') = 0 \quad ; \\
&\boldsymbol{D}\cdot\boldsymbol{V}_0' + \mathrm{S}\frac{\partial \omega_0'}{\partial \overline{p}_0} = 0 \quad ; \\
&\mathrm{S}\frac{\partial}{\partial t}\left(\frac{\partial \phi_2'}{\partial \overline{p}_0}\right) + \frac{\mathrm{S}}{\mathrm{M}_\infty^2}K_0(\overline{p}_0)\frac{\omega_0'}{\overline{p}_0^2} = 0 \quad .
\end{aligned} \tag{7.106}$$

To the above system (7.106), a slip condition is added:

$$\frac{\partial \phi_2'}{\partial t} + \frac{\theta_0(1)}{K_0(1)}\frac{\partial}{\partial t}\left(\frac{\partial \phi_2'}{\partial \overline{p}_0}\right) = 0 \quad , \quad \text{on} \quad \overline{p}_0 = 1 \quad , \tag{7.107}$$

[12] To be sure, these values are closely related to the initial values of the velocities and, more especially, to the initial profile of the vertical component of the velocity. The results cited are relative to a vertical wind (caused by heat convection) which reaches its maximum at an altitude of 3 km where it is of the order of 1 m/s, i.e., fifty times *weaker* than the horizontal wind at the same altitude. The calculation results can be found in the second part (Chap. 3) of Outrebon's 1981 Thesis.

where $\overline{p}_0 = 1$ is once again the solution of the equation $\phi_0(\overline{p}_0) = 0$ according to the classical principle of linearization since V_0', ω_0' and ϕ_2' are supposed to be first order, infinitely small quantities and all infinitely small quantities of second order or more have been dropped. It has already been pointed out that one of the most negative consequences of resorting to the primitive equations is related to the loss of the hyperbolic character of these equations. It has also been seen that the hyperbolicity can, in a certain sense, be reinstated by favoring the vertical structure in system (7.106). To accomplish this, it is necessary to rigorously study the spectral properties of the operator:

$$\mathcal{A} = -\frac{d}{d\overline{p}_0}\left(\frac{\overline{p}_0^2}{K_0(\overline{p}_0)}\frac{d}{d\overline{p}_0}\right) \quad , \tag{7.108}$$

of the Sturm-Liouville type on the segment]0, 1[, where

$$K_0(\overline{p}_0) \equiv -\frac{\overline{p}_0}{\mathrm{Bo}}\left(\frac{d\theta_0}{d\overline{p}_0} - \frac{\gamma-1}{\gamma}\frac{\theta_0}{\overline{p}_0}\right) > 0 \quad .$$

Hence, the Sturm-Liouville equation:

$$\frac{d}{d\overline{p}_0}\left(\frac{\overline{p}_0}{K_0(\overline{p}_0)}\frac{d\mathcal{L}}{d\overline{p}_0}\right) + \lambda\mathcal{L} = 0 \tag{7.109}$$

generally presents a singularity at the origin $\overline{p}_0 = 0$!

In the following, we want to characterize the "finite energy" solutions of (7.109) and obtain the *necessary boundary conditions* so that the associated evolution problem can be well-posed.

In order to accomplish the above, our work with El Mabrouk (1984) has served as a guide. Only the following hypotheses are employed [for the definitions and classical functional analysis results, see the book by Triebel, (1972)]:

$$K_0(\overline{p}_0) \in \mathcal{C}^1(]0, 1[; \mathbb{R}_+^*) \quad ;$$

$K_0(\overline{p}_0) = O(\overline{p}_0^a)$, in the vicinity of $\overline{p}_0 = 0$ with $a \geq 0$ being a real scalar. It is observed that $\mu_0 = K_0(1)/\theta_0(1)$. The operator $\mathcal{B}$ is associated with the operator $\mathcal{A} - \lambda$ in such a way that:

$$\mathcal{D}(\mathcal{B}) = \Big\{\mathcal{L} \in \mathcal{C}^2(]0, 1[) \cap \mathcal{C}^1(]0, 1[) \quad ;$$

$$\left(\frac{d}{d\overline{p}_0} + \mu_0\right)\mathcal{L}(1) = 0 \quad , \quad \mathcal{L} \equiv 0 \text{ in the vicinity of } \overline{p}_0 = 0\Big\} \quad ,$$

$$\mathcal{B}\mathcal{L} = (\mathcal{A} - \lambda)\mathcal{L} \quad , \text{ for } \mathcal{L} \text{ in } \mathcal{D}(\mathcal{B}) \quad .$$

We then have:

Lemma 1. – *The operator $\mathcal{B}$ is symmetrical with a dense domain in $L^2(]0, 1[)$ and semi-bounded in a lower position.* The adjoint-operator $\mathcal{B}^*$ of $\mathcal{B}$, can now be defined by:

$$\mathcal{D}(\mathcal{B}^*) = \left\{ \mathcal{L} \in L^2(]0, 1[) \quad , \quad \left(\frac{d\mathcal{L}}{d\overline{p}_0} + \mu_0 \mathcal{L} \right)(1) = 0 \quad ; \right.$$

$$\left. (\mathcal{A} - \lambda)\mathcal{L} \text{ "distribution" } \in L^2(]0, 1[) \right\} \quad ;$$

$\mathcal{B}^*\mathcal{L} = (\mathcal{A} - \lambda)\mathcal{L}$, as defined by the distributions $\forall \mathcal{L} \in \mathcal{D}(\mathcal{B}^*)$. The operator $\mathcal{B}$ then canonically admits a self-adjoint extension of Friedriechs $\mathcal{B}_F$ and we have:

$$\mathcal{B}_F = \begin{cases} \mathcal{B}_F\mathcal{L} = (\mathcal{A} - \lambda)\mathcal{L} \quad , \text{ distribution} \quad ; \\ \mathcal{D}(\mathcal{B}_F) = H_{\mathcal{B}} \cap \mathcal{D}(\mathcal{B}^*) \quad , \end{cases}$$

where $H_{\mathcal{B}}$ is the "energy space" associated with $\mathcal{B}$.

We define then: $\Omega_{\mathcal{B}} = \{\mathcal{L} \in H_{\mathcal{B}}, \|\mathcal{L}\|_{\mathcal{B}} \leq 1\}$, as the unit sphere of $H_{\mathcal{B}}$. This leads to:

Lemma 2. – *$\Omega_{\mathcal{B}}$ is precompact in $L^2(]0, 1[)$.*

Finally, lemmas 1 and 2 result in:

Theorem 1. *The operator $\mathcal{B}_F$ has the following properties:*

a) *its spectrum is purely punctual*

b) *its eigenvalues can be ordered as an enumerable increasing sequence by repeating each value as many times as its multiplicity*

$$\lambda \leq \lambda_0 \leq \lambda_1 \leq \ldots \leq \lambda_n \Rightarrow +\infty \quad ,$$

c) *if $(X_j)_{j=0,1,\ldots,+\infty}$ are the associated orthonormed eigenvectors, then the system $\{X_j\}$ is complete and we have:*

$$\mathcal{D}(\mathcal{B}_F) = \left\{ \mathcal{L} \in L^2(]0, 1[) \quad , \quad \sum_0^{+\infty} \lambda_j^2 |(\mathcal{L}, X_j)|^2 < +\infty \right\}$$

$$\text{and} \quad \mathcal{B}_F\mathcal{L} = \sum_0^{+\infty} \lambda_j(\mathcal{L}, X_j)X_j \quad ,$$

d) *$\mathcal{B}_F^{-1}$ is a compact operator of $L^2(]0, 1[)$ in itself and in $\mathcal{D}(\mathcal{B}_F)$.*

It is now necessary to characterize the behavior in $\overline{p}_0 = 0$ which is deduced from:

Theorem 2. *We have the following equivalence:*

$$\mathcal{L} \in \mathcal{D}(\mathcal{B}_F) \Leftrightarrow \begin{cases} \mathcal{L} \in \mathcal{D}(\mathcal{B}^*) \Leftrightarrow \begin{cases} \mathcal{L} \in L^2(]0, 1[); \\ \dfrac{d}{d\overline{p}_0}\left(\dfrac{\overline{p}_0^2}{K_0(\overline{p}_0)} \dfrac{d\mathcal{L}}{d\overline{p}_0} \right) \in L^2(]0, 1[); \\ \left(\dfrac{d}{d\overline{p}_0} + \mu_0 \right) \mathcal{L}(1) = 0 \quad , \end{cases} \\ \|\mathcal{L}\|_{\mathcal{B}} < +\infty \quad , \\ \lim\limits_{\overline{p}_0 \to 0} \left\{ \dfrac{\overline{p}_0}{K_0(\overline{p}_0)} \mathcal{L} \dfrac{d\mathcal{L}}{d\overline{p}_0} \right\} = 0 \quad . \end{cases}$$

We thus obtain a classical condition at the edge in $\overline{p}_0 = 1$ which is simply the slip condition along the wall $\overline{p}_0 = 1$ where $\overline{p}_0 = 1$ (which in linear theory simulates the surface of the ground assumed to be flat) is a solution to the equation:

$$\phi_0(\overline{p}_0) = -\frac{1}{\mathrm{Bo}} \int_1^{\overline{p}_0} \frac{\theta_0(p)}{p} dp = 0 \quad .$$

On the other hand, in $\overline{p}_0 = 0$, a growth condition exists both on $\mathcal{L}$ and on its derivative $d\mathcal{L}/d\overline{p}_0$. In particular, if the continuity of $\mathcal{L}$ is imposed all the way to the edge $\overline{p}_0 = 0$, a classical physical condition is again encountered [see (7.65)].

However, we can specify the condition

$$\lim_{\overline{p}_0 \to 0} \left(\frac{\overline{p}_0^2}{K_0(\overline{p}_0)} \mathcal{L} \frac{d\mathcal{L}}{d\overline{p}_0} \right) = 0 \quad , \tag{7.110}$$

according to the convergence of the integral:

$$J_0 = \int_0^1 \frac{K_0(p)}{p^2} dp \quad . \tag{7.111}$$

1st case:

$$J_0 < +\infty \Leftrightarrow K_0(p) = O(p^a) \quad , \quad a = 1 + \delta \quad , \quad \delta > 0 \quad ;$$

a) if $0 > \delta > 1$, then $[p^2/K_0(p)]d\mathcal{L}/dp \in H^1(]0,1[)$ and therefore in $\mathcal{C}(]0,1[)$, i.e., in particular,

$$\lim_{p \to 0} \left(\frac{p^2}{K_0(p)} \frac{d\mathcal{L}}{dp} \right) < +\infty \quad . \tag{7.112}$$

However, the trace of $\mathcal{L}$ in $p = 0$ is not always defined. Moreover, if it is supposed that

$$\frac{p}{\sqrt{K_0(p)}} \frac{d\mathcal{L}}{dp} = \mathrm{O}(p^{-\alpha}) \quad , \quad \text{then} \quad \alpha \leq \frac{1-\delta}{2} \quad ;$$

b) if $1 < \delta < +\infty$, then $H_{\mathcal{B}}$ is injected continually into $H^1(]0,1[)$ and the trace of $\mathcal{L}$ in $p = 0$ is equal to zero; but the trace of $[p^2/K_0(p)]d\mathcal{L}/dp$ is not always defined in $p = 0$;

c) if $\delta = 1$, then $\mathcal{L} \in H^2(]0,1[)$. The traces of $\mathcal{L}$ and of $[p^2/K_0(p)]d\mathcal{L}/dp$ exist in $p = 0$ and we have $\mathcal{L}(0) = 0$.

2nd case:

$$J_0 = +\infty \Leftrightarrow K_0(p) = \mathrm{O}(p^b) \quad , \quad b = 1 - \delta \quad .$$

$0 < \delta < 1$; then $[p^2/K_0(p)](d\mathcal{L}/dp) \in H^1(]0,1[)$, therefore we have the relation (7.112); the trace of $\mathcal{L}$ in zero, however, is not always defined. In fact, this is compensated for by the fact that then:

$$\frac{p^2}{K_0(p)}\frac{d\mathcal{L}}{dp} \to 0 \quad , \quad \text{when} \quad p \to 0^+ \quad ,$$

since

$$\frac{p}{\sqrt{K_0(p)}} = \mathrm{O}\left(p^{1/2+\delta/2}\right) \text{ and } \frac{p}{\sqrt{K_0(p)}}\frac{d\mathcal{L}}{dp} \text{ is, at most, of the order } -\frac{1}{2} \, .$$

In conclusion, the construction of the operator $\mathcal{B}_F$ makes it possible to reduce the linearized primitive equations to a family of first order linear hyperbolic systems. Theorem 2, which is an equivalence, gives the *conditions* at the edge which are *necessary* for the corresponding evolution problem to be well-posed. Nonetheless, the problem of the sufficient conditions remains unsolved. Concerning the mathematical "characterization" of the quasi-linear primitive equations which do not form a hyperbolic system and for which the theories (based essentially on the energy inequalities) developed by Kreiss (1970), and Majda and Osher (1975) are of no help, to our knowledge, practically nothing is known at present.

2. We return now to the end of Sect. 7.5 to specify what was said regarding the horizontal scale of the initial data which is assumed very large compared to the vertical scale – exactly like in the quasi-static approximation. In fact, a more general adjustment process is called for in the case of initial data whose horizontal length scale remains of the same order as the vertical scale. Such an adjustment process is described by the equations of three-dimensional unsteady motions; tackling this very difficult problem is beyond the scope of the present work. It is simply pointed out that the equations mentioned above are obtained from (7.33) by introducing simultaneously with $\hat{t} = t/\varepsilon$ and $z = Z/\varepsilon$, the variables:

$$\hat{x} = \frac{x}{\varepsilon} \quad , \quad \hat{y} = \frac{y}{\varepsilon} \Rightarrow \boldsymbol{D} = \frac{1}{\varepsilon}\hat{\boldsymbol{D}} \quad , \tag{7.113}$$

and also by carrying out the limiting process:

$$\varepsilon \to 0 \quad , \quad \text{at} \quad \hat{t}, \hat{x}, \hat{y} \quad \text{and} \quad z \quad \text{fixed} \quad , \tag{7.114}$$

in the Euler equations (7.33). The following adjustment equations thus result for[13] $\hat{\boldsymbol{v}}_0$, $\hat{w}_0$, $\hat{\varrho}_0$, $\hat{p}_0$ and $\hat{T}_0$:

$$\begin{aligned}
&\hat{\varrho}_0\left\{\mathrm{S}\frac{\partial\hat{\boldsymbol{v}}_0}{\partial\hat{t}} + \left(\hat{\boldsymbol{v}}_0\cdot\hat{\boldsymbol{D}}\right)\hat{\boldsymbol{v}}_0 + \hat{w}_0\frac{\partial\hat{\boldsymbol{v}}_0}{\partial z}\right\} + \frac{1}{\gamma\mathrm{M}_\infty^2}\hat{\boldsymbol{D}}\hat{p}_0 = 0 \quad ; \\
&\hat{\varrho}_0\left\{\mathrm{S}\frac{\partial\hat{w}_0}{\partial\hat{t}} + \hat{\boldsymbol{v}}_0\cdot\hat{\boldsymbol{D}}\hat{w}_0 + \hat{w}_0\frac{\partial\hat{w}_0}{\partial z}\right\} + \frac{1}{\gamma\mathrm{M}_\infty^2}\left(\frac{\partial\hat{p}_0}{\partial z} + \mathrm{Bo}\hat{\varrho}_0\right) = 0 \quad ; \\
&\mathrm{S}\frac{\partial\hat{\varrho}_0}{\partial\hat{t}} + \hat{\boldsymbol{D}}\cdot(\hat{\varrho}_0\hat{w}_0) + \frac{\partial\hat{\varrho}_0\hat{w}_0}{\partial z} = 0 \quad ; \\
&\hat{p}_0 = \hat{\varrho}_0\hat{T}_0 \quad ;
\end{aligned} \tag{7.115a}$$

[13] These are functions of the "short" variables $\hat{t}$, $\hat{x}$, $\hat{y}$, z; the index zero indicates that the limiting process (7.114) has been carried out.

$$\hat{\varrho}_0\left(S\frac{\partial\hat{T}_0}{\partial\hat{t}}+\hat{\boldsymbol{v}}_0\cdot\hat{\boldsymbol{D}}\hat{T}_0\right)-\frac{\gamma-1}{\gamma}\left(S\frac{\partial\hat{p}_0}{\partial\hat{t}}+\hat{\boldsymbol{v}}_0\cdot\hat{\boldsymbol{D}}\hat{p}_0\right)$$

$$+\hat{w}_0\left(\hat{\varrho}_0\frac{\partial\hat{T}_0}{\partial z}-\frac{\gamma-1}{\gamma}\frac{\partial\hat{p}_0}{\partial z}\right)=0\quad . \tag{7.115b}$$

The three-dimensional unsteady adjustment equations are thus obtained. It will be noted that when

$$\hat{t}\to+\infty\quad,\quad|\hat{x}|\to\infty\quad,\quad|\hat{y}|\to\infty\quad,\quad z\to+\infty\quad,$$

the behavior of the solution of system (7.115) must be compatible with the solution of the Euler equation (7.33) when:

$$t\to 0\quad,\quad x\to 0\quad,\quad ;y\to 0\quad,\quad Z\to 0\quad .$$

It is in this sense that the quasi-linear adjustment equations (7.115) can in fact be said to describe a "local" atmospheric phenomenon in the vicinity of the observation point $\mathcal{P}_0$ where the synoptic forecast was carried out from the primitive equations [as explained at the end of Sect. 7.1, when $Z\to 0$, the Euler equations (7.33) identify with the primitive equations].

It is emphasized that in the hydrostatic balance adjustment problem analyzed in Sect. 7.5:

$$\lim_{z\to\infty}\equiv\lim_{Z\to 0}$$

which means that at infinity in z, a hydrostatic equilibrium behavior must be imposed in (7.71). This was, moreover, the choice made by Outrebon (1981) in his numerical solution of the problem associated with (7.71).

To conclude this section, some bibliographical references are called for. Concerning the primitive equations, the reader might consult the reports by Hinkelmann, Bykov and Shuman in the book *Lectures on Numerical Short-Range Weather Prediction*, edited by Hydrometeoizdat, Leningrad (1969). The review papers by Phillips (1970) and Kibel (1970) are also cited as are the books by Monin (1969, Chaps. 4–9), and Houghton (1977, Chap. 11).

On the topic of initial conditions for the primitive equations, the work by Phillips (1960) is recommended above all; the reader is also oriented towards Wiin-Nielsen (1978) and Baer (1977), but it is pointed out that the adjustment problem considered in Sect. 7.5 is very different from the initialization problem of the primitive equations from the viewpoint of meteorological specialists in short-range numerical weather forecasting. This initialization problem is important from the forecasting point of view because it aims at eliminating the relatively short gravity waves which are still present in the primitive equations.[14] It should be noted that for the primitive equations, the adjustment problem considered in Sect. 7.5 is, in fact, analogous to what Rossby (1938) did for the quasi-geostrophic

[14] Some very worthwhile information concerning the problem of initialization can be found in the book by Bengtsson, Ghil and Källen (1981). In particular, see pages 111–138.

model equation (see Chap. 11) when he formulated the problem of adjustment to geostrophy. The difference lies in the fact that the latter is linear; it was solved in its entirety by Kibel (1955). Bolin (1953) and Blumen (1972) also dealt with this topic.

Finally, it is pointed out that many articles concerning the analysis and resolution of the primitive equations by numerical methods can be found in the "Journal of the Atmospheric Sciences" (in English) and also "Izvestiya of Acad. of Sci. USSR, Atmospheric and Oceanic physics" (in Russian but with an English version).

7.7 Complementary Remarks 2

Let us go back to the basic quasi-static equations obtained in Sect. 7.1, i.e., (7.14–18). It is obvious that the initial conditions to be imposed on this system cannot be deduced from the starting initial conditions (7.7). This is a direct result of the loss of the term $\partial w_0/\partial t$ in the limiting momentum equation in the vertical direction (which, after the limiting process $\lim_{q-s}{}^P$, identifies with the classical hydrostatic relation). In the vicinity of $t = 0$, there is, therefore, a singular phenomenon:

$$\lim_{t\to 0}\left(\lim_{q-s}{}^P\right) \neq \lim_{q-s}{}^P\left(\lim_{t\to 0}\right) \quad . \tag{7.116}$$

This singular phenomenon was analyzed by Zeytounian (1980) for the general case of the unsteady Navier-Stokes equations. He demonstrated that in the vicinity of $t = 0$, a new "doubly" local limiting process must be considered for the starting equations (7.1–4). It is thus necessary to specify *the limiting form of* (7.1–4) *when*:

$$\varepsilon \to 0 \quad \text{and} \quad \mathrm{Re} \to \infty \quad , \quad \text{with} \quad \varepsilon^2\,\mathrm{Re} = \mathrm{Re}_\perp = \mathrm{O}(1) \quad ,$$

in the *vicinity of the ground's surface* $z = 0$, and at the *initial time* $t = 0$.

It is pointed out that by examining the behavior of the solution of the quasi-static equations (7.14–18) in the vicinity of the initial time it can be seen that w_0 behaves like $t^{-1/2}$. This implies then that the quasi-static approximation cannot be valid "too close" to $t = 0$. This singularity is, of course, intrinsically linked to the compressibility of the fluid.[15]

In order to obtain an asymptotic representation of the solutions of the basic equations (7.1–4) in the vicinity of $t = 0$ and $z = 0$, the following local limiting process must be considered:

$$\lim_{q-s}{}^l \equiv \Big\{ \varepsilon \to 0 \;,\; \mathrm{Re} \to \infty \;;\; \varepsilon^2\,\mathrm{Re} = \mathrm{Re}_\perp = \mathrm{O}(1) \;; \quad \mathrm{S}, \gamma, \mathrm{M}_\infty, \mathrm{Pr}, \mathrm{Bo}, \sigma_{00}, \frac{t}{\varepsilon^a}, x, y \text{ and } \frac{z}{\varepsilon^b} \text{ fixed} \Big\} \quad , \tag{7.117}$$

[15] A theoretical analysis of unsteady laminar compressible boundary layers can be found in our "Note Technique", O.N.E.R.A. (Zeytounian, 1970).

where a and b are real numbers to be determined. In this case, for the limiting functions:

$$(\tilde{\boldsymbol{v}}_0, \tilde{w}_0, \tilde{p}_0, \tilde{\varrho}_0, \tilde{T}_0) = \lim_{q-s}{}^{l}(\boldsymbol{v}, \varepsilon w, p, \varrho, T) \tag{7.118}$$

once it has been determined that the correct choice is[16]:

$$a = 2 \quad , \quad b = 1 \quad , \tag{7.119}$$

the following system of equations results:

$$\tilde{\varrho}_0 \left(\mathrm{S}\frac{\partial \tilde{\boldsymbol{v}}_0}{\partial \tilde{t}} + w_0 \frac{\partial \tilde{\boldsymbol{v}}_0}{\partial \tilde{z}} \right) = \frac{1}{\mathrm{Re}_\perp} \frac{\partial^2 \tilde{\boldsymbol{v}}_0}{\partial \tilde{z}^2} \quad ; \tag{7.120a}$$

$$\tilde{\varrho}_0 \left(\mathrm{S}\frac{\partial \tilde{w}_0}{\partial \tilde{t}} + \tilde{w}_0 \frac{\partial \tilde{w}_0}{\partial \tilde{z}} \right) + \frac{1}{\gamma \mathrm{M}_\infty^2} \frac{\partial \tilde{p}_0}{\partial \tilde{z}} = \frac{1}{\mathrm{Re}_\perp} \left[\frac{\partial}{\partial \tilde{z}} \left(\tilde{\mu}_0 \frac{\partial \tilde{w}_0}{\partial \tilde{z}} \right) + \frac{1}{3} \frac{\partial^2 \tilde{w}_0}{\partial \tilde{z}^2} \right] \quad ; \tag{7.120b}$$

$$\mathrm{S}\frac{\partial \tilde{\varrho}_0}{\partial \tilde{t}} + \tilde{w}_0 \frac{\partial \tilde{\varrho}_0}{\partial \tilde{z}} + \tilde{\varrho}_0 \frac{\partial \tilde{w}_0}{\partial \tilde{z}} = 0; \tag{7.120c}$$

$$\tilde{p}_0 = \tilde{\varrho}_0 \tilde{T}_0 \quad , \tag{7.120d}$$

$$\tilde{\varrho}_0 \left(\mathrm{S}\frac{\partial \tilde{T}_0}{\partial \tilde{t}} + \tilde{w}_0 \frac{\partial \tilde{T}_0}{\partial \tilde{z}} \right) - \frac{\gamma - 1}{\gamma} \left(\tilde{S}\frac{\partial \tilde{p}_0}{\partial \tilde{t}} + \tilde{w}_0 \frac{\partial \tilde{p}_0}{\partial \tilde{z}} \right)$$
$$= \frac{1}{\mathrm{Pr}} \frac{1}{\mathrm{Re}_\perp} \frac{\partial}{\partial \tilde{z}} \left(\tilde{k}_0 \frac{\partial \tilde{T}_0}{\partial \tilde{z}} \right) + \frac{\gamma - 1}{\mathrm{Re}_\perp} \tilde{\mu}_0 \mathrm{M}_\infty^2 \left[\left| \frac{\partial \tilde{\boldsymbol{v}}_0}{\partial \tilde{z}} \right|^2 + \frac{4}{3} \left(\frac{\partial \tilde{w}_0}{\partial \tilde{z}} \right)^2 \right] , \tag{7.120e}$$

which is the so-called "Rayleigh" system (Howart, 1951).

The above Rayleigh equations (7.120) govern the class of unsteady motions of a heat conducting and compressible viscous fluid which only depend on the time $\tilde{t}$ and a space variable orthogonal to the ground's surface $\tilde{z}$ (simulated by the plane $\tilde{z} = 0$). It will be observed that the horizontal variables x and y play the role of parameters in (7.120).

The following ground conditions must be superimposed on (7.120):

$$\tilde{\boldsymbol{v}}_0 = \tilde{w}_0 = 0 \quad \text{and} \quad \frac{\partial \tilde{T}_0}{\partial \tilde{z}} = 0 \quad , \quad \text{for} \quad \tilde{z} = 0 \quad . \tag{7.121}$$

The above conditions result directly from (7.8).

If the initial data $\boldsymbol{v}^0$, w^0, p^0 and ϱ^0 which come into play in (7.7) do not have a Rayleigh-type vertical structure (i.e., that they are only functions of x, y

---

[16] It is necessary, on the one hand, that: $a = 2b$ and, on the other hand, that $2b = 2$. It will be noted that $\tilde{t} \equiv t/\varepsilon^2$ and $\tilde{z} \equiv z/\varepsilon$. Finally, it is observed that

$$\hat{t} = \varepsilon \tilde{t} \quad \text{and} \quad Z = \varepsilon^2 \tilde{z} \quad ,$$

where $\hat{t}$ and Z are defined by (7.68).

and z *but not* of $\tilde{z} = z/\varepsilon$), then the following initial conditions must be written for (7.120):

$$\tilde{\boldsymbol{v}}_0 = \boldsymbol{v}|^0_{z=0} \quad , \quad \tilde{w}_0 = v|^0_{z=0} ,$$
$$\tilde{p}_0 = p^0_{z=0} \quad , \quad \tilde{\varrho}_0 = \varrho^0_{z=0} \quad , \quad \text{for} \quad t = 0 \quad . \tag{7.122}$$

Naturally, matching conditions must also be imposed on the solution of the Rayleigh problem (7.120–122). Firstly, it should be noted that if in the limiting process (7.117), the particular choice of $a = 1$ and $b = 0$ is made, then we again find the hydrostatic adjustment equations for (7.67). If, however $1 < a < 2$ and $0 < b < 1$, then for the limiting functions:

$$(\boldsymbol{v}_I, w_I, p_I, \varrho_I, T_I) = \lim_{q-s}{}^{l_I}(\boldsymbol{v}, \varepsilon w, p, \varrho, T) \quad , \tag{7.123}$$

we find the equations usually employed in gas dynamics for describing the one-dimensional unsteady motions of a perfect fluid:

$$S\frac{\partial \boldsymbol{v}_I}{\partial t_I} + w_I\frac{\partial \boldsymbol{v}_I}{\partial z_I} = 0 \quad ; \tag{7.124a}$$

$$\gamma M^2_\infty \varrho_I\left(S\frac{\partial w_I}{\partial t_I} + w_I\frac{\partial w_I}{\partial z_I}\right) + \frac{\partial p_I}{\partial z_I} = 0 \quad ;$$
$$S\frac{\partial \varrho_I}{\partial t_I} + w_I\frac{\partial \varrho_I}{\partial z_I} = 0 \quad ; \quad p_I = \varrho_I T_I \quad ; \tag{7.124b}$$
$$\varrho_I\left(S\frac{\partial T_I}{\partial t_I} + w_I\frac{\partial T_I}{\partial z_I}\right) - \frac{\gamma-1}{\gamma}\left(S\frac{\partial p_I}{\partial t_I} + w_I\frac{\partial p_I}{\partial z_I}\right) = 0 \quad ,$$

where

$$t_I = \frac{t}{\varepsilon^a} \quad , \quad z_I = \frac{z}{\varepsilon^{a-1}} \quad , \quad 1 < a < 2 \quad . \tag{7.125}$$

Equations (7.124) are intermediate between the hydrostatic balance adjustment equations (7.70, 71), and the Rayleigh equations (7.120). They are, in fact, the degenerate forms of (7.120) where the second members (non-adiabatic) have been dropped. It is noted that $t_I \to \infty$ must correspond to $\hat{t} \to 0$, and $t_I \to 0$ to $\tilde{t} \to \infty$. Moreover, $z_I \to \infty$ corresponds to $z \to 0$ and $z_I \to 0$ to $\tilde{z} \to \infty$. Therefore, in the vicinity of $t = 0$ and in a region close to the ground whose thickness is of the order $H \sim (L\nu_0/U_0)^{1/2}$, the limiting form of the basic equations (7.1–4) gives rise to two different systems of equations, (7.120) and (7.124) which both govern one-dimensional unsteady motions. Obviously, the domains of validity of these two systems are distinct. Equations (7.120) which remain valid in the vicinity of the ground's surface of the order ε^2 (with respect to the vertical variable Z), degenerate when $\tilde{z} \to \infty$ into the one-dimensional unsteady perfect fluid equations. The order ε^2 is smaller than that of the boundary layer associated with the quasi-static equations (7.14–18) whose order is ε with respect to Z.

In order to study the compatibility of the Rayleigh equations (7.120) with the quasi-static equations (7.14–18) which are valid for $t = O(1)$, both the asymptotic structures of (7.120) in the vicinity of $\tilde{t} = \infty$ and $\tilde{z} = \infty$ and of (7.14–18) in the vicinity of $t = 0$ and $z = 0$ must be analyzed. Such an analysis brings to light an intermediate domain common to both systems which is characterized by the change of scales:

$$t = \delta t_* \quad , \quad z = \delta^{1/2} z_* \quad , \quad w = \delta^{-1/2} w_* \quad , \qquad (7.126)$$

with $\varepsilon^2 < \delta(\varepsilon) < 1$. In this intermediate domain, the limiting equations [see (7.127)] are those which are obtained by carrying out on the quasi-static equations (7.14–18), the simplifications leading to the Rayleigh equations when the exact equations (7.1–4) are used as the starting point.

The asymptotic structure of (7.124) still needs to be clarified when $t_I \to \infty$ and $z_I \to \infty$ (which represents a kind of compensation equations). They remain valid in a domain whose vertical thickness in the vicinity of $t = 0$ is equal to the deficit of thickness with respect to the thickness of the boundary layer associated with the quasi-static equations. In addition, (7.124) match with the linear primitive equations of the *second* approximation when $t \to 0$. In the compensation domain under consideration, a "small perturbation" is present which corresponds to the displacement thickness correction. Considered from the viewpoint of the boundary layer associated with the quasi-static equations, this correction tends to infinity as $t^{-1/2}$, when $t \to 0$; but from the viewpoint of the Rayleigh unsteady viscous layer, it tends to zero as $\tilde{t}^{-1/2}$ when $\tilde{t} \to \infty$. It is pointed out that for very large $\tilde{t}$, the classical Rayleigh problem can be split up into a boundary layer problem and a perfect fluid problem of second (linearized) approximation.

An important question arises quite naturally concerning the initialization of the quasi-static equations (7.14–18). From the above phenomenological analysis, it is obvious that this initialization must be carried out in such a way that *the behavior of the solution of the corresponding unsteady quasi-static equations for* $t \to 0$ *and* $z \to 0$ *must be compatible with the behavior of the solution of the associated Rayleigh equations when* $\tilde{t} \to \infty$ *and* $\tilde{z} \to \infty$. It can therefore be seen that *a priori, we cannot be completely certain of prescribing the "right" initial conditions for the quasi-static equations (7.14–18)* – we must make sure that the above-mentioned compatibility has been carefully verified.

Let us return to the quasi-static equations (7.14–18) written with the variables t, x, y, $z = Z/\varepsilon$ with respect to $\boldsymbol{v}_0$, $w_0 = \tilde{w}_0/\varepsilon$, p_0, ϱ_0, T_0. We then carry out for these equations, the change in scales which is valid in the vicinity of $t = 0$ and $z = 0$,

$$t = \varepsilon^{\varphi} t_* \quad , \quad z = \varepsilon^{\sigma} z_*$$

and we postulate the following proximal asymptotic representation:

$$\boldsymbol{v}_0 = \boldsymbol{v}_{0*} + \dots \quad , \quad w_0 = \varepsilon^{\psi} w_{0*} + \dots \quad , \quad p_0 = p_{0*} + \dots \quad ,$$
$$\varrho_0 = \varrho_{0*} + \dots \quad , \quad T_0 = T_{0*} + \dots \quad .$$

In order to obtain the significant degeneracy of (7.14–18) in the vicinity of $t = 0$ and $z = 0$, the following must be adopted:

$$\varphi = 2\sigma \quad \text{and} \quad \psi = -\sigma \quad ,$$

with $\sigma > 0$ remaining undetermined. In this way, for any

$$\delta \equiv \varepsilon^{2\sigma} \to 0 \quad , \quad \text{with} \quad \varepsilon \to 0 \quad ,$$

the change in scales (7.126) leads to the intermediate equations:

$$\begin{aligned}
&\varrho_{0*}\left(\mathrm{S}\frac{\partial \boldsymbol{v}_{0*}}{\partial t_*} + w_{0*}\frac{\partial \boldsymbol{v}_{0*}}{\partial z_*}\right) = \frac{1}{\mathrm{Re}_\perp}\frac{\partial}{\partial z_*}\left(\mu_{0*}\frac{\partial \boldsymbol{v}_{0*}}{\partial z_*}\right) \quad ; \\
&\frac{\partial p_{0*}}{\partial z_*} = 0 \quad ; \\
&\mathrm{S}\frac{\partial \varrho_{0*}}{\partial t_*} + w_{0*}\frac{\partial \varrho_{0*}}{\partial z_*} + \varrho_{0*}\frac{\partial w_{0*}}{\partial z_*} = 0 \quad ; \\
&p_{0*} = \varrho_{0*}T_{0*} \quad ; \\
&\varrho_{0*}\left(\mathrm{S}\frac{\partial T_{0*}}{\partial t_*} + w_{0*}\frac{\partial T_{0*}}{\partial z_*}\right) - \frac{\gamma - 1}{\gamma}\mathrm{S}\frac{\partial p_{0*}}{\partial t_*} \\
&\quad = \frac{1}{\mathrm{Pr}}\frac{1}{\mathrm{Re}_\perp}\frac{\partial}{\partial z_*}\left(k_{0*}\frac{\partial T_{0*}}{\partial z_*}\right) + \frac{\gamma - 1}{\mathrm{Re}_\perp}\mathrm{M}_\infty^2\mu_{0*}\left|\frac{\partial \boldsymbol{v}_{0*}}{\partial z_*}\right|^2 \quad .
\end{aligned} \tag{7.127}$$

We now return to the Rayleigh equations (7.120) written with the variables $\tilde{t}$ and $\tilde{z}$ for the funtions $\tilde{\boldsymbol{v}}_0$, $\tilde{w}_0 \equiv \varepsilon w_0$, $\tilde{p}_0$, $\tilde{\varrho}_0$ and $\tilde{T}_0$. To obtain the equations governing the behavior of the solution of (7.120) for very large values of $\tilde{t}$ and $\tilde{z}$, the following change in scales must be introduced:

$$\tilde{t} = \frac{t_{**}}{\Delta} \quad \text{and} \quad \tilde{z} = \frac{z_{**}}{\Delta^{1/2}} \quad ,$$

with $\Delta(\varepsilon) > 0$ remaining undetermined. If we now postulate the new distal asymptotic representation:

$$\begin{aligned}
&\tilde{\boldsymbol{v}}_0 = \boldsymbol{v}_{0**} + \dots \quad , \quad \tilde{w}_0 = \Delta^{1/2} w_{0**} + \dots \quad , \quad \tilde{p}_0 = p_{0**} + \dots \quad , \\
&\tilde{\varrho}_0 = \varrho_{0**} + \dots \quad , \quad \tilde{T}_0 = T_{0**} + \dots
\end{aligned}$$

from (7.120), for $\boldsymbol{v}_{0**}$, w_{0**}, p_{0**}, ϱ_{0**} and T_{0**} in the variables t_{**} and z_{**}, we again find the intermediate equations (7.127) which are the boundary layer form of the Rayleigh equations (7.120).

It is thus seen that *matching* indeed exists between the quasi-static equations [valid for $t = O(1)$] and the Rayleigh equations (valid for small t of order ε^2).

So as to have compatibility between the two behaviors:

$$\tilde{t} \to \infty \quad , \quad \tilde{z} \to \infty \quad \text{and} \quad t \to 0 \quad , \quad z \to 0 \quad ,$$

it is necessary that the changes in scales:

$$\begin{aligned}
&t = \frac{\varepsilon^2}{\Delta}t_{**} \quad , \quad z = \frac{\varepsilon}{\Delta^{1/2}}z_{**} \quad , \quad w = \frac{\Delta^{1/2}}{\varepsilon}w_{**} \quad \text{and} \\
&t = \delta t_* \quad , \quad z = \delta^{1/2}z_* \quad , \quad w = \delta^{-1/2}w_* \quad ,
\end{aligned}$$

be *identical*. This leads to the imposing on the scales δ and Δ, of the constraint:

$$\varepsilon^2 = \Delta\delta \quad . \tag{7.128}$$

This constraint must be verified to make sure there is matching between the solutions of (7.14–18) and (7.120). This matching takes place simultaneously in t and z in an intermediate domain which is characterized by $O(\delta)$, for t, and by $O(\varepsilon\delta^{1/2})$ for Z. It will be observed that naturally:

$$\varepsilon\delta^{1/2} \ll \varepsilon \quad ,$$

and thus, in the vicinity of $t = 0$ in the boundary layer, there necessarily exists a perfect fluid region which is the compensation layer in which (7.124) remain valid. Moreover, in the outer fringes of the matching intermediate layer which is governed by (7.127) (it is in this layer that the matching to the compensation flow of the pefect fluid has to take place), *we must* naturally find equations like (7.127) but *without* the second members related to non-adiabatic effects. Consequently, this explains why there is in this region *a small perturbation* which corresponds to the *displacement correction thickness*: from the viewpoint of the quasi-static equations, this correction tends towards infinity as $t^{-1/2}$ when $t \to 0$ since:

$$t = \delta t_* \quad , \quad \text{with} \quad \varepsilon^2 \ll \delta \ll 1 \quad ; \quad t \to 0 \sim t_* \to \infty \quad ,$$

whereas from the viewpoint of the Rayleigh equations, it tends towards zero as $\tilde{t}^{-1/2}$ when $\tilde{t} \to \infty$, since:

$$\tilde{t} = \frac{\delta}{\varepsilon^2} t_* = \frac{t_*}{\Delta} \quad , \quad \text{with} \quad \Delta > 0 \quad ; \quad \tilde{t} \to \infty \sim t_* \to 0 \quad .$$

Thus is implied that the solution in the domain of compensation "can" only match with the second approximation of a perfect fluid in $t \to 0$. Finally, it is recalled that $t_* = t/\delta$ is the time variable which characterizes the matching domain (in time) governed by the intermediate equations (7.127).

8. The Boussinesq Approximation

In the analysis of atmospheric flows, it is usual to consider the atmosphere as an incompressible fluid medium but which is stratified in altitude. We will see that this is the consequence of the characteristic velocity U_0 being always very small compared to the characteristic speed of sound $c^0_\infty = [\gamma R T_\infty(0)]^{1/2}$ (flows also called "hyposonic"). The so-called Boussinesq[1] approximation consists in neglecting the variations in density – except where they are coupled with the magnitude of the gravity g. Otherwise stated, this means that the influence of the stratification on the inertia terms is ignored in comparison to the influence related to the gravity.

For a thorough understanding of the simplification obtained from this Boussinesq approximation, it is worthwhile to consider the simple case of a so-called isochoric flow[2] which is both two-dimensional and unsteady and for which the only external force comes from the action of gravity of magnitude g. If the Coriolis and viscosity forces are ignored, along with the heat conduction, there results with the usual notations, the following dimensional equations:

$$\begin{aligned}
&\frac{\partial \boldsymbol{u}}{\partial t} + (\boldsymbol{u} \cdot \nabla_2)\boldsymbol{u} + \frac{1}{\varrho}\nabla_2 p = -g\boldsymbol{k} \quad ; \\
&\nabla_2 \cdot \boldsymbol{u} = 0 \quad ; \\
&\frac{\partial \varrho}{\partial t} + \boldsymbol{u} \cdot \nabla_2 \varrho = 0 \quad ,
\end{aligned} \tag{8.1}$$

where

$$\boldsymbol{u} = u\boldsymbol{i} + w\boldsymbol{k} \quad \text{and} \quad \nabla_2 = \frac{\partial}{\partial x}\boldsymbol{i} + \frac{\partial}{\partial z}\boldsymbol{k} \quad .$$

If from (8.1), the pressure p is eliminated and if the plane stream function $\psi(t, x, z)$ is introduced such that:

$$u = -\frac{\partial \psi}{\partial z} \quad , \quad w = \frac{\partial \psi}{\partial x} \quad ,$$

[1] This approximation is, in fact, originally due to Boussinesq (1903, pp. 157–176), but only for the case of a non-homogeneous incompressible fluid. For a physical discussion of this subject, the classic book by Landau and Lifshitz (1959, Sect. 56) is recommended; herein, the approximation is termed "free convection". The reader is likewise oriented towards the books by Phillips (1977, Sect. 2.4) and by Kochin, Kibel and Roze (1963, Sects. 36 and 38, Vol. 1).

[2] In Chap. 9, we will return to isochoric flows in detail. It is pointed out here that for an adiabatic isochoric flow, the conservation equation for the specific entropy is replaced by the conservation equation for the density: the isochoric flow is thus (kinematically) incompressible.

then we obtain (see our Notes, Zeytounian 1974, Chap. 2):

$$\left(\frac{\partial}{\partial t} + \boldsymbol{u} \cdot \nabla_2\right) \{\nabla_2 \cdot (\varrho \nabla_2 \psi) + \mathcal{J}(\mathcal{H}, \varrho)\} = 0 \quad , \tag{8.2}$$

where

$$\mathcal{H} = \int_0^t \left[\frac{|\nabla_2 \psi|^2}{2} - gz\right] dt \tag{8.3}$$

is the mechanical action defined for each fluid trajectory element and $\mathcal{J}(\alpha, \beta)$ is the Jacobian defined by:

$$\mathcal{J}(\alpha, \beta) = \frac{D(\alpha, \beta)}{D(x, z)} \equiv \frac{\partial \alpha}{\partial x}\frac{\partial \beta}{\partial z} - \frac{\partial \alpha}{\partial z}\frac{\partial \beta}{\partial x} \quad .$$

With the equation of isochoricity

$$\left(\frac{\partial}{\partial t} + \boldsymbol{u} \cdot \nabla_2\right) \varrho = 0 \quad ,$$

(8.2) forms a closed system allowing ψ and ϱ to be defined once adequate initial and boundary conditions have been added.

In a Boussinesq approximation, ϱ can be "taken out" of the derivation except when it is coupled with g. The following Boussinesq-type equation then replaces (8.2):

$$\left\{\frac{\partial}{\partial t} + \frac{\partial \psi}{\partial x}\frac{\partial}{\partial z} - \frac{\partial \psi}{\partial z}\frac{\partial}{\partial x}\right\} \left(\frac{\partial^2 \psi}{\partial x^2} + \frac{\partial^2 \psi}{\partial z^2}\right) + g\frac{\partial \log \varrho}{\partial x} = 0 \quad . \tag{8.4}$$

For the case of the compressible fluid, the Boussinesq approximation is justifiable for vertically confined flows (in the opposite direction of g) at a low altitude (for example, see Mihaljan, 1962). In the simulation of atmospheric flows in the tangent plane, the constraint of vertical confinement must be removed. Under certain conditions which are defined in Sect. 8.1, we have shown (Zeytounian, 1974) that the first approximation in asymptotic theory does indeed correspond to the Boussinesq equations. This result, however, is valid only locally (see discussion in Set. 8.2). The following authors had previously reduced the compressible fluid equations to Boussinesq equations: Spiegel and Veronis (1960), Dutton and Fichtl (1969), Ogura and Phillips (1962) and Phillips (1967). The methods employed (which do not constitute a formal development procedure) did not make it possible to obtain an "extension" of these equations. Moreover, in order to obtain a version of Boussinesq-type equations which are valid everywhere, it is necessary to introduce parallel to the dimensionless altitude variable $\overline{z}$, the slow altitude variable[3]

$$\overline{z}_\infty = \mathrm{Bo}\overline{z} \quad ,$$

as explained in Sect. 2.4.1. We will return to this question in Sect. 8.4.

[3] This idea was recently exploited by Bois in his thesis (1979a, Chap. 3). The reader might also consult Bois (1976 and 1979b).

In the forthcoming Sect. 8.1, a justification of the Boussinesq equations will be found within the framework of asymptotic theory. Section 8.2 aims at bringing to light various "singular" aspects of the Boussinesq approximation. In Sect. 8.3, three new forms of these equations are given. In Sect. 8.4, a linear theory is expounded on the so-called Boussinesq waves. This theory, inspired by the works of Bois (1979a), generalizes the one given in Sect. 6.2. Section 8.5 is devoted to the problem of adjustment to the Boussinesq state. Finally, in Sect. 8.6, miscellaneous complementary information can be found.

The basic equations in the present chapter are the full non-dimensional Navier-Stokes equations which can be written as follows:[4]

$$(1+\omega)\left\{\mathrm{S}\frac{D\boldsymbol{u}}{Dt}+\frac{1}{\mathrm{Ro}}(\boldsymbol{e}\times\boldsymbol{u})\right\}+\frac{T_\infty}{\gamma \mathrm{M}_\infty^2}\nabla\pi-(1+\omega)\frac{\mathrm{Bo}}{\varepsilon}\frac{1}{\gamma\mathrm{M}_\infty^2}\vartheta\boldsymbol{k}$$
$$=\frac{1}{\varrho_\infty}\frac{1}{\mathrm{Re}}\left\{\frac{1}{\varepsilon^2}\frac{\partial^2\boldsymbol{u}}{\partial z^2}+\boldsymbol{D}^2\boldsymbol{u}+\frac{1}{3}\nabla(\nabla\cdot\boldsymbol{u})\right\}\quad ; \tag{8.5}$$

$$\mathrm{S}\frac{D\omega}{Dt}+(1+\omega)(\nabla\cdot\boldsymbol{u})=(1+\omega)\frac{\mathrm{Bo}}{T_\infty}\left(1+\frac{dT_\infty}{dz_\infty}\right)w\quad ; \tag{8.6}$$

$$(1+\omega)\mathrm{S}\frac{D\vartheta}{Dt}-\frac{\gamma-1}{\gamma}\mathrm{S}\frac{D\pi}{Dt}+(1+\pi)\frac{\mathrm{Bo}}{T_\infty}\left(\frac{\gamma-1}{\gamma}+\frac{dT_\infty}{dz_\infty}\right)w$$
$$=\frac{1}{\varrho_\infty}\frac{1}{\mathrm{Pr}}\frac{1}{\mathrm{Re}}\left\{\boldsymbol{D}^2\vartheta+\frac{1}{\varepsilon^2}\left[\frac{\partial^2\vartheta}{\partial z^2}+2\frac{\mathrm{Bo}}{T_\infty}\frac{dT_\infty}{dz_\infty}\frac{\partial\vartheta}{\partial z}+\frac{Bo^2}{T_\infty}\frac{d^2T_\infty}{dz_\infty^2}\vartheta\right]\right\}$$
$$+\frac{\gamma-1}{\varrho_\infty T_\infty}\frac{\mathrm{M}_\infty^2}{\varepsilon^2\mathrm{Re}}\phi\quad ; \tag{8.7}$$

$$\pi=\omega+\vartheta+\omega\vartheta\quad . \tag{8.8}$$

In particular, the following boundary condition must be imposed on (8.5–8):

$$\frac{\partial\vartheta}{\partial z}+\mathrm{Bo}\frac{dT_\infty}{dz_\infty}\vartheta=0\quad ,\quad \text{on}\quad z=0\quad . \tag{8.9}$$

It is recalled that the phenomena associated with the radiations are supposed as only intervening in the definition of the standard temperature $T_\infty(z_\infty)$. On this subject, the reader is directed to (2.110) and its related boundary condition.

8.1 The Boussinesq Equations

Let us now consider (8.5–8). According to Zeytounian (1974), the Boussinesq approximation consists in seeking out the limiting form of (8.5–8) by taking into

[4] Once again, the bars have been omitted on the dimensionless quantities. The definition of $\boldsymbol{e}$, ∇, $\boldsymbol{D}$ and $\boldsymbol{u}$ is the same as in Chap. 7 [see (7.5) and (7.6)]. The perturbations π, ω and ϑ are defined in (2.70). Lastly, (2.110) will be taken into account and it will be noted that ϱ_∞ and T_∞ are functions of $z_\infty \equiv \mathrm{Bo}z$.

account the ground condition (8.9) when simultaneously:

$$\mathrm{M}_\infty \to 0 \quad , \quad \mathrm{Bo} \to 0 \quad , \tag{8.10}$$

in such a way that the similarity relation:

$$\frac{\mathrm{Bo}}{\mathrm{M}_\infty} = \hat{B} = \mathrm{O}(1) \quad , \tag{8.11}$$

is satisfied. It is implicitly understood as usual that in the limiting process (8.10, 11), the variables t, x, y and z remain fixed of the order unity just as do the dimensionless parameters S, Ro, ε, γ, Pr, Re. With the limiting process (8.10) which satisfies (8.11), the following asymptotic representation is associated:

$$\boldsymbol{u} = \boldsymbol{u}_0 + \dots \quad , \quad \omega = \mathrm{M}_\infty \omega_1 + \dots \quad , \quad \vartheta = \mathrm{M}_\infty \vartheta_1 + \dots \quad ,$$
$$\pi = \mathrm{M}_\infty^2 \pi_2 + \dots \quad . \tag{8.12}$$

In this case, the so-called Boussinesq equations result for the functions $\boldsymbol{u}_0$, ω_1, ϑ_1 and π_2:

$$\mathrm{S}\frac{D\boldsymbol{u}_0}{Dt} + \frac{1}{\mathrm{Ro}}(\boldsymbol{e} \times \boldsymbol{u}_0) + \frac{1}{\gamma}\nabla\pi_2 - \frac{\hat{B}}{\gamma\varepsilon}\vartheta_1\hat{\boldsymbol{k}}$$
$$= \frac{1}{\mathrm{Re}}\left\{\frac{1}{\varepsilon^2}\frac{\partial^2 \boldsymbol{u}_0}{\partial z^2} + \boldsymbol{D}^2\boldsymbol{u}_0\right\} \quad ; \tag{8.13}$$

$$\nabla \cdot \boldsymbol{u}_0 = 0 \quad ; \tag{8.14}$$

$$\mathrm{S}\frac{D\vartheta_1}{Dt} + \hat{B}\left[\frac{\gamma - 1}{\gamma} + \left(\frac{dT_\infty}{dz_\infty}\right)_0\right] w_0 = \frac{1}{\mathrm{Pr}}\frac{1}{\mathrm{Re}}\left\{\frac{1}{\varepsilon^2}\frac{\partial^2\vartheta_1}{\partial z^2} + \boldsymbol{D}^2\vartheta_1\right\} \; ; \tag{8.15}$$

$$\omega_1 = -\vartheta_1 \quad . \tag{8.16}$$

The following "athermanous wall" condition[5] must be assigned to the Boussinesq equations (8.13–16):

$$\frac{\partial\vartheta_1}{\partial z} = 0 \quad , \quad \text{on} \quad z = 0 \quad . \tag{8.17}$$

It is noted that the relation $z_\infty = \mathrm{Bo}z$ was used to obtain (8.13–16) and thus:

$$\lim_{\mathrm{Bo}\to 0}\begin{pmatrix}\varrho_\infty(z_\infty) \\ T_\infty(z_\infty)\end{pmatrix} = \begin{pmatrix}\varrho_\infty(0) \\ T_\infty(0)\end{pmatrix} \equiv \begin{pmatrix}1 \\ 1\end{pmatrix} \quad ,$$

since we are working with dimensionless quantities.

At the ground's surface, the velocity no-slip condition must be imposed on (8.13–16):

[5] It is naturally assumed that: $-(dT_\infty/dz_\infty)_0$ remains bounded and different from $(\gamma - 1)/\gamma$.

$$u_0 = 0 \quad , \quad \text{on} \quad z = 0 \quad . \tag{8.18}$$

It is important to emphasize right away that (8.13–16) remain valid only in an atmospheric layer whose thickness $H \equiv H_B$ satisfies the following constraint:

$$\frac{RT_\infty(0)}{g} \gg H = \hat{B}\frac{U_0}{g}\sqrt{\frac{RT_\infty(0)}{\gamma}} \equiv H_B \cong 200\,\mathrm{m} \quad \text{to} \quad 300\,\mathrm{m} \quad , \tag{8.19}$$

when $\hat{B}$, in fact, remains of the order unity.

Stated otherwise, it can be said that the asymptotic representation (8.12) which leads to the Boussinesq equations (8.13–16) via the hypotheses (8.10, 11) is valid only in atmospheric layers of a thickness H_B which is of the order of a few hundred meters. If, in particular, the wall conditions (8.17, 18) are prescribed, then the (local) asymptotic representation (8.12) can describe the atmospheric flow in first approximation only in the boundary layer in the vicinity of the wall $z = 0$, which has a thickness of the order of

$$H_B \equiv \hat{B}\frac{U_0}{g}\sqrt{\frac{RT_\infty(0)}{\gamma}} \quad . \tag{8.20}$$

This thickness is, of course, a function of the similarity parameter $\hat{B}$.

It is pointed out that because of (8.14) and (8.16), we can no longer impose an initial condition on ω_1 for $t = 0$ in the Boussinesq equations. This of course is linked to the fact that the Boussinesq approximation unfortunately filters out the internal acoustic waves having a frequency σ_a, as well as the acoustic-type plane waves having a frequency σ_p. It will be seen in Sect. 8.5 that a time adjustment problem to the Boussinesq state must be analyzed starting from a given initial situation (for $t = 0$).

One last remark is called for concerning the condition at the wall (8.17). In usual mesometeorological applications, it is most often benificial to suppose that the distribution of the temperature fluctuation on the ground is a given function of t, x and y. Let then $\Xi(t, x, y)$ be this function which is supposed non-dimensional, and let ΔT_0 be a characteristic temperature associated with $\Xi(t, x, y)$. In this case, the following condition must be assigned to (8.5–8) unstead of (8.9):

$$\vartheta = \tau_0 \Xi(t, x, y) \quad , \quad \text{on} \quad z = 0 \tag{8.21}$$

where

$$\tau_0 = \Delta t_0 / T_\infty(0) \tag{8.22}$$

is generally much smaller than unity. The Boussinesq approximation implies that:

$$\tau_0 \to 0 \quad , \quad \text{with} \quad \mathrm{M}_\infty \to 0 \tag{8.23}$$

such that

$$\frac{\tau_0}{\mathrm{M}_\infty} = \Lambda_0 = \mathrm{O}(1) \quad . \tag{8.24}$$

The following condition on ϑ_1 then replaces (8.17):

$$\vartheta_1 = \Lambda_0 \Xi(t, x, y) \quad , \quad \text{on} \quad z = 0 \quad . \tag{8.25}$$

A relatively general analysis of the case corresponding to (8.23) can be found in our work Zeytounian (1977).

8.2 Some Considerations concerning the Singular Nature of the Boussinesq Approximation

For atmospheric flows, the Boussinesq approximation must be considered only as a local representation which is valid in the vicinity of the ground. The limiting process (8.10, 11) is thus obviously singular. If we want ε to remain of order unity then the singular nature of this limiting process is magnified. As a matter of fact, in this case

$$L \cong H \equiv H_{\mathrm{B}}$$

and the Boussinesq equations (8.13–16) are valid only locally in a domain having a characteristic diameter of the order of $2H_{\mathrm{B}}$[6]. We are thus led to clarify the behavior "at infinity" of the solution of the Boussinesq equations (8.13–16). For the case of an adiabatic atmosphere, Re $\equiv \infty$, in steady evolution, this question was recently considered by Guiraud (1979) who from a multiple-scale technique obtained (see Sect. 8.6) a radiation condition for the so-called lee waves in (steady) three-dimensional flow relevant to the steady adiabatic Boussinesq equations.

However, let us return for the moment to (3.17). If the limiting process (8.10, 11) is applied to this equation, we obtain the following:

$$\left\{ \frac{\partial^2}{\partial \tau^2} + \frac{\lambda_\mu^{\mathrm{B}}/\mathrm{Ro}^2 + K^2 \hat{B}^2(\gamma - 1)/\gamma^2}{\varepsilon^2 K^2 + \lambda_\mu^B} \right\} \chi_{nm}^{(2)} = 0 \quad , \tag{8.26}$$

once it has been postulated that:

$$\chi_{nm} = \mathrm{M}_\infty^2 \chi_{nm}^{(2)} + \dots \quad ,$$

in agreement with the last relation of (8.12).

A strong degeneracy is observed. Instead of (3.17) of fourth order in τ, we obtain only a Boussinesq-type equation of second order in τ as a limit. Hence, given the constraint (8.11), the characteristic Mach number M_∞ is a small

---

[6] It will be noticed that when the problems being studied contain no boundary conditions (which is the case in a periodic wave propagation analysis), the characteristic quantities of the waves considered (frequency, wave length, etc.) are the same as those which define the characteristic length L and the characteristic velocity U_0 of the problem. Consequently, the wave length of the waves which are relevant to the Boussinesq approximation must be small compared to the characteristic scale of the atmosphere $RT_\infty(0)/g$.

singular perturbation parameter for the Cauchy problem (3.17, 18). Like in the case of the quasi-static approximation considered in Chap. 7, only internal gravity waves are obtained whose modified frequency $\sigma_{\mathrm{g\,B}}$ is given by the relation:

$$\sigma^2_{\mathrm{g\,B}} = \frac{\lambda^{\mathrm{B}}_{\mu}/\mathrm{Ro}^2 + K^2\hat{B}^2(\gamma-1)/\gamma^2}{\varepsilon^2K^2 + \lambda^{\mathrm{B}}_{\mu}} \quad . \tag{8.27}$$

From (3.21–23), it is confirmed that:

$$\lim_{\substack{\mathrm{Bo}=\hat{B}\mathrm{M}_\infty \\ \hat{B}=\mathrm{O}(1),\, \mathrm{M}_\infty \to 0}} \sigma^2_{\mathrm{g}} \equiv \sigma^2_{\mathrm{g\,B}} \quad .$$

In the same way, it is observed that:

$$\lim_{\substack{\mathrm{Bo}\to 0, \\ \mathrm{M}_\infty \to 0}} \sigma^2_{\mathrm{a}} \equiv \infty \quad , \tag{8.28}$$

even when $\mathrm{Bo} = \hat{B}M_\infty$, with $\hat{B} = \mathrm{O}(1)$.

The Boussinesq approximation thus filters out internal acoustic waves of frequency σ_{a}. However, this same approximation also filters out acoustic-type plane waves whose frequency σ_{p} satisfies relation (3.25). Thus, the acoustic oscillations in the plane $(x,\, y)$ are filtered out and π_2' satisfies the equation:

$$\overline{\Delta}_2 \frac{\partial \pi_2'}{\partial \tau} = 0 \quad .$$

Since the Boussinesq approximation is capable only of describing the waves developing in a thin layer within the troposphere, it is obvious that the corresponding eigenvalues $\lambda^{\mathrm{B}}_{\mu}$ are modified. For example, if we consider the Sturm-Liouville equation (3.10) in the troposphere, then by simulating the tropopause by the rigid plane $\overline{z} = 1/\mathrm{Bo}$ (in dimensionless variables), it is observed that when $\mathrm{Bo} \to 0$, this plane goes to infinity. Therefore, we are indeed dealing with a typical boundary layer phenomenon which is related to the singular nature of the Boussinesq approximation as concerns the vertical structure of the waves. The main outer flow for this boundary layer which must necessarily verify the boundary condition on $\overline{z}_\infty = \mathrm{Bo}\overline{z} \equiv 1$, has a double scale vertical structure. Guiraud and Zeytounian (1979) brought to light this double scale phenomenon for the case of a plane steady flow of a heavy perfect gas with constant specific heats but without viscosity and thermal conductivity. The flow in question took place above an obstacle and was bounded at the top by a fixed rigid wall. We will return to this matter in Chap. 13.

Lastly, it is pointed out that the Boussinesq approximation which leads to the Boussinesq equations (8.13–16) is deduced from the complete equations (8.5–8) with the following Boussinesq limiting process as the starting point:

$$\lim_{B} = \lim\left\{\mathrm{Bo}\to 0,\ \mathrm{M}_\infty \to 0;\ \frac{\mathrm{Bo}}{\mathrm{M}_\infty} = \hat{B} \quad , \right.$$

$$\left. \mathrm{S},\ \gamma,\ \mathrm{Pr},\ \mathrm{Re},\ \varepsilon,\ \mathrm{Ro},\ t,\ x,\ y,\ z\ \text{fixed}\right\} \quad . \tag{8.29}$$

These Boussinesq equations (8.13–16) govern an incompressible flow which consequently causes the vicinity of $t = 0$ to be singular:

$$\lim_{t\to 0}\lim_{B} \neq \lim_{B}\lim_{t\to 0} \quad .$$

A detailed analysis can be found in Sect. 8.5 of the vicinity of $t = 0$ which leads to the (adiabatic) problem of the adjustment to the Boussinesq state.

8.3 Three New Forms of the Boussinesq Equations

8.3.1 Taking into Account the Shearing of a Basic Wind; the So-called Long Equation

The classical theory of the so-called lee waves which appear downstream of the relief in the baroclinic adiabatic atmosphere in "small scale" steady evolution is usually made with the following hypotheses:

$$\varepsilon \equiv 1 \quad , \quad \mathrm{Ro} \equiv \infty \quad , \quad \mathrm{Re} \equiv \infty \quad , \quad \mathrm{S} \equiv 0 \quad . \tag{8.30}$$

Moreover, it is assumed that at upstream infinity where the relief disappears, a basic wind $\boldsymbol{v}_\infty(z_\infty)$ exists which is a function only of the standard altitude z_∞ such that:

$$\boldsymbol{v}_\infty(z_\infty) = U_\infty(z_\infty)\boldsymbol{i} + V_\infty(z_\infty)\boldsymbol{j} \quad . \tag{8.31}$$

By taking into account (8.31), we can see that it is a good idea to introduce the perturbation velocity $\boldsymbol{v}'$ in place of the horizontal velocity $\boldsymbol{v}$ in such a way that:

$$\boldsymbol{v} = \boldsymbol{v}_\infty(z_\infty) + \boldsymbol{v}' \quad . \tag{8.32}$$

We do not a priori suppose that $\boldsymbol{v}'$ is small compared to $\boldsymbol{v}_\infty(z_\infty)$. Let μ now be the dimensionless parameter which measures the magnitude of the shearing of the basic wind with the altitude. We then find:[7]

$$\frac{d\boldsymbol{v}^*(z^*_\infty)}{dz^*_\infty} = \frac{U_0/H}{\mu}\frac{d\boldsymbol{v}(z_\infty)}{dz_\infty} \quad , \tag{8.33}$$

where $\boldsymbol{v}^*(z^*_\infty)$ designates the dimensional basic wind (in m/s). If we suppose that:

$$\mathrm{Bo} \to 0 \quad , \quad \mu \to 0 \quad , \quad \text{with} \quad \mathrm{M}_\infty \to 0 \quad , \tag{8.34}$$

such that

$$\frac{\mathrm{Bo}}{\mathrm{M}_\infty} = \hat{B} = \mathrm{O}(1) \quad , \quad \frac{\mu}{\mathrm{M}_\infty} = \hat{\mu} = \mathrm{O}(1) \quad , \tag{8.35}$$

[7] The superscript asterisk characterizes in this instance a quantity *with* dimensions.

then like in Sect. 8.1, the following Boussinesq equations can be easily obtained:

$$\{[\boldsymbol{v}_\infty(0)+\boldsymbol{v}_0']\cdot\boldsymbol{D}\}\boldsymbol{v}_0' + w_0\frac{\partial \boldsymbol{v}_0'}{\partial z} + \frac{\hat{B}}{\hat{\mu}}\left(\frac{d\boldsymbol{v}_\infty}{dz_\infty}\right)_0 w_0 + \frac{T_\infty(0)}{\gamma}\boldsymbol{D}\pi_2 = 0 \quad ;$$

$$[\boldsymbol{v}_\infty(0)+\boldsymbol{v}_0']\cdot\boldsymbol{D}w_0 + w_0\frac{\partial w_0}{\partial z} + \frac{T_\infty(0)}{\gamma}\frac{\partial \pi_2}{\partial z} - \frac{\hat{B}}{\gamma}\vartheta_1 = 0 \quad ;$$

$$\boldsymbol{D}\cdot\boldsymbol{v}_0' + \frac{\partial w_0}{\partial z} = 0 \quad ; \tag{8.36}$$

$$\omega_1 = -\vartheta_1 \quad ;$$

$$[\boldsymbol{v}_\infty(0)+\boldsymbol{v}_0']\cdot\boldsymbol{D}\vartheta_1 + w_0\frac{\partial \vartheta_1}{\partial z} + \frac{\hat{B}}{T_\infty(0)}\left[\frac{\gamma-1}{\gamma} + \left(\frac{dT_\infty}{dz_\infty}\right)_0\right] w_0 = 0 \quad .$$

In the two-dimensional case $[\boldsymbol{v}_\infty(0) \equiv U_\infty(0)\boldsymbol{i}, \boldsymbol{v}_0' \equiv u_0'\boldsymbol{i}$ and $\boldsymbol{D} \equiv (\partial/\partial x)\boldsymbol{i}]$, we have:

$$\frac{\partial u_0'}{\partial x} + \frac{\partial w_0}{\partial z} = 0 \;\Rightarrow\; d\psi_0' = u_0' dz - w_0\, dx \quad , \tag{8.37}$$

where $\psi_0'(x, z)$ is the plane stream funtion.

By eliminating π_2, the following system results for ψ_0' and ϑ_1:

$$J\left\{\frac{\partial^2\psi_0'}{\partial x^2} + \frac{\partial^2\psi_0'}{\partial z^2}\right\} + \frac{\hat{B}}{\gamma}\frac{\partial\vartheta_1}{\partial x} - \frac{\hat{B}}{\hat{\mu}}\left(\frac{dU_\infty}{dz_\infty}\right)_0 \frac{\partial\psi_0'}{\partial x} = 0 \quad ;$$

$$J\{\vartheta_1\} - \frac{\hat{B}}{T_\infty(0)}\left[\frac{\gamma-1}{\gamma} + \left(\frac{dT_\infty}{dz_\infty}\right)_0\right]\frac{\partial\psi_0'}{\partial x} = 0 \quad , \tag{8.38}$$

where

$$J\{f\} \equiv \left(U_\infty(0) + \frac{\partial\psi_0'}{\partial z}\right)\frac{\partial f}{\partial x} - \frac{\partial\psi_0'}{\partial x}\frac{\partial f}{\partial z} \quad .$$

But

$$J\{\psi_0'\} \equiv U_\infty(0)\frac{\partial\psi_0'}{\partial x} \quad , \tag{8.39}$$

and therefore the second equation in (8.38) takes the form:

$$J\left\{\vartheta_1 - \frac{\hat{B}}{T_\infty(0)}\frac{1}{U_\infty(0)}\left[\frac{\gamma-1}{\gamma} + \left(\frac{dT_\infty}{dz_\infty}\right)_0\right]\psi_0'\right\} = 0 \quad . \tag{8.40}$$

Given the fact that ϑ_1 and ψ_0' cancel out at upstream infinity in the basic flow (when $x = -\infty$), we have the following under the hypothesis that the rotors are absent downstream[8]

[8] The problem of the rotors (closed streamlines) appearing downstream of the relief within the system of steady waves will be dealt with in Chap. 13. The review paper by Miles (1969) can be consulted on this topic.

$$\vartheta_1 = \frac{\hat{B}}{T_\infty(0)U_\infty(0)}\left[\frac{\gamma-1}{\gamma} + \left(\frac{dT_\infty}{dz_\infty}\right)_0\right]\psi_0' \quad . \tag{8.41}$$

If we now simply take into account (8.39) and (8.41) in the first equation of (8.38), we arrive at the following Helmoltz-type equation:

$$\frac{\partial^2\psi_0'}{\partial x^2} + \frac{\partial^2\psi_0'}{\partial z^2} + \Phi(0)\psi_0' = 0 \quad , \tag{8.42}$$

where

$$\Phi(z_\infty) = \frac{\hat{B}^2}{\gamma}\left\{\frac{(\gamma-1)/\gamma + (dT_\infty/dz_\infty)}{T_\infty(z_\infty)U_\infty^2(z_\infty)} - \frac{\gamma}{\hat{B}\hat{\mu}}\frac{1}{U_\infty(z_\infty)}\frac{dU_\infty}{dz_\infty}\right\} \quad . \tag{8.43}$$

In reference to Long (1953), (8.42) is often called "Long's equation". We will come back to this matter in Chap. 9. In Chap. 13, we will also see how (8.42) is interpreted from the viewpoint of the MMAE. For the time being, the problem of formulating the boundary conditions for (8.42) will not be dealt with. We will return to this question in detail in Chap. 13.

Backtracking to (8.36), it can be said that these equations represent the steady three-dimensional version of the Long equation (8.42) for the lee waves.

The solutions of (8.42) strongly depend on the parameter[9]

$$\Phi(0) = \frac{\hat{B}^2}{\gamma}\left\{\frac{\gamma-1}{\gamma} + \left(\frac{dT_\infty}{dz_\infty}\right)_0 - \frac{\gamma}{\hat{B}\hat{\mu}}\left(\frac{dU_\infty}{dz_\infty}\right)_0\right\} \quad , \tag{8.44}$$

that we will call the Dorodnitsyn-Scorer parameter [in reference to Dorodnitsyn (1950) and Scorer (1949)]. This parameter characterizes both the stratification of the standard atmosphere and the shearing influence of the basic wind at upstream infinity.

8.3.2 Generalized Boussinesq-Type Equations

Let us return to (8.5–8) where it is supposed that $\mathrm{Re} \equiv \infty$. According to the preceding, we can write that:

$$\pi = \mathrm{M}_\infty^2\overline{\pi} \quad , \quad \omega = \mathrm{M}_\infty\overline{\omega} \quad , \quad \vartheta = \mathrm{M}_\infty\overline{\vartheta} \quad , \tag{8.45}$$

and set

$$\mathrm{Bo} = \hat{B}\mathrm{M}_\infty \quad , \quad \zeta \equiv \frac{z_\infty}{\hat{B}} = \mathrm{M}_\infty z \quad . \tag{8.46}$$

Let us then take advantage of (8.45, 46) to replace (8.5–8) by the following generalized Boussinesq equations:

$$\mathrm{S}\frac{D\boldsymbol{u}}{Dt} + \frac{1}{\mathrm{Ro}}(\boldsymbol{e} \times \boldsymbol{u}) + \frac{T_\infty(\hat{B}\zeta)}{\gamma}\nabla\overline{\pi} - \frac{\hat{B}}{\varepsilon\gamma}\overline{\vartheta}\boldsymbol{k}$$
$$= -\mathrm{M}_\infty\overline{\omega}\left[\mathrm{S}\frac{D\boldsymbol{u}}{Dt} + \frac{1}{\mathrm{Ro}}(\boldsymbol{e} \times \boldsymbol{u}) - \frac{\hat{B}}{\varepsilon\gamma}\overline{\vartheta}\boldsymbol{k}\right] \quad ; \tag{8.47a}$$

---

9 It can always be assumed that $U_\infty^*(0)$ coincides with the reference velocity U_0 introduced in (2.44). U_∞^* is a velocity *with* dimensions according to our conventions given at the beginning of Chap. 7 and recalled again in (8.33).

$$\nabla \cdot \boldsymbol{u} = -\mathrm{M}_\infty \left[\mathrm{S}\frac{D\overline{\omega}}{Dt} - \overline{\omega}(\nabla \cdot \boldsymbol{u}) - \left(\hat{B} + \frac{dT_\infty(\hat{B}\zeta)}{d\zeta} \right) \frac{w}{T_\infty(\hat{B}\zeta)} \right]$$

$$+ \mathrm{M}_\infty^2 \overline{\omega} \left(\hat{B} + \frac{dT_\infty(\hat{B}\zeta)}{d\zeta} \right) \frac{w}{T_\infty(\hat{B}\zeta)} \quad ; \tag{8.47b}$$

$$\overline{\omega} + \overline{\vartheta} = \mathrm{M}_\infty (\overline{\pi} - \overline{\omega}\overline{\vartheta}) \quad ; \tag{8.47c}$$

$$\mathrm{S}\frac{D\overline{\vartheta}}{Dt} + \left[\hat{B}\frac{\gamma - 1}{\gamma} + \frac{dT_\infty(\hat{B}\zeta)}{d\zeta} \right] \frac{w}{T_\infty(\hat{B}\zeta)}$$

$$= -\mathrm{M}_\infty \left[\overline{\omega}\mathrm{S}\frac{D\overline{\vartheta}}{Dt} - \frac{\gamma - 1}{\gamma}\mathrm{S}\frac{D\overline{\pi}}{Dt} \right]$$

$$- \mathrm{M}_\infty^2 \overline{\pi} \left[\hat{B}\frac{\gamma - 1}{\gamma} + \frac{dT_\infty(\hat{B}\zeta)}{d\zeta} \right] \frac{w}{T_\infty(\hat{B}\zeta)} \quad . \tag{8.47d}$$

Equations (8.47) "generalize" the Boussinesq equations in as much as they remain valid throughout the entire adiabatic atmosphere but on the condition that the "fast" vertical variable z and the "slow" vertical variable $\zeta = \mathrm{M}_\infty z$ be considered simultaneously.

When

$$\mathrm{M}_\infty \to 0 \quad , \quad \text{at} \quad t, x, y, z \quad \text{fixed} \quad , \tag{8.48}$$

the classical Boussinesq equations in an adiabatic atmosphere again result. Their generalized form (8.47) can be useful when considering the propagation of periodic waves in the adiabatic atmosphere. This form necessarily allows certain results of Bois (1976) to be generalized in the case where the Coriolis force ($\mathrm{Ro} \neq \infty$) plays as significant a role as the stratification. This matter will come up again in Sect. 8.4.

8.3.3 Quasi-static Boussinesq Equations; the Problem of Meso-scale Circulations

As a matter of fact, when the boundary condition (8.21) is imposed on (8.5–8), an intrinsic vertical scale appears which is:

$$h_0 = \frac{R\Delta T_0}{g} \ll \frac{RT_\infty(0)}{g} \quad , \tag{8.49}$$

given the fact that the very physics of the atmospheric problems considered implies that

$$\Delta T_0 \ll T_\infty(0) \Rightarrow \tau_0 \ll 1 \quad .$$

Therefore, if during the non-dimensionalization of the equations of the atmosphere, h_0 is chosen instead of H as the vertical length scale, it is observed that:

$$\mathrm{Bo} = \frac{h_0}{RT_\infty(0)/g} \equiv \tau_0 \ll 1 \quad . \tag{8.50}$$

In this case, it also follows that:

$$\hat{B} \equiv \frac{\Delta T_0}{U_0}\sqrt{\frac{\gamma R}{T_\infty(0)}} \quad . \tag{8.51}$$

In addition, it must be imagined that in the Boussinesq equations (8.13–16), we have:

$$\varepsilon = \frac{h_0}{L} \equiv \frac{R\Delta T_0}{gL} \tag{8.52}$$

and that instead of the condition (8.25), we must write:

$$\vartheta_1 = \hat{B}\Xi(t, x, y) \quad , \quad \text{on} \quad z = 0 \quad . \tag{8.53}$$

In this form, the Boussinesq equations are well-adapted to describe circulation phenomena above a (flat) site which is thermically non-homogeneous [simulated by the boundary condition (8.53)]. Directly related to this problem, two different cases must be considered:

(i)

$$\varepsilon \cong 1 \Rightarrow L \cong h_0 = \frac{R\Delta T_0}{g} \cong 10^3\,\mathrm{m} \quad ,$$

$$\mathrm{Ro} \cong \frac{U_0 g}{l_0 R\Delta T_0} \cong 10^2 \gg 1 \quad ;$$

(ii)

$$\varepsilon \ll 1 \Rightarrow L \gg h_0 \quad .$$

In the first case (i), the terms associated with the Coriolis force proportional to 1/Ro can be neglected in the Boussinesq equations. On the contrary, the terms associated with 1/Ro must be retained in the second case (ii) which implies that $L \cong 10^5$ m; the quasi-static approximation discussed in Chap. 7 can, however, be carried out.

If, in fact, the quasi-static approximation is effected in the Boussinesq equations, then it must be kept in mind that the similarity relation (7.10) implies the constraint:

$$h_0 = \mathrm{Re}_\perp^{1/2}\frac{L}{\mathrm{Re}^{1/2}} = \mathrm{Re}_\perp^{1/2}\sqrt{\frac{L\nu_0}{U_0}} \quad ,$$

with $\mathrm{Re}_\perp \equiv h_0 W_0/\nu_0$. However, according to (8.49), which yields h_0, the following expression results for L:

$$L = \frac{1}{\mathrm{Re}_\perp}\frac{U_0}{g^2}\frac{(R\Delta T_0)^2}{\nu_0} \cong 10^5\,\mathrm{m} \quad , \tag{8.54}$$

which confirms that in the case (ii), the terms associated with 1/Ro must indeed be retained. Thus, on the condition (8.54), the following *quasi-static Boussinesq* equations can be written:

$$\mathrm{S}\frac{\partial \boldsymbol{v}_{00}}{\partial t} + (\boldsymbol{v}_{00} \cdot \boldsymbol{D})\boldsymbol{v}_{00} + w_{00}\frac{\partial \boldsymbol{v}_{00}}{\partial z} + \frac{1}{\mathrm{Ro}}\frac{\sin\varphi}{\sin\varphi_0}(\boldsymbol{k} \times \boldsymbol{v}_{00}) + \frac{1}{\gamma}\boldsymbol{D}\pi_{20} = \frac{1}{\mathrm{Re}_\perp}\frac{\partial^2 \boldsymbol{v}_{00}}{\partial z^2} \quad ;$$

$$\frac{\partial \pi_{20}}{\partial z} = \hat{B}\vartheta_{10} \quad ; \tag{8.55}$$

$$\boldsymbol{D} \cdot \boldsymbol{v}_{00} + \frac{\partial w_{00}}{\partial z} = 0 \quad ;$$

$$\mathrm{S}\frac{\partial \vartheta_{10}}{\partial t} + \boldsymbol{v}_{00} \cdot \boldsymbol{D}\vartheta_{10} + w_{00}\frac{\partial \vartheta_{10}}{\partial z} + \hat{B}\left[\frac{\gamma - 1}{\gamma} + \left(\frac{dT_\infty}{dz_\infty}\right)_0\right] w_{00} = \frac{1}{\mathrm{Pr}}\frac{1}{\mathrm{Re}_\perp}\frac{\partial^2 \vartheta_{10}}{\partial z^2} \quad .$$

The following boundary conditions must be superimposed on system (8.55) at the ground (supposed flat):

$$z = 0: \quad \boldsymbol{v}_{00} = 0 \quad , \quad w_{00} = 0 \quad , \quad \vartheta_{10} = \hat{B}\Xi(t, x, y) \quad . \tag{8.56}$$

Since the phenomenon of (free) circulation is independent of outside fields, we can impose the following:

$$\boldsymbol{v}_{00} = w_{00} = \pi_{20} = \vartheta_{10} \Rightarrow 0 \quad , \quad \text{with} \quad |x^2 + y^2| \to \infty \quad . \tag{8.57}$$

Finally, beyond the convective layer governed by (8.55), we have in first approximation the state of rest which means that the following behavior conditions must be imposed on system (8.55):

$$\boldsymbol{v}_{00} = w_{00} = \pi_{20} = \vartheta_{10} \Rightarrow 0 \quad , \quad \text{when} \quad z \to +\infty \quad . \tag{8.58}$$

In connection with (8.58), it must be emphasized that the order of system (8.55) with respect to z does not, in fact, allow the imposing of any condition on w_{00} when $z \to +\infty$. The equation for ϑ_{10}, however, shows that if:

$$\left(\frac{dT_\infty}{dz_\infty}\right)_0 \not\equiv \frac{\gamma - 1}{\gamma} \quad , \tag{8.59}$$

then it is absolutely necessary that:

$$\lim_{z \to +\infty} \vartheta_{10} = 0 \Rightarrow w_{00} = 0 \quad , \quad \text{for} \quad z = \infty \quad . \tag{8.60}$$

It must be clearly understood that the Boussinesq-type system (8.55) is not the same as the classical Prandtl boundary-layer system. The problem (8.55–58) with adequate initial conditions is one which describes in mesometeorology the circulations above a flat site without relief but having some thermal heterogeneities – these are the so-called breeze phenomena which are characterized by contrasts in ground temperatures. In particular, the presence of the term $\hat{B}[(\gamma-1)/\gamma+(dT_\infty/dz_\infty)_0]w_{00}$ in the equation for ϑ_{10}, brings about the formation of a compensating breeze (antibreeze) above the main breeze when this term is positive. One of the difficulties encountered in solving this breeze problem is linked to the presence of this term. However, only the "correct" taking into account of the latter can lead to a solution of the breeze problem which satisfies the constraint (8.60). On this topic, the reader might consult Chap. 7 in the book by Gutman (1969).

It will be remarked that thanks to (8.58), the first equation of system (8.55) can be rewritten in the following form:

$$\mathrm{S}\frac{\partial \boldsymbol{v}_{00}}{\partial t}+(\boldsymbol{v}_{00}\cdot \boldsymbol{D})\boldsymbol{v}_{00}+w_{00}\frac{\partial \boldsymbol{v}_{00}}{\partial z}+\frac{1}{\mathrm{Ro}}\frac{\sin\varphi}{\sin\varphi_0}(\boldsymbol{k}\times \boldsymbol{v}_{00})$$
$$+\frac{\hat{B}}{\gamma}\int_\infty^z \boldsymbol{D}\vartheta_{10}\,dz=\frac{1}{\mathrm{Re}_\perp}\frac{\partial^2 \boldsymbol{v}_{00}}{\partial z^2} \quad . \tag{8.61}$$

The second equation of (8.55) then permits the determining of π_{20}. We will return to this circulation problem in Chap. 13. As concerns the initial conditions, a special analysis is called for which should convince us that the state of rest is the right condition to prescribe:

$$t=0:\quad \boldsymbol{v}_{00}=0 \Rightarrow w_{00}=0 \quad , \quad \vartheta_{10}=0 \Rightarrow \pi_{20}=0 \quad . \tag{8.62}$$

It is pointed out that if the Strouhal number S must remain of the order unity, then $t_0 \cong L/U_0$ which shows that starting from the state of rest, system (8.55) makes it possible to analyze the evolution and the formation of the breeze during a period of several hours (as a function of the value of U_0 which is a few meters/second). Finally, the characteristic velocity U_0 remains to be defined. To this end, we can define an intrinsic characteristic velocity:

$$U_0=\tau_0\sqrt{gh_0} \quad , \tag{8.63}$$

and in this case, we can replace the Reynolds number Re in (8.5) and (8.7) by the Grashof number (see Chap. 13):

$$\mathrm{Gr}\equiv \mathrm{Re}^2=\varepsilon^2\frac{gh_0^3}{\nu_0^2}\tau_0^2 \quad . \tag{8.64}$$

8.4 Concerning a Linear Theory of the Boussinesq Waves ($\mathrm{Ro}\not\equiv\infty$)

Let us go back to (8.47) and assume that in the expression (7.5) of $\boldsymbol{e}$, we introduced $\varphi\equiv\varphi_0$ in such a way that $\boldsymbol{e}=\boldsymbol{k}+\operatorname{cotg}\varphi_0\boldsymbol{j}$. In the forthcoming, we will take into consideration the expressions (7.6) of $\boldsymbol{u}$, $\boldsymbol{\nabla}$ and $\mathrm{S}(D/Dt)$.

The fundamental hypothesis is that $\mathrm{M}_\infty \ll 1$,

$$\frac{\zeta}{z}\equiv \mathrm{M}_\infty \ll 1 \quad . \tag{8.65}$$

We note that if $\varepsilon = \mathrm{O}(1)$, then in fact $\mathrm{Ro} \gg 1$. On the other hand, when $\varepsilon \ll 1$, then the hypothesis that $\mathrm{Ro} = \mathrm{O}(1)$ can be made on the condition that $L \cong U_0/l_0 \cong 10^5\,\mathrm{m}$. If $L \gg U_0/l_0$ of the order of $10^6\,\mathrm{m}$, then $\mathrm{Ro} \ll 1$ with $\varepsilon \ll 1$, but

$$1 \gg \varepsilon \ll \mathrm{Ro} \ll 1 \quad .$$

Hypothesis (8.65) leads to the using of a multiple sale method (MSM) which is suggested by the two scales ζ and z which come into play simultaneously in (8.47)[10]. Nonetheless, experience has shown that a priori, z cannot be taken as a fast variable which generally speaking, must depend non-linearly on the slow variable. We will thus set:

$$\zeta = \mathrm{M}_\infty z \quad , \quad \xi = \frac{1}{\mathrm{M}_\infty}\varphi(\zeta) \quad , \tag{8.66}$$

where the function $\varphi(\zeta)$ will have to be determined during the construction of the double scale solution.

Let us designate by u^*, v^*, w^*, $\overline{\pi}^*$, $\overline{\omega}^*$ and $\overline{\vartheta}^*$ the unknowns of our problem supposed as functions of t, x, y, ζ, ξ and M_∞ in such a way that:

$$\frac{\partial}{\partial z} = \frac{d\varphi}{d\zeta}\frac{\partial}{\partial \xi} + \mathrm{M}_\infty \frac{\partial}{\partial \zeta} \quad . \tag{8.67}$$

Since we are here limiting our study to linear theory, we write the following expansions which are assumed uniformly valid:

$$\begin{pmatrix} u^* \\ v^* \\ w^* \\ \overline{\pi}^* \\ \overline{\omega}^* \\ \overline{\vartheta}^* \end{pmatrix} \equiv \mathcal{U}^* = \eta\left\{\mathcal{U}^*_{00} + \mathrm{M}_\infty \mathcal{U}^*_{01} + \mathrm{O}(\eta)\right\} \quad , \tag{8.68}$$

where $\mathcal{U}^*_{00}$ and $\mathcal{U}^*_{01}$ are functions of t, x, y, ζ and ξ. In (8.68), the second small parameter has been designated by η which characterizes the amplitude (supposed small) of the Boussinesq waves.

If (8.68) is substituted into (8.47), while taking into account (8.67) and if the terms proportional first to η, and then to $\eta\mathrm{M}_\infty$ are identified, two limiting systems of equations are obtained with respect to $\mathcal{U}^*_{00}$ and $\mathcal{U}^*_{01}$. It is pointed out that for the following linear study, it is not necessary to write the terms [in (8.68)] whose order in M_∞ is greater than one; the same holds for terms of order η^2 and $\eta^2\mathrm{M}_\infty$.

To order (0,0), the limiting equations for u^*_{00}, v^*_{00}, w^*_{00}, $\overline{\pi}^*_{00}$, $\overline{\vartheta}^*_{00}$ and $\overline{\omega}^*_{00}$ take the form:

[10] It must be kept in mind that according to the expressions (7.6), we have:

$$\mathrm{S}\frac{D}{Dt} \equiv \mathrm{S}\frac{\partial}{\partial t} + u\frac{\partial}{\partial x} + v\frac{\partial}{\partial y} + w\frac{\partial}{\partial z} \quad ,$$

with u, v and w the components of the relative velocity $\boldsymbol{u}$ in the direction of the axes P_0x, P_0y and P_0z.

$$\mathrm{S}\frac{\partial u_{00}^*}{\partial t} - \frac{1}{\mathrm{Ro}}v_{00}^* + \frac{\varepsilon}{\mathrm{Ro}}\operatorname{cotg}\varphi_0 w_{00}^* + \frac{T_\infty(\hat{B}\zeta)}{\gamma}\frac{\partial \pi_{00}^*}{\partial x} = 0 \quad ; \tag{8.69a}$$

$$\mathrm{S}\frac{\partial v_{00}^*}{\partial t} + \frac{1}{\mathrm{Ro}}u_{00}^* + \frac{T_\infty(\hat{B}\zeta)}{\gamma}\frac{\partial \pi_{00}^*}{\partial y} = 0 \quad ; \tag{8.69b}$$

$$\varepsilon^2\mathrm{S}\frac{\partial w_{00}^*}{\partial t} - \frac{\varepsilon}{\mathrm{Ro}}\operatorname{cotg}\varphi_0 u_{00}^* + \frac{T_\infty(\hat{B}\zeta)}{\gamma}\frac{d\varphi}{d\zeta}\frac{\partial \pi_{00}^*}{\partial \xi} - \frac{\hat{B}}{\gamma}\overline{\vartheta}_{00}^* = 0 \quad ; \tag{8.69c}$$

$$\frac{\partial u_{00}^*}{\partial x} + \frac{\partial v_{00}^*}{\partial y} + \frac{d\varphi}{d\zeta}\frac{\partial w_{00}^*}{\partial \xi} = 0 \quad , \tag{8.69d}$$

$$\mathrm{S}\frac{\partial \overline{\vartheta}_{00}^*}{\partial t} + \left[\hat{B}\frac{\gamma-1}{\gamma} + \frac{dT_\infty(\hat{B}\zeta)}{d\zeta}\right]\frac{w_{00}^*}{T_\infty(\hat{B}\zeta)} = 0 \quad ; \tag{8.69e}$$

$$\overline{\omega}_{00}^* = -\overline{\vartheta}_{00}^* \quad . \tag{8.69f}$$

The solution to (8.69) can be sought in the form of a progressive wave in a horizontal plane, i.e.:

$$\mathcal{U}_{00}^* = \tilde{\mathcal{U}}_{00}^*(\xi,\zeta)\exp(\mathrm{i}\theta) \quad , \tag{8.70}$$

with[11]

$$\theta = \sigma\frac{t}{\mathrm{S}} - kx - ly \quad ; \quad (\mathrm{i} = \sqrt{-1}) \quad . \tag{8.71}$$

For $\tilde{u}_{00}^*$, $\tilde{v}_{00}^*$, $\tilde{w}_{00}^*$, $\tilde{\overline{\pi}}_{00}^*$, $\tilde{\overline{\vartheta}}_{00}^*$ which depend solely on ξ and on ζ, the following equations then replace (8.69):

$$\begin{aligned}
&\mathrm{i}\sigma\tilde{u}_{00}^* - \frac{1}{\mathrm{Ro}}\tilde{v}_{00}^* + \frac{\varepsilon}{\mathrm{Ro}}\operatorname{cotg}\varphi_0\tilde{w}_{00}^* - \mathrm{i}\frac{k}{\gamma}T_\infty(\hat{B}\zeta)\tilde{\overline{\pi}}_{00}^* = 0 \quad ;\\
&\mathrm{i}\sigma\tilde{v}_{00}^* + \frac{1}{\mathrm{Ro}}\tilde{u}_{00}^* - \mathrm{i}\frac{l}{\gamma}T_\infty(\hat{B}\zeta)\tilde{\overline{\pi}}_{00}^* = 0;\\
&\varepsilon^2\mathrm{i}\sigma\tilde{w}_{00}^* - \frac{\varepsilon}{\mathrm{Ro}}\operatorname{cotg}\varphi_0\tilde{u}_{00}^* + \frac{T_\infty(\hat{B}\zeta)}{\gamma}\frac{d\varphi}{d\zeta}\frac{\partial\tilde{\overline{\pi}}_{00}^*}{\partial\xi} = \frac{\hat{B}}{\gamma}\tilde{\overline{\vartheta}}_{00}^* \quad ;\\
&-\mathrm{i}k\tilde{u}_{00}^* - \mathrm{i}l\tilde{v}_{00}^* + \frac{d\varphi}{d\zeta}\frac{\partial\tilde{w}_{00}^*}{\partial\xi} = 0 \quad ;\\
&\mathrm{i}\sigma\tilde{\overline{\vartheta}}_{00}^* + \left[\hat{B}\frac{\gamma-1}{\gamma} + \frac{dT_\infty(\hat{B}\zeta)}{d\zeta}\right]\frac{1}{T_\infty(\hat{B}\zeta)}\tilde{w}_{00}^* = 0 \quad .
\end{aligned} \tag{8.72}$$

First of all, from the first two equations of (8.72), the following is determined:

[11] Since we are taking the Coriolis force (Ro $\not\equiv \infty$) into account, we can no longer (like Bois, 1976) consider a two-dimensional flow with x and $z \Rightarrow (\xi,\zeta)$ as space variables.

$$\tilde{u}^*_{00} = \frac{1}{1/\mathrm{Ro}^2 - \sigma^2}\left\{\frac{T_\infty(\hat{B}\zeta)}{\gamma}\left(\mathrm{i}\frac{l}{\mathrm{Ro}} - \sigma k\right)\tilde{\bar{\pi}}^*_{00} - \mathrm{i}\frac{\sigma\varepsilon}{\mathrm{Ro}}\operatorname{cotg}\varphi_0\tilde{w}^*_{00}\right\} \quad ;$$

$$\tilde{v}^*_{00} = \frac{1}{1/\mathrm{Ro}^2 - \sigma^2}\left\{-\frac{T_\infty(\hat{B}\zeta)}{\gamma}\left(\mathrm{i}\frac{k}{\mathrm{Ro}} + \sigma l\right)\tilde{\bar{\pi}}^*_{00} + \frac{\varepsilon}{\mathrm{Ro}^2}\operatorname{cotg}\varphi_0\tilde{w}^*_{00}\right\} \quad , \tag{8.73}$$

as a function of $\tilde{\bar{\pi}}^*_{00}$ and $\tilde{w}^*_{00}$. Then the fifth equation of (8.72) determines:

$$\tilde{\bar{\vartheta}}^*_{00} = \frac{\mathrm{i}}{\sigma}\left(\hat{B}\frac{\gamma-1}{\gamma} + \frac{dT_\infty(\hat{B}\zeta)}{d\zeta}\right)\frac{1}{T_\infty(\hat{B}\zeta)}\tilde{w}^*_{00} \quad , \tag{8.74}$$

as a function of $\tilde{w}^*_{00}$.

By making use of (8.73, 74), instead of the fourth and third equations, we have respectively:

$$\left(\sigma^2 - \frac{1}{\mathrm{Ro}^2}\right)\frac{d\varphi}{d\zeta}\frac{\partial\tilde{w}^*_{00}}{\partial\xi} + \frac{\varepsilon}{\mathrm{Ro}}\operatorname{cotg}\varphi_0\left(\sigma k + \mathrm{i}\frac{l}{\mathrm{Ro}}\right)\tilde{w}^*_{00}$$
$$= \mathrm{i}\frac{\sigma}{\gamma}(k^2 + l^2)T_\infty(\hat{B}\zeta)\tilde{\bar{\pi}}^*_{00} \quad ; \tag{8.75}$$

$$\left(\sigma^2 - \frac{1}{\mathrm{Ro}^2}\right)\frac{T_\infty(\hat{B}\zeta)}{\gamma}\frac{d\varphi}{d\zeta}\frac{\partial\tilde{\bar{\pi}}^*_{00}}{\partial\xi} + \frac{\varepsilon}{\mathrm{Ro}}\operatorname{cotg}\varphi_0\frac{T_\infty(\hat{B}\zeta)}{\gamma}\left(\mathrm{i}\frac{l}{\mathrm{Ro}} - \sigma k\right)\tilde{\bar{\pi}}^*_{00}$$
$$= \mathrm{i}\left\{\left(\sigma^2 - \frac{1}{\mathrm{Ro}^2}\right)\frac{\Gamma_\infty(\zeta) - \varepsilon^2\sigma^2}{\sigma} + \frac{\varepsilon^2\sigma}{\mathrm{Ro}^2}\operatorname{cotg}^2\varphi_0\right\}\tilde{w}^*_{00} \quad , \tag{8.76}$$

where

$$\Gamma_\infty(\zeta) = \frac{\hat{B}}{\gamma}\left[\hat{B}\frac{\gamma-1}{\gamma} + \frac{dT_\infty(\hat{B}\zeta)}{d\zeta}\right]\frac{1}{T_\infty(\hat{B}\zeta)} \quad . \tag{8.77}$$

By eliminating $\tilde{\bar{\pi}}^*_{00}$ from (8.76) and (8.75), the following equation[12] is obtained for $\tilde{w}^*_{00}$

$$\left(\sigma^2 - \frac{1}{\mathrm{Ro}^2}\right)\left(\frac{d\varphi}{d\zeta}\right)^2\frac{\partial^2\tilde{w}^*_{00}}{\partial\xi^2} + 2\mathrm{i}\frac{l\varepsilon}{\mathrm{Ro}^2}\operatorname{cotg}\varphi_0\frac{d\varphi}{d\zeta}\frac{\partial\tilde{w}^*_{00}}{\partial\xi}$$
$$+\left\{(k^2 + l^2)\left[\Gamma_\infty(\zeta) - \varepsilon^2\sigma^2\right] + \frac{l^2\varepsilon^2}{\mathrm{Ro}^2}\operatorname{cotg}^2\varphi_0\right\}\tilde{w}^*_{00} = 0 \quad . \tag{8.78}$$

The case considered by Bois (1976) corresponds to:

$$l \equiv 0 \quad , \quad \varepsilon \equiv 1 \quad ; \quad \mathrm{Ro} \equiv \infty \Rightarrow L \cong H = 10^4\,\mathrm{m} \quad ,$$

and here, the following replaces (8.78):

$$\sigma^2\left(\frac{d\varphi}{d\zeta}\right)^2\frac{\partial^2\tilde{w}^*_{00}}{\partial\xi^2} + k^2\left[\Gamma_\infty(\zeta) - \sigma^2\right]\tilde{w}^*_{00} = 0 \quad , \tag{8.79}$$

---

[12] In all that follows, it is supposed that $\sigma^2 \not\equiv \mathrm{Ro}^{-2}$, or more precisely that $\sigma^2 > 1/\mathrm{Ro}^2$. To take into account the case where σ^2 is close to $1/\mathrm{Ro}^2$, another analysis would be necessary.

which is, in fact, (6.31).

Another interesting case is associated with:

$$\varepsilon \to 0 \quad , \quad \text{Ro} = O(1) \Rightarrow L \cong 10^5\,\text{m} \quad ,$$

which leads to the equation:

$$\left(\frac{d\varphi}{d\zeta}\right)^2 \frac{\partial^2 \tilde{w}_{00}^*}{\partial \xi^2} + \frac{k^2 + l^2}{\sigma^2 - 1/\text{Ro}^2} \Gamma_\infty(\zeta) \tilde{w}_{00}^* = 0 \quad . \tag{8.80}$$

The fundamental difference between (8.79) and (8.80) is connected to the cut-off phenomenon which is absent in (8.80)[13]. This phenomenon is present when at a given point $\zeta = \zeta_0$, we have:

$$\Gamma_\infty(\zeta_0) = \sigma^2 \quad , \tag{8.81}$$

which leads to a turning point problem [see, for example, Nayfeh (1973)]. In (8.79), the solution is oscillatory when $\Gamma_\infty(\zeta) > \sigma^2$, and non-periodic when $\Gamma_\infty(\zeta) < \sigma^2$. The problem thus arises of knowing how the transition is effected between these two types of solutions valid on either side of the point $\zeta = \zeta_0$ such that (8.81) is verified.

In (8.79), it is necessary to introduce the wave number ω_B defined by:

$$\omega_B^2 = \frac{k^2}{(d\varphi/d\zeta)^2} \left\{ \frac{\Gamma_\infty(\zeta)}{\sigma^2} - 1 \right\} \quad , \tag{8.82}$$

whereas in (8.80), the wave number is ω_c which is defined by:

$$\omega_c^2 = \frac{k^2 + l^2}{(d\varphi/d\zeta)^2} \left[\frac{\text{Ro}^2 \Gamma_\infty(\zeta)}{(\sigma\text{Ro})^2 - 1} \right] \quad . \tag{8.83}$$

In the general case associated with (8.78), the latter must return to a canonical form which does not contain a first derivative in ξ in order for the wave number ω_g to be introduced. In order to accomplish this, we introduce the new function $\tilde{W}_{00}^*(\xi, \zeta)$ such that

$$\tilde{W}_{00}^* = \exp\left\{ \mathrm{i} \frac{l\varepsilon \operatorname{cotg} \varphi_0}{(\sigma^2\text{Ro}^2 - 1) d\varphi/d\zeta} \xi \right\} \tilde{w}_{00}^* \quad . \tag{8.84}$$

The following equation then results for $\tilde{W}_{00}^*$

$$\frac{\partial^2 \tilde{W}_{00}^*}{\partial \xi^2} + \omega_g^2 \tilde{W}_{00}^* = 0 \quad , \tag{8.85}$$

where

---

[13] It is pointed out that if at a given point $\zeta = \zeta_a$, we have:

$$-dT_\infty(\hat{B}\zeta)/d\zeta \equiv \hat{B}\frac{\gamma - 1}{\gamma} \quad ,$$

then there is also a cut-off phenomenon in (8.80) which is related to the dry adiabatic temperature gradient.

$$\omega_g^2 = (k^2 + l^2)\frac{\Gamma_\infty(\zeta) - \varepsilon^2\sigma^2}{(\sigma^2 - 1/\mathrm{Ro}^2)(d\varphi/d\zeta)^2} + \frac{l^2\varepsilon^2 \cot g^2 \varphi_0}{(d\varphi/d\zeta)^2(\sigma^2\mathrm{Ro}^2 - 1)}\left[1 + \frac{1}{\sigma^2\mathrm{Ro}^2 - 1}\right] \quad . \tag{8.86}$$

The observation that ω_g can arbitrarily be chosen constant [otherwise, in (8.66), it would be enough to simply replace ξ by $\tilde{\xi} = \omega_g\xi$) shows that ω_g can be taken equal to 1. Therefore, (8.86) is the dispersion equation of (8.85) with the following for solution:

$$\left(\frac{d\varphi}{d\zeta}\right)^2 = (k^2 + l^2)\frac{\Gamma_\infty(\zeta) - \varepsilon^2\sigma^2}{\sigma^2 - 1/\mathrm{Ro}^2} + \frac{l^2\varepsilon^2 \cot g^2 \varphi_0}{(\sigma^2 - 1/\mathrm{Ro}^2)\mathrm{Ro}^2}\left[1 + \frac{1}{\sigma^2\mathrm{Ro}^2 - 1}\right] \quad . \tag{8.87}$$

Formula (8.87) demonstrates just how the fast variable $\xi = \varphi(\zeta)/\mathrm{M}_\infty$ depends on the slow variable $\zeta = \mathrm{M}_\infty z$. When $\mathrm{Ro} \equiv \infty$, $\varepsilon \equiv 1$ and $l = 0$, (8.87) reduces to the classical formula:

$$\left(\frac{d\varphi}{d\zeta}\right)^2 = k^2\left[\frac{\Gamma_\infty(\zeta)}{\sigma^2} - 1\right] \quad . \tag{8.88}$$

In the general case, it can be seen from (8.87) that the cut-off phenomenon appears when at the point ζ_c, we have

$$\Gamma_\infty(\zeta_c) = \varepsilon^2\sigma^2 - \frac{l^2\varepsilon^2 \cot g^2 \varphi_0}{(k^2 + l^2)\mathrm{Ro}^2}\left[1 + \frac{1}{\sigma^2\mathrm{Ro}^2 - 1}\right] \quad . \tag{8.89}$$

In the vicinity of this point$\zeta = \zeta_c$, the function $\Gamma_\infty(\zeta)$ is written in the following form:

$$\Gamma_\infty(\zeta) = \varepsilon^2\sigma^2 - \frac{l^2\varepsilon^2 \cot g^2 \varphi_0}{(k^2 + l^2)\mathrm{Ro}^2}\left[1 + \frac{1}{\sigma^2\mathrm{Ro}^2 - 1}\right] + (\zeta - \zeta_c)\left.\frac{d\Gamma_\infty(\zeta)}{d\zeta}\right|_{\zeta=\zeta_c} \quad . \tag{8.90}$$

To form a clearer idea, we can suppose that $(d\Gamma_\infty(\zeta)/d\zeta)_{\zeta=\zeta_c} < 0$.

But let us return to (8.85) for the moment. When $\zeta < \zeta_c$, this equation has an oscillatory solution of the form:

$$\tilde{W}_{00}^* = \mathcal{A}_{00}(\zeta)\,\mathrm{e}^{\mathrm{i}\xi} + \mathcal{B}_{00}(\zeta)\,\mathrm{e}^{-\mathrm{i}\xi} \quad . \tag{8.91}$$

However, when $\zeta > \zeta_c$, the solution of (8.85) is:

$$\tilde{W}_{00}^* = \mathcal{C}_{00}(\zeta)\,\mathrm{e}^{\xi} + \mathcal{D}_{00}(\zeta)\,\mathrm{e}^{-\xi} \quad . \tag{8.92}$$

The functions $\mathcal{A}_{00}(\zeta)$, $\mathcal{B}_{00}(\zeta)$, $\mathcal{C}_{00}(\zeta)$ and $\mathcal{D}_{00}(\zeta)$ are undetermined at this stage. The manner in which they depend on ζ is obtained by eliminating from the limiting equations of order (0,1), the terms which give rise to secular terms.

To order (0,1), the limiting equations for u^*_{01}, v^*_{01}, w^*_{01}, $\overline{\pi}^*_{01}$, $\overline{\vartheta}^*_{01}$, and $\overline{\omega}^*_{01}$ are:

$$\mathrm{S}\frac{\partial u^*_{01}}{\partial t} - \frac{1}{\mathrm{Ro}}v^*_{01} + \frac{\varepsilon}{\mathrm{Ro}}\operatorname{cotg}\varphi_0 w^*_{01} + \frac{T_\infty(\hat{B}\zeta)}{\gamma}\frac{\partial \overline{\pi}^*_{01}}{\partial x} = 0 \quad ;$$

$$\mathrm{S}\frac{\partial v^*_{01}}{\partial t} + \frac{1}{\mathrm{Ro}}u^*_{01} + \frac{T_\infty(\hat{B}\zeta)}{\gamma}\frac{\partial \overline{\pi}^*_{01}}{\partial y} = 0 \quad ;$$

$$\varepsilon^2\mathrm{S}\frac{\partial w^*_{01}}{\partial t} - \frac{\varepsilon}{\mathrm{Ro}}\operatorname{cotg}\varphi_0 u^*_{01} + \frac{T_\infty(\hat{B}\zeta)}{\gamma}\frac{d\varphi}{d\zeta}\frac{\partial \overline{\pi}^*_{01}}{\partial \xi} - \frac{\hat{B}}{\gamma}\overline{\vartheta}^*_{01} = -\frac{T_\infty(\hat{B}\zeta)}{\gamma}\frac{\partial \overline{\pi}^*_{00}}{\partial \zeta} \quad ; \tag{8.93}$$

$$\frac{\partial u^*_{01}}{\partial x} + \frac{\partial v^*_{01}}{\partial y} + \frac{d\varphi}{d\zeta}\frac{\partial w^*_{01}}{\partial \xi} = \left[\hat{B} + \frac{dT_\infty(\hat{B}\zeta)}{d\zeta}\right]\frac{1}{T_\infty(\hat{B}\zeta)}w^*_{00} - \left(\mathrm{S}\frac{\partial \overline{\omega}^*_{00}}{\partial t} + \frac{\partial w^*_{00}}{\partial \zeta}\right) \quad ;$$

$$\overline{\omega}^*_{01} + \overline{\vartheta}^*_{01} = \overline{\pi}^*_{00} \quad ;$$

$$\mathrm{S}\frac{\partial \overline{\vartheta}^*_{01}}{\partial t} + \left[\hat{B}\frac{\gamma - 1}{\gamma} + \frac{dT_\infty(\hat{B}\zeta)}{d\zeta}\right]\frac{1}{T_\infty(\hat{B}\zeta)}w^*_{01} = \frac{\gamma - 1}{\gamma}\mathrm{S}\frac{\partial \overline{\pi}^*_{00}}{\partial t}.$$

The solution to the limiting system (8.93) of order (0,1) is sought in a form analogous to (8.70):

$$\mathcal{U}^*_{01} = \tilde{\mathcal{U}}^*_{01}(\xi, \zeta)\exp\left[\mathrm{i}\left(\sigma\frac{t}{\mathrm{S}} - kx - ly\right)\right] \quad . \tag{8.94}$$

For the functions $\tilde{u}^*_{01}$, $\tilde{v}^*_{01}$, $\tilde{w}^*_{01}$, $\tilde{\overline{\pi}}^*_{01}$ and $\tilde{\overline{\vartheta}}^*_{01}$, there then results a system analogous to (8.72) but whose right hand members are respectively:

$$\left\{\begin{matrix} 0 \\ 0 \\ -\dfrac{T_\infty(\hat{B}\zeta)}{\gamma}\dfrac{\partial \tilde{\overline{\pi}}^*_{00}}{\partial \zeta} \\ \left(\hat{B} + \dfrac{dT_\infty(\hat{B}\zeta)}{d\zeta}\right)\dfrac{\tilde{w}^*_{00}}{T_\infty(\hat{B}\zeta)} - \mathrm{i}\sigma\tilde{\overline{\omega}}^*_{00} - \dfrac{\partial \tilde{w}^*_{00}}{\partial \zeta} \\ \mathrm{i}\sigma\dfrac{\gamma - 1}{\gamma}\tilde{\overline{\pi}}^*_{00} \end{matrix}\right\}$$

where $\tilde{\overline{\omega}}^*_{00} \equiv -\tilde{\overline{\vartheta}}^*_{00}$ and $\tilde{\overline{\omega}}^*_{01} = \tilde{\overline{\pi}}^*_{00} - \tilde{\overline{\vartheta}}^*_{01}$. Thus, the following replaces (8.75, 76):

$$\left(\sigma^2 - \frac{1}{\mathrm{Ro}^2}\right)\frac{d\varphi}{d\zeta}\frac{\partial \tilde{w}^*_{01}}{\partial \xi} + \frac{\varepsilon}{\mathrm{Ro}}\operatorname{cotg}\varphi_0\left(\sigma k + \mathrm{i}\frac{l}{\mathrm{Ro}}\right)\tilde{w}^*_{01} - \mathrm{i}\frac{\sigma}{\gamma}(k^2 + l^2)T_\infty(\hat{B}\zeta)\tilde{\overline{\pi}}^*_{01} = \left(\sigma^2 - \frac{1}{\mathrm{Ro}^2}\right)\tilde{\mathcal{G}}^*_{00} \quad ; \tag{8.95a}$$

$$\left(\sigma^2 - \frac{1}{\mathrm{Ro}^2}\right) \frac{T_\infty(\hat{B}\zeta)}{\gamma} \frac{d\varphi}{d\zeta} \frac{\partial \tilde{\bar{\pi}}^*_{01}}{\partial \xi} + \frac{\varepsilon}{\mathrm{Ro}} \operatorname{cotg} \varphi_0 \frac{T_\infty(\hat{B}\zeta)}{\gamma} \left(\frac{\mathrm{i}l}{\mathrm{Ro}} - \sigma k\right) \tilde{\bar{\pi}}^*_{01}$$
$$-\mathrm{i}\left\{\left(\sigma^2 - \frac{1}{\mathrm{Ro}^2}\right) \frac{\Gamma_\infty(\zeta) - \varepsilon^2\sigma^2}{\sigma} + \frac{\varepsilon^2\sigma}{\mathrm{Ro}^2} \operatorname{cotg}^2 \varphi_0\right\} \tilde{w}^*_{01}$$
$$= \left(\sigma^2 - \frac{1}{\mathrm{Ro}^2}\right) \tilde{f}^*_{00} \quad , \tag{8.95b}$$

where

$$\tilde{\mathcal{G}}^*_{00} = \mathrm{i}\sigma \tilde{\bar{\vartheta}}^*_{00} - \frac{\partial \tilde{w}^*_{00}}{\partial \zeta} + \left(\hat{B} + \frac{dT_\infty(\hat{B}\zeta)}{d\zeta}\right) \frac{\tilde{w}^*_{00}}{T_\infty(\hat{B}\zeta)} \quad ;$$
$$\tilde{f}^*_{00} = \frac{\hat{B}}{\gamma} \frac{\gamma - 1}{\gamma} \tilde{\bar{\pi}}^*_{00} - \frac{T_\infty(\hat{B}\zeta)}{\gamma} \frac{\partial \tilde{\bar{\pi}}^*_{00}}{\partial \zeta} \quad . \tag{8.96}$$

By eliminating once again $\tilde{\bar{\pi}}^*_{01}$ from (8.95), the following equation is obtained which is analogous to (8.78) but has a known right hand member:

$$\left(\sigma^2 - \frac{1}{\mathrm{Ro}^2}\right) \left(\frac{d\varphi}{d\zeta}\right)^2 \frac{\partial^2 \tilde{w}^*_{01}}{\partial \xi^2} + 2\mathrm{i} \frac{l\varepsilon}{\mathrm{Ro}^2} \operatorname{cotg} \varphi_0 \frac{d\varphi}{d\zeta} \frac{\partial \tilde{w}^*_{01}}{\partial \xi}$$
$$+ \left\{(k^2 + l^2)\left[\Gamma_\infty(\zeta) - \varepsilon^2\sigma^2\right] + \frac{l^2\varepsilon^2}{\mathrm{Ro}^2} \operatorname{cotg}^2 \varphi_0\right\} \tilde{w}^*_{01}$$
$$= \mathrm{i}\sigma(k^2 + l^2)\tilde{f}^*_{00} + \left[\left(\sigma^2 - \frac{1}{\mathrm{Ro}^2}\right) \frac{d\varphi}{d\zeta} \frac{\partial}{\partial \xi}\right.$$
$$\left. + \frac{\varepsilon}{\mathrm{Ro}} \operatorname{cotg} \varphi_0 \left(\mathrm{i}\frac{l}{\mathrm{Ro}} - \sigma k\right)\right] \tilde{\mathcal{G}}^*_{00} \quad . \tag{8.97}$$

It is now necessary to express the right hand member of (8.97) solely as a function of $\tilde{w}^*_{00}$ and then introduce a new unknown function:

$$\tilde{W}^*_{01} = \exp\left\{\mathrm{i} \frac{l\varepsilon \operatorname{cotg} \varphi_0}{(\sigma^2\mathrm{Ro}^2 - 1)d\varphi/d\zeta} \xi\right\} \tilde{w}^*_{01} \quad . \tag{8.98}$$

If (8.84) and (8.86) are now taken into account, the following non-homogeneous equation results for $\tilde{W}^*_{01}$:

$$\frac{\partial^2 \tilde{W}^*_{01}}{\partial \xi^2} + \tilde{W}^*_{01} = \frac{1}{(\sigma^2 - 1/\mathrm{Ro}^2)(d\varphi/d\zeta)^2} \left\{A \frac{\partial^2 \tilde{W}^*_{00}}{\partial \xi \partial \zeta}\right.$$
$$\left. + B \frac{\partial \tilde{W}^*_{00}}{\partial \zeta} + C \frac{\partial \tilde{W}^*_{00}}{\partial \xi} + D\tilde{W}^*_{00}\right\} \quad , \tag{8.99}$$

where

$$A = -2\left(\sigma^2 - \frac{1}{\mathrm{Ro}^2}\right) \frac{d\varphi}{d\zeta} \quad ;$$

$$B = -\frac{\varepsilon}{\mathrm{Ro}}\operatorname{cotg}\varphi_0\left(\sigma k + \frac{\mathrm{i}l}{\mathrm{Ro}}\right)\left\{1 + \frac{\hat{B}}{\gamma T_\infty(\hat{B}\zeta)}\frac{\mathrm{i}l/\mathrm{Ro} - \sigma k}{\mathrm{i}l/\mathrm{Ro} + \sigma k} - 2\frac{\mathrm{i}l/\mathrm{Ro}}{\sigma k + \mathrm{i}l/\mathrm{Ro}}\right\} \quad ;$$

$$C = \left(\sigma^2 - \frac{1}{\mathrm{Ro}^2}\right)\frac{d\varphi}{d\zeta}\left\{\left(\hat{B} + \frac{dT_\infty(\hat{B}\zeta)}{d\zeta}\right)\frac{1}{T_\infty(\hat{B}\zeta)} - \frac{d^2\varphi/d\zeta^2}{d\varphi/d\zeta}\right\} \quad ;$$

$$D = \frac{\varepsilon}{\mathrm{Ro}}\operatorname{cotg}\varphi_0\left(\sigma k + \frac{\mathrm{i}l}{\mathrm{Ro}}\right)\left\{\frac{\gamma}{\hat{B}}\Gamma_\infty(\zeta) + \frac{\hat{B}}{\gamma T_\infty(\hat{B}\zeta)}\frac{\mathrm{i}l/\mathrm{Ro} - \sigma k}{\mathrm{i}l/\mathrm{Ro} + \sigma k}\right.$$
$$\left. - \frac{\mathrm{i}l/\mathrm{Ro}}{\mathrm{i}l/\mathrm{Ro} + \sigma k}\left[\left(\hat{B} + \frac{dT_\infty(\hat{B}\zeta)}{d\zeta}\right)\frac{1}{T_\infty(\hat{B}\zeta)} - \frac{d^2\varphi/d\zeta^2}{d\varphi/d\zeta}\right]\right\} \quad .$$

It is now necessary to make use of the solution (8.91) for $\tilde{W}^*_{00}$ [or (8.92)] and, as explained in Sect. 6.2, eliminate the secular terms which will appear in the right-hand side of (8.99). After a standard calculation, we will obtain the following relation for determining the function $\mathcal{A}_{00}(\zeta)$:

$$-\frac{d\log\mathcal{A}_{00}(\zeta)}{d\zeta} = -\frac{\frac{\varepsilon}{\mathrm{Ro}}\operatorname{cotg}\varphi_0\left\{\frac{\gamma}{\hat{B}}\Gamma_\infty(\zeta)\frac{\sigma k + \mathrm{i}l/\mathrm{Ro}}{(k^2+l^2)\sigma} - \frac{\sigma k - \mathrm{i}l/\mathrm{Ro}}{(k^2+l^2)\sigma}\frac{\hat{B}}{\gamma T_\infty(\zeta)}\right\}}{2\left[\mathrm{i}\frac{l/\mathrm{Ro}}{\sigma(k^2+l^2)}\frac{\varepsilon\operatorname{cotg}\varphi_0}{\mathrm{Ro}} + \frac{\sigma^2 - 1/\mathrm{Ro}^2}{\sigma(k^2+l^2)}\frac{d\varphi}{d\zeta}\right]}$$
$$-\frac{\sigma^2 - 1/\mathrm{Ro}^2}{\sigma(k^2+l^2)}\frac{\left(\hat{B} + \frac{dT_\infty(\zeta)}{d\zeta}\right)\frac{1}{T_\infty(\zeta)}\frac{d\varphi}{d\zeta} - \frac{d^2\varphi}{d\zeta^2}}{2\left[\mathrm{i}\frac{l/\mathrm{Ro}}{\sigma(k^2+l^2)}\frac{\varepsilon\operatorname{cotg}\varphi_0}{\mathrm{Ro}} + \frac{\sigma^2 - 1/\mathrm{Ro}^2}{\sigma(k^2+l^2)}\frac{d\varphi}{d\zeta}\right]} \quad , \tag{8.100}$$

and an analogous equation for determining $\mathcal{B}_{00}(\zeta)$. When the standard temperature $T_\infty(\zeta)$ remains *bounded* with $\zeta \to +\infty$, it follows that $\mathcal{A}_{00}(\zeta)$ and $\mathcal{B}_{00}(\zeta)$ do not cancel out and remain well defined. In addition if $T_\infty(\zeta)$ is a *positive function* with continuous derivatives to the nth order whose behavior, for rather large ζ is of the form: ζ^α, $\alpha \geq 0$, then the functions $\mathcal{A}_{00}(\zeta)$ and $\mathcal{B}_{00}(\zeta)$ continue to be well-defined.

In conclusion, given the hypotheses concerning $T_\infty(\zeta)$ it is observed that any solution of (8.78) having the form:

$$\tilde{w}^*_{00} = \mathcal{A}_{00}(\zeta)\exp(\chi)\,; \quad \chi = \frac{\varphi(\zeta)}{\mathrm{M}_\infty}$$

with the phases:

$$\varphi_1(\zeta) = -\varphi_2(\zeta) = \mathrm{i}\frac{k^2+l^2}{\sigma}\int\sqrt{\Gamma_\infty(\zeta) - \sigma^2\varepsilon^2}\,d\zeta \quad ,$$

is damped when $\zeta \to +\infty$. Therefore, all Boussinesq waves are damped when ζ increases.

In the thesis[15] of M. Hasnaoui, the following case is studied:

$\varepsilon = \mathrm{O}(1)$ and $\mathrm{Ro} \gg 1$.

The above case study makes it possible to analyze the weak – but not nul-effect on the Coriolis force on the amplitude of the linear Boussinesq waves.

8.5 The Problem of Adjustment to the Boussinesq State

We will now return to the basic equations (8.5–8) and assume that in (8.5) and (8.7), Re $\equiv \infty$: this means that we are considering the Euler equations for an adiabatic atmosphere.

Let us see if a significant degeneracy of these Euler equations can be found which is related to the change in time scale:

$$t = \mathrm{M}_\infty^\alpha \hat{t} \quad , \tag{8.101}$$

with $\alpha > 0$ still to be determined, it being understood that the similarity relation (8.11) is supposed satisfied in such a way that:

$$\mathrm{Bo} \equiv \hat{B}\mathrm{M}_\infty \to 0 \quad , \quad \text{with} \quad \mathrm{M}_\infty \to 0 \quad ,$$
$$x, y, z \quad \text{and} \quad \hat{t} \quad \text{remaining fixed} \quad . \tag{8.102}$$

To the above (8.102), the following change in functions is assigned:

$$\boldsymbol{u} = \hat{\boldsymbol{u}} \quad , \quad \pi = \mathrm{M}_\infty^a \hat{\pi} \quad , \quad \omega = \mathrm{M}_\infty^b \hat{\omega} \quad \text{and} \quad \vartheta = \mathrm{M}_\infty^c \hat{\vartheta} \quad ,$$

where $\hat{\boldsymbol{u}}$, $\hat{\pi}$, $\hat{\omega}$ and $\hat{\vartheta}$ are functions of x, y, z and $\hat{t}$.

It is now easy to demonstrate that with the limiting process (8.102), there is just one possible significant degeneracy which corresponds to:

$$\alpha = a = b = c = 1 \quad . \tag{8.103}$$

As a consequence of (8.101–103), the following adjustment equations are obtained:

$$\begin{aligned} &\mathrm{S}\frac{\partial \hat{\boldsymbol{u}}_0}{\partial \hat{t}} + \frac{1}{\gamma}\nabla\hat{\pi}_1 = 0 \quad , \quad \mathrm{S}\frac{\partial \hat{\omega}_1}{\partial \hat{t}} + \nabla \cdot \hat{\boldsymbol{u}}_0 = 0 \\ &\mathrm{S}\frac{\partial}{\partial \hat{t}}\left(\hat{\vartheta}_1 - \frac{\gamma - 1}{\gamma}\hat{\pi}_1\right) = 0 \quad , \quad \hat{\pi}_1 = \hat{\omega}_1 + \hat{\vartheta}_1 \quad . \end{aligned} \tag{8.104}$$

The following (flat earth) slip condition must be added to the system of adjustment to the Boussinesq state:

$$\hat{\boldsymbol{u}}_0 \cdot \boldsymbol{k} \equiv \varepsilon\hat{w}_0 = 0 \quad , \quad \text{on} \quad z = 0 \quad . \tag{8.105}$$

Therefore, when $t = \mathrm{M}_\infty \hat{t} = \mathrm{O}(\mathrm{M}_\infty)$, we must consider the following local asymptotic representation:

---

[15] This thesis has been defended the 12 september 1986 at the University of Lille I (no.-d'ordre 1352).

$$u = \hat{u}_0 + \dots \quad , \quad \pi = M_\infty \hat{\pi}_1 + \dots \quad ,$$
$$\omega = M_\infty \hat{\omega}_1 + \dots \quad , \quad \vartheta = M_\infty \hat{\vartheta}_1 + \dots \quad , \tag{8.106}$$

which leads to the local limiting system (8.104) governing the unsteady phenomenon of adiabatic adjustment to the Boussinesq state. It must be remembered that the Boussinesq state is associated with the main asymptotic representation:

$$u = u_0 + \dots \quad , \quad \pi = M_\infty^2 \pi_2 + \dots \quad ,$$
$$\omega = M_\infty \omega_1 + \dots \quad , \quad \vartheta = M_\infty \vartheta_1 + \dots \tag{8.107}$$

which, in turn, is associated with $t = O(1)$. The resulting *adiabatic* Boussinesq equations are:

$$S \frac{Du_0}{Dt} + \frac{1}{Ro}(e \times u_0) + \frac{1}{\gamma} \nabla \pi_2 - \frac{\hat{B}}{\gamma \varepsilon} \vartheta_1 k = 0 \quad ;$$

$$\nabla \cdot u_0 = 0 \quad , \quad \omega_1 = -\vartheta_1 \quad ; \tag{8.108}$$

$$S \frac{D\vartheta_1}{Dt} + \hat{B} \left[\frac{\gamma - 1}{\gamma} + \left(\frac{dT_\infty}{dz_\infty} \right)_0 \right] w_0 = 0 \quad .$$

The application mechanism of the initial conditions in $t = 0$ to the Boussinesq system (8.108) is rather subtle and deserves special attention. As we already know, according to the MMAE (discussed in Sect. 5.1), these initial conditions must be applied to the local adjustment system (8.104). Let us suppose then that from the start, the following initial conditions were added to the Euler equations written for u, π, ω and ϑ:

$$t = 0 : \quad u = u^0 \quad , \quad \pi = \omega = \vartheta = 0 \quad ; \tag{8.109}$$

i.e., that the initial thermodynamic state is assumed to be the one corresponding to the standard atmosphere discussed at the end of Chap. 1 and which is solely a function of the standard altitude $z_\infty \equiv Bo z$ (with dimensionless variables and with bars omitted).

In this case, the following initial conditions must be imposed on the local adjustment system (8.104):

$$\hat{t} = 0 : \quad \hat{u}_0 = u^0 \quad , \quad \hat{\pi}_1 = \hat{\omega}_1 = \hat{\vartheta}_1 = 0 \quad . \tag{8.110}$$

On the contrary, we do not know for the time being what initial conditions must be assigned to the adiabatic Boussinesq equations (8.108) since the latter are not valid in the vicinity of $t = 0$ where, in fact, they must be replaced by the adjustment equations (8.104).

If the behavior of the solution of the local adjustment system (8.104) for $\hat{t} \to \infty$ is indeed decreasing such that the local perturbations do not persist for $t = O(1)$, then according to the MMAE, the initial conditions for (8.108) are matching conditions between the two asymptotic representations – the main one (8.107) and the local one (8.106).

It is further emphasized that we have implicitly admitted the initial conditions (8.109) for π, ω and ϑ during the derivation of the local system (8.104). This made it possible for us to impose the following:

$$\hat{\pi}_0 = \hat{\omega}_0 = \hat{\vartheta}_0 \equiv 0 \quad .$$

Thanks to the initial conditions (8.110), the local system (8.104) can now be replaced by the following equivalent local system:

$$\mathrm{S}\frac{\partial \hat{\boldsymbol{u}}_0}{\partial \hat{t}} + \nabla \hat{\omega}_1 = 0 \quad ; \quad \mathrm{S}\frac{\partial \hat{\omega}_1}{\partial \hat{t}} + \nabla \cdot \hat{\boldsymbol{u}}_0 = 0 \quad ; \tag{8.111a}$$

$$\hat{\vartheta}_1 = (\gamma - 1)\hat{\omega}_1 \quad ; \tag{8.111b}$$

$$\hat{\pi}_1 = \gamma \hat{\omega}_1 \quad . \tag{8.111c}$$

Thus, for $\hat{\boldsymbol{u}}_0$ and $\hat{\omega}_1$, we obtain the following boundary and initial values problem:

$$\begin{aligned} &\mathrm{S}\frac{\partial \hat{\boldsymbol{u}}_0}{\partial \hat{t}} + \nabla \hat{\omega}_1 = 0 \quad , \quad \mathrm{S}\frac{\partial \hat{\omega}_1}{\partial \hat{t}} + \nabla \cdot \hat{\boldsymbol{u}}_0 = 0 \quad ; \\ &\hat{t} = 0 : \quad \hat{\boldsymbol{u}}_0 = \boldsymbol{u}^0 \quad , \quad \hat{\omega}_1 = 0 \quad ; \\ &z = 0 : \quad \hat{\boldsymbol{u}}_0 \cdot \boldsymbol{k} = 0 \quad . \end{aligned} \tag{8.112}$$

Let us assume that the initial data $\boldsymbol{u}^0$ can be expressed in the relatively general form:

$$\boldsymbol{u}^0 = \nabla \varphi^0 + \nabla \times \boldsymbol{\psi}^0 \quad , \tag{8.113}$$

and let us seek out the solution for $\hat{\boldsymbol{u}}_0$ which satisfies (8.112) in the analogous form:

$$\hat{\boldsymbol{u}}_0 = \nabla \hat{\varphi}_0 + \nabla \times \hat{\boldsymbol{\psi}}_0 \quad . \tag{8.114}$$

We then find that:

$$\nabla \left(\mathrm{S}\frac{\partial \hat{\varphi}_0}{\partial \hat{t}} + \hat{\omega}_1 \right) + \nabla \times \mathrm{S}\frac{\partial \hat{\boldsymbol{\psi}}_0}{\partial \hat{t}} = 0 \quad . \tag{8.115}$$

Therefore

$$\nabla^2 \left(\mathrm{S}\frac{\partial \hat{\varphi}_0}{\partial \hat{t}} + \hat{\omega}_1 \right) \equiv \Delta \left(\mathrm{S}\frac{\partial \hat{\varphi}_0}{\partial \hat{t}} + \hat{\omega}_1 \right) = 0 \quad ,$$

which means that: $\mathrm{S}(\partial \hat{\varphi}_0 / \partial \hat{t}) + \hat{\omega}_1$ is a harmonic function. However, we can take into account the slip condition on $z = 0$ by extending $\hat{\boldsymbol{u}}_0$ and $\hat{\omega}_1$ throughout the entire physical space (x, y, z). To accomplish this, the horizontal components of the velocity $\hat{\boldsymbol{u}}_0$ and $\hat{\omega}_1$ must be even functions of z, whereas the vertical component of $\hat{\boldsymbol{u}}_0$ must be uneven in z. This would imply that the function $\hat{\varphi}_0$ be even in z as well as the vertical component of the vector $\hat{\boldsymbol{\psi}}_0$. On the other hand, the horizontal components of this same vector are uneven. These conditions are naturally assumed to be satisfied for the initial values φ^0 and $\boldsymbol{\psi}^0$.

We thus arrive at the conclusion that the expression: $S(\partial\hat{\varphi}_0/\partial\hat{t}) + \hat{\omega}_1$ is a harmonic function of x, y and z throughout the entire physical space (x, y, z). If this harmonic function must be regular throughout the entire space, then it is necessarily identically zero, i.e.,

$$\hat{\omega}_1 = -S\frac{\partial\hat{\varphi}_0}{\partial\hat{t}} \quad , \tag{8.116}$$

and

$$S\frac{\partial}{\partial\hat{t}}(\nabla \times \hat{\psi}_0) = 0 \Rightarrow \hat{\psi}_0 \equiv \psi^0 \quad . \tag{8.117}$$

Therefore

$$\hat{u}_0 = \nabla\hat{\varphi}_0 + \nabla \times \psi^0$$

where the initial vector $\psi^0\ (x, y, z)$ takes into consideration the rotational character of the initial velocity field u^0.

Finally, for the potential $\hat{\varphi}_0$, the following initial values problem results:

$$\begin{aligned} &S^2\frac{\partial^2\hat{\varphi}_0}{\partial\hat{t}^2} - \Delta\hat{\varphi}_0 = 0 \quad , \\ &\hat{t} = 0: \quad \hat{\varphi}_0 = \varphi^0 \quad , \quad S\frac{\partial\hat{\varphi}_0}{\partial\hat{t}} = 0 \quad . \end{aligned} \tag{8.118}$$

The solution of (8.118) is standard and demonstrates that when

$$\hat{t} \Rightarrow \infty: \quad |\nabla\hat{\varphi}_0| \Rightarrow 0 \quad , \quad \left|\frac{\partial\hat{\varphi}_0}{\partial\hat{t}}\right| \Rightarrow 0 \quad . \tag{8.119}$$

This means that when:

$$\hat{t} \Rightarrow \infty: \quad \hat{u}_0 \Rightarrow \nabla \times \psi^0 \quad \text{and} \quad \hat{\omega}_1 \Rightarrow 0 \quad , \tag{8.120}$$

which justifies the use of the MMAE:

$$\lim_{\hat{t}\to\infty} \hat{u}_0 = \lim_{t\to 0} u_0 = \nabla \times \psi^0 \quad , \tag{8.121a}$$

$$\lim_{t\to 0}(\omega_1, \vartheta_1) = 0 \quad . \tag{8.121b}$$

It will be remarked that there is no initial condition to be imposed on π_2 in the Boussinesq equations. This initial value can be calculated from the initial value of u_0.

Therefore, *based on the hypothesis* (8.109), *the following initial conditions must be imposed on the Boussinesq equations* (8.108):

$$t = 0: \quad u_0 = \nabla \times \psi^0 \quad \text{and} \quad \vartheta_1 = 0 \quad , \tag{8.122}$$

once it has been postulated that the initial data u^0 *is expressed in the form* (8.113).

8.6 Complementary Remarks

8.6.1. Let us return to the problem of adjustment to the Boussinesq state which was considered in the preceding section. We want to clarify whether or not it is possible from the start to adopt initial conditions for π, ω and ϑ, at $t = 0$ which are more general than those given by (8.109).

Let us therefore assume that:

$$\boldsymbol{u} = \boldsymbol{u}^0 \quad , \quad \pi = \pi^0 \quad , \quad \omega = \omega^0 \quad , \quad \vartheta = \vartheta^0 \quad , \quad \text{for} \quad t = 0 \quad , \tag{8.123}$$

in such a way that $\pi^0 = \omega^0 + \vartheta^0 + \omega^0\vartheta^0$. Generally speaking, the given quantities π^0, ω^0 and ϑ^0 are functions of x, y and z.

While taking into account (8.123), the following new local asymptotic representation [instead of (8.106)] must be postulated:

$$\begin{aligned} &\boldsymbol{u} = \hat{\boldsymbol{u}}_0 + \dots \quad , \quad \pi = \hat{\pi}_0 + \mathrm{M}_\infty \hat{\pi}_1 + \dots \quad , \quad \omega = \hat{\omega}_0 + \mathrm{M}_\infty \hat{\omega}_1 + \dots \quad , \\ &\vartheta = \hat{\vartheta}_0 + \mathrm{M}_\infty \hat{\vartheta}_1 + \dots \quad , \end{aligned} \tag{8.124}$$

where all the "local" functions with hats are dependent on $\hat{t} \equiv t/\mathrm{M}_\infty$ and on x, y and z.

From the general equations (8.5–8), where Re $\equiv \infty$[14] and $t = \mathrm{M}_\infty \hat{t}$, we can now obtain to the zeroth order for $\hat{\pi}_0$, $\hat{\omega}_0$ and $\hat{\vartheta}_0$, the following strongly degenerated system:

$$\begin{aligned} &\nabla \hat{\pi}_0 = 0 \quad , \quad \frac{\partial \hat{\omega}_0}{\partial \hat{t}} = 0 \quad , \quad \frac{\partial}{\partial \hat{t}}\left(\hat{\vartheta}_0 - \frac{\gamma - 1}{\gamma}\hat{\pi}_0\right) = 0 \quad , \\ &\hat{\pi}_0 = \hat{\omega}_0 + \hat{\vartheta}_0 + \hat{\omega}_0\hat{\vartheta}_0 \quad . \end{aligned} \tag{8.125}$$

Therefore, thanks to (8.123):

$$\hat{\omega}_0 \equiv \omega_0 \quad \text{and also} \quad \hat{\vartheta}_0 \equiv \vartheta^0 \Rightarrow \hat{\pi}_0 \equiv \pi_0^0 \quad , \tag{8.126}$$

where $\pi_0^0 \equiv$ const, since:

$$\frac{\partial \hat{\pi}_0}{\partial \hat{t}} = (1 + \omega^0)\frac{\partial \hat{\vartheta}_0}{\partial \hat{t}} \Rightarrow \left[1 - \frac{\gamma - 1}{\gamma}(1 + \omega^0)\right]\frac{\partial \hat{\pi}_0}{\partial \hat{t}} = 0 \quad ,$$

and if it is supposed that for all the values of x, y and z we have $\omega^0 \neq 1/(\gamma - 1)$.

Hence, the zero order terms in the local asymptotic representation (8.124) are "at most" functions of x, y and z ($\hat{\pi}_0$ being constant). If it is now taken into account that when $\hat{t} \to \infty$, (8.124) must match with the main asymptotic representation (8.107) which leads to the Boussinesq limiting equations, then of necessity:

[14] As a matter of fact, it is easy to see that the analysis from Sect. 8.5 along with the present one remain valid as long as Re $\sim \mathrm{M}_\infty$ (not realistic, however, for atmospheric flows). This means that the adjustment phenomenon is a purely adiabatic phenomenon during which the dissipative terms (proportional to 1/Re) do not come into play. Of course, in a boundary layer having a thickness of the order of $(\mathrm{M}_\infty/\mathrm{Re})^{1/2}$, there is an adjustment phenomenon (in the vicinity of $\hat{t} = 0$) for which the dissipative effects do play a role.

$$\hat{\pi}_0 = \hat{\omega}_0 = \hat{\vartheta}_0 \equiv 0 \quad ,$$

in order for this matching (according to the MMAE) to be possible.

It thus becomes imperative to suppose for the initial conditions (8.123) that:

$$\pi = \mathrm{M}_\infty \pi_1^0 \quad , \quad \omega = \mathrm{M}_\infty \omega_1^0, \quad \vartheta = \mathrm{M}_\infty \vartheta_1^0 \quad , \tag{8.127}$$

when $t = 0$, π_1^0, ω_1^0 and ϑ_1^0 being known functions of x, y and z.

In this case to order one, the same adjustment system (8.104) is obtained for $\hat{\pi}_1$, $\hat{\omega}_1$ and $\hat{\vartheta}_1$. The third equation of this system implies the following relation:

$$\hat{\vartheta}_1 - \frac{\gamma - 1}{\gamma}\hat{\pi}_1 \equiv \vartheta_1^0 - \frac{\gamma - 1}{\gamma}\hat{\pi}_1^0 \quad ,$$

once the following initial conditions are taken into account:

$$\hat{t} = 0 : \quad \hat{\pi}_1 = \pi_1^0 \quad , \quad \hat{\omega}_1 = \omega_1^0 \quad , \quad \hat{\vartheta}_1 = \vartheta_1^0$$

which ensue from (8.127). But

$$\hat{\pi}_1 = \hat{\omega}_1 + \hat{\vartheta}_1 \Rightarrow \frac{\hat{\pi}_1}{\gamma} = \hat{\omega}_1 + \frac{\pi_1^0}{\gamma} - \omega_1^0$$

and the following system results for $\hat{\boldsymbol{u}}_0$ and $\hat{\omega}_1$ instead of (8.112):

$$\begin{aligned}
&\mathrm{S}\frac{\partial \hat{\boldsymbol{u}}_0}{\partial \hat{t}} + \boldsymbol{\nabla}\left(\hat{\omega}_1 + \frac{\pi_1^0}{\gamma} - \omega_1^0\right) = 0 \quad ; \\
&\mathrm{S}\frac{\partial \hat{\omega}_1}{\partial \hat{t}} + \boldsymbol{\nabla} \cdot \hat{\boldsymbol{u}}_0 = 0 \quad , \\
&\hat{t} = 0 : \quad \hat{\boldsymbol{u}}_0 = \boldsymbol{u}^0 \quad , \quad \hat{\omega}_1 = \omega_1^0 \quad , \\
&z = 0 : \quad \hat{\boldsymbol{u}}_0 \cdot \boldsymbol{k} = 0 \quad .
\end{aligned} \tag{8.128}$$

By repeating the same procedure applied in Sect. 8.5, we are led to set:

$$\hat{\omega}_1 = -\mathrm{S}\frac{\partial \hat{\varphi}_0}{\partial \hat{t}} + \omega_1^0 - \frac{\pi_1^0}{\gamma} \tag{8.129}$$

and we thus arrive at the following problem with initial values:

$$\begin{aligned}
&\mathrm{S}^2 \frac{\partial^2 \hat{\varphi}_0}{\partial \hat{t}^2} - \Delta \hat{\varphi}_0 = 0 \\
&\hat{t} = 0 : \quad \hat{\varphi}_0 = \varphi^0 \quad , \quad \mathrm{S}\frac{\partial \hat{\varphi}_0}{\partial \hat{t}} = -\frac{\pi_1^0}{\gamma} \quad .
\end{aligned} \tag{8.130}$$

In conclusion, the only difference with the case studied in Sect. 8.5 is that the following initial conditions must be imposed on the Boussinesq equations [valid for $t = \mathrm{O}(1)$]:

$$t = 0 : \quad \boldsymbol{u}_0 = \boldsymbol{\nabla} \times \boldsymbol{\psi}^0 \quad \text{and} \quad \vartheta_1 = \frac{\pi_1^0}{\gamma} - \omega_1^0 \quad . \tag{8.131}$$

8.6.2. The second point that we wish to clarify in this section concerns the behavior of the solution of the Boussinesq equations at infinity. Like Guiraud (1979), we will consider here the steady, adiabatic, but three-dimensional case:

$$\mathrm{S} \equiv 0 \quad , \quad \mathrm{Re} \equiv \infty \quad .$$

Moreover, we will assume that:

$$\varepsilon \equiv 1 \quad \text{and} \quad \mathrm{Ro} \equiv \infty \quad .$$

The basic equations are those which were considered in Subsect. 8.3.1 and which are related to the "lee wave" problem. We introduce the convective operator:

$$\frac{\delta_0}{\delta t} = \left(U_\infty(0) + u_0'\right)\frac{\partial}{\partial x} + \left(V_\infty(0) + v_0'\right)\frac{\partial}{\partial y} + w_0\frac{\partial}{\partial z}$$

and note that:

$$\Gamma_\infty(0) \equiv \frac{\hat{B}}{T_\infty(0)}\left[\frac{\gamma - 1}{\gamma} + \left(\frac{dT_\infty}{dz}\right)_0\right] \quad .$$

This being the case, the system of *steady* adiabatic Boussinesq equations can be written in the form of a single matrix equation:

$$\mathcal{A}(\mathcal{U})\frac{\partial \mathcal{U}}{\partial x} + \mathcal{B}(\mathcal{U})\frac{\partial \mathcal{U}}{\partial y} + \mathcal{C}(\mathcal{U})\frac{\partial \mathcal{U}}{\partial z} + \mathcal{D}\mathcal{U} = 0 \quad , \tag{8.132}$$

where $\mathcal{A}$, $\mathcal{B}$, $\mathcal{C}$ and $\mathcal{D}$ are four 5×5 matrices. It is observed that $\mathcal{D}$ does not depend on $\mathcal{U}$,

$$\text{with} \quad \mathcal{U} = \left(u_0', v_0', w_0, \pi_2, \vartheta_1\right)^{\mathrm{T}} \equiv \left(\mathcal{U}_1, \mathcal{U}_2, \mathcal{U}_3, \mathcal{U}_4, \mathcal{U}_5\right)^{\mathrm{T}} \quad ,$$

where the superscript "T" indicates that we transpose a row in a column. Finally, it is pointed out that in the equations for u_0' and v_0', it has been assumed that $\hat{\mu} \equiv \infty$ (i.e., that $\mu \gg \mathrm{M}_\infty$).

For the lee wave problem (that we will consider in detail in Chap. 13), the steady adiabatic Boussinesq equations must be solved with (a) the slip condition on the ground, (b) the perturbation cancellation condition at infinity far upstream of the relief where the basic flow is found, (c) the condition that these perturbations remain bounded at downstream infinity, and (d) a radiation condition when $\sqrt{x^2 + y^2 + z^2} \Rightarrow \infty$. This last condition has to be formulated by studying the behavior of the solutions of (8.132) at infinity. Equation (8.132) is quasi-linear but since

$$\|\mathcal{U}\| = \sqrt{\sum_{k=1}^{5} \mathcal{U}_k^2}$$

obviously tends towards zero at infinity, we can take advantage of this fact to simplify the asymptotic analysis.

Experience in two-dimensional studies (see Chap. 13) prompts us to seek $\mathcal{U}$ in the form:

$$\mathcal{U}(x, y, z) = \tilde{\mathcal{U}}\left(\varphi(x, y, z), \overline{x}, \overline{y}, \overline{z}; \alpha\right) \quad , \tag{8.133}$$

where $\alpha \ll 1$ is used to characterize the order of magnitude of $r = \sqrt{x^2+y^2+z^2}$:

$$\overline{x} = \alpha x \quad , \quad \overline{y} = \alpha y \quad , \quad \overline{z} = \alpha z \quad , \quad \overline{r} = \alpha r \quad .$$

Then, we postulate that:

$$\tilde{\mathcal{U}} = \eta_1(\alpha)\tilde{\mathcal{U}}^{(1)} + \eta_2(\alpha)\tilde{\mathcal{U}}^{(2)} + \dots \quad , \qquad \lim_{\alpha\to 0}\frac{\eta_2}{\eta_1} = 0 \quad , \tag{8.134}$$

and by substitution, we obtain:

$$\eta_1\left\{(a\mathcal{A}_0 + b\mathcal{B}_0 + c\,\mathcal{C}_0)\frac{\partial\tilde{\mathcal{U}}^{(1)}}{\partial\varphi} + \mathcal{D}\tilde{\mathcal{U}}^{(1)}\right\} + \eta_2\{\quad\}$$
$$+\eta_1^2\{\quad\} + \eta_1\alpha\left\{\mathcal{A}_0\frac{\partial\tilde{\mathcal{U}}^{(1)}}{\partial\overline{x}} + \mathcal{B}_0\frac{\partial\tilde{\mathcal{U}}^{(1)}}{\partial\overline{y}} + \mathcal{C}_0\frac{\partial\tilde{\mathcal{U}}^{(1)}}{\partial\overline{z}}\right\} + \dots = 0 \quad , \tag{8.135}$$

where

$$a = \frac{\partial\varphi}{\partial x} \quad , \quad b = \frac{\partial\varphi}{\partial y} \quad , \quad c = \frac{\partial\varphi}{\partial z} \quad ;$$
$$\mathcal{A}_0 \equiv \mathcal{A}(0) \quad , \quad \mathcal{B}_0 \equiv \mathcal{B}(0) \quad , \quad \mathcal{C}_0 \equiv \mathcal{C}(0) \quad .$$

For the time being, it is not known how η_1^2 compares to η_2, or to $\eta_1\alpha$ but this fact is unimportant for our purposes. What is important, however, is that when $\alpha \Rightarrow 0$:

$$\frac{\eta_1^2}{\eta_1} \Rightarrow 0 \quad , \quad \frac{\eta_2}{\eta_1} \Rightarrow 0 \quad , \quad \frac{\eta_1\alpha}{\eta_1} \Rightarrow 0$$

and by carrying out a limiting process, we can deduce that we must have the following:

$$(a\mathcal{A}_0 + b\mathcal{B}_0 + c\,\mathcal{C}_0)\frac{\partial\tilde{\mathcal{U}}^{(1)}}{\partial\varphi} + \mathcal{D}\tilde{\mathcal{U}}^{(1)} = 0 \quad . \tag{8.136}$$

Quite naturally, we are led to seek out $\tilde{\mathcal{U}}^{(1)}$ in the form: $\tilde{\mathcal{U}}^{(1)} = K\,\mathrm{e}^{\mathrm{i}\varphi}\,D = \hat{\mathcal{U}}^{(1)}\,\mathrm{e}^{\mathrm{i}\varphi}$ and D is seen to be the right-hand proper solution of the matrix $\mathcal{L} = \mathrm{i}(a\mathcal{A}_0 + b\mathcal{B}_0 + c\,\mathcal{C}_0) + \mathcal{D}$. This means that $\mathcal{L}D = 0$, whereas K is an arbitrary scalar. We set:

$$\Delta(a,\, b,\, c) = \text{determinant of } \mathcal{L} \quad .$$

It is thus obvious that the phase φ must verify the first order partial derivative equation:

$$\Delta\left(\frac{\partial\varphi}{\partial x} \; , \; \frac{\partial\varphi}{\partial y} \; , \; \frac{\partial\varphi}{\partial z}\right) = 0 \quad . \tag{8.137}$$

Let us now set

$$D = (\hat{u},\, \hat{v},\, \hat{w},\, \hat{\pi},\, \hat{\vartheta})^{\mathrm{T}}$$

we have (by writing U_∞^0 V_∞^0, ... instead of $U_\infty(0)$, $V_\infty(0)$, ...)

$$\gamma(U_\infty^0 a + V_\infty^0 b)\hat{u} + aT_\infty^0\hat{\pi} = 0;$$

$$\gamma(U^0_\infty a + V^0_\infty b)\hat{v} + bT^0_\infty \hat{\pi} = 0 \quad ;$$
$$\gamma(U^0_\infty a + V^0_\infty b)\hat{w} + cT^0_\infty \hat{\pi} - \frac{\hat{B}}{\mathrm{i}}\hat{\vartheta} = 0 \quad ;$$
$$a\hat{u} + b\hat{v} + c\hat{w} = 0 \quad ;$$
$$\mathrm{i}(U^0_\infty a + V^0_\infty b)\hat{\vartheta} + \Gamma^0_\infty \hat{w} = 0 \quad ,$$

and we will want to know under what conditions this homogeneous system admits of a non-trivial solution. Two cases stand out: in the first: $U^0_\infty a + V^0_\infty b = 0$ and thus $\hat{\pi} = \hat{\vartheta} = \hat{w} = 0$ and $a\hat{u} + b\hat{v} = 0$. This solution corresponds to the wake of the obstacle which we will not study further. The second case corresponds to:

$$U^0_\infty a + V^0_\infty b \neq 0$$

and so we find that the following condition must be satisfied:

$$\left(U^0_\infty a + V^0_\infty b\right)^2 \left(a^2 + b^2 + c^2\right) = \frac{\hat{B}}{\gamma}\Gamma^0_\infty \left(a^2 + b^2\right) \quad . \tag{8.138}$$

We thus have:

$$D = \left(ac,\ bc,\ -\left(a^2 + b^2\right),\ -\frac{\gamma}{T^0_\infty}\left(U^0_\infty a + V^0_\infty b\right) c \quad , \right.$$
$$\left. -\frac{\mathrm{i}\gamma}{\hat{B}}\left(U^0_\infty a + V^0_\infty b\right)\left(a^2 + b^2 + c^2\right)\right)^{\mathrm{T}} \quad .$$

It is observed that

$$\Delta(a, b, c) = \frac{\mathrm{i}T^0_\infty}{\gamma}\left(U^0_\infty a + V^0_\infty b\right)\left\{\frac{\hat{B}}{\gamma}\Gamma^0_\infty\left(a^2 + b^2\right)\right.$$
$$\left. - \left(U^0_\infty a + V^0_\infty b\right)^2\left(a^2 + b^2 + c^2\right)\right\} \quad .$$

Since we know that $\Delta = 0$, then there is also a left-hand proper solution G of $\mathcal{L}$, such that: $G\mathcal{L} = 0$ and G is a row matrix. Furthermore[16]

$$\langle G, \mathcal{L}U\rangle = 0 \quad , \quad \text{for any} \quad U \quad . \tag{8.139}$$

Thus we find that:

$$G = \left(ac, bc,\ -\left(a^2 + b^2\right),\ -\left(U^0_\infty a + V^0_\infty b\right) c \quad ,\right.$$
$$\left.\frac{\mathrm{i}}{\Gamma^0_\infty}\left(U^0_\infty a + V^0_\infty b\right)\left(a^2 + b^2 + c^2\right)\right) \quad .$$

A direct calculation makes it possible to establish the relation:

$$\langle G, \mathcal{A}_0 \partial D\rangle = \langle \partial G, \mathcal{A}_0 D\rangle \quad ,$$

[16] The scalar product associated with the following norm || || is used:

$$\langle \mathcal{U}, U\rangle = \sum_{k=1}^{5} \mathcal{U}_k U_k \quad .$$

as well as the two analogous relations with $\mathcal{B}_0$ and $\mathcal{C}_0$ in the place of $\mathcal{A}_0$ respectively. It is understood that the differentiation is carried out while maintaining relation (8.138).

Let us go back to (8.135). In the sequence η_n, there is one denoted η_{n0}, such that $\eta_{n0} = \alpha\eta_1$. Under these conditions, we have:

$$\{a\mathcal{A}_0 + b\mathcal{B}_0 + c\,\mathcal{C}_0\}\frac{\partial\tilde{\mathcal{U}}^{(n0)}}{\partial\varphi} + \mathcal{D}\tilde{\mathcal{U}}^{(n0)}$$

$$= \left(\mathcal{A}_0\frac{\partial\hat{\mathcal{U}}^{(1)}}{\partial x} + \mathcal{B}_0\frac{\partial\hat{\mathcal{U}}^{(1)}}{\partial y} + \mathcal{C}_0\frac{\partial\hat{\mathcal{U}}^{(1)}}{\partial z}\right)\mathrm{e}^{\mathrm{i}\varphi} \quad , \quad \text{if} \quad \eta_1^2 \neq \eta_1\alpha \quad .$$

If such were not the case – and this is actually the situation as will be verified a posteriori – it would be necessary to add to the right-hand side of the above equation, certain terms whose dependence in φ is represented by $\mathrm{e}^{2\mathrm{i}\varphi}$ as factor.

Let us therefore consider the equation

$$\left(a\mathcal{A}_0 + b\mathcal{B}_0 + c\,\mathcal{C}_0\right)\frac{\partial\tilde{v}}{\partial\varphi} + \mathcal{D}\tilde{v} = F_n\,\mathrm{e}^{\mathrm{i}n\varphi} \quad . \tag{8.140}$$

By superposition, we can deal with the case where in the right-hand member of our equation for $\tilde{\mathcal{U}}^{(n0)}$, there is a sum of terms like in the right-hand member of (8.140). If $n \neq 1$, the solution can be sought in the form

$$\tilde{v} = \hat{v}\,\mathrm{e}^{\mathrm{i}n\varphi} \Rightarrow \left[\mathrm{i}n\left(a\mathcal{A}_0 + b\mathcal{B}_0 + c\,\mathcal{C}_0\right) + \mathcal{D}\right]\hat{v} = F_n \quad . \tag{8.141}$$

Since the determinant of this system is not zero for $n \neq 1$, a unique solution can be found for $\hat{v}$. If, on the contrary, $n \equiv 1$, there will not, in general, be a solution in the form sought except if F_1 verifies a compatibility condition that we are now going to formulate. Equation (8.139) shows that if $\hat{v}$ is a solution of (8.141), then for $n = 1$

$$\langle G,\, F_1\rangle = \langle G,\, \mathcal{L}\hat{v}\rangle = 0 \quad , \tag{8.142}$$

and such is the compatibility condition. If the latter is not satisfied, the solution to (8.140) should be sought in the form

$$\tilde{v} = \varphi\,\mathrm{e}^{\mathrm{i}\varphi}\,\tilde{V} + \chi\,\mathrm{e}^{\mathrm{i}\varphi}D, \tag{8.143}$$

where χ is an arbitrary constant. Thus, for $\tilde{V}$, we have the equation

$$\left[\mathcal{L} + a\mathcal{A}_0 + b\mathcal{B}_0 + c\,\mathcal{C}_0\right]\tilde{V} = F_1 \quad , \tag{8.144}$$

and this time, the determinant is not zero. Nonetheless, (8.143) remains unacceptable because the right hand member is not bounded. We are thus induced to prescribe the compatibility condition (8.142) which implies the cancelling of the secular term in $\varphi\,\mathrm{e}^{\mathrm{i}\varphi}$ in (8.143). Returning to the equation for $\tilde{\mathcal{U}}^{(n0)}$, this compatibility condition is written:

$$\left\langle G,\, \mathcal{A}_0\frac{\partial\hat{\mathcal{U}}^{(1)}}{\partial x} + \mathcal{B}_0\frac{\partial\hat{\mathcal{U}}^{(1)}}{\partial y} + \mathcal{C}_0\frac{\partial\hat{\mathcal{U}}^{(1)}}{\partial z}\right\rangle = 0 \quad ,$$

i.e., if the form of $\hat{\mathcal{U}}^{(1)}$ is taken into account:

$$\frac{\partial}{\partial \overline{x}}\left(K^2\langle G, \mathcal{A}_0 D\rangle\right) + \frac{\partial}{\partial \overline{y}}\left(K^2\langle G, \mathcal{B}_0 D\rangle\right) + \frac{\partial}{\partial \overline{z}}\left(K^2\langle G, \mathcal{C}_0 D\rangle\right) = 0 \quad . \tag{8.145}$$

The above relation can be given a form which is easily interpretable: Firstly, it is simple to demonstrate that:

$$\langle G, \mathcal{L}D\rangle = \sigma\Delta \quad , \quad \sigma = \frac{\gamma}{\hat{B}T^0_\infty \Gamma^0_\infty}\left(a^2+b^2+c^2\right) \quad .$$

After derivation, with respect to a, b and c, we have

$$\frac{\partial}{\partial \overline{x}}\left(\sigma K^2 \frac{\partial\Delta}{\partial a}\right) + \frac{\partial}{\partial \overline{y}}\left(\sigma K^2 \frac{\partial\Delta}{\partial b}\right) + \frac{\partial}{\partial \overline{z}}\left(\sigma K^2 \frac{\partial\Delta}{\partial c}\right) = 0 \quad .$$

Moreover, the radii associated with (8.137) are defined by

$$\frac{d\overline{x}}{(\partial\Delta/\partial a)} = \frac{d\overline{y}}{(\partial\Delta/\partial b)} = \frac{d\overline{z}}{(\partial\Delta/\partial c)} = \frac{da}{0} = \frac{db}{0} = \frac{dc}{0} = d\hat{B} \quad ,$$

in such a way that the following relation is obtained

$$\nabla\cdot\left(\sigma\frac{K^2}{\hat{B}}\overline{\boldsymbol{x}}\right) = 0 \quad , \tag{8.146}$$

where $\overline{\boldsymbol{x}}$ designates the vector of components $\overline{x}$, $\overline{y}$, $\overline{z}$, and ∇, the associated gradient operator vector.

The radii are rectilinear and all go through the origin which as a consequence of (8.146), implies that:

$$\frac{\sigma}{\hat{B}}K^2\overline{r}^3 = \text{const, on each radius} \quad .$$

In addition, since $\sigma\overline{r}/\hat{B} = \text{const}$, on each radius, it is finally seen that

$$K\overline{r} = \text{const, on each radius} \quad . \tag{8.147}$$

We now introduce:

$$\boldsymbol{\xi} = \overline{\boldsymbol{x}}/\overline{r} \quad , \quad |\boldsymbol{\xi}| = 1 \quad ;$$

it is obvious that:

$$\boldsymbol{\xi} = \left(\frac{\partial\Delta}{\partial a}\boldsymbol{i} + \frac{\partial\Delta}{\partial b}\boldsymbol{j} + \frac{\partial\Delta}{\partial c}\boldsymbol{k}\right) \Bigg/ \sqrt{\left(\frac{\partial\Delta}{\partial a}\right)^2 + \left(\frac{\partial\Delta}{\partial b}\right)^2 + \left(\frac{\partial\Delta}{\partial c}\right)^2}$$

and it can be deduced that:

$$a = a(\boldsymbol{\xi}) \quad , \quad b = b(\boldsymbol{\xi}) \quad , \quad c = c(\boldsymbol{\xi}) \quad .$$

Therefore:

$$K\overline{r} = \mathcal{F}(\boldsymbol{\xi}) \quad , \quad D = D(\boldsymbol{\xi}) \quad ,$$

such that:

$$\hat{\mathcal{U}}^{(1)} = \frac{\mathcal{F}(\overline{\boldsymbol{x}}/\overline{r})}{\overline{r}} D\left(\frac{\overline{\boldsymbol{x}}}{\overline{r}}\right) \quad . \tag{8.148}$$

Returning to (8.134) and the expression of $\tilde{\mathcal{U}}^{(1)}$, as a function of $\hat{\mathcal{U}}^{(1)}$, we have for $\mathcal{U}(x, y, z)$:

$$\mathcal{U} = \frac{\eta_1}{\alpha} \frac{\mathcal{F}(\boldsymbol{x}/r)}{r} \mathrm{e}^{\mathrm{i}\varphi} D\left(\frac{\boldsymbol{x}}{r}\right) + \mathrm{O}(\eta_1^2) + \mathrm{O}(\eta_1 \alpha) \quad .$$

Since the result must be independent of the choice of α, we find that $\eta_1 \equiv \alpha$, and hence:

$$\mathcal{U} = \frac{\mathcal{F}(\boldsymbol{x}/r)}{r} \mathrm{e}^{\mathrm{i}\varphi} D\left(\frac{\boldsymbol{x}}{r}\right) + \mathrm{O}\left(\frac{1}{r^2}\right), \tag{8.149}$$

which is the result sought after. To be sure, the function $\mathcal{F}$ cannot be determined by a local study in the vicinity of infinity! The phase φ is constructed in a standard way from (8.137) and from rectilinear radii emanating from the origin. The matrix column D is indeed determined by the condition $\mathcal{L}D = 0$ give or take a normalization which must be defined. Under these conditions, D is constant on each radius and thus only depends on the direction of the vector $\boldsymbol{x}$. The condition (8.149) generalizes to the three-dimensional case, the radiation condition applied to the Long equation (8.42). This condition will be taken up again in Chap. 13.

8.6.3. Let us now backtrack to the solution (8.91). When Ro $\equiv \infty$, $\varepsilon \equiv 1$ and $l = 0$, we know that $d\varphi/d\zeta$ is one of the roots of (8.88), and ξ is given by:

$$\xi = \frac{1}{\mathrm{M}_\infty} \int_0^\zeta \frac{d\varphi(u)}{du} du \quad , \quad \varphi(0) \equiv 0 \quad .$$

The problem which then arises is to know how to discern among the two waves composing solution (8.91), the one having an amplitude which tends towards zero when z goes to infinity. To do this, it is necessary to take into account (like Bois, 1979) the dissipative effects in the basic equations [instead of (6.21)]. In this case, for the function $(d\varphi/d\zeta)^2$, there results a third degree equation where the Reynolds number Re comes into play. If we set:

$$\frac{d\varphi}{d\zeta}(\zeta; \mathrm{Re}) = \frac{d\varphi^0}{d\zeta} + \frac{1}{\mathrm{Re}} \frac{d\varphi^1}{d\zeta} + \dots \quad ,$$

we again find the expression (6.34) for $d\varphi^0/d\zeta$, whereas the imaginary i appears in the expression of $d\varphi^1/d\zeta$. The presence of this imaginary i in $d\varphi^1/d\zeta$ shows that if φ^0 is real (rapid oscillation), φ^1 indicates a slow damping or a slow amplification. If φ^0 is imaginary, φ^1 is real and indicates a slow oscillation. In the case where $d\varphi^0/d\zeta$ is real, only one of the two waves is damped – the one corresponding to the positive root of (6.34). Thus is obtained a means of discerning among the two waves which make up solution (6.36), the one whose amplitude tends towards zero for infinite z, and the one which has an unbounded amplitude when the viscosity is taken into account at high altitude. The taking into account of (evanescent) dissipative effects at high altitude therefore makes it possible to solve the problem of the boundary condition at infinity in altitude.

To conclude, the lectures by Bois (1984) given at C.I.S.M. in Udine (Italy) in October 1983 deal with an asymptotic theory of the atmospheric Boussinesq waves which covers both the lee waves (see Chap. 13) and the problem of convective instability (which is that of Rayleigh-Bénard). In the last section of Chap. 13, the reader will find an asymptotic formulation of the Rayleigh-Bénard problem via the Boussinesq approximation for dilatable liquids (Zeytounian, 1983). For all questions concerned with the so-called "hydrodynamic" stability, we recommend the recent book by Drazin and Reid (1981).

9. The Isochoric Approximation

As was already explained in Chap. 3 (see Subsect. 3.3.4), when a so-called isochoric approximation[1] is carried out, quasi-incompressible atmospheric phenomena which are non-homogeneous (stratified with the altitude) are, in fact, considered. This isochoric approximation leads to the conservation of the density (the so-called equation of isochoricity):

$$\frac{D\varrho}{Dt} = 0 \quad , \tag{9.1}$$

when the adiabatic atmosphere ($\mathrm{Re} \equiv \infty$) is considered.

More precisely, if we consider the non-dimensional Euler equations [for example, in the form (2.69)], then the isochoric approximation consists in carrying out the limiting process:

$$\gamma \to \infty \quad , \quad \mathrm{M}_\infty \to 0 \quad ; \quad \gamma \mathrm{M}_\infty^2 = \hat{M}^2 \equiv \frac{\mathrm{Bo}}{\varepsilon}\mathrm{Fr}^2 = \mathrm{O}(1) \quad , \tag{9.2}$$

with

$$\varepsilon = \frac{H}{L} \quad , \quad \mathrm{Bo} = \frac{H}{RT_\infty(0)/g} \quad , \quad \mathrm{Fr}^2 = \frac{U_0^2}{gL} \quad .$$

It must be specified that $\gamma \to \infty$ means precisely

$$c_v \to 0 \quad , \quad c_p = \mathrm{O}(1) \quad . \tag{9.3}$$

It will be remarked that the isochoric approximation can also be effected "formally" from the *dimensional* Euler equations [see for instance, (2.15)] by simply causing $\gamma = c_p/c_v$ to tend to infinity $\big(c_p = \mathrm{O}(1),\ c_v \to 0\big)$. In this case, when $\gamma \to \infty$, the following limiting value results for the specific entropy $s = c_v \log p/\varrho^\gamma$ (for a perfect gas with constant c_p and c_v, with $c_v \to 0$):

$$s_0 = -c_p \log \varrho_0 \quad , \tag{9.4}$$

where the subscript "0" corresponds to the limit $\gamma \to \infty$. Thus the conservation of the specific entropy $Ds/Dt = 0$ in an (adiabatic) isochoric approximation is replaced by the conservation of ϱ_0 along the isochoric trajectories in an adiabatic atmosphere. Consequently, the continuity equation becomes:

$$\nabla \cdot \boldsymbol{u}_0 = 0 \quad . \tag{9.5}$$

[1] *Isochoric* flow, flow for which the *volume* remains *constant*.

The isochoric approximation is usually employed for the study of stratified heavy fluid flows [see, for example, the book by Yih (1980) and our thesis (Zeytounian, 1969)].

It is important to understand that the *incompressibility* of the flow, which is expressed by (9.5), is not a consequence of the physical properties of the fluid (which remains compressible), but rather is related solely to the *kinematic* properties of the (adiabatic) isochoric flow being considered.

Occasionally, there is a tendency to identify the isochoric approximation with the Boussinesq approximation considered in the last chapter. We will see in the present chapter that this is valid only if the *quasi-linear terms in the expression of D/Dt are not taken into account.*

In Sect. 9.1, we establish the isochoric equations and discuss under what conditions they are valid. Section 9.2 is devoted to the singular character of the isochoric approximation. In Sect. 9.3, the link between the isochoric and Boussinesq approximations is elucidated. Section 9.4 outlines the wave phenomena in isochoric flows. Finally, Sect. 9.5 ends this chapter with some complementary remarks.

9.1 The Isochoric Equations

Let us consider the complete equations which govern atmospheric flows: these are the non-dimensional equations (7.1–4). We now carry out the limiting process (9.2) in these equations. It is immediately observed that the momentum equation (7.1) does not change as long as $\gamma \mathrm{M}_\infty^2$ is replaced by

$$\hat{M}^2 = \frac{\mathrm{Bo}}{\varepsilon}\mathrm{Fr}^2$$

which is supposed of the order unity. The continuity equation (7.2) remains unchanged as does (7.3). Equation (7.4) becomes:

$$-T_0 \mathrm{S}\frac{D\varrho_0}{Dt} = \frac{1}{\mathrm{Pr}}\frac{1}{\mathrm{Re}}\left\{\frac{1}{\varepsilon^2}\frac{\partial}{\partial z}\left(k_0\frac{\partial T_0}{\partial z}\right) + \boldsymbol{D}^2 T_0\right\}$$
$$+\frac{\mathrm{Bo}\,\mathrm{Fr}^2}{\varepsilon^3\,\mathrm{Re}}\phi_0 + \frac{1}{\mathrm{Pr}}\frac{1}{\mathrm{Re}}\frac{\mathrm{Bo}\sigma_{00}}{\varepsilon^2}\frac{\partial \mathcal{R}_0}{\partial z} \quad , \tag{9.6}$$

where $\sigma_{00} = (c_p/g)\left[\mathcal{R}_\infty(T_\infty(0))/k_0\right]$.

The limiting equation (9.6) indicates that we must assume that $\mathrm{Re} \equiv \infty^2$ in order to recover the equation of isochoricity:

$$\mathrm{S}\frac{D\varrho_0}{Dt} = 0 \quad . \tag{9.7}$$

Therefore, when $\mathrm{Re} \equiv \infty$, the following adiabatic isochoric equations are obtaind

[2] It is noted that the cases of $\mathrm{Pr} \equiv \infty$ and $\mathrm{Fr}^2 \equiv 0$ lead to degeneracies that are far too strong in the momentum equation.

$$\hat{M}^2\left\{S\frac{D\boldsymbol{u}_0}{Dt}+\frac{1}{\mathrm{Ro}}(\boldsymbol{e}\times\boldsymbol{u}_0)\right\}+\frac{1}{\varrho_0}\nabla p_0+\frac{\mathrm{Bo}}{\varepsilon}\boldsymbol{k}=0 \quad ;$$
$$S\frac{D\varrho_0}{Dt}=0 \quad ; \quad \nabla\cdot\boldsymbol{u}_0=0 \quad ; \quad T_0=\frac{p_0}{\varrho_0} \quad . \tag{9.8}$$

In the above isochoric equations (9.8), the continuity equation has a totally incompressible form whereas the energy equation takes the form of the equation of isochoricity. The system (9.8) is very often used to simulate local phenomena (in this case $\varepsilon \equiv 1$ and $\mathrm{Ro} \equiv \infty$) which are then considered as stratified incompressible flows.

It is again recalled that in order to obtain (9.8) which is valid in an adiabatic atmosphere, the following limiting process must be carried out

$$\mathrm{Re}\equiv\infty, \quad \text{then} \quad \mathrm{M}_\infty\to 0 \quad \text{and} \quad \gamma\to\infty; \quad \mathrm{M}_\infty^2\gamma=\hat{M}^2=\mathrm{O}(1) \tag{9.9}$$

It is understood that when carrying out (9.9), t, x, y and z, as well as the other dimensionless parameters remain fixed of the order unity.

Since we have $c_v \equiv 0$ in an isochoric flow, it is thus obvious that the internal specific energy $e = e_0 \equiv 0$ and the specific enthalpy is $h_0 \equiv T_0$. For isochoric flows, (2.23) becomes:

$$\frac{D}{Dt}\{(\boldsymbol{\omega}_0+2\boldsymbol{\Omega})\cdot\nabla\log\varrho_0\}=0 \quad , \tag{9.10}$$

with dimensioned values.

Finally, the dimensioned potential temperature takes the form:

$$\theta_0=\frac{p_\infty(0)}{c_p}\frac{1}{\varrho_0} \quad . \tag{9.11}$$

If we now restrict our attention to the case of a steady isochoric flow, then the two first integrals (2.29) can be replaced by:

$$(\boldsymbol{\omega}_0+2\boldsymbol{\Omega})\cdot\nabla\psi_0=\frac{\partial I_0}{\partial\chi_0}+\frac{p_0}{\varrho_0^2}\frac{\partial\varrho_0}{\partial\chi_0} \quad ;$$
$$(\boldsymbol{\omega}_0+2\boldsymbol{\Omega})\cdot\nabla\chi_0=-\frac{\partial I_0}{\partial\psi_0}-\frac{p_0}{\varrho_0^2}\frac{\partial\varrho_0}{\partial\psi_0} \quad , \tag{9.12}$$

where

$$I_0=\frac{1}{2}|\boldsymbol{u}_0|^2+\frac{p_0}{\varrho_0}+gz=I_0(\psi_0,\chi_0) \quad ;$$
$$\varrho_0=\varrho_0(\psi_0,\chi_0), \quad \boldsymbol{u}_0=\nabla\psi_0\times\nabla\chi_0 \quad . \tag{9.13}$$

The relations (9.12) were exploited most particularly by Yih (1967) and Zeytounian (1969).

It is pointed out that (9.12) is a system of two equations for ψ_0 and χ_0 given the fact that the relative vorticity $\boldsymbol{\omega}_0$ can be expressed by using ψ_0 and χ_0:

$$\begin{aligned}\boldsymbol{\omega}_0&=\nabla\times\boldsymbol{u}_0=\nabla\times(\nabla\psi_0\times\nabla\chi_0)\\&=(\nabla\chi_0\cdot\nabla)\nabla\psi_0-(\nabla\psi_0\cdot\nabla)\nabla\chi_0+\Delta\chi_0\nabla\psi_0-\Delta\psi_0\nabla\chi_0 \quad ,\end{aligned} \tag{9.14}$$

where Δ is the three-dimensional Laplacian $\Delta \equiv \nabla^2$. On this subject, the thesis by Ionescu-Bujor (1961) and Zeytounian (1971) are both recommended.

9.2 Some Considerations concerning the Singular Nature of the Isochoric Approximation

If the limiting process (9.2) is applied to the Cauchy problem (3.17, 18), instead of (3.18), we obtain a second order in τ limiting equation:

$$\left\{\frac{\partial^2}{\partial\tau^2}\left[\frac{\lambda_\mu^{\mathrm{is}}/\mathrm{Ro}^2 + \varepsilon(\mathrm{Bo}/\mathrm{Fr}^2)K^2}{\varepsilon^2K^2 + \lambda_\mu^{\mathrm{is}}}\right]\right\}\chi_{nm}^{\mathrm{is}} = 0 \quad . \tag{9.15}$$

If (9.15) is compared to the equation obtained from the Boussinesq approximation [see (8.26)], it will be seen that the two equations coincide when $K^2\varepsilon\mathrm{Bo}/\mathrm{Fr}^2$ and $K^2[(\gamma-1)/\gamma]\hat{B}^2/\gamma$ are identified. Because of this, everything that has been said concerning the filtering role of the Boussinesq approximation is valid for the isochoric approximation in linear theory. In Sect. 9.3, we will see that such is not the case in an exact theory which takes into account the quasi-linear terms since the isochoric approximation is less "simplifying" than the Boussinesq approximation. Thus, unlike the latter, the isochoric approximation barely changes the vertical structure of the short internal waves (see the end of Sect. 3.2). This difference is related to the constraint (8.19). To be sure, the isochoric approximation filters out the short acoustical waves (as was pointed out in Sect. 3.3) since the short internal gravity waves have a frequeny $\sigma_{\mathrm{g\,is}}$ directly proportional to the intensity of the gravitational force g:

$$\sigma_{\mathrm{g\,is}} = \pm\frac{gH/U_0^2}{\left(1 + \lambda_\mu^{\mathrm{is}}/K^2\right)^{1/2}} \quad , \tag{9.16}$$

once it is supposed that $\varepsilon \equiv 1\,(H \equiv L)$ and $\mathrm{Ro} \equiv \infty$. It is essential to understand that both the Boussinesq and isochoric approximations are obtained from the limiting process $\mathrm{M}_\infty \to 0$ to which must be added in one case:

$$\mathrm{Bo} \to 0 \quad , \quad \text{with} \quad \mathrm{Bo} = \hat{B}\mathrm{M}_\infty \quad , \quad \hat{B} = \mathrm{O}(1)$$

and in the other case:

$$\gamma \to \infty \quad , \quad \text{with} \quad \gamma\mathrm{M}_\infty^2 = \hat{M}^2 = \mathrm{O}(1) \quad .$$

The singular nature of these approximations is above all due to the limiting process $\mathrm{M}_\infty \to 0$ although the singular nature of the latter is attenuated thanks to the similarity relations:

$$\frac{\mathrm{Bo}}{\mathrm{M}_\infty} = \hat{B} = \mathrm{O}(1) \quad \text{and} \quad \gamma\mathrm{M}_\infty^2 = \hat{M}^2 = \mathrm{O}(1) \quad .$$

Therefore, we again emphasize the importance of understanding that the central problem is related to the extremely singular nature of the limiting process $M_\infty \to 0$ for atmospheric flows.

In conclusion, it is pointed out that the isochoric approximation again leads to an adjustment problem in the vicinity of $t = 0$. This problem has not yet, however, been analyzed in detail.

9.3 The Relation Between the Isochoric and Boussinesq Approximations

In order to elucidate the difference between the isochoric and Boussinesq approximation, we are going to consider the simple case of the steady two-dimensional isochoric (adiabatic) flow while ignoring the Coriolis force, the fluid being a heavy perfect gas (with c_p constant and $c_v = 0$). The equations are then written in the following *dimensional* form

$$u\frac{\partial u}{\partial x} + w\frac{\partial u}{\partial z} + \frac{1}{\varrho}\frac{\partial p}{\partial x} = 0 \quad ; \quad u\frac{\partial w}{\partial x} + w\frac{\partial w}{\partial z} + \frac{1}{\varrho}\frac{\partial p}{\partial z} + g = 0 \quad ;$$
$$u\frac{\partial \varrho}{\partial x} + w\frac{\partial \varrho}{\partial z} = 0 \quad ; \quad \frac{\partial u}{\partial x} + \frac{\partial w}{\partial z} = 0 \quad . \tag{9.17}$$

The isochoric flow governed by (9.17) is supposed to take place in a duct having a curvilinear bottom. $U_\infty(z_\infty)\boldsymbol{i}$ and $\varrho_\infty(z_\infty)$ designate respectively the velocity and density at far upstream infinity where the duct is bounded by two parallel walls placed respectively in $z \equiv z_\infty = 0$ and $z \equiv z_\infty = H_0$, where z_∞ is the altitude of a streamline at an abscissa $x \equiv x_\infty = -\infty$. The curvilinear portion of the bottom of the duct is assumed as being confined to the vicinity of $x = 0$ between the abscisses $x = -L_0/2$ and $x = +L_0/2$.

$\Delta = z - z_\infty$ designates the variation (with dimensions) in altitude of the streamline in the flow perturbed by the curvilinear bottom of the channel with respect to its position in the flow at upstream infinity.

The equation for the curvilinear part of the bottom of the duct is assumed to be of the form:

$$z = h_0 h\left(\frac{x}{L_0}\right) \quad , \quad h_0 = \max_{-L_0/2 \le x \le L_0/2} \left| h\left(\frac{x}{L_0}\right)\right| \quad . \tag{9.18}$$

From system (9.17), we can construct[3] a second order partial differential equation for Δ:

$$\frac{\partial^2 \Delta}{\partial x^2} + \frac{\partial^2 \Delta}{\partial z^2} - \frac{g}{U_\infty^2 \varrho_\infty}\frac{d\varrho_\infty}{dz_\infty}\Delta$$
$$= \frac{1}{2}\left\{\left(\frac{\partial \Delta}{\partial x}\right)^2 + \left(\frac{\partial \Delta}{\partial z}\right)^2 - 2\frac{\partial \Delta}{\partial z}\right\}\frac{d\log(U_\infty^2 \varrho_\infty)}{dz_\infty} \quad . \tag{9.19}$$

[3] In Chap. 13, a general equation is obtained for the steady two-dimensional flow of a heavy (compressible) perfect gas with constant c_p and c_v. Equation (9.19) corresponds to a particular case of this equation.

The following boundary conditions must be assigned to (9.19)

$$\Delta\left(x,\, h_0 h\left(\frac{x}{L_0}\right)\right) = h_0 h\left(\frac{x}{L_0}\right) \quad , \quad -\frac{L_0}{2} \le x \le +\frac{L_0}{2} \quad ;$$
$$\Delta(-\infty,\, z_\infty) = 0 \quad , \quad \lim_{x \to -\infty} h\left(\frac{x}{L_0}\right) \equiv 0 \quad ;$$
$$\Delta(x,\, H_0) = 0 \quad ; \tag{9.20}$$
$$\lim_{x \to +\infty} \left\{ \left|\frac{\partial \Delta}{\partial x}\right| + \left|\frac{\partial \Delta}{\partial z}\right| \right\} < \infty \quad ,$$

this last condition being the only one that can be imposed due to the appearance of waves downstream.

A simple case is the following:

$$U_\infty \equiv U_0 = \text{const} \quad , \quad \varrho_\infty(z_\infty) \equiv \varrho_\infty(0)\exp\left(-\frac{g}{RT_\infty(0)} z_\infty \right) , \tag{9.21}$$

with

$$N^2_{\infty,\,\text{is}}(z_\infty) = -\frac{g}{\varrho_\infty}\frac{d\varrho_\infty}{dz_\infty} \equiv N^2_{\infty,\,\text{is}}(0) = \frac{g^2}{RT_\infty(0)} \quad , \tag{9.22}$$

in dimensional variables.

In this case, (9.19) becomes an equation with constant coefficients:

$$\frac{\partial^2 \Delta}{\partial x^2} + \frac{\partial^2 \Delta}{\partial z^2} + \frac{(g/U_0)^2}{RT_\infty(0)}\Delta$$
$$+\frac{1}{2}\frac{g}{RT_\infty(0)}\left[\left(\frac{\partial \Delta}{\partial x}\right)^2 + \left(\frac{\partial \Delta}{\partial z}\right)^2 - 2\frac{\partial \Delta}{\partial z}\right] = 0 \quad . \tag{9.23}$$

In the above-considered problem, various characteristic length scales are introduced: three outer scales related to the geometry of the problem – L_0, H_0 and h_0, and two inner scales related to the steady waves which appear downstream – $RT_\infty(0)/g$ and $RT_\infty(0)/(g/U_0)^2$. Let λ_0 be the wave length of these steady internal waves (which are mainly gravity waves since the isochoric approximation filters out the short internal acoustic waves). We will take h_0 as the characteristic amplitude of these waves. The following non-dimensionalization can then be carried out:

$$\xi = \frac{x}{\lambda_0} \quad , \quad \zeta = \frac{z}{H_0} \quad , \quad \delta = \frac{\Delta}{h_0} \quad , \tag{9.24}$$

and (9.23) takes on the following non-dimensional form:

$$\varepsilon^2\beta^2\frac{\partial^2 \delta}{\partial \xi^2} + \frac{\partial^2 \delta}{\partial \zeta^2} + \frac{\text{Bo}^2}{\hat{M}^2}\delta + \frac{1}{2}\text{Bo}\nu\left[\varepsilon^2\beta^2\left(\frac{\partial \delta}{\partial \xi}\right)^2 + \left(\frac{\partial \delta}{\partial \zeta}\right)^2\right] - \text{Bo}\frac{\partial \delta}{\partial \zeta} = 0, \tag{9.25}$$

where

$$\beta = \frac{L_0}{\lambda_0} \quad , \quad \varepsilon\beta = \frac{H_0}{\lambda_0} \quad , \quad \nu = \frac{h_0}{H_0} \quad . \tag{9.26}$$

The following (dimensionless) boundary conditions must be prescribed for (9.25):

$$\delta\left(\xi,\, \nu h\left(\frac{\xi}{\beta}\right)\right) = h(\xi/\beta) \quad , \quad -\frac{1}{2} \le \frac{\xi}{\beta} \le +\frac{1}{2} \quad ;$$

$$\lim_{\xi \to -\infty} \delta = 0 \quad , \quad \delta(\xi,\, 1) = 0 \quad , \quad h(-\infty) \equiv 0 \quad ; \tag{9.27}$$

$$\lim_{\xi \to +\infty} \left\{ \varepsilon\beta \left| \frac{\partial \delta}{\partial \xi} \right| + \left| \frac{\partial \delta}{\partial \zeta} \right| \right\} < \infty \quad .$$

If the Boussinesq equations (see Chap. 8) are now substituted for (9.17) while making the same hypotheses as found above, then the following equation is obtained instead of (9.25) [this could easily be confirmed via reasoning such as that which led to (8.42)]:

$$\varepsilon^2 \beta^2 \frac{\partial^2 \delta}{\partial \xi^2} + \frac{\partial^2 \delta}{\partial \zeta^2} + \frac{\gamma - 1}{\gamma} \frac{\mathrm{Bo}^2}{\gamma \mathrm{M}_\infty^2} \delta = 0 \quad . \tag{9.28}$$

Equation (9.28) can obviously be obtained immediately from (9.25) by making $\mathrm{Bo} \to 0$ and $\mathrm{M}_\infty \to 0$ such that $\mathrm{Bo}/\mathrm{M}_\infty = \hat{B} = \mathrm{O}(1)$ and also by making in (9.28) $\gamma \to \infty$ and $\mathrm{M}_\infty \to 0$ in such a way that $\gamma \mathrm{M}_\infty^2 = \hat{M}^2 = \mathrm{O}(1)$ since $\hat{B}^2/\gamma \equiv \mathrm{Bo}^2/\hat{M}^2$. Therefore:

$$\lim_{\text{Boussinesq}} (\text{isochoric eqs.}) = \lim_{\text{isochoric}} (\text{Boussinesq eqs.}) \quad .$$

It is readily observed that the isochoric equations lead to an equation [(9.25)] which is less degenerated than (9.28), which results from the Boussinesq equations. Moreover, it is obvious that if the Boussinesq approximation is carried out, then it becomes futile to effect the isochoric approximation. On the contrary, the Boussinesq approximation can be carried out following the isochoric approximation as a first approach allowing the resolution of the isochoric equation (9.25).

To finish up this discussion, it is emphasized that if we start with the general equation governing the steady two-dimensional flow of a heavy (compressible) perfect gas with constant c_p and c_v (see Chap. 13), then by respecting the hypotheses (9.21) and by setting $\varepsilon = \beta = \nu \equiv 1$, we arrive at the dimensionless equation:

$$\begin{aligned} &\frac{\partial^2 \delta}{\partial \xi^2} + \frac{\partial^2 \delta}{\partial \zeta^2} + (1 + \omega)^2 \frac{\gamma - 1}{\gamma} K_0^2 \delta \\ &= \frac{\gamma - 1}{\gamma} \frac{\mathrm{Bo}}{2} \left\{ (2 + \omega)\omega - \left[\left(\frac{\partial \delta}{\partial \xi} \right)^2 + \left(\frac{\partial \delta}{\partial \zeta} \right)^2 - 2 \frac{\partial \delta}{\partial \zeta} \right] \right\} \\ &+ \frac{1}{1 + \omega} \left\{ \frac{\partial \omega}{\partial \xi} \frac{\partial \delta}{\partial \xi} + \frac{\partial \omega}{\partial \zeta} \frac{\partial \delta}{\partial \zeta} - \frac{\partial \omega}{\partial \zeta} \right\} \quad ; \end{aligned} \tag{9.29}$$

with

$$K_0^2 \equiv \frac{\mathrm{Bo}^2}{\gamma \mathrm{M}_\infty^2} \quad ,$$

on which must be associated the following relation

$$(1+\omega)^{\gamma-1} = 1 + \frac{1}{2}\frac{\gamma-1}{\gamma}\frac{\text{Bo}^2}{K_0^2}\frac{1}{(1+\omega)^2}\left[\left(\frac{\partial\delta}{\partial\xi}\right)^2 + \left(\frac{\partial\delta}{\partial\zeta}\right)^2 - 2\frac{\partial\delta}{\partial\zeta} + 1\right]$$
$$+ \frac{\gamma-1}{\gamma}\text{Bo}\delta - \frac{1}{2}\frac{\gamma-1}{\gamma}\frac{\text{Bo}^2}{K_0^2} \quad . \tag{9.30}$$

We now assume that K_0^2 will always remain of the order unity while carrying out the limiting processes described below (which is obviously always the case within the framework of the isochoric and Boussinesq approximations).

If in (9.29, 30), $\text{Bo} \to 0$ with fixed γ, then it follows that

$$\omega \equiv 0 \quad , \quad \frac{\partial^2\delta_B}{\partial\xi^2} + \frac{\partial^2\delta_B}{\partial\zeta^2} + \frac{\gamma-1}{\gamma}K_0^2\delta_B = 0 \quad , \tag{9.31}$$

and thus, we once again arrive at the equation corresponding to the Boussinesq approximation (in this case $K_0^2 \equiv \hat{B}^2/\gamma$).

On the other hand, if in (9.29, 30), the limiting process $\gamma \to \infty$ is carried out with fixed Bo, then we have:

$$\omega \equiv 0 \quad ,$$
$$\frac{\partial^2\delta_{is}}{\partial\xi^2} + \frac{\partial^2\delta_{is}}{\partial\zeta^2} + K_0^2\delta_{is} + \frac{\text{Bo}}{2}\left[\left(\frac{\partial\delta_{is}}{\partial\xi}\right)^2 + \left(\frac{\partial\delta_{is}}{\partial\zeta}\right)^2 - 2\frac{\partial\delta_{is}}{\partial\zeta}\right] = 0 \quad , \tag{9.32}$$

with $K_0^2 \equiv \text{Bo}^2/\hat{M}^2$.

If in (9.31), we make $\gamma \to \infty$, and in (9.32), $\text{Bo} \to 0$, we arrive at the same Helmholtz equation:

$$\frac{\partial^2\delta_0}{\partial\xi^2} + \frac{\partial^2\delta_0}{\partial\zeta^2} + K_0^2\delta_0 = 0 \quad , \tag{9.33}$$

since

$$\lim_{\gamma\to\infty}\frac{\gamma-1}{\gamma}\frac{\text{Bo}^2}{\hat{M}^2} \equiv \frac{\text{Bo}^2}{\hat{M}^2} = \lim_{\text{Bo}\to 0}\frac{\hat{B}^2}{\gamma} \quad ,$$

and thus

$$K_0^2 = \frac{\text{Bo}^2}{\hat{M}^2} = \frac{\hat{B}^2}{\gamma} \equiv \frac{\text{Bo}^2}{\gamma\text{M}_\infty^2} = \text{O}(1) \quad .$$

It is thus obvious that the following limiting processes are equivalent:

$$\gamma \to \infty \quad , \text{ then } \text{Bo} \to 0 \text{ and } \text{Bo} \to 0 \quad , \text{ then } \gamma \to \infty \quad .$$

Nonetheless, it must be stressed that alone, the limiting process $\text{Bo} \to 0$ [of Boussinesq with $K_0^2 = \text{O}(1)$] which leads to the linear equation (9.31) yields a much stronger degeneracy of the complete equations (9.29, 30) than does the limiting process $\gamma \to \infty$ alone [isochoric and again with $K_0^2 = \text{O}(1)$] which leads to the non-linear equation (9.32). Hence, it can be said that the adiabatic Boussinesq approximation is a particular case of the isochoric approximation when $\text{Bo} \to 0$ as long as $(\gamma-1)/\gamma$ is identified with unity.

Finally, it is pointed out that the derivation of (9.33) [the so-called Long equation (1953)] is independent of the manner in which the parameters Bo and γ^{-1} tend towards zero, it being understood that K_0^2 remains fixed while carrying out these limiting processes.

All statements in the present section are relative to the case defined in (9.21). However, it is obvious that this hypothesis is not essential. Let us return to (9.19) and suppose only that $U_\infty \equiv U_0 = \text{const}$, with the function $\varrho_\infty(z_\infty)$ remaining arbitrary (no wind shear upstream but the stratification may be in any proportion). If we take into account the definition of $N^2_{\infty,\text{is}}(z_\infty)$ where $z_\infty = z - \Delta$,

$$N^2_{\infty,\text{is}}(z-\Delta) \equiv -\frac{g}{\varrho_\infty}\frac{d\varrho_\infty}{dz_\infty} \quad , \tag{9.34}$$

we can rewrite (9.19) in the following *dimensional form*:

$$\frac{\partial^2\Delta}{\partial x^2} + \frac{\partial^2\Delta}{\partial z^2} + N^2_{\infty,\text{is}}(z-\Delta)\left\{\frac{\Delta}{U_0^2} + \frac{1}{2g}\left[\left(\frac{\partial\Delta}{\partial x}\right)^2 + \left(\frac{\partial\Delta}{\partial z}\right)^2 - 2\frac{\partial\Delta}{\partial z}\right]\right\} = 0 \tag{9.35}$$

We now turn our attention to the *dimensionless* variables. According to (9.24):

$$x = \lambda_0\xi \quad , \quad z = H_0\zeta \quad , \quad \Delta = h_0\delta \quad , \quad z_\infty = \frac{RT_\infty(0)}{g}\zeta_\infty \quad .$$

We introduce the dimensionless function:

$$\mathcal{B}_\infty(\zeta_\infty) \equiv \frac{N^2_{\infty,\text{is}}\{[RT_\infty(0)/g]\zeta_\infty\}}{N^2_{\infty,\text{is}}(0)} \quad , \tag{9.36}$$

with $N^2_{\infty,\text{is}}(0) \equiv g^2/RT_\infty(0)$. We naturally have:

$$\zeta_\infty = \text{Bo}(\zeta - \nu\delta) \quad , \tag{9.37}$$

with $\nu = h_0/H_0$.

It thus follows for the non-dimensional variation of the streamline $\delta(\xi, \zeta)$, the dimensionless equation

$$\varepsilon^2\beta^2\frac{\partial^2\delta}{\partial\xi^2} + \frac{\partial^2\delta}{\partial\zeta^2} + \mathcal{B}_\infty\big(\text{Bo}(\zeta-\nu\delta)\big)\Bigg\{K_0^2\delta$$
$$+\frac{1}{2}\text{Bo}\nu\left[\varepsilon^2\beta^2\left(\frac{\partial\delta}{\partial\xi}\right)^2 + \left(\frac{\partial\delta}{\partial\zeta}\right)^2\right] - \text{Bo}\frac{\partial\delta}{\partial\zeta}\Bigg\} = 0 \quad , \tag{9.38}$$

with $K_0^2 \equiv \text{Bo}^2/\hat{M}^2$.

When $\text{Bo} \to 0$, the following Boussinesq equation results:

$$\varepsilon^2\beta^2\frac{\partial^2\delta_\text{B}}{\partial\xi^2} + \frac{\partial^2\delta_\text{B}}{\partial\zeta^2} + \text{Ko}^2\delta_\text{B} = 0 \quad ,$$

with $\text{Ko}^2 = \hat{B}^2/\gamma$, since $\mathcal{B}_\infty(0) \equiv 1$.

Equation (9.38) can serve as the starting point [with the boundary conditions (9.27)] of an asymptotic theory for isochoric waves in a duct. The reader is referred to the two articles by Leonov and Miropolsky (1975a,b) for further details. For our purposes here, we will simply consider the equation with constant coefficients (9.25) in order to outline a theory of isochoric waves.

9.4 Wave Phenomena in the Isochoric Flows

Equation (9.25) is then our departure point. Let it be specified from the start that at the present time, it is not known (at least, to our knowledge) how to insert a local theory of waves (both long and short) into a boundary-value problem which is posed with conditions of the type found in (9.27). In Chap. 13, we will come back to this important research problem which deserves a detailed analysis. For the time being, we will only consider either a duct with parallel walls $\zeta = 0$ and $\zeta = 1$, or a medium which is "locally infinite" both horizontally and vertically. Given these hypotheses, we can of course suppose that $\varepsilon \equiv 1$, i.e., that $L_0 \equiv H_0$ with H_0 being the width of the duct. Our basic equation is then:

$$\beta^2\frac{\partial^2\delta}{\partial\xi^2} + \frac{\partial^2\delta}{\partial\zeta^2} + \sigma^2\delta + \frac{1}{2}\mathrm{Bo}\nu\left[\beta^2\left(\frac{\partial\delta}{\partial\xi}\right)^2 + \left(\frac{\partial\delta}{\partial\zeta}\right)^2\right] - \mathrm{Bo}\frac{\partial\delta}{\partial\zeta} = 0 \quad , \tag{9.39}$$

where $\sigma^2 \equiv \mathrm{Bo}^2/\hat{M}^2$, $\beta \equiv H_0/\lambda_0$.

In what follows, we will first consider two limiting cases which correspond on the one hand to long waves, and on the other hand, to short waves.

9.4.1 The Long Wave Theory

The long wave theory is usually based on the hypothesis:

$$\beta \ll 1 \Rightarrow \lambda_0 \gg H_0 \quad . \tag{9.40}$$

Since, generally speaking, Bo = O(1), it turns out that in a duct with parallel walls having a width $H_0 \cong RT_\infty(0)/g \cong 10^4$ m, λ_0, the wave length of the long internal waves must be "much greater" than 10 km.

To (9.39), the following rectilinear wall conditions are added:

$$\delta(\xi, 0) = 0 \quad , \quad \delta(\xi, 1) = 0 \quad . \tag{9.41}$$

For the time being, no hypothesis will be made concerning $\nu = h_0/H_0$ which characterizes the non-dimensional amplitude of the wave which is assumed to be long.

When $\beta \to 0$, we postulate the asymptotic expansions:

$$\delta = \beta^a\delta_a + \beta^b\delta_b + \ldots \quad ; \quad \sigma^2 = \sigma_0^2 + \beta^c\sigma_c^2 + \ldots \quad . \tag{9.42}$$

After carrying out a few rather standard calculations, it can be confirmed that in order to obtain in the right hand member of the equation for δ_b the maximum number of terms associated with δ_a, and in particular, a term related to the "non-linearity", it is necessary that:

$$b = 2 + a \quad , \quad c = 2 \quad , \quad 2 + a = 2a \Rightarrow a = 2 \quad , \quad b = 4 \quad .$$

For δ_2 and δ_4, the following limiting equations thus result:

$$\frac{\partial^2 \delta_2}{\partial \zeta^2} - \mathrm{Bo}\frac{\partial \delta_2}{\partial \zeta} + \sigma_0^2 \delta_2 = 0 \quad , \tag{9.43}$$

$$\frac{\partial^2 \delta_4}{\partial \zeta^2} - \mathrm{Bo}\frac{\partial \delta_4}{\partial \zeta} + \sigma_0^2 \delta_4 = -\mathcal{L}_2(\delta_2) \quad , \tag{9.44}$$

where

$$\mathcal{L}_2(\delta_2) = \frac{d^2 \delta_2}{\partial \xi^2} + \sigma_2^2 \delta_2 + \frac{1}{2}\mathrm{Bo}\nu\left(\frac{\partial \delta_2}{\partial \zeta}\right)^2 \quad . \tag{9.45}$$

The following boundary conditions must be imposed on (9.43, 44)

$$\delta_2(\xi, 0) = \delta_4(\xi, 0) = 0 \quad , \quad \delta_2(\xi, 1) = \delta_4(\xi, 1) = 0 \quad . \tag{9.46}$$

The following is a solution to (9.43):

$$\delta_2(\xi, \zeta) = U_2(\xi) W_2(\zeta) \quad , \tag{9.47}$$

where the function $W_2(\zeta)$ must satisfy the following Sturm-Liouville eigenvalue problem:

$$\begin{aligned} &\frac{d^2 W_2}{d\zeta^2} - \mathrm{Bo}\frac{dW_2}{d\zeta} + \sigma_0^2 W_2 = 0 \quad ; \\ &W_2(0) = W_2(1) = 0 \quad . \end{aligned} \tag{9.48}$$

This eigenvalue problem have an enumerable set of eigenfunctions:

$$W_2^{(n)}(\zeta) = \mathcal{A}_n \, \mathrm{e}^{(\mathrm{Bo}/2)\zeta} \sin(n\pi\zeta) \tag{9.49}$$

and of eigenvalues:

$$\sigma_{0,n}^2 = n^2\pi^2 + \frac{\mathrm{Bo}^2}{4} \quad , \tag{9.50}$$

which correspond to the normal modes of the internal waves. It is observed that all the $\sigma_{0,n}^2$ are real and form an ascending sequence (which is enumerable) of which all the terms are positive and which tends to $+\infty$ with $n \to +\infty$. To determine the coefficient $\mathcal{A}_n$, it is supposed that the eigenfunctions (9.49) are normalized in such a way that:

$$\int_0^1 \left[W_2^{(n)}(\zeta)\right]^2 d\zeta = 1 \quad . \tag{9.51}$$

This leads to the following relation for $\mathcal{A}_n$:

$$|\mathcal{A}_n| = \sqrt{2} \tag{9.52}$$

and we take $\mathcal{A}_n > 0$.

It is pointed out that it is necessary to define in the space of solutions a product associated with the norm. Therefore, let:

$$\langle f, g\rangle = \int_0^1 e^{-\mathrm{Bo}\zeta} f(\zeta)g(\zeta)d\zeta \quad . \tag{9.53}$$

This scalar product is clearly a bilinear symmetrical form and, in addition, it is positive non-degenerated because:

$$\langle f, f\rangle \geq e^{-\mathrm{Bo}} \|f\|^2_{L^2(0,1)} \quad .$$

To be sure:

$$\langle W_2^{(p)}, W_2^{(q)}\rangle = 0 \quad , \quad \text{when} \quad p \neq q \quad .$$

Let us now return to the problem (9.44) with (9.45)

$$\frac{\partial^2 \delta_4}{\partial \zeta^2} - \mathrm{Bo}\frac{\partial \delta_4}{\partial \zeta} + \sigma_0^2 \delta_4 = -\left\{\frac{\partial^2 \delta_2}{\partial \xi^2} + \sigma_2^2 \delta_2 + \frac{1}{2}\mathrm{Bo}\nu\left(\frac{\partial \delta_2}{\partial \zeta}\right)^2\right\} \quad , \tag{9.54a}$$

and the conditions

$$\delta_4(\xi, 0) = \delta_4(\xi, 1) = 0 \quad , \tag{9.54b}$$

where δ_2 is the solution of (9.43) with (9.46) which is associated with the eigenvalue $\sigma_{0,n}^2 = \frac{1}{4}\mathrm{Bo}^2 + n^2\pi^2$.

We seek $\delta_4(\xi, \zeta)$ in the form of the product $U_4(\xi)W_4(\zeta)$ and it follows for $W_4(\zeta)$ the following equivalent problem (since $U_4(\xi) \neq 0$) which we write in the form:

$$\mathcal{M}W_4 + \sigma_0^2 W_4 = -\frac{1}{U_4(\xi)}\mathcal{L}_2(\delta_2) \quad , \quad W_4(0) = W_4(1) = 0 \quad , \tag{9.55}$$

where $\mathcal{M} \equiv (d^2/d\zeta^2) - \mathrm{Bo}(d/d\zeta)$ is an auto-adjoint operator which respect to the scalar product (9.53). Thus, (9.55) can only possess solutions if we have the following compatibility condition[4]:

$$\frac{1}{U_2(\xi)}\left\langle \delta_2^{(n)}, \frac{1}{U_4(\xi)}\mathcal{L}_2\left(\delta_2^{(n)}\right)\right\rangle = 0 \quad , \tag{9.56}$$

when $\sigma_{0,n}^2 = \frac{1}{4}\mathrm{Bo}^2 + n^2\pi^2$.

We express:

[4] On this subject, the reader is referred to Courant and Hilbert (1966; see pages 358–362 of Volume 1) or else Zeytounian (1986; see pages 59–74).

$$\mathcal{L}_2\left(\delta_2^{(n)}\right) = \frac{\partial^2}{\partial \xi^2}\left(\delta_2^{(n)}\right) + \sigma_2^2\delta_2^n + \frac{1}{2}\mathrm{Bo}\nu\left(\frac{\partial \delta_2^{(n)}}{\partial \zeta}\right)^2$$

$$= \sqrt{2}\,\mathrm{e}^{(\mathrm{Bo}/2)\zeta}\sin(n\pi\zeta)\left[\frac{d^2U_2(\xi)}{d\xi^2} + \sigma_2^2 U_2(\xi)\right]$$

$$+\,\mathrm{Bo}\nu\,\mathrm{e}^{\mathrm{Bo}\zeta}\left[\frac{\mathrm{Bo}}{2}\sin(n\pi\zeta) + n\pi\cos(n\pi\zeta)\right]^2 U_2^2(\xi) \quad .$$

In this case, the condition (9.56) yields the following relation:

$$2\left[\frac{d^2U_2(\xi)}{d\xi^2} + \sigma_2^2U_2(\xi)\right]\int_0^1 \sin^2(n\pi\zeta)d\zeta$$

$$+\sqrt{2}\mathrm{Bo}\nu U_2^2(\xi)\int_0^1 \mathrm{e}^{(\mathrm{Bo}/2)\zeta}\sin(n\pi\zeta)\left[\frac{\mathrm{Bo}}{2}\sin(n\pi\zeta)\right.$$

$$\left.+n\pi\cos(n\pi\zeta)\right]^2 d\zeta = 0 \quad . \tag{9.57}$$

After a rather tedious calculation, the following equation is obtained for determining the function $U_2^{(n)}(\xi)$:

$$\frac{d^2U_2^{(n)}(\xi)}{d\xi^2} + \sigma_2^2 U_2^{(n)}(\xi) + \alpha_n\left[U_2^{(n)}(\xi)\right]^2 = 0 \quad , \tag{9.58}$$

where

$$\alpha_n = \frac{3}{\sqrt{2}}\mathrm{Bo}\nu n^3\pi^3\left[1-(-1)^n\mathrm{e}^{\mathrm{Bo}/2}\right]\frac{(3/4)\mathrm{Bo}^2 + 2n^2\pi^2}{[\mathrm{Bo}^2/4 + 9n^2\pi^2][\mathrm{Bo}^2/4 + n^2\pi^2]} \ . \tag{9.59}$$

the sign of α being determined by the natural whole number n. If $n = 2p$, then $\alpha_n < 0$, whereas if $n = 2p+1$, then $\alpha_n > 0$.

Equation (9.58) is of the type known as "Korteweg-de Vries" which is a steady constant coefficient equation integrated once with respect to ξ. When $\alpha_n > 0$, (9.58) can be integrated once and yields

$$\frac{3}{2\alpha_n}\left(\frac{dU_2^{(n)}}{d\xi}\right)^2 = -\left[\left(U_2^{(n)}\right)^3 + \frac{3\sigma_2^2}{2\alpha_n}\left(U_2^{(n)}\right)^2 - \beta_0\right] \quad , \tag{9.60}$$

where $\beta_0 > 0$ is an integration constant which can be interpreted as the potential energy of the long steady waves. If $\alpha_n < 0$, a first integral can still be obtained as long as the invariance of (9.58) with respect to the following transformation is taken into account:

$$\xi \to -\xi \quad , \quad U_2^{(n)} \to -U_2^{(n)} \quad \text{and} \quad \alpha_n \to -\alpha_n \quad .$$

In a particular case, when $U_2^{(n)}(\xi)$ as well as its derivatives tend to zero with $\xi \to \infty$, $\beta_0 \equiv 0$ results and (9.60) can be written as follows

$$\frac{3}{2\alpha_n}\left[\frac{dU_2^{(n)}}{d\xi}\right]^2 = \left[U_2^{(n)}\right]^2 \left\{\gamma_n - U_2^{(n)}\right\} \quad , \tag{9.61}$$

with $\gamma_n = -(3/2\alpha_n)\sigma_2^2$. When $\gamma_n > 0$, i.e., when $\sigma_2^2 < 0$, the exact solution of (9.61) takes on the form:

$$U_2^{(n)}(\xi) = \gamma_n \operatorname{sech}^2\left\{\sqrt{\frac{\gamma_n \alpha_n}{6}}\xi\right\} \quad , \tag{9.62}$$

which gives us a solitary wave having a maximum height of γ_n.

Before turning our attention to the short wave theory, it is important to note that when $\beta \ll 1$, it is also necessary to assume that $\nu \ll 1$ in (9.39). Otherwise stated, for *long, quasi non-dispersive isochoric waves which are slightly non-linear*, the following hypothesis must be made:

$$\nu = \hat{\nu}\beta^2 \quad , \quad \text{with} \quad \hat{\nu} = \mathrm{O}(1) \quad . \tag{9.63}$$

This means that:

$$h_0 \ll H_0 \ll \lambda_0 \quad \text{and}$$

$$\lambda_0 \cong H_0\sqrt{\frac{H_0}{h_0}} \quad . \tag{9.64}$$

Relation (9.64) specifies the order of magnitude of the wave length λ_0 versus the scales H_0 and h_0. According to (9.63), in order to take into account a weak non-linearity, it is necessary to carry out the expansion (9.42) for δ up to the sixth order in β. This is due to the fact that the equation corresponding to (9.58) is then linear [the quadratic term does not appear since in this case, α_n is proportional to β^2 according to (9.59, 63)].

9.4.2 The Short Wave Theory

Let us return to (9.39) from the first part of Sect. 9.4. The short wave theory is fundamentally based on the hypothesis that:

$$\beta \gg 1 \Rightarrow \lambda_0 \ll H_0 \quad . \tag{9.65}$$

The above is, however, insufficient since for the short internal isochoric waves resulting from (9.9), we necessarily have:

$$\sigma^2 \gg 1 \Rightarrow H_0 \gg \frac{U_0}{N^2_{\infty,\mathrm{is}}(0)} \equiv \frac{U_0}{g}\sqrt{RT_\infty(0)} \approx \lambda_0 \quad . \tag{9.66}$$

Therefore, $\beta \gg 1$ and $\sigma^2 \gg 1$ and the significant situation corresponds to:

$$\frac{\sigma^2}{\beta^2} \equiv \frac{N^2_{\infty,\mathrm{is}}(0)}{U_0^2}\lambda_0^2 = \chi^2 = \mathrm{O}(1) \quad . \tag{9.67}$$

Following the above hypotheses, it seems a good idea to introduce into (9.39) the new variable:

$$Z = \beta\zeta = \frac{\zeta}{\mu} \equiv \frac{z}{\lambda_0} \quad , \quad \mu = \frac{1}{\beta} \quad . \tag{9.68}$$

In this way, both x and Z are reduced with respect to λ_0. A consequence of all of the preceding, is that instead of (9.39), the following equation is obtained for the function $\delta^*(\xi, Z) \equiv \delta(\xi, \mu Z)$

$$\frac{\partial^2\delta^*}{\partial\xi^2} + \frac{\partial^2\delta^*}{\partial Z^2} + \chi^2\delta^* + \frac{1}{2}\mathrm{Bo}\nu\left[\left(\frac{\partial\delta^*}{\partial\xi}\right)^2 + \left(\frac{\partial\delta^*}{\partial Z}\right)^2\right]$$
$$-\mathrm{Bo}\mu\frac{\partial\delta^*}{\partial Z} = 0 \quad . \tag{9.69}$$

But from (9.69), it is noticed that the most significant short wave theory corresponds to the following limiting situation:

$$\mathrm{Bo} \to \infty \quad \text{and} \quad \nu = \mu \to 0 \quad , \quad \text{such that} \quad \mathrm{Bo}\mu \equiv \hat{\mu} = \mathrm{O}(1) \quad . \tag{9.70}$$

Stated otherwise, this means that:

$$H_0 \gg \frac{RT_\infty(0)}{g} \quad \text{and} \quad h_0 = \lambda_0 \approx \frac{RT_\infty(0)}{g}$$

or, according to (9.66)

$$U_0^2 \approx RT_\infty(0) \Longleftrightarrow \gamma\mathrm{M}_\infty^2 \equiv \hat{M}^2 = \mathrm{O}(1) \quad .$$

We thus again encounter the condition of isochoricity (9.9) which ensures the formal coherence of the above hypotheses.

Finally, the following model equation is obtained for analyzing the short internal waves:

$$\frac{\partial^2\delta^*}{\partial\xi^2} + \frac{\partial^2\delta^*}{\partial Z^2} + \chi^2\delta^* + \frac{\hat{\mu}}{2}\left[\left(\frac{\partial\delta^*}{\partial\xi}\right)^2 + \left(\frac{\partial\delta^*}{\partial Z}\right)^2 - 2\frac{\partial\delta^*}{\partial Z}\right] = 0 \quad . \tag{9.71}$$

When $\hat{\mu} \to 0$, we find ourselves within the framework of the Boussinesq theory (i.e., Bo is fixed, $\mu \to 0$, and then Bo $\to \infty$).

A difficult and as yet unsolved problem concerns the reformulation of the boundary conditions (9.27) which we will not attempt to tackle here. As $\hat{\mu} \ll 1$, the "slow" space variables appear:

$$\hat{\mu}\xi \equiv \hat{\xi} \quad \hat{\mu}Z = \hat{Z} \quad , \tag{9.72}$$

and thus a Luke (1966) and Whitham (1970)–type double scale technique can be applied. To the end, we must introduce a fast phase:

$$\phi = \frac{1}{\hat{\mu}}\hat{\theta}(\hat{\xi}, \hat{Z}) \quad , \tag{9.73a}$$

and define the horizontal and vertical wave numbers k and l, respectively, as follows

$$\frac{\partial\phi}{\partial\hat{\xi}} = \frac{k(\hat{\xi},\,\hat{Z})}{\hat{\mu}} \quad \text{and} \quad \frac{\partial\phi}{\partial\hat{Z}} = -\frac{l(\hat{\xi},\,\hat{Z})}{\hat{\mu}} \quad . \tag{9.73b}$$

We thus have:

$$\frac{\partial k}{\partial\hat{Z}} + \frac{\partial l}{\partial\hat{\xi}} = 0 \quad . \tag{9.73c}$$

We now postulate the following ("free") solution of (9.71)

$$\delta^* = \hat{\delta}(\phi;\ \hat{\xi},\ \hat{Z};\ \hat{\mu})$$

and seek $\hat{\delta}$ in the form of an "adiabatic approximation" which corresponds to a modulated wave solution with slowly varying parameters

$$\hat{\delta} = \hat{\delta}_0(\phi;\ \hat{\xi},\ \hat{Z}) + \hat{\mu}\hat{\delta}_1(\phi;\ \hat{\xi},\ \hat{Z}) + \ldots \tag{9.74}$$

We have:

$$\frac{\partial\delta^*}{\partial\xi} = \frac{\partial\hat{\delta}_0}{\partial\phi}k + \hat{\mu}\left(\frac{\partial\hat{\delta}_0}{\partial\hat{\xi}} + \frac{\partial\hat{\delta}_1}{\partial\phi}k\right) + \mathrm{O}(\hat{\mu}^2) \quad ;$$

$$\frac{\partial\delta^*}{\partial Z} = -\frac{\partial\hat{\delta}_0}{\partial\phi}l + \hat{\mu}\left(\frac{\partial\hat{\delta}_0}{\partial\hat{Z}} - \frac{\partial\hat{\delta}_1}{\partial\phi}l\right) + \mathrm{O}(\hat{\mu}^2) \quad ;$$

$$\frac{\partial^2\delta^*}{\partial\xi^2} = \frac{\partial^2\hat{\delta}_0}{\partial\phi^2}k^2 + \hat{\mu}\left(\frac{\partial\hat{\delta}_0}{\partial\phi}\frac{\partial k}{\partial\hat{\xi}} + 2\frac{\partial^2\hat{\delta}_0}{\partial\phi\partial\hat{\xi}}k + \frac{\partial^2\hat{\delta}_1}{\partial\phi^2}k^2\right) + \mathrm{O}(\hat{\mu}^2) \quad ;$$

$$\frac{\partial^2\delta^*}{\partial Z^2} = \frac{\partial^2\hat{\delta}_0}{\partial\phi^2}l^2 + \hat{\mu}\left(-\frac{\partial\hat{\delta}_0}{\partial\phi}\frac{\partial l}{\partial\hat{Z}} - 2\frac{\partial^2\hat{\delta}_0}{\partial\phi\partial\hat{Z}}l + \frac{\partial^2\hat{\delta}_1}{\partial\phi^2}l^2\right) + \mathrm{O}(\hat{\mu}^2).$$

Thus, by retaining only the terms in $\hat{\mu}^0$ and $\hat{\mu}^1$ respectively, the following equations result for $\hat{\delta}_0$ and $\hat{\delta}_1$:

$$(k^2 + l^2)\frac{\partial^2\hat{\delta}_0}{\partial\phi^2} + \chi^2\hat{\delta}_0 = 0 \quad , \tag{9.75a}$$

$$(k^2 + l^2)\frac{\partial^2\hat{\delta}_1}{\partial\phi^2} + \chi^2\hat{\delta}_1 = \mathcal{G}(\hat{\delta}_0) \quad , \tag{9.75b}$$

with

$$\mathcal{G}(\hat{\delta}_0) = -\left\{\left(\frac{\partial k}{\partial\hat{\xi}} - \frac{\partial l}{\partial\hat{Z}}\right)\frac{\partial\hat{\delta}_0}{\partial\phi}\right.$$
$$\left. + 2\left[\frac{\partial^2\hat{\delta}_0}{\partial\phi\partial\hat{\xi}}k - \frac{\partial^2\hat{\delta}_0}{\partial\phi\partial\hat{Z}}l\right] + \frac{k^2 + l^2}{2}\left(\frac{\partial\hat{\delta}_0}{\partial\phi}\right)^2 + \frac{\partial\hat{\delta}_0}{\partial\phi}l\right\} \quad . \tag{9.76}$$

Equation (9.75a), which is, in fact, an ordinary differential equation with respect to the variable ϕ, can have the classical solution:

$$\hat{\delta}_0 = \hat{\mathcal{A}}_0(\hat{\xi},\ \hat{Z})\,\mathrm{e}^{\mathrm{i}\phi} + \hat{\mathcal{B}}_0(\hat{\xi},\ \hat{Z})\,\mathrm{e}^{-\mathrm{i}\phi} \tag{9.77}$$

once the following dispersion relation has been imposed:

$$k^2 + l^2 = \chi^2 \quad . \tag{9.78}$$

In order to obtain a closed system making it possible to determine the values k, l, $\hat{\mathcal{A}}_0$ and $\hat{\mathcal{B}}_0$, we must make use of (9.73c, 78), as well as the relations associated with $\mathcal{G}(\hat{\delta}_0)$ which express the absence of the secular terms in the solution $\hat{\delta}_1$ of (9.75b). Hence, the procedure to follow is exactly the same as in Sect. 6.2.

9.4.3 Solitary Internal Waves

At the end of Subsect. 9.4.1 which dealt with the case of long isochoric internal waves, it was seen that the order zero solution (proportional to β^2) made a solitary wave appear with respect to the horizontal variable ξ. With Long's work (1965) serving as our source of inspiration, we wish here to expound a more systematic theory which will bring to light solitary internal waves.

Let us go back to (9.39) and seek out its solution in the form:

$$\delta = \delta_{00} + \nu\delta_{10} + \mathrm{Bo}\delta_{01} + \nu\mathrm{Bo}\delta_{11} + \dots \quad , \tag{9.79a}$$

on the hypothesis that

$$\sigma^2 = \sigma_{00}^2 + \nu\sigma_{10}^2 + \mathrm{Bo}\sigma_{01}^2 + \nu\mathrm{Bo}\sigma_{11}^2 + \dots \tag{9.79b}$$

Various similarity relations can be written between the parameters β^2, Bo and ν. However, if we impose that the solution be zero for $\zeta = 0$ and $\zeta = 1$, and also that it tend to zero when $\xi \to \pm\infty$, then two cases must be considered:

$$\beta^2 = \nu\mathrm{Bo} \quad \text{and} \quad \beta^2 = \nu\mathrm{Bo}^2 \quad . \tag{9.80}$$

In the forthcoming, we consider the case $\beta^2 = \nu\mathrm{Bo}$ which means that

$$\frac{H_0^2}{\lambda_0^2} = \frac{h_0 g}{RT_\infty(0)} \quad \text{and if} \quad H_0 \equiv \frac{RT_\infty(0)}{g} \Rightarrow \lambda_0 = H_0\sqrt{\frac{H_0}{h_0}} \quad .$$

The above corresponds to the case where the non-linear and dispersion effects are weak [see (9.64)].

The solution to the equation for δ_{00} is trivial:

$$\delta_{00} = f(\xi)\sin(n\pi\zeta) \quad , \quad \sigma_{00}^2 = n^2\pi^2 \quad .$$

The following term of the expansion for δ_{10} satisfies the equation:

$$\frac{\partial^2\delta_{10}}{\partial\zeta^2} + n^2\pi^2\delta_{10} = -\sigma_{10}^2\delta_{00} \equiv -f(\xi)\sigma_{10}^2\sin(n\pi\zeta) \quad ,$$

and the solution which satisfies the condition $\delta_{10} = 0$ on $\zeta = 0$ is of the form:

$$\delta_{10} = g_1(\xi)\sin(n\pi\zeta) + \frac{\sigma_{10}^2}{2n\pi}f(\xi)\zeta\cos(n\pi\zeta) \quad .$$

However, if this latter solution must also satisfy the condition $\delta_{10} = 0$, on $\zeta = 1$, it is necessary that:

$$\sigma_{10}^2 = 0 \quad \text{and thus} \quad \delta_{10} = g_1(\xi)\,\sin(n\pi\zeta) \quad .$$

The equation for δ_{01} is:

$$\frac{\partial^2 \delta_{01}}{\partial \zeta^2} + n^2\pi^2 \delta_{01} = -\sigma_{01}^2 + \frac{\partial \delta_{00}}{\partial \zeta}$$

$$= -\sigma_{01}^2 f(\xi)\,\sin(n\pi\zeta) + n\pi f(\xi)\,\cos(n\pi\zeta)$$

and the conditions $\delta_{01} = 0$ on $\zeta = 0$ and $\zeta = 1$ yield:

$$\sigma_{01}^2 \equiv 0 \quad \text{and} \quad \delta_{01} = f_1(\xi)\,\sin(n\pi\zeta) + \tfrac{1}{2} f(\xi)\zeta\,\sin(n\pi\zeta) \quad .$$

Finally, the equation for δ_{11} is:

$$\frac{\partial^2 \delta_{11}}{\partial \zeta^2} + n^2\pi^2 \delta_{11} = -\frac{\partial^2 \delta_{00}}{\partial \xi^2} - \frac{1}{2}\left(\frac{\partial \delta_{00}}{\partial \zeta}\right)^2 + \frac{\partial \delta_{10}}{\partial \zeta} - \sigma_{11}^2 \delta_{00}$$

$$= -\left(\frac{d^2 f}{d\xi^2} + \sigma_{11}^2 f\right)\,\sin(n\pi\zeta) + n\pi g_1(\xi)\cos(n\pi\zeta)$$

$$-\frac{1}{4} f^2(\xi) n^2\pi^2 \left(1 + \cos(2n\pi\zeta)\right) \quad .$$

The solution to this last equation is:

$$\delta_{11} = f_2\,\sin(n\pi\zeta) + \frac{f^2}{6}\,\cos(n\pi\zeta)$$

$$+ \frac{1}{2n\pi}\left(\frac{d^2 f}{d\xi^2} + \sigma_{11}^2 f\right)\,\zeta\,\cos(n\pi\zeta)$$

$$+ \frac{1}{2} g_1 \zeta\,\sin(n\pi\zeta) - \frac{1}{4} f^2 + \frac{f^2}{12}\,\cos(2n\pi\zeta) \quad .$$

However, since $\delta_{11} = 0$ on $\zeta = 1$, we must have:

$$[(-1)^n - 1]\frac{f^2}{6} + \frac{(-1)^n}{2n\pi}\left[\frac{d^2 f}{d\xi^2} + \sigma_{11}^2 f\right] = 0 \quad . \tag{9.81}$$

If n is even, the first term of (9.81) disappears and the resulting equation does not have a solution which disappears for $\xi \to \pm\infty$. When n is odd, we find the following:

$$\frac{d^2 f}{d\xi^2} + \sigma_{11}^2 f + \frac{2}{3} n\pi f^2 = 0 \quad , \quad n = 1, 3, \ldots, \tag{9.82}$$

whose solution is:

$$f(\xi) = -\frac{9\sigma_{11}^2}{4n\pi}\,\text{sech}^2\left(\frac{\xi}{2} i\sigma_{11}\right) \quad , \quad n = 1, 3, \ldots \tag{9.83}$$

which disappears for $|\xi| \to \infty$ if $\sigma_{11}^2 < 0$. We can determine σ_{11}^2 by imposing the following:

$$\frac{-9\sigma_{11}^2}{4n\pi} = 1 \Rightarrow \sigma_{11}^2 = -\frac{4n\pi}{9} \quad , \quad n = 1, 3, \ldots \quad ,$$

seeing that in (9.79a), 1 can be taken as the dimensionless measure of the maximum amplitude of the perturbation.

Thus, we again encounter, to order zero, the following solitary wave solution

$$\delta = \mathrm{sech}^2\left(\frac{\sqrt{n\pi}}{3}\xi\right)\sin(n\pi\zeta) + \mathrm{O}(\nu) \quad , \quad n = 1, 3, \ldots \quad ;$$

$$\sigma^2 = n^2\pi^2 - \frac{4n\pi}{9}\nu\mathrm{Bo} + \ldots \quad . \tag{9.84}$$

A few comments concerning the case $\beta^2 = \nu\mathrm{Bo}^2$ are called for. In order to arrive at a solitary wave, in this case, it is necessary in expansions (9.79) to include the terms proportional to ν^2, Bo^2 and $\nu\mathrm{Bo}^2$ with the indexes (20), (02) and (12), respectively. It then follows that:

$$\sigma_{00}^2 = n^2\pi^2 \quad , \quad \sigma_{10}^2 = \sigma_{20}^2 = \sigma_{01}^2 = \sigma_{11}^2 = 0$$

$$\text{and} \quad \sigma_{02}^2 = \frac{1}{4} \quad , \quad \sigma_{12}^2 = -\frac{n\pi}{9} \quad , \quad n = 2, 4, \ldots$$

and instead of (9.84), we have:

$$\delta = -\mathrm{sech}^2\left(\frac{\sqrt{n\pi}}{6}\xi\right)\sin(n\pi\zeta) + \mathrm{O}(\nu) \quad , \quad n = 2, 4, \ldots \quad ,$$

$$\text{and} \quad \sigma^2 = n^2\pi^2 + \frac{1}{4}\mathrm{Bo}^2 - \frac{n\pi}{9}\nu\mathrm{Bo}^2 + \ldots \tag{9.85}$$

It will be noticed that the solitary waves (9.84, 85) do not appear when we drop the terms related to the Boussinesq number Bo in (9.79), i.e., when a Boussinesq approximation is carried out. We thus have an example of a physical phenomenon which is brought to light only when we take into account small terms which are neglected in the Boussinesq approximation. The preceding once again demonstrates the profound difference existing between the isochoric and Boussinesq approximations.

Finally, the reader is referred to the book by Dodd, Eilbeck, Gibbon and Morris (1982) which in a systematic manner deals with "solitons and non-linear wave equations".

9.5 Complementary Remarks

9.5.1. Firstly, we wish to demonstrate how the results of Subsect. 9.4.2 concerning short waves can be generalized. For this we return to the dimensional equation (9.35) which is written for the perturbation $\Delta(x, z)$ and we introduce the following dimensionless values:

$$\xi = \frac{x}{\lambda_0} \quad , \quad \eta = \frac{z}{\lambda_0} \quad \text{and} \quad \Delta^* = \frac{\Delta}{H_0} \quad , \tag{9.86}$$

where $\lambda_0 = U_0/N_{\infty,\,\mathrm{is}}(0)$. In this case, instead of (9.35), we obtain the following non-dimensional equation for the function $\Delta^*(\xi, \eta; \kappa, \alpha)$

$$\frac{\partial^2 \Delta^*}{\partial \xi^2} + \frac{\partial^2 \Delta^*}{\partial \eta^2} + \mathcal{N}^2(\kappa\eta - \Delta^*)\left\{\Delta^* + \frac{\alpha}{2}\left[\left(\frac{\partial \Delta^*}{\partial \xi}\right)^2 + \left(\frac{\partial \Delta^*}{\partial \eta}\right)^2\right] - \kappa\alpha\frac{\partial \Delta^*}{\partial \eta}\right\} = 0 \quad , \tag{9.87}$$

where

$$\kappa = \frac{U_0}{H_0 N_{\infty,\,\mathrm{is}}(0)} \quad , \quad \alpha = \frac{H_0}{g} N^2_{\infty,\,\mathrm{is}}(0)$$

$$\mathcal{N}^2(\kappa\eta - \Delta^*) \equiv \frac{N^2_{\infty,\,\mathrm{is}}[(U_0/N_{\infty,\,\mathrm{is}}(0))\eta - H_0\Delta^*]}{N^2_{\infty,\,\mathrm{is}}(0)} \tag{9.88}$$

We now suppose that $\kappa \ll 1$ which corresponds to slowly propagating internal waves. Let ω_0 be the frequency, k_0, the horizontal wave number and λ_0, the length of an internal wave. We then have:

$$U_0 = \frac{\omega_0}{k_0} = \frac{\omega_0 \lambda_0}{2\pi} \Rightarrow \kappa = \frac{1}{2\pi}\frac{\omega_0}{N_{\infty,\,\mathrm{is}}(0)}\frac{\lambda_0}{H_0}$$

and since $\omega_0 \approx N_{\infty,\,\mathrm{is}}(0)$, it is observed that:

$$\kappa \ll 1 \Rightarrow \lambda_0 \ll H_0 \gg U_0/\omega_0 \quad .$$

Once again it is emphasized that when $\kappa \ll 1$, a purely local study is made which means analyzing the propagation of short internal waves in an infinite medium both in ξ and in η. The walls of the duct are therefore ignored as are the conditions in ξ.

If we follow the procedure used in Subsect. 9.4.2, instead of (9.75a), we arrive at the equation:

$$(k^2 + l^2)\frac{\partial^2 \hat{\Delta}_0}{\partial \phi^2} + \frac{\partial \hat{\chi}_0}{\partial \hat{\Delta}_0} = 0 \quad , \tag{9.89}$$

where

$$\hat{\chi}_0(\hat{\eta}, \hat{\Delta}_0) = \int \mathcal{N}^2(\hat{\eta} - \hat{\Delta}_0)\hat{\Delta}_0 d\hat{\Delta}_0 \quad . \tag{9.90}$$

The equation for $\hat{\Delta}_0(\phi; \hat{\xi}, \hat{\eta})$ with $\hat{\xi} = \kappa\xi$ and $\hat{\eta} = \kappa\eta$ is an ordinary differential equation with respect to ϕ. This equation has as first integral:

$$\frac{1}{2}(k^2 + l^2)\left(\frac{\partial \hat{\Delta}_0}{\partial \phi}\right)^2 + \hat{\chi}_0(\hat{\eta}, \hat{\Delta}_0) = \hat{\varepsilon}_0(\hat{\xi}, \hat{\eta}) \quad ,$$

where $\hat{\varepsilon}_0$ is the integration "constant". The solution to this latter equation is:

$$\phi(\hat{\xi}, \hat{\eta}) = (l^2 + k^2)^{1/2} \int \left\{2[\hat{\varepsilon}_0 - \hat{\chi}_0(\hat{\eta}, \hat{\Delta}_0)]\right\}^{-1/2} d\hat{\Delta}_0 - \phi_0(\hat{\varepsilon}, \hat{\eta}) \quad , \tag{9.91}$$

with $\phi_0(\hat{\xi}, \hat{\eta})$ being a second integration "constant". By inversing (9.91), we find the solution:

$$\hat{\Delta}_0(\phi;\, \hat{\xi}, \hat{\eta}) = \mathcal{F}\left(\phi + \phi_0,\ \ \hat{\varepsilon}_0,\ \ l^2 + k^2\right) \quad , \tag{9.92}$$

which satisfies (9.89).

In this instance, however, we are interested in periodic oscillations. The function $\hat{\Delta}_0$ must oscillate between two zeros of the function $\hat{\varepsilon}_0 - \hat{\chi}_0$ and must be periodic in ϕ. This means that:

$$\hat{\Delta}_0(\phi;\, \hat{\xi}, \hat{\eta}) = \hat{\Delta}_0(\phi + n\mathcal{P};\, \hat{\xi}, \hat{\eta}) \quad ,$$

where $\mathcal{P}$ is the period, and n is an arbitrary whole number. In addition, it turns out that in order for $\partial\hat{\Delta}_0/\partial\phi$ to remain bounded, $\mathcal{P}$ must be independent of $\hat{\xi}$ and $\hat{\eta}$. In this case, from (9.91), the following dispersion relation results:

$$(l^2 + k^2)^{1/2} \oint \frac{d\hat{\Delta}_0}{\{2[\hat{\varepsilon}_0 - \hat{\chi}_0(\eta, \hat{\Delta}_0)]\}^{1/2}} = \mathcal{P} \quad . \tag{9.93}$$

Thus are obtained relations (9.93) and the following:

$$\frac{\partial\phi}{\partial\hat{\xi}} = \frac{k(\hat{\xi}, \hat{\eta})}{\kappa} \quad , \quad \frac{\partial\phi}{\partial\hat{\eta}} = \frac{l(\hat{\xi}, \hat{\eta})}{\kappa} \quad , \tag{9.94}$$

for our three unknowns $k(\hat{\xi}, \hat{\eta})$, $l(\hat{\xi}, \hat{\eta})$ and $\hat{\varepsilon}_0(\hat{\xi}, \hat{\eta})$. One additional relation is necessary to obtain a closed system which would permit us to determine k, l and $\hat{\varepsilon}_0$. This relation is obtained by taking advantage of the rank one equation for $\hat{\Delta}_1$, and by writing the condition which eliminates the secular terms from the solution of this equation.

In the particular case where $\alpha = O(\kappa)$, the above-mentioned elimination leads to the following:

$$\frac{\partial}{\partial\hat{\xi}}\left\{k \int_{-\mathcal{P}/2}^{+\mathcal{P}/2} \left(\frac{\partial\hat{\Delta}_0}{\partial\phi}\right)^2 d\phi\right\} + \frac{\partial}{\partial\hat{\eta}}\left\{l \int_{-\mathcal{P}/2}^{+\mathcal{P}/2} \left(\frac{\partial\hat{\Delta}_0}{\partial\phi}\right)^2 d\phi\right\} = 0 \quad . \tag{9.95}$$

We will not go into further detail here, but the interested reader is referred to the work of Leonov and Miropolski (1975).

9.5.2. It is much more difficult to effect theoretical analyses for *three-dimensional* flows. Nonetheless, such analyses are possible if a *linearization* is carried out and if the perturbations induced (e.g., by an obstacle) are supposed "small" compared to the values which characterize the undisturbed basic flow.

Let us assume then that in the undisturbed area, the basic flow is characterized by the given quantities

$$U_\infty(z) \quad , \quad V_\infty(z) \quad , \ 0,\ p_\infty(z) \quad \text{and} \quad \varrho_\infty(z) \quad ,$$

of the velocity, pressure and density. Let ε be the small parameter of our problem. The solution to the steady three-dimensional adiabatic isochoric equations will be sought in the form:

$$\begin{aligned} u &= U_\infty(z) + \varepsilon u'(x, y, z) \quad ; \quad v = V_\infty(z) + \varepsilon v'(x, y, z) \quad ; \\ w &= \varepsilon w'(x, y, z) \quad ; \\ p &= p_\infty(z) + \varepsilon p'(x, y, z) \quad ; \\ \varrho &= \varrho_\infty(z) + \varepsilon \varrho'(x, y, z) \quad . \end{aligned} \tag{9.96}$$

By retaining only the linear terms, the following system of linear equations results:

$$\begin{aligned} \varrho_\infty\left(U_\infty \frac{\partial u'}{\partial x} + V_\infty \frac{\partial u'}{\partial y} + \frac{dU_\infty}{dz} w'\right) &= -\frac{\partial p'}{\partial x} \quad ; \\ \varrho_\infty\left(U_\infty \frac{\partial v'}{\partial x} + V_\infty \frac{\partial v'}{\partial y} + \frac{dV_\infty}{dz} w'\right) &= -\frac{\partial p'}{\partial y} \quad ; \\ \varrho_\infty\left(U_\infty \frac{\partial w'}{\partial x} + V_\infty \frac{\partial w'}{\partial y}\right) &= -\frac{\partial p'}{\partial z} - g\varrho' \quad , \end{aligned} \tag{9.97}$$

$$U_\infty \frac{\partial \varrho'}{\partial x} + V_\infty \frac{\partial \varrho'}{\partial y} + \frac{d \log \varrho_\infty}{dz} \varrho_\infty w' = 0 \quad ;$$

$$\frac{\partial u'}{\partial x} + \frac{\partial v'}{\partial y} + \frac{\partial w'}{\partial z} = 0 \quad .$$

In order to obtain just one equation instead of system (9.97), the new velocity components can be introduced:

$$\overline{u} = \varrho_\infty u' \quad , \quad \overline{v} = \varrho_\infty v' \quad \text{and} \quad \overline{w} = \varrho_\infty w' \quad . \tag{9.98}$$

In this case, the following equation for $\overline{w}$ and p' can be deduced from the first two and the last equations of (9.97)

$$\begin{aligned} &\left(U_\infty \frac{\partial}{\partial x} + V_\infty \frac{\partial}{\partial y}\right)\left(\frac{d \log \varrho_\infty}{dz} \overline{w} - \frac{\partial \overline{w}}{\partial z}\right) + \frac{dU_\infty}{dz} \frac{\partial \overline{w}}{\partial x} \\ &+ \frac{dV_\infty}{dz} \frac{\partial \overline{w}}{\partial y} + \left(\frac{\partial^2}{\partial x^2} + \frac{\partial^2}{\partial y^2}\right) p' = 0 \quad . \end{aligned} \tag{9.99}$$

A second equation between $\overline{w}$ and p' is obtained from the third and fourth equations of (9.97)

$$\begin{aligned} &\left\{ g \frac{d \log \varrho_\infty}{dz} - \left(U_\infty \frac{\partial}{\partial x} + V_\infty \frac{\partial}{\partial y}\right)^2 \right\} \overline{w} \\ &= \left(U_\infty \frac{\partial}{\partial x} + V_\infty \frac{\partial}{\partial y}\right) \frac{\partial p'}{\partial z} \quad . \end{aligned} \tag{9.100}$$

Now, we need only elimnate p' from (9.99, 100) in order to arrive at the equation sought with respect to $\overline{w} = \varrho_\infty w'$

$$\left(U_\infty\frac{\partial}{\partial x}+V_\infty\frac{\partial}{\partial y}\right)^2\left(\frac{\partial^2\overline{w}}{\partial x^2}+\frac{\partial^2\overline{w}}{\partial y^2}+\frac{\partial^2\overline{w}}{\partial z^2}\right)-g\frac{d\log\varrho_\infty}{dz}\left(\frac{\partial^2\overline{w}}{\partial x^2}+\frac{\partial^2\overline{w}}{\partial y^2}\right)$$
$$-\left(U_\infty\frac{\partial}{\partial x}+V_\infty\frac{\partial}{\partial y}\right)\left(\frac{d^2U_\infty}{dz^2}\frac{\partial\overline{w}}{\partial x}+\frac{d^2V_\infty}{dz^2}\frac{\partial\overline{w}}{\partial y}\right)$$
$$+\left(U_\infty\frac{\partial}{\partial x}+V_\infty\frac{\partial}{\partial y}\right)\frac{\partial}{\partial z}\left\{\left(U_\infty\frac{\partial}{\partial x}+V_\infty\frac{\partial}{\partial y}\right)\frac{d\log\varrho_\infty}{dz}\overline{w}\right\}=0 \quad (9.101)$$

We limit ourselves to the particular case

$$\varrho_\infty(z)=\varrho_\infty(0)\exp\left(-\frac{N^2_{\infty,\,\mathrm{is}}(0)}{g}z\right)$$

and introduce the following dimensionless values:

$$\xi=\frac{x}{L_0}\quad,\quad\eta=\frac{y}{L_0}\quad,\quad\zeta=\frac{z}{L_0}\quad,\quad\omega=\frac{\overline{w}}{\varrho_\infty(0)U_\infty(0)}\quad,$$
$$\overline{U}_\infty=\frac{U_\infty}{U_\infty(0)}\quad,\quad\overline{V}_\infty=\frac{V_\infty}{U_\infty(0)}\quad,$$

where L_0 is a characteristic length scale. In this case, instead of (9.101), the following non-dimensional equation is obtained for $\omega(\xi,\eta,\zeta)$:

$$\left(\overline{U}_\infty\frac{\partial}{\partial\xi}+\overline{V}_\infty\frac{\partial}{\partial\eta}\right)^2\left(\frac{\partial^2\omega}{\partial\xi^2}+\frac{\partial^2\omega}{\partial\eta^2}+\frac{\partial^2\omega}{\partial\zeta^2}\right)+K_0^2\left(\frac{\partial^2\omega}{\partial\xi^2}+\frac{\partial^2\omega}{\partial\eta^2}\right)$$
$$-\left(\overline{U}_\infty\frac{\partial}{\partial\xi}+\overline{V}_\infty\frac{\partial}{\partial\eta}\right)\left\{\frac{d^2\overline{U}_\infty}{d\zeta^2}\frac{\partial\omega}{\partial\xi}+\frac{d^2\overline{V}_\infty}{d\zeta^2}\frac{\partial\omega}{\partial\eta}\right\}$$
$$=\alpha_0\left(\overline{U}_\infty\frac{\partial}{\partial\xi}+\overline{V}_\infty\frac{\partial}{\partial\eta}\right)\frac{\partial}{\partial\zeta}\left(\overline{U}_\infty\frac{\partial\omega}{\partial\xi}+\overline{V}_\infty\frac{\partial\omega}{\partial\eta}\right)\quad, \quad (9.102)$$

where

$$K_0^2\equiv\frac{N^2_{\infty,\,\mathrm{is}}(0)L_0^2}{U^2_\infty(0)}\quad\text{and}\quad\alpha_0=\frac{L_0}{g}N^2_{\infty,\,\mathrm{is}}(0)\quad.$$

To be sure, when $\alpha\to 0$, we again find the equation corresponding to the Boussinesq approximation. Moreover, when $\overline{U}_\infty\equiv 1$ $[U_\infty\equiv U_\infty(0)]$ and $\overline{V}_\infty\equiv 0$, we have the equation:

$$\left\{\frac{\partial^2}{\partial\xi^2}+\frac{\partial^2}{\partial\eta^2}+\frac{\partial^2}{\partial\zeta^2}-\alpha_0\frac{\partial}{\partial\zeta}\right\}\frac{\partial^2\omega}{\partial\xi^2}+K_0^2\left(\frac{\partial^2}{\partial\xi^2}+\frac{\partial^2}{\partial\eta^2}\right)\omega=0\quad. \quad (9.103)$$

The solution to the general equation (9.102) can again be sought in the form:

$$\omega(\xi,\eta,\zeta)=\omega_0(\zeta)\exp\{\mathrm{i}(k\xi+l\eta)\}\quad,$$

which yields for $\omega_0(\zeta)$, the following ordinary differential equation

$$\frac{d^2\omega_0}{d\zeta^2}-\mathrm{i}\alpha_0\frac{d\omega_0}{d\zeta}+\Lambda_0(\zeta)\omega_0=0\quad, \quad (9.104)$$

with

$$\Lambda_0(\zeta) = K_0^2 \frac{k^2 + l^2}{(\overline{U}_\infty k + \overline{V}_\infty l)^2} - \frac{(d^2\overline{U}_\infty/d\zeta^2)k + (d^2\overline{V}_\infty/d\zeta^2)l}{\overline{U}_\infty k + \overline{V}_\infty l}$$
$$- (k^2 + l^2) - \mathrm{i}\alpha_0 \frac{(d\overline{U}_\infty/d\zeta)k + d\overline{V}_\infty/d\zeta)l}{\overline{U}_\infty k + \overline{V}_\infty l} \quad , \tag{9.105}$$

where $\mathrm{i} = \sqrt{-1}$.

Equation (9.103) was analyzed by Kibel (1955), Wurtele (1957) and Zeytounian (1964) for the case $\alpha_0 = 0$. For the same case, (9.104) was analyzed by Sawyer (1962) and Veltichev (1965). Finally, the general equation (9.103) was analyzed for $\alpha_0 = 0$ by Zeytounian (1965). On this same subject, our thesis (1969) may be consulted.

Let us backtrack to (9.102) and suppose that $\alpha_0 = 0$. If we introduce the following values into this equation:

$$\mathcal{G}_\infty(\zeta) = \sqrt{\overline{U}^2_\infty(\zeta) + \overline{V}^2_\infty(\zeta)} \quad ; \quad \mathrm{tg}\, \Omega_\infty(\zeta) = \overline{V}_\infty(\zeta)/\overline{U}_\infty(\zeta) \quad , \tag{9.106}$$

we obtain the following

$$\left(\cos \Omega_\infty \frac{\partial}{\partial \xi} + \sin \Omega_\infty \frac{\partial}{\partial \eta}\right)^2 \left\{\frac{\partial^2 \omega}{\partial \xi^2} + \frac{\partial^2 \omega}{\partial \eta^2} + \frac{\partial^2 \omega}{\partial \zeta^2} + \mathcal{A}_0(\zeta)\omega\right\}$$
$$+\mathcal{B}_0(\zeta)\left(\cos \Omega_\infty \frac{\partial}{\partial \xi} + \sin \Omega_\infty \frac{\partial}{\partial \eta}\right)\left(\sin \Omega_\infty \frac{\partial \omega}{\partial \xi} - \cos \Omega_\infty \frac{\partial \omega}{\partial \eta}\right)$$
$$+\mathcal{C}_0(\zeta)\left(\frac{\partial^2 \omega}{\partial \xi^2} + \frac{\partial^2 \omega}{\partial \eta^2}\right) = 0 \quad . \tag{9.107}$$

The following three parameters are thus seen to appear

$$\mathcal{A}_0(\zeta) = \left(\frac{d\Omega_\infty}{d\zeta}\right)^2 - \frac{1}{\mathcal{G}_\infty} \frac{d^2\mathcal{G}_\infty}{d\zeta^2} \quad ;$$
$$\mathcal{B}_0(\zeta) = \frac{d^2\Omega_\infty}{d\zeta^2} + 2\frac{1}{\mathcal{G}_\infty} \frac{d\mathcal{G}_\infty}{d\zeta} \frac{d\Omega_\infty}{d\zeta} \quad ; \tag{9.108}$$
$$\mathcal{C}_0(\zeta) = \frac{K_0^2}{\mathcal{G}^2_\infty(\zeta)} \quad .$$

Therefore, the solution of (9.102) (where $\alpha_0 \equiv 0$) will strongly depend on the coefficients $\mathcal{A}_0$, $\mathcal{B}_0$ and $\mathcal{C}_0$. The first two of the latter coefficients reflect the influence of the velocity profile of the basic undisturbed flow both as a function of its algebraic value $\mathcal{G}_\infty(\zeta)$ and of its direction $\Omega_\infty(\zeta)$. The latter takes into account the rotary effect of the basic velocity with altitude. Additional information concerning (9.107, 108) can be found in Zeytounian (1965).

One final remark concerns the coefficients (9.108). Since we made $\alpha_0 \equiv 0$ (Boussinesq approximation), the values of $\mathcal{A}_0$, $\mathcal{B}_0$ and $\mathcal{C}_0$ should have been set as given by (9.108) with $\zeta = 0$. This means that in reality, $\mathcal{A}_0$, $\mathcal{B}_0$ and $\mathcal{C}_0$ are functions of $\zeta_\infty \equiv \alpha_0\zeta$ and when $\alpha_0 \ll 1$, a double-scale theory analogous to the one in Sect. 6.2 must be applied in (9.102) or possibly even in (9.104). Although such a problem remains unsolved for the present, an attempt of this kind was nonetheless made by Bois (1984).

9.5.3. To conclude the present chapter, a few references are called for. Many applications of the isochoric equations can be found in the book by Yih (1980). In particular, Chap. 4 is devoted to the problem of hydrodynamic stability. The isochoric equations also have many applications in ocean dynamics. The reader is referred to the thesis of Stéphane Godts (1985) which contains a rigorous asymptotic modelling of ocean motions in long wave theory and under the quasi-geostrophic approximation while taking into account the coupling with the atmosphere via the free surface and the Ekman layer. As concerns the waves in the ocean, the monography of LeBlond and Mysak (1978) may be consulted.

10. The Deep Convection Approximation

As has already been noted several times, atmospheric flows are low Mach number flows: $M_\infty^2 = U_0^2/\gamma RT_\infty(0) \ll 1$. Because of this, if we wish to "avoid" the constraint (8.19) imposed in the Boussinesq approximation, it becomes necessary to analyze flows with very low Froude numbers since [see (2.63) and (2.65)]

$$\mathrm{Bo} = \varepsilon \frac{\gamma M_\infty^2}{\mathrm{Fr}^2} \Rightarrow \mathrm{Fr} \approx M_\infty \ll 1 \quad , \tag{10.1}$$

if it is assumed that: $\mathrm{Bo} = gH/RT_\infty(0)$, $\gamma = c_p/c_v$ and $\varepsilon = H/L$ are of the order unity.

However, the limiting process $\mathrm{Fr} \to 0$, i.e., in fact, $M_\infty \to 0$, in the atmospheric equations [e.g., (8.5–8) where the expression $\mathrm{Bo}\,\mathrm{Fr}^2/\varepsilon$ must be substituted for γM_∞^2] leads to a very strong degeneracy of these equations to order zero.[1]

It soon became clear that it should be assumed (Zeytounian, 1974) that when $\mathrm{Fr} \to 0$, the term[2]

$$\mathrm{Bo}\left(\frac{\gamma - 1}{\gamma} + \frac{dT_\infty}{dz_\infty}\right)\frac{1}{T_\infty(z_\infty)} \equiv \alpha_\infty^0 N_\infty^2(z_\infty) \to 0 \quad ;$$

which means that we must consider the double limiting process:

$$\mathrm{Fr} \to 0 \quad , \quad \alpha_\infty^0 \to 0 \quad ; \quad \frac{\alpha_\infty^0}{\mathrm{Fr}^2} = O(1) \quad . \tag{10.2}$$

In order to confirm (10.2), it is sufficient to understand that when $\mathrm{Fr} \to 0$, (8.5) implies that $\vartheta \sim \mathrm{Fr}^2$. Because of this, (8.7) in fact implies the constraint written in (10.2). If the latter is not satisfied, then [when $\alpha_\infty^0 = O(1)$]: $w \to 0$ with $\mathrm{Fr}^2 \to 0$! This is precisely what brings on a very strong degeneracy of the basic equations at order zero.

Hence it is necessary that the characteristic value of the Väisälä internal frequency (with dimensions) satisfy the relation:

$$N_\infty^*(0) \cong \frac{U_0}{H} \approx 10^{-3}\,\mathrm{s}^{-1}, \tag{10.3}$$

since $\mathrm{Bo} \cong 1$ implies that $H \cong (RT_\infty(0)/g) \approx 10^4\,\mathrm{m}$.

[1] This degeneracy is related to the so-called quasi-solenoidal model which is analyzed in Chap. 12.

[2] In all of the forthcoming, although all of the values are dimensionless, we have omitted the bars on these non-dimensional values. If necessary, we will use a superscript asterisk to designate a *dimensional* value.

The constraint (10.3) is, in fact, the one imposed by Ogura and Phillips (1962) in order to obtain the so-called "anelastic" equations via the system (2.83).

In Sect. 10.1, the "anelastic" equations of Ogura and Phillips (1962) are obtained. Section 10.2 is devoted to a generalization of these "anelastic" equations which leads to the deep convection equations according to Zeytounian (1974). In Sect. 10.3, the rendering of the Boussinesq equations (8.13–16) from the deep convection equations is demonstrated. Finally, Sect. 10.4 contains some complementary remarks.

10.1 The "Anelastic" Equations of Ogura and Phillips

Let us return to (2.83) and replace M_∞^2 by $\mathrm{Bo}\,\mathrm{Fr}^2/\gamma\varepsilon$. When $\mathrm{Bo}/\gamma\varepsilon = O(1)$ and $\mathrm{Fr} \to 0$, the limiting form of these equations can be sought by postulating the following asymptotic representation

$$\begin{aligned} u &= u_0 + \dots \;;\quad v = v_0 + \dots \;;\quad w = w_0 + \dots \;; \\ \Pi &= \Pi_0 + \mathrm{Fr}^2 \Pi_2 + \dots \;; \\ \theta &= \theta_0 + \mathrm{Fr}^2 \theta_2 + \dots \;. \end{aligned} \tag{10.4}$$

To order zero, we have the following from the first three equations of (2.83)

$$\theta_0 \frac{\partial \Pi_0}{\partial x} = 0 \quad , \quad \theta_0 \frac{\partial \Pi_0}{\partial y} = 0 \quad ; \quad \theta_0 \frac{\partial \Pi_0}{\partial z} + \frac{\gamma - 1}{\gamma}\mathrm{Bo} = 0 \quad ,$$

i.e.,

$$\Pi_0 = \Pi_0(t, z) \quad , \quad \theta_0 = \theta_0(t, z) \quad .$$

The fourth equation in this system (2.83) shows, however, that

$$\frac{\partial \theta_0}{\partial t} + w_0 \frac{\partial \theta_0}{\partial z} = 0$$

and if at the initial instant $\theta_0 = \theta_{00}(z)$, then:[3]

$$\theta_0 \equiv \theta_{00}(z) \quad , \quad \Pi_0 = \Pi_{00}(z) \quad . \tag{10.5}$$

Again from the first three equations in (2.83), to order Fr^2, we obtain (with $\varepsilon \equiv 1$)[4]:

$$\begin{aligned} &\frac{\gamma - 1}{\gamma}\mathrm{Bo}\frac{Du_0}{Dt} + \theta_{00}\frac{\partial \Pi_2}{\partial x} = 0 \quad ; \qquad \frac{\gamma - 1}{\gamma}\mathrm{Bo}\frac{Dv_0}{Dt} + \theta_{00}\frac{\partial \Pi_2}{\partial y} = 0 \quad ; \\ &\frac{\gamma - 1}{\gamma}\mathrm{Bo}\frac{Dw_0}{Dt} + \theta_{00}\frac{\partial \Pi_2}{\partial z} = \frac{\gamma - 1}{\gamma}\mathrm{Bo}\frac{\theta_2}{\theta_{00}} \quad , \end{aligned} \tag{10.6}$$

[3] It might easily be thought that, in fact, $\theta_{00}(z) \equiv 1$ since $d\theta_{00}/dz = 0$ if $w_0 \not\equiv 0$. It will be seen farther on, however, that because of the similarity relation (10.8), we have $\theta_{00}(z) = 1 + \mathrm{Fr}^2\theta_{02}(z)$.

[4] In all that follows, we have:

$$\frac{D}{Dt} = \frac{\partial}{\partial t} + u_0\frac{\partial}{\partial x} + v_0\frac{\partial}{\partial y} + w_0\frac{\partial}{\partial z} \quad .$$

given the fact that

$$\theta_2 \frac{d\Pi_{00}}{dz} \equiv -\frac{\gamma-1}{\gamma}\mathrm{Bo}\frac{\theta_2}{\theta_{00}} \quad .$$

To order zero, the last equation of system (2.83) yields:

$$\frac{\partial u_0}{\partial x} + \frac{\partial v_0}{\partial y} + \frac{\partial w_0}{\partial z} = w_0 \frac{d}{dz}\left[\log \frac{\theta_{00}}{\Pi_{00}^{1/\gamma-1}}\right] = -w_0 \frac{1}{\varrho_{00}}\frac{d\varrho_{00}}{dz} \quad ,$$

since $\varrho \equiv \Pi^{1/(\gamma-1)}/\theta$. Therefore, to order zero, the following continuity equation results:

$$\frac{\partial \varrho_{00} u_0}{\partial x} + \frac{\partial \varrho_{00} v_0}{\partial y} + \frac{\partial \varrho_{00} w_0}{\partial z} = 0 \quad , \tag{10.7}$$

where $\varrho_{00}(z) = \Pi_{00}^{1/(\gamma-1)}/\theta_{00}$.

Let us consider the fourth equation of system (2.83). It gives:

$$w_0 \frac{d\theta_{00}}{dz} + \mathrm{Fr}^2 \frac{D\theta_2}{Dt} + \ldots = 0 \quad ,$$

and it is observed that the following similarity relation must be imposed:

$$\frac{d\theta_{00}}{dz} = \lambda_{00}\Gamma_{02}(z)\mathrm{Fr}^2 \quad , \tag{10.8}$$

where $\lambda_{00} = \text{const}$ and $\Gamma_{02}(z)$ is a function of order unity which takes into account the reference stratification. As a matter of fact, (10.8) necessarily implies that: $\partial\theta_{00}/\partial t \equiv 0$. On the hypothesis (10.8), for θ_2 it follows that

$$\frac{D\theta_2}{Dt} + \lambda_{00}\Gamma_{02}(z)w_0 = 0 \quad . \tag{10.9}$$

Let us now return to the zeroth order relation:

$$\theta_{00}\frac{d\Pi_{00}}{dz} + \frac{\gamma-1}{\gamma}\mathrm{Bo} = 0$$

and derive it with respect to z. There results:

$$\theta_{00}\frac{d^2\Pi_{00}}{dz^2} + \frac{d\theta_{00}}{dz}\frac{d\Pi_{00}}{dz} \equiv \theta_{00}\frac{d^2\Pi_{00}}{dz^2} + \lambda_{00}\Gamma_{02}(z)\mathrm{Fr}^2\frac{d\Pi_{00}}{dz} = 0 \quad ,$$

i.e., to order zero, when $\mathrm{Fr} \to 0$ we have:

$$\theta_{00}\frac{d^2\Pi_{00}}{dz^2} = 0 \Rightarrow \Pi_{00} = 1 + c_{00}z$$

or even:

$$c_{00}\theta_{00} + \frac{\gamma-1}{\gamma}\mathrm{Bo} = 0 \Rightarrow c_{00} = -\frac{\gamma-1}{\gamma}\mathrm{Bo}$$

since

$$\lim_{\mathrm{Fr}\to 0}\theta_{00} = 1 \quad .$$

Therefore:

$$\theta_{00}(z) = 1 + \underbrace{\lambda_{00} \int \Gamma_{02}(z)dz}_{\theta_{02}(z)} \, \mathrm{Fr}^2 \quad ,$$

$$\Pi_{00}(z) = 1 - \frac{\gamma - 1}{\gamma} \mathrm{Bo} z + \mathrm{Fr}^2 P_{02}(z) \quad , \quad \text{with} \tag{10.10}$$

$$\frac{dP_{02}}{dz} = \frac{\gamma - 1}{\gamma} \mathrm{Bo} \lambda_{00} \int \Gamma_{02}(z)dz \quad . \tag{10.11}$$

Finally, we arrive at the conclusion that the following asymptotic representation must be postulated

$$u = u_0 + \ldots \quad , \quad v = v_0 + \ldots \quad , \quad w = w_0 + \ldots \quad ,$$

$$\Pi = 1 - \frac{\gamma - 1}{\gamma} \mathrm{Bo} z + \mathrm{Fr}^2 \{ P_{02}(z) + \Pi_2 \} + \ldots \quad , \tag{10.12}$$

$$\theta = 1 + \mathrm{Fr}^2 \{ \theta_{02}(z) + \theta_2 \} + \ldots$$

to obtain the Ogura and Phillips (1962) type "anelastic" equations for u_0, v_0, w_0, Π_2 and θ_2

$$\frac{\gamma - 1}{\gamma} \mathrm{Bo} \frac{Du_0}{Dt} + \frac{\partial \Pi_2}{\partial x} = 0 \quad ;$$

$$\frac{\gamma - 1}{\gamma} \mathrm{Bo} \frac{Dv_0}{Dt} + \frac{\partial \Pi_2}{\partial y} = 0 \quad ;$$

$$\frac{\gamma - 1}{\gamma} \mathrm{Bo} \frac{Dw_0}{Dt} + \frac{\partial \Pi_2}{\partial z} = \frac{\gamma - 1}{\gamma} \mathrm{Bo} \left\{ \theta_2 + \lambda_{00} \int \Gamma_{02}(z)dz \right\} \quad ; \tag{10.13}$$

$$\frac{\partial \varrho_{00} u_0}{\partial x} + \frac{\partial \varrho_{00} v_0}{\partial y} + \frac{\partial \varrho_{00} w_0}{\partial z} = 0 \quad ;$$

$$\varrho_{00} = \left(1 - \frac{\gamma - 1}{\gamma} \mathrm{Bo} z \right)^{1/(\gamma - 1)} \quad ;$$

$$\frac{D\theta_2}{Dt} + \lambda_{00} \Gamma_{02}(z) w_0 = 0 \quad .$$

It will be observed that, in fact, it is the function

$$\mathcal{L}_2 \equiv \theta_2 + \lambda_{00} \int \Gamma_{02}(z)dz \tag{10.14}$$

which comes into play in the system (10.13).

10.2 The Deep Convection Equations According to Zeytounian

Let us now go back to the general equations (8.5–8) but with the ground condition (8.21) in place of (8.9) where [see (8.49)]

$$\tau_0 = \frac{\Delta T_0}{T_\infty(0)} \equiv \frac{\text{Bo}}{\nu_0} \quad , \quad \nu_0 \equiv \frac{gH}{R\Delta T_0} = \frac{H}{h_0} \quad . \tag{10.15}$$

Let:

$$\delta^* = z^* - z^*_\infty$$

be the vertical dimensional displacement of a fluid particle (at a fixed instant t^*) in the considered atmospheric phenomenon with respect to its position z^*_∞ in the standard situation. In dimensionless variables, we have:

$$z_\infty = \text{Bo}\left(z - \frac{\Delta}{\nu_0}\right) \tag{10.16}$$

where $\Delta = \delta^*/(R\Delta T_0/g)$.

In the forthcoming, we are going to consider the double limiting process:

$$\text{Fr} \to 0 \quad , \quad \nu_0 \to \infty \quad , \tag{10.17}$$

with t, x, y and z fixed (the parameter Bo is assumed to be of the order unity). It becomes obvious that (10.17) should be carried out under the following similarity relation:

$$\frac{1}{\varepsilon}\frac{\text{Fr}^2}{1/\nu_0} \equiv \frac{U_0^2}{R\Delta T_0} = \gamma\mu_{00}^2 = \text{O}(1) \quad , \tag{10.18}$$

where μ_{00} plays the part of a Mach number related to the temperature fluctuation ΔT_0 on the ground. It is pointed out that the hypothesis $\nu_0 \gg 1$ implies that $H \gg R\Delta T_0/g \equiv h_0$ which is, in fact, always the case when Bo $\cong 1$, i.e., $H \cong RT_\infty(0)/g$ since for the atmosphere, the following always holds true:

$$T_\infty(0) \gg \Delta T_0 \Rightarrow \tau_0 \ll 1 \quad .$$

Respecting the hypotheses (10.17, 18), the solution to (8.5–8) can be sought with the ground condition (8.21) by postulating the following asymptotic expansions (it will be remarked that $\text{Fr}^2/\varepsilon \equiv U_0^2/gH$ is the square of the Froude number constructed from H):

$$\begin{aligned} \boldsymbol{u} &= \boldsymbol{u}_0 + \dots \quad , \\ \pi &= \frac{\text{Fr}^2}{\varepsilon}\pi_2 + \dots \quad , \quad \omega = \frac{\text{Fr}^2}{\varepsilon}\omega_2 + \dots \quad , \quad \vartheta = \frac{\text{Fr}^2}{\varepsilon}\vartheta_2 + \dots \quad . \end{aligned} \tag{10.19}$$

After carrying out the limiting process, the following equation then replaces (8.5):

$$\begin{aligned} &\text{S}\frac{D\boldsymbol{u}_0}{Dt} + \frac{1}{\text{Ro}}(\boldsymbol{e} \times \boldsymbol{u}_0) + \frac{T_\infty^{(0)}}{\text{Bo}}\nabla\pi_2 - \frac{1}{\varepsilon}\vartheta_2\boldsymbol{k} \\ &= \frac{1}{\varrho_\infty^{(0)}}\frac{1}{\text{Re}}\left\{\frac{1}{\varepsilon^2}\frac{\partial^2\boldsymbol{u}_0}{\partial z^2} + \boldsymbol{D}^2\boldsymbol{u}_0 + \frac{\text{Bo}}{3\gamma}\nabla\left(\frac{w_0}{T_\infty^{(0)}}\right)\right\} \quad , \end{aligned} \tag{10.20}$$

once it has been understood that the continuity equation (8.6) takes on the limiting form:

$$\nabla\cdot\boldsymbol{u}_0 \equiv \frac{\partial u_0}{\partial x} + \frac{\partial v_0}{\partial y} + \frac{\partial w_0}{\partial z} = \frac{\text{Bo}}{\gamma}\frac{w_0}{T_\infty^{(0)}} \quad . \tag{10.21}$$

The above limiting form is the natural result of the similarity hypothesis which implies that:

$$\frac{dT_\infty}{dz_\infty} = \lambda_{00} \Gamma_\infty(z_\infty) \frac{1}{\varepsilon} \mathrm{Fr}^2 - \frac{\gamma - 1}{\gamma} \quad , \tag{10.22}$$

where λ_{00} = const and $\Gamma_\infty(z_\infty)$ is some arbitrary function of the order unity which takes into account a "weak" stratification with the altitude, of the standard atmosphere.

As we well know, the similarity hypothesis (10.22) is necessary in order to obtain a "valid" limiting form of the energy equation (8.7). First of all, it is pointed out that (10.22) makes it possible to determine $\varrho_\infty(z_\infty)$ from the equation:

$$\frac{d \log \varrho_\infty}{dz_\infty} = \left\{ \lambda_{00} \Gamma_\infty(z_\infty) \frac{1}{\varepsilon} \mathrm{Fr}^2 - \frac{1}{\gamma} \right\} \frac{1}{T_\infty(z_\infty)} \quad .$$

However, when $\mathrm{Fr} \to 0$, then $\nu_0 \to \infty$ and thus, from (10.16), it is obvious that:

$$z_\infty \to \mathrm{Bo}\, z \quad .$$

Therefore, after carrying out the limiting process, the following expressions are obtained

$$\begin{aligned} T_\infty^{(0)} &\equiv 1 - \frac{\gamma - 1}{\gamma} \mathrm{Bo}\, z \quad ; \\ \varrho_\infty^{(0)} &\equiv \left\{ T_\infty^{(0)} \right\}^{1/(\gamma - 1)} \quad \Rightarrow \quad \frac{d \log \varrho_\infty^{(0)}}{dz} \equiv -\frac{\mathrm{Bo}}{\gamma} \frac{1}{T_\infty^{(0)}} \quad . \end{aligned} \tag{10.23}$$

It is thus seen that (8.7) has the following limiting form:

$$\begin{aligned} &\mathrm{S} \frac{D\vartheta_2}{Dt} - \frac{\gamma - 1}{\gamma} \mathrm{S} \frac{D\pi_2}{Dt} + \lambda_{00}\, \mathrm{Bo} \frac{\Gamma_\infty^{(0)}(\mathrm{Bo}\, z)}{T_\infty^{(0)}(\mathrm{Bo}\, z)} w_0 \\ &= \frac{1}{\varrho_\infty^{(0)}(\mathrm{Bo}\, z)} \frac{1}{\mathrm{Pr}} \frac{1}{\mathrm{Re}} \left\{ \boldsymbol{D}^2 \vartheta_2 + \frac{1}{\varepsilon^2} \left[\frac{\partial^2 \vartheta_2}{\partial z^2} - 2 \frac{\gamma - 1}{\gamma} \frac{\mathrm{Bo}}{T_\infty^{(0)}(\mathrm{Bo}\, z)} \frac{\partial \vartheta_2}{\partial z} \right] \right\} \\ &\quad + \frac{(\gamma - 1)/\gamma}{\varrho_\infty^{(0)}(\mathrm{Bo}\, z) T_\infty^{(0)}(\mathrm{Bo}\, z)} \frac{\mathrm{Bo}}{\varepsilon^2 \mathrm{Re}} \phi_0 \end{aligned} \tag{10.24}$$

where ϕ_0 is given by the relation (2.102) in which u_0, v_0 and w_0 would replace $\overline{u}$, $\overline{v}$ and $\overline{w}$.

Naturally, in the *deep convection equations* (10.20, 21, 24), the following holds:

$$\mathrm{S} \frac{D}{Dt} \equiv \mathrm{S} \frac{\partial}{\partial t} + \boldsymbol{u}_0 \cdot \boldsymbol{\nabla}$$

and $\boldsymbol{u}_0 \equiv u_0 \boldsymbol{i} + v_0 \boldsymbol{j} + \varepsilon w_0 \boldsymbol{k}$, $\boldsymbol{\nabla} = \boldsymbol{D} + (1/\varepsilon)(\partial/\partial z)\boldsymbol{k}$ according to (7.6).

To these deep convection equations, the following ground condition must be added:

$$\vartheta_2 = \frac{\text{Bo}}{\gamma\mu_{00}^2}\Xi(t,\, x,\, y), \quad \text{on} \quad z = 0 \quad . \tag{10.25}$$

It is observed that $\text{Bo}/\gamma\mu_{00}^2 \equiv \varepsilon\tau_0/\text{Fr}^2 = \text{O}(1)$.

In (10.24), the term proportional to

$$\Gamma_\infty^{(0)}(\text{Bo}\, z) \equiv \lim_{z_\infty \to \text{Bo}\, z} \Gamma_\infty(z_\infty)$$

ensures the taking into account of the stratification of the standard atmosphere when $\text{Fr} \to 0$, for it must not be forgotten that $-dT_\infty^0/dz \equiv (\gamma - 1)/\gamma\text{Bo}$ corresponds only to a dry adiabatic temperature gradient. Finally, equation (8.8) yields the relation:

$$\pi_2 = \omega_2 + \vartheta_2 \quad , \tag{10.26}$$

for $\text{Fr} \to 0$, which makes it possible to calculate ω_2 once π_2 and ϑ_2 have been determined from the system of deep convection equations (10.20, 21, 24).

10.2.1 The Quasi-static Deep Convection Equations

If in the limiting equation (10.20), we *actually* want to take into account the influence of the Coriolis force, i.e., if:

$$\text{Ro} \cong 1 \Rightarrow L \cong \frac{U_0}{l_0} \cong 10^5\,\text{m} \quad ,$$

then

$$\varepsilon = \frac{H}{L} \equiv \text{Bo}\frac{RT_\infty(0)/g}{L} \cong 10^{-1} \ll 1$$

when $\text{Bo} \cong 1$.

It thus comes to light that in (10.20, 24), it is possible to carry out the quasi-static approximation which was discussed in Chap. 7. The resulting equations are the *quasi-static deep convection equations*. Let us then define:

$$\lim_{\substack{\varepsilon \to 0, \text{Re} \to \infty \\ \varepsilon^2\text{Re} \equiv \text{Re}_\perp = \text{O}(1)}} \begin{pmatrix} u_0 \\ v_0 \\ w_0 \\ \pi_2 \\ \vartheta_2 \end{pmatrix} \equiv \begin{pmatrix} u_{00} \\ v_{00} \\ w_{00} \\ \pi_{20} \\ \vartheta_{20} \end{pmatrix} \tag{10.27}$$

and

$$\text{S}\frac{D_0}{Dt} = \text{S}\frac{\partial}{\partial t} + u_{00}\frac{\partial}{\partial x} + v_{00}\frac{\partial}{\partial y} + w_{00}\frac{\partial}{\partial z} \quad . \tag{10.28}$$

We thus obtain the following equations

$$\text{S}\frac{D_0 u_{00}}{Dt} - \frac{1}{\text{Ro}}\frac{\sin\varphi}{\sin\varphi_0}v_{00} + \frac{T_\infty^0(\text{Bo}\, z)}{\text{Bo}}\frac{\partial\pi_{20}}{\partial x} = \frac{1}{\varrho_\infty^{(0)}(\text{Bo}\, z)}\frac{1}{\text{Re}_\perp}\frac{\partial^2 u_{00}}{\partial z^2} \quad ; \tag{10.29a}$$

$$\mathrm{S}\frac{D_0 v_{00}}{Dt} + \frac{1}{\mathrm{Ro}}\frac{\sin\varphi}{\sin\varphi_0}u_{00} + \frac{T^0_\infty(\mathrm{Bo}\,z)}{\mathrm{Bo}}\frac{\partial \pi_{20}}{\partial y} = \frac{1}{\varrho^{(0)}_\infty(\mathrm{Bo}\,z)}\frac{1}{\mathrm{Re}_\perp}\frac{\partial^2 v_{00}}{\partial z^2} \quad ; \tag{10.29b}$$

$$T^{(0)}_\infty(\mathrm{Bo}\,z)\frac{\partial \pi_{20}}{\partial z} = \mathrm{Bo}\,\vartheta_{20} \quad ; \tag{10.29c}$$

$$\frac{\partial u_{00}}{\partial x} + \frac{\partial v_{00}}{\partial y} + \frac{\partial w_{00}}{\partial z} = \frac{\mathrm{Bo}}{\gamma}\frac{1}{T^{(0)}_\infty(\mathrm{Bo}\,z)}w_{00} \quad ; \tag{10.29d}$$

$$\begin{aligned}\mathrm{S}\frac{D_0\vartheta_{20}}{Dt} &- \frac{\gamma-1}{\gamma}\mathrm{S}\frac{D_0\pi_{20}}{Dt} + \lambda_{00}\,\mathrm{Bo}\frac{\Gamma^{(0)}_\infty(\mathrm{Bo}\,z)}{T^{(0)}_\infty(\mathrm{Bo}\,z)}w_{00} \\ &= \frac{1}{\varrho^{(0)}_\infty(\mathrm{Bo}\,z)}\frac{1}{\mathrm{Pr}}\frac{1}{\mathrm{Re}_\perp}\left[\frac{\partial^2\vartheta_{20}}{\partial z^2} - 2\frac{\gamma-1}{\gamma}\frac{\mathrm{Bo}}{T^{(0)}_\infty(\mathrm{Bo}\,z)}\frac{\partial\vartheta_{20}}{\partial z}\right] \\ &+ \frac{\mathrm{Bo}(\gamma-1)/\gamma}{\varrho^{(0)}_\infty(\mathrm{Bo}\,z)T^{(0)}_\infty(\mathrm{Bo}\,z)}\frac{1}{\mathrm{Re}_\perp}\left[\left(\frac{\partial u_{00}}{\partial z}\right)^2 + \left(\frac{\partial v_{00}}{\partial z}\right)^2\right] \quad .\end{aligned} \tag{10.29e}$$

Once again, the following ground condition must be assigned to these quasi-static deep convection equations:

$$\vartheta_{20} = \frac{\mathrm{Bo}}{\gamma\mu^2_{00}}\Xi(t,\,x,\,y) \quad , \quad \text{on} \quad z = 0 \quad . \tag{10.30}$$

The above equations can serve as model equations for analyzing the breeze phenomena which are essentially associated with the specification of the thermal field $\Xi(t,\,x,\,y)$ on the ground (in $z = 0$). In Chap. 13, we will return to these problems which come under the heading of mesometeorology. On this subject, the reader is referred to the lectures by Zeytounian (1968) edited by the "Direction de la Météorologie Nationale de Paris."

10.2.2 A New Approach for the Derivation of the Deep Convection Equations (Case of the Adiabatic Atmosphere)

In this section, our starting point is the system of Euler equations written without dimensions – for example, (8.5–8) with Re $\equiv \infty$ in (8.5, 7). We will consider the limiting process $\mathrm{M}_\infty \to 0$ with Bo fixed at the order unity. The variables t, x, y, z, as well as all the other parameters, S, Ro, ε, γ, remain fixed at the order unity when $\mathrm{M}_\infty \to 0$. To this limiting process, we join the following asymptotic representation

$$\begin{aligned}&\boldsymbol{u} = \boldsymbol{u}_\mathrm{a} + \ldots \quad , \quad \omega = \mathrm{M}^2_\infty\omega_\mathrm{a} + \ldots \quad , \quad \vartheta = \mathrm{M}^2_\infty\vartheta_\mathrm{a} + \ldots \quad , \\ &\pi = \mathrm{M}^2_\infty\pi_\mathrm{a} + \ldots \quad .\end{aligned} \tag{10.31}$$

In this case, we have the following adiabatic deep convection equations for the functions $\boldsymbol{u}_\mathrm{a}$, ω_a, ϑ_a and π_a

$$
\begin{aligned}
&\mathrm{S}\frac{D\boldsymbol{u}_\mathrm{a}}{Dt} + \frac{1}{\mathrm{Ro}}(\boldsymbol{e} \times \boldsymbol{u}_\mathrm{a}) + \frac{T_\infty^{(0)}}{\gamma}\nabla\pi_\mathrm{a} = \frac{\mathrm{Bo}}{\varepsilon\gamma}\vartheta_\mathrm{a}\boldsymbol{k} \quad ; \\
&\nabla \cdot \boldsymbol{u}_\mathrm{a} = \frac{\mathrm{Bo}}{\gamma T_\infty^{(0)}} w_\mathrm{a} \quad ; \\
&\mathrm{S}\frac{D\vartheta_\mathrm{a}}{Dt} - \frac{\gamma-1}{\gamma}\mathrm{S}\frac{D\pi_\mathrm{a}}{Dt} + \frac{\mathrm{Bo}}{T_\infty^{(0)}}\chi_\infty(z_\infty)w_\mathrm{a} = 0 \quad ; \\
&\pi_\mathrm{a} = \omega_\mathrm{a} + \vartheta_\mathrm{a} \quad , \\
&\mathrm{S}\frac{D}{Dt} \equiv \mathrm{S}\frac{\partial}{\partial t} + \boldsymbol{u}_\mathrm{a} \cdot \nabla \quad ,
\end{aligned}
\tag{10.32}
$$

once the following hypothesis is made:

$$
\frac{dT_\infty}{dz_\infty} = -\frac{\gamma-1}{\gamma} + \chi_\infty(z_\infty)\mathrm{M}_\infty^2 \quad , \tag{10.33}
$$

$\chi_\infty(z_\infty)$ being a function which takes into account a weak stratification, with the altitude, of the standard atmosphere and which is assumed of the order unity in absolute values. In (10.32), we still have:

$$
T_\infty^{(0)} \equiv 1 - \frac{\gamma-1}{\gamma} z_\infty \quad ; \quad z_\infty \equiv \mathrm{Bo}\, z \quad . \tag{10.34}
$$

It is again pointed out that if (10.32) is to remain "asymptotically" valid, then according to (10.33), the temperature gradient $-dT_\infty/dz_\infty$ must be very close to $(\gamma-1)/\gamma$.

The approach given here is, in principle, similar to that of Batchelor (1953).

Before going on, we want to cite Gough's (1969) analysis of the anelastic approximation which was carried out with a view to thermal convection applications. His analysis was based on an approximation of the non-adiabatic terms.

10.3 The Relation Between the Boussinesq and the Deep Convection Approximations

Firstly let us consider the deep convection equations (10.20, 21, 24) with the ground condition (10.25). We sense right away that the Boussinesq equations of Chap. 8 [see (8.13, 15)] are obtainable from the deep convection equations via the limiting process $\mathrm{Bo} \to 0$. It can thus be stated that these Boussinesq equations are "*shallow*" *convection* equations.

However, when $\mathrm{Bo} \to 0$, it is obvious from (10.25) that the following must also be imposed:

$$
\gamma\mu_{00}^2 \to 0, \quad \text{with} \quad \mathrm{Bo} \to 0; \quad \frac{\mathrm{Bo}}{\gamma\mu_{00}^2} \equiv \frac{\Delta T_0/T_\infty(0)}{\mathrm{Fr}^2/\varepsilon} = \hat{\mu} = O(1)\,. \tag{10.35}
$$

Moreover, (10.24) implies that when $\mathrm{Bo} \to 0$

$$
\lambda_{00} \to \infty \quad , \quad \text{with} \quad \mathrm{Bo} \to 0 \quad ; \quad \lambda_{00}\mathrm{Bo} \equiv \varepsilon\frac{\mathrm{Bo}}{\mathrm{Fr}^2} = \hat{\lambda} = \mathrm{O}(1) \quad . \tag{10.36}
$$

Hence, with (10.35, 36), when $\mathrm{Bo} \to 0$, the limiting form of the deep convection equations (10.20, 21, 24) can be sought in the following form

$$\boldsymbol{u}_0 = \boldsymbol{u}_\mathrm{B} + \dots \quad , \quad \pi_2 = \mathrm{Bo}\pi_\mathrm{B} + \dots \quad , \quad \vartheta_2 = \vartheta_\mathrm{B} + \dots, \tag{10.37}$$

the limiting process $\mathrm{Bo} \to 0$ having been carried out at fixed t, x, y and z. Consequently, the system of Boussinesq equations in the following form is found for the functions $\boldsymbol{u}_\mathrm{B}$, π_B and ϑ_B

$$\mathrm{S}\frac{D\boldsymbol{u}_\mathrm{B}}{Dt} + \frac{1}{\mathrm{Ro}}(\boldsymbol{e} \times \boldsymbol{u}_\mathrm{B}) + \nabla\pi_\mathrm{B} - \frac{1}{\varepsilon}\vartheta_\mathrm{B}\boldsymbol{k} = \frac{1}{\mathrm{Re}}\left\{\frac{1}{\varepsilon^2}\frac{\partial^2 \boldsymbol{u}_\mathrm{B}}{\partial z^2} + \boldsymbol{D}^2\boldsymbol{u}_\mathrm{B}\right\} \quad ;$$

$$\nabla \cdot \boldsymbol{u}_\mathrm{B} = 0 \quad ; \tag{10.38}$$

$$\mathrm{S}\frac{D\vartheta_\mathrm{B}}{Dt} + \hat{\lambda}\Gamma_\infty^{(0)}(0)w_\mathrm{B} = \frac{1}{\mathrm{Pr}}\frac{1}{\mathrm{Re}}\left\{\frac{1}{\varepsilon^2}\frac{\partial^2\vartheta_\mathrm{B}}{\partial z^2} + \boldsymbol{D}^2\vartheta_\mathrm{B}\right\} \quad ,$$

$$\mathrm{S}\frac{D}{Dt} \equiv \mathrm{S}\frac{\partial}{\partial t} + \boldsymbol{u}_\mathrm{B} \cdot \nabla \quad ,$$

with the ground condition:

$$\vartheta_\mathrm{B} = \hat{\mu}\Xi(t, x, y) \quad , \quad \text{on} \quad z = 0 \quad . \tag{10.39}$$

It is observed that the contraints (10.35, 36) lead to:

$$\frac{\hat{\lambda}}{\hat{\mu}} \equiv \nu_0 \cong 1 \Rightarrow H \cong h_0 = \frac{R}{g}\Delta T_0 \quad . \tag{10.40}$$

If (10.40) on H is compared to the constraint obtained in Sect. 8.1 [see, e.g., (8.20)], it will be remarked that if

$$\Delta T_0 \approx \hat{B}U_0\sqrt{\frac{T_\infty(0)}{\gamma R}} \quad , \tag{10.41}$$

then the Boussinesq approximation remains valid. We will likewise see that (10.41) is equivalent to (8.24) when $\hat{B} \approx \Lambda_0 = \mathrm{O}(1)$.

Let us now consider the case of the adiabatic convection equations (10.32). On the one hand, we have the following asymptotic (Boussinesq) representation according to the results from Sect. 8.1:

$$\boldsymbol{u} = \boldsymbol{u}_\mathrm{B} + \dots \quad , \quad \omega = \mathrm{M}_\infty\omega_\mathrm{B} + \dots \quad , \quad \vartheta = \mathrm{M}_\infty\vartheta_\mathrm{B} + \dots \quad ,$$

$$\pi = \mathrm{M}_\infty^2\pi_\mathrm{B} + \dots \quad ,$$

with the hypotheses:

$$\mathrm{M}_\infty \to 0 \quad , \quad \mathrm{Bo} \to 0 \quad , \quad \frac{\mathrm{Bo}}{\mathrm{M}_\infty} = \hat{B} = \mathrm{O}(1) \quad . \tag{10.42}$$

On the other hand, within the framework of the Subsect. 10.2.2 theory, we have the asymptotic (deep convection) representation

$$\boldsymbol{u} = \boldsymbol{u}_a + \ldots \quad , \quad \omega = M_\infty^2 \omega_a + \ldots \quad ,$$
$$\vartheta = M_\infty^2 \vartheta_a + \ldots \quad , \quad \pi = M_\infty^2 \pi_a + \ldots \quad , \tag{10.43a}$$

with the hypotheses

$$M_\infty \to 0 \quad , \quad \mathrm{Bo} = O(1) \quad , \quad z_\infty \equiv \mathrm{Bo}\, z \quad , \tag{10.43b}$$

$$\frac{dT_\infty}{dz_\infty} = -\frac{\gamma - 1}{\gamma} + \chi_\infty(z_\infty) M_\infty^2 \quad . \tag{10.43c}$$

It is thus clear that

$$\boldsymbol{u}_a = \boldsymbol{u}_B \quad , \quad \pi_a = \pi_B, \quad \omega_a = \frac{\omega_B}{M_\infty} \quad , \quad \vartheta_a = \frac{\vartheta_B}{M_\infty} \quad ; \tag{10.44}$$

$$\chi_\infty(z_\infty) = \frac{\hat{B}}{\mathrm{Bo} M_\infty} \left(\frac{dT_\infty}{dz_\infty} + \frac{\gamma - 1}{\gamma} \right) \quad . \tag{10.45}$$

If we were now to take into account (10.44, 45) in the deep convection equations (10.32), then the limiting system which results when

$$M_\infty \to 0 \quad , \quad \mathrm{Bo} \to 0 \quad ; \quad \frac{\mathrm{Bo}}{M_\infty} = \hat{B} = O(1) \quad ,$$

is, in fact, the one which governs the "shallow" convection, i.e, the adiabatic Boussinesq equations [(8.13–15) with Re $\equiv \infty$]. It thus turns out that the Boussinesq equations are none other than the behavior of the deep convection equations when Bo $\to 0$. It is also interesting to note that (10.32) can be rewritten by replacing z and w_a with

$$z_\infty = \mathrm{Bo}\, z \quad , \quad \tilde{w}_a = \mathrm{Bo}\, w_a . \tag{10.46}$$

With the variables t, x, y, z_∞, the following deep convection equations are obtained for the functions $\boldsymbol{v}_a$, $\tilde{w}_a$, π_a and ϑ_a

$$\begin{aligned}
&\mathrm{S}\frac{\partial \boldsymbol{v}_a}{\partial t} + (\boldsymbol{v}_a \cdot \boldsymbol{D})\boldsymbol{v}_a + \tilde{w}_a \frac{\partial \boldsymbol{v}_a}{\partial z_\infty} + \frac{1}{\mathrm{Ro}} \frac{\sin\varphi}{\sin\varphi_0}(\boldsymbol{k} \times \boldsymbol{v}_a) \\
&\quad + \frac{1}{\mathrm{Ro}} \frac{\varepsilon}{\mathrm{Bo}} \frac{\cos\varphi}{\sin\varphi_0} \tilde{w}_a \boldsymbol{i} + \frac{1 - z_\infty(\gamma - 1)/\gamma}{\gamma} \boldsymbol{D}\pi_a = 0 \quad ; \\
&\mathrm{S}\frac{\partial \tilde{w}_a}{\partial t} + \boldsymbol{v}_a \cdot \boldsymbol{D}\tilde{w}_a + \tilde{w}_a \frac{\partial \tilde{w}_a}{\partial z_\infty} - \frac{1}{\mathrm{Ro}} \frac{\cos\varphi}{\sin\varphi_0} \boldsymbol{v}_a \cdot \boldsymbol{i} \\
&\quad + \frac{1}{\gamma}\left(\frac{\mathrm{Bo}}{\varepsilon}\right)^2 \left[\left(1 - \frac{\gamma - 1}{\gamma} z_\infty\right) \frac{\partial \pi_a}{\partial z_\infty} - \vartheta_a \right] = 0 \quad ; \\
&\boldsymbol{D} \cdot \boldsymbol{v}_a + \frac{\partial \tilde{w}_a}{\partial z_\infty} = \frac{\tilde{w}_a}{\gamma[1 - z_\infty(\gamma - 1)/\gamma]} \quad ; \\
&\mathrm{S}\frac{\partial \vartheta_a}{\partial t} + \boldsymbol{v}_a \cdot \boldsymbol{D}\vartheta_a - \frac{\gamma - 1}{\gamma}\left(\mathrm{S}\frac{\partial \pi_a}{\partial t} + \boldsymbol{v}_a \cdot \boldsymbol{D}\pi_a \right) \\
&\quad + \tilde{w}_a \left[\frac{\partial \vartheta_a}{\partial z_\infty} - \frac{\gamma - 1}{\gamma} \frac{\partial \pi_a}{\partial z_\infty} + \frac{\chi_\infty(z_\infty)}{1 - z_\infty(\gamma - 1)/\gamma} \right] = 0 \quad ,
\end{aligned} \tag{10.47}$$

which bring into play the parameters Bo/ε and $\chi_\infty(z_\infty)$.

Concerning the "shallow" convection Boussinesq equations, they can be rewritten by replacing z and w_{B} with:

$$\hat{z} = \hat{B}z = \frac{z_\infty}{\mathrm{M}_\infty} \quad , \quad \hat{w}_{\mathrm{B}} = \hat{B}w_{\mathrm{B}} = \frac{\tilde{w}_{\mathrm{a}}}{\mathrm{M}_\infty} \quad , \quad \text{respectively} \quad . \tag{10.48}$$

With the variables t, x, y, $\hat{z}$, we have the following Boussinesq equations for the functions $\boldsymbol{v}_{\mathrm{B}}$, $\hat{w}_{\mathrm{B}}$, π_{B} and ϑ_{B}

$$\mathrm{S}\frac{\partial \boldsymbol{v}_{\mathrm{B}}}{\partial t} + (\boldsymbol{v}_{\mathrm{B}} \cdot \boldsymbol{D})\boldsymbol{v}_{\mathrm{B}} + \hat{w}_{\mathrm{B}}\frac{\partial \boldsymbol{v}_{\mathrm{B}}}{\partial \hat{z}} + \frac{1}{\mathrm{Ro}}\frac{\sin\varphi}{\sin\varphi_0}(\boldsymbol{k} \times \boldsymbol{v}_{\mathrm{B}})$$
$$+\frac{1}{\mathrm{Ro}}\frac{\varepsilon}{\hat{B}}\frac{\cos\varphi}{\sin\varphi_0}\hat{w}_{\mathrm{B}}\boldsymbol{i} + \frac{1}{\gamma}\boldsymbol{D}\pi_{\mathrm{B}} = 0 \quad ; \tag{10.49a}$$

$$\mathrm{S}\frac{\partial \hat{w}_{\mathrm{B}}}{\partial t} + \boldsymbol{v}_{\mathrm{B}} \cdot \boldsymbol{D}\hat{w}_{\mathrm{B}} + \hat{w}_{\mathrm{B}}\frac{\partial \hat{w}_{\mathrm{B}}}{\partial \hat{z}} - \frac{1}{\mathrm{Ro}}\frac{\cos\varphi}{\sin\varphi_0}\boldsymbol{v}_{\mathrm{B}} \cdot \boldsymbol{i}$$
$$+\frac{1}{\gamma}\left(\frac{\hat{B}}{\varepsilon}\right)^2\left[\frac{\partial \pi_{\mathrm{B}}}{\partial \hat{z}} - \vartheta_{\mathrm{B}}\right] = 0 \quad ; \tag{10.49b}$$

$$\boldsymbol{D} \cdot \boldsymbol{v}_{\mathrm{B}} + \frac{\partial \hat{w}_{\mathrm{B}}}{\partial \hat{z}} = 0 \quad ; \tag{10.49c}$$

$$\mathrm{S}\frac{\partial \vartheta_{\mathrm{B}}}{\partial t} + \boldsymbol{v}_{\mathrm{B}} \cdot \boldsymbol{D}\vartheta_{\mathrm{B}} + \left[\frac{\gamma - 1}{\gamma} + \left(\frac{dT_\infty}{dz_\infty}\right)_0\right]\hat{w}_{\mathrm{B}} = 0 \quad , \tag{10.49d}$$

which bring into play the parameters $\hat{B}/\varepsilon$ and $(\gamma - 1)/\gamma + (dT_\infty/dz_\infty)_0$. We are thus led to consider the deep convection equations (10.47) as outer equations as defined by the MMAE, and the Boussinesq equations (10.49), as inner equations. The outer deep convection equations entirely encompass the inner Boussinesq equations, and are consequently uniformly valid in an atmospheric layer of thickness $H \cong RT_\infty(0)/g$. Otherwise stated, the Boussinesq equations can be "completed" by the deep convection equations; the latter would then contain the former which are valid in the vicinity of the ground.

10.4 Complementary Remarks

10.4.1. We will now turn back to the quasi-static deep convection equations (10.29). Let us suppose that $\mathrm{Ro} \equiv \mathrm{Re}_\perp = \infty$ and $(\partial/\partial x_2) = 0$ (two-dimensional case) and then let us linearize with respect to a state of rest. For the velocity components u', w', and also for the perturbations of pressure π', and of temperature ϑ', the following linear system results:

$$\mathrm{S}\frac{\partial u'}{\partial t} + \frac{T_\infty^{(0)}(\mathrm{Bo}\, z)}{\mathrm{Bo}}\frac{\partial \pi'}{\partial x} = 0 \quad ; \quad T_\infty^{(0)}(\mathrm{Bo}\, z)\frac{\partial \pi'}{\partial z} = \mathrm{Bo}\,\vartheta' \quad ; \tag{10.50a}$$

$$\varrho_\infty^{(0)}(\text{Bo}\,z)\frac{\partial u'}{\partial x} + \frac{\partial}{\partial z}\left[\varrho_\infty^{(0)}(\text{Bo}\,z)w'\right] = 0 \quad ;$$

$$\text{S}\left(\frac{\partial \vartheta'}{\partial t} - \frac{\gamma - 1}{\gamma}\frac{\partial \pi'}{\partial t}\right) + \lambda_{00}\text{Bo}\frac{\Gamma_\infty^{(0)}(\text{Bo}\,z)}{T_\infty^{(0)}(\text{Bo}\,z)}w' = 0 \quad ; \tag{10.50b}$$

$$T_\infty^{(0)}(\text{Bo}\,z) = 1 - \frac{\gamma-1}{\gamma}\text{Bo}\,z \quad , \quad \varrho_\infty^{(0)}(\text{Bo}\,z) = \left[T_\infty^{(0)}(\text{Bo}\,z)\right]^{1/(\gamma-1)} \quad .$$

The solution to (10.50) is sought in the form:

$$\begin{pmatrix} u' \\ w' \\ \pi' \\ \vartheta' \end{pmatrix} = \begin{pmatrix} U \\ W \\ \Pi \\ \Theta \end{pmatrix}(z)\exp\left[\mathrm{i}\left(\nu x + \frac{\sigma}{\text{S}}t\right)\right] \quad . \tag{10.51}$$

For $W(z)$, the above leads to the following ordinary second order differential equation:

$$\frac{d}{dz}\left[\frac{1}{\varrho_\infty^{(0)}}\frac{d\left(\varrho_\infty^{(0)}W\right)}{dz}\right] + \frac{\lambda_{00}\text{Bo}\nu^2}{\sigma^2}\frac{\Gamma_\infty^{(0)}}{T_\infty^{(0)}}W = 0 \quad . \tag{10.52}$$

One of the boundary conditions to be assigned to (10.52) is:

$$W(0) = 0 \quad . \tag{10.53}$$

If the flow is considered in a duct having a plane rigid surface at the top to simulate the tropopause (i.e., with dimensions, the surface $z^* = RT_\infty(0)/g$), then the second condition in z prescribed for (10.52) must be written in the form

$$W\left(\frac{1}{\text{Bo}}\right) = 0 \quad . \tag{10.54}$$

Let $\overline{W}$ be the conjugate complex of W; $\overline{W}(z)$ satisfies the same problem (10.52–54) as $W(z)$. We then multiply (10.52) by $\varrho_\infty^{(0)}\overline{W}$ and integrate the equation thus obtained once with respect to z, from $z = 0$ to $z = 1/\text{Bo}$. We obtain:

$$\int_0^{1/\text{Bo}} \frac{1}{\varrho_\infty^{(0)}}\left|\frac{d\left(\varrho_\infty^{(0)}W\right)}{dz}\right|^2 dz = \frac{\lambda_{00}\text{Bo}\nu^2}{\sigma^2}\int_0^{1/\text{Bo}} \frac{\Gamma_\infty^{(0)}}{T_\infty^{(0)}}\varrho_\infty^{(0)}|W|^2 dz \quad , \tag{10.55}$$

and it is remarked that if $\Gamma_\infty^{(0)}(\text{Bo}\,z) > 0$, then:

$$\omega_0^2 \equiv \frac{\lambda_{00}\text{Bo}\nu^2}{\sigma^2} > 0 \Rightarrow \sigma \cong \frac{\nu}{\omega_0} \leq 1 \quad ,$$

if, in fact, $\lambda_{00}\text{Bo} \cong 1$.

When Bo $\Rightarrow$ 0 (Boussinesq approximation) in such a way that $\lambda_{00}\text{Bo} \cong 1$, instead of (10.52), we have:

$$\frac{d^2W_\text{B}}{dz^2} + \omega_0^2\Gamma_\infty^{(0)}(0)W_\text{B} = 0 \quad , \tag{10.56}$$

whose solution is

$$W_B = \mathcal{A}_{00} \sin(\varphi_0 z) \quad , \tag{10.57}$$

where $\mathcal{A}_{00}$ = const and $\varphi_0^2 \equiv \omega_0^2 \Gamma_\infty^{(0)}(0) > 0$. This solution (10.57) satisfies the condition $W_B(0) = 0$ and has a sinusoidal behavior when $z \to +\infty$! Therefore, when Bo $\to 0$, a double scale wave phenomenon emerges in the duct of thickness 1/Bo.

In order to elucidate this phenomenon, we rewrite (10.52) for $W(z;\,\text{Bo})$ in the form:

$$\gamma T_\infty^{(0)}(\zeta)\frac{d^2W}{dz^2} - \text{Bo}\frac{dW}{dz} + \left[\omega_0^2 \gamma \Gamma_\infty^{(0)}(\zeta) - \text{Bo}^2\frac{\gamma-1}{\gamma}\frac{1}{T_\infty^{(0)}(\zeta)}\right] W = 0 \quad . \tag{10.58}$$

To the above equation, the following conditions are added:

$$W = 0 \quad , \quad \text{for} \quad z = 0\,(\zeta = 0) \quad \text{and} \quad \zeta = 1 \quad , \tag{10.59}$$

where $\zeta = \text{Bo}\, z$.

We now introduce the fast phase:

$$\phi(\zeta) = \frac{Z(\zeta)}{\text{Bo}} \quad , \quad \text{Bo} \ll 1 \quad , \tag{10.60}$$

and let:

$$W(z;\,\text{Bo}) = \tilde{W}(\phi,\,\zeta;\,\text{Bo}) \quad .$$

Hence:

$$\frac{dW}{dz} = \text{Bo}\frac{\partial \tilde{W}}{d\zeta} + \frac{dZ}{d\zeta}\frac{\partial \tilde{W}}{\partial \phi}$$

and also

$$\frac{d^2W}{dz^2} = \text{Bo}^2\frac{\partial^2 \tilde{W}}{\partial \zeta^2} + \left(\frac{dZ}{d\zeta}\right)^2\frac{\partial^2 \tilde{W}}{\partial \phi^2} + \text{Bo}\left[2\frac{dZ}{d\zeta}\frac{\partial^2 \tilde{W}}{\partial \zeta \partial \phi} + \frac{\partial^2 Z}{d\zeta^2}\frac{\partial \tilde{W}}{\partial \phi}\right] \quad .$$

If $\tilde{W}$ is now sought in the form of a uniformly valid expansion

$$\tilde{W} = \tilde{W}_0(\zeta,\,\phi) + \text{Bo}\,\tilde{W}_1(\zeta,\,\phi) + \dots \tag{10.61}$$

the following equations are obtained for $\tilde{W}_0$ and $\tilde{W}_1$

$$\gamma T_\infty^{(0)}(\zeta)\left(\frac{dZ}{d\zeta}\right)^2\frac{\partial^2 \tilde{W}_0}{\partial \phi^2} + \omega_0^2 \gamma \Gamma_\infty^{(0)}(\zeta)\tilde{W}_0 = 0 \quad , \tag{10.62a}$$

$$\gamma T_\infty^{(0)}(\zeta)\left(\frac{dZ}{d\zeta}\right)^2\frac{\partial^2 \tilde{W}_1}{\partial \phi^2} + \omega_0^2 \gamma \Gamma_\infty^{(0)}(\zeta)\tilde{W}_1$$

$$= \frac{dZ}{d\zeta}\frac{\partial \tilde{W}_0}{\partial \phi} - \gamma T_\infty^{(0)}(\zeta)\left\{2\frac{dZ}{d\zeta}\frac{\partial^2 \tilde{W}_0}{\partial \zeta \partial \phi} + \frac{d^2 Z}{d\zeta^2}\frac{\partial W_0}{\partial \phi}\right\} \quad . \tag{10.62b}$$

According to (10.62a), it is advisable [if we are to carry out the "matching" with (10.56)] to choose the following for the function $Z(\zeta)$

$$\gamma T_{\infty}^{(0)}(\zeta)\left(\frac{dZ}{d\zeta}\right)^2 = \gamma\frac{\Gamma_{\infty}^{(0)}(\zeta)}{\Gamma_{\infty}^{(0)}(0)}$$

or even

$$Z(\zeta) = \frac{1}{\Gamma_{\infty}^{(0)}(0)}\int_0^{\zeta}\sqrt{\frac{\Gamma_{\infty}^{(0)}(\zeta)}{T_{\infty}^{(0)}(\zeta)}}d\zeta, \tag{10.63}$$

provided that $Z(0) = 0$.

On the condition that $Z(\zeta)$ be given by (10.63), we find that:

$$\tilde{W}_0(\zeta, \phi) = \mathcal{A}_0(\zeta)\sin(\varphi_0\phi) + \mathcal{B}_0(\zeta)\cos(\varphi_0\phi) \quad .$$

However, in the vicinity of $\zeta \cong 0$, we have

$$Z(\zeta) \cong \zeta \quad \text{and} \quad \phi \cong z \quad ,$$

which ensures the matching with the proximal Boussinesq solution (10.57) once it is assumed that $\mathcal{B}_0(\zeta) \equiv 0$.

The coefficient $\mathcal{A}_0(\zeta)$ still remains to be determined. To this end, we make use of (10.62b) for $\tilde{W}_1(\zeta, \phi)$ by cancelling out the secular terms which appear in the right hand member of this equation. This leads to the following equation for $\mathcal{A}_0(\zeta)$:

$$\frac{d\mathcal{A}_0}{d\zeta} + \mathcal{Q}_0(\zeta)\mathcal{A}_0 = 0 \quad , \quad \mathcal{A}_0(1) = 0 \quad , \quad \text{where}$$

$$\mathcal{Q}_0(\zeta) = \frac{\gamma T_{\infty}^{(0)}(\zeta)d^2Z/d\zeta^2 - dZ/d\zeta}{2\gamma T_{\infty}^{(0)}(\zeta)(dZ/d\zeta)} \quad , \tag{10.64}$$

is a known function of ζ if (10.63) is taken into account.

At this point, we find

$$\mathcal{A}_0(\zeta) = \exp\left[-\int_1^{\zeta}\mathcal{Q}_0(\zeta)d\zeta\right] \quad . \tag{10.65}$$

Thus:

$$\tilde{W}_0(\zeta, \phi) = \exp\left[-\int_1^{\zeta}\mathcal{Q}_0(\zeta)d\zeta\right]\sin(\varphi_0\phi) \quad . \tag{10.66}$$

Finally, it is realized that in the proximal Boussinesq solution, we inevitably have:

$$\mathcal{A}_{00} = \exp\left[-\int_1^{0}\mathcal{Q}_0(\zeta)d\zeta)\right] \quad , \tag{10.67}$$

which completely solves the problem being considered to order zero.

10.4.2. It is noteworthy that the above analysis can be extended to the nonlinear case if for "exact" basic equations, the following steady two-dimensional adiabatic deep convection equations are adopted

$$u_0\frac{\partial u_0}{\partial x}+w_0\frac{\partial u_0}{\partial z}+\frac{T_\infty^{(0)}(\mathrm{Bo}\,z)}{\mathrm{Bo}}\frac{\partial \pi_2}{\partial x}=0\quad;$$

$$u_0\frac{\partial w_0}{\partial x}+w_0\frac{\partial w_0}{\partial z}+\frac{T_\infty^{(0)}(\mathrm{Bo}\,z)}{\mathrm{Bo}}\frac{\partial \pi_2}{\partial z}=\vartheta_2\quad;$$

$$\frac{\partial u_0}{\partial x}+\frac{\partial w_0}{\partial z}=\frac{\mathrm{Bo}}{\gamma}\frac{1}{T_\infty^{(0)}(\mathrm{Bo}\,z)}w_0\quad; \tag{10.68}$$

$$u_0\frac{\partial \vartheta_2}{\partial x}+w_0\frac{\partial \vartheta_2}{\partial z}-\frac{\gamma-1}{\gamma}\left(u_0\frac{\partial \pi_2}{\partial x}+w_0\frac{\partial \pi_2}{\partial z}\right)$$
$$+\lambda_{00}\mathrm{Bo}\frac{\Gamma_\infty^{(0)}(\mathrm{Bo}\,z)}{T_\infty^{(0)}(\mathrm{Bo}\,z)}w_0=0\quad;$$

$$T_\infty^{(0)}(\mathrm{Bo}\,z)=1-\frac{\gamma-1}{\gamma}\mathrm{Bo}\,z\quad;\qquad \frac{d\log \varrho_\infty^{(0)}(\mathrm{Bo}\,z)}{dz}=-\frac{\mathrm{Bo}}{\gamma}\frac{1}{T_\infty^{(0)}(\mathrm{Bo}\,z)}\quad, \tag{10.69}$$

once it has been assumed that $\varepsilon\equiv 1$, $\mathrm{Ro}\equiv\infty$ and $\mathrm{Re}\equiv\infty$. The continuity equation [third equation of system (10.68)] is integrated while taking into account the expression of the right hand member as a function of $\varrho_\infty^{(0)}(\mathrm{Bo}\,z)$, by introducing the generalized stream function $\psi_0(x,\,z;\,\mathrm{Bo})$:

$$u_0=-\exp\left[\frac{\mathrm{Bo}}{\gamma}\int\frac{dz}{T_\infty^{(0)}(\mathrm{Bo}\,z)}\right]\frac{\partial\psi_0}{\partial z}\quad;$$

$$w_0=\exp\left[\frac{\mathrm{Bo}}{\gamma}\int\frac{dz}{T_\infty^{(0)}(\mathrm{Bo}\,z)}\right]\frac{\partial\psi_0}{\partial x}\quad. \tag{10.70}$$

The fourth equation of (10.68) then leads to the following first integral:

$$\vartheta_2-\frac{\gamma-1}{\gamma}\pi_2+\lambda_{00}\mathrm{Bo}\int\frac{\Gamma_\infty^{(0)}(\mathrm{Bo}\,z)}{T_\infty^{(0)}(\mathrm{Bo}\,z)}dz=\theta(\psi_0)\quad. \tag{10.71}$$

Furthermore, the following vorticity equation is obtained from the first two equations of (10.68) when the third equation is also put to use

$$u_0\frac{\partial\Omega_0}{\partial x}+w_0\frac{\partial\Omega_0}{\partial z}+\frac{\mathrm{Bo}}{\gamma T_\infty^{(0)}(\mathrm{Bo}\,z)}w_0\Omega_0=\frac{\partial}{\partial x}\left(\vartheta_2-\frac{\gamma-1}{\gamma}\pi_2\right)\quad, \tag{10.72}$$

where $\Omega_0=(\partial w_0/\partial x)-(\partial u_0/\partial z)$. By making use of (10.70, 71), we obtain from (10.72), a second first integral

$$\exp\left[\frac{\mathrm{Bo}}{\gamma}\int\frac{dz}{T_\infty^{(0)}(\mathrm{Bo}\,z)}\right]\Omega_0-\frac{d\theta}{d\psi_0}z=\chi(\psi_0)\quad, \tag{10.73}$$

since

$$\left\{\frac{\partial\psi_0}{\partial z}\frac{\partial}{\partial x}-\frac{\partial\psi_0}{\partial x}\frac{\partial}{\partial z}\right\}\left(\frac{d\theta}{d\psi_0}\right)\equiv 0 \quad .$$

However,

$$\exp\left[\frac{\mathrm{Bo}}{\gamma}\int\frac{dz}{T_\infty^{(0)}(\mathrm{Bo}\,z)}\right]=\left(1-\frac{\gamma-1}{\gamma}\mathrm{Bo}\,z\right)^{-1/(\gamma-1)} \quad ,$$

and hence

$$\frac{\partial w_0}{\partial x}=\left(1-\frac{\gamma-1}{\gamma}\mathrm{Bo}\,z\right)^{-1/(\gamma-1)}\frac{\partial^2\psi_0}{\partial x^2} \quad ;$$

$$\frac{\partial u_0}{\partial z}=-\left(1-\frac{\gamma-1}{\gamma}\mathrm{Bo}\,z\right)^{-1/(\gamma-1)}\frac{\partial^2\psi_0}{\partial z^2}$$
$$-\frac{\mathrm{Bo}}{\gamma}\left(1-\frac{\gamma-1}{\gamma}\mathrm{Bo}\,z\right)^{-1/(\gamma-1)}\frac{\partial\psi_0}{\partial z} \quad .$$

Therefore, from (10.73), the following equation in ψ_0 is derived:

$$\frac{\partial^2\psi_0}{\partial x^2}+\frac{\partial^2\psi_0}{\partial z^2}+\frac{\mathrm{Bo}}{\gamma}\left(1-\frac{\gamma-1}{\gamma}\mathrm{Bo}\,z\right)^{-1}\frac{\partial\psi_0}{\partial z}$$
$$=\left(1-\frac{\gamma-1}{\gamma}\mathrm{Bo}\,z\right)^{2/(\gamma-1)}\left\{\chi(\psi_0)+\frac{d\theta}{d\psi_0}z\right\} \quad . \tag{10.74}$$

The arbitrary functions of ψ_0, $\chi(\psi_0)$ and $\theta(\psi_0)$, must be determined from the boundary conditions. This problem will be addressed in Chap. 13.

10.4.3. To be sure, in order to initialize the deep convection equations, strictly speaking, we shall consider an initial adjustment problem which is valid in the vicinity of $t = 0$. As a matter of fact, this adjustment problem can be considered as a particular case of the problem taken up in Chap. 12 which concerns the quasi non-divergent model (quasi-solenoidal).

We wisth to specify that in the *unsteady* two-dimensional adiabatic case, when $\mathrm{Re}\equiv\infty$, $\mathrm{Ro}\equiv\infty$ and $\varepsilon\equiv 1$, we have to consider in linear theory the following deep convection system instead of the complete system (3.30)

$$\mathrm{S}\frac{\partial u_0'}{\partial t}=-\frac{T_\infty^{(0)}(\mathrm{Bo}\,z)}{\mathrm{Bo}}\frac{\partial\pi_2'}{\partial x} \quad ; \quad \mathrm{S}\frac{\partial w_0'}{\partial t}=-\frac{T_\infty^{(0)}(\mathrm{Bo}\,z)}{\mathrm{Bo}}\frac{\partial\pi_2'}{\partial z}+\vartheta_2' \quad ;$$
$$\frac{\partial u_0'}{\partial x}+\frac{\partial w_0'}{\partial z}=\frac{\mathrm{Bo}}{\gamma}\frac{1}{T_\infty^{(0)}(\mathrm{Bo}\,z)}w_0' \quad ; \tag{10.75}$$
$$\mathrm{S}\left(\frac{\partial\vartheta_2'}{\partial t}-\frac{\gamma-1}{\gamma}\frac{\partial\pi_2'}{\partial t}\right)+\lambda_{00}\mathrm{Bo}\frac{\Gamma_\infty^{(0)}(\mathrm{Bo}\,z)}{T_\infty^{(0)}(\mathrm{Bo}\,z)}w_0'=0 \quad .$$

The above linear system (10.75) leads to a single equation for $\mathcal{D}_{20}'\equiv(\partial u_0'/\partial x)+(\partial w_0'/\partial z)$ which is written:

$$\left\{ S^2 \frac{\partial^2}{\partial t^2} \left[\frac{\partial^2}{\partial x^2} + \frac{\partial^2}{\partial z^2} - \frac{(2\gamma - 1)\mathrm{Bo}}{1 - (\gamma - 1)\mathrm{Bo}\, z/\gamma} \frac{\partial}{\partial z} \right] \right.$$

$$\left. + \lambda_{00}\, \mathrm{Bo} \frac{\Gamma_\infty^{(0)}(\mathrm{Bo}\, z)}{1 - (\gamma - 1)\mathrm{Bo}\, z/\gamma} \frac{\partial^2}{\partial x^2} \right\} \mathcal{D}'_{20} = 0 \quad . \tag{10.76}$$

The above equation is of second order in t, whereas (3.34) associated with the complete system (3.30) was of fourth order in t. Once again, there is a loss of the internal acoustic waves just like in the Boussinesq approximation. The difference is that in (10.76), there is a term in $\partial/\partial z$ which takes into account a weak compressibility.

To conclude, we wish to point out that various authors [in particular, Voulfson (1981)] have attempted to obtain forms of the deep convection equations which are "better" adapted for calculation so that the latter possesses integral conservation laws. Nonetheless, from the viewpoint of asymptotic methods, these various attempts cannot be considered coherent since they lead to the introduction of terms into the limiting equations which do not correlate.

11. The Quasi-geostrophic and Ageostrophic Models

As our starting point in this chapter, we will again use the exact quasi-static equations of Sect. 7.1. In this instance, however, a β-plane approximation will be carried out (see Sect. 2.3), and, like in Subsect. 2.4.3, the coordinate system τ, ξ, η, ζ, associated with isobaric surfaces will be used. This means that we will consider (4.3) to which we will add the right hand members of (7.14, 18) after rewriting them with respect to the variable ζ, while taking into account

$$\frac{\partial}{\partial z} = -\mathrm{Bo}\varrho\frac{\partial}{\partial \zeta} \quad .$$

If the following vertical Ekman number is introduced

$$\mathrm{Ek}_\perp \equiv \frac{\mathrm{Ro}}{\mathrm{Re}_\perp} = \frac{\nu_0}{l_0 H^2} \quad , \quad l_0 = 2\Omega_0 \sin\varphi_0 \quad , \tag{11.1}$$

and if it is assumed that the turbulent transfer coefficients μ_0 and k_0, as well as the radiation $\mathcal{R}$, are functions only of ζ such that[1]

$$-\frac{\mathrm{Bo}^2}{\mathrm{Pr}}\sigma_{00}\frac{d\mathcal{R}_0}{d\zeta} \equiv Q_0(\zeta) \quad , \tag{11.2}$$

then the following *dimensionless* equations (where bars have been omitted) can be written for U, V, W, $\mathcal{H}$, T and ϱ

$$\mathrm{Ki}\left\{\frac{\partial U}{\partial \tau} + \frac{1}{\mathrm{S}}\left(U\frac{\partial U}{\partial \xi} + V\frac{\partial U}{\partial \eta}\right) + W\frac{\partial U}{\partial \zeta}\right\} - \left(1 + \frac{\beta}{\mathrm{S}}\mathrm{Ki}\eta\right)V + \frac{\lambda_0}{\mathrm{Ki}}\frac{\partial \mathcal{H}}{\partial \xi}$$
$$= \mathrm{Bo}^2\mathrm{Ek}_\perp\frac{\partial}{\partial \zeta}\left(\varrho\mu_0\frac{\partial U}{\partial \zeta}\right) \quad ; \tag{11.3}$$

$$\mathrm{Ki}\left\{\frac{\partial V}{\partial \tau} + \frac{1}{\mathrm{S}}\left(U\frac{\partial V}{\partial \xi} + V\frac{\partial V}{\partial \eta}\right) + W\frac{\partial V}{\partial \zeta}\right\} + \left(1 + \frac{\beta}{\mathrm{S}}\mathrm{Ki}\eta\right)U$$
$$+\frac{\lambda_0}{\mathrm{Ki}}\frac{\partial \mathcal{H}}{\partial \eta} = \mathrm{Bo}^2\mathrm{Ek}_\perp\frac{\partial}{\partial \zeta}\left(\varrho\mu_0\frac{\partial V}{\partial \zeta}\right) \quad ; \tag{11.4}$$

$$\mathrm{Bo}\zeta\frac{\partial \mathcal{H}}{\partial \zeta} + T = 0 \quad , \quad \varrho = \frac{\zeta}{T} \quad ; \tag{11.5}$$

[1] In other words, $\mu_0(\zeta)$, $k_0(\zeta)$ and $Q_0(\zeta)$ are considered as being the values corresponding to the standard atmosphere which are functions only of ζ.

$$\frac{\partial U}{\partial \xi} + \frac{\partial V}{\partial \eta} + \mathrm{S}\frac{\partial W}{\partial \zeta} = 0 \quad ; \tag{11.6}$$

$$\mathrm{Ki}\left\{\frac{\partial T}{\partial \tau} + \frac{1}{\mathrm{S}}\left(U\frac{\partial T}{\partial \xi} + V\frac{\partial T}{\partial \eta}\right) + W\left(\frac{\partial T}{\partial \zeta} - \frac{\gamma - 1}{\gamma}\frac{T}{\zeta}\right)\right\}$$
$$= \frac{\mathrm{Bo}^2}{\mathrm{Pr}}\mathrm{Ek}_\perp \frac{\partial}{\partial \zeta}\left(\varrho k_0 \frac{\partial T}{\partial \zeta}\right)$$
$$+ \frac{\gamma - 1}{\gamma}\frac{\mathrm{Bo}^3}{\mathrm{S}\lambda_0}\mathrm{Ki}^2 \mathrm{Ek}_\perp \varrho \mu_0 \left[\left(\frac{\partial U}{\partial \zeta}\right)^2 + \left(\frac{\partial V}{\partial \zeta}\right)^2\right] + \mathrm{Ek}_\perp Q_0 \quad . \tag{11.7}$$

Let us recall that:

$$\lambda_0 \equiv \frac{\mathrm{Bo}}{\gamma \mathrm{S}}\left(\frac{\mathrm{Ki}}{\mathrm{M}_\infty}\right)^2 = \mathrm{O}(1) \quad .$$

The earth's surface will be assumed flat and thus the following boundary conditions can be assigned to (11.3, 4, 6, 7)

$$U = 0 \quad , \quad V = 0 \quad ; \quad W = \mathrm{Bo}\varrho\frac{\partial \mathcal{H}}{\partial \tau} \quad ; \quad k_0 \varrho \frac{\partial T}{\partial \zeta} = \sigma_{00}\mathcal{R}_0 \quad , \quad \text{on} \quad \mathcal{H} = 0 \quad . \tag{11.8}$$

Concerning the initial conditions, three are called for. First of all

$$U = U^0 \quad \text{and} \quad V = V^0 \quad , \quad \text{for} \quad \tau = 0 \quad , \tag{11.9}$$

and a condition on the temperature or on the altitude of the isobaric surface, again for $\tau = 0$. This condition will be specified in Sect. 11.2 when the analysis of the problem of adjustment to geostrophy is made. It will be noted that U^0 and V^0 are known functions of ξ, η and ζ.

With respect to ξ and η, we can assume that $U, V, T, \mathcal{H}$ and their derivatives remain bounded *at infinity* but the boundary conditions for $\zeta \to 0$ will not be given here. However, it will be seen in Sect. 11.1 how a reasonable behavior condition can be imposed in conjunction with the equation governing the so-called quasi-geostrophic model. In any case, it is important to note that the conditions in $\zeta \to 0$ play "practically" no role at all in the deriving of the models which constitute the focal point of the present chapter.

Throughout the forthcoming sections, we will be considering the double limiting process[2]:

$$\mathrm{Ki} \to 0 \quad \text{and} \quad \mathrm{Ek}_\perp \to 0 \quad . \tag{11.10}$$

It is easy to demonstrate that both Ki and $\mathrm{Ek}_\perp$ lead to a singular perturbation problem when they tend to zero if either one is fixed.

As a matter of fact, it has already been pointed out in Sect. 4.1 that when $\mathrm{Ki} \to 0$, the partial derivatives in τ disappear which means we can no longer

[2] To be precise, the situation considered here is: M_∞, Ki, $\mathrm{Ek}_\perp$ fixed; $\varepsilon \to 0$, $\mathrm{Re} \to \infty$, with $\varepsilon^2\mathrm{Re} = \mathrm{Re}_\perp \equiv \mathrm{Ki}/\mathrm{S}\,\mathrm{Ek}_\perp$ fixed; then $\mathrm{M}_\infty \to 0$, $\mathrm{Ki} \to 0$; λ_0 fixed, $\mathrm{Ek}_\perp \to 0$.

satisfy all the initial conditions: we are thus dealing with an initial layer phenomenon in the vicinity of $\tau = 0$. This in turn leads to the problem of adjustment to geostrophy which has already been considered in Sect. 6.1, and which will be analyzed in greater detail in Sect. 11.2.

Furthermore, when $\mathrm{Ek}_{\perp} \to 0$, the disappearance of the "non-adiabatic" terms from the right-hand members of (11.3.,4,7) is observed. The conditions (11.8) can thus no longer be applied and are then replaced by a single slip condition which is as follows:

$$W = \mathrm{Bo}_{\varrho}\left\{\frac{\partial \mathcal{H}}{\partial \tau} + \frac{1}{\mathrm{S}}\left(U\frac{\partial \mathcal{H}}{\partial \xi} + V\frac{\partial \mathcal{H}}{\partial \eta}\right)\right\} \quad , \quad \text{on} \quad \mathcal{H} = 0 \quad , \tag{11.11}$$

according to (2.92,93,95). Therefore, we are faced with a typical boundary layer phenomenon in the vicinity of the ground's surfae which, in first approximation, leads to the so-called Ackerblom problem which is analyzed in Sect. 11.3.

When $\mathrm{Ki} \to 0$ and $\mathrm{Ek}_{\perp} \to 0$, we thus have a "doubly" singular problem on our hands. However, we know that a special relationship exists between Ki and $\mathrm{Ek}_{\perp}$ such that when a description is obtained, all the others are contained therein [Darrozes (1972)].

Let then

$$\mathrm{Ek}_{\perp} = \sigma(\mathrm{Ki})\widehat{\mathrm{Ek}}_{\perp}, \tag{11.12}$$

where $\widehat{\mathrm{Ek}}_{\perp} = O(1)$ and $\sigma(\mathrm{Ki}) \to 0$, with $\mathrm{Ki} \to 0$.

The gauge function $\sigma(\mathrm{Ki})$ must be chosen in such a way that the maximum number of terms are retained in the system of local equations which is valid near the ground. This system is obtained in standard fashion via the complete equations (11.3–7) by introducing the (local) altitude variable

$$\hat{\zeta} = \frac{1 - \zeta}{\delta(\mathrm{Ki})} \quad , \quad \delta(\mathrm{Ki}) \to 0 \quad , \quad \text{with} \quad \mathrm{Ki} \to 0 \quad , \tag{11.13}$$

and by carrying out the *local* limiting process

$$\mathrm{Ki} \to 0 \quad , \quad \text{with} \quad \tau, \xi, \eta \quad \text{and} \quad \hat{\zeta} \quad \text{fixed} \quad . \tag{11.14}$$

It is emphasized from the start that in conjunction with the definition (11.13) of $\hat{\zeta}$, when $\mathrm{Ki} \to 0$ with τ, ζ, η and ζ fixed (*main limiting* process considered in Sect. 11.1), we are led to seek the solution of the function $\mathcal{H}$ in the form

$$\mathcal{H} = \mathcal{H}_0(\zeta) + \mathrm{Ki}\mathcal{H}_1 + \mathrm{Ki}^2\mathcal{H}_2 + \ldots \quad , \tag{11.15}$$

and thus $\mathcal{H} = 0$ means that

$$\mathcal{H}(\zeta) + \mathrm{Ki}\mathcal{H}_1 + \mathrm{Ki}^2\mathcal{H}_2 + \ldots = 0 \quad ,$$

or even

$$\zeta = \zeta_s(\xi, \eta, \tau; \mathrm{Ki}) = \zeta_{s0} + \mathrm{Ki}\zeta_{s1} + \ldots \quad . \tag{11.16}$$

However, the first term of (11.16) which is constant and a solution to $\mathcal{H}_0(\zeta) = 0$, can be taken equal to unity if it is assumed that in the dimensionless starting equations, the pressure variable was rendered dimensionless via the ground pressure corresponding to the standard atmosphere. In this case

$$\zeta_{s0} \equiv 1 \quad , \quad \zeta_{s1}(\xi, \eta, \tau) = -\frac{\mathcal{H}_1|_{\zeta=1}}{(d\mathcal{H}_0/d\zeta)_{\zeta=1}} \quad . \tag{11.17}$$

Therefore, when $\mathrm{Ki} \to 0$ with ζ fixed, the ground's surface (supposed flat) can, in fact, be simulated by the plane $\zeta = 1$. This is not an additional approximation, but rather a coherent conclusion of the fundamental hypothesis $\mathrm{Ki} \to 0$ in the MMAE.

From (11.12, 13), it is easy to demonstrate that

$$\sigma(\mathrm{Ki}) \equiv \delta^2(\mathrm{Ki}) \quad , \tag{11.18}$$

once the Van Dyke (1964) principle of the least local degeneracy has been applied. In order to determine $\delta(\mathrm{Ki})$, it is enough to impose the matching of the W components in the main and local representations. As it is assumed that

$$K_0(\zeta) \equiv \zeta \left\{ \frac{d}{d\zeta}\left(\zeta \frac{d\mathcal{H}_0}{d\zeta}\right) - \frac{\gamma - 1}{\gamma} \frac{d\mathcal{H}_0}{d\zeta} \right\} \not\equiv 0 \quad ,$$

which means that

$$W = \mathrm{Ki}\, w_1 + \dots \quad , \tag{11.19}$$

when $\mathrm{Ki} \to 0$ with τ, ξ, η, ζ fixed, we are led to the following relation

$$\mathrm{Ki} = \delta(\mathrm{Ki}) \quad , \tag{11.20}$$

since in the local representation, we have

$$W = \delta(\mathrm{Ki})\hat{w}_1 + \dots \quad , \tag{11.21}$$

when $\mathrm{Ki} \to 0$ with $\tau, \xi, \eta, \hat{\zeta}$ fixed.

Hence, according to (11.18, 20)

$$\sigma(\mathrm{Ki}) \equiv \mathrm{Ki}^2 \tag{11.22}$$

and the following similarity relation must be imposed

$$\mathrm{Ek}_\perp = \widehat{\mathrm{Ek}}_\perp \mathrm{Ki}^2, \quad \widehat{\mathrm{Ek}}_\perp = O(1) \quad , \tag{11.23}$$

and then

$$\mathrm{Re}_\perp \equiv \frac{1}{S\widehat{\mathrm{Ek}}_\perp} \frac{1}{\mathrm{Ki}} \gg 1 \quad .$$

It can thus be seen that $\mathrm{Ki} \to 0$ is the only small parameter of our problem. It is pointed out that (11.23) implies the following constraint on the characteristic time t_0

$$t_0 \cong \widehat{\mathrm{Ek}}_\perp^{1/2} \frac{H}{(l_0 \nu_0)^{1/2}} \tag{11.24}$$

which turns out to be of the order of 3–4 days for values of H, ν_0 and l_0 which are realistic from the viewpoint of synoptic phenomena. This corresponds very well to a classical short-range synoptic forecast.

Throughout this entire chapter it will be supposed that all the parameters γ, S, λ_0, Bo, Pr, $\widehat{\mathrm{Ek}}_\perp$ and σ_{00} are fixed and of the order unity when $\mathrm{Ki} \to 0$.

When $\mathrm{Ki} \to 0$, we must consider very carefully four regions in the plane (τ, ζ). They are as indicated below in the Fig. 3.

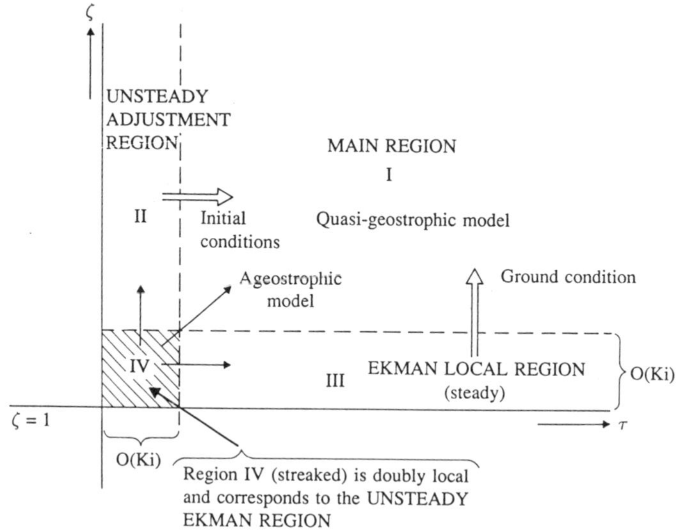

Fig. 3

Region I corresponds to $\mathrm{Ki} \to 0$ with τ and ζ fixed. To order zero, it leads to the classical geostrophic relations. By taking into account the order one equations, a first order evolution equation in τ can be formed afterwards. This equation governs the so-called quasi-geostrophic flow and is related to the geostrophic component $\mathcal{H}_1$ in the (main) expansion (11.15). This classical quasi-geostrophic theory is expounded in Sect. 11.1.

Region II corresponds to $\mathrm{Ki} \to 0$ with $\tilde{\tau} = \tau/\mathrm{Ki}$ and ζ fixed. To order zero, it leads to the problem which describes the unsteady phenomenon of adjustment to geostrophy. The analysis of this problem, which is carried out in detail in Sect. 11.2, clarifies the question of which initial condition to impose on the evolution equation which governs the quasi-geostrophic flow for $\mathcal{H}_1$ and is valid in region I.

Region III corresponds to $\mathrm{Ki} \to 0$ with τ and $\hat{\zeta} = (1 - \zeta)/\mathrm{Ki}$ fixed. To order zero, it leads to the so-called Ackerblom problem for the horizontal velocity components in the Ekman steady boundary layer. The solving of this problem makes it possible to determine the component $\hat{w}_1^\infty$ of the corresponding velocity which is induced by the Ekman boundary layer at infinity when $\hat{\zeta} \to \infty$. This, via the matching condition with the quasi-geostrophic flow of region I gives a boundary condition on $\zeta = 1$ for the equation governing $\mathcal{H}_1$ (which is of second order in ζ). The above elements concerning region III (for order zero in Ki) will be expounded in Sect. 11.3.

Region IV corresponds to the simultaneous introduction of the short variables $\tilde{\tau}$ and $\hat{\zeta}$. It is unnecessary to consider this region when dealing with the quasi-geostrophic asymptotic model (see Sects. 11.1–3). However, this region IV is significant for the so-called *ageostrophic* asymptotic model for the component $\mathcal{H}_2$ of the main expansion (11.15). Section 11.4 deals with this ageostrophic model: in Subsect. 11.4.1, we obtain the equation of the ageostrophic model; in Subsect. 11.4.2, the unsteady Ekman boundary layer (related to region IV) is analyzed; in Subsect. 11.4.3, the problem of adjustment to ageostrophy is exposed; in Subsect. 11.4.4, the problem of the second Ekman boundary layer approximation is analyzed (region III); finally in Sect. 11.5 can be found some complementary remarks as well as bibliographical references and a short history of the subject.

The main results exposed in the present chapter were published in an article by Guiraud and Zeytounian (1980). The details concerning the ageostrophic asymptotic model, however, are given for the first time in Sect. 11.4. That the present chapter is one of the longest in this book can be explained by the fact that the theory expounded herein is among the most exemplary since we have managed to put together a complete asymptotic theory which takes into account second order effects with respect to the Kibel number assumed small compared to unity. Such a situation is naturally quite exceptional and we will see, for example, in Chap. 12, which is devoted to models derived from low Mach number flows, that the situation is completely different. A coherent asymptotic theory, related to $M_\infty \to 0$ and comparable to the one expounded in the forthcoming sections, requires yet a considerable effort.

11.1 The Classical Quasi-geostrophic Model

We will be considering region I where τ and ζ are assumed of order unity. Let us focus on the main limiting process

$$\mathrm{Ki} \to 0 \quad , \quad \text{with} \quad \tau,\ \xi,\ \eta \quad \text{and} \quad \zeta \quad \text{fixed} \tag{11.25}$$

and postulate the main asymptotic expansion

$$\begin{pmatrix} U \\ V \\ W \\ \mathcal{H} \\ T \\ \varrho \end{pmatrix} = \begin{pmatrix} u_0 \\ v_0 \\ w_0 \\ \mathcal{H}_0 \\ T_0 \\ \varrho_0 \end{pmatrix} + \mathrm{Ki} \begin{pmatrix} u_1 \\ v_1 \\ w_1 \\ \mathcal{H}_1 \\ T_1 \\ \varrho_1 \end{pmatrix} + \mathrm{Ki}^2 \begin{pmatrix} u_2 \\ v_2 \\ w_2 \\ \mathcal{H}_2 \\ T_2 \\ \varrho_2 \end{pmatrix} + \dots \quad . \tag{11.26}$$

From (11.3–5), to order zero the following results

$$\frac{\partial \mathcal{H}_0}{\partial \xi} = \frac{\partial \mathcal{H}_0}{\partial \eta} = 0 \quad , \quad T_0 = -\mathrm{Bo}\varrho \frac{\partial \mathcal{H}_0}{\partial \zeta} \quad , \quad \varrho_0 = \frac{\zeta}{T_0} \quad . \tag{11.27}$$

Since we suppose that $Q_0 \equiv Q_0(\zeta)$ in (11.7), it is absolutely justified to also assume that $\mathcal{H}_0 \equiv \mathcal{H}_0(\zeta)$, $T_0 \equiv T_0(\zeta)$ and $\varrho_0 \equiv \varrho_0(\zeta)$ – which characterizes the standard atmosphere

$$\mathrm{Bo}\zeta\frac{d\mathcal{H}_0}{d\zeta} + T_0 = 0 \quad , \quad \varrho_0 = \frac{\zeta}{T_0} \quad . \tag{11.28}$$

We do not, however, know for the moment how T_0 depends on ζ. In the forthcoming, it will be assumed that

$$K_0(\zeta) \equiv -\frac{\zeta}{\mathrm{Bo}}\left(\frac{dT_0}{d\zeta} - \frac{\gamma-1}{\gamma}\frac{T_0}{\gamma}\right) \not\equiv 0 \quad , \tag{11.29}$$

which corresponds to the most realistic physical case.

According to the hypothesis (11.29), (11.7) degenerates strongly when $\mathrm{Ki} \to 0$ (with τ, ξ, η, ζ fixed). We have

$$w_0 \equiv 0 \tag{11.30}$$

and thus, (11.7) yields the following limiting equation for T_1

$$\begin{aligned} &\frac{\partial T_1}{\partial \tau} + \frac{1}{\mathrm{S}}\left(u_0\frac{\partial T_1}{\partial \xi} + v_0\frac{\partial T_1}{\partial \eta}\right) - \mathrm{Bo}\frac{K_0(\zeta)}{\zeta}w_1 \\ &= \widehat{\mathrm{Ek}}_\perp\left\{\frac{\mathrm{Bo}^2}{\mathrm{Pr}}\frac{d}{d\zeta}\left(\varrho_0 k_0\frac{dT_0}{d\zeta}\right) + Q_0(\zeta)\right\} \quad , \end{aligned} \tag{11.31}$$

where the right-hand member is a function only of ζ. Equations (11.3,4) give the geostrophic relations to order zero.

$$u_0 = -\lambda_0\frac{\partial \mathcal{H}_1}{\partial \eta} \quad \text{and} \quad v_0 = +\lambda_0\frac{\partial \mathcal{H}_1}{\partial \xi} \tag{11.32}$$

and the continuity equation (11.6) becomes

$$\frac{\partial u_0}{\partial \xi} + \frac{\partial v_0}{\partial \eta} = 0 \quad , \tag{11.33}$$

which is indeed coherent with (11.32) thanks to the fact that $\lambda_0 = \mathrm{const}$.

If, like in Chap. 7, we introduce the horizontal velocity $\boldsymbol{v}_\alpha = u_\alpha \boldsymbol{i} + v_\alpha \boldsymbol{j}$, and the operator $\boldsymbol{D} \equiv (\partial/\partial\xi)\boldsymbol{i} + (\partial/\partial\eta)\boldsymbol{j}$, $\boldsymbol{D}\cdot\boldsymbol{k} = 0$, we can write to the following order in Ki, the limiting equations

$$\begin{aligned} &\frac{\partial \boldsymbol{v}_0}{\partial \tau} + \frac{1}{\mathrm{S}}(\boldsymbol{v}_0\cdot\boldsymbol{D})\boldsymbol{v}_0 + \frac{\beta}{\mathrm{S}}\eta(\boldsymbol{k}\times\boldsymbol{v}_0) + \boldsymbol{k}\times\boldsymbol{v}_1 + \lambda_0\boldsymbol{D}\mathcal{H}_2 = 0 \quad , \\ &\boldsymbol{D}\cdot\boldsymbol{v}_1 + \mathrm{S}\frac{\partial w_1}{\partial \zeta} = 0 \quad ; \end{aligned} \tag{11.34}$$

$$\mathrm{Bo}\zeta\frac{\partial \mathcal{H}_1}{\partial \zeta} + T_1 = 0 \quad , \quad \varrho_1 = \frac{\zeta}{T_1} \quad , \tag{11.35}$$

where

$$\boldsymbol{v}_0 \equiv \lambda_0(\boldsymbol{k}\times\boldsymbol{D}\mathcal{H}_1) \quad . \tag{11.36}$$

Equations (11.31, 34, 35) form a system analogous to (4.12), the only difference being the presence of a right-hand member proportional to $\widehat{\mathrm{Ek}}_\perp$ in (11.31). Hence, if the procedure carried out in Set. 4.1 is repeated, the following equation is obtained for $\mathcal{H}_1$

$$\frac{D_0}{D\tau}(\wedge \mathcal{H}_1) + \lambda_0 \frac{\beta}{\mathrm{S}} \frac{\partial \mathcal{H}_1}{\partial \xi} = G_0(\zeta) \quad , \tag{11.37}$$

where the operators $D_0/D\tau$ and $\wedge$ are defined by (4.14, 16) respectively. The function $G_0(\zeta)$ is given by the relation

$$G_0(\zeta) = \frac{\mathrm{S}\widehat{\mathrm{Ek}}_\perp}{\mathrm{Bo}} \frac{d}{d\zeta} \left\{ \frac{\zeta}{K_0(\zeta)} \left[\frac{\mathrm{Bo}^2}{\mathrm{Pr}} \frac{d}{d\zeta} \left(\varrho_0 k_0 \frac{dT_0}{d\zeta} \right) + Q_0(\zeta) \right] \right\} \quad . \tag{11.38}$$

If it can then be proved that $G_0(\zeta) \equiv 0$, we will again find the classical equation (4.15) of the quasi-geostrophic model. To this end, let us consider a three-dimensional domain $\mathcal{D}$ in the space-time frame (τ, ξ, η) with $\zeta = \text{const}$. Let $\partial\mathcal{D}$ be its boundary and c, a, b be the direction cosines of the outward drawn unit normal to $\partial\mathcal{D}$ of which the element of area is denoted $d\Sigma$. Integrating the two sides of (11.37) in $\mathcal{D}$ yields

$$|\mathcal{D}| G_0(\zeta) = \iiint\limits_{\mathcal{D}} \left\{ \frac{D_0}{D\tau}(\wedge \mathcal{H}_1) + \lambda_0 \frac{\beta}{\mathrm{S}} \frac{\partial \mathcal{H}_1}{\partial \xi} \right\} d\tau \, d\xi \, d\eta \quad ,$$

where $|\mathcal{D}|$ designates the volume of the domain $\mathcal{D}$.

However, since we have [thanks to (11.33)]

$$\frac{D_0}{D\tau}(\wedge \mathcal{H}_1) \equiv \frac{\partial}{\partial \tau}(\wedge \mathcal{H}_1) + \frac{1}{\mathrm{S}} \left[\frac{\partial}{\partial \xi}(u_0 \wedge \mathcal{H}_1) + \frac{\partial}{\partial \eta}(v_0 \wedge \mathcal{H}_1) \right] \quad ,$$

it readily follows that

$$\iiint\limits_{\mathcal{D}} \left\{ \frac{D_0}{D\tau}(\wedge \mathcal{H}_1) + \lambda_0 \frac{\beta}{\mathrm{S}} \frac{\partial \mathcal{H}_1}{\partial \xi} \right\} d\tau \, d\xi \, d\eta$$

$$\equiv \iint\limits_{\partial\mathcal{D}} \left\{ \left[c + \frac{1}{\mathrm{S}}(a u_0 + b v_0) \right] \wedge \mathcal{H}_1 + \lambda_0 \frac{\beta}{\mathrm{S}} a \mathcal{H}_1 \right\} d\Sigma \quad ,$$

according to the classical Ostrogradski formula. Hence

$$G_0(\zeta) = \frac{1}{|\mathcal{D}|} \iint\limits_{\partial\mathcal{D}} \left\{ \frac{a}{\mathrm{S}}(\lambda_0 \beta \mathcal{H}_1 + u_0 \wedge \mathcal{H}_1) + \left(c + \frac{b}{\mathrm{S}} v_0 \right) \wedge \mathcal{H}_1 \right\} d\Sigma \tag{11.39}$$

To be sure, $\mathcal{H}_1$ remains bounded at infinity in τ, ξ, η, and if the following physical behavior condition

$$\lim_{|\mathcal{D}| \to \infty} (\wedge \mathcal{H}_1) < \infty \quad , \tag{11.40}$$

is imposed in the space-time frame (τ, ξ, η), and associated with equation (11.37), then from (11.39), it is found, taking the limit as $|\mathcal{D}| \to \infty$, that

$$G_0(\zeta) \equiv 0 \quad ,$$

since $G_0(\zeta)$ is not a function of τ, ξ and η (i.e., of $|\mathcal{D}|$). We thus arrive at the following "standard" relation

$$\frac{\zeta}{K_0(\zeta)}\left[\frac{\mathrm{Bo}^2}{\mathrm{Pr}}\frac{d}{d\zeta}\left(\varrho_0(\zeta)k_0(\zeta)\frac{dT_0}{d\zeta}\right)+Q_0(\zeta)\right] = \mathrm{const} \quad . \tag{11.41}$$

For the time being, nothing further can be said concerning (11.41). Nonetheless, we will see in Sect. 11.3 that the constant in (11.41) must mandatorily be equal to zero. In addition, we will see what condition is necessary on the ground for the "standard" equation

$$\frac{d}{d\zeta}\left[\varrho_0(\zeta)k_0(\zeta)\frac{dT_0}{d\zeta}\right]+\frac{\mathrm{Pr}}{\mathrm{Bo}^2}Q_0(\zeta) = 0 \quad , \tag{11.42}$$

which, in fact, appears as a non-secularity condition for the quasi-geostrophic model and which makes it possible to determine the standard distribution $T_0(\zeta)$ once adequate conditions in ζ (on the ground and for $\zeta \to 0$) have been prescribed.

The equation

$$\underbrace{\frac{D_0}{D\tau}(\wedge\mathcal{H}_1)+\frac{\lambda_0\beta}{\mathrm{S}}\frac{\partial\mathcal{H}_1}{\partial\xi} = 0}_{\sim\sim} \quad , \tag{11.43}$$

for the quasi-geostrophic asymptotic model is of first order in τ, and second order in ξ, η and ζ.

In order to determine $\mathcal{H}_1$, it is enough to know the initial value of $\mathcal{H}_1$ in $\tau = 0$. This value $\mathcal{H}_1^0$ (ξ,η,ζ) is obtained from the problem governing the adjustment to geostrophy considered in Sect. 11.2.

As concerns the conditions in ξ, η and ζ, it can be said for the time being that the density of the total energy

$$\frac{\zeta^2}{K_0(\zeta)}\left(\frac{\partial\mathcal{H}_1}{\partial\zeta}\right)^2+\frac{\lambda_0}{\mathrm{S}}|D\mathcal{H}_1|^2 \quad ,$$

can be required to decrease "at infinity" in the whole space ξ, η, ζ ($\xi^2+\eta^2\to\infty$ and $\zeta\to 0$) fast enough to ensure a "good behavior" of the solutions of (11.43). The ground condition for $\zeta = 1$ for (11.43) will be obtained in Sect. 11.3.

11.2 The Adjustment to Geostrophy

Let us return to (11.3–11.7). According to Sect. 6.1, in order to analyze the vicinity of $\tau = 0$ (region II), we have to introduce a short time $\tilde{\tau} = \tau/\mathrm{Ki}$, and then consider the local limiting process

$$\mathrm{Ki} \to 0 \quad \text{with} \quad \tilde{\tau}, \xi, \eta \quad \text{and} \quad \zeta \quad \text{fixed} \tag{11.44}$$

it being understood that in (11.3–6), the similarity relation (11.23) must be taken into account and that, in addition, $\lambda = O(1)$, $\mathrm{S} = O(1)$ and $\mathrm{Bo} = O(1)$. It is specified that

$$\tilde{\tau} = \frac{t^*}{1/l_0} \quad , \quad l_0 = 2\Omega_0 \sin\varphi_0 \quad , \tag{11.45}$$

where t^* is the dimensional time variable. Therefore, at a medium latitude, the characteristic short-time $1/l_0$ is of the order of 3 to 4 hours. It is precisely during this time lapse that the adjustment of the real wind (in a hydrostatic adiabatic atmosphere) to the geostrophic wind characterized by (11.32) takes place.

The following local asymptotic expansion is imposed on the local limiting process (11.44):

$$\begin{pmatrix} U \\ V \\ W \\ \mathcal{H} \\ T \\ \varrho \end{pmatrix} = \begin{pmatrix} \tilde{u}_0 \\ \tilde{v}_0 \\ \tilde{w}_0 \\ \tilde{\mathcal{H}}_1 \\ \tilde{T}_0 \\ \tilde{\varrho}_0 \end{pmatrix} + \mathrm{Ki}\begin{pmatrix} \tilde{u}_1 \\ \tilde{v}_1 \\ \tilde{w}_1 \\ \tilde{\mathcal{H}}_1 \\ \tilde{T}_1 \\ \tilde{\varrho}_1 \end{pmatrix} + \mathrm{Ki}^2\begin{pmatrix} \tilde{u}_2 \\ \tilde{v}_2 \\ \tilde{w}_2 \\ \tilde{\mathcal{H}}_2 \\ \tilde{T}_2 \\ \tilde{\varrho}_2 \end{pmatrix} + \dots \quad . \tag{11.46}$$

In this case, from (11.3–7), it follows that first of all:

$$\frac{\partial\tilde{\mathcal{H}}_0}{\partial\xi} = \frac{\partial\tilde{\mathcal{H}}_0}{\partial\eta} = 0 \quad , \quad \frac{\partial\tilde{T}_0}{\partial\tilde{\tau}} = 0 \quad , \quad \tilde{\varrho}_0 = \frac{\zeta}{\tilde{T}_0} \quad , \quad \mathrm{Bo}\zeta\frac{\partial\tilde{\mathcal{H}}_0}{\partial\zeta} + \tilde{T}_0 = 0 \quad ,$$

and it is remarked that to order zero, the thermodynamic values identify with those which characterize the standard atmosphere, and depend only on ζ (like in Sect. 11.1):

$$\tilde{\mathcal{H}}_0 \equiv \mathcal{H}_0(\zeta) \quad , \quad \tilde{T}_0 \equiv T_0(\zeta) \quad \text{and} \quad \tilde{\varrho}_0 = \frac{\zeta}{\tilde{T}_0} \equiv \varrho_0(\zeta) \quad . \tag{11.47}$$

To the next order, we obtain for $\tilde{u}_0$, $\tilde{v}_0$, $\tilde{w}_0$ and $\mathcal{H}_1$, a system of local adjustment analogous to the one obtained in Sect. 6.1. If we introduce the local horizontal velocity $\tilde{\boldsymbol{v}}_0 = \tilde{u}_0\boldsymbol{i} + \tilde{v}_0\boldsymbol{j}$, and the operator $\boldsymbol{D} \equiv (\partial/\partial\xi)\boldsymbol{i} + (\partial/\partial\eta)\boldsymbol{j}$, then this local system can be written as follows:

$$\frac{\partial\tilde{\boldsymbol{v}}_0}{\partial\tilde{\tau}} + \boldsymbol{k} \times \tilde{\boldsymbol{v}}_0 + \lambda_0\boldsymbol{D}\tilde{\mathcal{H}}_1 = 0 \quad ; \quad \boldsymbol{D}\cdot\tilde{\boldsymbol{v}}_0 - \mathrm{S}\frac{\partial}{\partial\zeta}\left(\frac{\zeta^2}{K_0(\zeta)}\frac{\partial^2\tilde{\mathcal{H}}_1}{\partial\tilde{\tau}\partial\zeta}\right) = 0 \quad , \tag{11.48}$$

which is a system of two equations for $\tilde{\boldsymbol{v}}_0$ and $\tilde{\mathcal{H}}_1$.

From (11.48), the following equation can be obtained:

$$\frac{\partial}{\partial\tilde{\tau}}\left\{\boldsymbol{k}\cdot(\boldsymbol{D}\times\tilde{\boldsymbol{v}}_0) + \mathrm{S}\frac{\partial}{\partial\zeta}\left(\frac{\zeta^2}{K_0(\zeta)}\frac{\partial\tilde{\mathcal{H}}_1}{\partial\zeta}\right)\right\} = 0 \quad . \tag{11.49}$$

Two initial conditions for $\tilde{\tau} = 0$ must be assigned to the system of local adjustment. Firstly, we have

$$\tilde{\boldsymbol{v}}_0 = \boldsymbol{V}^0 \quad , \quad \text{for} \quad \tilde{\tau} = 0 \quad , \tag{11.50a}$$

where $\boldsymbol{V}^0 = U^0\boldsymbol{i} + V^0\boldsymbol{j}$ according to (11.9).

Concerning the initial value for $\tilde{\mathcal{H}}_1$, it must first of all be supposed that in the system (11.3–7), the initial given value of $\mathcal{H}$, let us say H^0, can be decomposed in the form $H^0 = \mathcal{H}_0(\zeta) + \mathrm{Ki} H_1^0$. In this case, we can write:

$$\tilde{\mathcal{H}}_1 = H_1^0 \quad , \quad \text{for} \quad \tilde{\tau} = 0 \quad . \tag{11.50b}$$

It is pointed out that the initial given values $\boldsymbol{V}^0$ and H^0 are not arbitrary but come from the process of adjustment to hydrostatic balance (in an adiabatic atmosphere) analyzed in Sect. 7.5.

In a general context, the velocity field $\tilde{\boldsymbol{v}}_0$ can be represented in the form

$$\tilde{\boldsymbol{v}}_0 = \boldsymbol{D}\tilde{\varphi}_0 + \boldsymbol{k} \times \boldsymbol{D}\tilde{\psi}_0 \quad . \tag{11.51}$$

In this case, the first equation of (11.48) leads to

$$\boldsymbol{D}\left(\frac{\partial \tilde{\varphi}_0}{\partial \tilde{\tau}} + \lambda_0 \tilde{\mathcal{H}}_1 - \tilde{\psi}_0\right) + \boldsymbol{k} \times \boldsymbol{D}\left(\tilde{\varphi}_0 + \frac{\partial \tilde{\psi}_0}{\partial \tilde{\tau}}\right) = 0$$

which implies that the functions $J_0 \equiv (\partial\tilde{\varphi}_0/\partial\tilde{\tau}) + \lambda_0\tilde{\mathcal{H}}_1 - \tilde{\psi}_0$ and $\mathcal{L}_0 \equiv \tilde{\varphi}_0 + (\partial\tilde{\psi}_0/\partial\tilde{\tau})$ are harmonic functions throughout the entire plane (ξ, η). However, if we want these functions to be regular, at infinity in the plane (ξ, η), then, of necessity, $J_0 \equiv 0$ and $\mathcal{L}_0 \equiv 0$. Therefore, we obtain

$$\frac{\partial \tilde{\varphi}_0}{\partial \tilde{\tau}} + \lambda_0 \tilde{\mathcal{H}}_1 - \tilde{\psi}_0 = 0 \quad \text{and} \quad \frac{\partial \tilde{\psi}_0}{\partial \tilde{\tau}} + \tilde{\varphi}_0 = 0 \quad . \tag{11.52a}$$

If to (11.52a), we add the following equation

$$D^2\tilde{\varphi}_0 - \mathrm{S}\frac{\partial}{\partial \zeta}\left(\frac{\zeta^2}{K_0(\zeta)}\frac{\partial^2 \tilde{\mathcal{H}}_1}{\partial\tilde{\tau}\partial\zeta}\right) = 0 \quad , \tag{11.52b}$$

which results from (11.51) and the second equation of (11.48), a system of three equations for $\tilde{\varphi}_0$, $\tilde{\psi}_0$ and $\tilde{\mathcal{H}}_1$ is obtained. We can eliminate $\tilde{\psi}_0$ and $\tilde{\mathcal{H}}_1$ from (11.52) and thus obtain a single equation for $\tilde{\varphi}_0$. We have, in fact

$$\frac{\partial^2 \tilde{\varphi}_0}{\partial \tilde{\tau}^2} + \lambda_0 \frac{\partial \tilde{\mathcal{H}}_1}{\partial \tilde{\tau}} - \frac{\partial \tilde{\psi}_0}{\partial \tilde{\tau}} = 0 \quad , \quad \text{then} \quad \lambda_0 \frac{\partial \tilde{\mathcal{H}}_1}{\partial \tilde{\tau}} = -\left(\frac{\partial^2 \tilde{\varphi}_0}{\partial \tilde{\tau}^2} + \tilde{\varphi}_0\right) \quad ,$$

which gives us the following equation for $\tilde{\varphi}_0$:

$$\boldsymbol{D}^2\tilde{\varphi}_0 + \frac{\mathrm{S}}{\lambda_0}\frac{\partial}{\partial \zeta}\left\{\frac{\zeta^2}{K_0(\zeta)}\frac{\partial}{\partial \zeta}\left[\frac{\partial^2 \tilde{\varphi}_0}{\partial \tilde{\tau}^2} + \tilde{\varphi}_0\right]\right\} = 0 \quad . \tag{11.53}$$

The above equation (11.53) is identical to the one obtained by Kibel (1957) when it is supposed that $K_0(\zeta) \equiv \text{const}$. Two initial conditions in $\tilde{\tau}$, two boundary conditions in ζ, as well as behavior conditions in ξ and η must all be prescribed for (11.53).

When $K_0(\zeta) \equiv \text{const}$, Kibel (1957; Chap. IV, Sect. 4.2) demonstrated that $\tilde{\varphi}_0 \to 0$ and $(\partial\tilde{\varphi}_0/\partial\tilde{\tau}) \to 0$, when $\tilde{\tau} \to \infty$, as $\tilde{\tau}^{-1/2}\mathrm{osc}(\tilde{\tau})$, where $\mathrm{osc}(\tilde{\tau})$ designates bounded functions which oscillate like trigonometrical functions.

If $\partial\tilde{\varphi}_0/\partial\tilde{\tau}) \to 0$, then $\lambda_0\tilde{\mathcal{H}}_1 \to \tilde{\psi}_0$, and since $\tilde{\varphi}_0 \to 0$, according to (11.51), we again find the geostrophic relation (11.36):

$$\tilde{\boldsymbol{v}}_0^\infty = \lambda_0\big(\boldsymbol{k} \times \boldsymbol{D}\tilde{\mathcal{H}}_1^\infty\big) \quad , \tag{11.54}$$

where $\lim_{\tilde{\tau}\to\infty} \tilde{\boldsymbol{v}}_0 = \tilde{\boldsymbol{v}}_0^\infty$ and $\lim_{\tilde{\tau}\to\infty} \tilde{\mathcal{H}}_1 = \tilde{\mathcal{H}}_1^\infty$. When $K_0(\zeta) \not\equiv$ const, *Kibel*'s analysis again needs to be carried out, but it is clear that there is adjustment to geostrophy, i.e., that (11.54) is indeed satisfied.

It turns out, in fact, that an equation for $\tilde{\mathcal{H}}_1^\infty$ can be obtained. In order to accomplish this, we must backtrack to (11.49) and integrate it from $\tilde{\tau} = 0$ to $\tilde{\tau} = \infty$. If we then take into account (11.54) and the initial conditions (11.50), the following equation is obtained for $\tilde{\mathcal{H}}_1^\infty$:

$$\begin{aligned}\lambda_0 \boldsymbol{D}^2\tilde{\mathcal{H}}_1^\infty &+ \mathrm{S}\frac{\partial}{\partial\zeta}\left(\frac{\zeta^2}{K_0(\zeta)}\frac{\partial\tilde{\mathcal{H}}_1^\infty}{\partial\zeta}\right)\\ &= \boldsymbol{k}\cdot(\boldsymbol{D}\times\boldsymbol{V}^0) + \mathrm{S}\frac{\partial}{\partial\zeta}\left(\frac{\zeta^2}{K_0(\zeta)}\frac{\partial H_1^0}{\partial\zeta}\right) \quad .\end{aligned} \tag{11.55}$$

The matching condition between $\mathcal{H}_1$ and $\tilde{\mathcal{H}}_1$:

$$\mathcal{H}_1^0 \equiv \lim_{\tau\to 0}\mathcal{H}_1 = \lim_{\tilde{\tau}\to\infty}\tilde{\mathcal{H}}_1 \equiv \tilde{\mathcal{H}}_1^\infty \quad ,$$

now leads to the imposing of the following initial condition on (11.43) from the quasi-geostrophic model:

$$\wedge\,\mathcal{H}_1\big|_{\tau=0} = \boldsymbol{k}\cdot(\boldsymbol{D}\times\boldsymbol{V}^0) + \mathrm{S}\frac{\partial}{\partial\zeta}\left(\frac{\zeta^2}{K_0(\zeta)}\frac{\partial H_1^0}{\partial\zeta}\right) \quad . \tag{11.56}$$

It is of course necessary to know how to solve (11.55) in order to be in a position to assign the initial condition (11.56) to the equation:

$$\frac{D_0}{D\tau}(\wedge\,\mathcal{H}_1) + \lambda_0\frac{\beta}{\mathrm{S}}\frac{\partial\mathcal{H}_1}{\partial\xi} = 0 \quad .$$

To accomplish this, we must be able to formulate the boundary conditions associated with (11.55).

A moment's reflection reveals that the adjustment process to geostrophy is a phenomenon which takes place essentially outside of atmospheric boundary layers. There is no interaction, to this order, between region II and regions III and IV. This means that on the earth's surface (assumed flat), a condition must be imposed which results directly from the exact slip condition (11.11). However, when $\mathrm{Ki}\to 0$, $\mathcal{H} = 0$ signifies to order zero that $\zeta = 1$ and thus (11.11) becomes for $\tilde{\mathcal{H}}_1$:

$$\tilde{w}_0 = \mathrm{Bo}\varrho_0(1)\frac{\partial\tilde{\mathcal{H}}_1}{\partial\tilde{\tau}}\bigg|_{\zeta=1} \Rightarrow \frac{\partial}{\partial\tilde{\tau}}\left\{\frac{\partial\tilde{\mathcal{H}}_1}{\partial\zeta} + \mathrm{Bo}\frac{K_0(\zeta)}{T_0(\zeta)}\tilde{\mathcal{H}}_1\right\}\bigg|_{\zeta=1} = 0 \quad ,$$

or even

$$\left\{\frac{\partial}{\partial\zeta} + \mathrm{Bo}\frac{K_0(1)}{T_0(1)}\right\}(\tilde{\mathcal{H}}_1^\infty - H_1^0) = 0 \quad , \quad \text{on} \quad \zeta = 1 \quad . \tag{11.57}$$

The second condition in ζ, which is a behavior condition, must be written for $\zeta\to 0$. The results in Sect. 7.6 indicate that, generally speaking, it is necessary to impose that:

$$\frac{\zeta^2}{K_0(\zeta)} \tilde{\mathcal{H}}_1^\infty \frac{\partial \tilde{\mathcal{H}}_1^\infty}{\partial \zeta} \to 0 \quad , \quad \text{with} \quad \zeta \to 0 \quad . \tag{11.58}$$

Concerning the behavior condition for $\xi^2 + \eta^2 \to \infty$, the following must be imposed:

$$|D\tilde{\mathcal{H}}_1^\infty|^2 \to 0 \quad , \quad \text{with} \quad \xi^2 + \eta^2 \to \infty \quad . \tag{11.59}$$

The reader can find in Blumen (1972), a discussion concerning the adjustment to geostrophy where the various aspects of this problem [which go back to Rossby (1938)] are analyzed.

11.3 The Ekman Steady Boundary Layer and the Ackerblom Problem

Once again, we return to (11.3–7) and focus our attention on region III. A change in the vertical coordinate:

$$\hat{\zeta} = \frac{1-\zeta}{\text{Ki}} \quad , \tag{11.60}$$

is carried out, and the following local limiting process is considered:

$$\text{Ki} \to 0 \quad , \quad \text{with} \quad \tau, \xi, \eta \quad \text{and} \quad \hat{\zeta} \quad \text{fixed} \quad . \tag{11.61}$$

It is thus first necessary to rewrite the problem (11.3–7) with (11.8) with respect to $\hat{\zeta}$: $\partial/\partial\zeta = -(1/\text{Ki})(\partial/\partial\hat{\zeta})$, and then postulate the following local asymptotic expansion which is associated with (11.61):

$$\begin{pmatrix} U \\ V \\ W \\ \mathcal{H} \\ T \\ \varrho \end{pmatrix} = \begin{pmatrix} \hat{u}_0 \\ \hat{v}_0 \\ \hat{w}_0 \\ \hat{\mathcal{H}}_0 \\ \hat{T}_0 \\ \hat{\varrho}_0 \end{pmatrix} + \text{Ki} \begin{pmatrix} \hat{u}_1 \\ \hat{v}_1 \\ \hat{w}_1 \\ \hat{\mathcal{H}}_1 \\ \hat{T}_1 \\ \hat{\varrho}_1 \end{pmatrix} + \text{Ki}^2 \begin{pmatrix} \hat{u}_2 \\ \hat{v}_2 \\ \hat{w}_2 \\ \hat{\mathcal{H}}_2 \\ \hat{T}_2 \\ \hat{\varrho}_2 \end{pmatrix} + \dots \quad . \tag{11.62}$$

First of all, it results from (11.3–7) in this case that:

$$\begin{gathered} \frac{\partial \hat{\mathcal{H}}_0}{\partial \xi} = \frac{\partial \hat{\mathcal{H}}_0}{\partial \eta} = 0 \quad , \quad \frac{\partial \hat{w}_0}{\partial \hat{\zeta}} = 0 \quad , \quad \frac{\partial \hat{\mathcal{H}}_0}{\partial \hat{\zeta}} = 0 \quad , \quad \hat{\varrho}_0 \hat{T}_0 = 1 \quad , \\ \frac{B_0^2}{\text{Pr}} \widehat{\text{Ek}}_\perp \frac{\partial}{\partial \hat{\zeta}} \left(\hat{\varrho}_0 \hat{k}_0 \frac{\partial \hat{T}_0}{\partial \hat{\zeta}} \right) - \hat{w}_0 \frac{\partial \hat{T}_0}{\partial \hat{\zeta}} = 0 \quad . \end{gathered} \tag{11.63}$$

Then from the boundary conditions (11.8), we also have:

$$\frac{\partial \hat{T}_0}{\partial \hat{\zeta}} = 0 \quad , \quad \text{on} \quad \hat{\mathcal{H}}_0 = 0 \quad . \tag{11.64}$$

It will be remarked that k_0 is a given function of ζ. If this given function does not have an Ekman layer structure, then $\hat{k}_0 \equiv k_0(1)$ and it is possible not to take into account corrections of a higher order. If, on the other hand, this given function does have an Ekmann layer structure, then $\hat{k}_0 \equiv \hat{k}_0(\hat{\zeta})$ and it is unnecessary to consider higher order corrections.

The matching between region III (steady Ekman) and the main region I implies that $\hat{w}_0 \equiv 0$. Moreover, since $(\partial\hat{\mathcal{H}}_0/\partial\hat{\zeta}) = 0$ and $(\partial\hat{T}_0/\partial\hat{\zeta}) = 0$, the matching with region I leads to:

$$\lim_{\hat{\zeta}\to\infty} \hat{T}_0 = T_0|_{\zeta=1} \equiv T_0(1) \quad \text{and} \quad \lim_{\hat{\zeta}\to\infty} \hat{\mathcal{H}}_0 = \mathcal{H}_0|_{\zeta=1} = \mathcal{H}_0(1) \quad . \tag{11.65}$$

However, since $\hat{w}_0 \equiv 0$, we also have $(\partial\hat{\mathcal{H}}_0/\partial\tau) = 0$, on $\hat{\mathcal{H}}_0 = 0$, which means $\hat{\mathcal{H}}_0 = 0$. The latter is consistent with the condition $\mathcal{H}_0(1) = 0$.

Hence, for the time being, we have:

$$\hat{w}_0 = 0 \quad , \quad \hat{\mathcal{H}}_0 = 0 \quad \text{and} \quad \hat{T}_0 = T_0(1) = \frac{1}{\varrho_0} \quad . \tag{11.66}$$

To the following order, from (11.3–7) we obtain the system of equations:

$$\begin{aligned}
&\lambda_0 \boldsymbol{D}\hat{\mathcal{H}}_1 + \boldsymbol{k} \times \hat{\boldsymbol{v}}_0 - \mathrm{Bo}^2\widehat{\mathrm{Ek}}_\perp \frac{\partial}{\partial\hat{\zeta}}\left(\varrho_0(1)\hat{\mu}_0\frac{\partial\hat{\boldsymbol{v}}_0}{\partial\hat{\zeta}}\right) = 0 \quad ;\\
&\frac{\mathrm{Bo}^2}{\mathrm{Pr}}\widehat{\mathrm{Ek}}_\perp \frac{\partial}{\partial\hat{\zeta}}\left(\varrho_0(1)\hat{k}_0\frac{\partial\hat{T}_1}{\partial\hat{\zeta}}\right) = 0 \quad ;\\
&\mathrm{S}\frac{\partial\hat{w}_1}{\partial\hat{\zeta}} - \boldsymbol{D}\cdot\hat{\boldsymbol{v}}_0 = 0 \quad ;\\
&\mathrm{Bo}\frac{\partial\hat{\mathcal{H}}_1}{\partial\hat{\zeta}} - T_0(1) = 0 \quad ,
\end{aligned} \tag{11.67}$$

and (11.8) implies that:

$$\begin{aligned}
&\hat{\boldsymbol{v}}_0 = 0 \quad \hat{w}_1 = \frac{\mathrm{Bo}}{T_0(1)}\frac{\partial\hat{\mathcal{H}}_1}{\partial\tau} \quad , \quad \hat{k}_0\frac{1}{T_0(1)}\frac{\partial\hat{T}_1}{\partial\hat{\zeta}} = \sigma_{00}\mathcal{R}_0(1) \quad ,\\
&\text{on} \quad \hat{\mathcal{H}}_1 = 0
\end{aligned} \tag{11.68}$$

if it assumed that the radiation does not have an Ekman layer structure.

It is pointed out that in the steady Ekman problem, the ground is characterized by

$$\hat{\zeta} = \hat{\zeta}_{s0} + \mathrm{Ki}\hat{\zeta}_{s1} + \dots \quad . \tag{11.69}$$

The matching of the functions $\mathcal{H}$ and T must be considered for the calculation of $\hat{\zeta}_{s0}$ but for the present, we note that the last condition of (11.68) and the second equation of (11.67) lead to:

$$\varrho_0(1)\hat{k}_0\frac{\partial\hat{T}_1}{\partial\hat{\zeta}} \equiv \sigma_{00}\mathcal{R}_0(1) = \text{const} \quad . \tag{11.70}$$

The matching of the temperatures between regions III and I gives:

$$\hat{T}_1 = T_{1,1} + \left(\frac{dT_0}{d\zeta}\right)_{\zeta=1} \hat{\zeta} \quad , \tag{11.71}$$

where

$$T_{1,1} \equiv T_1(\tau, \xi, \eta, 1) = -\mathrm{Bo}\left(\frac{\partial \mathcal{H}_1}{\partial \zeta}\right)_{\zeta=1} \quad .$$

Thus, instead of (11.70), we have:

$$\varrho_0(1)\hat{k}_0\left(\frac{dT_0}{d\zeta}\right)_{\zeta=1} = \sigma_{00}\mathcal{R}_0(1) \quad .$$

However, by taking into account the expression (11.2) for $Q_0(\zeta)$, we also have from (11.42):

$$\varrho_0(\zeta)k_0(\zeta)\frac{dT_0}{d\zeta} - \sigma_{00}\mathcal{R}_0(\zeta) = \text{const} \quad .$$

Therefore, for the case $\hat{k}_0 \equiv k_0(1)$, the constant in the right-hand member is necessarily zero and the following equation results for the calculation of $T_0(\zeta)$:

$$k_0(\zeta)\frac{dT_0}{d\zeta} = \sigma_{00}\frac{\mathcal{R}_0(\zeta)}{\varrho_0(\zeta)} \quad . \tag{11.72}$$

One boundary condition for $\zeta \to 0$ is then sufficient for determining $T_0(\zeta)$ as a function of the radiation. In all of the forthcoming, we will suppose that $\hat{k}_0 \equiv k_0(1)$ and $\hat{\mu}_0 \equiv \mu_0(1)$. The last equation of (11.67) implies that:

$$\frac{\partial \hat{\mathcal{H}}_1}{\partial \hat{\zeta}} = \frac{T_0(1)}{\mathrm{Bo}} = \text{const} \Rightarrow \hat{\mathcal{H}}_1 = \mathcal{H}_{1,1} + \frac{T_0(1)}{\mathrm{Bo}}\hat{\zeta} \tag{11.73}$$

and hence, $\hat{\mathcal{H}}_1 = 0$ means that:

$$\hat{\zeta}_{s0} = -\frac{\mathrm{Bo}}{T_0(1)}\mathcal{H}_{1,1} \quad . \tag{11.74}$$

Let us go back to the first equation of (11.67) and write:

$$\hat{\boldsymbol{v}}_0 = \boldsymbol{v}_{0,1} + \hat{\boldsymbol{v}}_0' \quad , \quad \text{where} \quad \boldsymbol{v}_{0,1} \equiv \boldsymbol{v}_0(\tau, \xi, \eta, 1) \quad .$$

Because of the matching with the main region I, we should have:

$$\lim_{\hat{\zeta} \to \infty} \hat{\boldsymbol{v}}_0' = 0 \quad \text{and} \quad \lambda_0 \boldsymbol{D}\mathcal{H}_{1,1} + \boldsymbol{k} \times \boldsymbol{v}_{0,1} = 0 \quad ,$$

which, thanks to (11.73), implies that

$$\lambda_0 \boldsymbol{D}\hat{\mathcal{H}}_1 + \boldsymbol{k} \times \hat{\boldsymbol{v}}_0 \equiv \boldsymbol{k} \times \hat{\boldsymbol{v}}_0' \quad .$$

Therefore, we can formulate the following problem to determine the horizontal perturbation velocity $\hat{\boldsymbol{v}}_0'$:

$$\kappa_0 \frac{\partial^2 \hat{\boldsymbol{v}}_0'}{\partial \hat{\zeta}^2} - \boldsymbol{k} \times \hat{\boldsymbol{v}}_0' = 0 \quad ,$$

$$\hat{\boldsymbol{v}}_0' = -\boldsymbol{v}_{0,1} \quad , \quad \text{on} \quad \hat{\zeta} = -\frac{\text{Bo}}{T_0(1)} \mathcal{H}_{1,1} \quad ; \tag{11.75}$$

$$\hat{\boldsymbol{v}}_0' \to 0 \quad , \quad \text{when} \quad \hat{\zeta} \to \infty \quad ,$$

where $\kappa_0 = \text{Bo}^2 \widehat{\text{Ek}}_\perp [\mu_0(1)/T_0(1)]$. The above problem (11.75) is the so-called "Ackerblom" problem.

The solution to (11.75) is classical:

$$\hat{\boldsymbol{v}}_0' - \text{i}\boldsymbol{k} \times \hat{\boldsymbol{v}}_0' = -(\boldsymbol{v}_{0,1} - \text{i}\boldsymbol{k} \times \boldsymbol{v}_{0,1})\text{E} \quad ,$$

$$\text{E} \equiv \exp\left\{ -\frac{1+\text{i}}{\sqrt{2\kappa_0}} \left(\hat{\zeta} + \frac{\text{Bo}}{T_0(1)} \mathcal{H}_{1,1} \right) \right\} \quad ; \quad (\text{i} \equiv \sqrt{-1}) \quad . \tag{11.76}$$

We now have to calculate $\hat{w}_1$ by making use of the third equation of (11.67), and then clarify the behavior of $\hat{w}_1$ at infinity when $\hat{\zeta} \to \infty$. The matching condition:

$$\lim_{\hat{\zeta} \to \infty} \hat{w}_1 \equiv \hat{w}_1^\infty = w_{1,1} \equiv w_1(\tau, \xi, \eta, 1) \quad , \tag{11.77}$$

then gives the boundary condition in $\zeta = 1$ for the quasi-geostrophic model equation (11.43) once the following relation has been taken advantage of:

$$w_{1,1} = -\frac{1}{K_0(1)} \frac{D_0}{D\tau} \left(\frac{\partial \mathcal{H}_1}{\partial \zeta} \right)_{\zeta=1} \quad . \tag{11.78}$$

The above relation follows from the third relation of (4.13).

From the third equation in system (11.67), while taking into account the condition (11.68) for $\hat{w}_1$ on $\hat{\mathcal{H}}_1 = 0$ (i.e., on $\hat{\zeta} = \hat{\zeta}_{s0}$), we obtain:

$$\text{S}\hat{w}_1 = \int_{\hat{\zeta}_{s0}}^{\hat{\zeta}} (\boldsymbol{D} \cdot \hat{\boldsymbol{v}}_0) d\hat{\zeta} + \text{S} \frac{\text{Bo}}{T_0(1)} \left(\frac{\partial \hat{\mathcal{H}}_1}{\partial \tau} \right)_{\hat{\zeta} = \hat{\zeta}_{s0}} \quad , \quad \text{where}$$

$$\hat{\boldsymbol{v}}_0 = \Re\left\{ (\boldsymbol{v}_{0,1} - \text{i}\boldsymbol{k} \times \boldsymbol{v}_{0,1})(1 - \text{E}) \right\} \quad .$$

A simple calculation shows that:

$$\int_{\hat{\zeta}_{s0}}^{\hat{\zeta}} \hat{\boldsymbol{v}}_0 d\hat{\zeta} = (\hat{\zeta} - \hat{\zeta}_{s0})\boldsymbol{v}_{0,1} + \sqrt{\frac{\kappa_0}{2}} \boldsymbol{k} \times \boldsymbol{v}_{0,1} \quad ,$$

and when $\hat{\zeta} \to \infty$, it follows that:

$$\int_{\hat{\zeta}_{s0}}^{\infty} (\boldsymbol{D} \cdot \hat{\boldsymbol{v}}_0) d\hat{\zeta} = \boldsymbol{D} \cdot \int_{\hat{\zeta}_{s0}}^{\infty} \hat{\boldsymbol{v}}_0 d\hat{\zeta} = \sqrt{\frac{\kappa_0}{2}} \boldsymbol{D} \cdot (\boldsymbol{k} \times \boldsymbol{v}_{0,1}) \quad ,$$

given the fact that

$$\boldsymbol{D}\cdot\boldsymbol{v}_{0,1}=0 \quad , \quad \boldsymbol{v}_{0,1}\cdot\boldsymbol{D}\hat{\zeta}_{s0} \equiv -\frac{\mathrm{Bo}}{T_0(1)}\boldsymbol{v}_{0,1}\cdot\boldsymbol{D}\mathcal{H}_{1,1}=0 \quad .$$

Hence:

$$\hat{w}_1^{\infty}=\frac{\mathrm{Bo}}{T_0(1)}\frac{\partial\mathcal{H}_{1,1}}{\partial\tau}-\frac{\lambda_0}{\mathrm{S}}\sqrt{\frac{\kappa_0}{2}}\boldsymbol{D}^2\mathcal{H}_{1,1} \quad , \tag{11.79}$$

since according to (11.73), $\partial\hat{\mathcal{H}}_1/\partial\tau=\partial\mathcal{H}_{1,1}/\partial\tau$ is not a function of $\hat{\zeta}$.

Finally, from (11.77–79), the following boundary condition results:

$$\left\{\frac{\mathrm{Bo}}{T_0(1)}\frac{\partial}{\partial\tau}+\frac{1}{K_0(1)}\frac{D_0}{D\tau}\left(\frac{\partial}{\partial\zeta}\right)-\frac{\lambda_0}{\mathrm{S}}\sqrt{\frac{\kappa_0}{2}}\boldsymbol{D}^2\right\}\mathcal{H}_1=0 \quad ,$$
$$\text{on} \quad \zeta=1 \quad . \tag{11.80}$$

This condition (11.80) is the one which must be imposed on (11.43) which was obtained for the quasi-geostrophic model in Sect. 11.1. It is pointed out that the results given in the above Sects. 11.1–11.3 clarify those obtained by Zeytounian (1976; Chap. III).

11.4 The So-called "Ageostrophic" Model

The model referred to as "ageostrophic", which will be discussed in the present section, is relative to the ageostrophic component $\mathcal{H}_2$ from the main asymptotic expansion (11.15). Just like for the quasi-geostrophic model, it is first necessary to obtain the equation governing the ageostrophic asymptotic model – this will be the focal point of Subsect. 11.4.1. Once again, an initial condition in $\tau=0$ and a boundary condition in $\zeta=1$ are necessary for this ageostrophic model equation.

In order to obtain the initial condition in $\tau=0$, the problem of adjustment to ageostrophy in region II must be considered. Subsection 11.4.3 centers around this question. However, so as to be able to correctly formulate this problem of adjustment to ageostrophy, it is necessary to analyze the problem related to the unsteady Ekman boundary layer which develops in region IV. In addition, we must also elucidate the compatibility of the models between regions III and IV by analyzing the behavior of the unsteady Ekman boundary layer when $\tilde{\tau}\to\infty$. This analysis is carried out in Subsect. 11.4.2. Finally, in Subsect. 11.4.4, we will study the problem of the second approximation steady Ekman boundary layer. This will enable us to obtain the boundary condition in $\zeta=1$ for the equation of the ageostrophic model derived in Subsect. 11.4.1.

The results of the present section were published in an article jointly written with J. P. Guiraud [see Guiraud and Zeytounian (1980)]. Moreover, the entire theory which is presented in this chapter was expounded by Zeytounian and Guiraud at the ICMS[3] within the framework of an "Advanced School" (coordinated by R.Kh. Zeytounian) devoted to "Models for Atmospheric Flows".

[3] Footnote see opposite page

11.4.1 The Equation for the Ageostrophic Model

Let us return to the basic equations (11.3–7) and write the system of limiting equations derived from these equations to order Ki for $\boldsymbol{v}_1$, and to order Ki^2 for $\mathcal{H}_2$, T_2 and w_2:

$$\begin{aligned}\frac{\partial \boldsymbol{v}_1}{\partial \tau} &+ \frac{1}{\mathrm{S}}\left[(\boldsymbol{v}_1 \cdot \boldsymbol{D})\boldsymbol{v}_0 + (\boldsymbol{v}_0 \cdot \boldsymbol{D})\boldsymbol{v}_1\right] + w_1 \frac{\partial \boldsymbol{v}_0}{\partial \zeta} + \boldsymbol{k} \times \boldsymbol{v}_2 \\ &+ \frac{\beta}{\mathrm{S}}\eta(\boldsymbol{k} \times \boldsymbol{v}_1) + \lambda_0 \boldsymbol{D}\mathcal{H}_3 = \mathrm{Bo}^2 \widehat{\mathrm{Ek}}_\perp \frac{\partial}{\partial \zeta}\left(\varrho_0 \mu_0 \frac{\partial \boldsymbol{v}_0}{\partial \zeta}\right) \quad ,\end{aligned} \tag{11.81a}$$

$$\begin{aligned}\frac{\partial T_2}{\partial \tau} &+ \frac{1}{\mathrm{S}}(\boldsymbol{v}_1 \cdot \boldsymbol{D}T_1 + \boldsymbol{v}_0 \cdot \boldsymbol{D}T_2) - \mathrm{Bo}\frac{K_0(\zeta)}{\zeta} w_2 \\ &+ \left(\frac{\partial T_1}{\partial \zeta} - \frac{\gamma - 1}{\gamma}\frac{T_1}{\zeta}\right) w_1 \\ &= \frac{\mathrm{Bo}^2}{\mathrm{Pr}} \widehat{\mathrm{Ek}}_\perp \frac{\partial}{\partial \zeta}\left[\varrho_0 k_0 \frac{\partial T_1}{\partial \zeta} + \varrho_1 k_0 \frac{dT_0}{d\zeta}\right] \quad ;\end{aligned} \tag{11.81b}$$

$$\boldsymbol{D} \cdot \boldsymbol{v}_2 + \mathrm{S}\frac{\partial w_2}{\partial \zeta} = 0 \quad ; \tag{11.81c}$$

$$T_2 + \mathrm{Bo}\,\zeta \frac{\partial \mathcal{H}_2}{\partial \zeta} = 0 \quad ; \tag{11.81d}$$

According to the results in Sect. 11.1, we have the following expressions [see also (4.13) and (6.2)]:

$$\begin{aligned}\boldsymbol{v}_0 &= \lambda_0(\boldsymbol{k} \times \boldsymbol{D}\mathcal{H}_1) \quad ; \\ \boldsymbol{v}_1 &= \lambda_0(\boldsymbol{k} \times \boldsymbol{D}\mathcal{H}_2) + \boldsymbol{k} \times \left\{\frac{D_0}{D\tau}\boldsymbol{v}_0 + \frac{\beta}{\mathrm{S}}\eta(\boldsymbol{k} \times \boldsymbol{v}_0)\right\} \quad ; \\ w_1 &= -\frac{\zeta^2}{K_0(\zeta)}\frac{D_0}{D\tau}\left(\frac{\partial \mathcal{H}_1}{\partial \zeta}\right) \quad ; \\ T_1 &= -\mathrm{Bo}\,\zeta \frac{\partial \mathcal{H}_1}{\partial \zeta} \quad ; \\ \varrho_1 &= \mathrm{Bo}\frac{\varrho_0}{T_0}\zeta \frac{\partial \mathcal{H}_1}{\partial \zeta} \quad .\end{aligned} \tag{11.82}$$

Now, from (11.81a), we can derive the expression of $\boldsymbol{v}_2$ in the following form once (11.82) has also been taken into account:

---

[3] International Centre for Mechanical Sciences; session "Von Karman", Udine (Italy), October, 1983. The conference titles are: "General Introduction: Asymptotically consistent models for atmospheric flows", by R. Kh. Zeytounian, and "Some examples of applications of asymptotic techniques to the derivation of models for atmospheric flows", by J.P. Guiraud. During the writing of the present section, the conference notes written by Guiraud were put to very good use.

$$
\begin{aligned}
\boldsymbol{v}_2 &= \lambda_0(\boldsymbol{k} \times \boldsymbol{D}\mathcal{H}_3) + \boldsymbol{k} \times \left\{ \frac{D_0}{D\tau}\boldsymbol{v}_1 + \frac{1}{\mathrm{S}}(\boldsymbol{v}_1 \cdot \boldsymbol{D})\boldsymbol{v}_0 \right. \\
&\quad \left. + w_1 \frac{\partial \boldsymbol{v}_0}{\partial \zeta} + \frac{\beta}{\mathrm{S}}\eta(\boldsymbol{k} \times \boldsymbol{v}_1) - \mathrm{Bo}^2\widehat{\mathrm{Ek}}_\perp \frac{\partial}{\partial \zeta}\left(\varrho_0\mu_0 \frac{\partial \boldsymbol{v}_0}{\partial \zeta} \right) \right\} \\
&= \lambda_0(\boldsymbol{k} \times \boldsymbol{D}\mathcal{H}_3) - \lambda_0 \frac{D_0}{D\tau}(\boldsymbol{D}\mathcal{H}_2) - \frac{\lambda_0^2}{\mathrm{S}}[\boldsymbol{k} \times \boldsymbol{D}\mathcal{H}_2) \cdot \boldsymbol{D}](\boldsymbol{D}\mathcal{H}_1) \\
&\quad - \lambda_0 \frac{\beta}{\mathrm{S}}\eta(\boldsymbol{k} \times \boldsymbol{D}\mathcal{H}_2) + \boldsymbol{V}_1(\mathcal{H}_1) \quad ,
\end{aligned} \tag{11.83}
$$

where

$$
\begin{aligned}
\boldsymbol{V}_1 = &- \lambda_0 \boldsymbol{k} \times \left[\left(\frac{D_0}{D\tau} \right)^2 (\boldsymbol{D}\mathcal{H}_1) \right] + \frac{\lambda_0^2}{\mathrm{S}} \left[\frac{D_0}{D\tau}(\boldsymbol{D}\mathcal{H}_1) \cdot \boldsymbol{D} \right] (\boldsymbol{D}\mathcal{H}_1) \\
&+ \lambda_0 \frac{\zeta^2}{K_0(\zeta)} \left[\frac{D_0}{D\tau} \left(\frac{\partial \mathcal{H}_1}{\partial \zeta} \right) \right] \frac{\partial}{\partial \zeta}(\boldsymbol{D}\mathcal{H}_1) \\
&- \lambda_0 \mathrm{Bo}^2 \widehat{\mathrm{Ek}}_\perp \boldsymbol{k} \times \boldsymbol{D} \left[\frac{\partial}{\partial \zeta} \left(\varrho_0\mu_0 \frac{\partial \mathcal{H}_1}{\partial \zeta} \right) \right] + \frac{\beta}{\mathrm{S}}\eta\chi_1 \quad ,
\end{aligned} \tag{11.84}
$$

with

$$
\begin{aligned}
\chi_1 = &\lambda_0 \mathrm{S} \frac{D_0}{D\tau}(\boldsymbol{D}\mathcal{H}_1) + \frac{\lambda_0^2}{\mathrm{S}} \frac{1}{\eta} \frac{\partial \mathcal{H}_1}{\partial \xi} \boldsymbol{D}\mathcal{H}_1 \\
&+ \frac{\lambda_0^2}{\mathrm{S}}(\boldsymbol{k} \times \boldsymbol{D}\mathcal{H}_1) \cdot \boldsymbol{D}(\boldsymbol{D}\mathcal{H}_1) + \lambda_0 \frac{D_0}{D\tau}(\boldsymbol{D}\mathcal{H}_1) \\
&+ \frac{\lambda_0}{\mathrm{S}} \beta\eta \boldsymbol{k} \times \boldsymbol{D}\mathcal{H}_1 \quad ,
\end{aligned} \tag{11.85}
$$

being a collection of terms which disappear when $\beta \equiv 0$.

From (11.81b), we now deduce an expression for w_2. Once (11.81d, 82) have been taken into account, we have:

$$
\begin{aligned}
w_2 = &- \frac{\zeta^2}{K_0(\zeta)} \frac{D_0}{D\tau} \left(\frac{\partial \mathcal{H}_2}{\partial \zeta} \right) \\
&- \frac{\lambda_0}{\mathrm{S}} \frac{\zeta^2}{K_0(\zeta)} (\boldsymbol{k} \times \boldsymbol{D}\mathcal{H}_2) \cdot \boldsymbol{D} \left(\frac{\partial \mathcal{H}_1}{\partial \zeta} \right) + W_1(\mathcal{H}_1) \quad ,
\end{aligned} \tag{11.86}
$$

where

$$
\begin{aligned}
W_1 = &\lambda_0 \frac{\zeta^2}{K_0(\zeta)} \frac{D_0}{D\tau}(\boldsymbol{D}\mathcal{H}_1) \cdot \boldsymbol{D} \left(\frac{\partial \mathcal{H}_1}{\partial \zeta} \right) \\
&+ \frac{\zeta^3}{K_0^2(\zeta)} \frac{D_0}{D\tau} \left(\frac{\partial \mathcal{H}_1}{\partial \zeta} \right) \left[\zeta \frac{\partial^2 \mathcal{H}_1}{\partial \zeta^2} + \frac{1}{\gamma} \frac{\partial \mathcal{H}_1}{\partial \zeta} \right] \\
&+ \mathrm{Bo}^2 \frac{\zeta}{K_0(\zeta)} \frac{\widehat{\mathrm{Ek}}_\perp}{\mathrm{Pr}} \left\{ \frac{\partial}{\partial \zeta} \left[\varrho_0 k_0 \frac{\partial}{\partial \zeta} \left(\zeta \frac{\partial \mathcal{H}_1}{\partial \zeta} \right) \right] \right. \\
&\left. - \frac{\partial}{\partial \zeta} \left[\frac{\varrho_0 k_0}{T_0} \zeta \frac{dT_0}{d\zeta} \frac{\partial \mathcal{H}_1}{\partial \zeta} \right] \right\} + \frac{\beta}{\mathrm{S}} \eta (\boldsymbol{k} \times \boldsymbol{D}\mathcal{H}_1) \cdot \boldsymbol{D} \left(\frac{\partial \mathcal{H}_1}{\partial \zeta} \right) .
\end{aligned} \tag{11.87}
$$

We need only to take advantage of the continuity equation (11.81c) and we have:

$$D \cdot v_2 = -\lambda_0 \frac{D_0}{D\tau}(D^2\mathcal{H}_2) - \frac{\lambda_0^2}{S}(k \times D\mathcal{H}_2) \cdot D(D^2\mathcal{H}_1)$$
$$- \frac{\lambda_0}{S}\beta\frac{\partial \mathcal{H}_2}{\partial \xi} + D \cdot V_1$$

and

$$S\frac{\partial w_2}{\partial \zeta} = -S\frac{D_0}{D\tau}\left[\frac{\partial}{\partial \zeta}\left(\frac{\zeta^2}{K_0(\zeta)}\frac{\partial \mathcal{H}_2}{\partial \zeta}\right)\right]$$
$$- \lambda_0(k \times D\mathcal{H}_2) \cdot D\left[\frac{\partial}{\partial \zeta}\left(\frac{\zeta^2}{K_0(\zeta)}\frac{\partial \mathcal{H}_1}{\partial \zeta}\right)\right] + S\frac{\partial W_1}{\partial \zeta} \quad .$$

At this point, after some minor transformations and simplifications, the equation sought for $\mathcal{H}_2(\tau, \xi, \eta, \zeta)$ is obtained:

$$\frac{D_0}{D\tau}(\wedge \mathcal{H}_2) + \frac{\lambda_0}{S}(k \times D\mathcal{H}_2) \cdot D(\wedge \mathcal{H}_1)$$
$$+ \lambda_0\frac{\beta}{S}\frac{\partial \mathcal{H}_2}{\partial \xi} = D \cdot V_1 + S\frac{\partial W_1}{\partial \zeta} \quad , \tag{11.88}$$

with a right-hand member which is a function only of $\mathcal{H}_1$. The latter is a solution of the quasi-geostrophic model analyzed in Sects. 11.1–3.

11.4.2 The Problem of the Unsteady Ekman Boundary Layer. Adjustment to the Ackerblom Model

Region IV, which we want to consider now, corresponds to the simultaneous introduction of the short variables:

$$\tilde{\tau} = \frac{\tau}{\mathrm{Ki}} \quad \text{and} \quad \hat{\zeta} = \frac{1-\zeta}{\mathrm{Ki}} \quad .$$

If

$$\mathcal{U} = (U, V, W, \mathcal{H}, T, \varrho)^{\mathrm{T}} \quad , \quad \text{then}:$$
$$\mathcal{U}(\mathrm{Ki}\tilde{\tau}, \xi, \eta, 1 - \mathrm{Ki}\hat{\zeta}) \equiv \overline{\mathcal{U}}(\tilde{\tau}, \xi, \eta, \hat{\zeta}) \quad ,$$

and it is assumed that the ground is characterized by:

$$\hat{\zeta} = \overline{\zeta}_s(\tilde{\tau}, \xi, \eta) = \overline{\zeta}_{s0} + \mathrm{Ki}\overline{\zeta}_{s1} + \dots \quad . \tag{11.89}$$

Let us recall that:

in region I: $\mathcal{U}(\tau, \xi, \eta, \zeta) = \mathcal{U}_0 + \mathrm{Ki}\mathcal{U}_1 + \dots$;
in region II: $\tilde{\mathcal{U}}(\tilde{\tau}, \xi, \eta, \zeta) = \tilde{\mathcal{U}}_0 + \mathrm{Ki}\tilde{\mathcal{U}}_1 + \dots$;
in region III: $\hat{\mathcal{U}}(\tau, \xi, \eta, \hat{\zeta}) = \hat{\mathcal{U}}_0 + \mathrm{Ki}\hat{\mathcal{U}}_1 + \dots$;
in region IV: $\overline{\mathcal{U}}(\tilde{\tau}, \xi, \eta, \hat{\zeta}) = \overline{\mathcal{U}}_0 + \mathrm{Ki}\overline{\mathcal{U}}_1 + \dots$.

Finally, we note:

$$\overline{\mathcal{U}}|_{\hat{\zeta}=\overline{\zeta}_{s0}} \equiv \overline{\mathcal{U}}_s \quad ; \quad \text{also} \quad \tilde{\mathcal{U}}_s \equiv \tilde{\mathcal{U}}|_{\zeta=1} \quad .$$

First of all, we remark that:

$$\boldsymbol{D}\overline{\mathcal{H}}_0 = 0 \quad , \quad \frac{\partial \overline{\mathcal{H}}_0}{\partial \hat{\zeta}} = 0 \quad , \quad \frac{\partial \overline{w}_0}{\partial \hat{\zeta}} = 0 \quad \text{and} \quad \overline{\varrho}_0 \overline{T}_0 = 1 \quad , \tag{11.90}$$

and the following problem results for $\overline{T}_0$:

$$\begin{aligned}
&\frac{\partial \overline{T}_0}{\partial \bar{\tau}} - \overline{w}_0 \frac{\partial \overline{T}_0}{\partial \hat{\zeta}} = \frac{\mathrm{Bo}^2 \widehat{\mathrm{Ek}}_\perp}{\mathrm{Pr}} k_0(1) \frac{\partial}{\partial \hat{\zeta}} \left(\overline{\varrho}_0 \frac{\partial \overline{T}_0}{\partial \hat{\zeta}} \right) \quad ; \\
&\text{on} \quad \hat{\zeta} = \overline{\zeta}_{s0} : \; \overline{\varrho}_0 \frac{\partial \overline{T}_0}{\partial \hat{\zeta}} = 0 \quad , \\
&\text{when} \quad \hat{\zeta} \to \infty : \; \overline{T}_0 \to T_0(1) \quad .
\end{aligned} \tag{11.91}$$

A solution to (11.91) is $\overline{T}_0 \equiv T_0(1)$ and we can assume that it is, in fact, the only acceptable one as long as it is compatible with the initial condition corresponding to $\bar{\tau} = 0$.

After matching with region II, we find the following from (11.90):

$$\begin{aligned}
&\overline{\mathcal{H}}_0 \equiv \mathcal{H}_{0,1} = 0 \quad \text{and} \\
&\overline{w}_0 = -\frac{1}{K_0(1)} \left(\frac{\partial^2 \tilde{\mathcal{H}}_1}{\partial \bar{\tau} \partial \zeta} \right)_{\zeta=1} = \frac{\mathrm{Bo}}{T_0(1)} \frac{\partial \tilde{\mathcal{H}}_{1,1}}{\partial \bar{\tau}} \quad .
\end{aligned} \tag{11.92}$$

Concerning $\overline{T}_1$, it must satisfy the equation:

$$\frac{\partial \overline{T}_1}{\partial \bar{\tau}} - \left(\frac{\partial \overline{T}_1}{\partial \hat{\zeta}} + \frac{\gamma - 1}{\gamma} T_0(1) \right) \overline{w}_0 = \frac{\mathrm{Bo}^2 \widehat{\mathrm{Ek}}_\perp}{\mathrm{Pr}} \frac{k_0(1)}{T_0(1)} \frac{\partial^2 \overline{T}_1}{\partial \hat{\zeta}^2} \quad , \tag{11.93}$$

which must be solved for the following boundary conditions:

$$\begin{aligned}
&\frac{k_0(1)}{T_0(1)} \frac{\partial \overline{T}_1}{\partial \hat{\zeta}} = \sigma_{00} \mathcal{R}_0(1) \quad , \quad \text{on} \quad \hat{\zeta} = \overline{\zeta}_{s0} \quad ; \\
&\lim_{\hat{\zeta} \to \infty} \left[\overline{T}_1 - \tilde{T}_{1,1} + \hat{\zeta} \left(\frac{dT_0}{d\zeta} \right)_{\zeta=1} \right] = 0 \quad .
\end{aligned} \tag{11.94}$$

However, it was seen in Sect. 11.3 that the relation:

$$\frac{k_0(1)}{T_0(1)} \left(\frac{dT_0}{d\zeta} \right)_{\zeta=1} = \sigma_{00} \mathcal{R}_0(1)$$

has to be satisfied. Therefore, the only compatible solution of the problem (11.93,94) for $\overline{T}_1$ is:

$$\overline{T}_1 = \tilde{T}_{1,1} - \hat{\zeta} \left(\frac{dT_0}{d\zeta} \right)_{\zeta=1} \quad . \tag{11.95}$$

With (11.95), (11.93) becomes an equation for $\tilde{T}_{1,1}$ which is identically satisfied in region II on $\zeta = 1$. Once again, the compatibility with the initial condition corresponding to $\tilde{\tau} = 0$ must be satisfied.

Equation (11.5) leads to:

$$\mathrm{Bo}(1 - \mathrm{Ki}\hat{\zeta})\frac{\partial \overline{\mathcal{H}}}{\partial \hat{\zeta}} = \mathrm{Ki}\overline{T} \quad ,$$

and the matching with region II leads to:

$$\overline{\mathcal{H}}_1 = \tilde{\mathcal{H}}_{1,1} + \frac{T_0(1)}{\mathrm{Bo}}\hat{\zeta} \quad , \tag{11.96a}$$

$$\overline{\mathcal{H}}_2 = \tilde{\mathcal{H}}_{2,1} + \frac{1}{\mathrm{Bo}}\tilde{T}_{1,1}\hat{\zeta} + \frac{1}{2\mathrm{Bo}}\left[T_0(1) - \left(\frac{dT_0}{d\zeta}\right)_{\zeta=1}\right]\hat{\zeta}^2 \quad . \tag{11.96b}$$

The ground being in $\overline{\mathcal{H}} = \mathrm{Ki}\overline{\mathcal{H}}_1 + \mathrm{Ki}^2\overline{\mathcal{H}}_2 + \ldots = 0$, we find that:

$$\overline{\zeta}_{s0} = -\frac{\mathrm{Bo}}{T_0(1)}\tilde{\mathcal{H}}_{1,1} \quad ; \quad \overline{\zeta}_{s1} = -\frac{\mathrm{Bo}}{T_0(1)}\tilde{\mathcal{H}}_{2,1} + \frac{\mathrm{Bo}}{T_0^2(1)}\overline{\mathcal{H}}_{1,1}\tilde{T}_{1,1}$$
$$-\frac{1}{2}\left[\frac{\mathrm{Bo}}{T_0(1)}\right]^2\left\{1 - \frac{1}{T_0(1)}\left(\frac{dT_0}{d\zeta}\right)_{\zeta=1}\right\}(\tilde{\mathcal{H}}_{1,1})^2 \quad . \tag{11.97}$$

Now, according to the third condition in (11.8), on $\overline{\mathcal{H}} = 0$ we have:

$$\frac{\partial \overline{\mathcal{H}}}{\partial \tilde{\tau}} - \frac{\mathrm{Ki}}{\mathrm{Bo}}\frac{\overline{w}\overline{T}}{(1 - \mathrm{Ki}\hat{\zeta})} = 0 \quad , \quad \text{since} \quad \overline{\varrho} = \frac{1 - \mathrm{Ki}\hat{\zeta}}{\overline{T}} \quad .$$

Hence, to order Ki, it follows that:

$$\mathrm{Bo}\left.\frac{\partial \overline{\mathcal{H}}_1}{\partial \tilde{\tau}}\right|_s = T_0(1)\overline{w}_0|_s \quad ,$$

which is, in fact, equivalent to relation (11.92) for $\overline{w}_0$.

Therefore, we have:

$$\frac{\partial}{\partial \tilde{\tau}}\left\{\tilde{\mathcal{H}}_{1,1} + \frac{1}{\mathrm{Bo}}\frac{T_0(1)}{K_0(1)}\left(\frac{\partial \tilde{\mathcal{H}}_1}{\partial \zeta}\right)_{\zeta=1}\right\} = 0 \quad , \tag{11.98}$$

and we again encounter the slip condition of the problem of adjustment to geostrophy which led to condition (11.57) (see Sect. 11.2). It is observed that this slip condition for region II to order Ki (for $\tilde{\mathcal{H}}_1$) emerges naturally from the analysis of region IV and demonstrates that, in fact, *the process of adjustment to geostrophy is a purely adiabatic phenomenon.*

To order Ki^2, by making use of the expressions of $\overline{\mathcal{H}}_1$, $\overline{\mathcal{H}}_2$, $\overline{\zeta}_{s0}$ and $\overline{\zeta}_{s1}$, we find:

$$\mathrm{Bo}\left\{\frac{\partial \tilde{\mathcal{H}}_{2,1}}{\partial \tilde{\tau}} - \frac{1}{T_0(1)}\tilde{\mathcal{H}}_{1,1}\frac{\partial \tilde{T}_{1,1}}{\partial \tilde{\tau}} + \frac{\mathrm{Bo}}{T_0(1)}\tilde{\mathcal{H}}_{1,1}\frac{\partial \tilde{\mathcal{H}}_{1,1}}{\partial \tilde{\tau}}\right\}$$
$$= \tilde{w}_{0,1}\tilde{T}_{1,1} + \frac{\mathrm{Bo}}{T_0(1)}\tilde{w}_{0,1}\left(\frac{dT_0}{d\zeta}\right)_{\zeta=1}\tilde{\mathcal{H}}_{1,1} + T_0(1)\overline{w}_1|_s \quad . \tag{11.99}$$

The above relation makes it possible to calculate the value of $\overline{w}_1$ on the ground and in this case, the continuity equation makes it possible to calculate $\overline{w}_1$ according to the following:

$$\overline{w}_1 = \overline{w}_1\big|_s + \int\limits_{\hat{\zeta}=\overline{\zeta}_{s0}}^{\hat{\zeta}} (\boldsymbol{D}\cdot\overline{\boldsymbol{v}}_0)d\hat{\zeta} \Leftarrow \left(\frac{\partial \overline{w}_1}{\partial \hat{\zeta}} - \boldsymbol{D}\cdot\overline{\boldsymbol{v}}_0 = 0\right) \quad .$$

The matching with region II, however, implies that:

$$\lim_{\hat{\zeta}\to\infty} \left\{\overline{w}_1 - \tilde{w}_{1,1} + \left(\frac{\partial \tilde{w}_0}{\partial \zeta}\right)_{\zeta=1} \tilde{\zeta}\right\} = 0 \quad .$$

Naturally, it is necessary that:

$$\lim_{\hat{\zeta}\to\infty} \left(\overline{\boldsymbol{v}}_0 - \tilde{\boldsymbol{v}}_{0,1}\right) = 0 \quad .$$

So,

$$\tilde{w}_{1,1} = \overline{w}_1\big|_s - \overline{\zeta}_{s0}\boldsymbol{D}\cdot\tilde{\boldsymbol{v}}_{0,1} + \int\limits_{\hat{\zeta}=\overline{\zeta}_{s0}}^{+\infty} (\boldsymbol{D}\cdot\overline{\boldsymbol{v}}_0')d\hat{\zeta} \quad , \tag{11.100}$$

where $\overline{\boldsymbol{v}}_0 = \tilde{\boldsymbol{v}}_{0,1} + \overline{\boldsymbol{v}}_0'$.

Therefore, by taking advantage of (11.100), (11.99) may be rewritten as follows[4]:

$$\mathrm{Bo}\left\{\left[1 + \frac{T_0(1)}{\mathrm{Bo}\,K_0(1)}\frac{\partial}{\partial \zeta}\right]\frac{\partial \tilde{\mathcal{H}}_2}{\partial \tilde{\tau}}\right\}_{\zeta=1} = \tilde{\psi}_1 \quad , \tag{11.101}$$

where $\tilde{\psi}_1$ brings into play not only $\tilde{\mathcal{H}}_1$ and its derivatives taken in $\zeta = 1$, but also the integral $\int_{\hat{\zeta}=\overline{\zeta}_{s0}}^{+\infty}(\boldsymbol{D}\cdot\overline{\boldsymbol{v}}_0')d\hat{\zeta}$:

$$\begin{aligned}\tilde{\psi}_1 = &-\frac{\mathrm{Bo}^2}{T_0(1)}\left[\frac{\partial \tilde{\mathcal{H}}_{1,1}}{\partial \tilde{\tau}} + \left(\frac{\partial^2 \tilde{\mathcal{H}}_1}{\partial \tilde{\tau}\partial \zeta}\right)_{\zeta=1}\right]\tilde{\mathcal{H}}_{1,1} \\ &+ \mathrm{Bo}\left[-\left(\frac{\partial \tilde{\mathcal{H}}_1}{\partial \zeta}\right)_{\zeta=1} + \frac{1}{T_0(1)}\left(\frac{dT_0}{d\zeta}\right)_{\zeta=1}\tilde{\mathcal{H}}_{1,1}\right]\tilde{w}_{0,1} \\ &+ T_0(1)\left\{\frac{1}{\mathrm{S\,Bo}}\tilde{\boldsymbol{v}}_{0,1}\cdot\boldsymbol{D}\tilde{T}_{1,1} + \frac{1}{\mathrm{Bo}}\left[\left(\frac{\partial \tilde{T}_1}{\partial \zeta}\right)_{\zeta=1} - \frac{\gamma-1}{\gamma}\tilde{T}_{1,1}\right]\tilde{w}_{0,1}\right. \\ &\left.- \frac{\mathrm{Bo}}{T_0(1)}\tilde{\mathcal{H}}_{1,1}\boldsymbol{D}\cdot\tilde{\boldsymbol{v}}_{0,1} - \int\limits_{\hat{\zeta}=\overline{\zeta}_{s0}}^{+\infty}(\boldsymbol{D}\cdot\overline{\boldsymbol{v}}_0')d\hat{\zeta}\right\} \quad .\end{aligned} \tag{11.102}$$

[4] It is pointed out that in region II, we have the following (see Subsect. 11.4.3):

$$\begin{aligned}\tilde{w}_{1,1} = &-\frac{1}{K_0(1)}\frac{\partial}{\partial \tilde{\tau}}\left(\frac{\partial \tilde{\mathcal{H}}_2}{\partial \zeta}\right)_{\zeta=1} + \frac{1}{\mathrm{SBo}}\frac{1}{K_0(1)}\tilde{\boldsymbol{v}}_{0,1}\cdot\boldsymbol{D}\tilde{T}_{1,1} \\ &+ \frac{1}{\mathrm{Bo}}\frac{1}{K_0(1)}\left[\left(\frac{\partial \tilde{T}_1}{\partial \zeta}\right)_{\zeta=1} - \frac{\gamma-1}{\gamma}\tilde{T}_{1,1}\right]\tilde{w}_{0,1} \quad .\end{aligned}$$

The relation (11.101) is the boundary condition in $\zeta = 1$ to be assigned to the problem of adjustment to ageostrophy. The velocity $\overline{\boldsymbol{v}}_0'$ now remains to be calculated. In (11.102), $\tilde{\boldsymbol{v}}_{0,1}$ and $\tilde{w}_{0,1}$ are both expressed as functions of $\mathcal{H}_1$.

From (11.3, 4), it is readily found that $\overline{\boldsymbol{v}}_0$ has to satisfy the following equation:

$$\frac{\partial \overline{\boldsymbol{v}}_0}{\partial \tilde{\tau}} + \boldsymbol{k} \times \overline{\boldsymbol{v}}_0 - \frac{\partial \overline{\boldsymbol{v}}_0}{\partial \hat{\zeta}} \overline{w}_0 + \lambda_0 \boldsymbol{D} \tilde{\mathcal{H}}_{1,1} = \frac{\mathrm{Bo}^2 \widehat{\mathrm{Ek}}_\perp}{T_0(1)} \mu_0(1) \frac{\partial^2 \overline{\boldsymbol{v}}_0}{\partial \hat{\zeta}^2} \quad ,$$

on which must be superimposed the boundary conditions:

$$\overline{\boldsymbol{v}}_0 = 0 \quad , \quad \text{on} \quad \hat{\zeta} = \overline{\zeta}_{s0} \quad ;$$
$$\overline{\boldsymbol{v}}_0 \to \tilde{\boldsymbol{v}}_{0,1} \quad , \quad \text{when} \quad \hat{\zeta} \to \infty \quad .$$

However, according to the problem of adjustment to geostrophy [see (11.48)], we have:

$$\frac{\partial \tilde{\boldsymbol{v}}_{0,1}}{\partial \tilde{\tau}} + \boldsymbol{k} \times \tilde{\boldsymbol{v}}_{0,1} + \lambda_0 \boldsymbol{D} \tilde{\mathcal{H}}_{1,1} = 0 \quad ,$$

and, for $\overline{\boldsymbol{v}}_0' = \overline{\boldsymbol{v}}_0 - \tilde{\boldsymbol{v}}_{0,1}$, we obtain the following problem:

$$\begin{aligned} &\frac{\partial \overline{\boldsymbol{v}}_0'}{\partial \tilde{\tau}} + \boldsymbol{k} \times \overline{\boldsymbol{v}}_0' - \frac{\partial \overline{\boldsymbol{v}}_0'}{\partial \hat{\zeta}} \overline{w}_0 = \frac{\mathrm{Bo}^2 \widehat{\mathrm{Ek}}_\perp}{T_0(1)} \mu_0(1) \frac{\partial^2 \overline{\boldsymbol{v}}_0'}{\partial \hat{\zeta}^2} \quad , \\ &\overline{\boldsymbol{v}}_0'\big|_s = -\tilde{\boldsymbol{v}}_{0,1} \quad ; \\ &\lim_{\hat{\zeta} \to \infty} \overline{\boldsymbol{v}}_0' = 0 \quad , \quad \overline{\boldsymbol{v}}_0'\big|_{\tilde{\tau}=0} = \boldsymbol{V}^0 - \boldsymbol{V}^0\big|_{\zeta=1} \quad . \end{aligned} \tag{11.103}$$

In the preceding problem, the horizontal variables play the role of parameters and in the forthcoming, they will be ignored for the sake of simplification.

So as to give (11.103) a more classical form, we introduce the function:

$$\overline{\boldsymbol{v}}_0'(\tilde{\tau}, \hat{\zeta}) - \mathrm{i}\boldsymbol{k} \times \overline{\boldsymbol{v}}_0'(\tilde{\tau}, \hat{\zeta}) = \boldsymbol{X}\left(\tilde{\tau}, \hat{\zeta} + \int_0^{\tilde{\tau}} \overline{w}_0(\tilde{\tau}, \hat{\zeta}) d\tilde{\tau} \right) \quad .$$

Since according to (11.92),

$$\overline{w}_0 + \frac{1}{K_0(1)} \frac{\partial}{\partial \tilde{\tau}} \left(\frac{\partial \tilde{\mathcal{H}}_1}{\partial \zeta} \right)_{\zeta=1} = 0 \quad ,$$

we have:

$$\int_0^{\tilde{\tau}} \overline{w}_0(\tilde{\tau}, \hat{\zeta}) d\tilde{\tau} \equiv \frac{1}{K_0(1)} \left[\left(\frac{\partial H_1^0}{\partial \zeta} \right)_{\zeta=1} - \left(\frac{\partial \tilde{\mathcal{H}}_1}{\partial \zeta} \right)_{\zeta=1} \right] \quad ,$$

where H_1^0 is the initial value of $\tilde{\mathcal{H}}_1$ [see (11.50b)].

Hence, for $\boldsymbol{X}(\tilde{\tau}, \hat{\zeta})$, the following replaces (11.103)

$$\frac{\partial \boldsymbol{X}}{\partial \tilde{\tau}} + \mathrm{i}\boldsymbol{X} = \kappa_0 \frac{\partial^2 \boldsymbol{X}}{\partial \hat{\zeta}^2} \quad , \quad \kappa_0 \equiv \frac{\mathrm{Bo}^2 \widehat{\mathrm{Ek}}_\perp}{T_0(1)} \mu_0(1); \tag{11.104a}$$

$$
\begin{aligned}
&\boldsymbol{X}(\tilde{\tau}, \hat{\zeta} = \overline{\zeta}_{s0}(\tilde{\tau} = 0)) = -\tilde{\boldsymbol{v}}_{0,1} + \mathrm{i}\boldsymbol{k} \times \tilde{\boldsymbol{v}}_{0,1} \equiv \boldsymbol{X}_0(\tilde{\tau}) \quad ; \\
&\lim_{\hat{\zeta} \to \infty} \boldsymbol{X} = 0 \quad ; \\
&\boldsymbol{X}(0, \hat{\zeta}) = \boldsymbol{V}^0 - \boldsymbol{V}^0|_{\zeta=1} - \mathrm{i}\boldsymbol{k} \times \left(\boldsymbol{V}^0 - \boldsymbol{V}^0|_{\zeta=1}\right) \equiv \boldsymbol{X}^0(\hat{\zeta}) \quad ,
\end{aligned}
\tag{11.104b}
$$

once it is observed that:

$$
\overline{\zeta}_{s0} + \frac{1}{K_0(1)}\left[\left(\frac{\partial H_1^0}{\partial \zeta}\right)_{\zeta=1} - \left(\frac{\partial \tilde{\mathcal{H}}_1}{\partial \zeta}\right)_{\zeta=1}\right] = \overline{\zeta}_{s0}|_{\tilde{\tau}=0} \equiv \overline{\zeta}_{s0}(\tilde{\tau} = 0) \quad .
$$

Problem (11.104) can be solved by applying the Laplace transform:

$$
\boldsymbol{y}(\theta, \hat{\zeta}) = \int_0^\infty \mathrm{e}^{-\theta\tilde{\tau}}\, \boldsymbol{X}(\tilde{\tau}, \hat{\zeta}) d\tilde{\tau} \quad ,
$$

which leads to the problem:

$$
\begin{aligned}
&(\theta + \mathrm{i})\boldsymbol{y} - \kappa_0 \frac{\partial^2 \boldsymbol{y}}{\partial \hat{\zeta}^2} = \boldsymbol{X}^0(\hat{\zeta}) \quad ; \\
&\boldsymbol{y} = \boldsymbol{y}_0(\theta) \quad , \quad \text{for} \quad \hat{\zeta} = \overline{\zeta}_{s0}|_{\tilde{\tau}=0} \quad ; \\
&\lim_{\hat{\zeta} \to \infty} \boldsymbol{y} = 0 \quad .
\end{aligned}
\tag{11.105}
$$

The solution to this type of problem can be found in the book by Greenspan (1968; see Sect. 2.3).

The fundamental point is, in fact, related to the matching with the Ekman region III when $\tilde{\tau} \to \infty$! This matching imposes that:

$$
\lim_{\tilde{\tau} \to \infty} \overline{\boldsymbol{v}}_0 = \hat{\boldsymbol{v}}_0|_{\tau=0} = \left(\boldsymbol{v}_{0,1} + \hat{\boldsymbol{v}}_0'\right)|_{\tau=0} \quad , \tag{11.106}
$$

according to the notation of Sect. 11.3. However, the matching between regions I and II leads to:

$$
\lim_{\tilde{\tau} \to \infty} \tilde{\boldsymbol{v}}_0 = \boldsymbol{v}_0|_{\tau=0} \Rightarrow \boldsymbol{v}_{0,1}|_{\tau=0} \equiv \lim_{\tilde{\tau} \to \infty} \tilde{\boldsymbol{v}}_{0,1} \quad .
$$

Therefore, it is necessary that:

$$
\overline{\boldsymbol{v}}_0'^{\infty} \equiv \lim_{\tilde{\tau} \to \infty} \overline{\boldsymbol{v}}_0' = \hat{\boldsymbol{v}}_0'|_{\tau=0} \quad , \tag{11.107}
$$

in order for adjustment to the Ackerblom problem (11.75) to take place.

Problem (11.103) shows that $\overline{\boldsymbol{v}}_0'^{\infty}$ must satisfy the following boundary value problem:

$$
\begin{aligned}
&\boldsymbol{k} \times \overline{\boldsymbol{v}}_0'^{\infty} = \kappa_0 \frac{\partial^2 \overline{\boldsymbol{v}}_0'^{\infty}}{\partial \hat{\zeta}^2} \quad ; \\
&\overline{\boldsymbol{v}}_0'^{\infty} = -\boldsymbol{v}_{0,1} \quad , \quad \text{on} \quad \hat{\zeta} = \lim_{\tilde{\tau} \to \infty} \overline{\zeta}_{s0} \equiv -\frac{\mathrm{Bo}}{T_0(1)} H_{1,1}^0 \quad ; \\
&\lim_{\hat{\zeta} \to \infty} \overline{\boldsymbol{v}}_0'^{\infty} = 0 \quad .
\end{aligned}
\tag{11.108}
$$

If we now compare (11.108) to the Ackerblom problem (11.75), it will be seen that the matching condition (11.107) [i.e. (11.106)] does indeed take place since by using the notation in (11.50b):

$$\hat{\zeta}_{s0}|\mathcal{T} = 0 \equiv \overline{\zeta}_{s0}|_{\tilde{\tau} \to \infty} = -\frac{\text{Bo}}{T_0(1)} H^0_{1,1} \quad , \tag{11.109}$$

according to (11.74). Thus, to order zero, there is indeed adjustment to the Ackerblom model.

11.4.3 The Problem of Adjustment to Ageostrophy

We return again to (11.3–7) where $\tau \equiv \text{Ki}\tilde{\tau}$, and consider the local asymptotic expansion (11.46). For $\tilde{\boldsymbol{v}}_1$, $\tilde{\mathcal{H}}_2$, $\tilde{T}_2$ and $\tilde{w}_1$, the following local system should be considered:

$$\frac{\partial \tilde{\boldsymbol{v}}_1}{\partial \tilde{\tau}} + \boldsymbol{k} \times \tilde{\boldsymbol{v}}_1 + \lambda_0 \boldsymbol{D}\tilde{\mathcal{H}}_2 + \frac{1}{\text{S}}\left[(\tilde{\boldsymbol{v}}_0 \cdot \boldsymbol{D})\tilde{\boldsymbol{v}}_0 + \beta\eta\boldsymbol{k} \times \tilde{\boldsymbol{v}}_0\right] + \tilde{w}_0 \frac{\partial \tilde{\boldsymbol{v}}_0}{\partial \zeta} = 0 \quad ; \tag{11.110a}$$

$$\frac{\partial \tilde{T}_2}{\partial \tilde{\tau}} - \text{Bo}\frac{K_0(\zeta)}{\zeta}\tilde{w}_1 + \frac{1}{\text{S}}\tilde{\boldsymbol{v}}_0 \cdot \boldsymbol{D}\tilde{T}_1 + \left(\frac{\partial \tilde{T}_1}{\partial \zeta} - \frac{\gamma - 1}{\gamma}\frac{\tilde{T}_1}{\zeta}\right)\tilde{w}_0 = \frac{\text{Bo}^2}{\text{Pr}}\widehat{\text{Ek}}_\perp \left[\frac{\partial}{\partial \zeta}\left(\tilde{\varrho}_0 k_0 \frac{\partial \tilde{T}_0}{\partial \zeta}\right) + \sigma_{00}\frac{d\mathcal{R}_0}{d\zeta}\right] \quad ; \tag{11.110b}$$

$$\boldsymbol{D} \cdot \tilde{\boldsymbol{v}}_1 + \text{S}\frac{\partial \tilde{w}_1}{\partial \zeta} = 0 \quad ; \quad \text{Bo}\zeta\frac{\partial \tilde{\mathcal{H}}_2}{\partial \zeta} + \tilde{T}_2 = 0 \quad . \tag{11.110c}$$

Like in Sect. 11.2, we introduce the representation:

$$\tilde{\boldsymbol{v}}_1 = \boldsymbol{D}\tilde{\varphi}_1 + \boldsymbol{k} \times \boldsymbol{D}\tilde{\psi}_1 \quad , \tag{11.111}$$

which leads to the following relations instead of (11.52a):

$$\frac{\partial \tilde{\varphi}_1}{\partial \tilde{\tau}} + \lambda_0 \tilde{\mathcal{H}}_2 - \tilde{\psi}_1 = (\boldsymbol{D}^2)^{-1} A_1 \quad ; \quad \frac{\partial \tilde{\psi}_1}{\partial \tilde{\tau}} + \tilde{\varphi}_1 = (\boldsymbol{D}^2)^{-1} B_1 \quad , \tag{11.112}$$

where

$$\boldsymbol{D}^2\left\{(\boldsymbol{D}^2)^{-1}B_1\right\} \equiv -(\boldsymbol{k} \times \boldsymbol{D}) \cdot \left[\frac{1}{\text{S}}\left[(\tilde{\boldsymbol{v}}_0 \cdot \boldsymbol{D})\tilde{\boldsymbol{v}}_0 + \beta\eta\boldsymbol{k} \times \tilde{\boldsymbol{v}}_0\right] + \tilde{w}_0\frac{\partial \tilde{\boldsymbol{v}}_0}{\partial \zeta}\right] \quad ;$$

and

$$\boldsymbol{D}^2\left\{(\boldsymbol{D}^2)^{-1}A_1\right\} \equiv -\boldsymbol{D} \cdot \left[\frac{1}{\text{S}}\left[(\tilde{\boldsymbol{v}}_0 \cdot \boldsymbol{D})\tilde{\boldsymbol{v}}_0 + \beta\eta\boldsymbol{k} \times \tilde{\boldsymbol{v}}_0\right] + \tilde{w}_0\frac{\partial \tilde{\boldsymbol{v}}_0}{\partial \zeta}\right] \quad .$$

By using the second relation of (11.110c), while keeping in mind that we have (11.47,72) from (11.110b) we find $\tilde{w}_1$:

$$\tilde{w}_1 = -\frac{\zeta^2}{K_0(\zeta)}\frac{\partial^2 \tilde{\mathcal{H}}_2}{\partial \tilde{\tau}\partial \zeta} + \frac{\zeta}{\text{Bo}K_0(\zeta)}\left[\frac{\tilde{\boldsymbol{v}}_0 \cdot \boldsymbol{D}\tilde{T}_1}{\text{S}} + \left(\frac{\partial \tilde{T}_1}{\partial \zeta} - \frac{\gamma - 1}{\gamma}\frac{\tilde{T}_1}{\zeta}\right)\tilde{w}_0\right] \quad .$$

Finally, the following equation replaces (11.53) for $\tilde{\varphi}_1$:

$$\boldsymbol{D}^2\tilde{\varphi}_1 + \frac{\mathrm{S}}{\lambda_0}\left(\frac{\partial^2}{\partial\tilde{\tau}^2} + 1\right)\left[\frac{\partial}{\partial\zeta}\left(\frac{\zeta^2}{K_0(\zeta)}\frac{\partial\tilde{\varphi}_1}{\partial\zeta}\right)\right] = \mathcal{C}_1 \quad , \tag{11.113}$$

where

$$\begin{aligned}\mathcal{C}_1 = {} & \frac{\mathrm{S}}{\lambda_0}(\boldsymbol{D}^2)^{-1}\left\{\frac{\partial}{\partial\zeta}\left(\frac{\zeta^2}{K_0(\zeta)}(A_1 + B_1)\right)\right\} \\ & - \frac{\mathrm{S}}{\mathrm{Bo}}\frac{\partial}{\partial\zeta}\left\{\frac{\zeta^2}{K_0(\zeta)}\left[\frac{1}{\mathrm{S}}\tilde{\boldsymbol{v}}_0\cdot\boldsymbol{D}\tilde{T}_1 + \left(\frac{\partial\tilde{T}_1}{\partial\zeta} - \frac{\gamma-1}{\gamma}\frac{\tilde{T}_1}{\zeta}\right)\tilde{w}_0\right]\right\}. \end{aligned} \tag{11.114}$$

Equation (11.113) requires for $\tilde{\varphi}_1$ two initial conditions in $\tilde{\tau} = 0$ and one in $\zeta = 1$, as well as an adequate behavior condition when $\zeta \to 0$. The two initial conditions are written for $\tilde{\varphi}_1$ and $\partial\tilde{\varphi}_1/\partial\tilde{\tau}$ with known right-hand members, whereas the condition in $\zeta = 1$ is the one obtained in Subsect. 11.4.2 [see (11.101)] once $\tilde{\mathcal{H}}_2$ has been expressed as a function of $\tilde{\varphi}_1$ from (11.112). This leads to a complex problem which we will not take up here but which could possibly be treated numerically.

Let us now take a closer look at the matching phenomenon in τ between regions I and II. According to the results in Sect. 11.2 concerning the adjustment to geostrophy, when $\tilde{\tau} \to \infty$, we have:

$$\begin{aligned}&\tilde{\boldsymbol{v}}_0 = \tilde{\boldsymbol{v}}_0^\infty + \tilde{\boldsymbol{v}}_0' \quad , \quad \tilde{\mathcal{H}}_1 = \tilde{\mathcal{H}}_1^\infty + \tilde{\mathcal{H}}_1' \quad , \\ &\tilde{T}_1 = \tilde{T}_1^\infty + \tilde{T}_1' \quad , \quad \tilde{w}_0 = \tilde{w}_0' \quad . \end{aligned} \tag{11.115}$$

It is pointed out that in the vicinity of $\tilde{\tau} = \infty$, $\tilde{\mathcal{U}}$ has the following form:

$$\tilde{\mathcal{U}} = P\tilde{\mathcal{U}} + \tilde{\mathcal{U}}^\infty + \tilde{\mathcal{U}}' \quad ,$$

where $P\tilde{\mathcal{U}}$ is a polynomial in $\tilde{\tau}$, without constant terms, $\tilde{\mathcal{U}}^\infty$ is the limiting value of $\hat{\mathcal{U}} - P\tilde{\mathcal{U}}$, whereas:

$$\tilde{\mathcal{U}}' \to 0 \quad , \quad \text{with} \quad \tilde{\tau} \to \infty \quad .$$

Furthermore, by definition $(\mathcal{U})^0 \equiv \mathcal{U}(\tau = 0, \xi, \eta, \zeta)$.

The matching between the approximations in regions I and II yields:

$$\tilde{\boldsymbol{v}}_0^\infty = (\boldsymbol{v}_0)^0 \quad , \quad \tilde{\mathcal{H}}_1^\infty = (\mathcal{H}_1)^0 \quad \text{and} \quad \tilde{T}_1^\infty = (T_1)^0 \quad . \tag{11.116}$$

If (11.110) is now considered, the following behavior can be expected when $\tilde{\tau} \to \infty$:

$$\begin{aligned}&\tilde{\boldsymbol{v}}_1 = \tilde{\boldsymbol{v}}_1^1\tilde{\tau} + \tilde{\boldsymbol{v}}_1^\infty + \tilde{\boldsymbol{v}}_1' \quad , \quad \mathcal{H}_2 = \tilde{\mathcal{H}}_2^1\tilde{\tau} + \tilde{\mathcal{H}}_2^\infty + \tilde{\mathcal{H}}_2' \quad ; \\ &\tilde{w}_1 = \tilde{w}_1^\infty + \tilde{w}_1' \quad ; \quad \tilde{T}_2 = \tilde{T}_2^1\tilde{\tau} + \tilde{T}_2^\infty + \tilde{T}_2' \quad , \end{aligned}$$

and we obtain:

$$\begin{aligned}&\boldsymbol{k} \times \tilde{\boldsymbol{v}}_1^1 + \lambda_0\boldsymbol{D}\tilde{\mathcal{H}}_2^1 = 0 \quad ; \quad \boldsymbol{D}\cdot\tilde{\boldsymbol{v}}_1^1 = 0 \quad ; \\ &\mathrm{Bo}\zeta\frac{\partial\tilde{\mathcal{H}}_2^1}{\partial\zeta} + \tilde{T}_2^1 = 0 \quad ; \end{aligned}$$

$$\tilde{\boldsymbol{v}}_1^1 + \boldsymbol{k} \times \tilde{\boldsymbol{v}}_1^\infty + \lambda_0 \boldsymbol{D}\tilde{\mathcal{H}}_2^\infty + \frac{1}{\mathrm{S}}\left(\tilde{\boldsymbol{v}}_0^\infty \cdot \boldsymbol{D}\tilde{\boldsymbol{v}}_0^\infty + \beta\eta \boldsymbol{k} \times \tilde{\boldsymbol{v}}_0^\infty\right) = 0 \quad ;$$

$$\tilde{T}_2^1 + \frac{1}{\mathrm{S}}\tilde{\boldsymbol{v}}_0^\infty \cdot \boldsymbol{D}\tilde{T}_1^\infty - \mathrm{Bo}\frac{K_0(\zeta)}{\zeta}\tilde{w}_1^\infty = 0 \quad ;$$

$$\boldsymbol{D} \cdot \tilde{\boldsymbol{v}}_1^\infty + \mathrm{S}\frac{\partial \tilde{w}_1^\infty}{\partial \zeta} = 0 \quad ;$$

$$\mathrm{Bo}\zeta \frac{\partial \tilde{\mathcal{H}}_2^\infty}{\partial \zeta} + \tilde{T}_2^\infty = 0 \quad .$$

The matching implies then that:

$$\tilde{\boldsymbol{v}}_1^1 = \left(\frac{\partial \boldsymbol{v}_0}{\partial \tau}\right)^0 \quad , \quad \tilde{\mathcal{H}}_2^1 = \left(\frac{\partial \mathcal{H}_1}{\partial \tau}\right)^0 \quad , \quad \tilde{T}_2^1 = \left(\frac{\partial T_1}{\partial \tau}\right)^0 \quad , \tag{11.117}$$

and

$$\tilde{\boldsymbol{v}}_1^\infty = (\boldsymbol{v}_1)^0 \quad , \quad \tilde{\mathcal{H}}_2^\infty = (\mathcal{H}_2)^0 \quad , \quad \tilde{w}_1^\infty = (w_1)^0 \quad . \tag{11.118}$$

The matching conditions (11.116–118) result automatically from the comparison of the corresponding equations in region II (in the vicinity of $\tilde{\tau} = \infty$) and region I (in the vicinity of $\tau = 0$). It is thus remarked that matching must in fact take place between regions I and II both to order one and two (ageostrophy).

We now wish to derive an equation from $\tilde{\mathcal{H}}_2^\infty$ in the same way as (11.55) was obtained for $\tilde{\mathcal{H}}_1^\infty$ in Sect. 11.2. This will allow us to obtain an initial condition in $\tau = 0$ for the equation of the ageostrophic model (11.88) obtained in Subsect. 11.4.1 for $\mathcal{H}_2$. In this way, it will not be necessary to solve the problem of the adjustment to ageostrophy which is governed by (11.113) for $\tilde{\varphi}_1$.

Let us then consider (11.110a) and apply to it the operator $\boldsymbol{k} \cdot (\boldsymbol{D} \times \cdot)$. We obtain:

$$\frac{\partial}{\partial \tilde{\tau}}(\boldsymbol{k} \cdot (\boldsymbol{D} \times \tilde{\boldsymbol{v}}_1)) - \mathrm{S}\frac{\partial \tilde{w}_1}{\partial \zeta} + \boldsymbol{k} \cdot \left\{\boldsymbol{D} \times \left[\frac{1}{\mathrm{S}}\left(\tilde{\boldsymbol{v}}_0 \cdot \boldsymbol{D}\tilde{\boldsymbol{v}}_0 + \beta\eta \boldsymbol{k} \times \tilde{\boldsymbol{v}}_0\right) + \tilde{w}_0 \frac{\partial \tilde{\boldsymbol{v}}_0}{\partial \zeta}\right]\right\} = 0 \quad ,$$

once the first equation of (11.110c) has been made use of. If we now take advantage of the expression of $\tilde{w}_1$ obtained before (11.113), the above equation can be transformed into the following[5]:

[5] It will be observed that

$$\boldsymbol{k} \cdot \left[\boldsymbol{D} \times \left(\tilde{\boldsymbol{v}}_0 \cdot \boldsymbol{D}\tilde{\boldsymbol{v}}_0\right)\right] = \left(\tilde{\boldsymbol{v}}_0 \cdot \boldsymbol{D}\right)\left[\boldsymbol{k} \cdot \left(\boldsymbol{D} \times \tilde{\boldsymbol{v}}_0\right)\right] - \frac{\partial}{\partial \tilde{\tau}}\left(\frac{1}{2}|\boldsymbol{D} \times \tilde{\boldsymbol{v}}_0|^2\right) \quad ,$$

and

$$\boldsymbol{k} \cdot \left\{\boldsymbol{D} \times \left(\beta\eta \boldsymbol{k} \times \tilde{\boldsymbol{v}}_0\right]\right\} = \beta \tilde{\boldsymbol{v}}_0 \cdot \boldsymbol{j} - \frac{\partial}{\partial \tilde{\tau}}\left\{\beta\eta \boldsymbol{k} \cdot \left(\boldsymbol{D} \times \tilde{\boldsymbol{v}}_0\right)\right\} \quad .$$

$$\frac{\partial}{\partial \tilde{\tau}}\left\{ -\boldsymbol{D}\cdot(\boldsymbol{k}\times\tilde{\boldsymbol{v}}_1) + \mathrm{S}\frac{\partial}{\partial\zeta}\left(\frac{\zeta^2}{K_0(\zeta)}\frac{\partial\tilde{\mathcal{H}}_2}{\partial\zeta}\right) + \frac{1}{\mathrm{S}}\left[\frac{1}{2}|\boldsymbol{D}\times\tilde{\boldsymbol{v}}_0|^2 + \beta\eta\boldsymbol{k}\cdot(\boldsymbol{D}\times\tilde{\boldsymbol{v}}_0)\right]\right\} = \tilde{\mathcal{G}}_1 \quad , \tag{11.119}$$

where

$$\tilde{\mathcal{G}}_1 = \frac{\partial}{\partial\zeta}\left\{\frac{\zeta}{\mathrm{Bo}K_0(\zeta)}\left(\tilde{\boldsymbol{v}}_0\cdot\boldsymbol{D}\tilde{T}_1 + \left[\frac{\partial\tilde{T}_1}{\partial\zeta} - \frac{\gamma-1}{\gamma}\frac{\tilde{T}_1}{\zeta}\right]\tilde{w}_0\right)\right\} - \left[\frac{1}{\mathrm{S}}\tilde{\boldsymbol{v}}_0\cdot\boldsymbol{D} + \tilde{w}_0\frac{\partial}{\partial\zeta}\right]\boldsymbol{D}^2\tilde{\psi}_0 - \frac{\beta}{\mathrm{S}}\tilde{\boldsymbol{v}}_0\cdot\boldsymbol{j} \quad , \tag{11.120}$$

is a known functional of $\tilde{\mathcal{H}}_1$.

When $\tilde{\tau}\to\infty$, this functional tends to a definite limit and we have:

$$\lim_{\tilde{\tau}\to\infty}\tilde{\mathcal{G}}_1 = \tilde{\mathcal{G}}_1^\infty = -\frac{1}{\mathrm{S}}\left\{(\tilde{\boldsymbol{v}}_0^\infty\cdot\boldsymbol{D})(\wedge\tilde{\mathcal{H}}_1^\infty) + \frac{\beta}{\mathrm{S}}\tilde{\boldsymbol{v}}_0^\infty\cdot\boldsymbol{j}\right\} \quad . \tag{11.121}$$

It is pointed out that $\lim_{\tilde{\tau}\to\infty}\tilde{\psi}_0 = \tilde{\psi}_0^\infty \equiv \lambda_0\tilde{\mathcal{H}}_1^\infty$ according to the results of Sect. 11.2.

If we integrate (11.119) with respect to $\tilde{\tau}$ from 0 to $\tilde{\tau}$, then it is necessary to calculate the following integral:

$$\int_0^{\tilde{\tau}}\tilde{\mathcal{G}}_1 d\tilde{\tau} = \tilde{\tau}\tilde{\mathcal{G}}_1^\infty + \mathcal{PF}\int_0^\infty\tilde{\mathcal{G}}_1 d\tilde{\tau} \quad ,$$

where $\mathcal{PF}$ designates the finite part of the integral. Hence, in the vicinity of $\tilde{\tau}=\infty$, we find:

$$-\boldsymbol{D}\cdot(\boldsymbol{k}\times\tilde{\boldsymbol{v}}_1) + \mathrm{S}\frac{\partial}{\partial\zeta}\left(\frac{\zeta^2}{K_0(\zeta)}\frac{\partial\tilde{\mathcal{H}}_2}{\partial\zeta}\right) + \frac{1}{\mathrm{S}}\left[\frac{1}{2}|\boldsymbol{D}\times\tilde{\boldsymbol{v}}_0|^2 + \beta\eta\boldsymbol{k}\cdot(\boldsymbol{D}\times\tilde{\boldsymbol{v}}_0)\right]$$
$$= \left\{ -\boldsymbol{D}\cdot(\boldsymbol{k}\times\tilde{\boldsymbol{v}}_1) + \mathrm{S}\frac{\partial}{\partial\zeta}\left(\frac{\zeta^2}{K_0(\zeta)}\frac{\partial\tilde{\mathcal{H}}_2}{\partial\zeta}\right) + \frac{1}{\mathrm{S}}\left[\frac{1}{2}|\boldsymbol{D}\times\tilde{\boldsymbol{v}}_0|^2 + \beta\eta\boldsymbol{k}\cdot(\boldsymbol{D}\times\tilde{\boldsymbol{v}}_0)\right]\right\}_{\tilde{\tau}=0} + \tilde{\tau}\tilde{\mathcal{G}}_1^\infty + \mathcal{PF}\int_0^\infty\tilde{\mathcal{G}}_1 d\tilde{\tau} + \mathrm{O}\left(\frac{1}{\tilde{\tau}}\right) \quad , \tag{11.122}$$

where $\mathrm{O}(1/\tilde{\tau})$ are terms which tend towards zero with $\tilde{\tau}\to\infty$. The internal consistency of (11.122) with the preceding analysis implies that:

$$-\boldsymbol{D}\cdot(\boldsymbol{k}\times\tilde{\boldsymbol{v}}_1^1) + \mathrm{S}\frac{\partial}{\partial\zeta}\left(\frac{\zeta^2}{K_0(\zeta)}\frac{\partial\tilde{\mathcal{H}}_2^1}{\partial\zeta}\right) = \tilde{\mathcal{G}}_1^\infty \quad , \tag{11.123}$$

which also means that:

$$-\boldsymbol{D}\cdot(\boldsymbol{k}\times\tilde{\boldsymbol{v}}_1^{\infty})+\mathrm{S}\frac{\partial}{\partial\zeta}\left(\frac{\zeta^2}{K_0(\zeta)}\frac{\partial\tilde{\mathcal{H}}_2^{\infty}}{\partial\zeta}\right)=\Bigg\{-\boldsymbol{D}\cdot(\boldsymbol{k}\times\tilde{\boldsymbol{v}}_1)$$

$$+\mathrm{S}\frac{\partial}{\partial\zeta}\left(\frac{\zeta^2}{K_0(\zeta)}\frac{\partial\tilde{\mathcal{H}}_2}{\partial\zeta}\right)+\frac{1}{\mathrm{S}}\left[\frac{1}{2}|\boldsymbol{D}\times\tilde{\boldsymbol{v}}_0|^2+\beta\eta\boldsymbol{k}\cdot(\boldsymbol{D}\times\tilde{\boldsymbol{v}}_0)\right]\Bigg\}_{\tilde{\tau}=0}$$

$$+\mathcal{PF}\int_0^{\infty}\tilde{\mathcal{G}}_1 d\tilde{\tau}-\frac{1}{\mathrm{S}}\left[\frac{1}{2}|\boldsymbol{D}\times\tilde{\boldsymbol{v}}_0^{\infty}|^2+\beta\eta\boldsymbol{k}\cdot(\boldsymbol{D}\times\tilde{\boldsymbol{v}}_0^{\infty})\right] \quad . \qquad (11.124)$$

However, with the matching conditions (11.116, 117), and by taking into account (11.121), from (11.123) we have:

$$-\boldsymbol{D}\cdot\left[\boldsymbol{k}\times\left(\frac{\partial\boldsymbol{v}_0}{\partial\tau}\right)^0\right]+\mathrm{S}\frac{\partial}{\partial\zeta}\left[\frac{\zeta^2}{K_0(\zeta)}\frac{\partial}{\partial\zeta}\left(\frac{\partial\mathcal{H}_1}{\partial\tau}\right)^0\right]$$

$$=-\frac{1}{\mathrm{S}}\left\{(\boldsymbol{v}_0)^0\cdot\boldsymbol{D}(\wedge(\mathcal{H}_1)^0)+\beta(\boldsymbol{v}_0)^0\cdot\boldsymbol{j}\right\}$$

or, according to the adjustment to geostrophy:

$$\left[\frac{\partial}{\partial\tau}(\wedge\mathcal{H}_1)\right]^0=-\frac{1}{\mathrm{S}}\left\{(\boldsymbol{v}_0)^0\cdot\boldsymbol{D}(\wedge(\mathcal{H}_1)^0)+\beta(\boldsymbol{v}_0)^0\cdot\boldsymbol{j}\right\} \quad ,$$

i.e., we again encounter the equation of the quasi-geostrophic model (11.43) written for $\tau=0$.

For the moment, let us backtrack to (11.124). Taking advantage of (11.118), we obtain:

$$-\boldsymbol{D}\cdot(\boldsymbol{k}\times\tilde{\boldsymbol{v}}_1^{\infty})+\mathrm{S}\frac{\partial}{\partial\zeta}\left(\frac{\zeta^2}{K_0(\zeta)}\frac{\partial\tilde{\mathcal{H}}_2^{\infty}}{\partial\zeta}\right)$$

$$=-\boldsymbol{D}\cdot\left[\boldsymbol{k}\times(\tilde{\boldsymbol{v}}_1)^0\right]+\mathrm{S}\frac{\partial}{\partial\zeta}\left(\frac{\zeta^2}{K_0(\zeta)}\frac{\partial(\mathcal{H}_2)^0}{\partial\zeta}\right)$$

$$=\wedge(\mathcal{H}_2)^0+\frac{1}{\mathrm{S}}\boldsymbol{D}\cdot\left[(\boldsymbol{v}_0)^0\cdot\boldsymbol{D}(\boldsymbol{v}_0)^0+\beta\eta\boldsymbol{k}\times(\boldsymbol{v}_0)^0\right] \quad .$$

Finally, we find that the following initial condition must be imposed on the equation for $\mathcal{H}_2$ from the ageostrophic model [i.e., (11.88)]:

$$\wedge\mathcal{H}_2|_{\tau=0}=-\boldsymbol{D}\times(\boldsymbol{k}\times\tilde{\boldsymbol{v}}_1^0)+\mathrm{S}\frac{\partial}{\partial\zeta}\left(\frac{\zeta^2}{K_0(\zeta)}\frac{\partial\tilde{\mathcal{H}}_2^0}{\partial\zeta}\right)$$

$$+\frac{1}{\mathrm{S}}\left[\frac{1}{2}|\boldsymbol{D}\times\boldsymbol{V}^0|^2+\beta\eta\boldsymbol{k}\cdot(\boldsymbol{D}\times\boldsymbol{V}^0)\right]$$

$$-\frac{\lambda_0}{\mathrm{S}}\Bigg\{\boldsymbol{v}_0\cdot\boldsymbol{D}(\boldsymbol{D}^2\mathcal{H}_1)+\beta\eta\boldsymbol{D}^2\mathcal{H}_1+\beta\boldsymbol{j}\cdot\boldsymbol{D}\mathcal{H}_1$$

$$+\frac{\lambda_0}{2}|\boldsymbol{D}^2\mathcal{H}_1|^2+\beta\eta\boldsymbol{D}^2\mathcal{H}_1\Bigg\}_{\tau=0}+\mathcal{PF}\int_0^{\infty}\tilde{\mathcal{G}}_1 d\tilde{\tau} \quad , \qquad (11.125)$$

where $\tilde{\boldsymbol{v}}_1^0$ and $\tilde{\mathcal{H}}_2^0$ are the initial values of $\tilde{\boldsymbol{v}}_1$ and $\tilde{\mathcal{H}}_2$ which come into play in the problem of the unsteady adjustment to ageostrophy (11.110). To be sure, in order to calculate the last term in the right-hand member of (11.125), the problem of the adjustment to geostrophy must be resolved which means that $\tilde{\mathcal{H}}_1$ must be known!

11.4.4 The Second Approximation Steady Ekman Problem

Let us return to region III. From the expansion (11.62), the following system of equations is obtained for the terms $\hat{\boldsymbol{v}}_1$, $\hat{\mathcal{H}}_2$, and $\hat{w}_2$:

$$\begin{aligned}
&\lambda_0 \boldsymbol{D}\hat{\mathcal{H}}_2 + \boldsymbol{k}\times\hat{\boldsymbol{v}}_1 - \mathrm{Bo}^2\widehat{\mathrm{Ek}}_\perp \frac{\partial}{\partial\hat{\zeta}}\left(\varrho_0(1)\mu_0(1)\frac{\partial\hat{\boldsymbol{v}}_1}{\partial\hat{\zeta}} + \hat{\varrho}_1\mu_0(1)\frac{\partial\hat{\boldsymbol{v}}_0}{\partial\hat{\zeta}}\right)\\
&\quad -\frac{\partial\hat{\boldsymbol{v}}_0}{\partial\hat{\zeta}}\hat{w}_1 + \frac{\partial\hat{\boldsymbol{v}}_0}{\partial\tau} + \frac{1}{\mathrm{S}}\left(\hat{\boldsymbol{v}}_0\cdot\boldsymbol{D}\hat{\boldsymbol{v}}_0 + \beta\eta\boldsymbol{k}\times\hat{\boldsymbol{v}}_0\right) = 0 \quad ;\\
&\mathrm{S}\frac{\partial\hat{w}_2}{\partial\zeta} - \boldsymbol{D}\cdot\hat{\boldsymbol{v}}_1 = 0 \quad ;\\
&\mathrm{Bo}\left(\frac{\partial\hat{\mathcal{H}}_2}{\partial\hat{\zeta}} - \hat{\zeta}\frac{\partial\hat{\mathcal{H}}_1}{\partial\hat{\zeta}}\right) - \hat{T}_1 = 0 \quad .
\end{aligned} \tag{11.126}$$

To the above, the following boundary conditions on the ground

$$\hat{\zeta} = \hat{\zeta}_{s0} = -[\mathrm{Bo}/T_0(1)]\mathcal{H}_{1,1}$$

must be prescribed:

$$\begin{aligned}
&\hat{\mathcal{H}}_{2,s} + \hat{\zeta}_{s,1}\left(\frac{\partial\hat{\mathcal{H}}_1}{\partial\hat{\zeta}}\right)_s = 0 \quad ;\\
&\mathrm{Bo}\left[\left(\frac{\partial\hat{\mathcal{H}}_2}{\partial\tau}\right)_s + \hat{\zeta}_{s1}\left(\frac{\partial^2\hat{\mathcal{H}}_1}{\partial\tau\partial\hat{\zeta}}\right)_s - \hat{\zeta}_{s0}\left(\frac{\partial\hat{\mathcal{H}}_1}{\partial\tau}\right)_s\right]\\
&\qquad - \hat{w}_{2,s}T_0(1) - \hat{w}_{1,s}\hat{T}_{1,s} - \left(\frac{\partial\hat{w}_1}{\partial\hat{\zeta}}\right)_s T_0(1)\hat{\zeta}_{s1} = 0 \quad ;\\
&\hat{\boldsymbol{v}}_{1,s} + \hat{\zeta}_{s,1}\left(\frac{\partial\hat{\boldsymbol{v}}_0}{\partial\hat{\zeta}}\right)_s = 0 \quad ,
\end{aligned} \tag{11.127}$$

with the notation:

$$\hat{\mathcal{U}}_s = \hat{\mathcal{U}}|_{\hat{\zeta}=\hat{\zeta}_{s0}} \quad .$$

It will be observed that $\hat{\zeta} = \hat{\zeta}_{s0} + \mathrm{Ki}\hat{\zeta}_{s1} + \ldots$ characterizes the ground in the Ekman layer. We now can make use of (11.71, 73) for $\hat{T}_1$ and $\hat{\mathcal{H}}_1$, and from the last relation in (11.126) we find:

$$\begin{aligned}
\mathrm{Bo}\frac{\partial\hat{\mathcal{H}}_2}{\partial\hat{\zeta}} &= T_{1,1} + \left[T_0(1) - \left(\frac{dT_0}{d\zeta}\right)_{\zeta=1}\right]\hat{\zeta}\\
&= -\mathrm{Bo}\left(\frac{\partial\mathcal{H}_1}{\partial\zeta}\right)_{\zeta=1} + \mathrm{Bo}\left(\frac{\partial^2\mathcal{H}_0}{\partial\zeta^2}\right)_{\zeta=1}\hat{\zeta} \quad .
\end{aligned}$$

Hence, by matching with region I:

$$\hat{\mathcal{H}}_2 = \mathcal{H}_{2,1} - \left(\frac{\partial \mathcal{H}_1}{\partial \zeta}\right)_{\zeta=1} \hat{\zeta} + \frac{1}{2}\left(\frac{\partial^2 \mathcal{H}_0}{\partial \zeta^2}\right)_{\zeta=1} \hat{\zeta}^2 \quad . \tag{11.128}$$

Thus it is observed that:

$$\hat{\zeta}_{s1} = -\frac{\mathrm{Bo}}{T_0(1)}\left[\mathcal{H}_{2,1} + A_1(\mathcal{H}_1)\right] \quad , \tag{11.129}$$

where

$$A_1(\mathcal{H}_1) = \frac{\mathrm{Bo}}{T_0(1)}\mathcal{H}_{1,1}\left(\frac{\partial \mathcal{H}_1}{\partial \zeta}\right)_{\zeta=1} + \frac{1}{2}\left(\frac{\mathrm{Bo}}{T_0(1)}\mathcal{H}_{1,1}\right)^2\left(\frac{\partial^2 \mathcal{H}_0}{\partial \zeta^2}\right)_{\zeta=1} \quad .$$

A boundary-value problem for $\hat{\boldsymbol{v}}_1$ ("second approximation" Ackerblom problem) must now be formulated. Firstly, the matching with region I yields:

$$\hat{\boldsymbol{v}}_1 = \boldsymbol{v}_{1,1} - \left(\frac{\partial \boldsymbol{v}_0}{\partial \zeta}\right)_{\zeta=1} \hat{\zeta} + \hat{\boldsymbol{v}}_1' \quad ,$$

where $\hat{\boldsymbol{v}}_1'$ tends to zero with $\hat{\zeta} \to \infty$ and it is the solution of the equation:

$$\kappa_0 \frac{\partial^2 \hat{\boldsymbol{v}}_1'}{\partial \hat{\zeta}^2} - \boldsymbol{k} \times \hat{\boldsymbol{v}}_1' = \hat{\boldsymbol{Q}}_0' \quad , \tag{11, 130}$$

with the ground condition:

$$\hat{\boldsymbol{v}}_{1,s}' + \boldsymbol{v}_{1,1} - \hat{\zeta}_{s0}\left(\frac{\partial \boldsymbol{v}_0}{\partial \zeta}\right)_{\zeta=1} + \hat{\zeta}_{s1}\left(\frac{\partial \hat{\boldsymbol{v}}_0}{\partial \hat{\zeta}}\right)_s = \boldsymbol{0} \quad , \tag{11.131}$$

and where:

$$\hat{\boldsymbol{Q}}_0' = \frac{\partial \hat{\boldsymbol{v}}_0'}{\partial \tau} + \frac{1}{\mathrm{S}}\{(\boldsymbol{v}_{0,1} + \hat{\boldsymbol{v}}_0') \cdot \boldsymbol{D}(\boldsymbol{v}_{0,1} + \hat{\boldsymbol{v}}_0') - (\boldsymbol{v}_{0,1} \cdot \boldsymbol{D})\boldsymbol{v}_{0,1}$$
$$+ \beta\eta \boldsymbol{k} \times \hat{\boldsymbol{v}}_0'\} - \frac{\partial \hat{\boldsymbol{v}}_0}{\partial \hat{\zeta}}\hat{w}_1 - \kappa_0 \frac{\partial}{\partial \hat{\zeta}}\left(\frac{\hat{\varrho}_1}{\varrho_0(1)}\frac{\partial \hat{\boldsymbol{v}}_0'}{\partial \hat{\zeta}}\right) \quad .$$

It will be seen that we can express the term: $\hat{\boldsymbol{v}}_{1,s}' - \mathrm{i}\boldsymbol{k} \times \hat{\boldsymbol{v}}_{1,s}'$ as a function of $\mathcal{H}_{2,1}$ and $\mathcal{H}_{1,1}$. After a rather long but simple calculation, it is found that

$$\hat{\boldsymbol{v}}_{1,s}' - \mathrm{i}\boldsymbol{k} \times \hat{\boldsymbol{v}}_{1,s}' = -\lambda_0\left[\boldsymbol{k} \times \boldsymbol{D}\mathcal{H}_{2,1} + \mathrm{i}\boldsymbol{D}\mathcal{H}_{2,1}\right]$$
$$+\boldsymbol{\lambda}(\mathcal{H}_{1,1})\mathcal{H}_{2,1} + A_1(\mathcal{H}_1)\boldsymbol{\lambda}(\mathcal{H}_{1,1}) + \boldsymbol{\mu}(\mathcal{H}_1) \quad ,$$

which makes it possible to obtain the solution to (11.130) in the following form:

$$\hat{\boldsymbol{v}}_1' - \mathrm{i}\boldsymbol{k} \times \hat{\boldsymbol{v}}_1' = \{-\lambda_0\left[\boldsymbol{k} \times \boldsymbol{D}\mathcal{H}_{2,1} + \mathrm{i}\boldsymbol{D}\mathcal{H}_{2,1}\right]$$
$$+\left[A_1(\mathcal{H}_1) + \mathcal{H}_{2,1}\right]\boldsymbol{\lambda}(\mathcal{H}_{1,1}) + \boldsymbol{\mu}(\mathcal{H}_1)\}\mathrm{E}$$
$$+ \int_{\hat{\zeta}_{s0}}^{\hat{\zeta}} G(\hat{\zeta}, \hat{s})\left[\hat{\boldsymbol{Q}}_0'(\hat{s}) - \mathrm{i}\boldsymbol{k} \times \hat{\boldsymbol{Q}}_0'(\hat{s})\right] d\hat{s} + \int_{\hat{\zeta}}^{\infty} G(\hat{\zeta}, \hat{s})\left[\hat{\boldsymbol{Q}}_0'(\hat{s}) - \mathrm{i}\boldsymbol{k} \times \hat{\boldsymbol{Q}}_0'(\hat{s})\right] d\hat{s} \quad . \tag{11.132}$$

In the above formulas, we have introduced the following notations:

$$\boldsymbol{\lambda}(\mathcal{H}_{1,1}) = \lambda_0 \frac{\mathrm{Bo}}{T_0(1)} \frac{1+\mathrm{i}}{\sqrt{2\kappa_0}} \left[\boldsymbol{k} \times \boldsymbol{D}\mathcal{H}_{1,1} + \mathrm{i}\boldsymbol{D}\mathcal{H}_{1,1}\right] \quad ;$$

$$\begin{aligned}\boldsymbol{\mu}(\mathcal{H}_1) &= \lambda_0 \left[\frac{\partial}{\partial t} + \frac{\lambda_0}{\mathrm{S}}(\boldsymbol{k} \times \boldsymbol{D}\mathcal{H}_{1,1}) \cdot \boldsymbol{D} \right.\\ &\left. + \frac{\lambda_0}{\mathrm{S}}\beta\eta\boldsymbol{k} \times \right] (\boldsymbol{D}\mathcal{H}_{1,1} - \mathrm{i}\boldsymbol{k} \times \boldsymbol{D}\mathcal{H}_{1,1})\\ &- \frac{\lambda_0 \mathrm{Bo}}{T_0(1)}\mathcal{H}_{1,1}\left[\boldsymbol{k} \times \boldsymbol{D}\left(\frac{\partial \mathcal{H}_1}{\partial \zeta}\right)_{\zeta=1} + \mathrm{i}\boldsymbol{D}\left(\frac{\partial \mathcal{H}_1}{\partial \zeta}\right)_{\zeta=1}\right] \quad ,\end{aligned}$$

and

$$G(\hat{\zeta}, \hat{s}) = \begin{cases} -\dfrac{1-\mathrm{i}}{2\sqrt{2\kappa_0}}\left\{\exp\left[\dfrac{1+\mathrm{i}}{\sqrt{2\kappa_0}}(\hat{\zeta} - \hat{s})\right]\right. \\ \left. -\exp\left[\dfrac{1+\mathrm{i}}{\sqrt{2\kappa_0}}(2\hat{\zeta}_{s0} - \hat{\zeta} - \hat{s})\right]\right\} \quad , \quad \hat{\zeta} \leq \hat{s} \quad ; \\ -\dfrac{1-\mathrm{i}}{2\sqrt{2\kappa_0}}\left\{\exp\left[\dfrac{1+\mathrm{i}}{\sqrt{2\kappa_0}}(\hat{s} - \hat{\zeta})\right]\right. \\ \left. -\exp\left[\dfrac{1+\mathrm{i}}{\sqrt{2\kappa_0}}(2\hat{\zeta}_{s0} - \hat{s} - \hat{\zeta})\right]\right\} \quad , \quad 0 \leq \hat{s} \leq \hat{\zeta} \quad . \end{cases}$$

It is now necessary to put to good use the second equation of (11.126) to calculate $\hat{w}_2$. By taking advantage of the expression for $\hat{\boldsymbol{v}}_1$, we have:

$$\hat{w}_2 = (\hat{\zeta} - \hat{\zeta}_{s0})\boldsymbol{D} \cdot \boldsymbol{v}_{1,1} + \int_{\hat{\zeta}_{s0}}^{\hat{\zeta}} (\boldsymbol{D} \cdot \hat{\boldsymbol{v}}_1')\, d\hat{\zeta} + \hat{w}_{2,s} \quad ,$$

since

$$\boldsymbol{D} \cdot \left(\frac{\partial \boldsymbol{v}_0}{\partial \zeta}\right)_{\zeta=1} = 0 \quad .$$

Matching with region I yields:

$$\lim_{\hat{\zeta} \to \infty} \left(\hat{w}_2 - w_{2,1} + \hat{\zeta}\left(\frac{\partial w_1}{\partial \zeta}\right)_{\zeta=1}\right) = 0$$

and thus the following relation is deduced:

$$w_{2,1} = \hat{w}_{2,s} - \hat{\zeta}_{s0}\boldsymbol{D} \cdot \boldsymbol{v}_{1,1} + \hat{\boldsymbol{v}}_{1,s}' \cdot \boldsymbol{D}\hat{\zeta}_{s0} + \boldsymbol{D} \cdot \int_{\hat{\zeta}_{s0}}^{+\infty} \hat{\boldsymbol{v}}_1'\, d\hat{\zeta} \quad . \tag{11.133}$$

Relation (11.133) must enable us to obtain the boundary condition in $\zeta = 1$ for (11.88) of the ageostrophic model. The long and rather tedious calculation making it possible to obtain this condition was carried out by J.P. Guiraud in his Course Notes for the ICMS in Udine. Below are given the various steps of the procedure which leads to this condition.

Let us recall that:

$$\mathrm{E} = \exp\left\{ -\frac{1+\mathrm{i}}{\sqrt{2\kappa_0}}(\hat{\zeta} - \hat{\zeta}_{s0}) \right\} \quad \text{and hence:}$$

$$\frac{\partial \mathrm{E}}{\partial \tau} = -\frac{\mathrm{Bo}}{T_0(1)} \frac{1+\mathrm{i}}{\sqrt{2\kappa_0}} \frac{\partial \mathcal{H}_{1,1}}{\partial \tau}\mathrm{E} \quad ; \quad \boldsymbol{D}\mathrm{E} = -\frac{\mathrm{Bo}}{T_0(1)} \frac{1+\mathrm{i}}{\sqrt{2\kappa_0}}(\boldsymbol{D}\mathcal{H}_{1,1})\mathrm{E} \quad .$$

From (11.86) for w_2 we deduce:

$$w_{2,1} = -\frac{1}{K_0(1)}\left[\frac{\partial}{\partial \tau} + \frac{\lambda_0}{\mathrm{S}}(\boldsymbol{k} \times \boldsymbol{D}\mathcal{H}_{1,1}) \cdot \boldsymbol{D}\right]\left(\frac{\partial \mathcal{H}_2}{\partial \zeta}\right)_{\zeta=1}$$
$$-\frac{\lambda_0}{\mathrm{S}} \frac{1}{K_0(1)}(\boldsymbol{k} \times \boldsymbol{D}\mathcal{H}_{2,1}) \cdot \boldsymbol{D}\left(\frac{\partial \mathcal{H}_1}{\partial \zeta}\right)_{\zeta=1} + W_{1,1} \quad , \tag{11.134}$$

where $W_{1,1}$ is the expression (11.87) taken in $\zeta = 1$. Therefore, in (11.133) the left-hand term is expressed as a function of $\mathcal{H}_2$ and $\mathcal{H}_1$. Among the right-hand terms of (11.133), let us first consider the term $\hat{w}_{2,s}$. The latter intervenes in the second relation of (11.127) and because of this, we must calculate a whole sequence of terms. First of all, we have:

$$-\left(\frac{\partial \hat{w}_1}{\partial \hat{\zeta}}\right)_s T_0(1)\hat{\zeta}_{s1} = (\boldsymbol{D} \cdot \hat{\boldsymbol{v}}_0)_s \frac{\mathrm{Bo}}{\mathrm{S}}[\mathcal{H}_{2,1} + A_1(\mathcal{H}_1)] \quad , \tag{11.135}$$

according to the third equation of (11.67) and (11.129). The term $(\boldsymbol{D} \cdot \hat{\boldsymbol{v}}_0)_s$ can be calculated from the following relation according to the results of Sect. 11.3:

$$0 = \boldsymbol{D} \cdot \hat{\boldsymbol{v}}_{0,s} = (\boldsymbol{D} \cdot \hat{\boldsymbol{v}}_0)_s + \left(\frac{\partial \hat{\boldsymbol{v}}_0}{\partial \hat{\zeta}}\right)_s \cdot \boldsymbol{D}\hat{\zeta}_{s0}$$
$$= (\boldsymbol{D} \cdot \hat{\boldsymbol{v}}_0)_s - \Re\left\{ \frac{1+\mathrm{i}}{\sqrt{2\kappa_0}}(\boldsymbol{v}_{0,1} - \mathrm{i}\boldsymbol{k} \times \boldsymbol{v}_{0,1})\frac{\mathrm{Bo}}{T_0(1)}\boldsymbol{D}\mathcal{H}_{1,1} \right\}$$

and we find, instead of (11.135):

$$T_0(1)\hat{\zeta}_{s1}\left(\frac{\partial \hat{w}_1}{\partial \hat{\zeta}}\right)_s = \frac{\lambda_0}{\sqrt{2\kappa_0}} \frac{\mathrm{Bo}^2}{\mathrm{S}} \frac{1}{T_0(1)}[\mathcal{H}_{2,1} + A(\mathcal{H}_1)]|\boldsymbol{D}\mathcal{H}_{1,1}|^2 \quad . \tag{11.136}$$

We will now consider the term:

$$\hat{w}_{1,s}\hat{T}_{1,s} = -\frac{\mathrm{Bo}^2}{T_0(1)} \frac{\partial \mathcal{H}_{1,1}}{\partial \tau}\left(\frac{\partial \mathcal{H}_1}{\partial \zeta}\right)_{\zeta=1} - \frac{\mathrm{Bo}^2}{T_0^2(1)}\left(\frac{dT_0}{d\zeta}\right)_{\zeta=1}\mathcal{H}_{1,1}\frac{\partial \mathcal{H}_{1,1}}{\partial \tau} \quad ,$$

seeing that according to (11.71):

$$\hat{T}_{1,s} = -\mathrm{Bo}\left(\frac{\partial \mathcal{H}_1}{\partial \zeta}\right)_{\zeta=1} - \frac{\mathrm{Bo}}{T_0(1)}\left(\frac{dT_0}{d\zeta}\right)_{\zeta=1}\mathcal{H}_{1,1} \quad .$$

Then, we have

$$\mathrm{Bo}\left[\left(\frac{\partial \hat{\mathcal{H}}_2}{\partial \tau}\right)_s + \hat{\zeta}_{s1}\left(\frac{\partial^2 \hat{\mathcal{H}}_1}{\partial \tau \partial \hat{\zeta}}\right)_s - \hat{\zeta}_{s0}\left(\frac{\partial \hat{\mathcal{H}}_1}{\partial \tau}\right)_s\right]$$
$$= \mathrm{Bo}\left\{\frac{\partial \mathcal{H}_{2,1}}{\partial \tau} + \frac{\mathrm{Bo}}{T_0(1)}\mathcal{H}_{1,1}\left[\frac{\partial \mathcal{H}_{1,1}}{\partial \tau} + \frac{\partial}{\partial \tau}\left(\frac{\partial \mathcal{H}_1}{\partial \zeta}\right)_{\zeta=1}\right]\right\} \quad .$$

Finally, we find that:

$$\hat{w}_{2,s} = \frac{\text{Bo}}{T_0(1)} \frac{\partial \mathcal{H}_{2,1}}{\partial \tau} - \frac{\lambda_0}{\sqrt{2\kappa_0}} \frac{\text{Bo}^2}{\text{S}T_0^2(1)} |\boldsymbol{D}\mathcal{H}_{1,1}|^2 \mathcal{H}_{2,1} + \mathcal{N}(\mathcal{H}_1) , \qquad (11.137)$$

where

$$\mathcal{N}(\mathcal{H}_1) = \left(\frac{\text{Bo}}{T_0(1)}\right)^2 \left\{ - \frac{\lambda_0}{\text{S}\sqrt{2\kappa_0}} A(\mathcal{H}_1) |\boldsymbol{D}\mathcal{H}_{1,1}|^2 + \frac{\partial \mathcal{H}_{1,1}}{\partial \tau} \left(\frac{\partial \mathcal{H}_1}{\partial \zeta}\right)_{\zeta=1} + \left[1 + \frac{1}{T_0(1)} \left(\frac{dT_0}{d\zeta}\right)_{\zeta=1}\right] \frac{\partial \mathcal{H}_{1,1}}{\partial \tau} \mathcal{H}_{1,1} + \frac{\partial}{\partial \tau} \left(\frac{\partial \mathcal{H}_1}{\partial \zeta}\right)_{\zeta=1} \mathcal{H}_{1,1} \right\} .$$

The second and third terms of the right-hand members of (11.133) must now be expressed. Expressions (11.74, 129, 132) are made use of and it is found that:

$$-\hat{\zeta}_{s0} \boldsymbol{D} \cdot \boldsymbol{v}_{1,1} + \hat{\boldsymbol{v}}'_{1,s} \cdot \boldsymbol{D}\hat{\zeta}_{s0} = \frac{\lambda_0}{\sqrt{2\kappa_0}} \frac{1}{\text{S}} \left(\frac{\text{Bo}}{T_0(1)}\right)^2 |\boldsymbol{D}\mathcal{H}_{1,1}|^2 \mathcal{H}_{2,1} - \lambda_0 \frac{\text{Bo}}{T_0(1)} (\boldsymbol{k} \times \boldsymbol{D}\mathcal{H}_{1,1}) \cdot \boldsymbol{D}\mathcal{H}_{2,1} + \mathcal{P}(\mathcal{H}_1) \quad , \qquad (11.138)$$

where $\mathcal{P}(\mathcal{H}_1)$ is a group of terms which are functions only of $\mathcal{H}_1$ and its derivatives taken in $\zeta = 1$. We will not write here the expression of $\mathcal{P}(\mathcal{H}_1)$.

The term $\boldsymbol{D} \cdot \int_{\hat{\zeta}_{s0}}^{+\infty} \hat{\boldsymbol{v}}'_1 d\hat{\zeta}$ from (11.133) remains to be calculated. For this, we must take advantage of the solution of (11.132). To this end, we note that:

$$\int_{\hat{\zeta}_{s0}}^{+\infty} \text{E} d\hat{\zeta} = \frac{\sqrt{2\kappa_0}}{1+\text{i}} \quad , \quad \text{and} \quad \int_{\hat{\zeta}_{s0}}^{+\infty} G(\hat{\zeta}, \hat{s}) d\hat{s} = \text{i}\{1 - \text{E}\} \quad .$$

Then we find that

$$\boldsymbol{D} \cdot \int_{\hat{\zeta}_{s0}}^{+\infty} \hat{\boldsymbol{v}}'_1 d\hat{\zeta} = \boldsymbol{D} \cdot \left\{ \Re \left[\frac{\sqrt{2\kappa_0}}{1+\text{i}} \Big(- \lambda_0 [\boldsymbol{k} \times \boldsymbol{D}\mathcal{H}_{2,1} + \text{i}\boldsymbol{D}\mathcal{H}_{2,1}] + [A_1(\mathcal{H}_1) + \mathcal{H}_{2,1}] \boldsymbol{\lambda}(\mathcal{H}_{1,1}) + \boldsymbol{\mu}(\mathcal{H}_1) \Big) \right] \right\} - \boldsymbol{D} \cdot \left\{ \Im \int_{\hat{\zeta}_{s0}}^{+\infty} (1 - \text{E})(\hat{\boldsymbol{Q}}'_0 - \text{i}\boldsymbol{k} \times \hat{\boldsymbol{Q}}'_0) d\hat{\zeta} \right\} .$$

The following relation is thus obtained:

$$\boldsymbol{D} \cdot \int_{\hat{\zeta}_{s0}}^{+\infty} \hat{\boldsymbol{v}}'_1 d\hat{\zeta} = \lambda_0 \frac{\text{Bo}}{T_0(1)} (\boldsymbol{k} \times \boldsymbol{D}\mathcal{H}_{1,1}) \cdot \boldsymbol{D}\mathcal{H}_{2,1} - \frac{\lambda_0}{2} \sqrt{2\kappa_0} \boldsymbol{D}^2 \mathcal{H}_{2,1} + \mathcal{S}(\mathcal{H}_1) \quad , \qquad (11.139)$$

where once again $\mathcal{S}(\mathcal{H}_1)$ is a collection of terms which are functions only of $\mathcal{H}_1$ and its dervatives taken in $\zeta = 1$. The expression of $\mathcal{S}(\mathcal{H}_1)$ will not be given here. It is pointed out that the integral which appears in $\mathcal{S}(\mathcal{H}_1)$ can be calculated explicitly as a function of $\mathcal{H}_1$.

Therefore, if we go back to (11.133) and take into account (11.134, 137–139), we can obtain the following boundary condition in $\zeta = 1$ for $\mathcal{H}_2$:

$$\left\{\frac{\mathrm{Bo}}{T_0(1)}\frac{\partial}{\partial\tau} + \frac{1}{K_0(1)}\left[\frac{\partial}{\partial\tau} + \frac{\lambda_0}{\mathrm{S}}(\boldsymbol{k}\times\boldsymbol{D}\mathcal{H}_1)\cdot\boldsymbol{D}\right]\frac{\partial}{\partial\zeta} - \frac{\lambda_0}{\mathrm{S}}\sqrt{\frac{\kappa_0}{2}}\boldsymbol{D}^2 - \frac{\lambda_0}{\mathrm{S}}\frac{1}{K_0(1)}\left[\boldsymbol{k}\times\boldsymbol{D}\left(\frac{\partial\mathcal{H}_1}{\partial\zeta}\right)\right]\cdot\boldsymbol{D}\right\}\mathcal{H}_2 = \mathcal{M}(\mathcal{H}_1), \quad \text{on} \quad \zeta = 1\,, \tag{11.140}$$

where

$$\mathcal{M}(\mathcal{H}_1) = \mathcal{N}(\mathcal{H}_1) + \mathcal{P}(\mathcal{H}_1) + \mathcal{S}(\mathcal{H}_1) - W_{1,1} \quad ,$$

is a right-hand member which can be entirely expressed as a function of $\mathcal{H}_1$ and its derivatives in $\zeta = 1$.

The analysis of the ageostrophic model is thus completed but in Sect. 11.5 we will return to certain questions related to these quasi-geostrophic and ageostrophic asymptotic models.

11.5 Complementary Remarks

11.5.1. According to the results of Sects. 11.1–3, the following quasi-geostrophic model can be formulated:

$$\begin{aligned}
&\frac{D_0}{D\tau}(\wedge\,\mathcal{H}_1) + \lambda_0\frac{\beta}{\mathrm{S}}\frac{\partial\mathcal{H}_1}{\partial\xi} = 0 \quad , \\
&\wedge\,\mathcal{H}_1|_{\tau=0} = \boldsymbol{k}\cdot(\boldsymbol{D}\times\boldsymbol{V}^0) + \mathrm{S}\frac{\partial}{\partial\zeta}\left(\frac{\zeta^2}{K_0(\zeta)}\frac{\partial H_1^0}{\partial\zeta}\right) \quad ; \\
&\left\{\frac{\mathrm{Bo}}{T_0(1)}\frac{\partial}{\partial\tau} + \frac{1}{K_0(1)}\frac{D_0}{D\tau}\left(\frac{\partial}{\partial\zeta}\right) - \frac{\lambda_0}{\mathrm{S}}\sqrt{\frac{\kappa_0}{2}}\boldsymbol{D}^2\right\}\mathcal{H}_1 = 0 \quad , \quad \text{on} \quad \zeta = 1 \quad ; \\
&\frac{\zeta^2}{K_0(\zeta)}\left(\frac{\partial\mathcal{H}_1}{\partial\zeta}\right)^2 + \frac{\lambda_0}{\mathrm{S}}|\boldsymbol{D}\mathcal{H}_1|^2 \quad ,
\end{aligned} \tag{11.141}$$

decreases sufficiently rapidly at infinity when $|\xi^2 + \eta^2| \to \infty$ and $\zeta \to 0$.

In the model (11.141), we have the following notations:

$$\begin{aligned}
\frac{D_0}{D\tau} &= \frac{\partial}{\partial\tau} + \frac{\lambda_0}{\mathrm{S}}(\boldsymbol{k}\times\boldsymbol{D}\mathcal{H}_1)\cdot\boldsymbol{D} \quad ; \\
\wedge &= \lambda_0\boldsymbol{D}^2 + \mathrm{S}\frac{\partial}{\partial\zeta}\left(\frac{\zeta^2}{K_0(\zeta)}\frac{\partial}{\partial\zeta}\right) \quad ,
\end{aligned}$$

$$\text{with} \quad \lambda_0 \equiv \frac{\mathrm{Bo}}{\gamma\mathrm{S}}\left(\frac{\mathrm{Ki}}{\mathrm{M}_\infty}\right)^2 \quad \text{and} \quad K_0(\zeta) = -\frac{\zeta}{\mathrm{Bo}}\left(\frac{dT_0}{d\zeta} - \frac{\gamma-1}{\gamma}\frac{T_0}{\zeta}\right) \quad .$$

A first remark concerns the conditions at infinity with respect to ξ and η. Strictly speaking, only matching with a local solution valid in the vicinity of

$|\xi^2 + \eta^2| = \infty$ can enable us to choose lateral conditions in ξ and η which are coherent from the asymptotic viewpoint. The same is true of the boundary condition to be imposed for $\zeta \to 0$.

A second remark concerns the appearance of a term in $\partial/\partial\tau$ in the condition on $\zeta = 1$ for (11.141). An initial condition in $\tau = 0$ must be added to this condition on $\zeta = 1$. This initial condition is obtained from (11.57) (Sect. 11.2), once it is taken into account that according to the matching between regions I and II, we have: $\tilde{\mathcal{H}}_1^\infty \equiv \mathcal{H}_1|_{\tau=0}$. Hence, the initial condition:

$$\left[\left\{\frac{\partial \mathcal{H}_1}{\partial \zeta} + \frac{\mathrm{Bo}}{T_0(1)} K_0(1)\mathcal{H}_1\right\}_{\zeta=1}\right]_{\tau=0} = \left[\frac{\partial H_1^0}{\partial \tau} + \frac{\mathrm{Bo}}{T_0(1)} K_0(1) H_1^0\right]_{\zeta=1} , \tag{11.142}$$

must be superimposed on the condition in $\zeta = 1$ for (11.141). We will suppose now that $\mathcal{H}_1(\tau, \xi, \eta, \zeta)$ is defined in the domain:

$$\overline{Q}: \ |\xi| < \infty \ , \quad |\eta| < \infty \ , \quad : 0 \le \zeta \le 1 \ .$$

The equation of the quasi-geostrophic model (11.141) can be considered as a transport equation for $\wedge \mathcal{H}_1$ once the velocity field $\boldsymbol{v}_0$ is known. This equation can, in fact, be written as follows:

$$\frac{\partial}{\partial \tau}(\wedge \mathcal{H}_1) + \frac{1}{\mathrm{S}} \boldsymbol{v}_0 \cdot \boldsymbol{D}(\wedge \mathcal{H}_1) = -\frac{\beta}{\mathrm{S}} \boldsymbol{v}_0 \cdot \boldsymbol{j} \ , \tag{11.143}$$

with

$$\wedge \mathcal{H}_1|_{\tau=0} = f^0$$

where f^0 is an expression known from the initial conditions prescribed at the start.

Equation (11.143) with $\boldsymbol{v}_0$ known is solved in classic fashion by quadrature. However, the fundamental problem is to find $\boldsymbol{v}_0$ when $\wedge\mathcal{H}_1$ is known! To this end, we take advantage of the expression of $\boldsymbol{v}_0$ as a function of $\mathcal{H}_1$: $\boldsymbol{v}_0 = \lambda_0 \boldsymbol{k} \times \boldsymbol{D}\mathcal{H}_1$ and then we can construct the following numerical algorithm. We designate by

$$\mathcal{T}_{\boldsymbol{v}_0}(\tau, \Delta\tau) \wedge \mathcal{H}_1(\tau)$$

the solution of (11.143) at time $\tau + \Delta\tau$ obtained from the solution at time τ with $\boldsymbol{v}_0$ being assumed known. We then formulate an implicit numerical code by setting:

$$\begin{aligned} \mathcal{H}_1(\tau + \Delta\tau) &= \wedge^{-1} \mathcal{T}_{\boldsymbol{v}_{0,m}}(\tau, \Delta\tau) \wedge \mathcal{H}_1(\tau) \ ; \\ 2\boldsymbol{v}_{0,m} &= \lambda_0 \boldsymbol{k} \times \boldsymbol{D}\big(\mathcal{H}_1(\tau) + \mathcal{H}_1(\tau + \Delta\tau)\big) \ , \end{aligned} \tag{11.144}$$

on the condition that numerical codes are defined for the calculation of $\wedge^{-1}$, $\boldsymbol{D}$ and $\mathcal{T}_{\boldsymbol{v}_0}$ $(\tau, \Delta\tau)$.

It is thus necessary that we be able to find $\mathcal{H}_1$ when $\wedge \mathcal{H}_1$ is known. We will assume that $\wedge \mathcal{H}_1$ is known in the domain Q: $|\xi| < \infty$, $|\eta| < \infty$, $0 < \zeta < 1$. Let us consider the domains:

$$Q_R: \ \xi^2+\eta^2<R^2 \quad , \quad 0<\zeta<1 \quad ;$$
$$\Sigma_R: \ \xi^2+\eta^2=R^2 \quad , \quad 0<\zeta<1 \quad ;$$
$$D_R: \ \xi^2+\eta^2<R^2 \quad .$$

For any arbitrary "test" function ϕ, the following relation can be written:

$$\iiint_{Q_R} \phi \wedge \mathcal{H}_1 d\xi d\eta d\zeta = -\iiint_{Q_R} \Bigg(\lambda_0 \boldsymbol{D}\mathcal{H}_1 \cdot \boldsymbol{D}\phi + \mathrm{S}\frac{\zeta^2}{K_0(\zeta)} \frac{\partial \mathcal{H}_1}{\partial \zeta} \frac{\partial \phi}{\partial \zeta} \Bigg) d\xi\, d\eta\, d\zeta + \iint_{\Sigma_R} \lambda_0 \phi \frac{\partial \mathcal{H}_1}{\partial \nu} d\sigma\, d\zeta$$
$$+ \left\{ \iint_{D_R} \mathrm{S}\frac{\zeta^2}{K_0(\zeta)} \phi \frac{\partial \mathcal{H}_1}{\partial \zeta} d\xi\, d\eta \right\}_{\zeta=0}^{\zeta=1} \quad , \tag{11.145}$$

where $\partial/\partial\nu$ is the derivative along the normal to the circle $\xi^2+\eta^2=R^2$ in the plane $\zeta=$ const. Equation (11.145) indicates what "natural" boundary conditions are to be imposed for $|\xi^2+\eta^2| \to \infty$ and $\zeta \to 0$. For this purpose, we introduce a class $\mathcal{F}$ of functions which is defined as follows: the functions ϕ and ψ both belong to $\mathcal{F}$ if:

$$\lim_{R\to\infty} \iint_{\Sigma_R} \phi \frac{\partial \psi}{\partial \nu} d\sigma\, d\zeta = 0 \quad , \quad \lim_{\zeta \to 0} \iint_{-\infty}^{+\infty} \frac{\zeta^2}{K_0(\zeta)} \phi \frac{\partial \psi}{\partial \zeta} d\xi\, d\eta = 0 \quad ,$$

and naturally:

$$\iiint_{Q} \left\{ \lambda_0 |\boldsymbol{D}\phi|^2 + \mathrm{S}\frac{\zeta^2}{K_0(\zeta)} \left(\frac{\partial \phi}{\partial \zeta}\right)^2 \right\} d\xi\, d\eta\, d\zeta < +\infty \quad .$$

We thus again find a condition in $\zeta \to 0$ analogous to the one obtained by El Mabrouk and Zeytounian (1984). If we therefore consider the problem of seeking $\mathcal{H}_1$ when

$$\wedge \mathcal{H}_1 = \chi_0$$

is known, with a right condition in $\zeta=1$, we can formulate a variational problem in ξ, η, ζ by progressing in time with respect to τ: we must find $\mathcal{H}_1$ (ξ, η, ζ, τ) belonging to the above-defined class $\mathcal{F}$, which satisfies the right condition in $\zeta=1$ such that the following relation holds for any ϕ belonging to $\mathcal{F}$:

$$\iiint_{Q} \left\{ \lambda_0 \boldsymbol{D}\mathcal{H}_1 \cdot \boldsymbol{D}\phi + \mathrm{S}\frac{\zeta^2}{K_0(\zeta)} \frac{\partial \mathcal{H}_1}{\partial \zeta} \frac{\partial \phi}{\partial \zeta} + \phi \chi_0 \right\} d\xi\, d\eta\, d\zeta$$
$$= \iint_{-\infty}^{+\infty} \frac{\mathrm{S}}{K_0(1)} \left(\phi \frac{\partial \mathcal{H}_1}{\partial \zeta} \right)_{\zeta=1} d\xi\, d\eta \quad . \tag{11.146}$$

We can ask if this problem is well-posed? The answer will depend on the nature of the condition in $\zeta = 1$. The latter is homogenous with respect to $\mathcal{H}_1$ and, in addition, from (11.146), we find that $\mathcal{H}_1 \equiv 0$[6] when $\chi_0 = 0$ and $\phi \equiv \mathcal{H}_1$. This implies that unicity and existence can be reasonably expected.

11.5.2. Let us now turn our attention to (11.88) from the ageostrophic model. This equation, as well as (11.43) from the quasi-geostrophic model, can be obtained from the following more general equation:

$$\left\{\frac{\partial}{\partial\tau} + \frac{\lambda_0}{\mathrm{S}}(\boldsymbol{k} \times \boldsymbol{D}\mathcal{H}) \cdot \boldsymbol{D}\right\} \wedge \mathcal{H} + \lambda_0 \frac{\beta}{\mathrm{S}} \boldsymbol{i} \cdot \boldsymbol{D}\mathcal{H}$$
$$= \mathrm{Ki}\left(\boldsymbol{D} \cdot \boldsymbol{V}_1 + \mathrm{S}\frac{\partial W_1}{\partial\zeta}\right) + \mathrm{O}(\mathrm{Ki}^2) \qquad (11.147)$$

by setting:

$$\mathcal{H} = \mathcal{H}_1 + \mathrm{Ki}\mathcal{H}_2 + \mathrm{O}(\mathrm{Ki}^2) \quad .$$

In the same way, if we consider the following general condition in $\zeta = 1$:

$$\left\{\frac{\mathrm{Bo}}{T_0(1)}\frac{\partial}{\partial\tau} + \frac{1}{K_0(1)}\left[\frac{\partial}{\partial\tau} + \frac{\lambda_0}{\mathrm{S}}(\boldsymbol{k} \times \boldsymbol{D}\mathcal{H}) \cdot \boldsymbol{D}\right]\frac{\partial}{\partial\zeta}\right.$$
$$\left. - \frac{\lambda_0}{\mathrm{S}}\sqrt{\frac{\kappa_0}{2}}\boldsymbol{D}^2\right\}\mathcal{H}|_{\zeta=1} = \mathrm{Ki}\mathcal{M}(\mathcal{H}) + \mathrm{O}(\mathrm{Ki}^2) \quad , \qquad (11.148)$$

we again find with $\mathcal{H} = \mathcal{H}_1 + \mathrm{Ki}\mathcal{H}_2 + \mathrm{O}(\mathrm{Ki}^2)$, the conditions (11.80, 140).

From the mathematical point of view, the following problem must therefore be analyzed for the unknown function $\mathcal{H}$ (τ, ξ, η, ζ):

$$\left(\frac{\partial}{\partial\tau} + \frac{\lambda_0}{\mathrm{S}}(\boldsymbol{k} \times \boldsymbol{D}\mathcal{H}) \cdot \boldsymbol{D}\right)(\wedge \mathcal{H}) + \lambda_0 \frac{\beta}{\mathrm{S}} \boldsymbol{i} \cdot \boldsymbol{D}\mathcal{H} = \mathcal{A}_0 \quad ;$$
$$\wedge \mathcal{H}|_{\tau=0} = \mathcal{B}_0 \quad ,$$
$$\left\{\frac{\mathrm{Bo}}{T_0(1)}\frac{\partial}{\partial\tau} + \frac{1}{K_0(1)}\left[\frac{\partial}{\partial\tau} + \frac{\lambda_0}{\mathrm{S}}(\boldsymbol{k} \wedge \boldsymbol{D}\mathcal{H}) \cdot \boldsymbol{D}\right]\frac{\partial}{\partial\zeta}\right. \qquad (11.149)$$
$$\left. - \frac{\lambda_0}{\mathrm{S}}\sqrt{\frac{\kappa_0}{2}}\boldsymbol{D}^2\right\}\mathcal{H}|_{\zeta=1} = \mathcal{C}_0 \quad ,$$

$\mathcal{A}_0$, $\mathcal{B}_0$ and $\mathcal{C}_0$ being known functions. In addition, the solution $\mathcal{H}$ must belong to the class $\mathcal{F}$ which was defined above. We do not know if by using functional analysis techniques it is possible to demonstrate rigorous mathematical results concerning (11.149). Nonetheless, certain particular results were obtained by Dutton (1974).

---

[6] In this case, to $\wedge \mathcal{H}_1 = 0$, the condition

$$\left.\frac{\partial\mathcal{H}_1}{\partial\zeta}\right|_{\zeta=1} = -\frac{\mathrm{Bo}}{T_0(1)}K_0(1)\mathcal{H}_{1,1}$$

must be prescribed and the right-hand member of (11.146) becomes, with $\phi \equiv \mathcal{H}_1$,

$$-\iint_{-\infty}^{+\infty} \frac{\mathrm{BoS}}{T_0(1)}(\mathcal{H}_{1,1})^2 d\xi d\eta \quad .$$

11.5.3. Let us return now to the ageostrophic model of Sect. 11.4 and try to clarify the problem of the initial condition in $\tau = 0$ for the boundary condition (11.140). To this end, we must turn our attention to region IV and consider the relation (11.101). The matching between regions I and II implies that:

$$\lim_{\tau \to \infty} \left\{ \tilde{\mathcal{H}}_2 - \mathcal{H}_2|_{\tau=0} - \tilde{\tau}\left(\frac{\partial \mathcal{H}_1}{\partial \tau}\right)_{\tau=0} \right\} = 0 \quad ,$$

which allows us to calculate the left-hand member of (11.101). Therefore:

$$\lim_{\tilde{\tau} \to \infty} \left\{ \left[1 + \frac{T_0(1)}{\mathrm{Bo} K_0(1)} \frac{\partial}{\partial \zeta} \right] \frac{\partial \tilde{\mathcal{H}}_2}{\partial \tilde{\tau}} \right\} = \left(1 + \frac{T_0(1)}{\mathrm{Bo} K_0(1)} \frac{\partial}{\partial \zeta} \right) \frac{\partial \mathcal{H}_1}{\partial \tau} \bigg|_{\tau=0}$$
$$= \left\{ \frac{\partial}{\partial \tau} \left[\left(1 + \frac{T_0(1)}{\mathrm{Bo} K_0(1)} \frac{\partial}{\partial \zeta} \right) \mathcal{H}_1 \right] \right\}_{\tau=0} .$$

However, if we consider the condition (11.80), we can express:

$$\frac{\partial}{\partial \tau} \left\{ \mathcal{H}_1 + \frac{T_0(1)}{\mathrm{Bo} K_0(1)} \frac{\partial \mathcal{H}_1}{\partial \zeta} \right\}_{\zeta=1} = -\frac{T_0(1)}{\mathrm{Bo}} \left\{ \frac{1}{K_0(1)} \frac{1}{\mathrm{S}} \boldsymbol{v}_0 \cdot \boldsymbol{D}\left(\frac{\partial \mathcal{H}_1}{\partial \zeta} \right) \right.$$
$$\left. - \frac{\lambda_0}{\mathrm{S}} \sqrt{\frac{\kappa_0}{2}} \boldsymbol{D}^2 \mathcal{H}_1 \right\}_{\zeta=1} ,$$

which shows that the right-hand side of (11.101) tends towards a limit when $\tilde{\tau} \to \infty$:

$$\tilde{\psi}_1^\infty = \lim_{\tilde{\tau} \to \infty} \tilde{\psi}_1 \equiv -\frac{T_0(1)}{\mathrm{Bo}} \left\{ \frac{1}{\mathrm{S} K_0(1)} \boldsymbol{v}_{0,1} \cdot \boldsymbol{D}\left(\frac{\partial \mathcal{H}_1}{\partial \zeta} \right)_{\zeta=1} \right.$$
$$\left. - \frac{\lambda_0}{\mathrm{S}} \sqrt{\frac{\kappa_0}{2}} \boldsymbol{D}^2 \mathcal{H}_{1,1} \right\}_{\zeta=0} .$$

This last expression must be consistent with the expression (11.102) of $\tilde{\psi}_1$, but it does not appear to be easily verifiable (such a verification was not carried out by J.P. Guiraud). Another question which arises concerns the integrability in $\tilde{\tau}$ of the difference $\tilde{\psi}_1 - \tilde{\psi}_1^\infty$ If this difference is integrable, then the following condition is obtained:

$$\left[\left\{ \left(1 + \frac{T_0(1)}{\mathrm{Bo} K_0(1)} \frac{\partial}{\partial \zeta} \right) \mathcal{H}_2 \right\}_{\zeta=1} \right]_{\tau=0} = \frac{1}{\mathrm{Bo}} \int_0^\infty (\tilde{\psi}_1 - \tilde{\psi}_1^\infty) d\tilde{\tau} \quad , \qquad (11.150)$$

where $\tilde{\psi}_1^\infty$ can be expressed from the initial values $\boldsymbol{V}^0$ and H_1^0 (see Sect. 11.2) taken in $\zeta = 1$:

$$\tilde{\psi}_1^\infty = -\frac{T_0(1)}{\mathrm{SBo}} \left\{ \frac{1}{K_0(1)} \boldsymbol{V}^0 \cdot \boldsymbol{D}\left(\frac{\partial H_1^0}{\partial \zeta} \right) - \lambda_0 \sqrt{\frac{\kappa_0}{2}} \boldsymbol{D}^2 H_1^0 \right\}_{\zeta=1} . \qquad (11.151)$$

The initial condition (11.150) for $\mathcal{H}_2|_{\zeta=1}$ in $\tau = 0$ must be joined to the boundary condition (11.140). Furthermore, it is pointed out that according to (11.102), the calculation of $\tilde{\psi}_1^\infty$ necessitates the calculation of the following integral:

$$\int_0^{\infty}\left\{\int_{\hat{\zeta}=\overline{\zeta}_{s0}}^{+\infty}(\boldsymbol{D}\cdot\overline{\boldsymbol{v}}_0')d\hat{\zeta}-\lim_{\bar{\tau}\to\infty}\int_{\hat{\zeta}=\overline{\zeta}_{s0}}^{+\infty}(\boldsymbol{D}\cdot\overline{\boldsymbol{v}}_0')d\hat{\zeta}\right\}d\bar{\tau}\quad .$$

At the level of the ageostrophic model, a question which remains to be clarified concerns the possible appearance in the region II expansions of fractional powers of Ki, given the fact that the phenomenon of adjustment to geostrophy analyzed in Sect. 11.2 takes place via a decrease in $\bar{\tau}^{-1/2}$ when $\bar{\tau}\to\infty$. This means that terms in $\mathrm{Ki}^{1/2}$ could emerge in the expansions (11.46). In the same way, the solution of problem (11.103) when $\bar{\tau}\to\infty$ will surely bring into play terms in Ki with fractional powers. For the time being, we do not know just what role will be played by these terms when matching is carried out with the Ekman layer (region III).

Furthermore, the boundary condition for $\zeta\to 0$ associated with the quasi-geostrophic and ageostrophic models can be obtained by taking into account the fact that when $\zeta\to 0$, the basic quasi-static equations (11.3–7) must match with the adiabatic primitive equations of Chap. 7 (see Sects. 7.2, 3) written at the ground. For the quasi-geostrophic model associated with the search for $\mathcal{H}_1$, this leads to the imposing of the following boundary condition "at infinity":

$$\lim_{\zeta\to 0}\left\{\frac{\partial\mathcal{H}_1}{\partial\tau}+\frac{T_0(1)}{\mathrm{Bo}K_0(1)}\frac{D_0}{D\tau}\left(\frac{\partial\mathcal{H}_1}{\partial\zeta}\right)\right\}=0\quad . \tag{11.152}$$

11.5.4. We wish to finish the present chapter by making two remarks. The first concerns the obtaining by Monin (1958) and Charney (1962) of the so-called "balance" equation which appears in the quasi-non-divergent model (see Chap. 12) as a "second" approximation within the framework of the quasi-geostrophic expansion (11.15). Judging from the results given in the present chapter, the obtaining of the balance equation can be seen, in fact, to the erroneous. If the expression of $\boldsymbol{v}_1$ is put to good use, [see (11.82)] and if the expression of $\boldsymbol{v}_0$ is taken into account as a function of $\mathcal{H}_1$, the following relation can be obtained:

$$\begin{aligned}\boldsymbol{k}\cdot(\nabla\times\boldsymbol{v}_1)=\lambda_0\boldsymbol{D}^2\mathcal{H}_2+2\frac{\lambda_0^2}{\mathrm{S}}&\left[\left(\frac{\partial^2\mathcal{H}_1}{\partial\xi\partial\eta}\right)^2\right.\\&\left.-\frac{\partial^2\mathcal{H}_1}{\partial\xi^2}\frac{\partial^2\mathcal{H}_1}{\partial\eta^2}\right]-\frac{\beta}{\mathrm{S}}\lambda_0\frac{\partial\mathcal{H}_1}{\partial\eta}\quad .\end{aligned} \tag{11.153}$$

The above equation can be interpreted in two different ways. The first is consistent with the asymptotic theory expounded in this chapter and consists in considering (11.153) as a formula which permits the calculation of the vertical component of the eddy associated with the velocity vector $\boldsymbol{v}_1$, but only after $\mathcal{H}_1$ *and* $\mathcal{H}_2$ (which satisfy the quasi-geostrophic and ageostrophic models respectively) have been calculated.

The second manner (due to Monin and Charney) in which (11.153) can be interpreted consists in "mixing" (or confounding) $\mathcal{H}_1$ and $\mathcal{H}_2$ with a lone $\mathcal{H}$ and then in considering (11.153) as a Monge-Ampère-type equation for the calculation of this $\mathcal{H}$ when the horizontal velocity [the left-hand member of (11.153)] is known. Chapter 8 of Monin's book (1972) may be consulted on this topic. In fact, as was demonstrated by Monin (1961), the balance equation is obtained within the framework of a low Mach number M_∞ asymptotic theory which, in particular, leads to the so-called quasi-non-divergent (quasi-solenoidal) model. This asymptotic theory at $M_\infty \ll 1$ is still very incomplete despite the recent results obtained by Guiraud and Zeytounian (see Sect. 12.3) in the forthcoming chapter). One of the remaining problems is the insertion of the balance equation and the companion evolution equation (Monin's quasi-solenoidal model) into a hierarchy of consistent models. We will see that the results obtained by Guiraud and Zeytounian (see Chap. 12) afford, to a certain extent, a better understanding of the fundamental nature of this theory with $M_\infty \ll 1$.

The second remark that we wish to make as a conclusion to this chapter concerns the system of basic quasi-static equations (11.3–7). As a matter of fact, it is possible to carry out an asymptotic analysis for $\mathrm{Ki} \ll 1$ by using the complete Navier-Stokes equations [for example (7.1–4)]. However, if it is supposed that $\varepsilon = O(\mathrm{Ki}^2)$, which is a perfectly reasonable hypothesis for synoptic-scale phenomena, then the correction to be applied to the asymptotic theory carried out in this chapter concerns the term W_1 given by (11.87). This is dictated by the fact that to order Ki^2, the quasi-static equation (11.5) is "slightly" disturbed by a zero order term.

11.5.5. In conclusion, some bibliographical references along with their brief histories are given. Kibel (1940) appears to have been the first to introduce the notion of a small Ki flow [quasi-geostrophic expansion (11.15)] and to obtain an equation similar to the one derived in Sect. 11.1. Later on, this equation from the quasi-geostrophic model was obtained in various forms by Ertel (1941), Charney (1948, 1949) and then Oboukhov (1949). The obtaining of this equation as presented in Sect. 11.1 resembles the procedure described in Monin's book (1972).

An excellent exposé on geostrophy can be found in the article by Phillips (1963), as well as in the book by Pedlosky (1979) and the article by Hide (1971). It is pointed out that Chap. 8 of Pedlosky's book (1979) is devoted to "the ageostrophic motions", but the points discussed are not directly related to our report in Sect. 11.4 concerning the so-called ageostrophic model. Among others, the reader is directed towards the articles by Charney (1971), Kuo (1973), Hoskins (1975), White (1977) and Blumen (1968) which deal with the limiting equations for flows with low Rossby numbers. The article by Blumen (1968) was devoted to the stability of the quasi-geostrophic flow. On this same subject, the following are recommended: the articles by Hoskins, Blumen and Hocking in the book by Roberts and Soward (1978), the short book (in Russian) by Dikij (1976) and Chap. 7 in Pedlosky's book (1979).

On the subject of the existence and unicity of the solution of the equation for the quasi-geostrophic model in a bounded domain, the previously-cited article by Dutton (1974) is given as reference. As concerns the ageostrophic model, the results in Sect. 11.4 appear to be the first which enable the establishing of a rigorous asymptotic model as defined by the MMAE. In any case, research on this ageostrophic model can be mainly accredited to the Soviet school [Bykov (1962), Burtsev (1969) and Yudin (1963) among others] and to an article by Arakawa (1962). Unfortunately, the Soviets' results are erroneous since, once again, the approximations related to $\mathcal{H}_1$ and $\mathcal{H}_2$ have been confused. The result has thus not always been an elliptic equation for the ageostrophic model. In the article by Sadokov (1969), which has been included in the work "Lectures on Numerical Short-Range Weather Prediction", WMO, Leningrad (1969), the reader will find a representative example of the Soviet school of thought.

On the question of short-range weather prediction, the following may be consulted: the books by Kibel (1963) and Martchuk (1974), as well as the reviews by Phillips (1970), Kibel (1970) and Cullen (1983).

The problem of the adjustment to geostrophy considered in Sect. 11.2 has a long history beginning with the articles by Rossby (1938), then Cahn (1945) and Oboukhov (1949) whose works were relative to the so-called "barotropic" case. In the baroclinic case (but again with the quasi-static hypothesis), the problem was analyzed in various articles [Bolin (1953), Veronis (1956), Fjelstad (1958), Monin (1958)], but it appears that Kibel (1955) can be accredited with the complete solution of this problem of adjustment to geostrophy when the parameter $K_0(\zeta)$ is assumed constant. The discussion in Sect. 11.2 which gives the definitive asymptotic solution to this problem is strongly influenced by Zeytounian (1976; see Sect. 3.2). Finally, an original exposé of this same problem of adjustment to geostrophy can be found in Blumen (1972).

12. Models Derived from the Theory of Low Mach Number Flows

It soon became apparent that the quasi-geostrophic model considered in the previous chapter was too rough for many synoptic situations. It was thus necessary to devise a new approximation which would also lead to a simple limiting model, but a more efficient one.

It thus seemed reasonable to preserve the Mach number M_∞ as infinitely small (atmospheric motions are always low Mach number flows) but, on the other hand, to drop the idea of a small Kibel number. The latter would be assumed of the order unity.

The starting point once again consists of the primitive equations which we write in the following form in accordance with the results of Subsect. 2.4.3:

$$\mathrm{S}\frac{\partial \boldsymbol{v}}{\partial \tau} + (\boldsymbol{v}\cdot \boldsymbol{D})\boldsymbol{v} + w\frac{\partial \boldsymbol{v}}{\partial \zeta} + \left(\frac{1}{\mathrm{Ro}} + \beta\eta\right)\boldsymbol{k}\times\boldsymbol{v} + \frac{\mathrm{Bo}}{\gamma \mathrm{M}_\infty^2}\boldsymbol{D}\mathcal{H} = 0 \quad ; \tag{12.1a}$$

$$\left(\mathrm{S}\frac{\partial}{\partial \tau} + \boldsymbol{v}\cdot\boldsymbol{D}\right)\frac{\partial \mathcal{H}}{\partial \zeta} + \frac{w}{\zeta}\left\{\frac{\partial}{\partial \zeta}\left(\zeta\frac{\partial \mathcal{H}}{\partial \zeta}\right) - \frac{\gamma-1}{\gamma}\frac{\partial \mathcal{H}}{\partial \zeta}\right\} = 0 \quad ; \tag{12.1b}$$

$$\boldsymbol{D}\cdot\boldsymbol{v} + \frac{\partial w}{\partial \zeta} = 0 \quad . \tag{12.1c}$$

System (12.1) forms a closed system for:

$$\boldsymbol{v} = U\boldsymbol{i} + V\boldsymbol{j} \quad , \quad w = \mathrm{S}W \quad \text{and} \quad \mathcal{H} \quad ,$$

which are functions of τ, ξ, η and ζ.

For system (12.1), the following slip condition is prescribed:

$$\left(\mathrm{S}\frac{\partial}{\partial \tau} + \boldsymbol{v}\cdot\boldsymbol{D} + w\frac{\partial}{\partial \zeta}\right)\mathcal{H} = 0 \quad , \quad \text{on} \quad \mathcal{H} = 0 \quad . \tag{12.2}$$

If necessary, we can add the following non-adiabatic term to the right-hand side of (12.1a):

$$\frac{\mathrm{Bo}^2}{\mathrm{Re}_\perp}\frac{\partial}{\partial \zeta}\left(\varrho\mu_0\frac{\partial \boldsymbol{v}}{\partial \zeta}\right)$$

and by taking into account that

$$T = -\mathrm{Bo}\zeta\frac{\partial \mathcal{H}}{\partial \zeta} \quad ,$$

we can replace (12.1b) with the following energy equation for the temperature T written with dissipative terms:

$$\left(\mathrm{S}\frac{\partial}{\partial \tau}+\boldsymbol{v}\cdot\boldsymbol{D}\right)T+w\left(\frac{\partial T}{\partial \zeta}-\frac{\gamma-1}{\gamma}\frac{T}{\zeta}\right)$$
$$=\frac{\mathrm{Bo}^2}{\mathrm{Pr}}\frac{1}{\mathrm{Re}_\perp}\frac{\partial}{\partial \zeta}\left(\varrho k_0\frac{\partial T}{\partial \zeta}\right)+\frac{1}{\mathrm{Re}_\perp}Q_0$$
$$+(\gamma-1)\frac{\mathrm{Bo}^2 M_\infty^2}{\mathrm{Re}_\perp}\varrho\mu_0\left|\frac{\partial \boldsymbol{v}}{\partial \zeta}\right|^2 \quad . \tag{12.3}$$

For the moment, let us return to (12.1) and introduce the horizontal divergence:

$$\mathcal{D}\equiv\boldsymbol{D}\cdot\boldsymbol{v}=\frac{\partial U}{\partial \xi}+\frac{\partial V}{\partial \eta} \quad , \tag{12.4}$$

and the vertical component of the vorticity:

$$\Omega=\boldsymbol{k}\cdot(\boldsymbol{D}\times\boldsymbol{v})=\frac{\partial V}{\partial \xi}-\frac{\partial U}{\partial \eta} \quad . \tag{12.5}$$

It now becomes possible after a rather long but simple calculation to replace the vectorial equation (12.1a) with two scalar equations for $\mathcal{D}$ and Ω:

$$\left(\mathrm{S}\frac{\partial}{\partial \tau}+\boldsymbol{v}\cdot\boldsymbol{D}+w\frac{\partial}{\partial \zeta}\right)\mathcal{D}+\mathcal{D}^2+2\left(\frac{\partial V}{\partial \xi}\frac{\partial U}{\partial \eta}-\frac{\partial V}{\partial \eta}\frac{\partial U}{\partial \xi}\right)$$
$$+\frac{\partial U}{\partial \zeta}\frac{\partial w}{\partial \xi}+\frac{\partial V}{\partial \zeta}\frac{\partial w}{\partial \eta}-\left(\frac{1}{\mathrm{Ro}}+\beta\eta\right)\Omega+\beta U+\frac{\mathrm{Bo}}{\gamma \mathrm{M}_\infty^2}\boldsymbol{D}^2\mathcal{H}=0 \quad , \tag{12.6a}$$

$$\left(\mathrm{S}\frac{\partial}{\partial \tau}+\boldsymbol{v}\cdot\boldsymbol{D}+w\frac{\partial}{\partial \zeta}\right)\Omega+\mathcal{D}\Omega+\frac{\partial V}{\partial \zeta}\frac{\partial w}{\partial \xi}-\frac{\partial U}{\partial \zeta}\frac{\partial w}{\partial \eta}$$
$$+\left(\frac{1}{\mathrm{Ro}}+\beta\eta\right)\mathcal{D}+\beta V=0 \quad . \tag{12.6b}$$

It will be remarked that the slip condition (12.2) can be written for:

$$\zeta=\zeta_s(\tau,\,\xi,\,\eta) \quad \text{provided that} \quad \mathcal{H}(\tau,\,\xi,\,\eta,\,\zeta_s)\equiv 0 \quad .$$

According to the definition of W [see (2.88)], it is seen that $\zeta_s(\tau,\,\xi,\,\eta)$ must satisfy the equation:

$$\mathrm{S}\frac{\partial \zeta_s}{\partial \tau}+\boldsymbol{v}\cdot\boldsymbol{D}\zeta_s=w(\tau,\,\xi,\,\eta,\,\zeta_s) \quad . \tag{12.7}$$

Throughout the forthcoming, the main small parameter is the Mach number $\mathrm{M}_\infty \ll 1$. To be sure, if the non-adiabatic dissipative viscous terms are taken into account, then it can also be supposed that $\mathrm{Re}_\perp \gg 1$. In this case, various situations are possible:

a) $\mathrm{Re}_\perp\equiv\infty$, and then $\mathrm{M}_\infty\to 0$;

b) $\mathrm{Re}_\perp$ fixed , $\mathrm{M}_\infty\to 0$, and then $\mathrm{Re}_\perp\to\infty$;

c) M_∞ fixed , $Re_\perp \to \infty$, and then $Re_\perp \to \infty$;

d) $M_\infty \to 0$ and $Re_\perp \to \infty$ such that the similarity relation:

$$\frac{1/\sqrt{Re_\perp}}{M_\infty^\alpha} = \hat{S}_0 = O(1) \quad , \quad \alpha > 0 \quad , \tag{12.8}$$

is satisfied with α a real number to be determined.

In Sect. 12.1, we will first consider the classical *Monin-Charney* case which is based on the limiting process:

$$Re_\perp \equiv \infty \quad , \quad M_\infty \to 0 \quad \text{with} \quad \tau, \xi, \eta \text{ and } \zeta \text{ fixed} \quad . \tag{12.9}$$

Section 12.2 is devoted to various extensions of the classical case considered in Sect. 11.1. In particular, the similarity relation (12.8) will be clarified. Finally, certain recent results obtained by Guiraud and Zeytounian[1] will be reviewed in Sect. 12.3. These results were developed in the Notes written by J.P. Guiraud in October, 1983 for the ICMS Course in Udine (Italy).

Nevertheless, despite these recent results obtained via the systematic application of the MMAE, the development of an asymptotic model of atmospheric flows considered as low Mach number flows is still far from complete. A great deal of research is yet necessary and it is hoped that the modest contribution of the present chapter might enhance future progress in this domain.

In Sect. 12.4, the problem of adjustment to the classical quasi-nondivergent model is discussed and Sect. 12.5 contains various complementary remarks.

The ground will be considered flat in the following, but in Chap. 13, we will have the opportunity to return to low Mach number flows during the review of the models for lee waves which appear downstream of a mountain in a stratified baroclinic atmosphere. This review will be given *without* the quasi-static hypothesis [$\varepsilon = O(1)$] but the Coriolis force ($Ro \equiv \infty$) and the β effect will be ignored.

12.1 The So-called Classical "Quasi-nondivergent" Model and Its Limitations

In conjunction with the limiting process (12.9), let us consider the following main asymptotic expansion:

$$\begin{pmatrix} \boldsymbol{v} \\ w \\ \mathcal{H} \\ T \\ \zeta_s \end{pmatrix} = \begin{pmatrix} \boldsymbol{v}_0 \\ w_0 \\ \mathcal{H}_0 \\ T_0 \\ \zeta_{s0} \end{pmatrix} + M_\infty^2 \begin{pmatrix} \boldsymbol{v}_2 \\ w_2 \\ \mathcal{H}_2 \\ T_2 \\ \zeta_{s2} \end{pmatrix} + M_\infty^4 \begin{pmatrix} \boldsymbol{v}_4 \\ w_4 \\ \mathcal{H}_4 \\ T_4 \\ \zeta_{s4} \end{pmatrix} + \dots \quad . \tag{12.10}$$

[1] A part of these results were given in the paper prepared for the "Short Course" associated with the BAIL III Conference in Dublin (June, 1984). Section 4.2 of our "Recent Advances" to be published in the "International Journal of Engineering Sciences" also reviews these same results [see Zeytounian (1985)].

To order zero, (12.1a) leads to:

$$\frac{\partial \mathcal{H}_0}{\partial \xi} = 0 \quad , \quad \frac{\partial \mathcal{H}_0}{\partial \eta} = 0$$

and from (12.1b), it is noticed that:

$$\mathrm{S}\frac{\partial^2 \mathcal{H}_0}{\partial \tau \partial \zeta} + w_0\left(\frac{\partial^2 \mathcal{H}_0}{\partial \zeta^2} + \frac{1}{\gamma\zeta}\frac{\partial \mathcal{H}_0}{\partial \zeta}\right) = 0 \quad , \tag{12.11}$$

which means it can be assumed that $\mathcal{H}_0 \equiv \mathcal{H}_0(\tau, \zeta)$. In this case, we also have $w_0 \equiv w_0(\tau, \zeta)$, and from (12.1c) we find that $\mathcal{D}_0 = \boldsymbol{D} \cdot \boldsymbol{v}_0 \equiv \mathcal{D}_0(\tau, \zeta)$. Finally, according to (12.7), $\zeta_{s0} \equiv \zeta_{s0}(\tau)$ satisfies the following:

$$\mathrm{S}\frac{\partial \zeta_{s0}}{\partial \tau} = w_0(\tau, \zeta_{s0}). \tag{12.12}$$

It is obvious that the order zero is consistent only with a flat ground. In addition, the components U_0 and V_0 of $\boldsymbol{v}_0$ can be represented in the following form:

$$U_0 = \frac{1}{2}\mathcal{D}_0\xi + \frac{\partial \varphi_0^{\mathrm{h}}}{\partial \xi} - \frac{\partial \psi_0}{\partial \eta} \quad , \quad V_0 = \frac{1}{2}\mathcal{D}_0\eta + \frac{\partial \varphi_0^{\mathrm{h}}}{\partial \eta} + \frac{\partial \psi_0}{\partial \xi} \quad , \tag{12.13}$$

with

$$\psi_0(\tau, \xi, \eta, \zeta) = \frac{1}{2\pi}\iint_{-\infty}^{+\infty} \Omega_0(\tau, \xi, \eta, \zeta)\log\left\{(\xi - \xi')^2 + (\eta - \eta')^2\right\}^{1/2} d\xi' d\eta' \quad , \tag{12.13a}$$

where φ_0^{h} is an arbitrary function which is harmonic with respect to ξ and η.

The problem which now arises is to make sure that to order zero, our functions are independent of the time τ, i.e., that they characterize the standard state which is a function only of the altitude ζ. To this end, (12.3) must be used instead of (12.1b) and the expansion (12.10) must be extended to the term proportional to M_∞^α while taking into account the similarity relation (12.8) between $\mathrm{Re}_\perp$ and M_∞. Hence, if the following term is added to (12.10):

$$\ldots + \mathrm{M}_\infty^\alpha \begin{pmatrix} \tilde{\boldsymbol{v}} \\ \tilde{w} \\ \tilde{\mathcal{H}} \\ \tilde{T} \\ \tilde{\zeta}_s \end{pmatrix} + \mathrm{O}\left(\mathrm{M}_\infty^{\alpha+2}\right)$$

and if it is assumed that α is not an even whole number, then from (12.3), it is found that:

$$\mathrm{S}\frac{\partial \tilde{T}}{\partial \tau} + \boldsymbol{v}_0 \cdot \boldsymbol{D}\tilde{T} + w_0\left(\frac{\partial \tilde{T}}{\partial \zeta} - \frac{\gamma - 1}{\gamma}\frac{\tilde{T}}{\zeta}\right) + \tilde{w}\left(\frac{\partial T_0}{\partial \zeta} - \frac{\gamma - 1}{\gamma}\frac{T_0}{\zeta}\right)$$
$$= \frac{\mathrm{Bo}^2}{\mathrm{Pr}}\hat{S}_0^2\left[\frac{\partial}{\partial \zeta}\left(\varrho_0 k_0 \frac{\partial T_0}{\partial \zeta}\right) + \frac{\mathrm{Pr}}{\mathrm{Bo}^2}Q_0\right] \quad . \tag{12.14}$$

However, it is easy to see that the left-hand side of (12.3) can be written in the form $S\zeta^{(\gamma-1)/\gamma}(D/D\tau)[T\zeta^{-(\gamma-1)/\gamma}]$ and therefore, (12.14) can be integrated with respect to τ along the trajectories associated with $\boldsymbol{v}_0$. In this case, so that secularities do not appear during a sufficiently long period of time (since the temperature $\tilde{T}$ must remain bounded), it is necessary that the right-hand member be zero:

$$\frac{\mathrm{Bo}^2}{\mathrm{Pr}}\frac{d}{d\zeta}\left[\varrho_0(\zeta)k_0(\zeta)\frac{dT_0}{d\zeta}\right]+Q_0(\zeta)=0 \quad . \tag{12.15}$$

We then recover the standard atmosphere:

$$\mathcal{H}_0=\mathcal{H}_0(\zeta) \quad , \quad T_0=T_0(\zeta) \quad , \quad w_0=\frac{d\zeta_{s0}}{d\tau}=0 \quad , \tag{12.16}$$

associated with the thermal balance equation (12.15). In this case, the following system is derived from (12.1a, c) for $\boldsymbol{v}_0$ and $\Pi_2 \equiv (\mathrm{Bo}/\gamma)\mathcal{H}_2$:

$$\begin{aligned}&\mathrm{S}\frac{\partial \boldsymbol{v}_0}{\partial\tau}+(\boldsymbol{v}_0\cdot\boldsymbol{D})\boldsymbol{v}_0+\left(\frac{1}{\mathrm{Ro}}+\beta\eta\right)\boldsymbol{k}\times\boldsymbol{v}_0+\boldsymbol{D}\Pi_2=0 \quad ,\\ &\boldsymbol{D}\cdot\boldsymbol{v}_0=0 \quad .\end{aligned} \tag{12.17}$$

System (12.17) describes the flow of an incompressible fluid along isobaric surfaces ζ = const. This plane flow was projected onto the β plane and is strongly uncoupled with respect to the altitude. The only means of obtaining a coupling with respect to ζ is to impose on this flow initial conditions and lateral conditions in ξ,η!

Before carrying on, it should be noted that since $\mathcal{D}_0 \equiv 0$ and $w_0 \equiv 0$, from (12.6a, b), we find to order zero the following system for Ω_0 and $\mathcal{H}_2$ which is equivalent to (12.17):

$$\begin{aligned}&\boldsymbol{D}^2\mathcal{H}_2=\frac{\gamma}{\mathrm{Bo}}\left\{\left(\frac{1}{\mathrm{Ro}}+\beta\eta\right)\boldsymbol{D}^2\psi_0+\beta\frac{\partial\psi_0}{\partial\eta}+2\mathcal{J}\left(\frac{\partial\psi_0}{\partial\xi},\frac{\partial\psi_0}{\partial\eta}\right)\right\} \quad ;\\ &\left\{\mathrm{S}\frac{\partial}{\partial\tau}+\mathcal{J}(\psi_0,\cdot)\right\}\boldsymbol{D}^2\psi_0+\beta\frac{\partial\psi_0}{\partial\xi}=0 \quad ,\end{aligned} \tag{12.18}$$

where:

$$\mathcal{J}\equiv\frac{\partial\psi_0}{\partial\xi}\frac{\partial}{\partial\eta}-\frac{\partial\psi_0}{\partial\eta}\frac{\partial}{\partial\xi}=\mathcal{J}(\psi_0,\cdot) \quad .$$

System (12.18) forms the so-called classical "quasi-nondivergent" model (or quasi-solenoidal model). The first equation in (12.18) is the so-called "balance" equation whereas the second one is an evolution equation for ψ_0.

Certain remarks can now be made. Firstly, system (12.18) is of first order in τ and necessitates an initial condition $\psi_0|_{\tau=0}=\psi_0^0$. How can ψ_0^0 be found? In Sect. 12.4, we will see that the solution resides in posing the problem of the unsteady adjustment by introducing a short time $\tilde{\tau}=\tau/\mathrm{M}_\infty$ in order to take into account the internal gravity waves (which exist at the level of the primitive equations) which were filtered out during the limiting process (12.9). Secondly, in the quasi-nondivergent model (12.18), why is there no derivation with respect to the altitude ζ. As has already been pointed out, this quasi-nondivergent model describes the motion of an incompressible (barotropic) atmosphere stratified in horizontal layers in the planes $\zeta = \text{const}$, the latter being totally independent of each other. Hence, to order zero, the cancelling out of w_0 leads to the undesirable consequence of forcing the motion into the horizontal planes $\zeta = \text{const}$. The problem might be expected to be remedied if the expansion is carried out to the next following order. If $\alpha > 1$ in (12.8), then from (12.3) it follows that:

$$w_2 = \frac{1}{\text{Bo}} \frac{\zeta}{K_0(\zeta)} \left(\text{S} \frac{\partial T_2}{\partial \tau} + \boldsymbol{v}_0 \cdot \boldsymbol{D} T_2 \right) \quad , \tag{12.19}$$

where $K_0(\zeta)$ is given by (11.29), and T_2 is directly related to Π_2 satisfying (12.17):

$$T_2 = -\gamma \zeta \frac{\partial \Pi_2}{\partial \zeta} \quad . \tag{12.20}$$

Next, we have:

$$\boldsymbol{D} \cdot \boldsymbol{v}_2 \equiv \mathcal{D}_2 = -\frac{\partial w_2}{\partial \zeta} \quad , \tag{12.21}$$

and then:

$$\text{S} \frac{\partial \boldsymbol{v}_2}{\partial \tau} + (\boldsymbol{v}_0 \cdot \boldsymbol{D}) \boldsymbol{v}_2 + (\boldsymbol{v}_2 \cdot \boldsymbol{D}) \boldsymbol{v}_0 + w_2 \frac{\partial \boldsymbol{v}_0}{\partial \zeta} + \boldsymbol{D} \Pi_4 = 0 \quad . \tag{12.22}$$

It will be remarked that a coupling with the altitude exists thanks to the terms $\partial w_2 / \partial \zeta$ and $w_2 (\partial \boldsymbol{v}_0 / \partial \zeta)$. However, this is not enough and this main representation remains highly degenerated!

Let us now take a look at what happens to the slip condition on the ground. First of all, we have the following:

$$\mathcal{H}_0(\zeta_{s0}) = 0 \quad \Rightarrow \quad \zeta_{s0} = 1 \quad . \tag{12.23}$$

We denote $f_s \equiv f(\tau, \xi, \eta, \zeta_{s0})$ and we then have:

$$\left(\frac{\partial \mathcal{H}_0}{\partial \zeta} \right)_s \zeta_{s2} + \mathcal{H}_{2s} = 0 \; \Rightarrow \; \zeta_{s2} = -\left(\frac{\mathcal{H}_2}{d\mathcal{H}_0 / d\zeta} \right)_s \quad . \tag{12.24}$$

The above relation (12.24) is indeed compatible with the one which results from (12.7):

$$\text{S} \frac{\partial \zeta_{s2}}{\partial \tau} + \boldsymbol{v}_0 \cdot \boldsymbol{D} \zeta_{s2} = w_{2s} \quad , \tag{12.25}$$

given the fact that:

$$\left(\text{S} \frac{\partial \mathcal{H}}{\partial \tau} + \boldsymbol{v} \cdot \boldsymbol{D} \mathcal{H} + w \frac{\partial \mathcal{H}}{\partial \zeta} \right)_{\zeta = \zeta_s} = 0 \quad .$$

12.1.1 Analysis of Singularities Related to the Monin-Charney Limiting Process

Let us backtrack to (12.1b, c) and (12.6a, b) and seek the solution to these equations in the form:

$$\mathcal{H} = \mathcal{H}_0(\zeta) + M_\infty^2 \chi \quad ; \quad w = M_\infty^2 \omega \quad ; \quad \mathcal{D} = M_\infty^2 \delta \quad , \quad \text{and} \tag{12.26}$$

$$U = -\frac{\partial \psi}{\partial \eta} + M_\infty^2 \frac{\partial \varphi}{\partial \xi} \quad , \quad V = +\frac{\partial \psi}{\partial \xi} + M_\infty^2 \frac{\partial \varphi}{\partial \eta} \quad . \tag{12.27}$$

In this case:

$$\Omega = \boldsymbol{D}^2 \psi \quad \text{and} \quad \delta = \boldsymbol{D}^2 \varphi \quad , \tag{12.28}$$

and in a quasi-static adiabatic atmosphere, we have the following system of equations for δ, Ω, χ and ω:

$$\begin{aligned} &\mathrm{S} M_\infty^2 \frac{\partial \delta}{\partial \tau} - \left(\frac{1}{\mathrm{Ro}} + \beta\eta\right)\Omega + \frac{\mathrm{Bo}}{\gamma}\boldsymbol{D}^2\chi = -\mathcal{A}_\delta \quad ; \\ &\mathrm{S}\frac{\partial \Omega}{\partial \tau} + M_\infty^2\left(\frac{1}{\mathrm{Ro}} + \beta\eta\right)\delta = -\mathcal{A}_\Omega \quad ; \\ &\delta + \frac{\partial \omega}{\partial \zeta} = 0 \quad ; \\ &\mathrm{S}\frac{\partial}{\partial \tau}\left(\frac{\partial \chi}{\partial \zeta}\right) + \frac{K_0(\zeta)}{\zeta^2}\omega = -\mathcal{A}_{\mathcal{H}} \quad , \end{aligned} \tag{12.28}$$

where the right-hand members are given by:

$$\begin{aligned} \mathcal{A}_\delta &= 2\left(\frac{\partial V}{\partial \xi}\frac{\partial U}{\partial \eta} - \frac{\partial V}{\partial \eta}\frac{\partial U}{\partial \xi}\right) + \beta U + M_\infty^2\left\{\boldsymbol{v}\cdot\boldsymbol{D}\delta + \frac{\partial U}{\partial \zeta}\frac{\partial \omega}{\partial \xi} + \frac{\partial V}{\partial \zeta}\frac{\partial \omega}{\partial \eta}\right\} \\ &\quad + M_\infty^4\left(\omega\frac{\partial \delta}{\partial \zeta} + \delta^2\right); \end{aligned} \tag{12.29a}$$

$$\mathcal{A}_\Omega = \boldsymbol{v}\cdot\boldsymbol{D}\Omega + \beta V + M_\infty^2\left\{\frac{\partial V}{\partial \zeta}\frac{\partial \omega}{\partial \xi} - \frac{\partial U}{\partial \zeta}\frac{\partial \omega}{\partial \eta} + \delta\Omega + \omega\frac{\partial \Omega}{\partial \zeta}\right\} \quad ; \tag{12.29b}$$

$$\mathcal{A}_{\mathcal{H}} = \boldsymbol{v}\cdot\boldsymbol{D}\left(\frac{\partial \chi}{\partial \zeta}\right) + M_\infty^2\frac{\omega}{\zeta}\left\{\frac{1}{\gamma}\frac{\partial \chi}{\partial \zeta} + \zeta\frac{\partial^2 \chi}{\partial \zeta^2}\right\} \quad . \tag{12.29c}$$

The function ω can easily be eliminated from (12.28) and then for Ω, χ and δ, the following system of three evolution equations results:

$$\begin{aligned} &\mathrm{S}\frac{\partial \Omega}{\partial \tau} + M_\infty^2\left(\frac{1}{\mathrm{Ro}} + \beta\eta\right)\delta + \mathcal{A}_\Omega = 0 \quad ; \\ &M_\infty^2 \mathrm{S}\frac{\partial \delta}{\partial \tau} - \left(\frac{1}{\mathrm{Ro}} + \beta\eta\right)\Omega + \frac{\mathrm{Bo}}{\gamma}\boldsymbol{D}^2\chi + \mathcal{A}_\delta = 0 \quad ; \\ &\mathrm{S}\frac{\partial}{\partial \tau}\left[\frac{\partial}{\partial \zeta}\left(\frac{\zeta^2}{K_0(\zeta)}\frac{\partial \chi}{\partial \zeta}\right)\right] - \delta + \frac{\partial}{\partial \zeta}\left(\frac{\zeta^2}{K_0(\zeta)}\mathcal{A}_{\mathcal{H}}\right) = 0 \quad . \end{aligned} \tag{12.30}$$

The following three initial conditions must be prescribed for (12.30):

$$\tau = 0: \quad \delta = \delta^0 \quad , \quad \Omega = \Omega^0 \quad \text{and} \quad \chi = \chi^0 \quad , \tag{12.31}$$

as well as the slip condition:

$$\left[1 + \frac{1}{\text{Bo}} \frac{T_0(\zeta)}{K_0(\zeta)} \zeta \frac{\partial}{\partial \zeta}\right] \text{S} \frac{\partial \chi}{\partial \tau} = -\mathcal{A}_\chi \quad , \quad \text{on} \quad \mathcal{H}_0 + \text{M}_\infty^2 \chi = 0 \quad , \tag{12.32}$$

with

$$\mathcal{A}_\chi = \boldsymbol{v} \cdot \boldsymbol{D}\chi + \frac{1}{\text{Bo}} \frac{T_0(\zeta)}{K_0(\zeta)} \zeta \mathcal{A}_{\mathcal{H}} + \text{M}_\infty^2 \omega \frac{\partial \chi}{\partial \zeta} \quad .$$

From (12.30), a single equation is formed whose left-hand member contains only one of the three unknown functions. In particular, we can write for Ω:

$$\text{S}\left\{\boldsymbol{D}^2 + \frac{\gamma \text{M}_\infty^2}{\text{Bo}} \left[\text{S}^2 \frac{\partial^2}{\partial \tau^2} + \left(\frac{1}{\text{Ro}} + \beta\eta\right)^2\right] \frac{\partial}{\partial \zeta}\left(\frac{\zeta^2}{K_0(\zeta)} \frac{\partial}{\partial \zeta}\right)\right\} \frac{\partial \Omega}{\partial \tau}$$
$$= -\boldsymbol{D}^2 \mathcal{A}_\Omega + \frac{\gamma \text{M}_\infty^2}{\text{Bo}} \text{S} \left\{\frac{\partial}{\partial \zeta}\left(\frac{\zeta^2}{K_0(\zeta)} \frac{\partial}{\partial \zeta}\right) \frac{\partial}{\partial \tau}\left[-\frac{\partial \mathcal{A}_\Omega}{\partial \tau} + \left(\frac{1}{\text{Ro}} + \beta\eta\right) \mathcal{A}_\delta\right]\right.$$
$$\left. - \frac{\text{Bo}}{\gamma}\left(\frac{1}{\text{Ro}} + \beta\eta\right) \boldsymbol{D}^2 \frac{\partial}{\partial \zeta}\left(\frac{\zeta^2}{K_0(\zeta)} \mathcal{A}_{\mathcal{H}}\right)\right\} \quad . \tag{12.33}$$

The following three initial conditions are necessary for (12.33):

$$\Omega = \Omega^0 \quad ;$$

$$\text{S} \frac{\partial \Omega}{\partial \tau} = -\left(\mathcal{A}_\Omega^0 + \text{M}_\infty^2 \left(\frac{1}{\text{Ro}} + \beta\eta\right) \delta^0\right) \quad ;$$

$$\text{S}^2 \frac{\partial^2 \Omega}{\partial \tau^2} = -\left[\left(\frac{1}{\text{Ro}} + \beta\eta\right)^2 \Omega^0 - \frac{\text{Bo}}{\gamma}\left(\frac{1}{\text{Ro}} + \beta\eta\right) \boldsymbol{D}^2 \chi^0\right.$$
$$\left. - \left(\frac{1}{\text{Ro}} + \beta\eta\right) \mathcal{A}_\delta^0 + \text{S}\left(\frac{\partial \mathcal{A}_\Omega}{\partial \tau}\right)^0\right] \quad , \quad \text{in} \quad \tau = 0 \quad . \tag{12.34}$$

A slip condition deduced from (12.32) and written for Ω may also be assigned to (12.33). However, we will not prescribe one here since we will be considering a linear theory which corresponds to the "small" motions of the adiabatic atmosphere with respect to the standard atmosphere (with zero velocity). If we suppose that the β effect is negligible, then the terms related to $\mathcal{A}_\Omega$, $\mathcal{A}_\delta$, $\mathcal{A}_{\mathcal{H}}$ and $\mathcal{A}_\chi$ can be cancelled and the following linear problem can be written for the function ψ (since $\boldsymbol{D}^2\psi = \Omega$):

$$\text{S}\boldsymbol{D}^2 \frac{\partial \psi}{\partial \tau} + \frac{\gamma \text{M}_\infty^2}{\text{Bo}} \text{S}\left[\text{S}^2 \frac{\partial^2}{\partial \tau^2} + \frac{1}{\text{Ro}^2}\right] \frac{\partial}{\partial \zeta}\left(\frac{\zeta^2}{K_0(\zeta)} \frac{\partial^2 \psi}{\partial \tau \partial \zeta}\right) = 0 \quad , \tag{12.35a}$$

$$\tau = 0: \begin{cases} \psi = \psi^0 \quad ; \\ \text{S} \dfrac{\partial \psi}{\partial \tau} = -\dfrac{\text{M}_\infty^2}{\text{Ro}} \varphi^0 ; \\ \text{S}^2 \dfrac{\partial^2 \psi}{\partial \tau^2} = \dfrac{\text{Bo}}{\gamma \text{Ro}} \chi^0 - \dfrac{1}{\text{Ro}^2} \psi^0 \quad , \end{cases} \tag{12.35b}$$

$$\left[1+\frac{1}{\mathrm{Bo}}\frac{T_0(1)}{K_0(1)}\frac{\partial}{\partial\zeta}\right]\left(\mathrm{S}^2\frac{\partial^2}{\partial\tau^2}+\frac{1}{\mathrm{Ro}^2}\right)\mathrm{S}\frac{\partial\psi}{\partial\tau}=0\,,\quad \text{on } \zeta=1\,. \tag{12.35c}$$

A behavior condition on ψ when $\zeta\to 0$ and also when $|\xi^2+\eta^2|\to\infty$ needs to be formulated. These conditions must be chosen in such a way that the boundary and initial values problem (12.35) is *well-posed.*

In any case, we believe that an asymptotic analysis of the linear problem (12.35) when $\mathrm{M}_\infty\to 0$ with Ro, γ, Bo and S fixed, and also $K_0(\zeta)>0$ should make it possible to clearly define the domain of validity of the quasi-nondivergent model which is obtained when $\mathrm{M}_\infty\to 0$ with τ, ξ, η and ζ fixed. In this case, instead of (12.35a), we have:

$$\boldsymbol{D}^2\left(\frac{\partial\psi_0}{\partial\tau}\right)=0\quad . \tag{12.36}$$

This quasi-nondivergent model could possibly be completed by its related local models. Of course, it is first essential to know if such an analysis could be carried out by the MMAE.

While carrying out the main limiting process (12.9), the higher order derivatives in τ and in ζ are lost and thus, the vicinities of $\tau=0$, $\zeta=1$ and $\zeta=0$ are certainly singular.

As a matter of fact, it is very instructive to note that three transformations are possible on the variables τ, ζ and ξ, η respectively, such that when $\mathrm{M}_\infty\to 0$ (the new variables remaining fixed), we obtain three local limiting equations which are far less degenerate than the main limiting equation (12.36) written for $\psi_0\equiv\overset{P}{\lim}\,\psi$ with $\overset{P}{\lim}=\{\mathrm{M}_\infty\to 0$ with τ, ξ, η, ζ fixed$\}$.

First of all, we introduce the short-time $\tilde{\tau}=\tau/\mathrm{M}_\infty$ in the vicinity of $\tau=0$. In this case, when:

$$\overset{l^\tau}{\lim}\,\psi\equiv\tilde{\psi}_0\quad\text{with}$$

$$\overset{l^\tau}{\lim}=\left\{\mathrm{M}_\infty\to 0\quad,\quad\text{with } \tilde{\tau}\equiv\frac{\tau}{\mathrm{M}_\infty},\ \xi,\ \eta\ \text{and}\ \zeta\ \text{ fixed}\right\}\quad,$$

the following equation is obtained for $\tilde{\psi}_0(\tilde{\tau},\xi,\eta,\zeta)$:

$$\mathrm{S}\boldsymbol{D}^2\left(\frac{\partial\tilde{\psi}_0}{\partial\tilde{\tau}}\right)+\frac{\gamma}{\mathrm{Bo}}\mathrm{S}^3\frac{\partial^3}{\partial\tilde{\tau}^3}\left\{\frac{\partial}{\partial\zeta}\left(\frac{\zeta^2}{K_0(\zeta)}\frac{\partial\tilde{\psi}_0}{\partial\zeta}\right)\right\}=0\quad . \tag{12.37}$$

We next introduce the following in the vicinity of the ground $\zeta=1$:

$$\hat{\zeta}=\frac{1-\zeta}{\mathrm{M}_\infty}\quad,$$

$$\overset{l^\zeta}{\lim}\,\psi=\hat{\psi}_0\quad\text{with}\quad\overset{l^\zeta}{\lim}=\left\{\mathrm{M}_\infty\to 0\ \text{with}\ \tau,\ \xi,\ \eta\ \text{and}\ \hat{\zeta}\ \text{fixed}\right\}\quad,$$

and the following equation results for $\hat{\psi}(\tau,\xi,\eta,\hat{\zeta})$

$$\mathrm{S}\boldsymbol{D}^2\left(\frac{\partial\hat{\psi}_0}{\partial\tau}\right)+\frac{\gamma\mathrm{S}}{\mathrm{Bo}}\frac{1}{K_0(1)}\left(\mathrm{S}^2\frac{\partial^2}{\partial\tau^2}+\frac{1}{\mathrm{Ro}^2}\right)\frac{\partial^2}{\partial\hat{\zeta}^2}\left(\frac{\partial\hat{\psi}_0}{\partial\tau}\right)=0\quad . \tag{12.38}$$

Finally, in the horizontal planes ζ = const in the vicinity of infinity, we introduce:

$$\overline{\xi} = \mathrm{M}_\infty \xi \quad , \quad \overline{\eta} = \mathrm{M}_\infty \eta \quad , \quad \overset{l_\infty}{\lim}\, \psi = \overline{\psi}_0 \quad ,$$

$$\overset{l_\infty}{\lim} = \left\{ \mathrm{M}_\infty \to 0 \quad , \text{ with } \tau, \overline{\xi}, \overline{\eta} \text{ and } \zeta \text{ fixed} \right\} \quad ,$$

then the following limiting equation is obtained for $\overline{\psi}_0(\tau, \overline{\xi}, \overline{\eta}, \zeta)$:

$$\mathrm{S}\left(\frac{\partial^2}{\partial \overline{\xi}^2} + \frac{\partial^2}{\partial \overline{\eta}^2} \right) \frac{\partial \overline{\psi}_0}{\partial \tau} + \frac{\gamma S}{\mathrm{Bo}} \left(\mathrm{S}^2 \frac{\partial^2}{\partial \tau^2} + \frac{1}{\mathrm{Ro}^2} \right) \frac{\partial}{\partial \zeta} \left(\frac{\zeta^2}{K_0(\zeta)} \frac{\partial^2 \overline{\psi}_0}{\partial \zeta \partial \tau} \right) = 0 \quad . \tag{12.39}$$

It is thus remarked that when $\mathrm{M}_\infty \to 0$, the main representation (which leads to the classical quasi-nondivergent model) must be considered conjointly with three local representations in the vicinities of $\tau = 0$, $\zeta = 1$ and $\xi^2 + \eta^2 = \infty$, respectively. Therefore, the analysis of the behavior of the solutions of (12.37–39) [with adequate initial and boundary conditions resulting from those of (12.35)] when $\tilde{\tau} \to \infty$, $\hat{\zeta} \to \infty$ and $\overline{\xi}^2 + \overline{\eta}^2 \to 0$, will solve the problem of which asymptotic method (MMAE or MSM) should be used. Let us recall that the use of the MMAE is legitimate when the behavior of the local solution towards the main solution does not involve any cumulative effects which could affect this main representation (in our case, the classical quasi-nondivergent model which emerges from this main representation).

In Sect. 12.4, the above question is resolved for the vicinity of $\tau = 0$. An analysis is given in Sect. 12.3 of the vicinity of $\xi^2 + \eta^2 = \infty$ in the planes ζ = const. The problems related to the vicinity of $\zeta = 1$ and to the behavior in the vicinity of $\zeta = 0$ remain as yet unsettled.[2]

Concerning this latter problem which would make it possible to find a condition when $\zeta \to 0$, it can once again be said to be related to the eigenvalue problem considered in Sect. 7.6. So let us look for the solution to (12.35a) in the following form:

$$\psi(\tau, \xi, \eta, \zeta) = \sum_{n=1}^{\infty} \chi_n(\zeta) F_n(\tau, \xi, \eta) \quad , \tag{12.40}$$

where the $\chi_n(\zeta)$ issue from a spectral problem associated with the operator $(\partial/\partial\zeta)\left[(\zeta^2/K_0(\zeta))\partial/\partial\zeta\right]$, each eigenfunctions $\chi_n(\zeta)$, $n = 1, 2, \ldots$ having its corresponding eigenvalue μ_n:

$$\frac{d}{d\zeta}\left(\frac{\zeta^2}{K_0(\zeta)} \frac{d\chi_n}{d\zeta} \right) + \mu_n \chi_n = 0 \quad ; \tag{12.41}$$

$$\zeta = 1: \quad \chi_n + \frac{1}{\mathrm{Bo}} \frac{T_0(1)}{K_0(1)} \frac{d\chi_n}{d\zeta} = 0 \quad , \tag{12.42}$$

[2] As concerns the vicinity of $\zeta = 1$ in the adiabatic case, see the Subsect. 12.5.3 of the Complementary Remarks (Sect. 12.5).

$$\zeta \to 0: \quad \frac{\zeta^2}{K_0(\zeta)} \chi_n \frac{d\chi_n}{d\zeta} \to 0 \quad . \tag{12.43}$$

The functions $F_n(\tau, \xi, \eta)$ are themselves solutions of the following second order partial differential equation:

$$\left\{ \frac{\partial^2}{\partial \xi^2} + \frac{\partial^2}{\partial \eta^2} - \frac{\gamma \mathrm{M}_\infty^2}{\mathrm{Bo}} \mu_n \left(\mathrm{S}^2 \frac{\partial^2}{\partial \tau^2} + \frac{1}{\mathrm{Ro}^2} \right) \right\} \mathrm{S} \frac{\partial F_n}{\partial \tau} = 0 \quad . \tag{12.44}$$

Initial conditions and behavior conditions in ξ and η must be prescribed for (12.44). It thus appears reasonable to assign to (12.35a) the following boundary condition:

$$\frac{\zeta^2}{K_0(\zeta)} \psi \frac{\partial \psi}{\partial \zeta} \to 0 \quad , \quad \text{with} \quad \zeta \to 0 \quad .$$

12.2 The Generalized Quasi-nondivergent Model and its Limitations

12.2.1. Let us now consider the non-adiabatic case and suppose, first of all, that $\mathrm{Re}_\perp = \mathrm{O}(1)$. In this case, the limiting process:

$$\mathrm{Re}_\perp = \mathrm{O}(1) \quad ; \quad \mathrm{M}_\infty \to 0 \quad \text{with} \quad \tau, \xi, \eta \quad \text{and} \quad \zeta \quad \text{fixed} \quad , \tag{12.45}$$

again leads to:

$$w_0 = 0 \Rightarrow \mathcal{D}_0 = 0 \tag{12.46}$$

due to the thermal equilibrium (12.15). The balance equation [the first equation of (12.18)] remains the same and its is remarked that the derivation of this equation is independent of the hypothesis made concerning $\mathrm{Re}_\perp$ [either $\mathrm{Re}_\perp \equiv \infty$, or, $\mathrm{Re}_\perp = \mathrm{O}(1)$]. On the other hand, instead of the evolution equation for ψ_0 [the second equation of (12.18)], a parabolic-type equation is found (with respect to τ and ζ):

$$\left\{ \mathrm{S} \frac{\partial}{\partial \tau} + \mathcal{J}(\psi_0, \cdot) \right\} \boldsymbol{D}^2 \psi_0 + \beta \frac{\partial \psi_0}{\partial \xi} = \frac{\mathrm{Bo}^2}{\mathrm{Re}_\perp} \frac{\partial}{\partial \zeta} \left(\frac{\mu_0(\zeta)}{T_0(\zeta)} \frac{\partial \boldsymbol{D}^2 \psi_0}{\partial \zeta} \right) \quad . \tag{12.47}$$

An initial condition in $\tau = 0$ and two conditions in ζ for $\zeta = 1$ and $\zeta \to 0$ must be prescribed for (12.47). Obviously, we have:

$$\psi_0 = 0 \quad , \quad \text{on} \quad \zeta = 1 \quad , \tag{12.48}$$

since we have $\boldsymbol{v}_0 = \boldsymbol{k} \times \boldsymbol{D}\psi_0$ and $\Omega_0 \equiv \boldsymbol{D}^2 \psi_0$ when $\mathcal{D}_0 \equiv 0$. However, it is also found that from (12.48) by using the balance equation:

$$\mathcal{H}_2 = 0 \quad , \quad \text{on} \quad \zeta = 1 \quad , \tag{12.49}$$

which implies that $\zeta_{s2} \equiv 0$ according to (12.24). Even so, on $\zeta = 1$, it is also necessary to write the condition which results from the exact condition $w = \mathrm{SBo}\varrho(\partial\mathcal{H}/\partial\tau)$ on $\mathcal{H} = 0$. By taking into account (12.49) and given the fact that $\mathcal{H}_0(1) = 0$ this leads to imposing that:

$$w_2 = 0 \quad , \quad \text{on} \quad \zeta = 1 \quad . \tag{12.50}$$

Again on $\zeta = 1$, we must write the condition resulting from the exact condition $k_0\varrho(\partial T/\partial\zeta) = \sigma_{00}\mathcal{R}_0$ on $\mathcal{H} = 0$. To order zero, this leads to:

$$\frac{k_0(1)}{T_0(1)}\frac{dT_0}{d\zeta}\bigg|_{\zeta=1} = \sigma_{00}\mathcal{R}_0(1)$$

which shows that the integration constant in (12.15) is zero [(the identity (11.2) is used]. Therefore:

$$\frac{k_0(\zeta)}{T_0(\zeta)}\frac{dT_0}{d\zeta} = \sigma_{00}\mathcal{R}_0(\zeta) \quad , \tag{12.51}$$

for all ζ.

Let us now turn back to (12.3). To the order M_∞^2, the following limiting equation is obtained for T_2:

$$\mathrm{S}\frac{\partial T_2}{\partial\tau} + \mathcal{J}(\psi_0, T_2) - \mathrm{Bo}\frac{K_0(\zeta)}{\zeta}w_2 - (\gamma-1)\frac{\mathrm{Bo}^2}{\mathrm{Re}_\perp}\frac{\mu_0(\zeta)}{T_0(\zeta)}\left|\boldsymbol{k}\times\boldsymbol{D}\frac{\partial\psi_0}{\partial\zeta}\right|^2$$
$$= \frac{\mathrm{Bo}^2}{\mathrm{Pr}}\frac{1}{\mathrm{Re}_\perp}\left[\frac{\partial}{\partial\zeta}\left(\frac{k_0(\zeta)}{T_0(\zeta)}\frac{\partial T_2}{\partial\zeta}\right) - \sigma_{00}\frac{\partial}{\partial\zeta}\left(\frac{\zeta}{T_0(\zeta)}\mathcal{R}_0(\zeta)T_2\right)\right] \quad , \tag{12.52}$$

since $\varrho_2 = -[\zeta/T_0^2(\zeta)]T_2$. However, as:

$$T_2 = -\mathrm{Bo}\zeta\frac{\partial\mathcal{H}_2}{\partial\zeta} \quad , \tag{12.53}$$

(12.52) is in fact an equation which determines w_2 since ψ_0 and $\mathcal{H}_2$ satisfy the generalized quasi-nondivergent model [$\mathrm{Re}_\perp = \mathrm{O}(1)$] which is made up of (12.47) and the first equation of (12.18). We can then calculate:

$$\mathcal{D}_2 = -\frac{\partial w_2}{\partial\zeta} \quad . \tag{12.54}$$

We do not know what role is played by the condition (12.50) in $\zeta = 1$ for w_2, but what we do know is that the above generalized quasi-nondivergent model is certainly not valid in the vicinity of $\zeta = 1$.

It is remarked that for T_2, we have the following ground condition:

$$\frac{k_0(1)}{T_0(1)}\frac{\partial T_2}{\partial\zeta} - \sigma_{00}\frac{\mathcal{R}_0(1)}{T_0(1)}T_2 = 0 \quad , \quad \text{on} \quad \zeta = 1 \quad . \tag{12.54}$$

If we now impose on (12.52), the "naturally" associated boundary condition:

$$\lim_{\zeta\to 0}\left\{\frac{k_0(\zeta)}{T_0(\zeta)}\frac{\partial T_2}{\partial\zeta} - \sigma_{00}\frac{\zeta}{T_0(\zeta)}\mathcal{R}_0(\zeta)T_2\right\} = 0 \quad , \tag{12.55}$$

after the integration of (12.52) from $\zeta = 1$ to $\zeta = 0$, the following mean relation can be obtained:

$$\lim_{\zeta \to 0} \frac{1}{|\zeta - 1|} \int_1^\zeta \left\{ \mathrm{S}\frac{\partial T_2}{\partial \tau} + \mathcal{J}(\psi_0, T_2) - \mathrm{Bo}\frac{K_0(\zeta)}{\zeta} w_2 \right.$$

$$\left. -(\gamma - 1)\frac{\mathrm{Bo}^2}{\mathrm{Re}_\perp} \frac{\mu_0(\zeta)}{T_0(\zeta)} \left| \boldsymbol{k} \times \boldsymbol{D}\frac{\partial \psi_0}{\partial \zeta} \right|^2 \right\} d\zeta = 0 \quad . \tag{12.56}$$

From (12.56), it can be sensed that the seeking of an approximate solution which is uniformly valid in ζ must be related to a mean principle and a double-scale phenomenon in ζ can thus be expected when $\mathrm{M}_\infty \to 0$!

12.2.2. When $\mathrm{Re}_\perp \to \infty$ with $\mathrm{M}_\infty \to 0$ such that the similarity relation (12.8) is satisfied with $\alpha > 0$ an arbitrary real number to be determined, two limiting processes must be considered. The first one is a "local" limiting process which is valid in the vicinity of $\zeta = 1$ (which simulates the ground when $\mathrm{M}_\infty \to 0$). This first limiting process brings into play the short vertical variable:

$$\hat{\zeta} = \frac{1 - \zeta}{\mathrm{M}_\infty^\beta} \quad , \quad \beta > 0 \quad . \tag{12.57}$$

In order for the local degeneracy associated with the local limiting process:

$$\begin{aligned} &\mathrm{Re}_\perp \to \infty \quad , \quad \mathrm{M}_\infty \to 0 \quad , \\ &\frac{1/\sqrt{\mathrm{Re}_\perp}}{\mathrm{M}_\infty^\alpha} = \hat{\mathcal{S}}_0 = \mathrm{O}(1) \quad , \quad \text{with} \quad \tau, \xi, \eta \quad \text{and} \quad \hat{\zeta} \quad \text{fixed} \quad , \end{aligned} \tag{12.58}$$

to be the *most* significant, it is necessary to impose that:

$$2(\alpha - \beta) = 0 \Rightarrow \alpha \equiv \beta \quad .$$

$\beta > 0$ is then determined by imposing the matching of the vertical velocities in the local representation [associated with (12.58)], and in the main representation (associated with the main limiting process:

$$\frac{1}{\sqrt{\mathrm{Re}_\perp}} \equiv \hat{\mathcal{S}}_0 \mathrm{M}_\infty^\alpha \quad , \quad \mathrm{M}_\infty \to 0 \quad \text{with} \quad \tau, \xi, \eta \quad \text{and} \quad \zeta \quad \text{fixed}) \quad .$$

We then have:

$$\beta = 2 \Rightarrow \alpha = 2 \quad .$$

Hence, the following similarity relation must be imposed:

$$\frac{1/\sqrt{\mathrm{Re}_\perp}}{\mathrm{M}_\infty^2} = \hat{\mathcal{S}}_0 = \mathrm{O}(1) \quad . \tag{12.58a}$$

In first approximation, the main limiting process leads to the classical quasi-nondivergent model (12.18), also called the "quasi-solenoidal" model.

Let us consider then the local limiting process (12.58) and assign to it the following local asymptotic expansion:

$$\begin{pmatrix} \boldsymbol{v} \\ w \\ \mathcal{H} \\ T \\ \Omega \\ \mathcal{D} \end{pmatrix} = \begin{pmatrix} \hat{\boldsymbol{v}}_0 \\ \hat{w}_0 \\ \hat{\mathcal{H}}_0 \\ \hat{T}_0 \\ \hat{\Omega}_0 \\ \hat{\mathcal{D}}_0 \end{pmatrix} + \mathrm{M}_\infty^2 \begin{pmatrix} \hat{\boldsymbol{v}}_2 \\ \hat{w}_2 \\ \hat{\mathcal{H}}_2 \\ \hat{T}_2 \\ \hat{\Omega}_2 \\ \hat{\mathcal{D}}_2 \end{pmatrix} + \dots \quad ., , \tag{12.59}$$

Once again, (12.1a) implies that $\boldsymbol{D}\hat{\mathcal{H}}_0 = 0$, and the relation $T = -\mathrm{Bo}\zeta(\partial\mathcal{H}/\partial\zeta)$, which is written [with $\hat{\zeta} = (1-\zeta)/\mathrm{M}_\infty^2$]:

$$\hat{T} = \mathrm{Bo}\frac{1-\mathrm{M}_\infty^2\hat{\zeta}}{\mathrm{M}_\infty^2}\frac{\partial\hat{\mathcal{H}}}{\partial\hat{\zeta}} \quad , \tag{12.60}$$

yields

$$\frac{\partial\hat{\mathcal{H}}_0}{\partial\hat{\zeta}} = 0 \Rightarrow \hat{\mathcal{H}}_0 \equiv \hat{\mathcal{H}}_0(\tau) \quad \text{and}$$

$$\hat{T}_0 = \mathrm{Bo}\frac{\partial\hat{\mathcal{H}}_2}{\partial\hat{\zeta}} \quad . \tag{12.61}$$

The matching of the $\mathcal{H}$, however, leads to the choosing of:

$$\hat{\mathcal{H}}_0 \equiv \mathcal{H}_0(1) = 0 \quad . \tag{12.62}$$

In addition, the continuity equation (12.1c) becomes:

$$\mathrm{M}_\infty^2\mathcal{D} = \frac{\partial\hat{w}}{\partial\hat{\zeta}} \Rightarrow \frac{\partial\hat{w}_0}{\partial\hat{\zeta}} = 0$$

and the matching with the main representation (in which $w_0 = 0$) leads to:

$$\hat{w}_0 = 0 \quad . \tag{12.63}$$

In this case, the energy equation (12.3) gives the following local limiting equation for $\hat{T}_0$:

$$\mathrm{S}\frac{\partial\hat{T}_0}{\partial\tau} + \hat{\boldsymbol{v}}_0\cdot\boldsymbol{D}\hat{T}_0 - \hat{w}_2\frac{\partial\hat{T}_0}{\partial\hat{\zeta}} = \frac{\mathrm{Bo}^2}{\mathrm{Pr}}\hat{\mathcal{S}}_0^2 k_0(1)\frac{\partial}{\partial\hat{\zeta}}\left(\frac{1}{\hat{T}_0}\frac{\partial\hat{T}_0}{\partial\hat{\zeta}}\right) \quad , \tag{12.64}$$

since $\hat{\varrho}_0 \equiv 1/\hat{T}_0$.

The ground being in $\hat{\mathcal{H}} = 0 + \mathrm{M}_\infty^2\hat{\mathcal{H}}_2 + \dots = 0$, is also simulated by the equation:

$$\hat{\zeta} = \hat{\zeta}_{s0} + \mathrm{M}_\infty^2\hat{\zeta}_{s2} + \dots \quad , \tag{12.65}$$

with $\hat{\mathcal{H}}_2(\hat{\zeta}_{s0}) = 0$. The balance condition on the ground, $k_0\varrho(\partial T/\partial\zeta) = \sigma_{00}\mathcal{R}_0$ on $\mathcal{H} = 0$, becomes:

$$k_0(1)\frac{1}{\hat{T}_0}\frac{\partial\hat{T}_0}{\partial\hat{\zeta}} = 0 \quad , \quad \text{on} \quad \hat{\zeta} = \hat{\zeta}_{s0} \quad . \tag{12.66}$$

The second condition in $\hat{\zeta}$, which must be assigned to (12.64) for $\hat{T}_0$, is a matching condition:

$$\lim_{\hat{\zeta}\to\infty} \hat{T}_0 = T_0(1) \quad . \tag{12.67}$$

It seems reasonable to postulate that the homogeneous problem (12.64, 66, 67) for $\hat{T}_0 - T_0(1)$ has the trivial solution:

$$\hat{T}_0 - T_0(1) \equiv 0 \Rightarrow T_0(1) = \mathrm{Bo}\frac{\partial \hat{\mathcal{H}}_2}{\partial \hat{\zeta}} \quad . \tag{12.68}$$

We have to acknowledge that it is the only admissible solution.

Hence, it is observed from (12.68) that:

$$\frac{\partial \hat{\mathcal{H}}_2}{\partial \hat{\zeta}} = \frac{T_0(1)}{\mathrm{Bo}} = -\left(\frac{d\mathcal{H}_0}{d\zeta}\right)_{\zeta=1} \quad .$$

The matching with $\mathcal{H}_2$ (from the main representation) shows that:

$$\hat{\mathcal{H}}_2 = \mathcal{H}_{2,1} + \frac{T_0(1)}{\mathrm{Bo}}\hat{\zeta} = \mathcal{H}_{2,1} - \left(\frac{d\mathcal{H}_0}{d\zeta}\right)_{\zeta=1}\hat{\zeta} \quad , \tag{12.69}$$

which also implies that:

$$\hat{\zeta}_{s0} \equiv \frac{\mathcal{H}_{2,1}}{(d\mathcal{H}_0/d\zeta)|_{\zeta=1}} \quad . \tag{12.70}$$

It will be observed that $\mathcal{H}_{2,1} \equiv \mathcal{H}_2(\tau, \xi, \eta, 1)$.

Finally, for $\hat{\boldsymbol{v}}_0$ and $\hat{w}_2$, the following system of boundary layer equations is obtained:

$$\mathrm{S}\frac{\partial \hat{\boldsymbol{v}}_0}{\partial \tau} + (\hat{\boldsymbol{v}}_0 \cdot \boldsymbol{D})\hat{\boldsymbol{v}}_0 - \hat{w}_2\frac{\partial \hat{\boldsymbol{v}}_0}{\partial \hat{\zeta}} + \left(\frac{1}{\mathrm{Ro}} + \beta\eta\right)\boldsymbol{k} \times \hat{\boldsymbol{v}}_0$$

$$+\frac{\mathrm{Bo}}{\gamma}\boldsymbol{D}\mathcal{H}_{2,1} = \nu_0\frac{\partial^2 \hat{\boldsymbol{v}}_0}{\partial \hat{\zeta}^2} \quad ; \tag{12.71a}$$

$$\boldsymbol{D} \cdot \hat{\boldsymbol{v}}_0 = \frac{\partial \hat{w}_2}{\partial \hat{\zeta}} \quad , \tag{12.71b}$$

where $\nu_0 \equiv [\mathrm{Bo}^2/T_0(1)]\mu_0(1)\hat{S}_0^2$. The following boundary conditions in $\hat{\zeta}$ must be added to the system (12.71):

$$\boldsymbol{v}_0 = 0 \quad , \quad \hat{w}_2 = \mathrm{S}\frac{\mathrm{Bo}}{T_0(1)}\frac{\partial \mathcal{H}_{2,1}}{\partial \tau} \quad , \quad \text{on} \quad \hat{\zeta} = \hat{\zeta}_{s0} = -\frac{\mathrm{Bo}}{T_0(1)}\mathcal{H}_{2,1} \quad . \tag{12.72a}$$

$$\hat{\boldsymbol{v}}_0 \to \boldsymbol{v}_{0,1} \quad , \quad \text{when} \quad \hat{\zeta} \to \infty \quad . \tag{12.72b}$$

It is important to point out here that the matching of the local and main representations implies that:

$$\lim_{\hat{\zeta}\to\infty} \hat{w}_2 = w_{2,1} \equiv w_2(\tau, \xi, \eta, 1) \quad , \tag{12.73}$$

since $\lim_{\hat{\zeta} \to \infty} \boldsymbol{D} \cdot \hat{\boldsymbol{v}}_0 = \boldsymbol{D} \cdot \boldsymbol{v}_{0,1} \equiv 0$, and the term proportional to $\hat{\zeta}$ in the behavior of $\hat{w}_2$ at infinity is zero. In the main representation, however, it is noticed that according to (12.19):

$$\left(S\frac{\partial}{\partial \tau} + \boldsymbol{v}_{0,1} \cdot \boldsymbol{D} \right) T_{2,1} = \mathrm{Bo} K_0(1) \lim_{\hat{\zeta} \to \infty} \hat{w}_2 \quad ,$$

which makes it possible to determine $T_{2,1}$ once adequate conditions have been specified. In this case, the following constraint (in ζ) results for $\mathcal{H}_2$:

$$\left. \frac{\partial \mathcal{H}_2}{\partial \zeta} \right|_{\zeta=1} = -\mathrm{Bo} T_{2,1} \quad .$$

The above constraint leads us to think that an intermediary layer must no doubt be introduced between the main layer and the local boundary layer (at least if our singular perturbation problem falls within the framework of the MMAE which is not at all sure). In this case, the behaviors (12.72b, 73) may no longer hold and the classical quasi-nondivergent model, which makes it possible to determine $\mathcal{H}_{2,1}$ and $\boldsymbol{v}_{0,1}$, is no longer valid for calculating the boundary layer associated with the problem (12.71, 72). This important question remains for the time being unresolved.

From a conceptual point of view, the above problem is very interesting since it demonstrates the application limitations of the MMAE. Until "all the terms" of the local and main expansions are known, we cannot be sure that at a certain order, we will not have to introduce an intermediary expansion in order to deal with possible incompatibilities.

We thus end this section which is merely a sketch of a more detailed asymptotic study aiming at bringing to light the structure in ζ of low Mach number flows. We will have the opportunity in Chap. 13 to return to these questions which are only beginning to receive answers thanks the works by Guiraud and Zeytounian (see Zeytounian, 1985).

12.3 Analysis of Guiraud and Zeytounian's Recent Results

When no relief is present, the classical quasi-nondivergent model of Sect. 12.1 enables the study of the evolution of an eddy spot. This model cannot, however, elucidate the origin of such a spot. This is explained by the fact that each isobaric surface evolves independently. Only a weak coupling is achieved by differential iteration due to higher order approximations. It turns out that the most apparent – if not the most efficient–coupling takes place in the horizontal plane at infinity. In order to carry out the corresponding asymptotic analysis[3] with the least number

[3] Once again, we will make use of the "Course Notes" of J.P. Guiraud written for the conferences at the "Advanced School" in October, 1983 at the ICMS in Udine (Italy) devoted to "Models for Atmospheric Flows".

of problems, we will limit our attention to the case of Ro $\equiv \infty$ (neglecting the Coriolis force) and $\beta \equiv 0$ (ignoring the beta effect). In the horizontal direction at infinity, we suppose that:

$$\lim_{|\xi^2+\eta^2|\to\infty} \left(\boldsymbol{v}, w, \mathcal{H}, T, \zeta_s\right) = \left(\boldsymbol{v}^{\infty}(\zeta), 0, \mathcal{H}_0(\zeta), T_0(\zeta), 1\right) \quad . \tag{12.74}$$

It is pointed out that the basic flow characterized by:

$$\boldsymbol{v} = \boldsymbol{v}^{\infty}(\zeta) \quad , \quad \mathcal{H} = \mathcal{H}_0(\zeta) \quad , \quad T = T_0(\zeta) \quad \text{and} \quad w = 0$$

is an exact solution of the basic equations:

$$\begin{aligned}
&\mathrm{S}\frac{\partial \boldsymbol{v}}{\partial \tau} + (\boldsymbol{v}\cdot\boldsymbol{D})\boldsymbol{v} + w\frac{\partial \boldsymbol{v}}{\partial \zeta} + \frac{\mathrm{Bo}}{\gamma \mathrm{M}_{\infty}^2}\boldsymbol{D}\mathcal{H} = 0 \quad ; \\
&\mathrm{S}\frac{\partial T}{\partial \tau} + \boldsymbol{v}\cdot\boldsymbol{D}T + w\left(\frac{\partial T}{\partial \zeta} - \frac{\gamma-1}{\gamma}\frac{T}{\zeta}\right) = 0 \quad ; \\
&\boldsymbol{D}\cdot\boldsymbol{v} + \frac{\partial w}{\partial \zeta} = 0 \quad ; \\
&T = -\mathrm{Bo}\,\zeta\frac{\partial \mathcal{H}}{\partial \zeta} \quad .
\end{aligned} \tag{12.75}$$

Let us return to (12.13) where $\mathcal{D}_0 \equiv 0$. The harmonic function φ_0^{h} can be taken in the form:

$$\varphi_0^{\mathrm{h}} = \boldsymbol{v}^{\infty}\cdot\boldsymbol{\xi} \quad , \quad \text{where} \quad \boldsymbol{\xi} = \xi\boldsymbol{i} + \eta\boldsymbol{j}$$

and the following is obtained:

$$\boldsymbol{v}_0 = \boldsymbol{v}_{\infty}(\zeta) + \boldsymbol{k}\times\boldsymbol{D}\psi_0 \quad , \tag{12.76}$$

where ψ_0 can be calculated from (12.13a) once Ω_0 is determined. In the forthcoming, we will focus our attention on the behavior at infinity when $|\boldsymbol{\xi}|\to\infty$ and we will assume that Ω_0 cancels out beyond a certain bounded domain. From (12.13a), the following asymptotic behavior is obtained:

$$\psi_0 = \frac{\overline{\Omega_0}}{2\pi}\log|\boldsymbol{\xi}| - \frac{\boldsymbol{\mu}_0\cdot\boldsymbol{\xi}}{2\pi|\boldsymbol{\xi}|^2} + \frac{\boldsymbol{\xi}\cdot\boldsymbol{N}_0\cdot\boldsymbol{\xi}}{2\pi|\boldsymbol{\xi}|^4} + \mathrm{O}\left(\frac{1}{|\boldsymbol{\xi}|^3}\right) \quad , \tag{12.77}$$

where

$$\begin{aligned}
&\overline{\Omega_0}(\zeta) = \iint_{\mathbf{R}^2} \Omega_0(\tau,\xi,\eta,\zeta)\,d\xi\,d\eta \quad ; \\
&\boldsymbol{\mu}_0(\tau,\zeta) = \iint_{\mathbf{R}^2} \boldsymbol{\xi}\Omega_0(\tau,\xi,\eta,\zeta)\,d\xi\,d\eta \quad ; \\
&\boldsymbol{N}_0(\tau,\zeta) = \frac{1}{2}\iint_{\mathbf{R}^2} \Omega_0(\tau,\xi,\eta,\zeta)\left[|\boldsymbol{\xi}|^2\mathbb{1} - 2\boldsymbol{\xi}\otimes\boldsymbol{\xi}\right]d\xi\,d\eta \quad ,
\end{aligned} \tag{12.78}$$

where $\mathbb{1}$ is the second order unit tensor and $\boldsymbol{\xi}\otimes\boldsymbol{\xi}$ is the tensor product of the vector $\boldsymbol{\xi}$ with the vector $\boldsymbol{\xi}$. As a consequence of (12.18), we have:

$$\frac{\partial \overline{\Omega_0}}{\partial \tau} = 0 \quad \text{and} \quad \mathrm{S}\frac{\partial \boldsymbol{\mu}_0}{\partial \tau} = \overline{\Omega_0}\boldsymbol{v}^\infty \quad .$$

Thanks to (12.77), the following behavior is derived for $\boldsymbol{v}_0$ from (12.76):

$$\boldsymbol{v}_0 = \boldsymbol{v}^\infty(\zeta) + \frac{\overline{\Omega_0}}{2\pi}\frac{\boldsymbol{k}\times\boldsymbol{\xi}}{|\boldsymbol{\xi}|^2} - \frac{1}{2\pi|\boldsymbol{\xi}|^2}\left\{\boldsymbol{k}\times\boldsymbol{\mu}_0 - 2\frac{(\boldsymbol{\mu}_0\cdot\boldsymbol{\xi})(\boldsymbol{k}\times\boldsymbol{\xi})}{|\boldsymbol{\xi}|^2}\right\} + \mathrm{O}\left(\frac{1}{|\boldsymbol{\xi}|^3}\right) \quad . \tag{12.79}$$

The behavior of $\mathcal{H}_2$ when $|\boldsymbol{\xi}| \to \infty$ must now be obtained. To accomplish this, we will take advantage of the following relation:

$$\boldsymbol{D}\left(\frac{\mathcal{H}_2}{\gamma/\mathrm{Bo}}\right) + \mathrm{S}\boldsymbol{k}\times\boldsymbol{D}\frac{\partial\psi_0}{\partial\tau} + \left[(\boldsymbol{v}^\infty + \boldsymbol{k}\times\boldsymbol{D}\psi_0)\cdot\boldsymbol{D}\right](\boldsymbol{k}\times\boldsymbol{D}\psi_0) = 0 \quad ,$$

which results trivially from (12.76) and the first equation in (12.17). However, in the vicinity of infinity ($|\boldsymbol{\xi}| \approx \infty$), the function ψ_0 is harmonic. If the latter's conjugate harmonic χ_0 is considered, we find: $\boldsymbol{k}\times\boldsymbol{D}\psi_0 = \boldsymbol{D}\chi_0$. The following is thus obtained:

$$\boldsymbol{D}\left\{\frac{\mathcal{H}_2}{\gamma/\mathrm{Bo}} + \mathrm{S}\frac{\partial\chi_0}{\partial\tau} + \left(\boldsymbol{v}_\infty + \frac{1}{2}\boldsymbol{D}\chi_0\right)\cdot\boldsymbol{D}\chi_0\right\} = 0 \quad . \tag{12.80}$$

From (12.77, 80), it then follows that:

$$\chi_0 = \frac{\overline{\Omega_0}}{2\pi}\mathrm{arctg}\frac{\eta}{\xi} + \frac{(\boldsymbol{k}\times\boldsymbol{\mu}_0)\cdot\boldsymbol{\xi}}{2\pi|\boldsymbol{\xi}|^2} + \frac{\boldsymbol{\xi}\cdot\boldsymbol{N}_0\cdot(\boldsymbol{k}\times\boldsymbol{\xi}) + (\boldsymbol{k}\times\boldsymbol{\xi})\cdot\boldsymbol{\mu}_0\cdot\boldsymbol{\xi}}{4\pi|\boldsymbol{\xi}|^4} + \ldots$$

and the following behavior can be found for $\mathcal{H}_2$:

$$\mathcal{H}_2 = -\frac{\gamma/\mathrm{Bo}}{2\pi|\boldsymbol{\xi}|^2}\left\{\frac{(\overline{\Omega_0})^2}{4\pi} + \boldsymbol{v}^\infty\cdot(\boldsymbol{k}\times\boldsymbol{\mu}_0) + \frac{\boldsymbol{\xi}\cdot\mathrm{S}(\partial\boldsymbol{N}_0/\partial\tau)\cdot\boldsymbol{k}\times\boldsymbol{\xi}}{|\boldsymbol{\xi}|^2} - \frac{2}{|\boldsymbol{\xi}|^2}\left[(\boldsymbol{k}\times\boldsymbol{\mu}_0)\cdot\boldsymbol{\xi}\right]\left[\boldsymbol{v}^\infty\cdot\boldsymbol{\xi}\right]\right\} + \mathrm{O}\left(\frac{1}{|\boldsymbol{\xi}|^3}\right) \quad . \tag{12.81}$$

When looking at higher order terms, we find that in the expansion (12.10) (which leads to the quasi-nondivergent model in particular), the term $\mathcal{H}_4$ is not bounded when $|\boldsymbol{\xi}| \to \infty$. This indicates that the main expansion (12.10) is certainly not valid in the vicinity of infinity (with respect to the horizontal variables ξ and η). In order to shed light on the nature of this non-uniformity in $\boldsymbol{\xi}$ when $|\boldsymbol{\xi}| \to \infty$, let us consider, as did Zeytounian (1976), the following local limiting process:

$$\mathrm{M}_\infty \to 0 \quad , \quad \tilde{\xi} = \mathrm{M}_\infty\xi \quad , \quad \tilde{\eta} = \mathrm{M}_\infty\eta \quad , \quad \tau \text{ and } \xi \text{ fixed} \quad . \tag{12.82}$$

In the starting system (12.75), we introduce the following relations:

$$\begin{aligned} &\boldsymbol{v} = \boldsymbol{v}^\infty(\zeta) + \mathrm{M}_\infty^a\tilde{\boldsymbol{v}} \quad , \quad w = \mathrm{M}_\infty^{a+1}\tilde{w} \quad , \quad \mathcal{H} = \mathcal{H}_0(\zeta) + \mathrm{M}_\infty^{a+1}\tilde{\mathcal{H}} \quad , \\ &T = T_0(\zeta) + \mathrm{M}_\infty^{a+1}\tilde{T} \quad , \quad \zeta_s = 1 + \mathrm{M}_\infty^{a+1}\tilde{\xi}_s \quad . \end{aligned} \tag{12.83}$$

The following system of equations is then obtained:

$$\mathrm{S}\frac{\partial \tilde{\boldsymbol{v}}}{\partial \tau} + \tilde{\boldsymbol{D}}\left(\frac{\tilde{\mathcal{H}}}{\gamma/\mathrm{Bo}}\right) + \mathrm{M}_\infty \left\{ \left[\left(\boldsymbol{v}^\infty + \mathrm{M}_\infty^a \tilde{\boldsymbol{v}} \right) \cdot \tilde{\boldsymbol{D}} \right] \tilde{\boldsymbol{v}} \right.$$

$$\left. + \tilde{w} \left(\frac{d\boldsymbol{v}^\infty}{d\zeta} + \mathrm{M}_\infty^a \frac{\partial \tilde{\boldsymbol{v}}}{\partial \zeta} \right) \right\} = \mathrm{O}\left(\mathrm{M}_\infty^{2\alpha - a} \right) \quad ; \tag{12.84a}$$

$$\tilde{\boldsymbol{D}} \cdot \tilde{\boldsymbol{v}} + \frac{\partial \tilde{w}}{\partial \zeta} = 0 \quad ; \tag{12.84b}$$

$$\mathrm{S}\frac{\partial \tilde{T}}{\partial \tau} - \mathrm{Bo}\frac{K_0(\zeta)}{\zeta}\tilde{w} + \mathrm{M}_\infty \left(\boldsymbol{v}^\infty + \mathrm{M}_\infty^a \tilde{\boldsymbol{v}} \right) \cdot \tilde{\boldsymbol{D}}\tilde{T}$$

$$+ \mathrm{M}_\infty^{a+1} \tilde{w} \left(\frac{\partial \tilde{T}}{\partial \zeta} - \frac{\gamma - 1}{\gamma} \frac{\tilde{T}}{\zeta} \right) = \mathrm{O}\left(\mathrm{M}_\infty^{2\alpha - a - 1} \right) \quad ; \tag{12.84c}$$

$$\tilde{T} + \mathrm{Bo}\zeta \frac{\partial \tilde{\mathcal{H}}}{\partial \zeta} = 0 \quad , \tag{12.84d}$$

where

$$\tilde{\boldsymbol{D}} \equiv \frac{\partial}{\partial \tilde{\xi}}\boldsymbol{i} + \frac{\partial}{\partial \tilde{\eta}}\boldsymbol{j} \equiv \frac{1}{\mathrm{M}_\infty}\boldsymbol{D} \quad .$$

In the right-hand members of (12.84a, c), the exponent 2α issues from the dissipative terms when the similarity relation (12.8) is taken into account. It will be assumed that α is such that these dissipation terms do not enter into the asymptotic analysis related to the limiting process (12.82).

Concerning $\tilde{\zeta}_s$, it can be said to satisfy the equation:

$$\mathrm{S}\frac{\partial \tilde{\zeta}_s}{\partial \tau} - \tilde{w} + \mathrm{M}_\infty \left(\boldsymbol{v}^\infty + \mathrm{M}_\infty^a \tilde{\boldsymbol{v}} \right) \cdot \tilde{\boldsymbol{D}}\tilde{\zeta}_s = 0 \quad , \quad \text{on} \quad \zeta = 1 + \mathrm{M}_\infty^{a+1}\tilde{\zeta}_s \quad . \tag{12.85}$$

If we now introduce into the behaviors (12.79, 81), the new variables $\tilde{\xi} = \mathrm{M}_\infty \xi$ and $\tilde{\eta} = \mathrm{M}_\infty \eta$, i.e.,

$$\tilde{\boldsymbol{\xi}} \equiv \mathrm{M}_\infty \boldsymbol{\xi} \quad ,$$

we see that:

$$\boldsymbol{v}_0 = \boldsymbol{v}^\infty(\zeta) + \mathrm{M}_\infty \frac{\overline{\Omega_0}}{2\pi} \frac{\boldsymbol{k} \times \tilde{\boldsymbol{\xi}}}{|\tilde{\boldsymbol{\xi}}|^2} + \ldots \quad ;$$

$$\mathcal{H}_0(\zeta) + \mathrm{M}_\infty^2 \mathcal{H}_2 + \ldots = \mathcal{H}_0(\zeta) + \mathrm{M}_\infty^4 \mathrm{O}\left(\frac{1}{|\tilde{\boldsymbol{\xi}}|^2} \right) + \ldots \quad .$$

The above indicates that the correct choice is:

$$a = 1 \quad . \tag{12.86}$$

Therefore, at the limit, when (12.82) takes place, the following local system results:

$$
\begin{aligned}
&\mathrm{S}\frac{\partial \tilde{\boldsymbol{v}}_0}{\partial \tau} + \tilde{\boldsymbol{D}}\left(\frac{\tilde{\mathcal{H}}}{\gamma/\mathrm{Bo}}\right) = 0 \quad ; \\
&\mathrm{S}\frac{\partial \tilde{T}_0}{\partial \tau} - \mathrm{Bo}\frac{K_0(\zeta)}{\zeta}\tilde{w} = 0 \quad ; \\
&\tilde{\boldsymbol{D}} \cdot \tilde{\boldsymbol{v}}_0 + \frac{\partial \tilde{w}_0}{\partial \zeta} = 0 \quad ; \\
&\tilde{T}_0 + \mathrm{Bo}\zeta\frac{\partial \tilde{\mathcal{H}}_0}{\partial \zeta} = 0 \quad ,
\end{aligned}
\tag{12.87}
$$

once the following local expansion has been carried out:

$$
\begin{pmatrix} \tilde{\boldsymbol{v}} \\ \tilde{w} \\ \tilde{\mathcal{H}} \\ \tilde{T} \end{pmatrix} = \begin{pmatrix} \tilde{\boldsymbol{v}}_0 \\ \tilde{w}_0 \\ \tilde{\mathcal{H}}_0 \\ \tilde{T}_0 \end{pmatrix} + \mathrm{M}_\infty \begin{pmatrix} \tilde{\boldsymbol{v}}_1 \\ \tilde{w}_1 \\ \tilde{\mathcal{H}}_1 \\ \tilde{T}_1 \end{pmatrix} + \ldots \quad . \tag{12.88}
$$

The matching with the main quasi-nondivergent expansion when $|\boldsymbol{\xi}| \to \infty$ leads us to choose the following:

$$
\tilde{\boldsymbol{v}}_0 \equiv \frac{\overline{\Omega_0}}{2\pi}\frac{\boldsymbol{k} \times \tilde{\boldsymbol{\xi}}}{|\tilde{\boldsymbol{\xi}}|^2} \quad \text{and} \quad \tilde{w}_0 = \tilde{\mathcal{H}}_0 = \tilde{T}_0 \equiv 0 \quad . \tag{12.89}
$$

To the following order, we obtain for $\tilde{\boldsymbol{v}}_1$, $\tilde{\mathcal{H}}_1$, $\tilde{T}_1$ and $\tilde{w}_1$, the local equations:

$$
\begin{aligned}
&\mathrm{S}\frac{\partial \tilde{\boldsymbol{v}}_1}{\partial \tau} + \tilde{\boldsymbol{D}}\left(\frac{\tilde{\mathcal{H}}_1}{\gamma/\mathrm{Bo}}\right) + (\boldsymbol{v}^\infty \cdot \tilde{\boldsymbol{D}})\tilde{\boldsymbol{v}}_0 = 0 \quad ; \\
&\mathrm{S}\frac{\partial \tilde{T}_1}{\partial \tau} - \mathrm{Bo}\frac{K_0(\zeta)}{\zeta}\tilde{w}_1 = 0 \quad ; \\
&\tilde{\boldsymbol{D}} \cdot \tilde{\boldsymbol{v}}_1 + \frac{\partial \tilde{w}_1}{\partial \zeta} = 0 \quad ; \\
&\tilde{T}_1 + \mathrm{Bo}\zeta\frac{\partial \tilde{\mathcal{H}}_1}{\partial \zeta} = 0 \quad ,
\end{aligned}
\tag{12.90}
$$

and the matching imposes that:

$$
\tilde{\boldsymbol{v}}_1 \cong -\frac{\boldsymbol{k} \times \boldsymbol{\mu}_0}{2\pi|\tilde{\boldsymbol{\xi}}|^2} + \frac{(\boldsymbol{\mu}_0 \cdot \boldsymbol{\xi})\boldsymbol{k} \times \boldsymbol{\xi}}{\pi|\tilde{\boldsymbol{\xi}}|^4} + \ldots, \quad \tilde{\mathcal{H}}_1 \to 0 \ \text{ when } \ |\boldsymbol{\xi}| \to 0. \tag{12.91}
$$

From the linear system (12.90), however, a homogeneous equation for $\tilde{\mathcal{H}}_1$ is extracted:

$$
\tilde{\boldsymbol{D}}^2\tilde{\mathcal{H}}_1 + \gamma \mathrm{S}^2\frac{\partial^2}{\partial \tau^2}\left[\frac{\partial}{\partial \zeta}\left(\frac{\zeta^2}{K_0(\zeta)}\frac{\partial \tilde{\mathcal{H}}_1}{\partial \zeta}\right)\right] = 0 \quad .
$$

It is now noticed that the solution:

$$
\tilde{\boldsymbol{v}}_1 = -\frac{\boldsymbol{k} \times \boldsymbol{\mu}_0}{2\pi|\tilde{\boldsymbol{\xi}}|^2} + \frac{(\boldsymbol{\mu}_0 \cdot \boldsymbol{\xi})\boldsymbol{k} \times \boldsymbol{\xi}}{\pi|\tilde{\boldsymbol{\xi}}|^4} \quad ; \tag{12.92a}
$$

$$\tilde{w}_1 = 0 \quad ;$$
$$\tilde{T}_1 = \tilde{\mathcal{H}}_1 = 0 \quad , \tag{12.92b}$$

does indeed satisfy (as concerns $\tilde{\boldsymbol{v}}_1$) the relations:

$$\tilde{\boldsymbol{D}} \cdot \tilde{\boldsymbol{v}}_1 = 0 \quad \text{and} \quad \mathrm{S}\frac{\partial \tilde{\boldsymbol{v}}_1}{\partial \tau} + \left(\boldsymbol{v}^\infty \cdot \tilde{\boldsymbol{D}}\right)\tilde{\boldsymbol{v}}_0 = 0 \quad .$$

It can thus be stated that (12.92) is indeed the solution of (12.90).

The first two terms of the asymptotic expansion (12.88), which is valid in the vicinity of $|\boldsymbol{\xi}| = \infty$, are thus obtained from the behavior of the solution of the classical quasi-nondivergent model when $|\boldsymbol{\xi}| \to \infty$. It will be remarked that $\overline{\Omega_0}(\zeta)$ is the total eddy intensity contained within the isobaric surface of level ζ. It is thus seen that $\tilde{\boldsymbol{v}}_0$ is an eddy potential, whereas $\tilde{\boldsymbol{v}}_1$ is an eddy potential doublet. These two terms from (12.88) for $\tilde{\boldsymbol{v}}(\tau, \tilde{\boldsymbol{\xi}}, \zeta)$ do not then give any information which is not already contained in the classical quasi-nondivergent model governing $\boldsymbol{v}_0$ and $\mathcal{H}_2$. To prove that $\mathcal{H}_4$ does not remain bounded when $|\boldsymbol{\xi}| \to \infty$, it is necessary to return to the second approximation equations (12.19–25).

By taking into account (12.20), we have from (12.19):

$$w_2 = -\frac{\zeta^2}{K_0(\zeta)}\left(\mathrm{S}\frac{\partial}{\partial \tau} + \boldsymbol{v}_0 \cdot \boldsymbol{D}\right)\frac{\partial \mathcal{H}_2}{\partial \zeta} \quad . \tag{12.93}$$

Then from (12.22), the following equation results for $\Omega_2 = \boldsymbol{k} \cdot (\boldsymbol{D} \times \boldsymbol{v}_2)$:

$$\mathrm{S}\frac{\partial \Omega_2}{\partial \tau} + \boldsymbol{v}_0 \cdot \boldsymbol{D}\Omega_2 + \boldsymbol{v}_2 \cdot \boldsymbol{D}\Omega_0 + w_2\frac{\partial \Omega_0}{\partial \zeta} + \boldsymbol{k} \cdot \left(\boldsymbol{D}w_2 \times \frac{\partial \boldsymbol{v}_0}{\partial \zeta}\right) = 0 \quad .$$

Since the analysis is being carried out in the vicinity of infinity in $\boldsymbol{\xi}$, the term $\Omega_0(\equiv 0$ when $|\boldsymbol{\xi}| \to \infty)$ can be neglected in the above equation. Therefore:

$$\left(\mathrm{S}\frac{\partial}{\partial \tau} + \boldsymbol{v}^\infty \cdot \boldsymbol{D}\right)\left\{\Omega_2 - \frac{\zeta^2}{K_0(\zeta)}\boldsymbol{k} \cdot \left(\boldsymbol{D}\frac{\partial \mathcal{H}_2}{\partial \zeta} \times \frac{d\boldsymbol{v}^\infty}{d\zeta}\right)\right\} + \frac{\overline{\Omega_0}}{2\pi}\frac{\boldsymbol{k} \times \boldsymbol{\xi}}{|\boldsymbol{\xi}|^2} \cdot \boldsymbol{D}\Omega_2$$
$$-\frac{\zeta^2}{K_0(\zeta)}\boldsymbol{k} \cdot \left\{\left[\boldsymbol{D}\left(\frac{\overline{\Omega_0}}{2\pi}\frac{(\boldsymbol{k} \times \boldsymbol{\xi})}{|\boldsymbol{\xi}|^2} \cdot \boldsymbol{D}\left(\frac{\partial \mathcal{H}_2}{\partial \zeta}\right)\right] \times \frac{d\boldsymbol{v}^\infty}{d\zeta}\right\} = 0 \quad ,$$

once we have made use of (12.93) and the behavior (12.79) for $\boldsymbol{v}_0$ when $|\boldsymbol{\xi}| \to \infty$. If we take into account (12.81) (which gives us the behavior of $\mathcal{H}_2$ when $|\boldsymbol{\xi}| \to \infty$), it is found that:

$$\Omega_2 \cong \frac{\zeta^2}{K_0(\zeta)}\left(\frac{d\boldsymbol{v}^\infty}{d\zeta} \times \boldsymbol{k}\right) \cdot \boldsymbol{D}\frac{\partial \mathcal{H}_2}{\partial \zeta} = \mathrm{O}\left(\frac{1}{|\boldsymbol{\xi}|^3}\right) \quad , \tag{12.94}$$

when $|\boldsymbol{\xi}| \to \infty$.

The following representation can help us find the behavior of $\boldsymbol{v}_2$ when $|\boldsymbol{\xi}| \to \infty$:

$$\boldsymbol{v}_2 = \boldsymbol{D}\varphi_2 + \boldsymbol{k} \times \boldsymbol{D}\psi_2 \quad , \tag{12.95}$$

and we have:

$$D^2\varphi_2 = -\frac{\partial w_2}{\partial \zeta} \quad , \quad D^2\psi_2 = \Omega_2 \quad .$$

In the vicinity of infinity in $\boldsymbol{\xi}$ we have the behavior (12.94) for Ω_2, and it is thus found that:

$$\psi_2 = \frac{\overline{\Omega_2}}{2\pi}\log|\boldsymbol{\xi}| + \mathrm{O}\left(\frac{1}{|\boldsymbol{\xi}|}\right) \quad , \tag{12.96}$$

where

$$\overline{\Omega_2} = \iint_{R^2} \Omega_2(\tau, \boldsymbol{\xi}, \zeta)d\boldsymbol{\xi} \quad \text{and} \quad \frac{\partial \overline{\Omega_2}}{\partial \tau} = 0 \quad .$$

A particular solution in the vicinity of infinity in $\boldsymbol{\xi}$ of the equation

$$D^2\varphi_2 = -\frac{\partial w_2}{\partial \zeta}$$

is denoted φ_2^P. We thus find that:

$$\varphi_2 = \varphi_2^P - \frac{1}{2\pi}\frac{\partial \overline{w_2}}{\partial \zeta}\log|\boldsymbol{\xi}| + \mathrm{O}\left(\frac{1}{|\boldsymbol{\xi}|}\right) \quad , \tag{12.97}$$

where

$$\overline{w_2} = \lim_{R\to\infty} \iint_{|\boldsymbol{\xi}|^2 < R^2} w_2(\tau, \boldsymbol{\xi}, \zeta)d\boldsymbol{\xi} \equiv \overline{w_2}(\tau, \zeta) \quad . \tag{12.98}$$

The behavior of $\mathcal{H}_4$ in the vicinity of infinity when $|\boldsymbol{\xi}| \to \infty$ must now be obtained. Equation (12.22) is used to this end and is written in the vicinity of infinity in $\boldsymbol{\xi}$ in the following form:

$$\mathrm{S}\frac{\partial \boldsymbol{v}_2}{\partial \tau} + (\boldsymbol{v}^\infty \cdot \boldsymbol{D})\boldsymbol{v}_2 + w_2\frac{d\boldsymbol{v}_\infty}{d\zeta} + \boldsymbol{D}\left(\frac{\mathcal{H}_4}{\gamma/\mathrm{Bo}}\right) = 0.$$

By analogy with (12.80), the following is obtained:

$$\boldsymbol{D}\left\{\frac{\mathcal{H}_4}{\gamma/\mathrm{Bo}} + \left(\mathrm{S}\frac{\partial}{\partial \tau} + \boldsymbol{v}^\infty \cdot \boldsymbol{D}\right)\left(\varphi_2 + \frac{\overline{\Omega_2}}{2\pi}\mathrm{arctg}\left(\frac{\eta}{\xi}\right)\right)\right\} + \mathrm{O}\left(\frac{1}{|\boldsymbol{\xi}|^2}\right) = 0 \quad .$$

However, since $\partial\overline{\Omega_2}/\partial\tau = 0$, it is found that:

$$\mathcal{H}_4 = -\frac{\gamma \mathrm{S}}{\mathrm{Bo}}\frac{\partial \varphi_2}{\partial \tau} + \mathrm{O}\left(\frac{1}{|\boldsymbol{\xi}|}\right) \quad , \quad \text{when} \quad |\boldsymbol{\xi}| \to \infty \quad . \tag{12.99}$$

Seeing that

$$\mathrm{S}\frac{\partial \varphi_2^P}{\partial \tau} = \frac{\gamma}{8\pi}\frac{\partial}{\partial \xi}\left\{\frac{\zeta^2}{K_0(\zeta)}\frac{\partial}{\partial \zeta}\left[\frac{\boldsymbol{\xi}\cdot \mathrm{S}^3(\partial^3 \boldsymbol{N}_0/\partial\tau^3)\cdot(\boldsymbol{k}\times\boldsymbol{\xi})}{|\boldsymbol{\xi}|^2}\right]\right\} \quad ,$$

we find that:

$$\mathcal{H}_4 = \frac{\gamma \mathrm{S}}{\mathrm{Bo}}\frac{1}{2\pi}\frac{\partial^2 \overline{w_2}}{\partial\tau\partial\zeta}\log|\boldsymbol{\xi}|$$
$$- \frac{\gamma^2}{8\pi \mathrm{Bo}}\frac{\partial}{\partial \zeta}\left\{\frac{\zeta^2}{K_0(\zeta)}\frac{\partial}{\partial \zeta}\left[\frac{\boldsymbol{\xi}\cdot \mathrm{S}^3(\partial^3 \boldsymbol{N}_0/\partial\tau^3)\cdot(\boldsymbol{k}\times\boldsymbol{\xi})}{|\boldsymbol{\xi}|^2}\right]\right\} + \ldots \tag{12.100}$$

which confirms that $\mathcal{H}_4$ is not bounded at infinity in $\boldsymbol{\xi}$. But, in the vicinity of infinity $\boldsymbol{\xi} = (1/\mathrm{M}_\infty)\tilde{\boldsymbol{\xi}}$, and we remark that the main quasi-nondivergent expansion (12.10) must be completed by a term $\mathrm{M}_\infty^4 \log \mathrm{M}_\infty$. In particular, the following main expansion is written for $\mathcal{H}$:

$$\mathcal{H} = \mathcal{H}_0(\zeta) + \mathrm{M}_\infty^2 \mathcal{H}_2 + \left(\mathrm{M}_\infty^4 \log \mathrm{M}_\infty\right) \frac{\gamma \mathrm{S}}{2\pi \mathrm{Bo}} \frac{\partial^2 \overline{w_2}}{\partial \tau \partial \zeta} + \mathrm{M}_\infty^4 \mathcal{H}_4 + \dots . \tag{12.101}$$

In addition, the following local expansion which is valid in the vicinity of infinity in $\boldsymbol{\xi}$ is written for $\mathcal{H}$:

$$\mathcal{H} = \mathcal{H}_0(\zeta) + \mathrm{M}_\infty^4 \tilde{\mathcal{H}}_2 + \dots \quad . \tag{12.102}$$

When $|\tilde{\boldsymbol{\xi}}| \to 0$, matching imposes that $\tilde{\mathcal{H}}_2$ behaves as follows:

$$\begin{aligned}
\tilde{\mathcal{H}}_2 \cong \; & -\frac{\gamma}{8\pi^2 \mathrm{Bo}} (\overline{\Omega}_0)^2 \frac{1}{|\tilde{\boldsymbol{\xi}}|^2} - \frac{\gamma}{2\pi \mathrm{Bo}} \frac{1}{|\tilde{\boldsymbol{\xi}}|^2} \Bigg\{ \boldsymbol{v}^\infty \cdot (\boldsymbol{k} \times \boldsymbol{\mu}_0) \\
& + \frac{\tilde{\boldsymbol{\xi}} \cdot \mathrm{S}(\partial \boldsymbol{N}_0/\partial \tau) \cdot \boldsymbol{k} \times \tilde{\boldsymbol{\xi}}}{|\tilde{\boldsymbol{\xi}}|^2} - \frac{2}{|\tilde{\boldsymbol{\xi}}|^2} [(\boldsymbol{k} \times \boldsymbol{\mu}_0) \cdot \tilde{\boldsymbol{\xi}}]\, [\boldsymbol{v}^\infty \cdot \tilde{\boldsymbol{\xi}}] \Bigg\} \\
& - \frac{\gamma^2}{8\pi \mathrm{Bo}} \frac{\partial}{\partial \zeta} \left\{ \frac{\zeta^2}{K_0(\zeta)} \frac{\partial}{\partial \zeta} \left[\frac{\tilde{\boldsymbol{\xi}} \cdot \mathrm{S}^3 (\partial^3 \vec{N}_0/\partial \tau^3) \cdot (\boldsymbol{k} \times \tilde{\boldsymbol{\xi}})}{|\tilde{\boldsymbol{\xi}}|^2} \right] \right\} \\
& + \frac{\gamma \mathrm{S}}{\mathrm{Bo}} \frac{1}{2\pi} \frac{\partial^2 \overline{w_2}}{\partial \tau \partial \zeta} \log |\tilde{\boldsymbol{\xi}}| + \dots \equiv \tilde{\mathcal{H}}_2^\infty \quad .
\end{aligned} \tag{12.103}$$

We now want to obtain the equation which governs $\tilde{\mathcal{H}}_2(\tau, \tilde{\boldsymbol{\xi}}, \zeta)$. The following local equations are used for this purpose:

$$\begin{aligned}
& \mathrm{S} \frac{\partial \tilde{\boldsymbol{v}}_2}{\partial \tau} + \tilde{\boldsymbol{D}} \left(\frac{\tilde{\mathcal{H}}_2}{\gamma/\mathrm{Bo}} \right) + (\boldsymbol{v}^\infty \cdot \tilde{\boldsymbol{D}}) \tilde{\boldsymbol{v}}_1 + (\tilde{\boldsymbol{v}}_0 \cdot \tilde{\boldsymbol{D}}) \tilde{\boldsymbol{v}}_0 = 0 \quad ; \\
& \tilde{\boldsymbol{D}} \cdot \tilde{\boldsymbol{v}}_2 + \frac{\partial \tilde{w}_2}{\partial \zeta} = 0 \quad ; \\
& \mathrm{S} \frac{\partial \tilde{T}_2}{\partial \tau} - \mathrm{Bo} \frac{K_0(\zeta)}{\zeta} \tilde{w}_2 = 0 \quad ; \\
& \mathrm{Bo}\, \zeta \frac{\partial \tilde{\mathcal{H}}_2}{\partial \zeta} = -\tilde{T}_2 \quad .
\end{aligned} \tag{12.104}$$

From the above system, the following equation can be extracted for $\tilde{\mathcal{H}}_2$:

$$\begin{aligned}
& \left\{ \frac{\partial^2}{\partial \tilde{\xi}^2} + \frac{\partial^2}{\partial \tilde{\eta}^2} + \gamma \mathrm{S}^2 \frac{\partial^2}{\partial \tau^2} \left[\frac{\partial}{\partial \zeta} \left(\frac{\zeta^2}{K_0(\zeta)} \frac{\partial}{\partial \zeta} \right) \right] \right\} \tilde{\mathcal{H}}_2 \\
& + \frac{2\gamma}{\mathrm{Bo}} \frac{1}{|\tilde{\boldsymbol{\xi}}|^4} \left(\frac{\overline{\Omega}_0}{2\pi} \right)^2 = 0 \quad .
\end{aligned} \tag{12.105}$$

Equation (12.105) is the analogue of the wave equation for the theory of Viviand (1970) and Crow (1970) in unsteady external aerodynamics. It is pointed out that in this case, the acoustics emerge from the limiting process:

$$\mathrm{S} = \mathrm{O}(1) \quad , \quad \mathrm{M}_\infty \to 0 \quad , \quad \text{with} \quad t\,,\ \tilde{\boldsymbol{x}} = \mathrm{M}_\infty \boldsymbol{x} \ \text{fixed} \quad .$$

Equation (12.105) emerges from the same limiting process starting from the primitive equations. It will also be seen that the acoustics likewise result from the following limiting process [see Zeytounian and Guiraud (1984), in particular]:

$$\mathrm{S} = \mathrm{O}(1) \quad , \quad \mathrm{M}_\infty \to 0 \quad , \quad \text{with} \quad \hat{t} = t/\mathrm{M}_\infty \quad \text{and} \quad \boldsymbol{x} \ \text{fixed} \quad .$$

In Sect. 12.4 which is devoted to the analysis of the problem of unsteady adjustment, the above will be shown to be true also for atmospheric phenomena.

First of all, the following boundary condition on the (flat) ground must be prescribed for (12.105):

$$\tilde{\mathcal{H}}_2 + \frac{1}{\mathrm{Bo}} \frac{T_0(1)}{K_0(1)} \frac{\partial \tilde{\mathcal{H}}_2}{\partial \zeta} = 0 \quad , \quad \text{on} \quad \zeta = 1 \quad . \tag{12.106}$$

Next, according to the result obtained by El Mabrouk and Zeytounian (1984), the following far field behavior can be imposed in ζ:

$$\lim_{\zeta \to 0} \left\{ \frac{\zeta^2}{K_0(\zeta)} \tilde{\mathcal{H}}_2 \frac{\partial \tilde{\mathcal{H}}_2}{\partial \zeta} \right\} = 0 \quad . \tag{12.107}$$

Then at infinity in $\tilde{\boldsymbol{\xi}}$, the following must hold:

$$\tilde{\mathcal{H}}_2 \to 0 \quad \text{when} \quad |\tilde{\boldsymbol{\xi}}| \to \infty \quad . \tag{12.108}$$

Finally, the matching imposes:

$$\lim_{|\tilde{\xi}| \to 0} \tilde{\mathcal{H}}_2 = \tilde{\mathcal{H}}_2^\infty \quad , \tag{12.109}$$

where $\tilde{\mathcal{H}}_2^\infty$ is the expression (12.103).

In order to solve the problem (12.105–109), the following decomposition is first introduced:

$$\begin{aligned} \tilde{\mathcal{H}}_2 = &- \frac{\gamma}{8\pi^2 \mathrm{Bo}} (\overline{\Omega}_0)^2 \frac{1}{|\tilde{\xi}|^2} - \frac{\gamma}{2\pi \mathrm{Bo}} \frac{1}{|\tilde{\xi}|^2} \Big\{ \boldsymbol{v}^\infty \cdot (\boldsymbol{k} \times \boldsymbol{\mu}_0) \\ &- \frac{2}{|\tilde{\xi}|^2} [(\boldsymbol{k} \times \boldsymbol{\mu}_0) \cdot \tilde{\boldsymbol{\xi}}] (\boldsymbol{v}^\infty \cdot \tilde{\boldsymbol{\xi}}) \Big\} + \tilde{\mathcal{H}}_2^* \quad . \end{aligned} \tag{12.110}$$

We then represent $\tilde{\mathcal{H}}_2^*$ in the form:

$$\tilde{\mathcal{H}}_2^* = \sum_{n=1}^{\infty} \chi_n(\zeta) H_n(\tau, \tilde{\boldsymbol{\xi}}) \quad , \tag{12.111}$$

where the $\chi_n(\zeta)$ come from a spectral problem related to the operator

$$\frac{\partial}{\partial \zeta} \left[\frac{\zeta^2}{K_0(\zeta)} \frac{\partial}{\partial \zeta} \right] \quad ,$$

with the enumerable sequence μ_n^2 of eigenvalues and under the conditions in $\zeta = 1$ and $\zeta \to 0$ corresponding to (12.106, 107).

The functions $H_n(\tau, \tilde{\boldsymbol{\xi}})$ satisfy an equation of the type (12.44) where $1/\mathrm{Ro} \equiv \infty$, with the following condition:

$$H_n \to 0 \quad , \quad \text{with} \quad |\tilde{\boldsymbol{\xi}}| \to \infty \quad ,$$

as well as a behavior condition issued from (12.103, 110) when $|\tilde{\boldsymbol{\xi}}| \to 0$. It then turns out that the terms H_n can be expressed with the help of the integrals:

$$J_n(\tau, \tilde{\boldsymbol{\xi}}) = \int_{-\infty}^{\tau - \mathrm{S}\mu_n\sqrt{\gamma}|\tilde{\boldsymbol{\xi}}|} \frac{\lambda_n(\tau_1) d\tau_1}{\left[(\tau - \tau_1)^2 - \mathrm{S}^2\mu_n^2\gamma|\tilde{\boldsymbol{\xi}}|^2\right]^{1/2}} d\tau_1 \quad , \tag{12.112}$$

where the function λ_n comes into play in the behavior of H_n when $|\tilde{\boldsymbol{\xi}}| \to 0$.

It can thus be seen that the problem (12.105-109) is well-posed and has a unique solution which means that the above asymptotic analysis based on the MMAE is consistent.

In conclusion, the local expansion (12.102) which is valid in the vicinity of infinity in $\tilde{\boldsymbol{\xi}}$ can be ignored as long as the term proportional to $\mathrm{M}_\infty^4 \log \mathrm{M}_\infty$ in the main expansion (12.101) is taken into account. In this case, the so-called quasi-nondivergent approximation is included in a consistent expansion scheme and the vertical uncoupling is no longer a mystery given the apparition of an unsuspected term proportional to $\mathrm{M}_\infty^4 \log \mathrm{M}_\infty$. The coupling in the direction of the vertical coordinate ζ is realized via a local approximation which is valid horizontally in the vicinity of infinity in $\boldsymbol{\xi}$. For the time being, it is not known whether or not a local approximation must *also* be considered in the vicinity of $\zeta = 1$ (which simulates the ground in the primitive equations) when $\mathrm{M}_\infty \to 0$. It would be most worthwhile to be able to resolve this question.

12.4 The Problem of Adjustment to the Quasi-nondivergent Flow

We now return to the basic primitive equations (12.1) and introduce a short adjustment time:

$$\hat{\tau} = \tau/\mathrm{M}_\infty \quad . \tag{12.113}$$

We denote by $\hat{\mathcal{U}}$ any function $\mathcal{U}$ considered to be dependent on $\hat{\tau}$ rather than τ. The following equations are obtained for $\hat{\boldsymbol{v}}$, $\hat{w}$ and $\hat{\mathcal{H}}$:

$$\mathrm{S}\frac{\partial \hat{\boldsymbol{v}}}{\partial \hat{\tau}} + \frac{\mathrm{Bo}}{\gamma \mathrm{M}_\infty} \boldsymbol{D}\hat{\mathcal{H}} + \mathrm{M}_\infty \left\{ (\hat{\boldsymbol{v}} \cdot \boldsymbol{D})\hat{\boldsymbol{v}} + \hat{w}\frac{\partial \hat{\boldsymbol{v}}}{\partial \zeta} + \left(\frac{1}{\mathrm{Ro}} + \beta\eta\right) \boldsymbol{k} \times \hat{\boldsymbol{v}} \right\} = 0 \quad ; \tag{12.114a}$$

$$\mathrm{S}\frac{\partial}{\partial \hat{\tau}}\left(\frac{\partial \hat{\mathcal{H}}}{\partial \zeta}\right) + \mathrm{M}_\infty \left\{ \hat{\boldsymbol{v}} \cdot \boldsymbol{D}\left(\frac{\partial \hat{\mathcal{H}}}{\partial \zeta}\right) + \frac{\hat{w}}{\zeta}\left[\frac{\partial}{\partial \zeta}\left(\zeta \frac{\partial \hat{\mathcal{H}}}{\partial \zeta}\right) - \frac{\gamma - 1}{\gamma}\frac{\partial \hat{\mathcal{H}}}{\partial \zeta}\right] \right\} = 0 \quad ;$$

$$\boldsymbol{D}\cdot\hat{\boldsymbol{v}}+\frac{\partial\hat{w}}{\partial\zeta}=0 \quad . \tag{12.114b}$$

The solution to (12.114) is sought in the form of the following local expansion:

$$\begin{pmatrix}\hat{\boldsymbol{v}}\\ \hat{\mathcal{H}}\\ \hat{w}\end{pmatrix}=\begin{pmatrix}\hat{\boldsymbol{v}}_0\\ \hat{\mathcal{H}}_0\\ \hat{w}_0\end{pmatrix}+\mathrm{M}_\infty\begin{pmatrix}\hat{\boldsymbol{v}}_1\\ \hat{\mathcal{H}}_1\\ \hat{w}_1\end{pmatrix}+\dots \quad . \tag{12.115}$$

To order zero, it is noticed that:

$$\boldsymbol{D}\hat{\mathcal{H}}_0=0 \quad\text{and}\quad \mathrm{S}\frac{\partial}{\partial\hat{\tau}}\left(\frac{\partial\hat{\mathcal{H}}_0}{\partial\zeta}\right)=0\Rightarrow\hat{\mathcal{H}}_0\equiv\mathcal{H}_0(\zeta) \quad ,$$

and we again find the standard atmosphere with:

$$\hat{T}_0\equiv T_0(\zeta)=-\mathrm{Bo}\zeta\frac{d\mathcal{H}_0}{d\zeta} \quad .$$

To the next order, the following adjustment equations result for the functions $\hat{\boldsymbol{v}}_0$, $\hat{\mathcal{H}}_1$ and $\hat{w}_0$:

$$\frac{\partial\hat{\boldsymbol{v}}_0}{\partial\hat{\tau}}+\frac{\mathrm{Bo}}{\gamma\mathrm{S}}\boldsymbol{D}\hat{\mathcal{H}}_1=0 \quad ; \quad \boldsymbol{D}\cdot\hat{\boldsymbol{v}}_0+\frac{\partial\hat{w}_0}{\partial\zeta}=0 \quad ; \tag{12.116a}$$

$$\mathrm{S}\frac{\partial}{\partial\hat{\tau}}\left(\frac{\partial\hat{\mathcal{H}}_1}{\partial\zeta}\right)+\frac{K_0(\zeta)}{\zeta^2}\hat{w}_0=0 \quad . \tag{12.116b}$$

From (12.116), an equation for $\hat{\mathcal{H}}_1$ can once again be derived:

$$\frac{\gamma\mathrm{S}^2}{\mathrm{Bo}}\frac{\partial^2}{\partial\hat{\tau}^2}\left[\frac{\partial}{\partial\zeta}\left(\frac{\zeta^2}{K_0(\zeta)}\frac{\partial\hat{\mathcal{H}}_1}{\partial\zeta}\right)\right]+\boldsymbol{D}^2\hat{\mathcal{H}}_1=0 \quad . \tag{12.117}$$

So as to be able to pursue the analysis, we will limit our attention to the case $K_0(\zeta)\equiv$ const. Let us introduce the following change in variables:

$$t=\frac{1}{\mathrm{S}}\sqrt{\frac{\mathrm{Bo}K_{00}}{\gamma}}\hat{\tau} \quad\text{and}\quad \log\frac{1}{\zeta}=z \quad , \tag{12.118}$$

where K_{00} is the constant value of $K_0(\zeta)$ [$K_{00}\equiv K_0(1)$ may be used.] Hence, instead of (12.117), we have:

$$\frac{\partial^2}{\partial t^2}\left\{\frac{\partial^2}{\partial z^2}-\frac{\partial}{\partial z}\right\}\hat{\mathcal{H}}_1+\boldsymbol{D}^2\hat{\mathcal{H}}_1=0 \quad ,$$

or, by introducing the new function:

$$\hat{\mathcal{G}}_1\equiv\exp\left(-\frac{z}{2}\right)\hat{\mathcal{H}}_1 \quad , \tag{12.119}$$

$$\frac{\partial^2}{\partial t^2}\left(\frac{\partial^2\hat{\mathcal{G}}_1}{\partial z^2}-\frac{1}{4}\hat{\mathcal{G}}_1\right)+\frac{\partial^2\hat{\mathcal{G}}_1}{\partial\xi^2}+\frac{\partial^2\hat{\mathcal{G}}_1}{\partial\eta^2}=0 \quad . \tag{12.120}$$

The general solution to (12.120) which is valid throughout the entire space (t, z, ξ, η) is given by the formula:

$$\hat{\mathcal{G}}_1 = \iiint_{-\infty}^{+\infty} \exp\{\mathrm{i}(kz + l\xi + m\eta)\}\left\{A\cos(\sigma t) + B\frac{\sin(\sigma t)}{\sigma}\right\}dk\,dl\,dm \quad , \tag{12.121}$$

where

$$\sigma = \sqrt{\frac{l^2 + \mathrm{M}^2}{k^2 + 1/4}} \quad .$$

The functions A and B are determined from the initial conditions. We must prove now that $\hat{\mathcal{G}}_1 \to 0$ with $t \to \infty$. To this end, the expression under the triple integral is first put into the form of a sum of $E \equiv \exp\{\mathrm{i}(kz + l\xi + m\eta) \pm \mathrm{i}\sigma t\}$. Then the steady phase method is applied in order to obtain an asymptotic behavior of $\hat{\mathcal{G}}_1$ when $t \to \infty$. This leads us to seek out k^*, l^* and m^* such that:

$$z \pm t\left(\frac{\partial\sigma}{\partial k}\right)^* = \xi \pm t\left(\frac{\partial\sigma}{\partial l}\right)^* = \eta \pm t\left(\frac{\partial\sigma}{\partial m}\right)^* = 0 \quad .$$

Let us then consider:

$$\sigma = \sigma^* + \tilde{L} + \tilde{\Delta} + \text{ higher order terms}$$

where σ^* is the value of σ in $k = k^*$, $l = l^*$ and $m = m^*$, whereas $\tilde{L}$ is linear, and $\tilde{\Delta}$ quadratic with respect to $\tilde{k} = k - k^*$, $\tilde{l} = l - l^*$ and $\tilde{m} = m - m^*$. In this case, asymptotically we obtain for $\hat{\mathcal{G}}_1$:

$$\hat{\mathcal{G}}_1 = \exp\{\mathrm{i}(k^* z + l^*\xi + m^*\eta)\}\left[A^* \iiint_{-\infty}^{+\infty} \cos(t\tilde{\Delta})d\tilde{k}\,d\tilde{l}\,d\tilde{m} + \frac{B^*}{\sigma^*}\iiint_{-\infty}^{+\infty} \sin(t\tilde{\Delta})d\tilde{k}\,d\tilde{l}\,d\tilde{m}\right] \quad . \tag{12.122}$$

It must now be kept in mind that $\tilde{\Delta}$ is a homogeneous quadratic polynomial in $\tilde{k}$, $\tilde{l}$ and $\tilde{m}$ and so, by setting:

$$t^{1/2}\tilde{k} = \hat{k} \quad , \quad t^{1/2}\tilde{l} = \hat{l} \quad \text{and} \quad t^{1/2}\tilde{m} = \hat{m} \Rightarrow t\tilde{\Delta} = \hat{\Delta} \quad ,$$

where $\hat{\Delta}$ is the same homogeneous quadratic polynomial but with $\hat{k}$, $\hat{l}$ and $\hat{m}$. Consequently, the following expression results:

$$\hat{\mathcal{G}}_1 = \frac{\varepsilon^*}{t^{3/2}}\left\{\alpha A^* + \beta\frac{B^*}{\sigma^*}\right\} \quad ,$$

where

$$\varepsilon^* = \exp\{\mathrm{i}(k^* z + l^*\xi + m^*\eta)\}$$

and

$$\alpha = \iiint_{-\infty}^{+\infty} \cos(\hat{\Delta})d\hat{k}\,d\hat{l}\,d\hat{m} \quad \text{and} \quad \beta = \iiint_{-\infty}^{+\infty} \sin(\hat{\Delta})d\hat{k}\,d\hat{l}\,d\hat{m} \quad .$$

The above well demonstrates the decrease of $\hat{\mathcal{G}}_1$ towards zero with $t \to +\infty$. Thus, there is indeed adjustment to the classical quasi-nondivergent flow when $t \to \infty$, i.e.:

$$\lim_{\hat{\tau}\to\infty} \hat{w}_0 \equiv 0 \quad , \quad \lim_{\hat{\tau}\to\infty} \hat{v}_0 \equiv \lim_{\tau\to 0} v_0 \quad , \quad \lim_{\hat{\tau}\to\infty} \hat{\mathcal{H}}_1 \equiv 0 \tag{12.123a}$$

$$\text{and} \quad \lim_{\hat{\tau}\to\infty} \hat{T}_1 \equiv 0 \quad , \quad \text{with} \quad (\boldsymbol{D}\cdot \boldsymbol{v}_0)_{\tau=0} = 0 \quad . \tag{12.123b}$$

In conclusion, it is observed that we can impose on the classical quasi-nondivergent model (12.17), the initial condition prescribed at the start for $\boldsymbol{v}$ according to (11.9):

$$\boldsymbol{v}_0 = \boldsymbol{V}^0 \quad , \quad \text{for} \quad \tau = 0 \quad , \tag{12.124}$$

where $\boldsymbol{V}^0 \equiv U^0\boldsymbol{i} + V^0\boldsymbol{j}$.

12.5 Complementary Remarks

12.5.1. Let us go back to the representation (12.111) for $\tilde{\mathcal{H}}_2^*$. First of all, it is specified that the set of eigenfunctions $\chi_n(\zeta)$ is *orthonormed*:

$$\int_0^1 \chi_n(\zeta)\chi_m(\zeta)d\zeta = \delta_{nm} = \begin{cases} 0 & n \neq m \\ 1 & n \equiv m \end{cases}$$

and forms a *complete system*:

for any function $f(\zeta) \subset L^2(0,1)$ (square integrable), we have:
$f(\zeta) = \sum_{n=1}^{\infty}(f,\chi_n)\chi_n$, where $(f,\chi_n) = \int_0^1 f(\zeta)\chi_n(\zeta)d\zeta$ and $\|f\|^2 = (f,f)$.
$H_n(\tau,\tilde{\boldsymbol{\xi}})$ must satisfy the following problem:[4]

$$\begin{aligned} &\frac{\partial^2 H_n}{\partial\tilde{\xi}^2} + \frac{\partial^2 H_n}{\partial\tilde{\eta}^2} - \frac{1}{\alpha_n^2}\frac{\partial^2 H_n}{\partial\tau^2} = 0 \quad ; \\ &|\tilde{\boldsymbol{\xi}}| \to 0 : \quad H_n \to 0 \quad ; \\ &H_n \equiv H_n^{(1)} + H_n^{(2)} \quad , \end{aligned} \tag{12.125a}$$

$$\begin{aligned} &H_n^{(1)} \cong -\frac{\gamma}{2\pi \mathrm{Bo}}\frac{1}{|\tilde{\boldsymbol{\xi}}|^2}\frac{\tilde{\boldsymbol{\xi}}\cdot(\mathrm{S}(\partial \boldsymbol{N}_0/\partial\tau),\chi_n)\cdot \boldsymbol{k}\times\tilde{\boldsymbol{\xi}}}{|\tilde{\boldsymbol{\xi}}|^2} \quad , \\ &\text{and } H_n^{(2)} \cong \frac{\gamma \mathrm{S}}{\mathrm{Bo}}\frac{1}{2\pi}\left(\frac{\partial^2\overline{w_2}}{\partial\tau\partial\zeta},\chi_n\right)\log|\tilde{\boldsymbol{\xi}}| \quad , \quad \text{when } |\tilde{\boldsymbol{\xi}}| \to 0 \quad . \end{aligned} \tag{12.125b}$$

[4] If we return to (12.103, 110), it will be remarked that in the behavior of $H_n^{(1)}$ when $|\tilde{\boldsymbol{\xi}}| \to 0$, we have (purposely) omitted the term related to $\mathrm{S}^3(\partial^3 \boldsymbol{N}_0/\partial\tau^3)$! This is due to the fact that in the vicinity of $|\tilde{\boldsymbol{\xi}}| = 0$, during the expansion of the term explicitly written, this term will appear naturally. In addition, it will be noted that in the equation for H_n, we have:

$$\alpha_n = \left[\mathrm{S}\mu_n\sqrt{\gamma}\right]^{-1}.$$

It then turns out that in (12.112), we also have two contributions related respectively to:

$$\lambda_n^{(1)} = \left(\mathrm{S}\frac{\partial N_0}{\partial \tau}, \chi_n\right) \Rightarrow J_n^{(1)}, \quad \text{and} \quad \lambda_n^{(2)} = \left(\frac{\partial^2 \overline{w_2}}{\partial \tau \partial \zeta}, \chi_n\right) \Rightarrow J_n^{(2)} \quad .$$

Therefore, the solution to problem (12.125) can be written in the following form:

$$H_n = \frac{\gamma}{4\pi \mathrm{Bo}} \frac{1}{|\tilde{\boldsymbol{\xi}}|^2} (\boldsymbol{k} \times \tilde{\boldsymbol{\xi}}) \cdot \frac{\partial^2 J_n^{(1)}}{\partial |\tilde{\boldsymbol{\xi}}|^2} \cdot \tilde{\boldsymbol{\xi}} - \frac{\gamma \mathrm{S}}{\mathrm{Bo}} \frac{1}{2\pi} J_n^{(2)}, \tag{12.126}$$

where $J_n^{(i)}$ is the integral (12.112) with $\lambda_n^{(i)}$, $i = 1, 2$.

12.5.2. Let us turn our attention back to the asymptotic expansions (12.101, 102), which are respectively the main and the local expansions. We wish to demonstrate that the inclusion of the term proportional to $\mathrm{M}_\infty^4 \log \mathrm{M}_\infty$ in the main expansion (12.101) is indeed the only possibility for taking into account this term within the framework of the asymptotic theory expounded in Sect. 12.3. In fact, if this term is included in the local expansion (12.102):

$$\mathcal{H} = \mathcal{H}_0(\zeta) + (\mathrm{M}_\infty^4 \log \mathrm{M}_\infty)\tilde{\mathcal{H}}_{20} + \mathrm{M}_\infty^4 \tilde{\mathcal{H}}_2 + \dots \tag{12.127}$$

we are then led to solve the following problem for $\tilde{\mathcal{H}}_{20}$:

$$\begin{aligned}
&\frac{\partial^2 \tilde{\mathcal{H}}_{20}}{\partial \tilde{\xi}^2} + \frac{\partial^2 \tilde{\mathcal{H}}_{20}}{\partial \tilde{\eta}^2} + \gamma \mathrm{S}^2 \frac{\partial^2}{\partial \tau^2}\left[\frac{\partial}{\partial \zeta}\left(\frac{\zeta^2}{K_0(\zeta)} \frac{\partial \tilde{\mathcal{H}}_{20}}{\partial \zeta}\right)\right] = 0 \quad , \\
&\lim_{\zeta \to 0}\left(\frac{\zeta^2}{K_0(\zeta)} \tilde{\mathcal{H}}_{20} \frac{\partial \tilde{\mathcal{H}}_{20}}{\partial \zeta}\right) = 0 \quad ; \\
&\left(\tilde{\mathcal{H}}_{20} + \frac{1}{\mathrm{Bo}} \frac{T_0(1)}{K_0(1)} \frac{\partial \tilde{\mathcal{H}}_{20}}{\partial \zeta}\right)_{\zeta=1} = 0 \quad ; \\
&\tilde{\mathcal{H}}_{20} \to 0 \quad , \quad \text{with} \quad |\tilde{\boldsymbol{\xi}}| \to \infty \quad ; \\
&\tilde{\mathcal{H}}_{20} \to -\frac{\gamma \mathrm{S}}{\mathrm{Bo}} \frac{1}{2\pi} \frac{\partial^2 \overline{w_2}}{\partial \tau \partial \zeta} \quad , \quad \text{when} \quad |\tilde{\boldsymbol{\xi}}| \to 0 \quad .
\end{aligned} \tag{12.128}$$

Like in Sect. 12.3, we set $\tilde{\mathcal{H}}_{20} = \sum_{n=1}^{\infty} A_n(\tau, \tilde{\xi}, \tilde{\eta})\chi_n(\zeta)$ and the following must be solved for A_n:

$$\begin{aligned}
&\frac{\partial^2 A_n}{\partial \tilde{\xi}^2} + \frac{\partial^2 A_n}{\partial \tilde{\eta}^2} - \frac{1}{\alpha_n^2} \frac{\partial^2 A_n}{\partial \tau^2} = 0 \quad , \\
&|\tilde{\boldsymbol{\xi}}| \to 0 : A_n \to -\frac{\gamma \mathrm{S}}{\mathrm{Bo}} \frac{1}{2\pi}\left(\frac{\partial^2 \overline{w_2}}{\partial \tau \partial \zeta}, \chi_n\right) \quad , \\
&|\tilde{\boldsymbol{\xi}}| \to \infty : A_n \to 0 \quad .
\end{aligned} \tag{12.129}$$

According to the results in Sect. 12.4, the following initial conditions can be prescribed for (12.129):

$$\tau = 0 : A_n = \frac{\partial A_n}{\partial \tau} = 0 \quad . \tag{12.130}$$

Let us demonstrate that this problem (12.129, 130) can only have the identically zero trivial solution. Firstly, we can write:

$$\begin{aligned}
0 &= \left(\frac{1}{\alpha_n^2} \frac{\partial^2 A_n}{\partial \tau^2} - \tilde{\boldsymbol{D}}^2 A_n \right) \frac{\partial A_n}{\partial \tau} = \frac{1}{\alpha_n^2} \frac{\partial^2 A_n}{\partial \tau^2} \frac{\partial A_n}{\partial \tau} + \tilde{\boldsymbol{D}} A_n \cdot \frac{\partial \tilde{\boldsymbol{D}} A_n}{\partial \tau} \\
&\quad - \frac{\partial A_n}{\partial \tau} \tilde{\boldsymbol{D}} \cdot (\tilde{\boldsymbol{D}} A_n) - \tilde{\boldsymbol{D}} A_n \cdot \frac{\partial \tilde{\boldsymbol{D}} A_n}{\partial \tau} \\
&= \frac{\partial}{\partial \tau} \left\{ \frac{1}{2} \left[\frac{1}{\alpha_n^2} \left(\frac{\partial A_n}{\partial \tau} \right)^2 + |\tilde{\boldsymbol{D}} A_n|^2 \right] \right\} - \tilde{\boldsymbol{D}} \cdot \left(\frac{\partial A_n}{\partial \tau} \tilde{\boldsymbol{D}} A_n \right) \quad .
\end{aligned}$$

Let us integrate this relation with respect to $\tilde{\xi}$ and $\tilde{\eta}$ in the domain $\tilde{\xi}^2 + \tilde{\eta}^2 > R^2$ (where R is an arbitrary constant). We obtain:

$$\begin{aligned}
&\frac{1}{2} \frac{\partial}{\partial \tau} \iint\limits_{|\tilde{\boldsymbol{\xi}}| > R} \left\{ \frac{1}{\alpha_n^2} \left(\frac{\partial A_n}{\partial \tau} \right)^2 + |\tilde{\boldsymbol{D}} A_n|^2 \right\} d\tilde{\xi}\, d\tilde{\eta} \\
&+ \int\limits_{|\tilde{\boldsymbol{\xi}}| = R} \frac{\partial A_n}{\partial \tau} \boldsymbol{n} \cdot \tilde{\boldsymbol{D}} A_n \, d\sigma = 0 \quad ,
\end{aligned}$$

while taking into account that for any bounded time, $A_n(\tau, \tilde{\xi}, \tilde{\eta})$ is certainly identically zero in the vicinity of $|\tilde{\boldsymbol{\xi}}| = \infty$ [which is obvious from (12.130)]. Thus, when $|\tilde{\boldsymbol{\xi}}| \to 0$, $\partial A_n / \partial \tau$ and $\tilde{\boldsymbol{D}} A_n$ tend to bounded values in such a way that:

$$\lim_{R \to 0} \int\limits_{|\tilde{\boldsymbol{\xi}}| = R} \frac{\partial A_n}{\partial \tau} \boldsymbol{n} \cdot \tilde{\boldsymbol{D}} A_n \, d\sigma = \lim_{|\tilde{\boldsymbol{\xi}}| \to 0} \left(\frac{\partial A_n}{\partial \tau} \tilde{\boldsymbol{D}} A_n \right) \cdot \int\limits_{|\tilde{\boldsymbol{\xi}}| = R} \boldsymbol{n} \, d\sigma = 0 \quad .$$

Hence,

$$\frac{\partial}{\partial \tau} \iint \left(\frac{1}{\alpha_n^2} \left(\frac{\partial A_n}{\partial \tau} \right)^2 + |\tilde{\boldsymbol{D}} A_n|^2 \right) d\tilde{\xi}\, d\tilde{\eta} = 0 \quad ,$$

where the integral is extended over the entire plane $(\tilde{\xi}, \tilde{\eta})$. As the initial conditions are zero, we finally obtain:

$$\iint \left(\frac{1}{\alpha_n^2} \left(\frac{\partial A_n}{\partial \tau} \right)^2 + |\tilde{\boldsymbol{D}} A_n|^2 \right) d\tilde{\xi}\, d\tilde{\eta} = 0 \quad , \tag{12.131}$$

which implies that the solution to (12.129, 130) is identically zero. It thus becomes obvious that the choice opted for in Sect. 12.3 concerning the inclusion of the term proportional to $\mathrm{M}_\infty^4 \log \mathrm{M}_\infty$ in the main expansion (12.101) is the only possibility.

12.5.3. To finish up this Chap. 12, let us return to (12.19) which allows w_2 to be determined once the quasi-nondivergent model (12.17) has yielded $\boldsymbol{v}_0$ and $\Pi_2 \equiv (\mathrm{Bo}/\gamma)\mathcal{H}_2$, then T_2, according to (12.20). In the main expansion (12.10), the ground (adiabatic and flat) is simulated by the equation $\zeta = 1$. It is thus necessary to verify whether or not $\lim\limits_{\zeta \to 1} w_2$ is compatible with the slip condition to order M_∞^2 in $\zeta = 1$. Since

$$\left.\frac{d\mathcal{H}_0}{d\zeta}\right|_{\zeta=1} = -\frac{T_0(1)}{\mathrm{Bo}} \quad \text{and} \quad \mathcal{H}_0(1) = 0 \quad ,$$

it is readily found that:

$$w_2 = \frac{\mathrm{Bo}}{T_0(1)}\left\{\mathrm{S}\frac{\partial \mathcal{H}_2}{\partial \tau} + \boldsymbol{v}_0 \cdot \boldsymbol{D}\mathcal{H}_2\right\} \quad , \quad \text{on} \quad \zeta = 1 \quad . \tag{12.132}$$

Moreover:

$$\zeta = \zeta_{\text{ground}} = 1 + \mathrm{M}_\infty^2 \zeta_{s2} + \ldots \quad , \quad \text{with} \quad \zeta_{s2} = \mathrm{Bo}\frac{\mathcal{H}_2|_{\zeta=1}}{T_0(1)} \quad . \tag{12.133}$$

Once again, it is emphasized that (12.132, 133) are direct consequences of the slip condition (12.2) when the main asymptotic expansion (12.10) with τ, ξ, η and ζ fixed is taken into account.

Let us now consider $\lim\limits_{\zeta \to 1} w_2$ where w_2 is calculated as a function of $T_2 = -\mathrm{Bo}\zeta(\partial \mathcal{H}_2/\partial \zeta)$ from (12.19). Hence:

$$\begin{aligned}\lim_{\zeta \to 1} w_2 &= \frac{1}{\mathrm{Bo}}\frac{1}{K_0(1)}\left(\mathrm{S}\frac{\partial T_2}{\partial \tau} + \boldsymbol{v}_0 \cdot \boldsymbol{D}T_2\right)_{\zeta=1} \\ &= -\frac{1}{K_0(1)}\left[\mathrm{S}\frac{\partial}{\partial \tau}\left(\frac{\partial \mathcal{H}_2}{\partial \zeta}\right)_{\zeta=1} + \boldsymbol{v}_0|_{\zeta=1} \cdot \boldsymbol{D}\left(\frac{\partial \mathcal{H}_2}{\partial \zeta}\right)_{\zeta=1}\right]\end{aligned} \tag{12.134}$$

We obviously have:

$$\lim_{\zeta \to 1} w_2 \neq w_2|_{\zeta=1}$$

where $w_2|_{\zeta=1}$ is calculated from (12.132) and $\lim\limits_{\zeta \to 1} w_2$, from (12.134).

In conclusion, it can be stated that *the vicinity of* $\zeta = 1$ *is certainly singular for the main asymptotic representation* (12.9, 10) which leads to the quasi-nondivergent model (12.17), and then to relations (12.19, 20) and to the "second approximation" equations (12.21, 22).

It thus becomes necessary to postulate the existence of a local asymptotic representation by introducing a local altitude variable $\tilde{\zeta} = (1 - \zeta)/\mathrm{M}_\infty^\alpha$, $\alpha > 0$:

$$\mathrm{Re}_\perp \equiv \infty \quad , \quad \mathrm{M}_\infty \to 0 \ \text{ with } \ \tau, \xi, \eta \ \text{ and } \ \tilde{\zeta} \equiv \frac{1-\zeta}{\mathrm{M}_\infty^\alpha}, \ a > 0 \ \text{ fixed} \quad ;$$

$$\begin{pmatrix} \boldsymbol{v} \\ w \\ \mathcal{H} \\ T \end{pmatrix} = \begin{pmatrix} \tilde{\boldsymbol{v}}_0 \\ \tilde{w}_0 \\ \tilde{\mathcal{H}}_0 \\ \tilde{T}_0 \end{pmatrix} + \mathrm{M}_\infty^\alpha \begin{pmatrix} \tilde{\boldsymbol{v}}_\alpha \\ \tilde{w}_\alpha \\ \tilde{\mathcal{H}}_\alpha \\ \tilde{T}_\alpha \end{pmatrix} + \ldots + \mathrm{M}_\infty^2 \begin{pmatrix} \tilde{\boldsymbol{v}}_2 \\ \tilde{w}_2 \\ \tilde{\mathcal{H}}_2 \\ \tilde{T}_2 \end{pmatrix} + \ldots \quad . \tag{12.135}$$

According to the analysis in Subsect. 12.1.1, it would appear that $\alpha = 1$ can be used.

If we now take into account the behavior of the functions characterizing the main asymptotic representation (which is valid far from the plane wall $\zeta = 1$) in the vicinity of $\zeta = 1$, i.e., when $\tilde{\zeta} = (1 - \zeta)/\mathrm{M}_\infty \to \infty$, we must set:

$$
\begin{aligned}
&\tilde{\boldsymbol{v}}_0 \equiv \boldsymbol{v}_0|_{\zeta=1}\,; \quad \tilde{\mathcal{H}}_0 \equiv \mathcal{H}_0(1) = 0\,; \quad \tilde{T}_0 \equiv T_0(1)\,; \quad \tilde{w}_0 = \tilde{w}_1 \equiv 0 \quad ; \\
&\tilde{\mathcal{H}}_1 \equiv \tilde{\zeta}\frac{T_0(1)}{\mathrm{Bo}} \quad ; \quad \tilde{T}_1 \equiv -\tilde{\zeta}\left.\frac{dT_0}{d\zeta}\right|_{\zeta=1} \quad ; \\
&\tilde{\mathcal{H}}_2 \equiv \mathcal{H}_2|_{\zeta=1} + \frac{\tilde{\zeta}^2}{2\mathrm{Bo}}\left[T_0(1) - \left.\frac{dT_0}{d\zeta}\right|_{\zeta=1}\right] \quad ; \\
&\tilde{T}_2 \equiv \tilde{\zeta}^2\left[\left.\frac{dT_0}{d\zeta}\right|_{\zeta=1} - T_0(1)\right] + \tilde{\theta}_2(\tau, \xi, \eta, \tilde{\zeta}) \quad .
\end{aligned}
\tag{12.136}
$$

In this case, the following system of local equations results for $\tilde{\boldsymbol{v}}_1$, $\tilde{w}_2$, $\tilde{\mathcal{H}}_3$ and $\tilde{\theta}_2$ which are functions of τ, ξ, η and $\tilde{\zeta}$:

$$
\begin{aligned}
&\mathrm{S}\frac{\partial \tilde{\boldsymbol{v}}_1}{\partial \tau} + \boldsymbol{v}_0|_{\zeta=1} \cdot \boldsymbol{D}\tilde{\boldsymbol{v}}_1 + \tilde{\boldsymbol{v}}_1 \cdot \boldsymbol{D}\boldsymbol{v}_0|_{\zeta=1} + \left(\frac{1}{\mathrm{Ro}} + \beta\eta\right)(\boldsymbol{k} \times \tilde{\boldsymbol{v}}_1) \\
&\quad + \frac{\mathrm{Bo}}{\gamma}\boldsymbol{D}\tilde{\mathcal{H}}_3 = 0 \quad ; \\
&\boldsymbol{D} \cdot \tilde{\boldsymbol{v}}_1 = \frac{\partial \tilde{w}_2}{\partial \tilde{\zeta}}; \quad \tilde{\theta}_2 = \mathrm{Bo}\frac{\partial \tilde{\mathcal{H}}_3}{\partial \tilde{\zeta}} \quad ; \\
&\mathrm{S}\frac{\partial \tilde{\theta}_2}{\partial \tau} + \boldsymbol{v}_0|_{\zeta=1} \cdot \boldsymbol{D}\tilde{\theta}_2 = \mathrm{Bo}K_0(1)\tilde{w}_2 \quad .
\end{aligned}
\tag{12.137}
$$

In the local representation, the ground is in:

$$
\tilde{\zeta} = \tilde{\zeta}_{\text{ground}} = \mathrm{M}_\infty \tilde{\zeta}_{s1} + \ldots, \quad \text{where} \quad \tilde{\zeta}_{s1} \equiv -\zeta_{s2} = \mathrm{Bo}\frac{\mathcal{H}_2|_{\zeta=1}}{T_0(1)}
$$

and the following slip condition must be imposed on (12.137):

$$
\tilde{w}_2 = \frac{\mathrm{Bo}}{T_0(1)}\left\{\mathrm{S}\frac{\partial \mathcal{H}_2|_{\zeta=1}}{\partial \tau} + \boldsymbol{v}_0|_{\zeta=1} \cdot \boldsymbol{D}\mathcal{H}_2|_{\zeta=1}\right\} \quad , \quad \text{on } \tilde{\zeta} = 0 \quad . \tag{12.138}
$$

Matching conditions between the main representation (12.9, 10) and the local representation (12.135) must still be written. In particular, for the w's, this yields:

$$
\lim_{\tilde{\zeta} \to \infty} \tilde{w}_2 = w_2(\tau, \xi, \eta, 1) \quad , \tag{12.139}
$$

or according to (12.134, 137):

$$
\lim_{\tilde{\zeta} \to \infty} \frac{\partial \tilde{\mathcal{H}}_3}{\partial \tilde{\zeta}} = -\left(\frac{\partial \mathcal{H}_2}{\partial \zeta}\right)_{\zeta=1} \quad . \tag{12.140}
$$

We will not, however, pursue this analysis any further. More research will be necessary in order to prove the internal coherency of the considered asymptotic model.

13. The Models for the Local and Regional Scales Atmospheric Flows

In this chapter, *two* privileged meteorological situations will be considered as a function of the hydrostatic parameter $\varepsilon = H/L$:

$$\varepsilon \sim 10^{-1} \Rightarrow \mathrm{Ro} \sim 1 \quad ; \tag{13.1}$$

and

$$\varepsilon \sim 1 \Rightarrow \mathrm{Ro} \gg 1 \quad . \tag{13.2}$$

The first situation characterized by (13.1) is related to *regional* type phenomena (or mesometeorological), such as the *breeze* phenomenon. In order to model the breeze, it is in fact necessary to take into account the intrinsic characteristic vertical scale $h_0 = R\Delta T_0/g$ where ΔT_0 is the characteristic temperature fluctuation related to the thermal field on the ground. In this case [see (8.50)], the significant Boussinesq number for our problem is:

$$\mathrm{Bo} = \frac{h_0 g}{RT_\infty(0)} = \frac{\Delta T_0}{T_\infty(0)} \equiv \tau_0 \ll 1 \quad . \tag{13.3}$$

Hence, the Boussinesq approximation according to Zeytounian (1974) can be carried out:

$$\tau_0 \Rightarrow 0 \quad , \quad \mathrm{M}_\infty \Rightarrow 0 \quad , \quad \frac{\tau_0}{\mathrm{M}_\infty} = \hat{\tau} = \mathrm{O}(1) \quad , \tag{13.4}$$

and then:

$$\varepsilon \to 0 \quad , \quad \mathrm{Re} \to \infty \quad \text{with} \quad \varepsilon^2 \mathrm{Re} = \mathrm{Re}_\perp = \mathrm{O}(1) \quad .$$

We will see, however, that it then becomes necessary to define a characteristic velocity U_0 intrinsically linked to the breeze phenomenon which also leads to the introducing of a Grashof number Gr instead of the Reynolds number Re.

But with the help of situation (13.1), the problem of the flow over a high relief in an adiabatic atmosphere can be also considered. In this case, we must be able to explain the implications of the following asymptotic situation:

$$\mathrm{Re} \equiv \infty \quad ; \quad \varepsilon \to 0 \quad \text{and} \quad \mathrm{M}_\infty \to 0 \quad , \quad \text{with} \quad \mathrm{M}_\infty = \hat{\varepsilon}\varepsilon^\beta \quad , \tag{13.5}$$

where $\hat{\varepsilon} = \mathrm{O}(1)$ and $\beta > 0$ is a real exponent to be determined.

When $\mathrm{Bo} = H/(RT_\infty(0)/g) = \mathrm{O}(1)$, the asymptotic model associated with the limiting process (13.5) remains to be constructed. It will be remarked that all asymptotic situations related to (13.1) are characterized by the filtering out of internal acoustic waves, as well as short internal gravity waves (in particular, those which propagate in nearly horizontal directions such as the so-called "lee waves").

On the other hand, (13.2) does not filter out the short internal gravity waves. This meteorological situation is representative of *local* scale phenomena for which *the Coriolis force is negligible* – these are precisely the *lee wave* type phenomena which form downstream of a mountain. The classical lee wave theory (see, for example, Zeytounian, 1969, and Yih's book, 1980) is often carried out within the framework of the adiabatic Boussinesq approximation:

$$\varepsilon \equiv 1 \quad , \quad \mathrm{Ro} \equiv \infty \quad , \quad \mathrm{Re} \equiv \infty \quad , \quad \mathrm{Bo} \to 0 \quad , \quad \mathrm{M}_\infty \to 0 \quad , \tag{13.6}$$

with $\mathrm{Bo} = \hat{B}\mathrm{M}_\infty$ and $\hat{B} = O(1)$.

Nevertheless, this Boussinesq-type model obtained from the limiting process (13.6) is valid only in a restricted vicinity of the mountain. And so it is necessary to complete this Boussinesq lee wave model ("inner" model) by an "outer" model which is valid beyond the immediate vicinity of the mountain. For the time being, it is not known how to obtain such an outer model for a three-dimensional flow (steady or unsteady). However, a result by Guiraud (1979) shows that, at infinity, far from the mountain, it is necessary to impose a classical Sommerfeld-type radiation condition for the three-dimensional model of classical steady lee waves (valid in the vicinity of the mountain). The above result suggests that the related outer asymptotic representation (valid outside the immediate vicinity of the mountain) must of necessity be in double scale. Just such a double scale representation was constructed by Guiraud and Zeytounian (1979) for the plane steady adiabatic flow. We will return to this matter later on (see Subsect. 13.2.4).

Concerning relevant bibliographical references, let us first cite the book by Gutmam (1969, English translation 1972), our Course edited by the "Direction de la Météorologie Nationale" in Paris (Zeytounian, 1968), the monograph edited under the direction of Paul Queney (1960) and also the relatively recent monograph by Atkinson (1981). Next are cited the books by Yih (1980) and Scorer (1978) which give a general review of the problems posed by the aerodynamics of the environment. On the subject of the fluid mechanics of the environment, the report by Hunt (1980) at the IUTAM Congress (1980) is most instructive. Finally, we give as reference the review by Smith (1979) in "Advances in Geophysics" concerning the influence of mountains on atmospheric flows.

In the above-cited works, the reader will find a sufficiently complete list of articles on the various subjects treated in the present chapter. Nonetheless, too few articles have been devoted to the consistent asymptotic modelling of these local and regional scales phenomena. A systematic review of this question remains to be carried out and the present work is a first attempt in this direction.

The organization of the present chapter is as follows: Section 13.1 is devoted entirely to a review of breeze-type *circulation* models referred to as *free* or *natural*. The various *lee wave models* are studied in Sect. 13.2. The *interaction phenomenon between free and forced* circulations is the subject of Sect. 13.3. "Complementary Remarks", Sect. 13.4, deals with related questions and problems to be resolved.

The various reviews which make up this final chapter have been directly inspired from the author's works as well as those carried out in collaboration with

J.P. Guiraud from the University of Paris VI. These studies were made public at the ICMS course in Udine, Italy (October, 1983) as well as in our review paper (Zeytounian 1985).

The basic equations are those of Chap. 8 written for the horizontal velocity $\boldsymbol{v}$, the vertical component of the velocity w and the thermodynamic perturbations π, w and ϑ. In non-dimensional form, they are written:[1]

$$(1+\omega)\left\{\mathrm{S}\frac{D\boldsymbol{v}}{Dt}+\frac{1}{\mathrm{Ro}}(\boldsymbol{k}\times\boldsymbol{v})+\frac{\varepsilon}{\operatorname{tg}\varphi_0}\frac{1}{\mathrm{Ro}}w\boldsymbol{i}\right\}+\frac{T_\infty}{\gamma\mathrm{M}_\infty^2}\boldsymbol{D}\pi$$
$$=\frac{1}{\varrho_\infty}\frac{1}{\mathrm{Re}}\left\{\boldsymbol{D}^2\boldsymbol{v}+\frac{1}{\varepsilon^2}\frac{\partial^2\boldsymbol{v}}{\partial z^2}+\frac{1}{3}\boldsymbol{D}(\nabla\cdot\boldsymbol{u})\right\} \quad ; \tag{13.7a}$$

$$(1+\omega)\left\{\varepsilon^2\mathrm{S}\frac{Dw}{Dt}-\frac{\varepsilon}{\operatorname{tg}\varphi_0}\frac{1}{\mathrm{Ro}}\boldsymbol{v}\cdot\boldsymbol{i}\right\}+\frac{T_\infty}{\gamma\mathrm{M}_\infty^2}\frac{\partial\pi}{\partial z}-(1+\omega)\frac{\mathrm{Bo}}{\gamma\mathrm{M}_\infty^2}\vartheta$$
$$=\frac{1}{\varrho_\infty}\frac{1}{\mathrm{Re}}\left\{\varepsilon^2\boldsymbol{D}^2w+\frac{\partial^2 w}{\partial z^2}+\frac{1}{3}\frac{\partial}{\partial z}(\nabla\cdot\boldsymbol{u})\right\} \quad ; \tag{13.7b}$$

$$\mathrm{S}\frac{D\omega}{Dt}+(1+\omega)(\nabla\cdot\boldsymbol{u})=(1+\omega)\frac{\mathrm{Bo}}{T_\infty}\left(1+\frac{dT_\infty}{dz_\infty}\right)w \quad ; \tag{13.7c}$$

$$(1+\omega)\mathrm{S}\frac{D\vartheta}{Dt}-\frac{\gamma-1}{\gamma}\mathrm{S}\frac{D\pi}{Dt}+(1+\pi)\frac{\mathrm{Bo}}{T_\infty}\left[\frac{\gamma-1}{\gamma}+\frac{dT_\infty}{dz_\infty}\right]w$$
$$=\frac{1}{\varrho_\infty}\frac{1}{\mathrm{Pr}}\frac{1}{\mathrm{Re}}\left\{\boldsymbol{D}^2\vartheta+\frac{1}{\varepsilon^2}\left[\frac{\partial^2\vartheta}{\partial z^2}+2\frac{\mathrm{Bo}}{T_\infty}\frac{dT_\infty}{dz_\infty}\frac{\partial\vartheta}{\partial z}\right.\right.$$
$$\left.\left.+\frac{\mathrm{Bo}^2}{T_\infty}\frac{d^2T_\infty}{dz_\infty^2}\vartheta\right]\right\}+\frac{\gamma-1}{\varrho_\infty T_\infty}\frac{\mathrm{M}_\infty^2}{\varepsilon^2\mathrm{Re}}\phi \quad ; \tag{13.7d}$$

$$\pi=\omega+\vartheta+\omega\vartheta \quad . \tag{13.7e}$$

In addition, in the energy equation (13.7d), the viscous dissipation function is given by the expression:

$$\phi=\left(\frac{\partial u}{\partial z}+\varepsilon^2\frac{\partial w}{\partial x}\right)^2+\left(\frac{\partial v}{\partial z}+\varepsilon^2\frac{\partial w}{\partial y}\right)^2+\varepsilon^2\left\{\left(\frac{\partial u}{\partial y}+\frac{\partial v}{\partial x}\right)^2\right.$$
$$\left.+2\left[\left(\frac{\partial u}{\partial x}\right)^2+\left(\frac{\partial v}{\partial y}\right)^2+\left(\frac{\partial w}{\partial z}\right)^2\right]-\frac{2}{3}(\nabla\cdot\boldsymbol{u})^2\right\} \quad , \tag{13.8}$$

where u and v are the components of $\boldsymbol{v}=u\boldsymbol{i}+v\boldsymbol{j}$.

---

[1] Once again:

$$\boldsymbol{u}=\boldsymbol{v}+\varepsilon w\boldsymbol{k} \quad , \quad \nabla=\boldsymbol{D}+\frac{1}{\varepsilon}\frac{\partial}{\partial z}\boldsymbol{k} \quad , \quad \mathrm{S}\frac{D}{Dt}=\mathrm{S}\frac{\partial}{\partial t}+\boldsymbol{v}\cdot\boldsymbol{D}+w\frac{\partial}{\partial z} \quad ;$$

T_∞ and ϱ_∞ are functions of $z_\infty=\mathrm{Bo}z$.

13.1 The Free Circulation Models

The so-called *free* – or *natural* – circulations which are very often the direct causes of regional "microclimates" are characterized essentially by local contrasts in ground temperature, for the mostpart *independent* of *any influence from outside fields*. One example is the breeze phenomena which result from a thermal contrast between the temperatures of the earth's surface and the sea. In our opinion, the bringing to light of free circulation models is a fundamental problem which would allow the establishment of a solid theoretical base for a regional weather prediction above a thermally non-homogeneous site. In the forthcoming, this site will be assumed to have no relief[2] and we will suppose that the temperature distribution on the ground (simulated by the equation $z = 0$) is known via the function $\Xi(t, x, y)$ [see, for example (8.21)]:

$$z = 0: \quad \vartheta = \tau_0 \Xi(t, x, y) \quad , \tag{13.9}$$

where $\tau_0 = \Delta T_0 / T_\infty(0)$ and ΔT_0 is the characteristic constant temperature fluctuation on the ground associated with Ξ.

In this section, we therefore assign to (13.7), the following ground conditions:

$$\boldsymbol{v} = 0 \quad , \quad w = 0 \quad \text{and} \quad T = 1 + \tau_0 \Xi(t, x, y) \quad , \quad \text{on} \quad z = 0 \quad . \tag{13.10}$$

As it is supposed here that there is no outside field, it is necessary to define an intrinsic characteristic velocity related to our free circulation phenomenon. From the results of Subsect. 8.3.3, we can introduce the characteristic velocity $(gh_0)^{1/2}\tau_0$ where $h_0 \equiv R\Delta T_0/g$. In this way, we arrive at (8.55) in which $\mathrm{Re}_\perp^2$ must be replaced by the Grashof number $\mathrm{Gr}_\perp \equiv \mathrm{Re}_\perp^2 = \varepsilon^2 (g h_0^3 / \nu_0^2) \tau_0^2$.

We can, however, also introduce the characteristic velocity (with dimensions) U_0 from:

$$U_0 = \sqrt{gL\tau_0} \quad , \tag{13.11}$$

where L is the characteristic horizontal length scale associated with the domain $\mathcal{D}_2$ of the plane $z = 0$ where the horizontal variables x and y are defined in the function Ξ. Thus, in the dimensionless parameters which come into play in (13.7), L must be imagined instead of H which implies that $\varepsilon \equiv 1$. We then have:

$$\mathrm{M}_\infty^2 = \frac{\mathrm{Bo}}{\gamma}\tau_0^2 \quad ; \quad \mathrm{Bo} = \frac{gL}{RT_\infty(0)} \quad ; \quad \mathrm{Re}^2 = \frac{gL^3}{\nu_0^2}\tau_0^2 \equiv \mathrm{Gr} \quad . \tag{13.12}$$

For the free convection problems in the atmosphere, naturally τ_0 and $\kappa_0 \equiv \mathrm{Gr}^{-1/2}$ are simultaneously small and must be compared. It is thus assumed that:

$$\kappa_0 = \hat{G}\tau_0^a \quad , \tag{13.13}$$

[2] In Sect. 13.4, a model is given for local winds on slopes and in valleys where the relief must be taken into account.

with $a > 0$ being a positive real number to be determined, and $\hat{G}$, a constant similarity parameter of the order unity for the limiting flow obtained when $\tau_0 \to 0$.

Just like in the classical case of low viscosity flows (with high Reynolds number), two different asymptotic representations must be sought which are related to significant degeneracies of problem (13.7, 10) while taking into account (13.12, 13). To this end, we introduce a change in vertical scale:

$$z = \tau_0^b \zeta \tag{13.14}$$

with $b \geq 0$ – just like $a > 0$ – being a real number to be determined.

We therefore consider the following degeneracy which is associated with (13.14):

$$\boldsymbol{v} = \tau_0^\alpha \boldsymbol{v}_\alpha + \dots \quad , \quad w = \tau_0^\beta w_\beta + \dots \quad , \tag{13.15a}$$

$$\pi = \tau_0^\sigma \pi_\sigma + \dots \quad , \quad \omega = \tau_0^\varphi \omega_\varphi + \dots \quad , \quad \vartheta = \tau_0^\psi \vartheta_\psi + \dots \tag{13.15b}$$

with the variables t, x and y remaining unchanged. The real numbers α, β, σ, ϑ and ψ are ≥ 0 and must be determined at the same time as a and b in such a way that the considered degeneracy (corresponding either to $b > 0$, or to $b = 0$) is significant according to the definition of the MMAE (see Sect. 5.1).

It is pointed out as a final remark that the asymptotic representations of the type (13.15) which are obtained in what follows are dependent on the choice (13.11) made for U_0.

13.1.1 Inner Degeneracies

1) Let us consider $b > 0$. First of all, we will suppose that Bo, as well as S, Ro, Pr and γ remain constant and of the order unity when $\tau_0 \to 0$ with t, x, y *and* $\zeta = z/\tau_0^b$, $b > 0$, fixed. In this case, a simple analysis shows that the corresponding significant degeneracy is obtained if:

$$a = 2 ; \quad b = 1 ; \quad \alpha = 0 \ \text{and} \ \beta = 1 , \quad \sigma = 2 , \quad \varphi = \psi = 1 . \tag{13.16}$$

Therefore, in this case:

$$\frac{\mathrm{Gr}^{-1/2}}{\tau_0^2} \equiv \hat{G} = \mathrm{O}(1) \quad ; \tag{13.17a}$$

$$\begin{aligned} &\boldsymbol{v} = \boldsymbol{v}_0(t, x, y, \zeta) + \dots \quad ; \quad w = \tau_0 w_1(t, x, y, \zeta) + \dots \quad ; \\ &\pi = \tau_0^2 \pi_2(t, x, y, \zeta) + \dots \quad ; \quad \vartheta = \tau_0 \vartheta_1(t, x, y, \zeta) + \dots \quad ; \\ &\omega = \tau_0 \omega_1(t, x, y, \zeta) + \dots \quad , \end{aligned} \tag{13.17b}$$

where $\zeta = z/\tau_0$.

The following limiting equations then result for the functions $\boldsymbol{v}_0$, w_1, π_2, ϑ_1 and ω_1:

$$\mathrm{S}\frac{\partial \boldsymbol{v}_0}{\partial t} + \boldsymbol{v}_0 \cdot \boldsymbol{D}\boldsymbol{v}_0 + w_1 \frac{\partial \boldsymbol{v}_0}{\partial \zeta} + \frac{1}{\mathrm{Ro}}\left(\boldsymbol{k} \times \boldsymbol{v}_0\right) + \frac{1}{\mathrm{Bo}} \boldsymbol{D}\pi_2 = \hat{G}\frac{\partial^2 \boldsymbol{v}_0}{\partial \zeta^2} ; \tag{13.18a}$$

$$\frac{\partial \pi_2}{\partial \zeta} - \mathrm{Bo}\vartheta_1 = 0 \quad ; \tag{13.18b}$$

$$\boldsymbol{D} \cdot \boldsymbol{v}_0 + \frac{\partial w_1}{\partial \zeta} = 0 \quad ; \tag{13.18c}$$

$$\mathrm{S}\frac{\partial \vartheta_1}{\partial t} + \boldsymbol{v}_0 \cdot \boldsymbol{D}\vartheta_1 + \mathrm{Bo}\left[\frac{\gamma-1}{\gamma} + \left(\frac{dT_\infty}{dz_\infty}\right)_0\right] w_1$$
$$+ w_1 \frac{\partial \vartheta_1}{\partial \zeta} = \frac{\hat{G}}{\mathrm{Pr}} \frac{\partial^2 \vartheta_1}{\partial \zeta^2} \quad ; \tag{13.18d}$$

$$\omega_1 = -\vartheta_1 \quad . \tag{13.18e}$$

It will be remarked that $z_\infty = \mathrm{Bo}z \equiv \mathrm{Bo}\tau_0\zeta$ tends to zero when $\tau_0 \to 0$. Let us point out that in (13.7), T_∞, dT_∞/dz_∞ and d^2T_∞/dz_∞^2 are given functions of z_∞. The same holds for ϱ_∞.

The limiting equations (13.18) remain valid in a boundary layer having a thickness of the order of τ_0 and which exists in the vicinity of $\zeta = 0$. On this surface, the following boundary conditions must be imposed:

$$\boldsymbol{v}_0 = 0 \quad , \quad w_1 = 0 \quad \text{and} \quad \vartheta_1 = \Xi(t, x, y) \quad , \quad \text{on} \quad \zeta = 0 \quad ,$$
$$t > 0 \quad , \quad x, y \in \mathcal{D}_2 \quad . \tag{13.19}$$

In (13.18, 19), we have:

$$\hat{G} = \frac{\nu_0}{\tau_0^3 g^{1/2} L^{3/2}} \quad . \tag{13.20}$$

Thus, if $\hat{G}$ is, in fact, to be of the order unity, then L must be chosen as follows:

$$L \sim \frac{\nu_0^{2/3}}{\tau_0^2 g^{1/3}} \quad . \tag{13.21}$$

However, as Bo is also supposed to be of the order unity, then it is necessary that $L \sim RT_\infty(0)/g$ and so:

$$\tau_0 \sim \frac{(g\nu_0)^{1/3}}{\left[RT_\infty(0)\right]^{1/2}} \quad . \tag{13.22}$$

The limiting problem (13.18, 19) thus remains valid with the constraint (13.22) which determines the order of magnitude of the characteristic temperature fluctuation on the ground:

$$\Delta T_0 \sim \left(\frac{T_\infty(0)}{R}\right)^{1/2} (g\nu_0)^{1/3} \quad . \tag{13.23}$$

2) Again let $b > 0$, but with $\mathrm{Bo} \gg 1$ $(L \gg RT_\infty(0)/g)$. More precisely:

$$\mathrm{Bo} = \frac{\hat{B}}{\tau_0^\delta} \quad , \quad \delta > 0 \quad , \quad \hat{B} = \mathrm{O}(1) \quad . \tag{13.24}$$

The other parameters S, Ro, Pr and γ remain fixed and of the order unity when $\tau_0 \to 0$ with t, x, y and $\zeta = z/\tau_0^b$ fixed. In this case, the significant degeneracy is obtained if:

$$a = 4 \quad ; \quad b = 2 \quad ; \quad \alpha = 1 \quad , \quad \beta = 3 \quad , \quad \sigma = 1 \quad , \quad \varphi = \psi = 1 \quad , \quad \delta = 2 \quad . \tag{13.25}$$

Therefore,

$$\frac{\mathrm{Gr}^{-1/2}}{\tau_0^4} \equiv \hat{G} = \mathrm{O}(1) \quad \text{and} \quad \mathrm{Bo}\tau_0^2 \equiv \hat{B} = \mathrm{O}(1) \quad ; \tag{13.26a}$$

$$\begin{aligned} &\boldsymbol{v} = \tau_0 \boldsymbol{v}_1(t,x,y,\zeta) + \dots \quad ; \quad w = \tau_0^3 w_3(t,x,y,\zeta) + \dots \quad ; \\ &\pi = \tau_0 \pi_1(t,x,y,\zeta) + \dots \quad ; \quad \vartheta = \tau_0 \vartheta_1(t,x,y,\zeta) + \dots \quad ; \\ &\omega = \tau_0 \omega_1(t,x,y,\zeta) + \dots \quad , \end{aligned} \tag{13.26b}$$

where $\zeta = z/\tau_0^2$ and the following linear limiting equations are obtained for the functions $\boldsymbol{v}_1$, w_3, π_1, ω_1 and ϑ_1:

$$\mathrm{S}\frac{\partial \boldsymbol{v}_1}{\partial t} + \frac{1}{\mathrm{Ro}}(\boldsymbol{k} \times \boldsymbol{v}_1) + \frac{T_\infty}{\hat{B}} \boldsymbol{D}\pi_1 = \frac{\hat{G}}{\varrho_\infty} \frac{\partial^2 \boldsymbol{v}_1}{\partial \zeta^2} \quad ; \tag{13.27a}$$

$$T_\infty \frac{\partial \pi_1}{\partial \zeta} - \hat{B}\vartheta_1 = 0 \quad ; \tag{13.27b}$$

$$\mathrm{S}\frac{\partial \omega_1}{\partial t} + \boldsymbol{D} \cdot \boldsymbol{v}_1 + \frac{\partial w_3}{\partial \zeta} = \frac{1}{T_\infty}\left(\hat{B} + \frac{dT_\infty}{d\zeta}\right) w_3 \quad ; \tag{13.27c}$$

$$\begin{aligned} &\mathrm{S}\left(\frac{\partial \vartheta_1}{\partial t} - \frac{\gamma - 1}{\gamma}\frac{\partial \pi_1}{\partial t}\right) + \frac{1}{T_\infty}\left[\frac{\gamma-1}{\gamma}\hat{B} + \frac{dT_\infty}{d\zeta}\right] w_3 \\ &\quad = \frac{\hat{G}}{\varrho_\infty \mathrm{Pr}}\left\{\frac{\partial^2 \vartheta_1}{\partial \zeta^2} + 2\frac{1}{T_\infty}\frac{dT_\infty}{d\zeta}\frac{\partial \vartheta_1}{\partial \zeta} + \frac{1}{T_\infty}\frac{d^2 T_\infty}{d\zeta^2}\vartheta_1\right\} \quad ; \end{aligned} \tag{13.27d}$$

$$\pi_1 = \omega_1 + \vartheta_1 \quad , \tag{13.27e}$$

where T_∞ and ϱ_∞ should be considered as functions of $\hat{B}\zeta$.

To the linear limiting system (13.27), the following boundary conditions must be assigned:

$$\begin{aligned} &\boldsymbol{v}_1 = 0 \quad , \quad w_3 = 0 \quad \text{and} \quad \vartheta_1 = \Xi(t,x,y) \quad , \quad \text{on} \quad \zeta = 0 \quad ; \\ &\qquad t > 0 \quad , \quad x,y \in \mathcal{D}_2 \quad . \end{aligned} \tag{13.28}$$

The problem (13.27, 28) remains valid in a boundary layer in the vicinity of $\zeta = 0$ having a thickness of the order of τ_0^2. The similarity parameters $\hat{B}$ and $\hat{G}$ are expressed in the following form:

$$\hat{B} = \frac{gL\tau_0^2}{RT_\infty(0)} \quad \text{and} \quad \hat{G} = \frac{\nu_0}{g^{1/2}L^{3/2}\tau_0^5} \quad . \tag{13.29}$$

From (13.29), the order of magnitude of ΔT_0 is readily determined:

$$\Delta T_0 \sim \left(\frac{T_\infty(0)}{R^3}\right)^{1/4} (g\nu_0)^{1/2} \quad . \tag{13.30}$$

Naturally, it is easy to demonstrate that the ΔT_0 obtained from (13.30) is much smaller (nearly ten times) than the ΔT_0 given by expression (13.23).

The following behavior conditions can be added to (13.18):

$$\boldsymbol{v}_0 = w_1 = \pi_2 = \vartheta_1 = \omega_1 \to 0 \quad , \quad \text{when} \quad |x^2 + y^2|^{1/2} \to \infty \quad , \tag{13.31}$$

whereas for (13.27), these behavior conditions are:

$$\boldsymbol{v}_1 = w_3 = \pi_1 = \vartheta_1 = \omega_1 \to 0 \quad , \quad \text{when} \quad |x^2 + y^2|^{1/2} \to \infty \quad . \tag{13.32}$$

The above conditions (13.31, 32) result naturally from the very physics of the free circulation phenomenon which takes place outside of all outer fields. Concerning the initial conditions, we have:[3]

$$\text{for system (13.18): } t = 0: \ \boldsymbol{v}_0 = 0 \ \text{ and } \ \vartheta_1 = 0 \ ; \tag{13.33a}$$

$$\text{for system (13.27): } t = 0: \ \boldsymbol{v}_1 = 0 \ , \ \vartheta_1 = 0 \ \text{ and } \ \omega_1 = 0 \ . \tag{13.33b}$$

Behavior conditions for $\zeta \to \infty$ are still required. To this end, we have to consider the outer degeneracies, i.e., those which correspond to $b = 0$.

13.1.2 Outer Degeneracies

We thus have $b = 0$ and hence, $z \equiv \zeta$. There is no change in vertical scale. In this case, (13.15) with the values (13.16) or (13.25) for α, β, σ, and φ and ψ yields from (13.76): $\vartheta_1 \equiv 0$ and so from (13.7d), we have either $w_1 = 0$, or $w_3 = 0$ when $-dT_\infty/dz_\infty \neq (\gamma - 1)/\gamma$. It is thus remarked that the outer degeneracies lead to the equations of an incompressible perfect fluid in two-dimensional unsteady flow in the plane x, y when $\tau_0 \to 0$.

The following outer system corresponds to (13.18):

$$\begin{aligned} &\mathrm{S}\frac{\partial \boldsymbol{v}_0}{\partial t} + \boldsymbol{v}_0 \cdot \boldsymbol{D}\boldsymbol{v}_0 + \frac{1}{\mathrm{Ro}}(\boldsymbol{k} \times \boldsymbol{v}_0) + \frac{T_\infty}{\mathrm{Bo}}\boldsymbol{D}\pi_2 = 0 \quad ; \\ &\boldsymbol{D} \cdot \boldsymbol{v}_0 = 0 \quad , \quad T_\infty \equiv T_\infty(\mathrm{Bo}z) \quad , \end{aligned} \tag{13.34}$$

whereas to (13.27) corresponds the outer system:

$$\begin{aligned} &\mathrm{S}\frac{\partial \boldsymbol{v}_1}{\partial t} + \frac{1}{\mathrm{Ro}}(\boldsymbol{k} \times \boldsymbol{v}_1) + \frac{T_\infty(\infty)}{\hat{B}}\boldsymbol{D}\pi_1 = 0 \quad , \\ &\boldsymbol{D} \cdot \boldsymbol{v}_1 = -\mathrm{S}\frac{\partial \pi_1}{\partial t} \quad . \end{aligned} \tag{13.35}$$

[3] Strictly speaking, the adjustment problems corresponding to the vicinity of $t = 0$ should be resolved and the matching with the (main) representations (13.17, 26) should be worked out. Nonetheless, it appears reasonable to postulate the initial conditions (13.33) by using "physics" common sense!

However, it is in fact necessary that $\pi_1 \equiv 0$.

Because of the initial and behavior conditions when $|x^2 + y^2| \to \infty$, the solutions to (13.34, 35) must be identically zero. In particular, they are zero for $z \to 0$.

13.1.3 Matching. Formulation of the Free Circulation Problem

By matching the outer asymptotic representations ($b = 0$) and the inner ones ($b > 0$), the following behaviors are found for the solutions of (13.8, 27) respectively, *when* $\zeta \to +\infty$:

$$v_0 = w_1 = \vartheta_1 = \pi_2 = \omega_1 \to 0 \quad , \quad \text{for (13.18),} \tag{13.36a}$$

and

$$v_1 = w_3 = \vartheta_1 = \pi_1 = \omega_1 \to 0 \quad , \quad \text{for (13.27)} \quad . \tag{13.36b}$$

It is, however, easy to demonstrate that the order of the inner systems of limiting equations (13.18, 27) in ζ, does not permit the imposing of conditions on w_1 or w_3 when $\zeta \to \infty$! As a matter of fact, the equation for ϑ_1 in the systems (13.18, 27) shows that with the condition $-dT_\infty/d\zeta \neq (\gamma - 1)/\gamma$:

$$\vartheta_1 \to 0 \quad , \quad \pi_1 \to 0 \quad \text{and} \quad \pi_2 \to 0 \Rightarrow w_1 \quad \text{or} \quad w_3 \to 0 \quad ,$$
$$\text{when} \quad \zeta \to +\infty \quad .$$

The following constraints:

$$w_1 = 0 \quad , \quad \text{when} \quad \zeta \to +\infty \quad , \quad \text{for (13.18)} \quad ,$$

and

$$w_3 = 0 \quad , \quad \text{when} \quad \zeta \to +\infty \quad , \quad \text{for (13.27)} \tag{13.37}$$

must therefore be satisfied.

It is important to fully understand that the inner limiting systems (13.18, 27) are not those of the classical Prandtl boundary layer. In particular, the quasi-linear problem constituted by equations (13.18) and conditions (13.19, 31, 33a, 36a) is the one which, in mesometeorology, describes the free circulations over a site perfectly flat, but having thermal non-uniformities. These are the "breeze"-type phenomena which are characterized by a) *contrasts in ground temperature* (related to the function Ξ in the condition (13.19) for ϑ_1), b) *the effect of the Coriolis force* (related to 1/Ro), and c) *the stratification* {the term $\mathrm{Bo}[(\gamma-1)/\gamma+(dT_\infty/dz_\infty)|_0]w_1$ in (13.18d)}. The presence, in particular, of this term: $\mathrm{Bo}[(\gamma-1)/\gamma+(dT_\infty/dz_\infty)|_0]w_1$ in (13.18d) for ϑ_1 brings about, when it is positive (standard stable stratification), the formation of a compensating breeze (called "anti-breeze") above the main breeze. One of the difficulties involved in solving this problem (13.18, 19, 31, 33a and 36a) is related precisely to the presence of this term in (13.18d). It is, however, only by "correctly" taking into account this term that a solution can be found which satisfies the constraint:

$$w_1 = 0 \quad , \quad \text{when} \quad \zeta = +\infty \tag{13.38}$$

on $w_1(t, x, y, \zeta)$. The reader is directed to the book by Gutman (1972, Chap. 7) for more details on this subject.

It is pointed out that thanks to (13.36a), (13.18a) can be written in the following form:

$$S\frac{\partial \boldsymbol{v}_0}{\partial t} + (\boldsymbol{v}_0 \cdot \boldsymbol{D})\boldsymbol{v}_0 + w_1\frac{\partial \boldsymbol{v}_0}{\partial \zeta} + \frac{1}{\mathrm{Ro}}(\boldsymbol{k} \times \boldsymbol{v}_0) + \int_{\infty}^{\zeta} \boldsymbol{D}\vartheta_1 \, d\zeta = \hat{G}\frac{\partial^2 \boldsymbol{v}_0}{\partial \zeta^2} , \tag{13.39}$$

which means that we have a system of three equations: (13.39, 18c, 18d) for the three unknowns: $\boldsymbol{v}_0$, w_1 and ϑ_1. Furthermore, from (13.18c) it is also found that:

$$\int_0^{\infty} (\boldsymbol{D} \cdot \boldsymbol{v}_0) d\zeta = 0 \quad , \tag{13.40}$$

which is one of the prerequisites for the existence of the anti-breeze.

13.2 The Models for the Asymptotic Analysis of Lee Waves

In Sect. 6.3, we brought to light via the local limiting process (6.90), the local quasi-steady equations (6.92) which follow from the complete Euler equations (6.76) written in non-dimensional form. These local equations, which are valid in a baroclinic and compressible adiabatic atmosphere, admit as a boundary condition on the relief, simulated by the non-dimensional equation:

$$z = \sigma h(\alpha x, \beta y) \quad , \tag{13.41}$$

the following slip condition:

$$w = \sigma \boldsymbol{v} \cdot \boldsymbol{D}h(\alpha x, \beta y) \quad . \tag{13.42}$$

Condition (13.42) is the one given by (6.93) once the hats and the zero subscripts have been eliminated from all the values.

If the local equations (6.92) are likewise written without the hats and zero subscripts, the following non-dimensional system is obtained for the functions $\boldsymbol{v}$, w, p, ϱ and T:

$$\begin{aligned}
&(\boldsymbol{v} \cdot \boldsymbol{D})\boldsymbol{v} + w\frac{\partial \boldsymbol{v}}{\partial z} + \frac{1}{\varrho}\frac{1}{\gamma \mathrm{M}_\infty^2}\boldsymbol{D}p = 0 \quad ; \\
&\boldsymbol{v} \cdot \boldsymbol{D}w + w\frac{\partial w}{\partial z} + \frac{1}{\gamma \mathrm{M}_\infty^2}\left[\frac{1}{\varrho}\frac{\partial p}{\partial z} + \mathrm{Bo}\right] = 0 \quad ; \\
&\boldsymbol{D} \cdot (\varrho \boldsymbol{v}) + \frac{\partial \varrho w}{\partial z} = 0 \quad ; \\
&p = \varrho T \quad ; \\
&\varrho \boldsymbol{v} \cdot \boldsymbol{D}T - \frac{\gamma - 1}{\gamma}\boldsymbol{v} \cdot \boldsymbol{D}p + \left[\varrho\frac{\partial T}{\partial z} - \frac{\gamma - 1}{\gamma}\frac{\partial p}{\partial z}\right]w = 0 \quad .
\end{aligned} \tag{13.43}$$

It is emphasized that these local equations are valid in the vicinity of the origin (x_0, y_0) of the plane $z = 0$. It was already seen in Sect. 6.3 that in order to carry out a quasi-steady dynamic local forecast above a site of origin (x_0, y_0) having a given relief (characterized by the function h), it is necessary to display at this point (x_0, y_0), and at the chosen time $t = t_0$, the short-range synoptic forecast obtained from the primitive equations (6.80). This short-range prediction thus plays the role of "conditions at infinity" in any plane z = const for the local equations (13.43). These conditions at infinity are as follows:

$$\begin{aligned}&\overline{\boldsymbol{v}}_0(t_0, x_0, y_0, z) \equiv \boldsymbol{V}_\infty(z) \quad ; \quad \overline{T}_0(t_0, x_0, y_0, z) \equiv T_\infty(z) \quad ;\\ &\overline{p}_0(t_0, x_0, y_0, z) \equiv p_\infty(z) \quad ; \quad \overline{\varrho}_0(t_0, x_0, y_0, z) \equiv \varrho_\infty(z) \quad ,\end{aligned} \tag{13.44}$$

such that

$$\frac{dp_\infty}{dz} + \mathrm{Bo}\,\varrho_\infty = 0 \quad \text{and} \quad p_\infty = \varrho_\infty T_\infty \quad . \tag{13.45}$$

To be sure, $\boldsymbol{V}_\infty(z)$, $T_\infty(z)$, $p_\infty(z)$ and $\varrho_\infty(z)$ must be assumed known in the local problem. Hence:

$$\lim_{\sqrt{x^2+y^2}\to\infty} (\boldsymbol{v}, w, p, \varrho, T) = \left(\boldsymbol{V}_\infty, 0, p_\infty, \varrho_\infty, T_\infty\right) \quad . \tag{13.46}$$

If we take $t_0 \equiv 0$, then on each plane z = const, we must display as conditions (13.46) for the problem (13.41, 42) the initial conditions obtained for the primitive equations in Sect. 7.5, from the analysis of the problem of adjustment to hydrostatic balance.

Quite naturally in (13.41, 42) it is supposed that:

$$h \equiv 0 \quad , \quad \text{when} \quad x^2 + y^2 > R^2 \quad , \tag{13.47}$$

where R is a bounded real scalar.

In order to obtain a consistent mathematical problem (since a theorem of unicity and existence is lacking), a behavior condition on w must be prescribed for (13.43) when $z \to +\infty$: this is the focal point of the forthcoming section.

13.2.1 Emergence of the Vertical Structure. Condition for $z \to +\infty$

We now return to (13.43) and arrange these equations in the following form which although not classical, proves advantageous for the forthcoming:

$$\begin{aligned}&\varrho\left[(\boldsymbol{v}\cdot\boldsymbol{D})\boldsymbol{v} + w\frac{\partial \boldsymbol{v}}{\partial z}\right] + \frac{1}{\gamma \mathrm{M}_\infty^2}\boldsymbol{D}p = 0 \quad ;\\ &\boldsymbol{v}\cdot\boldsymbol{D}p + w\frac{\partial p}{\partial z} - \frac{\gamma p}{\varrho}\left(\boldsymbol{v}\cdot\boldsymbol{D}\varrho + w\frac{\partial \varrho}{\partial z}\right) = 0 \quad ;\\ &\boldsymbol{D}\cdot(\varrho\boldsymbol{v}) + \frac{\partial \varrho w}{\partial z} = 0 \quad ;\\ &\varrho\left[\boldsymbol{v}\cdot\boldsymbol{D}w + w\frac{\partial w}{\partial z}\right] + \frac{1}{\gamma \mathrm{M}_\infty^2}\left(\frac{\partial p}{\partial z} + \mathrm{Bo}\,\varrho\right) = 0 \quad .\end{aligned} \tag{13.48}$$

We set: $\mathcal{U} = (u, v, \varrho, w, p)^T$, where u and v are the components of the horizontal velocity and $\boldsymbol{v} \cdot \boldsymbol{D} = u(\partial/\partial x) + v(\partial/\partial y)$. System (13.48) can then be put into the following matrix form:

$$\mathcal{A}(\mathcal{U})\frac{\partial \mathcal{U}}{\partial x} + \mathcal{B}(\mathcal{U})\frac{\partial \mathcal{U}}{\partial y} + \mathcal{C}(\mathcal{U})\frac{\partial \mathcal{U}}{\partial z} + \mathcal{D}_0\mathcal{U} = 0 \quad , \tag{13.49}$$

with $\mathcal{A}$, $\mathcal{B}$, $\mathcal{C}$ and $\mathcal{D}_0$ being 5×5 matrices. The matrix $\mathcal{D}_0$ is a constant matrix with just one element which is $\mathrm{Bo}/\gamma \mathrm{M}_\infty^2$. It is supposed that condition (13.46) is indeed satisfied and when $r = (x^2 + y^2)^{1/2} \to \infty$, the following far-field behavior is postulated:

$$\mathcal{U} = \mathcal{U}_\infty(z) + \tilde{\mathcal{U}}[\theta(x, y), z; x, y) + \ldots \tag{13.50}$$

where $\mathcal{U}_\infty(z) = (U_\infty, V_\infty, \varrho_\infty, 0, p_\infty)^\mathrm{T}$, U_∞ and V_∞ being the components of $V_\infty(z)$ with respect to x and y. Concerning the component $\tilde{\mathcal{U}}$, we suppose (in the spirit of the MSM) that:

$$|\tilde{\mathcal{U}}| \ll |\mathcal{U}_\infty| \quad .$$

It must be fully understood that the variation of $\mathcal{U}$ with respect to x and y comes into play via two different scales. One of them increases with $r = \left(x^2 + y^2\right)^{1/2}$, whereas the other corresponds to the internal waves which were discussed in Sects. 3.1 and 3.2. The scale related to these internal gravity waves remains of the order unity when $r \to \infty$. The existence of these internal gravity waves is due to the fact that when $r \to \infty$, the relief can be assumed flat and the perturbations satisfy the system of linear equations associated with (13.48) with a slip condition written on the plane surface $z = 0$. When $r \to \infty$, the horizontal wave length of these internal gravity waves becomes very small compared to the distance from the relief and these waves appear locally as plane waves which radiate afar (see Fig. 4).

In the forthcoming, we will characterize a quantity which is a function of $\mathcal{U}$, but which is evaluated in $\mathcal{U} = \mathcal{U}_\infty$ with the subscript "∞". It is obvious that:

$$\mathcal{C}_\infty \frac{d\mathcal{U}_\infty}{dz} + \mathcal{D}_0\mathcal{U}_\infty = 0$$

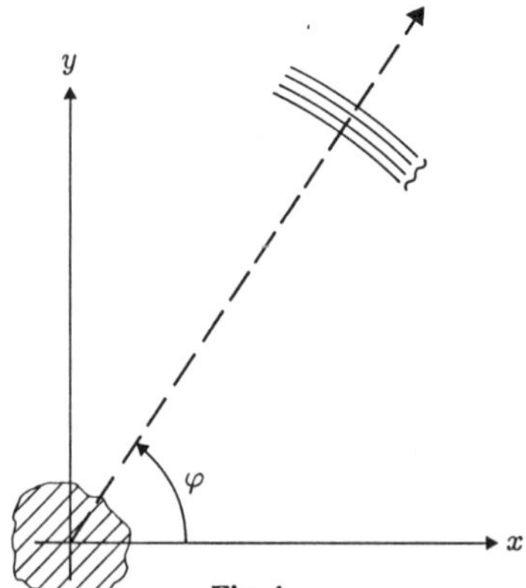

Fig. 4

and thus, the substitution of (13.50) into (13.49) after rearrangement yields:

$$\left(\frac{\partial\theta}{\partial x}\mathcal{A}_\infty + \frac{\partial\theta}{\partial y}\mathcal{B}_\infty\right)\frac{\partial\tilde{\mathcal{U}}}{\partial\theta} + \mathcal{C}_\infty\frac{\partial\tilde{\mathcal{U}}}{\partial z} + \left[\mathcal{D}_0 + \mathcal{M}\left(\mathcal{U}_\infty,\ \frac{d\mathcal{U}_\infty}{dz}\right)\right]\tilde{\mathcal{U}}$$
$$+\mathcal{A}_\infty\frac{\partial\tilde{\mathcal{U}}}{\partial x} + \mathcal{B}_\infty\frac{\partial\tilde{\mathcal{U}}}{\partial y} + \ldots = 0 \quad . \tag{13.51}$$

The matrix $\mathcal{M}$ will also be seen to be a 5×5 matrix. The term $\mathcal{M}[\mathcal{U}_\infty, (d\mathcal{U}_\infty/dz)]\tilde{\mathcal{U}}$ results from the term $\mathcal{C}(\mathcal{U})\partial\mathcal{U}/\partial z$ of (13.49) after taking into account (13.50).

Since it is supposed that:

$$\left|\frac{\partial\tilde{\mathcal{U}}}{\partial x}\right| + \left|\frac{\partial\tilde{\mathcal{U}}}{\partial y}\right| \ll \left\{\left|\frac{\partial\tilde{\mathcal{U}}}{\partial\theta}\right|,\ \left|\frac{\partial\tilde{\mathcal{U}}}{\partial z}\right|,\ |\tilde{\mathcal{U}}|\right\} \quad ,$$

the following *dominant* equation can be considered instead of (13.51):

$$\left(\frac{\partial\theta}{\partial x}\mathcal{A}_\infty + \frac{\partial\theta}{\partial y}\mathcal{B}_\infty\right)\frac{\partial\tilde{\mathcal{U}}}{\partial\theta} + \mathcal{C}_\infty\frac{\partial\tilde{\mathcal{U}}}{\partial z}$$
$$+\left[\mathcal{D}_0 + \mathcal{M}\left(\mathcal{U}_\infty,\ \frac{d\mathcal{U}_\infty}{dz}\right)\right]\tilde{\mathcal{U}} = 0 \quad . \tag{13.52}$$

A "plane wave" solution to (13.52) can be written:

$$\tilde{\mathcal{U}} = \mathrm{e}^{\mathrm{i}\theta}\hat{\mathcal{U}}\,,\quad \frac{\partial\theta}{\partial x} = k_1\,,\quad \frac{\partial\theta}{\partial y} = k_2$$

and it follows that:

$$(\mathcal{D}_0 + \mathcal{M} + \mathrm{i}k_1\mathcal{A}_\infty + \mathrm{i}k_2\mathcal{B}_\infty)\hat{\mathcal{U}} + \mathcal{C}_\infty\frac{d\hat{\mathcal{U}}}{dz} = 0 \tag{13.53}$$

where $\hat{\mathcal{U}} = \hat{\mathcal{U}}(z)$.

By decomposing the matrices $\mathcal{C}_\infty$ and $\mathcal{D}_0 + \mathcal{M} + \mathrm{i}(k_1\mathcal{A}_\infty + k_2\mathcal{B}_\infty)$, instead of (13.53), we arrive at the following:

$$\begin{aligned} &L\hat{X} + M\hat{Y} = 0 \quad ; \\ &N\hat{X} + S\hat{Y} + T\frac{d\hat{Y}}{dz} = 0 \quad , \end{aligned} \tag{13.54}$$

once the following decomposition has been introduced:

$$\hat{\mathcal{U}} = (\hat{X}|\hat{Y})^{\mathrm{T}} \quad , \quad \text{with} \quad \hat{X} = (\hat{u},\ \hat{v},\ \hat{\varrho})^{\mathrm{T}} \quad \text{and} \quad \hat{Y} = (\hat{w},\ \hat{p})^{\mathrm{T}} \quad .$$

The matrix L is a 3×3 matrix and M, N and S are respectively 2×3, 3×2, and 2×2 matrices:

$$\mathcal{D}_0 + \mathcal{M} + \mathrm{i}(k_1\mathcal{A}_\infty + k_2\mathcal{B}_\infty) = \left(\begin{array}{c|c} L & M \\ \hline \underbrace{N}_{3\text{ columns}} & \underbrace{S}_{2\text{ columns}} \end{array}\right)\begin{array}{l}\}\ 3\text{ rows} \\ \}\ 2\text{ rows}\end{array}$$

Finally,

$$T = \begin{pmatrix} \varrho_\infty(z) & 0 \\ 0 & 1/\gamma M_\infty^2 \end{pmatrix} \quad \text{and} \quad \mathcal{C}_\infty = \left(\begin{array}{c|c} 0 & 0 \\ \hline 0 & T \end{array}\right) .$$

The matrix L is invertible provided that:

$$\big(k_1 U_\infty + k_2 V_\infty\big)\left\{\frac{B_0}{\gamma M_\infty^2} - \frac{\gamma p_\infty}{\varrho_\infty}\mathrm{i}\big(k_1 U_\infty + k_2 V_\infty\big)\right\} \neq 0 \quad . \tag{13.55}$$

If it is assumed that condition (13.55) is fulfilled, then the second equation in (13.54) leads to:

$$T\frac{d\hat{Y}}{dz} + \{S - NL^{-1}M\}\hat{Y} = 0 \tag{13.56}$$

with the matrix:

$$NL^{-1}M = \left(\begin{array}{c|c} \lambda_\infty(z) & \mu_\infty(z) \\ \hline 0 & 0 \end{array}\right) \quad , \tag{13.57}$$

where

$$\lambda_\infty(z) = \frac{k_1(dU_\infty/dz) + k_2(dV_\infty/dz)}{\rho_\infty\big(k_1 U_\infty + k_2 V_\infty\big)} + \frac{\mathrm{i}(k_1 U_\infty + k_2 V_\infty)\big(dp_\infty/dz - (\gamma p_\infty/\varrho_\infty)(d\varrho_\infty/dz)\big)}{\mathrm{Bo}/\gamma M_\infty^2 - (\gamma p_\infty/\varrho_\infty)\mathrm{i}(k_1 U_\infty + k_2 V_\infty)}; \tag{13.58a}$$

$$\mu_\infty(z) = \frac{\mathrm{i}\big(k_1^2 + k_2^2\big)}{\gamma M_\infty^2(k_1 U_\infty + k_2 V_\infty)} - \frac{(k_1 U_\infty + k_2 V_\infty)^2}{\mathrm{Bo}/\gamma M_\infty^2 - \mathrm{i}(\gamma p_\infty/\varrho_\infty)(k_1 U_\infty + k_2 V_\infty)} \quad . \tag{13.58b}$$

Finally, for $\hat{w}$ and $\hat{p}$, the following system of two ordinary first order differential equations in z results:

$$\varrho_\infty\frac{d\hat{w}}{dz}\left(\frac{d\varrho_\infty}{dz} - \lambda_\infty(z)\right)\hat{w} - \mu_\infty(z)\hat{p} = 0 \quad ; \tag{13.59a}$$

$$\frac{1}{\gamma M_\infty^2}\frac{d\hat{p}}{dz} + \mathrm{i}\varrho_\infty\big(k_1 U_\infty + k_2 V_\infty\big)\hat{w} = 0 \quad , \tag{13.59b}$$

where U_∞, V_∞, ϱ_∞ and p_∞ are all functions of z, but $M_\infty^2 = U_0^2/\gamma R T_\infty(0) =$ const and we can take $U_0 \equiv U_\infty(0)$ and $T_\infty(0) \equiv p_\infty(0)/\varrho_\infty(0)$.

To system (13.59), the following conditions must be imposed on the ground:

$$\hat{w} = 0 \Rightarrow \frac{d\hat{p}}{dz} = 0 \quad , \quad \text{on} \quad z = 0 \quad . \tag{13.60}$$

A behavior condition on $\hat{w}$ when $z \to +\infty$ must still be prescribed. This condition must be such that it leads to a well-posed Sturm-Liouville-type problem for $\hat{w}$! This is a delicate question as yet not entirely resolved about which we will say no more for the moment.

Let us take note, however, that if we return to the matrix equation (13.52), then we can clarify it by introducing:

$$
\begin{aligned}
&\frac{\partial\theta}{\partial x} = \chi \sin\varphi \quad , \quad \frac{\partial\theta}{\partial y} = \chi\cos\varphi \quad ; \\
&x = \xi\cos\varphi - \eta\sin\varphi \quad , \quad y = \xi\sin\varphi + \eta\cos\varphi \quad ; \\
&\tilde{u}\cos\varphi + \tilde{v}\sin\varphi = \approx{u} \quad , \quad -\tilde{u}\sin\varphi + \tilde{v}\cos\varphi = \overset{\approx}{v} \quad , \quad \tilde{w} \equiv \overset{\approx}{w} \quad ; \\
&\boldsymbol{V}_\infty \cdot (\boldsymbol{i}\cos\varphi + \boldsymbol{j}\sin\varphi) \equiv U_\infty \cos\varphi + V_\infty\sin\varphi = \overset{\approx}{U}_\infty(z) \quad , \\
&\tilde{\varrho} \equiv \overset{\approx}{\varrho} \quad , \quad \tilde{p} \equiv \overset{\approx}{p} \quad .
\end{aligned}
\tag{13.61}
$$

For $\overset{\approx}{u}$, $\overset{\approx}{w}$, $\overset{\approx}{p}$ and $\overset{\approx}{\varrho}$, in fact, the following two-dimensional system results with respect to ξ and z:

$$
\begin{aligned}
&\varrho_\infty \overset{\approx}{U}_\infty \frac{\partial\overset{\approx}{u}}{\partial\xi} + \frac{1}{\gamma M_\infty^2}\frac{\partial\overset{\approx}{p}}{\partial\xi} + \varrho_\infty \frac{d\overset{\approx}{U}_\infty}{dz}\overset{\approx}{w} = 0 \quad ; \\
&\varrho_\infty \overset{\approx}{U}_\infty \frac{\partial\overset{\approx}{w}}{\partial\xi} + \frac{1}{\gamma M_\infty^2}\left(\frac{\partial\overset{\approx}{p}}{\partial z} + \mathrm{Bo}\,\overset{\approx}{\varrho}\right) = 0 \quad ; \\
&\varrho_\infty\left(\overset{\approx}{U}_\infty\frac{\partial\overset{\approx}{p}}{\partial\xi} + \frac{dp_\infty}{dz}\overset{\approx}{w}\right) - \gamma p_\infty\left(\overset{\approx}{U}_\infty\frac{\partial\overset{\approx}{\varrho}}{\partial\xi} + \frac{d\varrho_\infty}{dz}\overset{\approx}{w}\right) = 0 \quad ; \\
&\overset{\approx}{U}_\infty\frac{\partial\overset{\approx}{\varrho}}{\partial\xi} + \varrho_\infty\left(\frac{\partial\overset{\approx}{u}}{\partial\xi} + \frac{\partial\overset{\approx}{w}}{\partial z}\right) + \frac{d\varrho_\infty}{dz}\overset{\approx}{w} = 0 \quad ,
\end{aligned}
\tag{13.62}
$$

and then an equation for $\overset{\approx}{v}$:

$$
\overset{\approx}{U}_\infty\, \partial\overset{\approx}{v}/\partial\xi = 0 \quad . \tag{13.63}
$$

We can take $\overset{\approx}{v} \equiv 0$. It is pointed out that $\overset{\approx}{U}_\infty(z)$ is the velocity of a basic flow which is a function of z and parallel to the ξ-axis. The following condition must be prescribed for the two-dimensional linear system (13.62):

$$
\overset{\approx}{w}(\xi, 0) = 0 \quad , \tag{13.64}
$$

as well as an adequate condition in $z \to +\infty$. The problem (13.62, 64) resembles the one already dealt with in Sect. 3.2. A classical but rather long calculation makes it possible to extract from system (13.62) a second order partial derivative equation in ξ and z for $\overset{\approx}{w}(\xi, z)$:

$$
\begin{aligned}
&\frac{\partial^2\overset{\approx}{w}}{\partial\xi^2} + \frac{1}{1-\gamma M_\infty^2\mu_\infty^2}\frac{\partial^2\overset{\approx}{w}}{\partial z^2} + \frac{1}{\varrho_\infty}\frac{d}{dz}\left(\frac{\varrho_\infty}{1-\gamma M_\infty^2\mu_\infty^2}\right)\frac{\partial\overset{\approx}{w}}{\partial z} \\
&-\left\{\frac{\mathrm{Bo}}{\gamma M_\infty^2 \overset{\approx}{U}{}_\infty^2}\left[\frac{1}{\varrho_\infty}\frac{d\varrho_\infty}{dz} + \frac{\mathrm{Bo} + \gamma M_\infty^2 \overset{\approx}{U}_\infty (d\overset{\approx}{U}_\infty/dz)}{c_\infty^2\left(1-\gamma M_\infty^2\mu_\infty^2\right)}\right]\right. \\
&\left. + \frac{1}{\varrho_\infty\overset{\approx}{U}_\infty}\frac{d}{dz}\left[\frac{\varrho_\infty\overset{\approx}{U}_\infty\left(\mathrm{Bo} + (c_\infty^2/\overset{\approx}{U}_\infty)(d\overset{\approx}{U}_\infty/dz)\right)}{c_\infty^2(1-\gamma M_\infty^2\mu_\infty^2)}\right]\right\}\overset{\approx}{w} = 0 ,
\end{aligned}
\tag{13.65}
$$

where

$$\mu_\infty^2(z) \equiv \varrho_\infty \overset{\approx}{U}_\infty^2/\gamma p_\infty \quad \text{and} \quad c_\infty^2(z) = \gamma p_\infty/\varrho_\infty \quad ;$$

$$\mu_\infty^2 \equiv \overset{\approx}{U}_\infty^2/c_\infty^2 \quad .$$

The book by Yih (1980; see Sect. 16 in Chap. 2), proposes an equation which can be identified with (13.65) when dimensionless variables are used.

In all the mesometeorological applications, we have $\mathrm{M}_\infty \ll 1$ and thus the coefficients of $\partial^2\overset{\approx}{w}/\partial z^2$ and $\partial\overset{\approx}{w}/\partial z$ in (13.65) can with a very good approximation be replaced by 1 and $d\log\varrho_\infty/dz$ respectively. Concerning the coefficient of $\overset{\approx}{w}$ in (13.65), we have to proceed more carefully so we will leave it as is for now. In this case, the transformation:

$$\overset{\approx}{w} = \frac{1}{\varrho_\infty^{1/2}} W \quad , \tag{13.66}$$

leads to the following equation for $W(\xi, z)$:

$$\frac{\partial^2 W}{\partial \xi^2} + \frac{\partial^2 W}{\partial z^2} + L_\infty^{-2}(z) W = 0 \quad , \tag{13.67}$$

where

$$\begin{aligned} L_\infty^{-2}(z) = & -\left\{ \frac{\mathrm{Bo}}{\gamma \mathrm{M}_\infty^2 \overset{\approx}{U}_\infty^2} \left[\frac{1}{\varrho_\infty}\frac{d\varrho_\infty}{dz} + \frac{\mathrm{Bo} + \gamma \mathrm{M}_\infty^2 \overset{\approx}{U}_\infty (d\overset{\approx}{U}_\infty/dz)}{c_\infty^2(1-\gamma \mathrm{M}_\infty^2 \mu_\infty^2)} \right] \right. \\ & \left. + \frac{1}{\varrho_\infty \overset{\approx}{U}_\infty} \frac{d}{dz}\left[\frac{\varrho_\infty \overset{\approx}{U}_\infty \left(\mathrm{Bo} + (c_\infty^2/\overset{\approx}{U}_\infty)(d\overset{\approx}{U}_\infty/dz)\right)}{c_\infty^2(1-\gamma \mathrm{M}_\infty^2 \mu_\infty^2)} \right] \right\} \\ & - \frac{1}{2\varrho_\infty}\frac{d^2\varrho_\infty}{dz^2} + \frac{3}{4}\left(\frac{1}{\varrho_\infty}\frac{d\varrho_\infty}{dz}\right)^2 \quad . \end{aligned} \tag{13.68}$$

It will be remarked that $|L_\infty|$ is a vertical length scale for the internal gravity waves being considered. A simplified form of L_∞^{-2} is given by the expression:

$$\begin{aligned} L_\infty^{-2}(z) \sim & -\frac{\mathrm{Bo}}{\gamma \mathrm{M}_\infty^2}\frac{1}{\overset{\approx}{U}_\infty^2}\left(\frac{1}{\varrho_\infty}\frac{d\varrho_\infty}{dz} + \frac{\mathrm{Bo}}{c_\infty^2}\right) - \frac{1}{\overset{\approx}{U}_\infty}\frac{d^2\overset{\approx}{U}_\infty}{dz^2} \\ & + \frac{1}{\varrho_\infty}\frac{d\varrho_\infty}{dz}\frac{1}{\overset{\approx}{U}_\infty}\frac{d\overset{\approx}{U}_\infty}{dz} - \frac{1}{2\varrho_\infty}\frac{d^2\varrho_\infty}{dz^2} + \frac{3}{4}\left(\frac{1}{\varrho_\infty}\frac{d\varrho_\infty}{dz}\right)^2 \quad . \end{aligned} \tag{13.69}$$

Furthermore, for all mesometeorological situations encountered in practice, the following holds:

$$L_\infty^{-2}(z) > 0 \quad , \quad \text{for all} \quad z > 0 \quad .$$

Equation (13.67) for W, with the condition $W(\xi, 0)$ = a given function of ξ, has been analyzed by many authors.[4] What makes (13.67) interesting is that the following upper boundary condition can be imposed on W:

$$|W| < \infty \quad , \quad \text{when} \quad z \to +\infty \quad .$$

This means that $\approx\!\!\!\!w$ *must not increase towards infinity* faster than $\varrho_\infty^{-1/2}$. It is seen that when $T_\infty(z) \equiv 1$, then $\varrho_\infty(z)$ tends to zero as $\exp(-\mathrm{Bo}\,z)$ when $z \to +\infty$. In this case, the approximate expression (13.69) for L_∞^{-2} becomes:

$$L_\infty^{-2}(z) \sim \frac{\mathrm{Bo}^2}{4} + \frac{\gamma-1}{\gamma^2}\left(\frac{\mathrm{Bo}}{\mathrm{M}_\infty}\right)^2 \frac{1}{\overset{\approx}{U}{}^2_\infty} - \frac{\mathrm{Bo}}{\overset{\approx}{U}_\infty}\frac{d\overset{\approx}{U}_\infty}{dz} - \frac{1}{\overset{\approx}{U}_\infty}\frac{d^2\overset{\approx}{U}_\infty}{dz^2} \,. \tag{13.70}$$

From (13.70), we are better able to understand the significance of the Boussinesq approximation: $\mathrm{Bo} \equiv \hat{B}\mathrm{M}_\infty$, $\mathrm{M}_\infty \to 0$, $\hat{B} = \mathrm{O}(1)$, which leads to:

$$L_\infty^{-2}(z) \sim \frac{1}{\overset{\approx}{U}_\infty}\left\{(\gamma-1)\frac{\hat{B}^2}{\gamma^2}\frac{1}{\overset{\approx}{U}_\infty} - \frac{d^2\overset{\approx}{U}_\infty}{dz^2}\right\} \quad . \tag{13.71}$$

The above expression (13.71) appears in particular, in the first works of Dorodnitsyn (1940) and Scorer (1949).

When $\overset{\approx}{U}_\infty(z) \equiv 1$ (since we are working with dimensionless values), it is seen that:

$$L_\infty^{-2} \equiv L_{\infty,0}^{-2} = (\gamma-1)\left(\frac{\hat{B}}{\gamma}\right)^2 = \text{const} \quad .$$

In this last case, if we suppose that W is zero in $z = 0$ and at the altitude $z = Z_\infty$, then for (13.67), the following simple solution is found:

$$W(\xi, z) = A\,\sin(n\pi z/Z_\infty)\cos\left[\sqrt{L_{\infty,0}^{-2} - \frac{n^2\pi^2}{Z_\infty^2}}\,\xi + \text{const}\right] \quad ,$$

as long as:

$$n^2 < (\gamma-1)\left[Z_\infty\frac{\hat{B}}{\gamma\pi}\right]^2 \quad .$$

13.2.2 The General Requirement for Trapped Lee Waves

In a more general situation, a solution to (13.51) will exist as a so-called *trapped lee wave*, persisting indefinitely in the direction of ξ, if and only if $\partial\theta/\partial x$ and $\partial\theta/\partial y$ in (13.57) are solutions of the so-called *dispersion relation*:

$$\mathcal{F}\left(\frac{\partial\theta}{\partial x}, \frac{\partial\theta}{\partial y}\right) = 0 \quad . \tag{13.72}$$

[4] For example, Queney et al. (1960), Miles (1969), Scorer (1957), Sawyer (1962), Zeytounian (1969) and Smith (1979), as well as the books by Beer (1974) and by Tolstoy (1973) can all be cited. In the books by Scorer (1978) and Yih (1980), a relatively complete review of these lee waves can be found.

We now return to (13.51) and write:[5]

$$\tilde{\mathcal{U}} = \tilde{\mathcal{U}}_0 + \tilde{\mathcal{U}}_1 + \dots \quad , \tag{13.73}$$

where $|\tilde{\mathcal{U}}_1| \ll |\tilde{\mathcal{U}}_0|$. It is found that:

$$\left(\frac{\partial\theta}{\partial x}\mathcal{A}_\infty + \frac{\partial\theta}{\partial y}\mathcal{B}_\infty\right)\frac{\partial\tilde{\mathcal{U}}_0}{\partial\theta} + \mathcal{C}_\infty\frac{\partial\tilde{\mathcal{U}}_0}{\partial z} + (\mathcal{D}_0 + \mathcal{M})\tilde{\mathcal{U}}_0 = 0 \quad ; \tag{13.74a}$$

$$\left(\frac{\partial\theta}{\partial x}\mathcal{A}_\infty + \frac{\partial\theta}{\partial y}\mathcal{B}_\infty\right)\frac{\partial\tilde{\mathcal{U}}_1}{\partial\theta} + \mathcal{C}_\infty\frac{\partial\tilde{\mathcal{U}}_1}{\partial z} + (\mathcal{D}_0 + \mathcal{M})\tilde{\mathcal{U}}_1$$
$$+\mathcal{A}_\infty\frac{\partial\tilde{\mathcal{U}}_0}{\partial x} + \mathcal{B}_\infty\frac{\partial\tilde{\mathcal{U}}_0}{\partial y} = 0 \quad . \tag{13.74b}$$

It is assumed that $\tilde{\mathcal{U}}_0$ is given by:

$$\tilde{\mathcal{U}}_0 = \Re\left\{a_0(x,y)\hat{\mathcal{U}}_0\left(z, \frac{\partial\theta}{\partial x}, \frac{\partial\theta}{\partial y}\right)\mathrm{e}^{\mathrm{i}\theta}\right\} \quad , \tag{13.75}$$

where $a_0(x,y)$ is an amplitude function that we want to determine and $\hat{\mathcal{U}}_0(z, \partial\theta/\partial x, \partial\theta/\partial y)$ is a shape factor which can be found as a solution to

$$\left\{i\left(\frac{\partial\theta}{\partial x}\mathcal{A}_\infty + \frac{\partial\theta}{\partial y}\mathcal{B}_\infty\right) + \mathcal{D}_0 + \mathcal{M}\right\}\hat{\mathcal{U}}_0 + \mathcal{C}_\infty\frac{\partial\hat{\mathcal{U}}_0}{\partial z} = 0 \quad , \tag{13.76}$$

with proper boundary conditions on the ground and at an upper altitude. It will be assumed here that $\hat{\mathcal{U}}_0$ has been properly normalized so that only the amplitude function is unknown in (13.75).

The solution for $\tilde{\mathcal{U}}_0$ is now substituted into (13.74b) which gives:

$$\left(\frac{\partial\theta}{\partial x}\mathcal{A}_\infty + \frac{\partial\theta}{\partial y}\mathcal{B}_\infty\right)\frac{\partial\tilde{\mathcal{U}}_1}{\partial\theta} + \mathcal{C}_\infty\frac{\partial\tilde{\mathcal{U}}_1}{\partial z} + (\mathcal{D}_0 + \mathcal{M})\tilde{\mathcal{U}}_1$$
$$+\Re\left\{\left[\left(\frac{\partial a_0}{\partial x}\mathcal{A}_\infty + \frac{\partial a_0}{\partial y}\mathcal{B}_\infty\right)\hat{\mathcal{U}}_0\right.\right.$$
$$+a_0\left(\frac{\partial^2\theta}{\partial x^2}\mathcal{A}_\infty + \frac{\partial\theta}{\partial x\partial y}\mathcal{B}_\infty\right)\frac{\partial\hat{\mathcal{U}}_0}{\partial(\partial\theta/\partial x)}$$
$$\left.\left.+a_0\left(\frac{\partial^2\theta}{\partial x\partial y}\mathcal{A}_\infty + \frac{\partial^2\theta}{\partial y^2}\mathcal{B}_\infty\right)\frac{\partial\hat{\mathcal{U}}_0}{\partial(\partial\theta/\partial y)}\right]\mathrm{e}^{\mathrm{i}\theta}\right\} = 0 \quad .$$

We look for a solution to the following:

$$\tilde{\mathcal{U}}_1 = \Re\left\{\hat{\mathcal{U}}_1\,\mathrm{e}^{\mathrm{i}\theta}\right\} \quad , \tag{13.77}$$

and we have:

---

[5] Let us simply assume that there are trapped waves propagating away from the relief without considering how they have been created since the present analysis can be of no help in answering this question.

$$\mathrm{i}\left(\frac{\partial\theta}{\partial x}\mathcal{A}_\infty + \frac{\partial\theta}{\partial y}\mathcal{B}_\infty\right)\hat{\mathcal{U}}_1 + \mathcal{C}_\infty\frac{\partial\hat{\mathcal{U}}_1}{\partial z} + (\mathcal{D}_0 + \mathcal{M})\hat{\mathcal{U}}_1 + \hat{\phi}_0 = 0 \quad , \tag{13.78}$$

where

$$\hat{\phi}_0 \equiv \left(\frac{\partial a_0}{\partial x}\mathcal{A}_\infty + \frac{\partial a_0}{\partial y}\mathcal{B}_\infty\right)\hat{\mathcal{U}}_0 + a_0\left[\left(\frac{\partial^2\theta}{\partial x^2}\mathcal{A}_\infty + \frac{\partial^2\theta}{\partial x\partial y}\mathcal{B}_\infty\right)\frac{\partial\hat{\mathcal{U}}_0}{\partial(\partial\theta/\partial x)} + \left(\frac{\partial^2\theta}{\partial x\partial y}\mathcal{A}_\infty + \frac{\partial^2\theta}{\partial y^2}\mathcal{B}_\infty\right)\frac{\partial\hat{\mathcal{U}}_0}{\partial(\partial\theta/\partial y)}\right] \quad . \tag{13.79}$$

Let $\hat{W}_0$ be some five-element column matrix dependent on z. We will start from the obvious relation:[6]

$$\left\langle \hat{W}_0 , \left\{\mathrm{i}\left(\frac{\partial\theta}{\partial x}\mathcal{A}_\infty + \frac{\partial\theta}{\partial y}\mathcal{B}_\infty\right)\hat{\mathcal{U}}_1 + \mathcal{C}_\infty\frac{\partial\hat{\mathcal{U}}_1}{\partial z}(\mathcal{D}_0 + \mathcal{M})\hat{U}_1 \right\rangle + \langle \hat{W}_0 , \hat{\phi}_0\rangle = 0 \quad , \tag{13.80}$$

which may be written as:

$$\left\langle \hat{\mathcal{U}}_1^* , \mathrm{i}\left(\frac{\partial\theta}{\partial x}\mathcal{A}_\infty^{\mathrm{T}} + \frac{\partial\theta}{\partial y}\mathcal{B}_\infty^{\mathrm{T}}\right)\hat{W}_0^* - \frac{d}{dz}\left(\mathcal{C}_\infty^{\mathrm{T}}\hat{W}_0^*\right) + \left(\mathcal{D}_0^{\mathrm{T}} + \mathcal{M}^{\mathrm{T}}\right)\hat{W}_0^* \right\rangle + \frac{d}{dz}\langle \hat{W}_0, \mathcal{C}_\infty\hat{\mathcal{U}}_1\rangle + \langle \hat{W}_0, \hat{\phi}_0\rangle = 0 \quad . \tag{13.81}$$

We now choose $\hat{W}_0$ in such a way that:

$$-\mathrm{i}\left(\frac{\partial\theta}{\partial x}\mathcal{A}_\infty^{\mathrm{T}} + \frac{\partial\theta}{\partial y}\mathcal{B}_\infty^{\mathrm{T}}\right)\hat{W}_0 - \frac{d}{dz}\left(\mathcal{C}_\infty^{\mathrm{T}}\hat{W}_0\right) + \left(\mathcal{D}_0^{\mathrm{T}} + \mathcal{M}^{\mathrm{T}}\right)\hat{W}_0 = 0, \tag{13.82}$$

and that the following relation holds:

$$\langle \hat{W}_0, \mathcal{C}_\infty\hat{\mathcal{U}}_1\rangle = 0 \quad , \tag{13.83}$$

both at the ground and at the upper boundary. It is observed that the conditions defining $\hat{W}_0$ are precisely the adjoints of those which define $\hat{\mathcal{U}}_0$. Both have a non-trivial solution if, and only if, the same dispersion relation (13.72) holds [see, for instance, Coddington and Levinson (1955, Chap. 11)].

If we now integrate (13.80) over the whole of the altitude interval Z, we get:

$$\int_Z \langle \hat{W}_0, \hat{\phi}_0\rangle dz = 0 \quad , \tag{13.84}$$

as a consequence of (13.81–83). Taking (13.79) into account, we have:

---

[6] By using an asterisk for complex conjugate, we have:

$$\langle \hat{w}, \hat{f}\rangle = \sum_{k=1}^{5} \hat{w}_k^* \hat{f}_k \quad , \quad \text{where} \quad \hat{f} = \left(\hat{f}_1, \hat{f}_2, \hat{f}_3, \hat{f}_4, \hat{f}_5\right)^{\mathrm{T}} \quad .$$

$$\int_Z \left\{ \langle \hat{W}_0, \mathcal{A}_\infty \hat{\mathcal{U}}_0 \rangle \frac{\partial a_0}{\partial x} + \langle \hat{W}_0, \mathcal{B}_\infty \hat{\mathcal{U}}_0 \rangle \frac{\partial a_0}{\partial y} \right.$$

$$+ a_0 \left\langle \hat{W}_0, \left(\frac{\partial^2 \theta}{\partial x^2} \mathcal{A}_\infty + \frac{\partial^2 \theta}{\partial x \partial y} \mathcal{B}_\infty \right) \frac{\partial \hat{\mathcal{U}}_0}{\partial(\partial\theta/\partial x)} \right\rangle$$

$$\left. + a_0 \left\langle \hat{W}_0, \left(\frac{\partial^2 \theta}{\partial x \partial y} \mathcal{A}_\infty + \frac{\partial^2 \theta}{\partial y^2} \mathcal{B}_\infty \right) \frac{\partial \hat{\mathcal{U}}_0}{\partial(\partial\theta/\partial y)} \right\rangle \right\} dz = 0 \quad , \tag{13.85}$$

which is the required equation for computing the variation of the amplitude function $a_0(x, y)$.

Let us set ($k_1 \equiv \partial\theta/\partial x$ and $k_2 \equiv \partial\theta/\partial y$):

$$\mathcal{L} = \mathcal{C}_\infty \frac{\partial}{\partial z} + \mathcal{D}_0 + \mathcal{M} + \mathrm{i}(k_1 \mathcal{A}_\infty + k_2 \mathcal{B}_\infty) \quad ;$$

$$\mathcal{L}^* = -\mathcal{C}_\infty^{\mathrm{T}} \frac{\partial}{\partial z} + \mathcal{D}_0^{\mathrm{T}} + \mathcal{M}^{\mathrm{T}} - \frac{\partial \mathcal{C}_\infty^{\mathrm{T}}}{\partial z} - \mathrm{i}\left(k_1 \mathcal{A}_\infty^{\mathrm{T}} + k_2 \mathcal{B}_\infty^{\mathrm{T}}\right) \quad ,$$

and we have:

$$\mathcal{L}\hat{\mathcal{U}}_0 = \mathcal{L}^* \hat{W}_0 \quad . \tag{13.86}$$

The term $\hat{\mathcal{U}}_0$ is considered as a function of k_1 and k_2 which, of course, are not independent because of the dispersion relation (13.72).

We set

$$d\hat{\mathcal{U}}_0 = \frac{\partial \hat{\mathcal{U}}_0}{\partial k_1} dk_1 + \frac{\partial \hat{\mathcal{U}}_0}{\partial k_2} dk_2 \quad , \quad \text{and} \quad \frac{\partial \mathcal{F}}{\partial k_1} dk_1 + \frac{\partial \mathcal{F}}{\partial k_2} dk_2 = 0 \quad . \tag{13.87}$$

From the definition of $\hat{\mathcal{U}}_0$, we have:

$$\mathcal{L}\hat{\mathcal{U}}_0 = 0 \quad .$$

This relation may be differentiated which leads to:

$$\mathcal{L} d\hat{\mathcal{U}}_0 + \mathrm{i}\mathcal{A}_\infty \hat{\mathcal{U}}_0 dk_1 + \mathrm{i}\mathcal{B}_\infty \hat{\mathcal{U}}_0 dk_2 = 0 \quad .$$

From the above, we obtain

$$\int_Z \langle \hat{W}_0, \mathcal{L} d\hat{\mathcal{U}}_0 + \mathrm{i}\mathcal{A}_\infty \hat{\mathcal{U}}_0 dk_1 + \mathrm{i}\mathcal{B}_\infty \hat{\mathcal{U}}_0 dk_2 \rangle dz = 0 \quad .$$

But, we have:

$$\int_Z \langle \hat{W}_0, \mathcal{L} d\hat{\mathcal{U}}_0 \rangle dz = \int_Z \langle \mathcal{L}^* \hat{W}_0, d\hat{\mathcal{U}}_0 \rangle dz = 0 \quad ,$$

and, consequently,

$$\left\{ \int_Z \langle \hat{W}_0, \mathcal{A}_\infty \hat{\mathcal{U}}_0 \rangle dz \right\} dk_1 + \left\{ \int_Z \langle \hat{W}_0, \mathcal{B}_\infty \hat{\mathcal{U}}_0 \rangle dz \right\} dk_2 = 0 \quad . \tag{13.88}$$

By comparing (13.87) to (13.88), we find:

$$\int_Z \langle \hat{W}_0, \mathcal{A}_\infty \hat{\mathcal{U}}_0 \rangle dz = \Lambda_0 \frac{\partial \mathcal{F}}{\partial k_1} \quad ;$$
$$\int_Z \langle \hat{W}_0, \mathcal{B}_\infty \hat{\mathcal{U}}_0 \rangle dz = \Lambda_0 \frac{\partial \mathcal{F}}{\partial k_2} \quad . \tag{13.89}$$

Hence, (13.85) may be rewritten as:

$$\Lambda_0 \left\{ \frac{\partial \mathcal{F}}{\partial(\partial\theta/\partial x)} \frac{\partial a_0}{\partial x} + \frac{\partial \mathcal{F}}{\partial(\partial\theta/\partial y)} \frac{\partial a_0}{\partial y} \right\}$$
$$+ a_0 \left\{ \int_Z \left[\left\langle \hat{W}_0, \left(\frac{\partial^2\theta}{\partial x^2} \mathcal{A}_\infty + \frac{\partial^2\theta}{\partial x \partial y} \mathcal{B}_\infty \right) \frac{\partial \hat{\mathcal{U}}_0}{\partial(\partial\theta/\partial x)} \right\rangle \right.\right.$$
$$\left.\left. + \left\langle \hat{W}_0, \left(\frac{\partial^2\theta}{\partial x \partial y} \mathcal{A}_\infty + \frac{\partial^2\theta}{\partial y^2} \mathcal{B}_\infty \right) \frac{\partial \hat{\mathcal{U}}_0}{\partial(\partial\theta/\partial y)} \right\rangle \right] dz \right\} = 0 \quad , \tag{13.90}$$

which looks like an ordinary differential equation for $a_0(x, y)$ along the rays associated with the dispersion relation. Such rays are defined as the integrals of the first relation in the system:

$$\frac{dx}{\partial \mathcal{F}/\partial(\partial\theta/\partial x)} = \frac{dy}{\partial \mathcal{F}/\partial(\partial\theta/\partial y)} = \frac{\partial\theta/\partial x}{0} = \frac{\partial\theta/\partial y}{0} \quad .$$

Our purpose here was quite limited. We intended to show that the excited trapped lee waves travel along the rays and that their amplitudes may, at least in principle, be computed. It is almost obvious that this amplitude decays when travelling away from the relief although it would be difficult to prove formally. If the lee wave phenomenon were self-adjoint, and if the matrix $\mathcal{A}_\infty$, $\mathcal{B}_\infty$ were symmetric, (13.90) might be rewritten as:

$$\frac{\partial}{\partial x} \left\{ a_0^2(x, y) \langle \hat{W}_0, \mathcal{A}_\infty \hat{\mathcal{U}}_0 \rangle \right\} + \frac{\partial}{\partial y} \left\{ a_0^2(x, y) \langle \hat{W}_0, \mathcal{B}_\infty \hat{\mathcal{U}}_0 \rangle \right\} = 0 \quad . \tag{13.91}$$

Let us now consider two neighbouring rays and set $\Sigma\delta\omega$ as the distance between them. If they start at the relief with an angle $\delta\omega$ between them, then (13.91) tells us that the product $a_0^2\Sigma$ remains constant along each ray. If this were true, roughly speaking, "a_0" would decay as (distance)$^{-1/2}$, where distance means the arc length along the ray from the relief to the point under consideration.

13.2.3 Non-linear Models for Two-Dimensional Steady Lee Waves

1. When in (13.41) it can be supposed that $\beta \ll 1$ (a relief of "quasi" infinite length in the direction perpendicular to the ξ-axis [see (13.61)] parallel to the basic wind), then it is possible to only consider a two-dimensional problem in the plane (ξ, z).

Let us then return to *dimensional* values and consider the following classical Euler equations in the plane (x, z) for u, w, p, ϱ and T:

$$\varrho\left(u\frac{\partial u}{\partial x}+w\frac{\partial u}{\partial z}\right)+\frac{\partial p}{\partial x}=0 \quad ;$$
$$\varrho\left(u\frac{\partial w}{\partial x}+w\frac{\partial w}{\partial z}\right)+\frac{\partial p}{\partial z}+g\varrho=0 \quad ;$$
$$\frac{\partial \varrho u}{\partial x}+\frac{\partial \varrho w}{\partial z}=0 \quad ; \quad p=R\varrho T \tag{13.92}$$
$$\left(u\frac{\partial}{\partial x}+w\frac{\partial}{\partial z}\right)\left(\frac{p}{\varrho^\gamma}\right)=0 \quad .$$

We adopt the following as a relief:

$$z=f(x) \quad ; \quad f(\mp\infty)\equiv 0 \quad , \quad -\frac{l_0}{2}\leq x\leq +\frac{l_0}{2} \quad . \tag{13.93}$$

We can define a stream function:

$$\psi=\psi(x,z) \Rightarrow \varrho u=-\frac{\partial\psi}{\partial z} \quad \text{and} \quad \varrho w=+\frac{\partial\psi}{\partial x} \quad ,$$

and in this case, we know that the following first integrals can be obtained from (13.92):

$$\frac{p}{\varrho^\gamma}=\pi(\psi) \quad ;$$
$$\frac{u^2+w^2}{2}+\frac{\gamma}{\gamma-1}\varrho^{\gamma-1}\pi(\psi)+gz=J(\psi) \quad ; \tag{13.94}$$
$$\frac{\partial u}{\partial z}-\frac{\partial w}{\partial x}=-\varrho\left\{\frac{dJ}{d\psi}-\frac{1}{\gamma-1}\frac{p}{\varrho}\frac{d\log\pi}{d\psi}\right\} \quad .$$

The functions $\pi(\psi)$ and $J(\psi)$ (which depend on ψ alone) can be determined from the conditions to which the non-perturbed basic flow has been subjected far away at upstream infinity from the relief for $x\to-\infty$ (see Fig. 5)

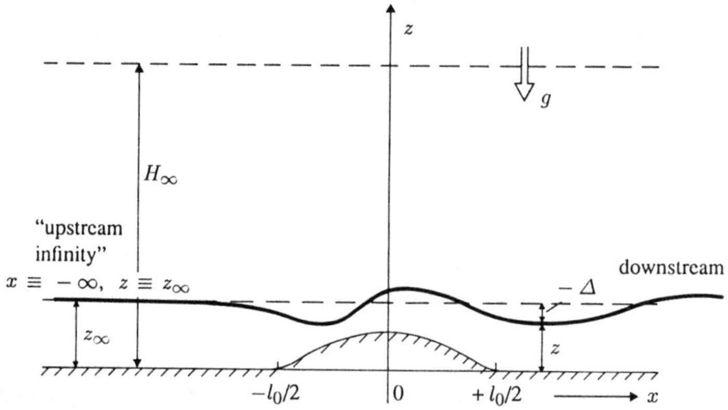

Fig. 5

In fact, if the functions

$$u_\infty = U_\infty \quad , \quad w_\infty \equiv 0 \quad , \quad p_\infty, \varrho_\infty, T_\infty \quad \text{and} \quad s_\infty \equiv c_v \log \frac{p_\infty}{\varrho_\infty^\gamma} \quad ,$$

defining the velocity and the thermodynamic state of the non-disturbed flow, are acknowledged to vary only with the altitude denoted z_∞, then we readily obtain:

$$\frac{dJ}{d\psi} - \frac{1}{\gamma - 1}\frac{p}{\varrho}\frac{d\log\pi}{d\psi} \equiv -\frac{1}{\varrho_\infty}\left\{\frac{dU_\infty}{dz_\infty} - \frac{\gamma R/c_p}{U_\infty(\gamma-1)}\frac{ds_\infty}{dz_\infty}(T - T_\infty)\right\} \quad , \tag{13.95}$$

where

$$\frac{ds_\infty}{dz_\infty} = \frac{c_p}{T_\infty}\left(\frac{dT_\infty}{dz_\infty} + \frac{\gamma - 1}{\gamma}\frac{g}{R}\right) \quad , \tag{13.96}$$

and it is assumed that $ds_\infty/dz_\infty > 0$.

By taking into account the conservation of the functions $\pi(\psi)$ and $J(\psi)$ along the streamlines, the temperature $T(x, z)$ can be expressed from the second first integral in (13.94) in the following form:

$$T = T_\infty(z_\infty) - \frac{\gamma - 1}{\gamma R}\left\{\frac{1}{2}\left(u^2 + w^2 - U_\infty^2\right) + g(z - z_\infty)\right\} \quad . \tag{13.97}$$

We then have for the density

$$\varrho = \varrho_\infty\left\{1 + \frac{T - T_\infty}{T_\infty}\right\}^{1/(\gamma-1)} \quad . \tag{13.98}$$

If we now take advantage of (13.94, 95, 97) and of the definition of ψ, after a few calculations, the following *exact* equation for $\psi(x, z)$ can be obtained:

$$\begin{aligned} &\frac{\partial^2\psi}{\partial x^2} + \frac{\partial^2\psi}{\partial z^2} + \varrho^2\frac{g}{c_p\varrho_\infty U_\infty}\frac{ds_\infty}{dz_\infty}(z - z_\infty) \\ &\quad = \frac{\partial\log\varrho}{\partial x}\frac{\partial\psi}{\partial x} + \frac{\partial\log\varrho}{\partial z}\frac{\partial\psi}{\partial z} - \frac{\varrho^2}{\varrho_\infty}\frac{dU_\infty}{dz_\infty} \\ &\quad\quad - \frac{1}{2}\frac{1}{c_p\varrho_\infty U_\infty}\frac{ds_\infty}{dz_\infty}\left[\left(\frac{\partial\psi}{\partial x}\right)^2 + \left(\frac{\partial\psi}{\partial z}\right)^2 - (\varrho U_\infty)^2\right] \quad . \end{aligned} \tag{13.99}$$

We must of course understand that in (13.99), ϱ is related to T via (13.98), whereas T is expressed in the form:

$$\begin{aligned} T = T_\infty(z_\infty) - \frac{\gamma - 1}{\gamma R}\frac{1}{\varrho^2}&\left\{\frac{1}{2}\left[\left(\frac{\partial\psi}{\partial x}\right)^2 + \left(\frac{\partial\psi}{\partial z}\right)^2 - (\varrho U_\infty)^2\right]\right. \\ &\left. + g\varrho^2(z - z_\infty)\right\} \quad . \end{aligned} \tag{13.100}$$

The above equation (13.99) generalizes the one obtained by Mrs. Dubreil-Jacotin (1937) and analyzed by Long (1953). It was discussed in Chap. 9 (see Sect. 9.3) which was devoted to isochoric flows.

Since our perfect gas (atmospheric air) was supposed inviscid, the slip condition along the relief surface is expressed by the relation:

$$\psi(x, f(x)) = 0 \quad , \tag{13.101}$$

once we have taken into account the fact that at upstream infinity (when $x \to -\infty$) we have:

$$\psi = -\int_0^{z_\infty} \varrho_\infty(z) U_\infty(z) dz \equiv \psi_\infty(z_\infty) \quad . \tag{13.102}$$

A second condition in z is yet necessary. In an infinite atmosphere, it must be a behavior condition for $z \to +\infty$.

From a mathematical point of view, the simplest case is that of a flow in a duct bounded at the top by the surface $z = H_\infty$; the bottom is simulated by the relief (13.93). We must then write the upper boundary condition in z as follows:

$$\psi = \psi_\infty(H_\infty) \quad , \quad \text{on} \quad z = H_\infty \quad . \tag{13.103}$$

Two boundary conditions in x are likewise necessary. The first is written:

$$\psi \to \psi_\infty(z_\infty) \quad , \quad T \to T_\infty(z_\infty) \quad , \quad \varrho \to \varrho_\infty(z_\infty) \quad , \quad \text{when} \quad x \to -\infty \quad . \tag{13.104}$$

The second condition is directly inspired by the physics of the problem:

$$\left|\frac{\partial \psi}{\partial x}\right| + \left|\frac{\partial \psi}{\partial z}\right| \quad \text{remains bounded when} \quad x \to +\infty \quad . \tag{13.105}$$

It is further pointed out that from (13.102), it follows that:

$$z_\infty = \psi_\infty^{-1}(\psi) \equiv z_\infty(\psi) \quad , \tag{13.106}$$

the function $z_\infty(\psi)$ being uniform if $U_\infty(z_\infty)$ is strictly positive when $z_\infty \in [0, H_\infty]$.

2. Let us now define (see Fig. 5):

$$z_\infty(\psi) + \Delta(x, z) = z \quad , \tag{13.107}$$

as the altitude of a streamline in the flow perturbed by the relief in such a way that the expression $z_\infty(\psi) = z - \Delta(x, z)$ remains constant along a streamline. By taking into account the obvious relation $d\psi = U_\infty \varrho_\infty d\Delta - U_\infty \varrho_\infty dz$, as well as the derivation formulas:

$$\frac{\partial z_\infty}{\partial x} = -\frac{\partial \Delta}{\partial x} \quad \text{and} \quad \frac{\partial z_\infty}{\partial z} = 1 - \frac{\partial \Delta}{\partial z} \quad ,$$

the following equation is obtained for $\Delta(x, z)$ instead of (13.99):

$$
\begin{aligned}
&\frac{\partial^2 \Delta}{\partial x^2} + \frac{\partial^2 \Delta}{\partial z^2} + \left(\frac{\varrho}{\varrho_\infty}\right)^2 \frac{g}{c_p U_\infty^2} \frac{d s_\infty}{d z_\infty} \Delta \\
&\quad = -\frac{1}{2}\left(\frac{\varrho}{\varrho_\infty}\right)^2 \frac{d}{d z_\infty}\left\{\log\left(U_\infty^2\, \mathrm{e}^{-s_\infty/c_p}\right)\right\} \\
&\quad + \frac{1}{2}\frac{d}{d z_\infty}\left\{\log\left(\varrho_\infty^2 U_\infty^2\, \mathrm{e}^{-s_\infty/c_p}\right)\right\}\left[\left(\frac{\partial \Delta}{\partial x}\right)^2 + \left(\frac{\partial \Delta}{\partial z}\right)^2 - 2\frac{\partial \Delta}{\partial z} + 1\right] \\
&\quad + \frac{\partial \log \varrho}{\partial x}\frac{\partial \Delta}{\partial x} + \frac{\partial \log \Delta}{\partial z}\left(\frac{\partial \Delta}{\partial z} - 1\right) \quad . \qquad (13.108)
\end{aligned}
$$

The following boundary conditions on Δ must be added to (13.108):

$$
\begin{aligned}
&\Delta\big(x, f(x)\big) = f(x) \quad , \quad -\frac{l_0}{2} \le x \le \frac{l_0}{2} \quad ; \\
&\Delta(x, H_\infty) = \Delta(-\infty, z_\infty) = 0 \quad , \quad f(-\infty) \equiv 0 \quad ; \qquad (13.109) \\
&\left|\frac{\partial \Delta}{\partial x}\right| + \left|\frac{\partial \Delta}{\partial y}\right| < \infty \quad , \quad \text{when} \quad x \to +\infty \quad .
\end{aligned}
$$

Two unknown functions, however, appear in (13.108). These are: $\Delta(x, z)$ and $\varrho(x, z)$. The functions U_∞, ϱ_∞ and s_∞ are given data in the problem and are expressed by Δ and z via the relation $z_\infty = z - \Delta$.

Equation (13.108) thus needs to be completed by a second formula. To this end, we will make use of (13.98, 100). After introducing Δ to replace ψ, the following results:

$$
\begin{aligned}
\varrho = \varrho_\infty &\left\{1 + \frac{U_\infty^2}{2 c_p T_\infty}\left(\frac{\varrho_\infty}{\varrho}\right)^2\left[\left(\frac{\partial \Delta}{\partial x}\right)^2 + \left(\frac{\partial \Delta}{\partial z}\right)^2 - 2\frac{\partial \Delta}{\partial z} + 1\right]\right. \\
&\left. + \frac{U_\infty^2}{2 c_p T_\infty}\left[\frac{2g}{U_\infty^2}\Delta - 1\right]\right\}^{1/(\gamma-1)} \quad , \qquad (13.110)
\end{aligned}
$$

and the condition:

$$
\varrho \to \varrho_\infty \quad , \quad \text{when} \quad x \to -\infty \qquad (13.111)
$$

is indeed satisfied since $\Delta(-\infty, z_\infty) \equiv 0$.

The problem which is made up of the fundamental equation (13.108), of relation (13.110) and the boundary conditions (13.109, 111) is closed and "seems" to be well-posed, mathematically speaking. In principle, it should enable the determining of the unknown functions Δ and ϱ of x and z. This problem (13.108–111) is *strongly non-linear* and only a numerical resolution can make it possible to find its solution in the general case. This numerical resolution gives rise, moreover, to various difficulties which are discussed in Zeytounian (1969a).

The non-linearities of problem (13.108–111) are, in fact, of three different kinds:

a. those related to the nature of the basic non-disturbed flow at upstream infinity, far from the relief; they disappear if the following constraint is imposed:

$$\varrho_\infty^2 U_\infty^2 \, \mathrm{e}^{-s_\infty/c_p} \equiv \text{const} \quad ;$$

b. those related to the fact that $\varrho = \chi(z_\infty, \Delta)$; they disappear if the function χ identifies with $\varrho_\infty(z_\infty)$ which happens in the isochoric flow;

c. those related to the condition $\Delta\left(x, f(x)\right) = f(x)$ imposed on a curvilinear contour; they disappear if the following linearization is carried out:

$$\Delta(x, 0) = f(x) \quad .$$

An asymptotic analysis of this non-linear problem (13.108–111) remains to be worked out. We will, however, restrict our attention here to a particular case.

3. Let us then consider the case of:

$$U_\infty(z) \equiv U_\infty^0 = \text{const} \quad \text{and} \quad -\frac{dT_\infty}{dz_\infty} \equiv \Gamma_\infty^0 = \text{const} \quad .$$

Therefore:

$$T_\infty(z_\infty) = T_\infty(0)\left\{1 - \frac{\Gamma_\infty^0}{T_\infty(0)} z_\infty\right\} \quad , \tag{13.112}$$

when $z_\infty \leq H_\infty \approx 11\,\text{km}$ and a typical value of Γ_∞^0 is 0.0065° C/m. In addition, as:

$$\frac{1}{\varrho_\infty}\frac{d\varrho_\infty}{dz_\infty} = -\frac{1}{T_\infty}\left\{\frac{dT_\infty}{dz} + \frac{g}{R}\right\}$$

it follows that:

$$\varrho_\infty(z_\infty) = \varrho_\infty(0)\left\{1 - \frac{\Gamma_\infty^0}{T_\infty(0)} z_\infty\right\}^{(g/R\Gamma_\infty^0)-1} \quad , \tag{13.113}$$

and also:

$$p_\infty(z_\infty) = p_\infty(0)\left\{1 - \frac{\Gamma_\infty^0}{T_\infty(0)} z_\infty\right\}^{g/R\Gamma_\infty^0} \quad , \tag{13.114}$$

since $p_\infty = R\varrho_\infty T_\infty$ and $dp_\infty/dz_\infty + g\varrho_\infty = 0$.

It is observed that the characteristic vertical scale

$$H_T = \frac{T_\infty(0)}{\Gamma_\infty^0} \approx 44\,308\,\text{m} \quad , \text{ appears when } T_\infty(0) \approx 288°\,\text{C} \quad .$$

A second characteristic vertical scale appears which is related to the following hydrostatic relation:

$$\frac{d\log p_\infty}{dz_\infty} + \frac{g}{RT_\infty} = 0 \Rightarrow H_p = \frac{RT_\infty(0)}{g} \approx 8\,434\,\text{m} \quad ,$$

which is the altitude of the so-called homogeneous atmosphere and it is remarked that:

$$R \approx 287\,\text{m}^2/\text{s}^2 \text{ degrees} \quad \text{and} \quad g \approx 9.8\,\text{m/s}^2 \quad .$$

Let us now introduce the density perturbation:

$$\omega = \frac{\varrho - \varrho_\infty}{\varrho_\infty} \tag{13.115}$$

and the following dimensionless values:

$$\xi = \frac{x}{l_0} \quad , \quad \zeta = \frac{z}{H_0} \quad , \quad \delta = \frac{\Delta}{h_0} \quad , \quad \tau_\infty = \frac{T_\infty}{T_\infty(0)} \quad , \tag{13.116}$$

where:

$$h_0 = \max_{-l_0/2 \leq x \leq l_0/2} |f(x)| \quad \text{and} \quad f(x) \equiv h_0 h(x/l_0) \quad .$$

The vertical scale H_0 is introduced assuming that what takes place at order H_0 vertical distances from the relief will be examined. Finally, instead of (13.108–110), we obtain the following reduced problem:

$$\tau_\infty \left\{ \varepsilon_0^2 \frac{\partial^2 \delta}{\partial \xi^2} + \frac{\partial^2 \delta}{\partial \zeta^2} - \frac{1}{1+\omega} \left[\varepsilon_0^2 \frac{\partial \omega}{\partial \xi} \frac{\partial \delta}{\partial \xi} + \frac{\partial \omega}{\partial \zeta} \frac{\partial \delta}{\partial \zeta} - \nu_0 \frac{\partial \omega}{\partial \zeta} \right] \right\} + (1+\omega)^2 K_0 \delta$$

$$= \frac{\nu_0}{2} \Lambda_0 \omega (2+\omega) - \frac{1}{2} \frac{\Lambda_0}{\nu_0} \left[\varepsilon_0^2 \left(\frac{\partial \delta}{\partial \xi} \right)^2 + \left(\frac{\partial \delta}{\partial \zeta} \right)^2 - 2\nu_0 \frac{\partial \delta}{\partial \zeta} \right] \quad ; \tag{13.117a}$$

$$(1+\omega)^{\gamma-1} = 1 - \frac{1}{\tau_\infty} \left\{ \frac{\gamma-1}{2} \frac{\mathrm{M}_\infty^2}{\nu_0^2} \frac{1}{(1+\omega)^2} \left[\varepsilon_0^2 \left(\frac{\partial \delta}{\partial \xi} \right)^2 + \left(\frac{\partial \delta}{\partial \zeta} \right)^2 \right. \right.$$

$$\left. \left. - 2\nu_0 \frac{\partial \delta}{\partial \zeta} + \nu_0^2 \right] + \frac{\gamma-1}{\gamma} \frac{\lambda_0}{\nu_0} \delta - \frac{\gamma-1}{2} \mathrm{M}_\infty^2 \right\} \quad ; \tag{13.117b}$$

$$\delta\left(\zeta, \frac{1}{\nu_0} h(\xi)\right) = h(\xi) \quad , \quad \xi \in \left[-\frac{1}{2}, +\frac{1}{2} \right] \quad ;$$

$$\delta(-\infty, \zeta_\infty) = \delta(\xi, \kappa_0) = 0 \quad ;$$

$$\left| \frac{\partial \delta}{\partial \xi} \right| + \left| \frac{\partial \delta}{\partial \zeta} \right| < \infty \quad , \quad \text{when} \quad \xi \to +\infty \quad , \tag{13.117c}$$

where

$$\tau_\infty(\xi, \zeta) \equiv 1 - \mu_0 \zeta + \frac{\mu_0}{\nu_0} \delta \quad . \tag{13.118}$$

In system (13.117) for $\delta(\xi, \zeta)$ and $\omega(\xi, \zeta)$, the following dimensionless parameters are introduced:

$$\varepsilon_0 = \frac{H_0}{L_0} \quad , \quad \nu_0 = \frac{H_0}{h_0} \quad , \quad \kappa_0 = \frac{H_\infty}{H_0} \quad ; \tag{13.119a}$$

$$\mu_0 = \frac{H_0}{H_T} \equiv \frac{\Gamma_\infty^0 H_0}{T_\infty(0)} \quad , \quad \lambda_0 = \frac{H_0}{H_p} \equiv \frac{g H_0}{R T_\infty(0)} \equiv \mathrm{Bo} \quad ; \tag{13.119b}$$

$$\mathrm{M}_\infty^2 = \frac{U_\infty^{02}}{\gamma R T_\infty(0)} \quad , \quad \Lambda_0 = \frac{\gamma-1}{\gamma}\lambda_0 - \mu_0 \equiv \mathrm{Bo}\left(\frac{\gamma-1}{\gamma} - \frac{R}{g}\Gamma_\infty^0\right) \quad ,$$

$$K_0 = \frac{\mathrm{Bo}}{\gamma \mathrm{M}_\infty^2}\Lambda_0 . \tag{13.119c}$$

Finally, we have made use of the relation:

$$\mathrm{Fr}_{H_0}^2 \equiv \frac{U_\infty^{02}}{g H_0} = \frac{\gamma \mathrm{M}_\infty^2}{\mathrm{Bo}} \quad .$$

It will be noted that:

$$\alpha_0 \equiv \frac{\gamma-1}{\gamma} - \frac{R}{g}\Gamma_\infty^0 \approx 0.28571 - 0.19037 \approx 0.09534 \quad ,$$

since $\gamma \approx 1.4$ for dry air. Furthermore, as $[\gamma R T_\infty(0)]^{1/2} \approx 340.17\,\mathrm{m/s}$ when $T_\infty(0) \approx 288°\,\mathrm{C}$, it is observed that when the characteristic velocity U_∞^0 varies between 34 m/s (i.e., 122.4 km/h) and 10 m/s (36 km/h), the characteristic Mach number M_∞ varies in the following way:

$$0.03 \leq \mathrm{M}_\infty \leq 0.1$$

and thus, we must take advantage of the limiting process $\mathrm{M}_\infty \to 0$ in order to model problem (13.117). However, when $\mathrm{M}_\infty \to 0$, we must also make sure that the parameter:

$$K_0 = \frac{\mathrm{Bo}^2}{\gamma \mathrm{M}_\infty^2}\alpha_0 \equiv \frac{\Gamma_A - \Gamma_\infty^0}{T_\infty(0)} g \left(\frac{H_0}{U_\infty^0}\right)^2 \tag{13.120}$$

remains of the order unity.[7] $\Gamma_A \equiv (\gamma-1)g/\gamma R$ is the dry adiabatic temperature gradient.

If we wish for $K_0 = \mathrm{O}(1)$, then it is necessary that:

$$H_0 \approx U_\infty^0 \sqrt{\frac{T_\infty(0)/g}{\Gamma_A - \Gamma_\infty^0}} \equiv \hat{H}_0 \quad , \tag{13.121}$$

and hence:

$$K_0 = \left(\frac{H_0}{\hat{H}_0}\right)^2 \quad .$$

The scale $\hat{H}_0$ is related to the steady waves which appear downstream of the relief.

The condition:

$$\nu_0 \gg 1 \Rightarrow h_0 \ll H_0 \quad , \tag{13.122}$$

is what makes it possible to linearize the first boundary condition in (13.117c). In this case:

[7] The case of $\mathrm{M}_\infty \to 0$ *alone* with all the other parameters remaining of the order unity is considered in Sect. 13.4. It is the quasi-nondivergent model for three-dimensional steady lee waves in an adiabatic atmosphere. In this case, $K_0 \to \infty$ with $\mathrm{M}_\infty \to 0$ and problem (13.117) becomes strongly degenerate. The asymptotic analysis of (13.117) remains to be carried out when *only* $\mathrm{M}_\infty \to 0$.

$$\delta(\xi,0) = h(\xi) \quad , \quad \xi \in [-\tfrac{1}{2}, +\tfrac{1}{2}] \quad . \tag{13.123}$$

In fact, since:

$$K_0 = \frac{\text{Bo}}{\gamma \text{M}_\infty^2} \Lambda_0 \equiv \frac{\text{Bo}^2}{\gamma \text{M}_\infty^2} \alpha_0 \quad ,$$

three cases can be considered:

a. $\text{Bo} = \text{O}(1)$ and $\Lambda_0 = \hat{\hat{\Lambda}}\text{M}_\infty^2$, $\hat{\hat{\Lambda}} = \text{O}(1)$ when $\text{M}_\infty \to 0$. This implies that:

$$\mu_0 \equiv \text{Bo}\frac{\gamma - 1}{\gamma} - \hat{\hat{\Lambda}}\text{M}_\infty^2 \quad ;$$

b. $\text{Bo} = \hat{B}\text{M}_\infty$ and $\Lambda_0 = \hat{\Lambda}\text{M}_\infty$ which implies that:

$$\mu_0 = \hat{\mu}\text{M}_\infty \quad ;$$

c. $\text{Bo} = \hat{\hat{B}}\text{M}_\infty^2$ and $\Lambda_0 = \text{O}(1)$ which implies that:

$$\mu_0 = \hat{\hat{B}}\frac{\gamma - 1}{\gamma}\text{M}_\infty^2 - \Lambda_0 \quad ;$$

$\hat{\Lambda}$, $\hat{\hat{\Lambda}}$, $\hat{B}$, $\hat{\hat{B}}$ and $\hat{\mu}$ are similarity parameters of order unity for the various limiting flows obtained when $\text{M}_\infty \to 0$.

Generally speaking, it will be supposed that $\varepsilon_0 = \text{O}(1)$ since it is only in this case that waves appear downstream of the relief. The case $\varepsilon_0 \ll 1$ which leads to the considering of the limiting case $\varepsilon_0 \to 0$ requires a special analysis which will not be taken up here (for the isochoric case, such an analysis was outlined in Sect. 9.4).

Since we are essentially interested in the lee waves generated in the troposphere, the following hypothesis can be made:

$$H_\infty \equiv H_p = RT_\infty(0)/g \Rightarrow \kappa_0 \equiv \frac{1}{\text{Bo}} \quad .$$

Finally, it is pointed out that when $\text{Bo} = \text{O}(1)$ and $\nu_0 = \text{O}(1)$ [which is the nonlinearized case (a)], the density perturbation ω remains finite when $\text{M}_\infty \to 0$, given the fact that, according to the first boundary condition in (13.117c), we still have $\delta = \text{O}(1)$ even when $\text{M}_\infty \to 0$. On the other hand, in cases (b) and (c), we necessarily have $\omega \ll 1$ when $\text{Bo} \to 0$ with $\text{M}_\infty \to 0$.

When $\omega \ll 1$, relation (13.117b) can be replaced by the expression:

$$\begin{aligned}
&\left(1 - \mu_0\zeta + \frac{\mu_0}{\nu_0}\delta\right)\left[\omega - \left(1 - \frac{\gamma}{2}\right)\omega^2 + \dots\right] \\
&= -\left\{\frac{1}{2}\left(\frac{\text{M}_\infty}{\nu_0}\right)^2\left[\varepsilon_0^2\left(\frac{\partial\delta}{\partial\xi}\right)^2 + \left(\frac{\partial\delta}{\partial\zeta}\right)^2 - 2\nu_0\frac{\partial\delta}{\partial\zeta}\right]\right. \\
&\quad + \frac{1}{\gamma}\frac{\text{Bo}}{\nu_0}\delta - \left(\frac{\text{M}_\infty}{\nu_0}\right)^2\left[\varepsilon_0^2\left(\frac{\partial\delta}{\partial\xi}\right)^2 + \left(\frac{\partial\delta}{\partial\zeta}\right)^2 - 2\nu_0\frac{\partial\delta}{\partial\zeta} + \nu_0^2\right]\omega \\
&\quad \left. + \frac{3}{2}\left(\frac{\text{M}_\infty}{\nu_0}\right)^2\left[\varepsilon_0^2\left(\frac{\partial\delta}{\partial\xi}\right)^2 + \left(\frac{\partial\delta}{\partial\zeta}\right)^2 - 2\nu_0\frac{\partial\delta}{\partial\zeta} + \nu_0^2\right]\omega^2 + \dots\right\} .
\end{aligned} \tag{13.124}$$

4. We will now examine case (a):

$$\mathrm{Bo} = \mathrm{O}(1) \quad , \quad \Lambda_0 = \hat{\Lambda}\mathrm{M}_\infty^2 \quad , \quad \mu_0 = \mathrm{Bo}\frac{\gamma - 1}{\gamma} - \hat{\Lambda}\mathrm{M}_\infty^2 \quad .$$

In addition $\kappa_0 \equiv 1/\mathrm{Bo}$.

All the other parameters (including ν_0 !) and the variables ξ, ζ remaining fixed, the following asymptotic expansions can be postulated for δ and ω:

$$\delta = \delta_0 + \mathrm{O}(\mathrm{M}_\infty) \quad , \quad \omega = \omega_0 + \mathrm{O}(\mathrm{M}_\infty) \quad , \tag{13.125}$$

when $\mathrm{M}_\infty \to 0$ with ξ and ζ fixed.

To order zero, the following explicit relation (function of δ_0) results from (13.124):

$$\omega_0 = \left\{ \frac{1 - \frac{\gamma-1}{\gamma}\mathrm{Bo}\zeta}{1 - \frac{\gamma-1}{\gamma}\mathrm{Bo}\left(\zeta - \frac{1}{\nu_0}\delta_0\right)} \right\}^{1/(\gamma-1)} - 1 \quad . \tag{13.126}$$

As for δ_0, it must satisfy the following limiting equation:

$$\begin{aligned} &\left\{1 - \frac{\gamma - 1}{\gamma}\mathrm{Bo}\left(\zeta - \frac{1}{\nu_0}\delta_0\right)\right\}\left[\varepsilon_0^2\frac{\partial^2\delta_0}{\partial\xi^2} + \frac{\partial^2\delta_0}{\partial\zeta^2}\right. \\ &\left. - \frac{1}{1+\omega_0}\left(\varepsilon_0^2\frac{\partial\omega_0}{\partial\xi}\frac{\partial\delta_0}{\partial\xi} + \frac{\partial\omega_0}{\partial\zeta}\frac{\partial\delta_0}{\partial\zeta} - \nu_0\frac{\partial\omega_0}{\partial\zeta}\right)\right] \\ &+(1+\omega_0)^2\frac{\mathrm{Bo}}{\gamma}\hat{\Lambda}\delta_0 = 0. \end{aligned} \tag{13.127}$$

Therefore, in this limiting case, a single, *non-linear* equation results for δ_0 $(\xi, \zeta, 1/\nu_0)$. We write this equation in the form:

$$\begin{aligned} &\varepsilon_0^2\frac{\partial^2\delta_0}{\partial\xi^2} + \frac{\partial^2\delta_0}{\partial\zeta^2} + \mathcal{A}_0\left(\delta_0 ; \frac{1}{\nu_0}, \hat{\Lambda}, \mathrm{Bo}\right) \\ &= \mathcal{B}_0\left(\frac{\partial\delta_0}{\partial\xi}, \frac{\partial\delta_0}{\partial\zeta}, \delta_0 ; \frac{1}{\nu_0}, \mathrm{Bo}\right) \quad , \end{aligned} \tag{13.128}$$

where the functions $\mathcal{A}_0$ and $\mathcal{B}_0$ can be readily obtained after introducing (13.126) into (13.127).

The following boundary conditions must be imposed on (13.128):

$$\begin{aligned} &\delta_0\left(\xi, \frac{1}{\nu_0}h(\xi)\right) = h(\xi) \quad , \quad \xi \in \left[-\frac{1}{2}, +\frac{1}{2}\right] \quad ; \\ &\delta_0(-\infty, \zeta_\infty) = \delta_0\left(\xi, \frac{1}{\mathrm{Bo}}\right) = 0 \quad ; \\ &\left|\frac{\partial\delta_0}{\partial\xi}\right| + \left|\frac{\partial\delta_0}{\partial\zeta}\right| < \infty \quad , \quad \text{for} \quad \xi \to +\infty \quad . \end{aligned} \tag{13.129}$$

Problem (13.128, 129) must be resolved numerically. The reader is oriented towards the work by Pekelis (1976) for information on this subject.

If we wish to linearize (13.128, 129), then we must take advantage of the fact that $\nu_0 \gg 1$. In particular, it will be observed that the smallness of the density perturbations ω is related to the order of magnitude of the parameter ν_0. It is therefore only in the case of:

$$h_0 \ll \frac{RT_\infty(0)}{g} \approx H_0 \quad ,$$

i.e.,

$$h_0 \approx U_\infty^0 \sqrt{\frac{T_\infty(0)/g}{\Gamma_A - \Gamma_\infty^0}} \quad , \tag{13.130}$$

that we have: $\omega \ll 1$ and that (13.128, 129) can be linearized for δ_0.

Let us now define:

$$\frac{1}{\nu_0} = \hat{\nu} \mathrm{M}_\infty^a \quad , \quad a > 0 \quad , \quad \hat{\nu} = \mathrm{O}(1) \quad . \tag{13.131}$$

When $\mathrm{M}_\infty \to 0$, the following results for ω:

$$\omega = -\frac{\mathrm{Bo}}{\gamma} \hat{\nu} \frac{\delta_0^*}{1 - \mathrm{Bo}(\gamma - 1)\zeta/\gamma} \mathrm{M}_\infty^a + \ldots \quad , \tag{13.132}$$

where $\delta_0^* = \lim_{\mathrm{M}_\infty \to 0} \delta$ satisfies the following linear problem:

$$\mathcal{H}\delta_0^* \equiv \varepsilon_0^2 \frac{\partial^2 \delta_0^*}{\partial \xi^2} + \frac{\partial^2 \delta_0^*}{\partial \zeta^2} - \frac{\mathrm{Bo}}{\gamma} \frac{1}{1 - \mathrm{Bo}(\gamma - 1)\zeta/\gamma} \frac{\partial \delta_0^*}{\partial \zeta}$$
$$\left[\hat{\Lambda} - \frac{(\gamma - 1)\mathrm{Bo}/\gamma}{1 - \mathrm{Bo}(\gamma - 1)\zeta/\gamma}\right] \frac{\mathrm{Bo}/\gamma}{1 - \mathrm{Bo}(\gamma - 1)\zeta/\gamma} \delta_0^* = 0 \quad ; \tag{13.133a}$$

$$\delta_0^*(\xi, 0) = h(\xi) \quad , \quad \xi \in [-\tfrac{1}{2}, +\tfrac{1}{2}] \quad ;$$
$$\delta_0^*\left(\xi, \frac{1}{\mathrm{Bo}}\right) = \delta_0^*(-\infty, \zeta_\infty) = 0; \tag{13.133b}$$
$$\left|\frac{\partial \delta_0^*}{\partial \xi}\right| + \left|\frac{\partial \delta_0^*}{\partial \zeta}\right| < \infty \quad , \quad \text{for} \quad \xi \to +\infty \quad .$$

Hence:

$$\lim_{1/\nu_0 \to 0} \mathcal{A}_0 \equiv \left[\hat{\Lambda} - \frac{(\gamma - 1)\mathrm{Bo}/\gamma}{1 - \mathrm{Bo}(\gamma - 1)\zeta/\gamma}\right] \frac{\mathrm{Bo}/\gamma}{1 - \mathrm{Bo}(\gamma - 1)\zeta/\gamma} \delta_0^* \quad ;$$

$$\lim_{1/\nu_0 \to 0} \mathcal{B}_0 \equiv \frac{\mathrm{Bo}}{\gamma} \frac{1}{1 - \mathrm{Bo}(\gamma - 1)\zeta/\gamma} \frac{\partial \delta_0^*}{\partial \zeta} \quad .$$

In order to find the value of a in (13.132), we have to go as far as the second approximation in (13.124). In this case, the calculation leads us to choose:

$$2a = 2 + a \Rightarrow a = 2 \quad ,$$

and it turns out that:

$$\omega = \omega_2^* \mathrm{M}_\infty^2 + \omega_4^* \mathrm{M}_\infty^4 + \dots \quad ,$$

with

$$\omega_2^* = -\frac{\mathrm{Bo}}{\gamma}\hat{\nu}\frac{\delta_0^*}{1-\mathrm{Bo}(\gamma-1)\zeta/\gamma} \quad ;$$

$$\omega_4^* = \frac{1}{1-\mathrm{Bo}(\gamma-1)\zeta/\gamma}\left\{\left[1-\hat{\Lambda}\zeta-\mathrm{Bo}\frac{\gamma-1}{\gamma}\hat{\nu}\delta_0^*\right]\omega_2^* + \hat{\nu}\frac{\partial\delta_0^*}{\partial\zeta}-\frac{\mathrm{Bo}}{\gamma}\hat{\nu}\delta_2^*\right\}+\left(1-\frac{\gamma}{2}\right)\omega_2^{*2} \quad ,$$

once we have taken into account the fact that the function δ can be asymptotically expanded as follows:

$$\delta = \delta_0^* + \mathrm{M}_\infty^2 \delta_2^* + \dots \quad .$$

To be sure, the corresponding linear problem for δ_2^* can be written if it is so desired.

The linear problem for $\delta_0^*(\xi,\zeta)$ is close to the one analyzed by Dorodnitsyn (1950) in a study which although not recent, is little known in the West.

In order to analyze the linear problem (13.133) for δ_0^*, a good approach is to introduce the new function:

$$\chi_0(\xi,\zeta) = \left(1-\frac{\gamma-1}{\gamma}\mathrm{Bo}\zeta\right)^{\beta}\delta_0^*(\xi,\zeta) \quad , \quad \beta \equiv \frac{1}{2(\gamma-1)} \quad .$$

The following problem then results for χ_0:

$$\begin{aligned}
&\varepsilon_0^2\frac{\partial^2\chi_0}{\partial\xi^2}+\frac{\partial^2\chi_0}{\partial\zeta^2}+\mathcal{D}_0(\mathrm{Bo}\zeta)\chi_0 = 0 \quad ;\\
&\chi_0(\xi,0) = h(\xi) \quad , \quad \xi\in[-\tfrac{1}{2},+\tfrac{1}{2}] \quad ,\\
&\chi_0\left(\xi,\frac{1}{\mathrm{Bo}}\right) = \chi_0(-\infty,\zeta_\infty) = 0 \quad ,\\
&\left|\frac{\partial\chi_0}{\partial\xi}\right|+\left|\frac{\partial\chi_0}{\partial\zeta}\right|<\infty \quad , \quad \xi\to+\infty \quad ,
\end{aligned} \tag{13.134}$$

with

$$\mathcal{D}_0(\mathrm{Bo}\zeta) = \frac{\hat{\Lambda}\mathrm{Bo}/\gamma}{1-\mathrm{Bo}(\gamma-1)\zeta/\gamma}-\frac{B_0^2(2\gamma-1)/4\gamma^2}{(1-\mathrm{Bo}(\gamma-1)\zeta/\gamma)^2} \quad , \tag{13.135}$$

and under normal mesometeorological conditions, $\mathcal{D}_0 > 0$ holds.

The solution to the linear canonical problem (13.134) can be put into the following form:

$$\chi_0(\xi,\zeta) = \sum_{n=1}^{\infty}\left(\frac{dL_n}{d\zeta}\right)_{\zeta=0} L_n(\zeta)\psi_n(\xi) \quad , \tag{13.136}$$

where $L_n(\zeta)$ and μ_n^2 designate respectively the eigenfunctions and eigenvalues of the associated Sturm-Liouville problem:

$$\frac{d^2 L_n}{d\zeta^2} + \left[\mu_n^2 + \mathcal{D}_0(\text{Bo}\lambda)\right] L_n = 0 \quad , \tag{13.137a}$$

$$L_n(0) = L_n\left(\frac{1}{\text{Bo}}\right) = 0 \quad . \tag{13.137b}$$

It is then found that:

$$\psi_n(\xi) = -\frac{1}{2\mu_n\varepsilon_0} \int_{-\infty}^{+\infty} \exp\left[-\frac{\mu_n}{\varepsilon_0}\,|\xi - \xi'|\right] h(\xi')d\xi' \quad ,$$

$$\text{when} \quad \mu_n^2 > 0 \quad ; \tag{13.138a}$$

$$\psi_n(\xi) = \frac{1}{2\gamma_n\varepsilon_0} \int_{-\infty}^{\xi} \sin\left[\frac{\gamma_n}{\varepsilon_0}\,(\xi - \xi')\right] h(\xi')d\xi' \quad ,$$

$$\text{when} \quad \mu_n^2 \equiv -\gamma_n^2 < 0 \quad . \tag{13.138b}$$

The set of $L_n(\zeta)$ constitutes an orthonormal base on the interval [0, 1/Bo]. While writing the solutions (13.138), we took into account the conditions in ξ for $\xi \to \pm\infty$. Nonetheless, we still must be able to determine the sign of μ_n^2. To this end, it can be demonstrated that the number of negative eigenvalues $\mu_n^2 = -\gamma_n^2 < 0$ of the eigenvalue problem (13.137) is equal to the number of zeros of the function:

$$Y(Z) = Z^{1/2}\left\{c_0 J_\beta\left(2\varphi_0 Z^{1/2}\right) + c_1 J_{-\beta}\left(2\varphi_0 Z^{1/2}\right)\right\} \quad ,$$

in the interval $]Z_0, 1[$. J_β and $J_{-\beta}$ are the first kind Bessel functions of order $+\beta$ and $-\beta$ with:

$$\beta \equiv \sqrt{1 + 4\sigma_0^2} \quad , \quad \sigma_0^2 = \frac{2\gamma - 1}{4(\gamma - 1)^2} \quad ; \quad \varphi_0^2 = \frac{\gamma}{(\gamma - 1)^2}\,\frac{\hat{\Lambda}}{\text{Bo}} \quad ,$$

we thus have:

$$Z \equiv 1 - \frac{\gamma - 1}{\gamma}\text{Bo}\zeta \quad , \quad Z_0 = 1 - \frac{\gamma - 1}{\gamma} \equiv \frac{1}{\gamma} \quad .$$

The integration constants c_0 and c_1 are readily determined from the conditions $Y(1/\gamma) = 0$ and $Y(1) = 1$. Under normal meteorological conditions, the following holds:

$$\varphi_0^2 \gg \sigma_0^2 \Rightarrow \frac{\hat{\Lambda}}{\text{Bo}} \gg \frac{2\gamma - 1}{4\gamma} \approx 0.32 \quad .$$

In this case, we can deduce that the number of zeros of the function $Y(Z)$ is equal to the whole part of the number:

$$\mathcal{N} = \frac{2\varphi_0}{\pi}\left(1 - Z_0^{1/2}\right) - \frac{1}{2} \quad ,$$

if $\mathcal{N}$, itself, is not "too close" to a whole number. The case $\varphi_0^2 \gg \sigma_0^2$ was analyzed in detail by Dorodnitsyn (1950).

5. Let us now consider the case (b):

$$\mathrm{Bo} = \hat{B}\mathrm{M}_\infty \quad , \quad \Lambda_0 = \hat{\Lambda}\mathrm{M}_\infty \quad \text{and} \quad \mu_0 = \hat{\mu}\mathrm{M}_\infty \quad .$$

This is the case corresponding to the Boussinesq approximation which was treated in detail in Chap. 8. In this case, we examine what happens at vertical distances from the relief which are of the order of:

$$H_0 \approx \frac{U_\infty^0}{g}\sqrt{\frac{RT_\infty(0)}{\gamma}} \quad .$$

It is no longer necessary to impose the constraint $\nu_0 \gg 1$ and it will be assumed that $\nu_0 = \mathrm{O}(1)$. Thererfore, when $\mathrm{M}_\infty \to 0$, the following results:

$$\omega = -\frac{\hat{B}}{\gamma\nu_0}\mathrm{M}_\infty\delta_\mathrm{B} \quad , \quad \delta_\mathrm{B} = \lim_{\mathrm{M}_\infty \to 0} \delta$$

and $\delta_\mathrm{B}(\xi, \zeta)$ satisfies the so-called Long equation (1953):

$$\varepsilon_0^2\frac{\partial^2\delta_\mathrm{B}}{\partial\xi^2} + \frac{\partial^2\delta_\mathrm{B}}{\partial\zeta^2} + \frac{\hat{B}\hat{\Lambda}}{\gamma}\delta_\mathrm{B} = 0 \quad , \tag{13.139}$$

to which must be assigned the boundary conditions:

$$\begin{aligned} &\delta_\mathrm{B}\left(\xi, \frac{1}{\nu_0}h(\xi)\right) = h(\xi) \quad , \quad \xi \in [-\tfrac{1}{2}, +\tfrac{1}{2}] \quad ; \\ &\delta_\mathrm{B}(-\infty, \zeta_\infty) = 0 \quad , \\ &\left|\frac{\partial\delta_\mathrm{B}}{\partial\xi}\right| + \left|\frac{\partial\delta_\mathrm{B}}{\partial\zeta}\right| < \infty \quad , \quad \xi \to +\infty \quad . \end{aligned} \tag{13.140}$$

For $\zeta \to +\infty$, we must join to problem (13.139, 140) (which is then called *inner*), an *outer* problem and then match the two! We will return to this situation in Subsect. 13.2.4.

6. Finally, let us consider the case (c):

$$\mathrm{Bo} \to 0 \quad , \quad \text{with} \quad \mathrm{M}_\infty \to 0 \quad , \quad \text{such that :}$$

$$\mathrm{Bo} = \hat{\hat{B}}\mathrm{M}_\infty^2 \quad , \quad \Lambda_0 = \mathrm{O}(1) \quad \text{and} \quad \mu_0 = \hat{\hat{B}}\frac{\gamma - 1}{\gamma}\mathrm{M}_\infty^2 - \Lambda_0 \quad .$$

We will assume that either $\nu_0 = \mathrm{O}(1)$ (non-linear model), or $\nu_0 \gg 1$ (linear model). It will be observed that:

$$\frac{\mathrm{Bo}}{\mathrm{M}_\infty^2} = \hat{\hat{B}} = \mathrm{O}(1) \Rightarrow \frac{\gamma}{\mathrm{Fr}_{H_0}^2} \approx 1 \Rightarrow H_0 \approx \gamma\frac{U_\infty^{02}}{g} \quad .$$

We will thus examine what happens in a *thin* layer near the relief (according to the vertical direction).

When $\nu_0 = O(1)$, the following representation results for ω when $M_\infty \to 0$:

$$\omega = -\frac{1}{2\nu_0^2}\frac{1}{1+\Lambda_0\zeta-(\Lambda_0/\nu_0)\delta_0}\left[\varepsilon_0^2\left(\frac{\partial\delta_0}{\partial\xi}\right)^2+\left(\frac{\partial\delta_0}{\partial\zeta}\right)^2-2\nu_0\frac{\partial\delta_0}{\partial\zeta}\right.$$
$$\left.+2\frac{\hat{\hat{B}}\nu_0}{\gamma}\delta_0\right]M_\infty^2+\dots\quad, \tag{13.141}$$

and $\lim_{M_\infty\to 0}\delta = \delta_0$ satisfies the equation:

$$\left(1+\Lambda_0\zeta-\frac{\Lambda_0}{\nu_0}\delta_0\right)\left[\varepsilon_0^2\frac{\partial^2\delta_0}{\partial\xi^2}+\frac{\partial^2\delta_0}{\partial\zeta^2}\right]$$
$$+\frac{\Lambda_0}{2\nu_0}\left[\varepsilon_0^2\left(\frac{\partial\delta_0}{\partial\xi}\right)^2+\left(\frac{\partial\delta_0}{\partial\zeta}\right)^2\right]-\Lambda_0\frac{\partial\delta_0}{\partial\zeta}+2\frac{\hat{\hat{B}}\Lambda_0}{\gamma}\delta_0=0\quad. \tag{13.142}$$

When $\hat{\hat{B}}\to\infty$ in such a way that $\Lambda_0\equiv\hat{\Lambda}\hat{B}/\hat{\hat{B}}\to 0$ but $\hat{\hat{B}}\Lambda_0\equiv\hat{B}\hat{\Lambda}=O(1)$, (13.142) brings us back to the Long equation (13.139).

The following boundary conditions must be added to (13.142):

$$\begin{aligned}&\delta_0\left(\xi,\frac{1}{\nu_0}h(\xi)\right)=h(\xi)\quad,\quad \xi\in[-\tfrac{1}{2},+\tfrac{1}{2}]\quad;\\&\delta_0(-\infty,\zeta_\infty)=0\quad,\\&\left|\frac{\partial\delta_0}{\partial\xi}\right|+\left|\frac{\partial\delta_0}{\partial\zeta}\right|<\infty\quad,\quad\xi\to+\infty\quad.\end{aligned} \tag{13.143}$$

Once again, a singular phenomenon is encountered when $\zeta\to+\infty$!

The non-linearity of (13.142) is related to the fact that $\Lambda_0 = O(1)$, i.e., that the static stability parameter is not assumed small in the thin layer considered above the relief (which is not small). The model (13.142, 143) does not appear to have as yet been analyzed.

This model (13.142, 143) can be linearized if the following complementary constraint is imposed:

$$\nu_0\gg 1\Rightarrow h_0\ll\gamma\frac{U_\infty^{02}}{g}$$

on the thickness h_0 of the relief.

In this case, the function $\delta_0^\infty = \lim_{\nu_0\to\infty}\delta_0$ satisfies the following linear problem:

$$\begin{aligned}&\varepsilon_0^2\frac{\partial^2\delta_0^\infty}{\partial\xi^2}+\frac{\partial^2\delta_0^\infty}{\partial\zeta^2}-\frac{\Lambda_0}{1+\Lambda_0\zeta}\left(\frac{\partial\delta_0^\infty}{\partial\zeta}-\frac{\hat{\hat{B}}}{\gamma}\delta_0^\infty\right)=0\quad,\\&\delta_0^\infty(\xi,0)=h(\xi)\quad,\quad\xi\in[-\tfrac{1}{2},+\tfrac{1}{2}]\quad;\\&\delta_0^\infty(-\infty,\zeta_\infty)=0\quad;\\&\left|\frac{\partial\delta_0^\infty}{\partial\xi}\right|+\left|\frac{\partial\delta_0^\infty}{\partial\zeta}\right|<\infty\quad,\quad\xi\to+\infty\quad.\end{aligned} \tag{13.144}$$

If instead of ζ, we now introduce the new variable $\eta = 1 + \Lambda_0\zeta$, and instead of the function δ_0^∞, we introduce the new function $\phi = \eta^{-1/2}\delta_0^\infty$, the following equation results for $\phi(\xi, \eta)$

$$\left(\frac{\varepsilon_0}{\Lambda_0}\right)^2 \frac{\partial^2 \phi}{\partial \xi^2} + \frac{\partial^2 \phi}{\partial \eta^2} + \left[\frac{\hat{B}/\gamma\Lambda_0}{\eta} - \frac{3/4}{\eta^2}\right]\phi = 0 \quad , \qquad (13.145)$$

which is of the same type as the equation for the linear problem (13.134).

We will end here this asymptotic analysis[8] and turn our attention in the forthcoming to the asymptotic interpretation of the Long model, i.e., the model (13.139, 140), with a behavior condition for $\zeta \to +\infty$.

13.2.4 Asymptotic Interpretation of the Long Model in the Troposphere[9]

1. Let us return to the problem (13.139, 140) where we set:

$$\frac{\hat{B}\hat{\Lambda}}{\gamma} \equiv \frac{\hat{B}^2}{\gamma}\alpha_0 = k_0^2 \quad .$$

We can no longer apply the condition $\delta(\xi, 1/\mathrm{Bo}) = 0$ to (13.139, 140): it must be replaced by a matching condition. More precisely, we know from Sommerfeld (see Wilcox, 1959) that for $r \to +\infty$, a so-called radiation condition must be imposed:

$$\delta_0 \sim \sqrt{\frac{2k_0}{\pi r}} \sin\theta\, \Re\left\{G(\cos\theta)\, e^{i(k_0 r - \pi/4)}\right\} \quad , \qquad (13.146)$$

where the function $G(\cos\theta)$ is arbitrary and depends on the form of the relief via the function $h(\xi)$. So as to satisfy the behavior at upstream infinity, the following condition must also be imposed:

$$G(\cos\theta) = 0 \quad , \quad \text{for} \quad \cos\theta < 0 \quad . \qquad (13.147)$$

It is pointed out that the polar coordinates r, θ in the upper half-plane $\zeta \geq 0$ are defined such that:

$$\varepsilon_0^{-1}\xi = r\cos\theta \quad \text{and} \quad \zeta = r\sin\theta \quad .$$

Problem (13.139, 140), while taking into account (13.146, 147), was first resolved by Kozhevnikov (1963) and then by Miles (1968) for a semi-circular obstacle. Huppert and Miles (1969) worked out the same problem for a semi-elliptical obstacle. On this subject, the book by Gutman (1972), as well as the review by Miles (1969) are recommended.

Thanks to (13.146), we will see that the classical Long model can, in fact, be considered as a zero order inner representation – in an asymptotic scheme

---

[8] The results of the present section were published in an article by Zeytounian (1979).

[9] According to the article by Guiraud and Zeytounian (1979).

of matched inner and outer representations (MMAE) – of the initial non-linear problem (13.117) when:

$$\mathrm{Bo} = \hat{B}\mathrm{M}_\infty \quad , \quad \Lambda_0 = \hat{\Lambda}\mathrm{M}_\infty \quad \text{and} \quad \mu_0 = \hat{\mu}\mathrm{M}_\infty \quad ; \quad \mathrm{M}_\infty \to 0 \quad .$$

2. The condition $\delta(\xi, 1/\mathrm{Bo}) = 0$ is beyond the domain of validity of the inner representation $\delta_0(\xi, \zeta)$ which leads to the Long-Miles problem discussed above. The following outer variables must first of all be introduced in the outer representation:

$$y = \mathrm{M}_\infty \zeta \quad \text{and} \quad x = \mathrm{M}_\infty \frac{\xi}{\varepsilon_0} \quad . \tag{13.148}$$

The radiation condition (13.146) then indicates the following as the correct choice:

$$\tilde{\delta} = \mathrm{M}_\infty^{-1/2} \delta(x, y; \mathrm{M}_\infty) \quad , \tag{13.149}$$

for carrying out the matching with the inner representation. In this case, (13.124) implies the introduction of:

$$\tilde{\omega} = \mathrm{M}_\infty^{-3/2} \omega(x, y; \mathrm{M}_\infty) \quad , \tag{13.150}$$

given the fact that $(1/\gamma)(\mathrm{Bo}/\nu_0)\delta_0 \equiv (\hat{B}/\gamma\nu_0)\mathrm{M}_\infty^{3/2}\tilde{\delta}$ is the dominant term in (13.124) when $\mathrm{M}_\infty \to 0$, with $\hat{B}$, $\hat{\Lambda}$ and $\hat{\mu}$ remaining fixed of the order unity.

From the inner approximation, the perturbations in the outer region can be expected to consist mainly of lee waves with a wavelength of the order of M_∞ [see (13.146)] to the scale of the outer region. Consequently, there is a double scale built into the solution for the outer region which must be dealt with.

According to (13.107), with dimensionless values, we have:

$$\zeta_\infty = \mathrm{Bo}\left(\zeta - \frac{1}{\nu_0}\delta\right) = \hat{B}\left(y - \frac{\mathrm{M}_\infty^{3/2}}{\nu_0}\tilde{\delta}\right) \quad .$$

Furthermore:

$$\begin{aligned} \tau_\infty &= 1 - \mu_0\left(\zeta - \frac{1}{\nu_0}\delta\right) = 1 - \frac{R}{g}\Gamma_\infty^0 \zeta_\infty \\ &= 1 - \frac{R}{g}\Gamma_\infty^0 \hat{B} y + \mathrm{M}_\infty^{3/2}\frac{R}{g}\Gamma_\infty^0 \frac{\hat{B}}{\nu_0}\tilde{\delta} \quad , \end{aligned}$$

since

$$\mu_0 = \frac{\gamma - 1}{\gamma}\mathrm{Bo} - \Lambda_0 = \mathrm{Bo}\left(\frac{\gamma - 1}{\gamma} - \alpha_0\right) \equiv \mathrm{Bo}\frac{R}{g}\Gamma_\infty^0 \quad .$$

If the expressions of ξ, ζ, δ, ω, τ_∞ and ζ_∞ are now substituted into the exact equation (13.117a), then the following outer equation is obtained for $\tilde{\delta}(x, y;\ \mathrm{M}_\infty)$:

$$\begin{aligned} &\frac{\partial^2 \tilde{\delta}}{\partial x^2} + \frac{\partial^2 \tilde{\delta}}{\partial y^2} + \nu_0 \frac{\partial \tilde{\omega}}{\partial y} - \frac{\gamma}{\hat{B}}\phi_\infty(y)\frac{\partial \tilde{\delta}}{\partial y} + \phi_\infty(y)\frac{1}{\mathrm{M}_\infty^2}\tilde{\delta} \\ &- \frac{\gamma\nu_0}{\hat{B}}\phi_\infty(y)\tilde{\omega} = \mathrm{O}\left(\frac{1}{\mathrm{M}_\infty^{1/2}}\right), \end{aligned} \tag{13.151}$$

where:

$$\phi_\infty(y) \equiv \frac{k_0^2}{1-(R/g)\Gamma_\infty^0 \hat{B} y} \quad , \quad \text{and thus} \quad \phi_\infty(0) \equiv k_0^2 \quad .$$

In (13.151), only the dominant terms in M_∞ have been written. The following must also be taken into consideration:

$$\tilde{\omega} = -\frac{\hat{B}}{\gamma\nu_0}\frac{\phi_\infty(y)}{k_0^2}\tilde{\delta} + \ldots \quad . \tag{13.152}$$

As a matter of fact, if we take into account (13.152), it is remarked that the last term of (13.151) is not a dominant term and so, for the *outer representation*, the following *dominant equation* in $\tilde{\delta}(x, y; M_\infty)$ results:

$$\frac{\partial^2\tilde{\delta}}{\partial x^2} + \frac{\partial^2\tilde{\delta}}{\partial y^2} + \frac{\phi_\infty(y)}{M_\infty^2}\tilde{\delta} = \frac{\hat{B}}{\gamma}\frac{\phi_\infty(y)}{k_0^2}\left\{\frac{d\log\phi_\infty(y)}{dy}\tilde{\delta} + \left[1+\left(\frac{\gamma k_0}{\hat{B}}\right)^2\right]\frac{\partial\tilde{\delta}}{\partial y}\right\} \quad . \tag{13.153}$$

This last equation must be associated with on the Long equation (13.139) for $\delta_B(\xi,\zeta)$ when $M_\infty \to 0$ with x and y fixed.

When the variables (13.148) are used, the condition $\delta(\xi, 1/\mathrm{Bo}) = 0$ becomes: $\tilde{\delta}(x, 1/\hat{B}; M_\infty) = 0$ and, in addition:

$$h\left(\varepsilon_0\frac{x}{M_\infty}\right) \Rightarrow 0 \quad , \quad \text{with} \quad M_\infty \to 0$$

except at point $x = 0$.

Furthermore, $\phi_\infty(y) > 0$ when $0 \le y \le 1/\hat{B}$ (since $(R/g)\Gamma_\infty^0 < 1$). Equation (13.153) with the boundary conditions:

$$\tilde{\delta}(x,0) = 0 \quad , \quad \tilde{\delta}\left(x, \frac{1}{\hat{B}}\right) = 0 \quad , \tag{13.154}$$

as well as the matching condition, at the origin $x = 0$, $y = 0$, with (13.146) was solved asymptotically for $M_\infty \to 0$ bringing to light a *double scale structure* from the solution [on account of the term $\phi_\infty(y)\tilde{\delta}/M_\infty^2$ in (13.153)].

As a matter of fact, we set the following according to the multiple scale technique (see Sect. 6.2):

$$\tilde{\delta} = \tilde{\delta}_0(x, y, \chi) + M_\infty\tilde{\delta}_1(x, y, \chi) + \ldots, \quad \tilde{\omega} = \tilde{\omega}_0(x, y, \chi) + \ldots \quad , \tag{13.155}$$

where:

$$\chi = \frac{\Theta(x, y)}{M_\infty} \quad .$$

From (13.152), we find that:

$$\tilde{\omega}_0 = -\frac{\hat{B}}{\gamma\nu_0}\phi_\infty(y)\frac{1}{k_0^2}\tilde{\delta}_0 \quad ,$$

$\tilde{\delta}_0$ being a solution of the equation [from (13.153)]:

$$\left[\left(\frac{\partial\Theta}{\partial x}\right)^2+\left(\frac{\partial\Theta}{\partial y}\right)^2\right]\frac{\partial^2\tilde{\delta}_0}{\partial\chi^2}+\phi_\infty(y)\tilde{\delta}_0=0 \quad . \tag{13.156}$$

We choose $\Theta(x, y)$ to be a solution of:

$$\left(\frac{\partial\Theta}{\partial x}\right)^2+\left(\frac{\partial\Theta}{\partial y}\right)^2=\frac{\phi_\infty(y)}{k_0^2} \quad , \tag{13.157}$$

and an obvious solution to:

$$\frac{\partial^2\tilde{\delta}_0}{\partial\chi^2}+k_0^2\tilde{\delta}_0=0$$

is then:

$$\tilde{\delta}_0(x,y,\chi)=\Re\left\{\mathcal{A}(x,y)\,\mathrm{e}^{\mathrm{i}(k_0\chi-\pi/4)}\right\} \quad , \tag{13.158}$$

where the phase difference $\pi/4$ is used for the matching with (13.146).

In order to obtain an equation for the amplitude $\mathcal{A}(x, y)$, the next order has to be considered and from (13.153), it is found for $\tilde{\delta}_1$:

$$\begin{aligned}\frac{\partial^2\tilde{\delta}_1}{\partial\chi^2}+k_0^2\tilde{\delta}_1 = &-\frac{k_0^2}{\phi_\infty(y)}\left\{2\left[\frac{\partial\Theta}{\partial x}\frac{\partial^2\tilde{\delta}_0}{\partial x\partial\chi}+\frac{\partial\Theta}{\partial y}\frac{\partial^2\tilde{\delta}_0}{\partial y\partial\chi}\right]\right.\\ &\left.+\left(\frac{\partial^2\Theta}{\partial x^2}+\frac{\partial^2\Theta}{\partial y^2}\right)\frac{\partial\tilde{\delta}_0}{\partial\chi}+\nu_0\frac{\partial\Theta}{\partial y}\frac{\partial\tilde{\omega}_0}{\partial\chi}\right\}\\ &+\frac{\gamma}{\hat{B}}k_0^2\frac{\partial\Theta}{\partial y}\frac{\partial\tilde{\delta}_0}{\partial\chi} \quad .\end{aligned} \tag{13.159}$$

Secular terms appear in the solution of (13.159) for $\tilde{\delta}_1$ which must then be removed.

3. For the moment, however, let us focus on (13.157). In the vicinity of $x = y = 0$, the latter is written $(\partial\Theta/\partial x)^2 + (\partial\Theta/\partial y)^2 = 1$ and only the solution $\Theta \equiv \tilde{r}$ with $x = \tilde{r}\cos\theta$ and $y = \tilde{r}\sin\theta$ makes it possible to carry out the matching ($\tilde{r} \equiv \mathrm{M}_\infty r$). In fact, it is necessary that:

$$\begin{aligned}\left(\mathrm{M}_\infty^{1/2}\tilde{\delta}_0\right) &\equiv \Re\left\{\mathrm{M}_\infty^{1/2}\mathcal{A}(\tilde{r}\cos\theta,\,\tilde{r}\sin\theta)\,\mathrm{e}^{\mathrm{i}(k_0\tilde{r}/\mathrm{M}_\infty-\pi/4)}\right\}\\ &\sim\Re\left\{\left(\frac{2k_0}{\pi r}\right)^{1/2}\sin\theta\,G(\cos\theta)\,\mathrm{e}^{\mathrm{i}(k_0r-\pi/4)}\right\} \quad ,\end{aligned}$$

when $\tilde{r}\to 0$, or:

$$\lim_{\mathrm{M}_\infty\to 0}\left[\mathrm{M}_\infty^{1/2}\mathcal{A}(\tilde{r}\cos\theta,\,\tilde{r}\sin\theta)\right]=\sqrt{\frac{2k_0}{\pi r}}\sin\theta\,G(\cos\theta) \quad .$$

The following matching condition can then be derived:

$$\mathcal{A}(\tilde{r}\cos\theta,\,\tilde{r}\sin\theta)\sim\sqrt{\frac{2k_0}{\pi\tilde{r}}}\sin\theta\,G(\cos\theta) \quad , \quad \text{when} \quad \tilde{r}\to 0 \quad . \tag{13.160}$$

The solution (13.158) of the equation $\partial^2\tilde{\delta}_0/\partial\chi^2 + k_0^2\tilde{\delta}_0 = 0$ cannot simultaneously satisfy the boundary conditions:

$$\tilde{\delta}_0(x,0) = 0 \quad \text{and} \quad \tilde{\delta}_0\left(x, \frac{1}{\hat{B}}\right) = 0 \quad , \quad \text{for all } x \neq 0 \quad , \tag{13.161}$$

(see footnote[10]). In order for them to be verified, it is necessary to superpose solutions of this type in the form:

$$\begin{aligned}\tilde{\delta}_0 &= \Re \sum_{n\in\mathbf{N}} \mathcal{A}_n(x,y) \exp\left\{ \mathrm{i}\left[\frac{k_0}{\mathbf{M}_\infty}\Theta_n(x,y) - \frac{\pi}{4} + \varphi_n \right] \right\} \\ &\equiv \Re \sum_{n\in\mathbf{N}} \mathcal{A}_n E_n,\end{aligned} \tag{13.162}$$

with φ_n a phase displacement, the usefulness of which will become obvious further on. We want to be sure that any solution $\mathcal{A}_n E_n$ propagates along the characteristic rays of (13.157) for $\Theta(x,y)$. At a given point $M(x,y)$, inside the duct formed by the rectilinear walls $y = 0$ and $y = 1/\hat{B}$, let us consider all the rays emanating from the origin $x = y = 0$ which arrive at this point M after n reflection on the walls. A solution $\mathcal{A}_n E_n$ corresponds to each ray and all these solutions are superposed on the set $\mathbf{N}$ of all the rays emanating from the origin and which arrive at the given point M after as many possible reflections on the walls.

Let us then consider a point close to the upper wall $y = 1/\hat{B}$. When n is even, it corresponds to the rays which either arrive directly, or which arrive after one last (nth) reflection against the lower wall $y = 0$. If the ray had been reflected for the last time against the upper wall, then n would necessarily be uneven.

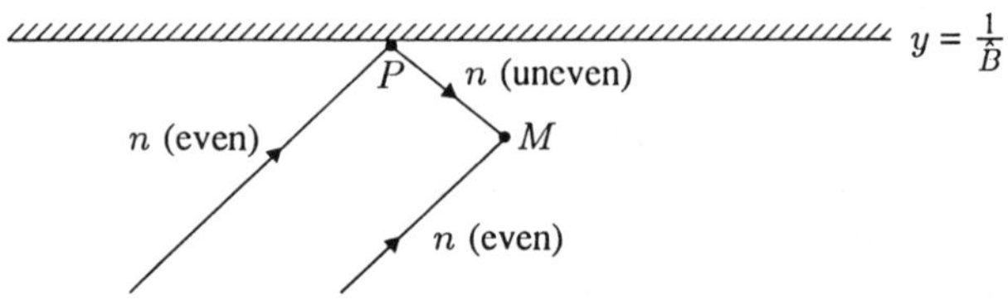

Fig. 6

Let us now consider the expression:

$$\psi_p = \mathcal{A}_{2p}E_{2p} + \mathcal{A}_{2p+1}E_{2p+1} \quad .$$

We will obtain $\lim_{y\to 1/\hat{B}} \psi_p = 0$ if the following is imposed:

$$\lim_{y\to 1/\hat{B}} \mathcal{A}_{2p} = \lim_{y\to 1/\hat{B}} \mathcal{A}_{2p+1} \quad \text{and} \quad \varphi_{2p+1} = \varphi_{2p} + \pi \quad , \quad \text{as well as}$$

$$\lim_{y\to 1/\hat{B}} \Theta_{2p} = \lim_{y\to 1/\hat{B}} \Theta_{2p+1} \quad .$$

[10] The matching condition (13.160) must be applied for $x \equiv 0$.

We can, of course, make an analogous observation: $\lim_{y \to 0} \psi_p$, for the lower wall (for all $x \neq 0$), by taking $n = 2p - 1$ and $n = 2p$.

Finally, we can satisfy the boundary conditions (13.161) on the walls by using the following solution:

$$\tilde{\delta}_0(x,y) = \Re \sum_{n \in \mathbf{N}} \mathcal{A}_n(x,y) \exp\left[\mathrm{i}\left(\frac{k_0}{\mathrm{M}_\infty} \Theta_n(x,y) - \frac{\pi}{4} + n\pi \right) \right] \quad , \qquad (13.163)$$

where:

$$\mathcal{A}_{2p}\left(x, \frac{1}{\hat{B}}\right) = \mathcal{A}_{2p+1}\left(x, \frac{1}{\hat{B}}\right) \quad ;$$
$$\Theta_{2p}\left(x, \frac{1}{\hat{B}}\right) = \Theta_{2p+1}\left(x, \frac{1}{\hat{B}}\right) \quad , \qquad (13.164a)$$

$$\mathcal{A}_{2p-1}(x,0) = \mathcal{A}_{2p}(x,0) \quad ; \quad \Theta_{2p-1}(x,0) = \Theta_{2p}(x,0) \quad . \qquad (13.164b)$$

It is thus remarked that by following the evolution of $\Theta_n(x,y)$ and $\mathcal{A}_n(x,y)$ on a given ray, these two quantities remain constant for each reflection.

Let us set in (13.157) [which is a first order, partial differential equation for $\Theta(x,y)$]:

$$p = \frac{\partial \Theta}{\partial x} \quad , \quad q = \frac{\partial \Theta}{\partial y} \Rightarrow p^2 + q^2 = \frac{\phi_\infty(y)}{k_0^2} \quad ,$$

or, along a ray (i.e., along a characteristic strip):

$$\frac{ds}{2} = \frac{dx}{2p} = \frac{dy}{2q} = \frac{dp}{0} = \frac{dq}{[d\phi_\infty(y)/dy]/k_0^2} = \frac{d\Theta}{2(p^2+q^2)} \quad , \qquad (13.165)$$

where s is the distance along an arbitrary ray from the origin ($x = 0$, $y = 0$) to the given point $M(x,y)$. If $x = s\sigma$ is the parametrical equation of a ray emanating from the origin:

$$p = \sigma = \text{const} \quad \text{on a ray} \quad \Rightarrow q = \pm \sqrt{\frac{\phi_\infty(y)}{k_0^2} - \sigma^2}$$

then:

$$ds = \frac{dy}{\pm \left[\phi_\infty(y)/k_0^2 - \sigma^2 \right]^{1/2}} \quad . \qquad (13.166)$$

Let us designate respectively by R_n, the portion of the ray between the reflections n and $n+1$, by R_{2p}, the ascending ray, and by R_{2p+1}, the descending ray. We then have:

$$\text{on } R_{2p}: \quad x = s\sigma \quad , \quad q = +\sqrt{\frac{\phi_\infty(y)}{k_0^2} - \sigma^2}, \quad ds = \frac{dy}{+\left(\frac{\phi_\infty(y)}{k_0^2} - \sigma^2\right)^{1/2}} \quad ;$$

$$\text{on } R_{2p+1}: \quad x = s\sigma \quad , \quad q = -\sqrt{\frac{\phi_\infty(y)}{k_0^2} - \sigma^2}, \quad ds = -\frac{dy}{+\left(\frac{\phi_\infty(y)}{k_0^2} - \sigma^2\right)^{1/2}} \cdot$$

Finally, we define $\alpha(n)$:

$$2\alpha(n) \equiv n \ , \text{ if } n \text{ is even, and } 2\alpha(n) - 1 \equiv n \ , \text{ if } n \text{ is uneven} \quad .$$

In this case, by a recurrent process for any n, it can be proven that on R_n, we have:

$$x = s\sigma \quad , \quad \text{and} \quad s = 2\alpha(n) \int_0^{1/\hat{B}} \frac{dt}{\left[\phi_\infty(t)/k_0^2 - \sigma^2\right]^{1/2}}$$

$$+(-1)^n \int_0^y \frac{dt}{\left[\phi_\infty(t)/k_0^2 - \sigma^2\right]^{1/2}} \quad , \tag{13.167}$$

and in the vicinity of the origin, we are readily brought back to:

$$s \equiv \tilde{r} \quad \text{and thus,} \quad \sigma \equiv \cos\theta \quad .$$

Nonetheless, on R_n, we also have:

$$ds = \frac{dy}{q} = \frac{d\Theta}{\phi_\infty(y)/k_0^2} = \frac{dy}{(-1)^n \left[\phi_\infty(y)/k_0^2 - \sigma^2\right]^{1/2}} \quad ,$$

and hence:

$$d\Theta = (-1)^n \frac{\phi_\infty(y)/k_0^2}{\left[\phi_\infty(y)/k_0^2 - \sigma^2\right]^{1/2}} dy$$

$$= (-1)^n \left\{ \frac{\sigma^2}{\left[\phi_\infty(y)/k_0^2 - \sigma^2\right]^{1/2}} dy + \sqrt{\frac{\phi_\infty(y)}{k_0^2} - \sigma^2} dy \right\} \quad .$$

But, we also have on R_n:

$$\frac{dx}{\sigma} = \frac{dy}{(-1)^n \left[\phi_\infty(y)/k_0^2 - \sigma^2\right]^{1/2}} \Rightarrow \sigma\, dx = \frac{(-1)^n \sigma^2}{\left[\phi_\infty(y)/k_0^2 - \sigma^2\right]^{1/2}} dy \quad .$$

Therefore, finally:

$$d\Theta = \sigma\, dx + (-1)^n \sqrt{\frac{\phi_\infty(y)}{k_0^2} - \sigma^2}\, dy \quad ,$$

and on R_n, it follows that:

$$\Theta = \Theta_n(x, y) = \sigma x + 2\alpha(n) \int_0^{1/\hat{B}} \sqrt{\frac{\phi_\infty(t)}{k_0^2} - \sigma^2}\, dt$$

$$+ (-1)^n \int_0^y \sqrt{\frac{\phi_\infty(t)}{k_0^2} - \sigma^2}\, dt \quad . \tag{13.168}$$

In particular, on R_0, we will have:

$$\Theta = \sigma x + \int_0^y \sqrt{\frac{\phi_\infty(t)}{k_0^2} - \sigma^2}\, dt \quad ,$$

and in the vicinity of the origin, it is again found that $\Theta \cong \tilde{r}$.

Formula (13.168) shows that:

$$\Theta_n(x,y) \equiv \vartheta_n\left[x, y;\ \sigma_n(x,y)\right]$$

and the relation:

$$\left.\frac{\partial \vartheta_n}{\partial \sigma}\right|_{\sigma=\sigma_n} = 0 \quad , \tag{13.169}$$

defines σ_n. If, moreover, (13.167) is taken advantage of, then (13.169) can be directly verified via (13.168).

4. The coeffcients $\mathcal{A}_n(x,y)$ must now be determined in solution (13.163). To accomplish this, we must return to (13.159) for $\tilde{\delta}_1$. If both (13.163) and the expression of $\tilde{\omega}_0$ (13.152) are put to good use, it is observed that in order to eliminate the secular terms which appear in the solution of $\tilde{\delta}_1$, we must impose the following condition which is an equation for the coefficients $\mathcal{A}_n(x,y)$:

$$2\left(\frac{\partial \Theta_n}{\partial x}\frac{\partial}{\partial x} + \frac{\partial \Theta_n}{\partial y}\frac{\partial}{\partial y}\right)\log \mathcal{A}_n(x,y)$$
$$= \mathcal{G}_\infty(y)\frac{\partial \Theta_n}{\partial y} - \left(\frac{\partial^2 \Theta_n}{\partial x^2} + \frac{\partial^2 \Theta_n}{\partial y^2}\right) \quad ,$$

where

$$\mathcal{G}_\infty(y) = \frac{\gamma}{\hat{B}}\left[1 + \left(\frac{\hat{B}}{\gamma k_0}\right)^2\right]\phi_\infty(y) \quad .$$

But, along a ray, we have:

$$\frac{\partial \Theta_n}{\partial x}\frac{\partial}{\partial x} + \frac{\partial \Theta_n}{\partial y}\frac{\partial}{\partial y} \equiv \frac{\partial \Theta_n}{\partial y}\left(\frac{\partial}{\partial y}\right)_{\sigma_n=\text{const}} \quad ,$$

and therefore, again along a ray:

$$2\frac{\partial \Theta_n}{\partial y}\left(\frac{d\log \mathcal{A}_n}{dy}\right)_{\sigma_n=\text{const}} + \frac{\partial^2 \Theta_n}{\partial x^2} + \frac{\partial^2 \Theta_n}{\partial y^2} - \mathcal{G}_\infty(y)\frac{\partial \Theta_n}{\partial y} = 0 \quad . \tag{13.170}$$

Let us calculate $\partial^2\Theta_n/\partial x^2 + \partial^2\Theta_n/\partial y^2$. Since $\partial\vartheta_n/\partial\sigma = 0$ along a ray, it is seen that:

$$\frac{\partial^2 \Theta_n}{\partial x^2} + \frac{\partial^2 \Theta_n}{\partial y^2} = \frac{\partial^2 \vartheta_n}{\partial x^2} + \frac{\partial^2 \vartheta_n}{\partial y^2} + \frac{\partial^2 \vartheta_n}{\partial \sigma \partial x}\frac{\partial \sigma_n}{\partial x} + \frac{\partial^2 \vartheta_n}{\partial \sigma \partial y}\frac{\partial \sigma_n}{\partial y} \quad ,$$

but

$$\frac{\partial \vartheta_n}{\partial x} \equiv \sigma_n \quad , \quad \frac{\partial^2 \vartheta_n}{\partial \sigma \partial x} = 1 \quad , \quad \frac{\partial^2 \vartheta_n}{\partial x^2} \equiv 0 \quad ,$$

$$\frac{\partial \vartheta_n}{\partial y} = (-1)^n \sqrt{\frac{\phi_\infty(y)}{k_0^2} - \sigma_n^2} \quad ,$$

$$\frac{\partial^2 \vartheta_n}{\partial \sigma \partial y} = -(-1)^n \frac{\sigma_n}{\left[\phi_\infty(y)/k_0^2 - \sigma_n^2\right]^{1/2}}$$

and

$$\frac{\partial^2 \vartheta_n}{\partial y^2} = (-1)^n \frac{(d\phi_\infty/dy)/k_0^2}{2\left[\phi_\infty(y)/k_0^2 - \sigma_n^2\right]^{1/2}} \quad .$$

Hence, along a ray:

$$\frac{\partial^2 \Theta_n}{\partial x^2} + \frac{\partial^2 \Theta_n}{\partial y^2} = \frac{\partial \sigma_n}{\partial x} - (-1)^n \frac{\partial \sigma_n}{\partial y}\left(\frac{\phi_\infty(y)}{k_0^2} - \sigma_n^2\right)^{-1/2} + (-1)^n \left(2k_0^2\right)^{-1} \frac{d\phi_\infty}{dy}\left(\frac{\phi_\infty(y)}{k_0^2} - \sigma_n^2\right)^{-1/2} \quad .$$

We must then still calculate $\partial\sigma_n/\partial x$ and $\partial\sigma_n/\partial y$. We know that $\sigma_n(x,y)$ is given by (13.169), i.e., by the relation which results from (13.168):

$$x - \sigma_n \left\{ 2\alpha(n) \int_0^{1/\hat{B}} \frac{dt}{\left(\frac{\phi_\infty(t)}{k_0^2} - \sigma_n^2\right)^{1/2}} + (-1)^n \int_0^y \frac{dt}{\left(\frac{\phi_\infty(t)}{k_0^2} - \sigma_n^2\right)^{1/2}} \right\} = 0 \quad ,$$

and if we introduce:

$$\Delta_n \equiv 2\alpha(n) \int_0^{1/\hat{B}} \frac{\phi_\infty(t)/k_0^2}{\left(\frac{\phi_\infty(y)}{k_0^2} - \sigma_n^2\right)^{3/2}} dt + (-1)^n \int_0^y \frac{\phi_\infty(t)/k_0^2}{\left(\frac{\phi_\infty(y)}{k_0^2} - \sigma_n^2\right)^{3/2}} dt \quad ,$$

then we can easily calculate:

$$\frac{\partial \sigma_n}{\partial x} = \frac{1}{\Delta_n} \quad \text{and} \quad \frac{\partial \sigma_n}{\partial y} = -(-1)^n \frac{\sigma_n}{\Delta_n\left(\frac{\phi_\infty(y)}{k_0^2} - \sigma_n^2\right)^{1/2}} \quad .$$

Therefore:

$$\frac{\partial \sigma_n}{\partial x} - (-1)^n \sigma_n \frac{\partial \sigma_n}{\partial y}\left(\frac{\phi_\infty(y)}{k_0^2} - \sigma_n^2\right)^{-1/2} \equiv \frac{\phi_\infty(y)/k_0^2}{\Delta_n\left(\frac{\phi_\infty(y)}{k_0^2} - \sigma_n^2\right)^{1/2}} \quad .$$

Finally, by taking into account the expression of $\partial\Theta_n/\partial y$, the following relation results on R_n:

$$\frac{(\partial^2\Theta_n/\partial x^2)+(\partial^2\Theta_n/\partial y^2)}{\partial\Theta_n/\partial y}=\frac{(d\phi_\infty/dy)/k_0^2}{2\left(\frac{\phi_\infty(y)}{k_0^2}-\sigma_n^2\right)}+\frac{(-1)^n\phi_\infty(y)/k_0^2}{\Delta_n\left(\frac{\phi_\infty(y)}{k_0^2}-\sigma_n^2\right)^{3/2}} .$$

But,

$$\frac{(d\phi_\infty/dy)/k_0^2}{2\left(\frac{\phi_\infty(y)}{k_0^2}-\sigma_n^2\right)}\equiv\left\{\frac{d}{dy}\left[\log\left(\frac{\phi_\infty(y)}{k_0^2}-\sigma_n^2\right)^{1/2}\right\}_{\sigma_n=\text{const}} ,$$

and thank to the definition of Δ_n:

$$\frac{(-1)^n\phi_\infty(y)/k_0^2}{\left(\frac{\phi_\infty(y)}{k_0^2}-\sigma_n^2\right)^{3/2}\Delta_n}\equiv\left\{\frac{d}{dy}(\log\Delta_n)\right\}_{\sigma_n=\text{const}} .$$

Thus the following relation can replace (13.170):

$$\left(\frac{d}{dy}\left[\log\left\{\mathcal{A}_n^2\sqrt{\frac{\phi_\infty(y)}{k_0^2}-\sigma_n^2}\Delta_n\right\}\right]\right)_{\sigma_n=\text{const}}=\mathcal{G}_\infty(y) . \tag{13.171}$$

If the following function is introduced:

$$\mathcal{F}_\infty(y)=\exp\left\{\frac{1}{2}\int_0^y\mathcal{G}_\infty(t)dt\right\}\Leftrightarrow\frac{1}{\mathcal{F}_\infty^2}\frac{d\mathcal{F}_\infty^2}{dy}\equiv\mathcal{G}_\infty(y) ,$$

it is then remarked that according to (13.171), the following necessarily results on the portion R_n:

$$\frac{\mathcal{A}_n\left[\phi_\infty(y)/k_0^2-\sigma_n^2\right]^{1/4}\Delta_n^{1/2}}{\mathcal{F}_\infty(y)}=\text{const} . \tag{13.172}$$

Relation (13.172) is verified over the entire ray, given the fact that:

$$\left(\frac{\phi_\infty(y)}{k_0^2}-\sigma_n^2\right)^{1/4}\Delta_n^{1/2}\mathcal{F}_\infty^{-1}$$

is continuous for each reflection.

At the outset, we have the following for the ray R_0 near the origin:

$$\left(\frac{\phi_\infty(y)}{k_0^2}-\sigma_n^2\right)^{1/4}\Delta_0^{1/2}\mathcal{F}_\infty^{-1}\cong\left(1-\sigma_0^2\right)^{1/4}\left(1-\sigma_0^2\right)^{-3/4}y^{1/2}$$
$$=\left(1-\sigma_0^2\right)^{-1/2}y^{1/2}=\frac{(\tilde{r}\sin\theta)^{1/2}}{\sin\theta} .$$

If we make use of the matching condition (13.160), we obtain:

$$\mathcal{A}_n(x,y)=\sqrt{\frac{2k_0}{\pi}}\frac{(1-\sigma^2)^{1/4}G(\sigma)}{\Delta_n^{1/2}\left(\frac{\phi_\infty(y)}{k_0^2}-\sigma^2\right)^{1/4}}\mathcal{F}_\infty(y). \tag{13.173}$$

5. Thus, on the one hand, we have clarified the role played by the Long-Miles model in an infinite atmosphere. On the other hand, we have demonstrated that the upper wall of the duct which simulates the tropopause has, in fact, no influence on the wave structure in the "inner" Long-Miles model which is valid only in the immediate vicinity of the relief. Far from this vicinity, the radiation condition induced by the Long-Miles model excites internal waves throughout the entire upper part of the troposphere. We have shown that it is possible to construct a double scale, asymptotic approximation for these short waves which propagate along the characteristic strips of (13.157) by alternately reflecting on the upper and lower walls of the duct simulating the troposphere. These waves propagate only in the downstream direction and do not perturb the structure of the inner (Long-Miles) flow near the relief – at least not in first approximation.

We have related the calculations in detail with the aim of pointing out the difficulties encountered when attempting to carry out the asymptotic theory of a problem by using the MSM. The mjaor difficulty here was due to the fact that (13.157, 170) are partial differential equations which, technically speaking, necessitate a more complex development than ordinary differential equations (like, for example, in Sect. 6.2).

13.2.5 Asymptotic Representation of Three-Dimensional Linearized Lee Waves in the Lower Atmosphere [11]

1. Let us now backtrack to the generalized Boussinesq equations discussed in Subsect. 8.3.2 and write them in the general case: viscous with heat conduction and viscous dissipation. We will neglect the Coriolis force ($\mathrm{Ro} \equiv \infty$) and suppose that $\varepsilon \equiv 1$ and $\mathrm{S} \equiv 1$ in the complete non-dimensional Navier-Stokes equations [e.g., (7.1–4)]. When $\mathrm{M}_\infty \to 0$ and $\mathrm{Bo} \to 0$ such that:

$$\mathrm{M}_\infty = \mathrm{Bo} \quad , \quad \text{i.e.,} \quad \hat{B} \equiv 1 \quad ; \quad \zeta \equiv \mathrm{M}_\infty z \quad ,$$

a solution is sought for these Navier-Stokes equations in the form:

$$\begin{aligned} &\boldsymbol{u} = \tilde{\boldsymbol{u}} + \mathrm{M}_\infty \tilde{\boldsymbol{u}}_1 \quad ; \quad p = p_0 + \mathrm{M}_\infty \tilde{p} \quad ; \\ &\varrho = \varrho_0 + \mathrm{M}_\infty \tilde{\varrho} \quad ; \quad T = T_0 + \mathrm{M}_\infty \tilde{T} \quad , \end{aligned} \tag{13.174}$$

$\varrho_0(\zeta)$, $T_0(\zeta)$ and $p_0(\zeta)$ are solutions of the standard atmosphere equations. Once the terms of order M_∞^2 have been neglected, the following system results for $\tilde{\boldsymbol{u}}$, $\tilde{p}$, $\tilde{\varrho}$ and $\tilde{T}$ which are functions of t, x, y and z:

$$\varrho_0(\zeta)\boldsymbol{\nabla}\cdot\tilde{\boldsymbol{u}} = -\mathrm{M}_\infty\left\{\frac{\partial\tilde{\varrho}}{\partial t} + \boldsymbol{\nabla}\cdot(\tilde{\varrho}\tilde{\boldsymbol{u}}) + \frac{d\varrho_0}{d\zeta}\tilde{w}\right\} \quad ; \tag{13.175a}$$

$$\gamma\varrho_0(\zeta)\frac{D\tilde{\boldsymbol{u}}}{Dt} + \boldsymbol{\nabla}\tilde{p} + \gamma\tilde{\varrho}\boldsymbol{k} = -\gamma\mathrm{M}_\infty\tilde{\varrho}\frac{D\tilde{\boldsymbol{u}}}{Dt} + \frac{\gamma}{\mathrm{Re}}\left[\Delta\tilde{\boldsymbol{u}} + \frac{1}{3}\boldsymbol{\nabla}(\boldsymbol{\nabla}\cdot\tilde{\boldsymbol{u}})\right] \quad ;$$

[11] According to the works by Bois (1982) and Darnaudet and Bois (1985). For details, the reader may consult the doctoral thesis of F. Darnaudet, "Calcul d'ondes de relief bidimensionnelles et tridimensionnelles par voie asymptotique et numérique" (Université de Paris VI, 1985).

$$\varrho_0(\zeta)\left[\frac{D\tilde{T}}{Dt} + \frac{dT_0}{d\zeta}\tilde{w}\right] - \frac{\gamma - 1}{\gamma}\frac{dp_0}{d\zeta}\tilde{w} - \frac{1}{\operatorname{Pr}\operatorname{Re}}\Delta\tilde{T}$$
$$= \mathrm{M}_\infty\left\{\frac{\gamma-1}{\gamma}\frac{D\tilde{p}}{Dt} - \tilde{\varrho}\tilde{w}\frac{dT_0}{d\zeta} - \tilde{\varrho}\frac{D\tilde{T}}{Dt} + \frac{\gamma-1}{\operatorname{Re}}\tilde{\phi}\right\} \quad ; \tag{13.175b}$$

$$\varrho_0(\zeta)\tilde{T} + T_0(\zeta)\tilde{\varrho} = \mathrm{M}_\infty(\tilde{p} - \tilde{\varrho}\tilde{T}) \quad ,$$

where $\tilde{w}$ designates the vertical component of $\tilde{\boldsymbol{u}}(\tilde{\boldsymbol{u}} \cdot \boldsymbol{k} = \tilde{w})$. The classical Boussinesq approximation according to Zeytounian (1974), appears as the first term (order 0 in M_∞) of an asymptotic expansion in M_∞ when $\mathrm{M}_\infty \to 0$.

2. Let us now linearize (13.175) around the basic solution (valid to the order M_∞, inclusive):

$$\tilde{\boldsymbol{u}} = U_\infty(\zeta)\boldsymbol{i} \quad , \quad \tilde{p} = \tilde{\varrho} = \tilde{T} = 0 \quad , \tag{13.176}$$

the vector $\boldsymbol{i}$ being the unit vector of the x-axis. We have by hypothesis $U_\infty(0) = 1$, which means that the characteristic velocity was taken equal to $U_\infty(0) \equiv U_0$ when the non-dimensionalization was carried out. The hypothesis that U_∞ depends significantly on $\zeta \equiv \mathrm{M}_\infty z$ has been confirmed by experience. The reader is referred to the paper by Palm and Foldvik (1959) on this topic.

After the linearization of (13.175), an equation is extracted for the perturbation of the vertical velocity w' in the steady case:[12]

$$\frac{\operatorname{Re}^{-2}}{\operatorname{Pr}\varrho_0(\zeta)}\Delta\Delta\Delta w' - \operatorname{Re}^{-1}U_\infty(\zeta)\left(1 + \frac{1}{\operatorname{Pr}}\right)\Delta\Delta\frac{\partial w'}{\partial x}$$
$$+ U_\infty^2(\zeta)\varrho_0(\zeta)\Delta\frac{\partial^2 w'}{\partial x^2} + \varrho_0(\zeta)\Gamma_\infty(\zeta)\boldsymbol{D}^2 w'$$
$$= -\mathrm{M}_\infty\left\{\frac{\operatorname{Re}^{-2}}{\operatorname{Pr}\varrho_0(\zeta)}\mathcal{A}(\zeta)\Delta\Delta\frac{\partial w'}{\partial z} - \operatorname{Re}^{-1}U_\infty(\zeta)\mathcal{B}(\zeta)\Delta\frac{\partial^2 w'}{\partial x \partial z}\right.$$
$$\left. + U_\infty^2(\zeta)\varrho_0(\zeta)\mathcal{C}(\zeta)\frac{\partial^3 w'}{\partial x^2 \partial z}\right\} \quad , \tag{13.177}$$

where:

$$\Gamma_\infty(\zeta) = \frac{1}{T_0(\zeta)}\left[\frac{dT_0}{d\zeta} + \gamma - 1\right]$$

plays the role of the Väisälä frequency. The functions $\mathcal{A}$, $\mathcal{B}$ and $\mathcal{C}$ can be expressed as follows:

$$\mathcal{A}(\zeta) = \frac{1}{T_0}\left(3\frac{dT_0}{d\zeta} + 2\gamma\right) \quad , \quad \mathcal{B}(\zeta) = \frac{1}{T_0}\left(\frac{\gamma}{\operatorname{Pr}} + 2\right) - \frac{2}{\operatorname{Pr}}\frac{1}{U_\infty}\frac{dU_\infty}{d\zeta} \quad ,$$

---

[12] It will be observed that:

$$\Delta = \frac{\partial^2}{\partial x^2} + \frac{\partial^2}{\partial y^2} + \frac{\partial^2}{\partial z^2} \equiv \boldsymbol{D}^2 + \frac{\partial^2}{\partial z^2} \quad .$$

and $\quad C(\zeta) = \dfrac{1}{T_0}\left(\dfrac{dT_0}{d\zeta} + \gamma\right) \equiv -\dfrac{1}{\varrho_0}\dfrac{d\varrho_0}{d\zeta}$.

In the inviscid adiabatic case Re $\equiv \infty$, the following replaces (13.177):

$$\Delta\frac{\partial^2 w'}{\partial x^2} + k(\zeta)\boldsymbol{D}^2 w' = -\mathrm{M}_\infty\frac{1}{\varrho_0}\frac{d\varrho_0}{d\zeta}\frac{\partial^3 w'}{\partial x^2\partial z} \quad , \tag{13.178}$$

where the Scorer parameter has been introduced:

$$k(\zeta) = \frac{\Gamma_\infty(\zeta)}{U_\infty^2(\zeta)} \equiv \frac{1}{U_\infty^2(\zeta)}\left\{\frac{d\log T_0}{d\zeta} + \frac{\gamma-1}{T_0(\zeta)}\right\} \quad .$$

In the forthcoming, our attention will be mainly focused on the solutions to (13.178). The general equation (13.177) will serve only as a support to (1), satisfy the boundary conditions at infinity, and (2), authorize certain expansions into a series. In any case, the effects of terms related to Re^{-1} are totally neglected near the relief.

3. Let us then consider the flow around a small height, three-dimensional obstacle. Because of the linearization, it is enough to study the flow induced by a Dirac mass placed at the origin. A convolution with respect to x and y will allow any obstacle to be treated ulteriorly.

Thus, for w', we have the conditions:

$$\begin{aligned} &w'(x,y,z;\mathrm{M}_\infty) \to 0 \quad , \quad \text{at infinity} \\ &w'(x,y,0;\mathrm{M}_\infty) = \frac{\partial\delta_0}{\partial x}(x,y) \quad , \end{aligned} \tag{13.179}$$

where $\delta_0(x,y)$ designates here the Dirac function.[13]

A Fourier transformation is now carried out on (13.178), with respect to x and y:

$$W(z,\mathrm{M}_\infty) = \iint_{-\infty}^{+\infty} w'(x,y,z;\mathrm{M}_\infty)\,\mathrm{e}^{-\mathrm{i}(kx+ly)}\,dx\,dy \quad ,$$

and W verifies the equation:

$$\frac{d^2W}{dz^2} + \left[\frac{k(\zeta)}{\cos^2\gamma} - m^2\right]W = -\mathrm{M}_\infty\frac{1}{\varrho_0}\frac{d\varrho_0}{d\zeta}\frac{\partial W}{\partial z} \quad , \tag{13.180}$$

where the following has been set:

$$m^2 = k^2 + l^2 \quad , \quad \gamma = \mathrm{arctg}\left(\frac{l}{k}\right) \quad , \quad \cos\gamma = \frac{k}{m} \quad .$$

---

[13] Dirac function (or impulsion): "Function" denoted δ_0, zero beyond the origin and with an "integral" equal to 1. At the origin, it is a unit impulsion:

$$\iint \varphi(x,y)\delta_0(x,y)dx\,dy \equiv \varphi(0,0) \quad .$$

The Dirac function δ (or more correctly, the *distribution*) appears as the limit of the density of a continuous distribution of very concentrated masses on a small region around the origin.

The general solution of (13.180) is:

$$W = \frac{1}{[\varrho_0(d\varphi/d\zeta)]^{1/2}}\left\{A\,e^{\xi} + B\,e^{-\xi}\right\} \quad , \tag{13.181a}$$

or

$$W = \frac{1}{[\varrho_0(d\varphi/d\zeta)]^{1/2}}\left\{C\,e^{i\xi} + D\,e^{-i\xi}\right\} \quad , \tag{13.181b}$$

with

$$\xi = \frac{1}{M_\infty}\varphi(\zeta, k, l) \quad , \quad \left(\frac{d\varphi}{d\zeta}\right)^2 = \left|\frac{k(\zeta)}{\cos^2\gamma} - m^2\right| \quad . \tag{13.182}$$

In case (a), $(d\varphi/d\zeta)^2 = m^2 - k(\zeta)/\cos^2\gamma$, and in case (b), $(d\varphi/d\zeta)^2 = k(\zeta)/\cos^2\gamma - m^2$. The turning points (points where $d\varphi/d\zeta = 0$) are determined by the equation $k(\zeta) = k^2$. When ζ goes beyond a turning point ζ^*, the solution takes the form (13.181a) on one side of the turning point, and the form (13.181b), on the other side. A localized study demonstrates that the constants A, B, C and D are linked by the relations:

$$2B\,e^{-\beta} + iA\,e^{\beta} = D\,e^{-i\alpha} \quad , \quad -2B\,e^{-\beta} + iA\,e^{\beta} = -C\,e^{i\alpha} \quad , \tag{13.183}$$

$$\text{where} \quad \alpha = \frac{1}{M_\infty}\varphi(\zeta^*) - \frac{\pi}{4} \quad , \quad \beta = \frac{1}{M_\infty}\varphi(\zeta^*) \quad .$$

4. We now want to determine the solution w' for a decreasing monotonic function $k(\zeta)$ (from k_0 to k_1 when ζ varies from 0 to $+\infty$). First of all, to determine W for a fixed value of ζ, four possibilities must be considered: a) $W = [\varrho_0(d\varphi/d\zeta)]^{-1/2}B\,e^{-\xi}$, B being determined by the ground condition which is written $W(0) = ik$ for a Dirac mass; b) $W = [\varrho_0(d\varphi/d\zeta)]^{-1/2}B\,e^{-\xi}$, but B is linked by a condition of the type (13.183) to two coefficients C and D. According to (13.183), these verify:

$$D\,e^{-i\alpha} = 2B\,e^{-\beta} \quad , \quad -C\,e^{i\alpha} = -2B\,e^{-\beta} \quad , \text{ and thus } \quad D = C\,e^{2i\alpha} \quad ,$$

whence C and D by writing that $W = ik$ in $\zeta = 0$, then B; c) analogous reasoning to b); d) in this case it is necessary to consider the viscous flow and, finally, the following four expressions are found:

$$\begin{aligned}
&1)\quad W = iG(\zeta, k, l)\,e^{-\xi} \quad ; \\
&2)\quad W = iG(\zeta, k, l)\frac{e^{\beta + i\alpha}}{2(1 + e^{2i\alpha})}e^{-\xi} \quad ; \\
&3)\quad W = iG(\zeta, k, l)\left[\frac{e^{i\xi}}{1 + e^{2i\alpha}} + \frac{e^{2i\alpha - i\xi}}{1 + e^{2i\alpha}}\right] \quad ; \\
&4)\quad W = iG(\zeta, k, l)\,e^{i\xi} \quad , \quad k > 0 \quad , \\
&\qquad W = iG(\zeta, k, l)\,e^{-i\xi} \quad , \quad k < 0 \quad ,
\end{aligned} \tag{13.184}$$

with

$$G(\zeta, k, l) \equiv k \sqrt{\left.\frac{d\varphi}{d\zeta}\right|_{\zeta=0} \Big/ \varrho_0(\zeta)\frac{d\varphi}{d\zeta}} \quad .$$

Once W is known, w' can be obtained by inverse Fourier transforms. In order to obtain an asymptotic estimate of the integrals which enter into the solution of w', the viscosity must be introduced.

We will consider, in particular, a flow whose Scorer parameter is given by:

$$k(\zeta) = k_0 - a\zeta, \quad \text{if} \quad \zeta \leq \zeta_0, \tag{13.185a}$$

$$k(\zeta) = k_0 - a\zeta_0 \equiv k_1 \quad , \quad \text{if } \zeta \geq \zeta_0 \quad . \tag{13.185b}$$

First of all, the following expression is found for w':

$$w' = J_1 + J_2$$

with:

$$J_1 = \frac{1}{\pi^2}\int_0^{+\infty}\int_0^{\sqrt{k_1}} H(\zeta, k, l)\left(\frac{d\varphi}{d\zeta}\right)^{-1/2} \sin(\xi + kx)\cos(ly)\,dk\,dl \quad ,$$

$$J_2 = \frac{1}{\pi^2}\int_0^{+\infty}\int_{\sqrt{k_1}}^{\sqrt{k(\zeta)}} H(\zeta, k, l)\left(\frac{d\varphi}{d\zeta}\right)^{-1/2} \left\{\sum_{n=0}^{+\infty}(-1)^n \sin(2n\alpha + \xi + kx)\cos(ly)\right.$$

$$\left. + \sum_{n=0}^{+\infty}(-1)^n \sin\big(2(n+1)\alpha - \xi + kx\big)\cos(ly)\right\} dk\,dl \quad ,$$

where:

$$H(\zeta, k, l) \equiv \sqrt{\frac{d\varphi}{d\zeta}}\, G(\zeta, k, l) \quad .$$

The contributions in w' of wave numbers k not belonging to the interval $[0, \sqrt{k(\zeta)}]$ are negligible. The function $H(\zeta, k, l)$ which appears in J_1 and J_2 is regular and α is defined by:

$$\alpha = \frac{1}{\mathrm{M}_\infty}\varphi(\zeta^*, k, l) \quad , \quad \text{with} \quad k(\zeta^*)/\cos^2\gamma = m^2 \quad .$$

The two integrals J_1 and J_2 are finite or infinite sums of integrals having the form:

$$J = \iint_{\mathcal{D}} G(\zeta, k, l)\, \mathrm{e}^{(\mathrm{i}/\mathrm{M}_\infty)\omega(k,l,\chi,\eta,\zeta)}\, dk\,dl \quad . \tag{13.186}$$

An estimate of (13.186) can be carried out by the steady phase method when the three variables $\chi \equiv \mathrm{M}_\infty x$, $\eta \equiv \mathrm{M}_\infty y$ and ζ are O(1). The integral J is in general $O(\mathrm{M}_\infty^2)$ unless if in the domain $\mathcal{D}$ exists some *steady points*, i.e., couples (k_i, l_i) which are solutions to the equations:

$$\frac{\partial\omega}{\partial k}(k_i, l_i) = 0 \quad , \quad \frac{\partial\omega}{\partial l}(k_i, l_i) = 0 \quad . \tag{13.187}$$

In this case, J, which is then $O(M_\infty)$, is expressed from the value of the integrand in (k_i, l_i) [see Queney (1977)]. The problem is thus to determine the solutions to (13.187) for the phases figuring in J_1 and J_2.

5. The writing of (13.187) for the function $k(\zeta)$ considered in (13.185) yields the following equations for k and l:

a) if $\zeta > \zeta_0$:

$$\chi = \frac{2\cos\gamma}{ak}\left\{\left(k_0 - k^2\right)^{1/2}\left[k^2\left(1+\frac{2}{3}\mathrm{tg}^2\gamma\right)+\frac{k_0}{3}\mathrm{tg}^2\gamma\right]\right.$$
$$\left. - \left(k_1 - k^2\right)^{1/2}\left[k^2\left(1+\frac{2}{3}\mathrm{tg}^2\gamma\right)+\frac{k_1}{3}\mathrm{tg}^2\gamma\right]\right\}$$
$$+\frac{(\zeta-\zeta_0)\left[k^2 + k_1(\mathrm{tg}^2\gamma)/3\right]\cos\gamma}{k\left(k_1-k^2\right)^{1/2}} \quad ;$$

$$\eta = -\frac{2\sin\gamma}{3ak}\left\{\left(k_0-k^2\right)^{3/2} - \left(k_1-k^2\right)^{3/2}\right\}$$
$$-(\zeta-\zeta_0)\frac{\left(k_1-k^2\right)^{1/2}\sin\gamma}{k\left(k_1-k^2\right)^{1/2}} \quad , \qquad (13.188a)$$

b) if $\zeta \le \zeta_0$:

$$\chi = \frac{2\cos\gamma}{ak}\left\{(2n+1)\left(k_0-k^2\right)^{1/2}\left[k^2\left(1+\frac{2}{3}\mathrm{tg}^2\gamma\right)+\frac{k_0}{3}\mathrm{tg}^2\gamma\right]\right.$$
$$\left. - \Lambda\left(k_0 - a\zeta - k^2\right)^{1/2}\left[k^2\left(1+\frac{2}{3}\mathrm{tg}^2\gamma\right)+\frac{1}{3}(k_0-a\zeta)\mathrm{tg}^2\gamma\right]\right\} \quad ;$$
$$\eta = -\frac{2\sin\gamma}{3ak}\left\{(2n+1)\left(k_0-k^2\right)^{3/2} - \Lambda\left(k_0-a\zeta-k^2\right)^{3/2}\right\} \quad ,(13.188b)$$

where $\Lambda = \pm 1$.

Rather than carry out a numerical resolution of (13.188a, b), Darnaudet (1985) considered for a fixed couple (k^*, l^*), the curves $\{\chi = \chi(k^*, l^*, \zeta), \eta = \eta(k^*, l^*, \zeta)\}$ which are defined by (13.188). These curves represent the locus of the points in the physical space where the perturbation of wave vector (k^*, l^*) exists, i.e., the propagation ray corresponding to this wave vector. The zone disturbed by the relief is the area of the physical space covered by the rays when (k, l) varies in $[0, \sqrt{k_0}] \times]-\infty, +\infty[$.

The following results can thus be established:

(1) The perturbation only exists downstream of the relief. The waves having a wave number k between 0 and $\sqrt{k_1}$ form a progressive wave which covers a quarter of the space ($\chi > 0$, $\zeta > 0$), not including a zone surrounding the half plane $\eta = 0$ which is limited in height by a plateau placed in ζ_0 and which is

shaped like a semi-infinite strip with a rounded edge (this zone is referred to as "zone Z" in the forthcoming).

(2) The waves with a wave number $k \in [\sqrt{k_1}, \sqrt{k_0}]$ form a system of lee waves confined within the area $\zeta \leq \zeta_0$. These waves are contained in a dihedron Δ whose plane $0\chi\zeta$ is bisecting. The opening of the dihedron depends on k_0 and k_1. The propagation rays are periodic figures, each ray having its own period. The set of lee waves forms a wake which is arranged nearly periodically near the relief and which slowly spreads out as one moves downstream.

(3) Unlike the two-dimensional case, the progressive wave and the lee waves exist simultaneously in the (non empty) zone $\Delta - Z$. Within the zone Z, however, there are areas which remain unperturbed. They form small "bubbles" which constitute the only non-perturbed zones of the flow. Moving downstream, they slowly disappear.

The above results demonstrate that the three-dimensional flow around (and above) a relief does not directly generalize the results of the two-dimensional flow. In particular, the system of lee waves appears to form a wake downstream of the relief which is analogous to that of ships. This latter property confirms not only results obtained by calculation (e.g., see Zeytounian, 1969), but also those obtained experimentally [see Gjevik and Marthinsen (1977), as well as Smith (1979) and Scorer (1978, p. 195)].

13.3 Modeling of the Interaction Phenomenon Between Free and Forced Circulations

Within the framework of a so-called "triple deck" structure in Sect. 6.4, an interaction phenomenon was modelled between an Ekman flow (forced) and a local free circulation due to a non-homogeneous thermal site.

Thus, we must first of all define a basic flow which is not disturbed by the local site considered. We will suppose here that this site is flat and that it is characterized only by the data of a thermal field. More precisely, we will consider at the start the Navier-Stokes equations (8.5–8) to which will be assigned the following boundary conditions on the ground:

$$\boldsymbol{u} = 0 \quad , \quad \vartheta = \tau_0 \Xi(t, x, y) \quad , \quad \text{on} \quad z = 0 \tag{13.189}$$

for $t > 0$ and $x, y \subset \mathcal{D}_2$ being a bounded domain in the plane $z = 0$. In this case, the length scale L is of course directly related to the diameter of domain $\mathcal{D}_2$. As concerns the length scale H which remains arbitrary, it could possibly be determined by a similarity relation related to the parameters of the problem.

Let us return to the characteristics of the basic flow. This flow must necessarily be a function of z in such a way that it takes into account a boundary layer structure (like the Ekman flow). We saw in Chap. 7 that to the primitive equations, it was necessary to join the boundary layer equations of Sect. 7.3 where the altitude is $\hat{z} = z/(1/\mathrm{Re}_{\perp}^{1/2})$.

It will thus be assumed that the basic flow is characterized by a horizontal velocity:

$$\boldsymbol{V}_\infty(z, \hat{z}) = U_\infty(z, \hat{z})\boldsymbol{i} + V_\infty(z, \hat{z})\boldsymbol{j} \quad , \tag{13.190}$$

such that:

$$\lim_{\hat{z} \to +\infty} \boldsymbol{V}_\infty = \boldsymbol{V}_{\infty,\mathrm{e}}(0) \quad \text{and} \quad \boldsymbol{V}_\infty \sim \hat{z} \quad , \quad \text{when} \quad \hat{z} \to 0 \quad .$$

It will be noted that $\boldsymbol{V}_{\infty,\mathrm{e}}(z) = U_{\infty,\mathrm{e}}(z)\boldsymbol{i} + V_{\infty,\mathrm{e}}(z)\boldsymbol{j}$ is the velocity of the basic flow beyond the boundary layer, i.e., in the region where the primitive equations remain valid.

Hence, the problem considered below is one of the interaction of two convection phenomena: a natural (free) convection which develops in the vicinity of a non-homogeneous thermal site [simulated by condition (13.189)], and a forced convection related to the basic flow having a velocity of $\boldsymbol{V}_\infty(z, \hat{z})$.

13.3.1 Formulation of the Regional Boundary Layer Problem

In the present case, the non-dimensional numbers M_∞, τ_0, 1/Re and ε are *simultaneously small*. We have already noted:

$$\varepsilon^2\,\mathrm{Re} = \mathrm{Re}_\perp \;\Rightarrow\; \frac{1}{\mathrm{Re}_\perp} \equiv \delta^2 \quad . \tag{13.191}$$

In the forthcoming, it will be supposed that:

$$\delta \ll 1 \quad \text{and} \quad \mathrm{M}_\infty = \delta^a \quad , \quad \tau_0 = \delta^b \quad , \tag{13.192}$$

where $a > 0$ and $b > 0$ are real scalars which must be determined in such a way that the limiting processes related to $\delta \to 0$ will be significant. The dimensionless numbers S, Bo, Ro and Pr remain of the order unity when $\delta \to 0$.

The first limiting process (called "outer") associated with $\delta \to 0$ is carried out with t, x, y and z fixed. In this case, a solution to the general Navier-Stokes equations (8.5–8) is sought in the form:

$$\begin{aligned} \boldsymbol{v} &= \overline{\boldsymbol{v}}_0 + \delta\overline{\boldsymbol{v}}_1 + \dots \quad ; \quad w = \overline{w}_0 + \delta\overline{w}_1 + \delta^2\overline{w}_2 + \dots \quad ; \\ \pi &= \delta^2\overline{\pi}_2 + \dots \quad ; \quad \omega = \delta\overline{\omega}_1 + \delta^2\overline{\omega}_2 + \dots \quad ; \quad \vartheta = \delta\overline{\vartheta}_1 + \delta^2\vartheta_2 + \dots \quad , \end{aligned} \tag{13.193}$$

once the correct choice of $a = 1$ has been made. A simple analysis then shows that:

$$\overline{\vartheta}_1 = 0 \quad , \quad \overline{\omega}_1 = 0 \quad , \quad \overline{w}_0 = 0 \quad ,$$

and the functions $\overline{\boldsymbol{v}}_0$ and $\overline{\pi}_2$ satisfy the equations:

$$\begin{aligned} &\boldsymbol{D} \cdot \overline{\boldsymbol{v}}_0 = 0 \quad ; \\ &\mathrm{S}\frac{\partial \overline{\boldsymbol{v}}_0}{\partial t} + (\overline{\boldsymbol{v}}_0 \cdot \boldsymbol{D})\overline{\boldsymbol{v}}_0 + \frac{1}{\mathrm{Ro}}\left[\boldsymbol{k} \times (\overline{\boldsymbol{v}}_0 - \boldsymbol{V}_{\infty,\mathrm{e}})\right] \\ &\qquad + \frac{T_\infty(\mathrm{Bo}z)}{\gamma}\boldsymbol{D}\overline{\pi}_2 = 0 \quad . \end{aligned} \tag{13.194}$$

Thus is the classical quasi non-divergent model of Chap. 12 recovered and the variations of $\overline{\boldsymbol{v}}_0$ and $\overline{\pi}_2$ with respect to z are parametrical via $T_\infty(\mathrm{Bo}z)$ and also the far-field boundary conditions:

$$\overline{\boldsymbol{v}}_0 \to \boldsymbol{V}_{\infty,\mathrm{e}}(z) \quad , \quad \overline{\pi}_2 \to 0 \quad , \tag{13.195}$$

at infinity in the planes $z = \mathrm{const}$.

The quasi-nondivergent system (13.194) can be completed by the relations:

$$\overline{\pi}_2 = \overline{\omega}_2 + \overline{\vartheta}_2 \quad , \quad \frac{\partial \overline{\pi}_2}{\partial z} = \frac{\mathrm{Bo}}{T_\infty(\mathrm{Bo}z)} \overline{\vartheta}_2 \quad ; \quad \overline{w}_1 \equiv 0 \quad . \tag{13.196}$$

$$\overline{w}_2 = \frac{T_\infty(\mathrm{Bo}z)}{\mathrm{Bo}(\gamma-1)/\gamma + dT_\infty/dz} \left(\mathrm{S}\frac{\partial}{\partial t} + \boldsymbol{v}_0 \cdot \boldsymbol{D} \right) \left(\overline{\vartheta}_2 - \frac{\gamma-1}{\gamma} \overline{\pi}_2 \right) \quad .$$

The second limiting process (called "inner") related to $\delta \to 0$ is carried out with t, x, y and $\hat{z} = z/\delta$ fixed. In this case, a solution to (8.5–8) is sought in the following form:

$$\begin{aligned}
\boldsymbol{v} &= \hat{\boldsymbol{v}}_0 + \delta \hat{\boldsymbol{v}}_1 + \dots \quad ; \\
w &= \delta \hat{w}_1 + \dots \quad ; \\
\pi &= \delta^2 \hat{\pi}_2 + \dots \quad , \\
\omega &= \delta \hat{\omega}_1 + \dots \quad , \\
\vartheta &= \delta \hat{\vartheta}_1 + \dots \quad ,
\end{aligned} \tag{13.197}$$

once we have chosen $b \equiv 1$: $\tau_0 = \delta$. Then, taking the limit, the functions $\hat{\boldsymbol{v}}_0$, $\hat{w}_1$, $\hat{\pi}_2$, $\hat{\vartheta}_1$ and $\hat{\omega}_1$ satisfy the following limiting equations:

$$\begin{aligned}
&\hat{\omega}_1 = -\hat{\vartheta}_1 \quad ; \\
&\boldsymbol{D} \cdot \hat{\boldsymbol{v}}_0 + \frac{\partial \hat{w}_1}{\partial \hat{z}} = 0 \quad ; \\
&\mathrm{S}\frac{\partial \hat{\boldsymbol{v}}_0}{\partial t} + (\hat{\boldsymbol{v}}_0 \cdot \boldsymbol{D})\hat{\boldsymbol{v}}_0 + \hat{w}_1 \frac{\partial \hat{\boldsymbol{v}}_0}{\partial \hat{z}} + \frac{1}{\mathrm{Ro}}[\boldsymbol{k} \times (\hat{\boldsymbol{v}}_0 - \boldsymbol{V}_{\infty,\mathrm{cl}})] \\
&\qquad + \frac{1}{\gamma}\boldsymbol{D}\hat{\pi}_2 = \frac{\partial^2 \hat{\boldsymbol{v}}_0}{\partial \hat{z}^2} \quad ; \\
&\frac{\partial \hat{\pi}_2}{\partial \hat{z}} = \mathrm{Bo}\hat{\vartheta}_1 \quad ; \\
&\mathrm{S}\frac{\partial \hat{\vartheta}_1}{\partial t} + (\hat{\boldsymbol{v}}_0 \cdot \boldsymbol{D})\hat{\vartheta}_1 + \hat{w}_1 \frac{\partial \hat{\vartheta}_1}{\partial \hat{z}} + \left[\mathrm{Bo}\frac{\gamma-1}{\gamma} + \left(\frac{dT_\infty(\mathrm{Bo}z)}{dz} \right)_{z=0} \right] \hat{w}_1 \\
&\qquad = \frac{1}{\mathrm{Pr}} \frac{\partial^2 \hat{\vartheta}_1}{\partial \hat{z}^2} \quad .
\end{aligned} \tag{13.198}$$

Since the surface $\hat{z} = 0$ lies within the domain of validity of the expansions (13.197), we can impose on (13.198) the following boundary conditions:

$$\hat{z} = 0: \quad \hat{\boldsymbol{v}}_0 = 0 \quad , \quad \hat{w}_1 = 0 \quad , \quad \hat{\vartheta}_1 = \Xi(t, x, y) \quad , \tag{13.199}$$

when $t > 0$ and x and $y \subset \mathcal{D}_2$.

Furthermore, *far-away* at *upstream infinity*, the following behavior conditions must be satisfied:

$$\hat{v}_0 \to V_{\infty,\mathrm{cl}}(\hat{z}) \quad , \quad \hat{\pi}_2 \to 0 \quad , \quad \hat{w}_1 \to 0 \quad , \quad \hat{\vartheta}_1 \to 0 \quad , \tag{13.200}$$

where $V_{\infty,\mathrm{cl}}(\hat{z})$ is the velocity of the basic flow in the boundary layer where the equations of Sect. 7.3 remain valid. The matching conditions between (13.193) and (13.197) must also be written. They are as follows:

$$\begin{aligned} \hat{z} \to +\infty : \quad & \hat{v}_0 \to \overline{v}_0(t,x,y,0) \quad , \quad \hat{\pi}_2 \to \overline{\pi}_2(t,x,y,0) \quad , \\ & \hat{\vartheta}_1 \to 0 \quad \text{and} \quad \hat{\omega}_1 \to 0 \quad . \end{aligned} \tag{13.201}$$

In addition, the last equation in (13.198) when:

$$\left(- \frac{dT_\infty(B_0 z)}{dz} \right)_{z=0} \neq \mathrm{Bo} \frac{\gamma - 1}{\gamma} \quad ,$$

leads to the following condition:

$$\lim_{\hat{z} \to +\infty} \hat{w}_1 = 0 \quad , \tag{13.202}$$

which is in good agreement with (13.196).

As initial conditions, the basic flow can be prescribed:[14]

$$t = 0 : \quad \hat{v}_0 = V_{\infty,\mathrm{cl}}(\hat{z}) \quad , \quad \text{and} \quad \hat{\vartheta}_1 = 0 \quad . \tag{13.203}$$

The regional boundary layer problem (13.198–203) remains valid with the constraint:

$$\varepsilon^2 \,\mathrm{Re}\, \mathrm{M}_\infty^2 = 1 \quad , \quad \text{when } \varepsilon \to 0 \;.\; \mathrm{M}_\infty \to 0 \text{ and } \mathrm{Re} \to \infty \quad . \tag{13.204}$$

This means that:

$$H^2 = L \frac{\gamma R \nu_0 T_\infty(0)}{U_0^3} \quad .$$

If we want $\mathrm{Ro} \approx 1$, then it is necessary that $L \approx 10^5$ m which implies that H is of the order of a kilometer.

In (13.198), $\hat{\pi}_2$ may of course be eliminated and replaced by $\hat{\vartheta}_1$:

$$\hat{\pi}_2 = \overline{\pi}_2(t,x,y,0) + \mathrm{Bo} \int_{\hat{z}}^{\infty} \hat{\vartheta}_1 \, d\hat{z} \quad ,$$

where $(D\overline{\pi}_2)_{z=0}$ is given by the second equation in (13.194) written for $z = 0$.

It is thus remarked that the classical quasi-nondivergent model analyzed in Chap. 12 is directly related to the regional boundary layer phenomenon (with L of the order of 100 km) above a non-homogeneous thermal site.

A final remark is called for concerning the outer solution which must satisfy (13.194, 195) and the initial condition: $t = 0$, $\overline{v}_0 = V_{\infty,\mathrm{e}}(z)$. An obvious trivial

---

[14] In this case, the following initial condition can be assigned to (13.194):

$$\overline{v}_0 = V_{\infty,\mathrm{e}}(z) \quad \text{for} \quad t = 0 \quad .$$

solution (since there is no relief) is:

$$\overline{v}_0 \equiv V_{\infty,e}(z) \quad \text{and} \quad \overline{\pi}_2 \equiv 0 \quad . \tag{13.204}$$

Thus, for the functions $\hat{v}_0$, $\hat{w}_1$ and $\hat{\vartheta}_1$, we can formulate the following regional boundary layer problem above a flat, thermally non-homogeneous site:

$$D \cdot \hat{v}_0 + \frac{\partial \hat{w}_1}{\partial \hat{z}} = 0 \quad ;$$

$$\mathrm{S}\frac{\partial \hat{v}_0}{\partial t} + (\hat{v}_0 \cdot D)\hat{v}_0 + \hat{w}_1 \frac{\partial \hat{v}_0}{\partial \hat{z}} + \frac{1}{\mathrm{Ro}}\left[k \times (\hat{v}_0 - V_{\infty,\mathrm{cl}})\right]$$

$$+ \frac{\mathrm{Bo}}{\gamma} D \int_{\infty}^{\hat{z}} \hat{\vartheta}_1 \, d\hat{z} = \frac{\partial^2 \hat{v}_0}{\partial \hat{z}^2} \quad ; \tag{13.205a}$$

$$\mathrm{S}\frac{\partial \hat{\vartheta}_1}{\partial t} + \hat{v}_0 \cdot D\hat{\vartheta}_1 + \left[\frac{\partial \hat{\vartheta}_1}{\partial \hat{z}} + \Gamma_\infty(0)\right]\hat{w}_1 = \frac{1}{\mathrm{Pr}}\frac{\partial^2 \hat{\vartheta}_1}{\partial \hat{z}^2} \quad ,$$

where:

$$\Gamma_\infty(0) \equiv \mathrm{Bo}\frac{\gamma - 1}{\gamma} + \left(\frac{dT_\infty(\mathrm{Bo}z)}{dz}\right)_{z=0} \quad ,$$

$$\hat{z} = 0: \ \hat{v}_0 = 0 \quad , \quad \hat{w}_1 = 0 \quad , \quad \hat{\vartheta}_1 = \Xi(t, x, y) \quad , \tag{13.205b}$$

when $t > 0$ and $x, y \subset \mathcal{D}_2$;

at upstream infinity:[15]

$$\hat{v}_0 \to V_{\infty,\mathrm{cl}}(\hat{z}) \quad , \quad \hat{w}_1 \to 0 \quad , \quad \hat{\vartheta}_1 \to 0 \quad ;$$

$$\hat{z} \to +\infty: \ \hat{v}_0 \to V_{\infty,e}(0) \quad , \quad \hat{\vartheta}_1 \to 0 \Rightarrow \hat{w}_1 \to 0 \quad ; \tag{13.205c}$$

$$t = 0: \ \hat{v}_0 = V_{\infty,\mathrm{cl}}(\hat{z}) \quad , \quad \hat{\vartheta}_1 = 0 \quad .$$

13.3.2 The Interaction Model

The regional boundary layer model of the preceding section does not interact directly with the outer flow in the sense that the latter must first be known, and then only afterwards, can the corresponding regional boundary layer problem be solved. We seek in the forthcoming to obtain a model for the direct interaction

[15] Given the parabolic nature of problem (13.205), it is obvious that as long as $\hat{\vartheta}_1 = 0$ on $\hat{z} = 0$, then the trivial solution to (13.205) is:

$$\hat{v}_0 \equiv V_{\infty,\mathrm{cl}}(\hat{z}) \quad , \quad \hat{w}_1 \equiv 0 \quad \text{and} \quad \hat{\vartheta}_1 \equiv 0 \quad .$$

For this it is necessary that

$$\frac{d^2 V_{\infty,\mathrm{cl}}}{d\hat{z}^2} - \frac{1}{\mathrm{Ro}}\left(k \times V_{\infty,\mathrm{cl}}\right) = 0 \quad ,$$

and

$$V_{\infty,\mathrm{cl}}(0) = 0 \quad , \quad \lim_{\hat{z} \to \infty} V_{\infty,\mathrm{cl}} = V_{\infty,e}(0) \quad .$$

phenomenon. We will limit our attention to the two-dimensional steady case and take for basic equations the Boussinesq equations of Sect. 8.1 in which $\mathrm{Ro} \equiv \infty$ ($L \ll U_0/l_0$ with $l_0 = 2\Omega_0 \cos\varphi_0$) and $\varepsilon \equiv 1$:

$$\begin{aligned}
&u_0\frac{\partial u_0}{\partial x} + w_0\frac{\partial u_0}{\partial z} + \frac{1}{\gamma}\frac{\partial \pi_2}{\partial x} = \frac{1}{\mathrm{Re}}\left(\frac{\partial^2 u_0}{\partial x^2} + \frac{\partial^2 u_0}{\partial z^2}\right) \quad ; \\
&u_0\frac{\partial w_0}{\partial x} + w_0\frac{\partial w_0}{\partial z} + \frac{1}{\gamma}\frac{\partial \pi_2}{\partial z} - \frac{\hat{B}}{\gamma}\vartheta_1 = \frac{1}{\mathrm{Re}}\left(\frac{\partial^2 w_0}{\partial x^2} + \frac{\partial^2 w_0}{\partial z^2}\right) \quad ; \\
&\frac{\partial u_0}{\partial x} + \frac{\partial w_0}{\partial z} = 0 \quad ; \\
&u_0\frac{\partial \vartheta_1}{\partial x} + w_0\frac{\partial \theta_1}{\partial z} + \hat{B}\mu_0 w_0 = \frac{1}{\mathrm{Pr}}\frac{1}{\mathrm{Re}}\left(\frac{\partial^2 \vartheta_1}{\partial x^2} + \frac{\partial^2 \vartheta_1}{\partial z^2}\right) \quad ,
\end{aligned} \tag{13.206}$$

where:

$$\mu_0 \equiv \frac{\gamma - 1}{\gamma} + \left.\frac{dT_\infty}{dz_\infty}\right|_{z_\infty = 0} \quad .$$

The following boundary conditions are imposed on (13.206):

$$\begin{aligned}
&z = 0: \quad u_0 = w_0 = 0 \quad , \quad \vartheta_1 = \Lambda_0 \Xi(x) \quad , \quad x \in [0, 1] \quad , \\
&x \to -\infty: \quad u_0 \to U_\infty(z, z/\Delta) \quad , \quad w_0 \to 0 \quad , \quad \vartheta_1 \to 0 \quad , \quad \pi_2 \to 0 \quad ,
\end{aligned} \tag{13.207}$$

where $\Delta \equiv 1/\sqrt{\mathrm{Re}}$. Once again, in the vicinity of $z = 0$ in the boundary layer, U_∞ is replaced by $U_{\infty,\mathrm{cl}}(\hat{z})$ with $\hat{z} = z/\Delta$. Beyond the boundary layer, $U_{\infty,\mathrm{e}}(z)$ is taken into account. Naturally: $U_{\infty,\mathrm{cl}}(\infty) = U_{\infty,\mathrm{e}}(0)$.

Let it be recalled that $\Lambda_0 = \tau_0/\mathrm{M}_\infty$. However, it turns out that to obtain a direct coupling between the boundary layer and the outer inviscid flow, the following must be employed:

$$\Lambda_0 \gg 1 \Rightarrow \Lambda_0\Delta \equiv \frac{\tau_0}{\mathrm{M}_\infty}\Delta = \sigma_0 = \mathrm{O}(1) \quad , \tag{13.208}$$

when simultaneously $\Delta \to 0$, $\tau_0 \to 0$ and $\mathrm{M}_\infty \to 0$ with σ_0 as a similarity parameter.

We will now consider a first limiting process:

$$\Delta \to 0 \quad , \quad \text{with} \quad x \quad \text{and} \quad z \quad \text{fixed} \quad , \tag{13.209a}$$

to which is added the following (outer) asymptotic representation:

$$\begin{aligned}
u_0 &= U_{\infty,\mathrm{e}}(z) + \Delta\overline{u}' + \ldots \quad ; \\
w_0 &= \Delta\overline{w}' + \ldots \quad ; \\
\pi_2 &= \Delta\overline{\pi}' + \ldots \quad ; \\
\vartheta_1 &= \Delta\overline{\vartheta}' + \ldots \quad .
\end{aligned} \tag{13.209b}$$

For $\overline{u}'$, $\overline{w}'$, $\overline{\pi}'$ and $\overline{\vartheta}'$, which are functions of x and z, the following equations are then obtained:

$$U_{\infty,e}\frac{\partial \overline{u}'}{\partial x}+\frac{dU_{\infty,e}}{dz}\overline{w}'+\frac{1}{\gamma}\frac{\partial \overline{\pi}'}{\partial x}=0 \quad ;$$
$$U_{\infty,e}\frac{\partial \overline{w}'}{\partial x}+\frac{1}{\gamma}\frac{\partial \overline{\pi}'}{\partial z}-\frac{\hat{B}}{\gamma}\overline{\vartheta}'=0 \quad ; \qquad (13.210)$$
$$\frac{\partial \overline{u}'}{\partial x}+\frac{\partial \overline{w}'}{\partial z}=0 \quad ;$$
$$U_{\infty,e}\frac{\partial \overline{\vartheta}'}{\partial x}+\hat{B}\mu_0\overline{w}'=0 \quad .$$

From (13.210), the following single equation is derived for $\overline{w}'$:

$$\frac{\partial^2\overline{w}'}{\partial x^2}+\frac{\partial^2\overline{w}'}{\partial z^2}+\left[\frac{\hat{B}^2}{\gamma}\frac{\mu_0}{U^2_{\infty,e}(z)}-\frac{1}{U_{\infty,e}(z)}\frac{d^2U_{\infty,e}}{dz^2}\right]\overline{w}'=0 \quad . \qquad (13.211)$$

On (13.211), the following conditions are imposed:

$$\overline{w}'\rightarrow 0 \quad , \quad \text{when} \quad x\rightarrow -\infty \quad \text{and} \quad \lim_{x\rightarrow +\infty}|\overline{w}'|<\infty \quad . \qquad (13.212)$$

Moreover, when $z\rightarrow +\infty$, a radiation condition must be written. A condition in $z=0$ also remains to be written.

We must therefore consider a second limiting process:

$$\Delta\rightarrow 0 \quad , \quad \text{with} \quad x \quad \text{and} \quad \hat{z}=z/\Delta \quad \text{fixed} \quad . \qquad (13.213a)$$

The following (inner) asymptotic representation is associated with (13.213a):

$$u_0=\hat{u}'+\ldots \quad ; \quad w_0=\Delta\hat{w}'+\ldots \quad ;$$
$$\pi_2=\hat{\pi}'+\ldots \quad ; \quad \vartheta_1=\frac{1}{\Delta}\hat{\vartheta}'+\ldots \quad . \qquad (13.213b)$$

For $\hat{u}'$, $\hat{w}'$, $\hat{\pi}'$ and $\hat{\vartheta}'$, which are functions of x and of $\hat{z}$, the following equations are obtained:

$$\hat{u}'\frac{\partial \hat{u}'}{\partial x}+\hat{w}'\frac{\partial \hat{u}'}{\partial \hat{z}}+\frac{1}{\gamma}\frac{\partial \hat{\pi}'}{\partial x}=\frac{\partial^2 u'}{\partial \hat{z}^2} \quad ; \quad \frac{\partial \hat{\pi}'}{\partial \hat{z}}=\hat{B}\hat{\vartheta}' \quad ;$$
$$\frac{\partial \hat{u}'}{\partial x}+\frac{\partial \hat{w}'}{\partial \hat{z}}=0 \quad ; \quad \hat{u}'\frac{\partial \hat{\vartheta}'}{\partial x}+\hat{w}'\frac{\partial \hat{\vartheta}'}{\partial \hat{z}}=\frac{1}{\mathrm{Pr}}\frac{\partial^2\hat{\vartheta}'}{\partial \hat{z}^2} \quad . \qquad (13.214)$$

Equations (13.214) form, in fact, a system for $\hat{u}'$, $\hat{w}'$ and $\hat{\vartheta}'$ since:

$$\hat{\pi}'=\hat{B}\int_{\infty}^{\hat{z}}\hat{\vartheta}'\,d\hat{z} \quad , \qquad (13.215)$$

as long as the matching of π_2 according to (13.209b, 213b) has been taken into account.

The following boundary conditions must be imposed on the boundary layer system (13.214) with (13.215):

$$\hat{z}=0: \ \hat{u}'=0 \ , \ \hat{w}'=0 \ , \ \hat{\vartheta}'=\sigma_0 \Xi(x) \ , \ x\in[0,1] \ ; \tag{13.216a}$$

$$x\to -\infty: \ \hat{u}'\to U_{\infty,\mathrm{cl}}(\hat{z}) \ , \ \hat{w}'\to 0 \ , \ \hat{\vartheta}'\to 0 \ ; \tag{13.216b}$$

$$\hat{z}\to +\infty: \ \hat{u}'\to U_{\infty,\mathrm{e}}(0) \ , \ \hat{\vartheta}'\to 0 \ . \tag{13.216c}$$

We still have to write a matching condition on the terms w_0:

$$\lim_{\hat{z}\to +\infty} \hat{w}' = \overline{w}'(x,0) \ . \tag{13.217}$$

It is precisely this condition which must be added to the outer equation (13.211) for $\overline{w}'$. Condition (13.217) takes into account the overall effect of the boundary layer in the outer flow which is governed by (13.211).

The continuity equation from (13.214) makes it possible to write the following relation:

$$\hat{w}' = -\int_0^{\hat{z}} \frac{\partial \hat{u}'}{\partial x} d\hat{z} \equiv \frac{\partial}{\partial x}\int_0^{\infty}\Big(U_{\infty,\mathrm{e}}(0)-\hat{u}'\Big)d\hat{z}$$
$$+\frac{\partial}{\partial x}\int_{\hat{z}}^{\infty}\Big(U_{\infty,\mathrm{e}}(0)-\hat{u}'\Big)d\hat{z} \ ,$$

where the last term is an infinitely small [o(1)] when $\hat{z}\to +\infty$. We can thus replace (13.217) with the following:

$$\overline{w}'(x,0) = U_{\infty,\mathrm{e}}(0)\frac{\partial}{\partial x}\left\{\int_0^{\infty}\left[1-\frac{\hat{u}'(x,\hat{z})}{U_{\infty,\mathrm{e}}(0)}\right]d\hat{z}\right\} \ . \tag{13.218}$$

Hence, it can be said in conclusion that (13.211) for $\overline{w}'(x,z)$ must be resolved subject to conditions (13.212, 218) and the radiation condition when $z\to +\infty$. When the coefficient:

$$\left|\frac{\hat{B}^2}{\gamma}\frac{\mu_0}{U_{\infty,\mathrm{e}}^2(z)} - \frac{1}{U_{\infty,\mathrm{e}}(z)}\frac{d^2U_{\infty,\mathrm{e}}}{dz^2}\right| > 0$$

is positive, this outer problem will cause a system of steady waves to appear downstream of the thermal spot where $x>1$. These waves are of the same type as those found downstream of a mountain (lee waves). Unfortunately, no calculation has as yet been carried out with the interaction model described in this section.

13.4 Complementary Remarks

13.4.1 A Model for the Local Winds of Slopes and Valleys

It has been observed that over the warm versant of a hill or mountain, an ascending local wind appears, whereas on the cold versant, this wind is descending.

Moreover, during the day, a zone of vertical ascending currents can be found near the crest of the slope. On the contrary, at the bottom of the valley, the currents are descending. In Chap. 6 of our Course on mesometeorology (Zeytounian, 1968), various indications are given on this topic. We wish to define here a simple asymptotic model which would allow the fundamental characteristics of the phenomenon to be taken into account. As a characteristic vertical length scale, we will take $h_0 = R\Delta T_0/g$. In this case, $\varepsilon = h_0/L \ll 1$ and we introduce the constant characteristic velocity $U_0 = (gh_0)^{1/2}\tau_0$ where $\tau_0 = \Delta T_0/T_\infty(0)$. In this situation:

$$\mathrm{Re}_\perp = \varepsilon^2\,\mathrm{Re} \equiv \frac{h_0^2}{L}\frac{\tau_0}{\nu_0}(gh_0)^{1/2} \quad \text{and thus}$$

$$\mathrm{Gr}_\perp^{-1/2} \equiv \frac{1}{\mathrm{Re}_\perp} = \frac{\nu_0^2}{\varepsilon^2 g h_0^3 \tau_0^2} \quad . \tag{13.219}$$

We can then carry out, on the one hand, the Boussinesq approximation, and, on the other hand, the quasi-static approximation. Thus, our basic equations are the quasi-static Boussinesq equations:

$$\begin{aligned}
&\mathrm{S}\frac{\partial \boldsymbol{v}}{\partial t} + (\boldsymbol{v}\cdot\boldsymbol{D})\boldsymbol{v} + w\frac{\partial \boldsymbol{v}}{\partial z} + \frac{1}{\mathrm{Ro}}(\boldsymbol{k}\times\boldsymbol{v}) + \frac{1}{\gamma}\boldsymbol{D}\pi = \mathrm{Gr}_\perp^{-1/2}\frac{\partial^2 \boldsymbol{v}}{\partial z^2} \quad ;\\
&\frac{\partial \pi}{\partial z} = \hat{B}\vartheta \quad ;\\
&\boldsymbol{D}\cdot\boldsymbol{v} + \frac{\partial w}{\partial z} = 0 \quad ;\\
&\mathrm{S}\frac{\partial \vartheta}{\partial t} + \boldsymbol{v}\cdot\boldsymbol{D}\vartheta + w\frac{\partial \vartheta}{\partial z} + \hat{B}\mu_0 w = \frac{1}{\mathrm{Pr}}\mathrm{Gr}_\perp^{-1/2}\frac{\partial^2 \vartheta}{\partial z^2} \quad .
\end{aligned} \tag{13.220}$$

The relief which simulates the slope or the valley is given in the following non-dimensional form:

$$z = \alpha_0\chi(x,y) \quad , \quad \text{with} \quad \alpha_0 = \frac{\chi_0}{h_0} \quad , \tag{13.221}$$

where $\chi_0 = \max|\chi|$ when $x, y, \subset \mathcal{D}_2$ ($\mathcal{D}_2$ is the plane form of the relief).

Instead of x, y and z, the following new independent variables are introduced:

$$x \equiv \xi \quad , \quad y = \eta \quad , \quad \zeta = z - \alpha_0\chi(\xi,\eta) \quad . \tag{13.222}$$

Naturally:

$$\frac{\partial}{\partial z} \equiv \frac{\partial}{\partial \zeta} \quad , \quad \frac{\partial}{\partial x} = \frac{\partial}{\partial \xi} - \alpha_0\frac{\partial \chi}{\partial \xi}\frac{\partial}{\partial \zeta} \quad , \quad \frac{\partial}{\partial y} = \frac{\partial}{\partial \eta} - \alpha_0\frac{\partial \chi}{\partial \eta}\frac{\partial}{\partial \zeta} \quad ,$$

and:

$$\boldsymbol{D}\cdot\boldsymbol{v} + \frac{\partial w}{\partial z} \equiv \frac{\partial u}{\partial x} + \frac{\partial v}{\partial y} + \frac{\partial w}{\partial z} = \frac{\partial u}{\partial \xi} + \frac{\partial v}{\partial \eta} + \frac{\partial W}{\partial \zeta} \quad ,$$

where:

$$W = w - \alpha_0\left(\frac{\partial \chi}{\partial \xi}u + \frac{\partial \chi}{\partial \eta}v\right) \quad .$$

We thus obtain with the new variables ξ, η, ζ, the following equations as replacements of (13.220) for the functions $\boldsymbol{v}$, W, π and ϑ which are dependent on t, ξ, η and ζ:

$$\begin{aligned}
&\mathrm{S}\frac{\partial \boldsymbol{v}}{\partial t} + (\boldsymbol{v}\cdot\boldsymbol{D})\boldsymbol{v} + W\frac{\partial \boldsymbol{v}}{\partial \zeta} + \frac{1}{\mathrm{Ro}}(\boldsymbol{k}\times\boldsymbol{v}) \\
&\qquad + \frac{1}{\gamma}\boldsymbol{D}\pi - \frac{\alpha_0}{\gamma}\hat{B}\boldsymbol{D}\chi\vartheta = \mathrm{Gr}_{\perp}^{-1/2}\frac{\partial^2 \boldsymbol{v}}{\partial \zeta^2}; \\
&\frac{\partial \pi}{\partial \zeta} = \hat{B}\vartheta \quad ; \\
&\boldsymbol{D}\cdot\boldsymbol{v} + \frac{\partial W}{\partial \zeta} = 0 \quad ; \qquad\qquad (13.223) \\
&\mathrm{S}\frac{\partial \vartheta}{\partial t} + \boldsymbol{v}\cdot\boldsymbol{D}\vartheta + W\frac{\partial \vartheta}{\partial \zeta} + \hat{B}\mu_0(W + \alpha_0\boldsymbol{v}\cdot\boldsymbol{D}\chi) \\
&\qquad = \frac{1}{\mathrm{Pr}}\mathrm{Gr}_{\perp}^{-1/2}\frac{\partial^2 \vartheta}{\partial \zeta^2} \quad ,
\end{aligned}$$

where

$$\boldsymbol{D} \equiv \frac{\partial}{\partial \xi}\boldsymbol{i} + \frac{\partial}{\partial \eta}\boldsymbol{j} \quad \text{and} \quad W = w - \alpha_0\boldsymbol{v}\cdot\boldsymbol{D}\chi \quad .$$

To the above equations, we can assign the following ground conditions:

$$\zeta = 0: \quad \boldsymbol{v} = 0 \quad , \quad W = 0 \quad , \quad \vartheta = \hat{B}\Xi(t, \xi, \eta) \quad , \qquad (13.224)$$

when $t > 0$ and ξ, $\eta \subset \mathcal{D}_2$. Of course:

$$\boldsymbol{v} \to 0 \; , \; W \to 0 \; , \; \pi \to 0 \text{ and } \vartheta \to 0 \quad , \quad \text{when} \quad |\xi^2 + \eta^2| \to \infty \quad . \qquad (13.225)$$

The fact that $\mathrm{Gr}_{\perp}^{-1/2} \ll 1$ can now be put to good use. When $\mathrm{Gr}_{\perp}^{-1/2} \to 0$ with t, ξ, η and ζ fixed, it is no longer possible to use conditions (13.224) for the limiting equations obtained from (13.223). However, the behavior conditions allow us to postulate the outer trivial solution:

$$\overline{\boldsymbol{v}}_0 \equiv 0 \quad , \quad \overline{W}_0 \equiv 0 \quad , \quad \overline{\pi}_0 \equiv 0 \quad \text{and} \quad \overline{\vartheta}_0 \equiv 0 \quad . \qquad (13.226)$$

Let us now introduce:

$$\hat{\zeta} = \frac{\zeta}{\mathrm{Gr}_{\perp}^{-1/4}} \quad , \quad \text{and} \quad \hat{W} = \frac{W}{\mathrm{Gr}_{\perp}^{-1/4}} \quad , \qquad (13.227)$$

and carry out the limiting process:

$$\mathrm{Gr}_{\perp}^{-1/2} \to 0 \quad , \quad \text{with} \quad t, \xi, \eta \quad \text{and} \quad \hat{\zeta} \quad \text{fixed} \quad , \qquad (13.228a)$$

with which is associated the following inner asymptotic representation:

$$\begin{aligned}
\boldsymbol{v} &= \hat{\boldsymbol{v}}_0 + \ldots \quad , \quad \hat{W} = \hat{W}_0 + \ldots \quad , \\
\pi &= \hat{\pi}_0 + \ldots \quad , \quad \vartheta = \hat{\vartheta}_0 + \ldots \quad .
\end{aligned} \qquad (13.228b)$$

For $\hat{\boldsymbol{v}}_0$, $\hat{W}_0$, $\hat{\pi}_0$ and $\hat{\vartheta}_0$, the following boundary layer system is then obtained:

$$
\begin{aligned}
&\mathrm{S}\frac{\partial \hat{\boldsymbol{v}}_0}{\partial t} + (\hat{\boldsymbol{v}}_0 \cdot \boldsymbol{D})\hat{\boldsymbol{v}}_0 + \hat{W}_0 \frac{\partial \hat{\boldsymbol{v}}_0}{\partial \hat{\zeta}} + \frac{1}{\mathrm{Ro}}(\boldsymbol{k} \times \hat{\boldsymbol{v}}_0) + \frac{1}{\gamma}\boldsymbol{D}\hat{\pi}_0 \\
&\qquad - \frac{\alpha_0}{\gamma}\hat{B}\boldsymbol{D}\chi\hat{\vartheta}_0 = \frac{\partial^2 \hat{\boldsymbol{v}}_0}{\partial \hat{\zeta}^2} \quad ; \\
&\frac{\partial \hat{\pi}_0}{\partial \hat{\zeta}} = 0 \quad ; \qquad\qquad (13.229) \\
&\boldsymbol{D} \cdot \hat{\boldsymbol{v}}_0 + \frac{\partial \hat{w}_0}{\partial \hat{\zeta}} = 0 \quad ; \\
&\mathrm{S}\frac{\partial \hat{\vartheta}_0}{\partial t} + \hat{\boldsymbol{v}}_0 \cdot \boldsymbol{D}\hat{\vartheta}_0 + \hat{W}_0 \frac{\partial \hat{\vartheta}_0}{\partial \hat{\zeta}} + \hat{B}\mu_0\alpha_0\hat{\boldsymbol{v}}_0 \cdot \boldsymbol{D}\chi = \frac{1}{\mathrm{Pr}}\frac{\partial^2 \hat{\vartheta}_0}{\partial \hat{\zeta}^2} \quad .
\end{aligned}
$$

First of all:

$$\hat{\pi}_0 \equiv \hat{\pi}_0(t, \xi, \eta) = \overline{\pi}_0(t, \xi, \eta, \zeta = 0) \equiv 0 \quad .$$

Then from the condition [which results from (13.219)]:

$$\mathrm{Gr}_{\perp}^{-1/2} \ll 1 \Rightarrow L^2 \ll \frac{g h_0^5 \tau_0^2}{\nu_0} \Rightarrow \mathrm{Ro} \ll 1 \quad ,$$

it is seen that the Coriolis term can be neglected in the first equation of (13.229). Thus, for the local wind phenomenon of slopes and valleys, the following model equations are finally derived:

$$
\begin{aligned}
&\mathrm{S}\frac{\partial \hat{\boldsymbol{v}}_0}{\partial t} + (\hat{\boldsymbol{v}}_0 \cdot \boldsymbol{D})\hat{\boldsymbol{v}}_0 + \hat{W}_0 \frac{\partial \hat{\boldsymbol{v}}_0}{\partial \hat{\zeta}} - \frac{\alpha_0}{\gamma}\hat{B}\boldsymbol{D}\chi\hat{\vartheta}_0 = \frac{\partial^2 \hat{\boldsymbol{v}}_0}{\partial \hat{\zeta}^2} \quad ; \\
&\boldsymbol{D} \cdot \hat{\boldsymbol{v}}_0 + \frac{\partial \hat{W}_0}{\partial \hat{\zeta}} = 0 \quad ; \qquad\qquad (13.230) \\
&\mathrm{S}\frac{\partial \hat{\vartheta}_0}{\partial t} + \hat{\boldsymbol{v}}_0 \cdot \boldsymbol{D}\hat{\vartheta}_0 + \hat{W}_0 \frac{\partial \hat{\vartheta}_0}{\partial \hat{\zeta}} + \hat{B}\mu_0\alpha_0\hat{\boldsymbol{v}}_0 \cdot \boldsymbol{D}\chi = \frac{1}{\mathrm{Pr}}\frac{\partial^2 \hat{\vartheta}_0}{\partial \hat{\zeta}^2} \quad ,
\end{aligned}
$$

on which the following conditions are imposed:

$$
\begin{aligned}
&\hat{\zeta} = 0: \ \hat{\boldsymbol{v}}_0 = 0 \quad , \quad \hat{W}_0 = 0 \quad , \quad \hat{\vartheta}_0 = \hat{B}\Xi(t, \xi, \eta) \quad , \\
&\text{when} \quad t > 0 \quad \text{and} \quad \xi, \eta \subset \mathcal{D}_2 \quad ; \\
&t = 0: \ \hat{\boldsymbol{v}}_0 = 0 \quad , \quad \hat{\vartheta}_0 = 0 \quad ; \qquad\qquad (13.231) \\
&|\xi^2 + \eta^2| \to \infty: \ \hat{\boldsymbol{v}}_0 \to 0 \quad , \quad \hat{W}_0 \to 0 \quad \text{and} \quad \hat{\vartheta}_0 \to 0 \quad ; \\
&\hat{\zeta} \to +\infty: \ \hat{\boldsymbol{v}}_0 \to 0 \quad , \quad \hat{\vartheta}_0 \to 0 \quad .
\end{aligned}
$$

13.4.2 Double Layer Periodic Slope (or Valley) Winds

Let us consider the problem (13.230, 231) as a two-dimensional unsteady case $(t, \xi, \hat{\zeta})$. We will assume that $\mathrm{S} \gg 1$ such that:

$$\frac{\alpha_0}{\mathrm{S}} = \mathrm{O}(1) \quad , \quad \mathrm{S}^{1/2}\hat{\zeta} \equiv Z = \mathrm{O}(1) \quad ,$$
$$\hat{W}_0 = \mathrm{S}^{1/2}\omega_0 \quad . \tag{12.232}$$

If the stream function $\hat{\psi}_0(t, \xi, Z)$ is introduced so that:

$$\frac{\partial \hat{u}_0}{\partial \xi} + \frac{\partial \omega_0}{\partial Z} = 0 \Rightarrow \hat{u}_0 = \frac{\partial \hat{\psi}_0}{\partial Z} \quad \text{and} \quad \omega_0 = -\frac{\partial \hat{\psi}_0}{\partial \xi} \quad ,$$

then the following system for ψ_0 and $\hat{\vartheta}_0$ replaces (13.230):

$$\frac{\partial^2 \hat{\psi}_0}{\partial t \partial Z} + \frac{1}{\mathrm{S}}\left(\frac{\partial \hat{\psi}_0}{\partial Z}\frac{\partial^2 \hat{\psi}_0}{\partial \xi \partial Z} - \frac{\partial \hat{\psi}_0}{\partial \xi}\frac{\partial^2 \hat{\psi}_0}{\partial Z^2}\right) = \Gamma(\xi)\hat{\vartheta}_0 + \frac{\partial^3 \hat{\psi}_0}{\partial Z^3} \quad ;$$
$$\frac{\partial \hat{\vartheta}_0}{\partial t} + \frac{1}{\mathrm{S}}\left(\frac{\partial \hat{\psi}_0}{\partial Z}\frac{\partial \hat{\vartheta}_0}{\partial \xi} - \frac{\partial \hat{\psi}_0}{\partial \xi}\frac{\partial \hat{\vartheta}_0}{\partial Z}\right) + \gamma\mu_0\Gamma(\xi)\frac{\partial \hat{\psi}_0}{\partial Z} = \frac{1}{\mathrm{Pr}}\frac{\partial^2 \hat{\vartheta}_0}{\partial Z^2} \quad , \tag{13.233}$$

where $\Gamma(\xi) = (\alpha_0/\mathrm{S})(\hat{B}/\gamma)(\partial\chi/\partial\xi)$. In the forthcoming, we will assume $\mathrm{Pr} \equiv 1$. To (13.233), where $1/\mathrm{S} \ll 1$ is a small parameter (high frequency periodic oscillations are the focal point here), the following boundary conditions[16] are added:

$$Z = 0: \ \frac{\partial \hat{\psi}_0}{\partial Z} = \frac{\partial \hat{\psi}_0}{\partial \xi} = 0 \quad , \quad \hat{\vartheta}_0 = \cos t + \frac{1}{\mathrm{S}}\mathcal{A}(\xi),$$
$$Z \to \infty: \ \frac{\partial \hat{\psi}_0}{\partial Z} = \frac{1}{\mathrm{S}}\mathcal{B}(\xi) \quad , \quad \hat{\vartheta}_0 \to 0 \quad . \tag{12.234}$$

Hence, there are no initial conditions (we have periodicity).

From the physics point of view, it is then a question of a slope wind engendered by a high frequency periodic temperature oscillation ($t_0^{-1} \gg U_0/L$) on a curved slope (or valley). When $\mathrm{S} \gg 1$, the velocity field is kept in check by the equilibrium between the vorticity diffusion via the viscosity effect and its creation via the Archimedes force (the relative effect of the convection being of the order of $1/\mathrm{S}$). The terms $\mathcal{A}(\xi)$ and $\mathcal{B}(\xi)$ which come into play in the boundary conditions (13.234) can be interpreted as a secondary effect of the convection due to the quasi-linear terms which appear in (13.233). As a matter of fact, $1/\mathrm{S} \ll 1$ is known to be a small singular perturbation parameter from the works by Riley (1965) and Stuart (1966).

For (13.233), the "real" physical conditions are, of course:

$$\frac{\partial \hat{\psi}_0}{\partial Z} = \frac{\partial \hat{\psi}_0}{\partial \xi} = 0 \quad ; \quad \hat{\vartheta}_0 = \cos t \quad , \quad \text{on} \quad Z = 0 \quad ;$$
$$\frac{\partial \hat{\psi}_0}{\partial Z} \to 0 \quad , \quad \hat{\vartheta}_0 \to 0 \quad , \quad \text{when} \quad Z \to +\infty \quad . \tag{13.235}$$

[16] The conditions imposed in our article (Zeytounian, 1968) are again used here. We will see further on, in the light of the results obtained by J.M. Noe in his doctoral thesis entitled "Sur une théorie asymptotique de la convection naturelle" (defended in March, 1981 at the University of Lille I), how these conditions can be correctly interpreted from the asymptotic point of view when $\mathrm{S} \to \infty$.

The fact that (13.235) *cannot be imposed* on the asymptotic solution of Zeytounian (1968):

$$\hat{\psi}_0 = \psi_0 + \frac{1}{S}\psi_1 + \dots \quad , \qquad \hat{\vartheta}_0 = \vartheta_0 + \frac{1}{S}\vartheta_1 + \dots \quad , \tag{13.236}$$

for (13.233) clearly shows that these expansions (13.236) are not uniformly valid throughout the convective layer considered. It turns out that (13.236) must be considered only as a proximal solution which is valid in the main Stokes layer which develops in the vicinity of the curved slope $Z = 0$ [expansions (13.236) where $S \to \infty$ must therefore be considered for any fixed ξ *and* Z)].

It should be noted, however, that the aim of our 1968 study (carried out in 1961 at the Hydrometeorology Center in Moscow) was to show that by taking into account the Archimedes force, a steady temperature increase was introduced on the surface of the slope $Z = 0$. This increase is characterized by $\mathcal{A}(\xi) = -(1/4)(d\Gamma/d\xi)$ and is due to the quasi-linear terms which generalized the classical Schlichting study (1932).

We are about to demonstrate that the introduction in the vicinity of $Z = \infty$ of an upper distal layer within the convective layer being considered makes it possible to recover the singular behavior of $\hat{\psi}_0$ and $\hat{\vartheta}_0$ according to (13.236) (when $Z \to +\infty$) at the upper boundary of the main Stokes layer when $\tilde{Z} = \delta(1/S)Z$ tends to zero. [$\delta(1/S) \to 0$ with $S \to \infty$ being a gauge which will be determined later.] This means that the present analysis permits us to confirm that the Riley-Stuart type[17] phenomenon of "double oscillating boundary layer" is also present in periodic natural convection motions. It is also pointed out that the seeking of a proximal sublayer in the vicinity of $Z = 0$ proved in vain and thus only a distal outer layer could be introduced.

First of all, we are going to assume that $\mu_0 = 0$ (neutral reference stratification). We will then specify in what ways the results must be changed when a stable reference stratification is to be taken into account.

If we consider solution (13.236), we will notice that to order zero, the solution ψ_0, ϑ_0 is uniformly valid over the entire convective layer and so it is not necessary to this order to introduce an outer distal layer. On the other hand, to the first order, if we impose on the solution ψ_1, ϑ_1 to satisfy the conditions:

$$\frac{\partial \psi_1}{\partial Z} = \frac{\partial \psi_1}{\partial \xi} = 0 \quad , \quad \vartheta_1 = 0 \quad , \quad \text{on} \quad Z = 0 \quad ,$$

which result from (13.235), then when $Z \to +\infty$, we are led to the following behaviors for $\hat{\vartheta}_0$ and $\hat{\psi}_0$:

[17] Rosenblat (1959) appears to be the first to have brought to light this double layer phenomenon within a boundary layer. This was accomplished during the resolution of the oscillating disk problem. By using a double scale technique, Benney (1964) analyzed this double layer phenomenon for the classical problem of a disk oscillating in its own plane.

$$\hat{\vartheta}_0 \sim \frac{1}{\mathrm{S}}\frac{1}{16}\frac{d\Gamma}{d\xi}(C_1 Z + 4) + \mathrm{O}\left(\frac{1}{\mathrm{S}^2}\right) \quad ;$$

$$\begin{aligned} \hat{\psi}_0 \sim\ & \frac{\Gamma}{2}\left[\frac{1+\mathrm{i}}{2\sqrt{2}}\mathrm{e}^{\mathrm{i}t} + \frac{1-\mathrm{i}}{2\sqrt{2}}\mathrm{e}^{-\mathrm{i}t}\right] \\ & + \frac{1}{\mathrm{S}}\Gamma\frac{d\Gamma}{d\xi}\left\{\frac{23\sqrt{2}-34}{128}\left[\frac{1+\mathrm{i}}{2}\mathrm{e}^{2\mathrm{i}t} + \frac{1-\mathrm{i}}{2}\mathrm{e}^{-2\mathrm{i}t}\right]\right. \\ & \left. - \frac{C_1}{24}Z^4 - \frac{1}{24}Z^3 + \frac{C_3}{2}Z^2 - \frac{7}{8}Z + \frac{41}{32\sqrt{2}}\right\} + \mathrm{O}\left(\frac{1}{\mathrm{S}^2}\right) \quad , \end{aligned} \tag{13.237}$$

with C_1 and C_2 being two arbitrary integration constants. However, on the one hand, it is easy to show by analogy with an argument given by Riley (see pages 168, 169 of his 1965 article) that the constant C_1 must necessarily take the value zero. On the other hand, the constant C_3 only comes into play as a parameter in the calculation of higher order terms. We can therefore take $C_3 \equiv 0$. Consequently, in (13.237) we set: $C_1 \equiv 0$ and $C_3 \equiv 0$.

Let us now define $\delta(1/\mathrm{S}) = 1/\mathrm{S}^{\gamma}$ with $\gamma > 0$ such that:

$$\tilde{Z} = \frac{1}{\mathrm{S}^{\gamma}} Z \quad . \tag{13.238}$$

For the outer distal layer, the functions $\tilde{\vartheta}_0(t, \xi, \tilde{Z}; 1/\mathrm{S})$ and $\tilde{\psi}_0(t, \xi, \tilde{Z}; 1/\mathrm{S})$ are introduced such that:

$$\hat{\vartheta}_0 = \frac{1}{\mathrm{S}^{\alpha}}\tilde{\vartheta}_0 \quad \text{and} \quad \hat{\psi}_0 - \frac{\Gamma}{2}\left[\frac{1+\mathrm{i}}{2\sqrt{2}}\mathrm{e}^{\mathrm{i}t} + \frac{1-\mathrm{i}}{2\sqrt{2}}\mathrm{e}^{-\mathrm{i}t}\right] = \frac{1}{\mathrm{S}^{\beta}}\tilde{\psi}_0 \quad .$$

The real members α, β and γ must now be determined. Firstly, since $C_1 = 0$, from (13.237), we have $\alpha = 1$. In addition, rewriting (13.233) with respect to the variable $\tilde{Z}$ and for the functions $\tilde{\vartheta}_0$ and $\tilde{\psi}_0$, we find that the following must hold:

$$1 + \beta + \gamma = 2\gamma \quad ,$$

if we want to retain the quasi-linear and viscous terms. Finally, from (13.237), when $\hat{\psi}_0$ is rewritten in the distal variable $\tilde{Z}$, and if it is taken into consideration that $C_1 = C_3 = 0$, we have (since $\gamma > 0$) according to Van Dyke's simplified matching rule (1964):

$$\beta = 1 - 3\gamma \quad .$$

Thus, the following must be taken:

$$\gamma = \tfrac{1}{2} \quad \text{and} \quad \beta = -\tfrac{1}{2} \quad .$$

The functions $\tilde{\vartheta}_0$ and $\tilde{\psi}_0$ must then satisfy the following problem:

$$\begin{aligned} & \frac{\partial\tilde{\vartheta}_0}{\partial t} + \frac{1}{\mathrm{S}}\left(\frac{\partial\tilde{\psi}_0}{\partial\tilde{Z}}\frac{\partial\tilde{\vartheta}_0}{\partial\xi} - \frac{\partial\tilde{\psi}_0}{\partial\xi}\frac{\partial\tilde{\vartheta}_0}{\partial\tilde{Z}}\right) = \frac{1}{\mathrm{S}}\frac{\partial^2\tilde{\vartheta}_0}{\partial\tilde{Z}^2} \\ & + \left(\frac{1}{\mathrm{S}}\right)^{3/2}\frac{1}{2}\frac{d\Gamma}{d\xi}\left[\frac{1+\mathrm{i}}{2\sqrt{2}}\mathrm{e}^{\mathrm{i}t} + \frac{1-\mathrm{i}}{2\sqrt{2}}\mathrm{e}^{-\mathrm{i}t}\right]\frac{\partial\tilde{\vartheta}_0}{\partial\tilde{Z}} \quad ; \end{aligned} \tag{13.239a}$$

$$\frac{\partial^2 \tilde{\psi}_0}{\partial t \partial \tilde{Z}} + \frac{1}{S}\left(\frac{\partial \tilde{\psi}_0}{\partial \tilde{Z}} \frac{\partial^2 \tilde{\psi}_0}{\partial \tilde{Z} \partial \xi} - \frac{\partial \tilde{\psi}_0}{\partial \xi} \frac{\partial \tilde{\psi}_0}{\partial \tilde{Z}^2}\right) = \frac{1}{S} \Gamma \tilde{\vartheta}_0 + \frac{1}{S} \frac{\partial^3 \tilde{\psi}_0}{\partial \tilde{Z}^3}$$
$$+ \left(\frac{1}{S}\right)^{3/2} \frac{1}{2} \frac{d\Gamma}{d\xi} \left[\frac{1+i}{2\sqrt{2}} e^{it} + \frac{1-i}{2\sqrt{2}} e^{-it}\right] \frac{\partial^2 \tilde{\psi}_0}{\partial \tilde{Z}^2} \quad , \tag{13.239b}$$

$$\tilde{Z} = 0: \begin{cases} \tilde{\vartheta}_0 = \frac{1}{4} \frac{d\Gamma}{d\xi} + O\left(\frac{1}{S}\right) \quad ; \\ \tilde{\psi}_0 = \Gamma \frac{d\Gamma}{d\xi} \left(\frac{1}{S}\right)^{3/2} \left[\frac{41}{32\sqrt{2}} + \frac{23\sqrt{2} - 34}{128} \left(\frac{1+i}{2} e^{2it} \right.\right. \\ \quad \left.\left. + \frac{1-i}{2} e^{-2it}\right)\right] + O\left(\frac{1}{S^{5/2}}\right) \quad ; \\ \frac{\partial \tilde{\psi}_0}{\partial \tilde{Z}} = -\frac{7}{8} \frac{1}{S} \Gamma \frac{d\Gamma}{d\xi} + O\left(\frac{1}{S^{5/2}}\right) \quad , \end{cases} \tag{13.240a}$$

$$\tilde{Z} \to +\infty: \quad \frac{\partial \tilde{\psi}_0}{\partial \tilde{Z}} \to 0 \quad , \quad \tilde{\vartheta}_0 \to 0 \quad . \tag{13.240b}$$

The solution to the distal problem (13.239, 240) must therefore be sought in the form of asymptotic distal expansions:

$$\begin{aligned} \tilde{\vartheta}_0 &= \tilde{\vartheta}^0 + \left(\frac{1}{S}\right)^{1/2} \tilde{\vartheta}^1 + \frac{1}{S} \tilde{\vartheta}^2 + O\left(\frac{1}{S^{3/2}}\right) \quad ; \\ \tilde{\psi}_0 &= \tilde{\psi}^0 + \left(\frac{1}{S}\right)^{1/2} \tilde{\psi}^1 + \frac{1}{S} \tilde{\psi}^2 + O\left(\frac{1}{S^{3/2}}\right) \quad , \end{aligned} \tag{13.241}$$

where only the terms introduced for obtaining the distal equations for $\tilde{\vartheta}^0$ and $\tilde{\psi}^0$ are specified. It is now easy to prove that both $\tilde{\vartheta}^0$ and $\tilde{\psi}^0$ can only be functions of ξ and $\tilde{Z}$; the same is true of $\tilde{\vartheta}^1$ and $\tilde{\psi}^1$. Hence, from (13.239) written to order 1/S, it results that the cancellation of the secular terms in t leads to:

$$\begin{aligned} \frac{\partial \tilde{\psi}^0}{\partial \tilde{Z}} \frac{\partial^2 \tilde{\psi}^0}{\partial \xi \partial \tilde{Z}} - \frac{\partial \tilde{\psi}^0}{\partial \xi} \frac{\partial^2 \tilde{\psi}^0}{\partial \tilde{Z}^2} &= \Gamma \tilde{\vartheta}^0 + \frac{\partial^3 \tilde{\psi}^0}{\partial \tilde{Z}^3} \quad ; \\ \frac{\partial \tilde{\psi}^0}{\partial \tilde{Z}} \frac{\partial \tilde{\vartheta}^0}{\partial \xi} - \frac{\partial \tilde{\psi}^0}{\partial \xi} \frac{\partial \tilde{\vartheta}^0}{\partial \tilde{Z}} &= \frac{\partial^2 \tilde{\vartheta}^0}{\partial \tilde{Z}^2} \quad . \end{aligned} \tag{13.242}$$

According to (13.240), the following boundary conditions must be prescribed for these steady equations (13.242):

$$\begin{aligned} \tilde{Z} = 0: \quad & \tilde{\psi}^0 = \frac{\partial \tilde{\psi}^0}{\partial \tilde{Z}} = 0 \quad , \quad \tilde{\vartheta}^0 = \frac{1}{4} \frac{d\Gamma}{d\xi} \quad ; \\ \tilde{Z} \to +\infty: \quad & \frac{\partial \tilde{\psi}^0}{\partial \tilde{Z}} \to 0 \quad , \quad \tilde{\vartheta}^0 \to 0 \quad . \end{aligned} \tag{13.243}$$

Problem (13.242, 243) is the dominant steady distal problem which, to order zero, governs the convective motion in the outer distal layer. We thus have in the latter, the following asymptotic representation:

$$\begin{aligned}\hat{\psi}_0 &= \frac{\Gamma}{2}\left[\frac{1+\mathrm{i}}{2\sqrt{2}}\mathrm{e}^{\mathrm{i}t} + \frac{1-\mathrm{i}}{2\sqrt{2}}\mathrm{e}^{-\mathrm{i}t}\right] + \mathrm{S}^{1/2}\tilde{\psi}^0\left(\xi, \frac{1}{\mathrm{S}^{1/2}}Z\right) + \dots;\\ \hat{\vartheta}_0 &= \frac{1}{\mathrm{S}}\tilde{\vartheta}^0\left(\xi, \frac{1}{\mathrm{S}^{1/2}}Z\right) + \dots \quad .\end{aligned} \tag{13.244}$$

Let us now assume that $\mu_0 \neq 0$ in (13.233). The following hypothesis must then be made:

$$\mu_0 = \hat{\mu}\frac{1}{\mathrm{S}^2} \quad , \quad \text{with} \quad \hat{\mu} = \mathrm{O}(1) \quad .$$

In this case, it is necessary to add the term $\gamma\hat{\mu}\Gamma(\partial\tilde{\psi}^0/\partial\tilde{Z})$ to the left-hand side of the second equation in (13.242). Various developments and certain calculation results concerning this problem can be found in the thesis by Noe (1981).

13.4.3 Low Mach Number Flow over a Relief[18]

1. Our purpose here is to investigate the scheme of Chap. 12 (see Sect. 12.3) when there is a relief defined by a function $h(\xi,\eta)$. It will be assumed that $h = \mathrm{O}(1)$. The notations of Chap. 12 will be used starting with (12.10). The following ground condition is written:

$$\mathcal{H}(\tau, \xi, \eta, \zeta_s(\tau, \xi, \eta)) = h(\xi, \eta) \quad ,$$

and the expansion is carried out. We obtain to leading order [remembering that (12.16) holds]:

$$\mathcal{H}_0(\zeta_{s0}(\xi,\eta)) = h(\xi,\eta) \quad , \quad \frac{\partial \zeta_{s0}}{\partial \tau} \equiv 0 \quad .$$

From (12.7), we get to leading order:

$$\boldsymbol{v}_0 \cdot \boldsymbol{D}\zeta_{s0} = 0 \quad , \tag{13.245}$$

and all along the relief we have:

$$\boldsymbol{v}_0 \cdot \boldsymbol{D}h(\xi,\eta) = 0 \quad , \quad \zeta = \zeta_{s0}(\xi,\eta) \quad . \tag{13.246}$$

For $\boldsymbol{v}_0$, $\mathcal{H}_2$, the following problem is obtained:

$$\begin{aligned}&S\frac{\partial \boldsymbol{v}_0}{\partial \tau} + (\boldsymbol{v}_0 \cdot \boldsymbol{D})\boldsymbol{v}_0 + \boldsymbol{D}\left(\frac{\mathcal{H}_2}{\gamma/\mathrm{Bo}}\right) = 0 \quad ;\\ &\boldsymbol{D} \cdot \boldsymbol{v}_0 = 0 \quad ,\\ &\boldsymbol{v}_0 \cdot \boldsymbol{D}h(\xi,\eta) = 0 \quad , \quad \text{for} \quad (\xi,\eta) \in \mathcal{C}_0(\zeta) \quad ,\\ &(\xi^2+\eta^2)^{1/2} \to \infty: \quad \boldsymbol{v}_0 \to \boldsymbol{v}^\infty(\zeta) \quad , \quad \mathcal{H}_2 \to 0 \quad .\end{aligned} \tag{13.247}$$

[18] Based on the Course given at the ICMS in Udine, October, 1983, by J.P. Guiraud.

Here $\mathcal{C}_0(\zeta)$ is defined by the intersection of the relief and the isobaric surface $\zeta = \text{const}$ according to $\mathcal{H} = \mathcal{H}_0(\zeta)$, i.e.,

$$(\xi, \eta) \in \mathcal{C}_0(\zeta) \Leftrightarrow \mathcal{H}_0(\zeta) = h(\xi, \eta) \quad .$$

In each isobaric surface which intersects the relief, there is a partial blocking phenomenon. As the flow is constrained to remain on its own isobaric surface, it cannot ride over the topography but must flow around it. For the isobaric surfaces which do not intersect the relief, the situation is exactly the same as in Sect. 12.3. One exception is that the model generates a flow which is disturbed with respect to (12.74) due to the contouring process, but only within the isobaric surfaces which intersect the relief. For those lying entirely above the relief, the model is unable to generate perturbations. The same holds concerning vertical vorticity.

Let us now underline the differences with respect to Sect. 12.3. Instead of (12.76), we have:

$$\boldsymbol{v}_0 = \boldsymbol{v}^{\infty}(\zeta) + \boldsymbol{k} \times \boldsymbol{D}\psi_0 + \boldsymbol{D}\varphi_0 \quad , \tag{13.248}$$

where φ_0 and ψ_0 have to be computed (for isobaric surfaces which intersect the relief) as follows:

$$\begin{aligned} &\frac{\partial^2 \psi_0}{\partial \xi^2} + \frac{\partial^2 \psi_0}{\partial \eta^2} = \Omega_0 \quad ; \quad \psi_0 = 0 \quad , \quad \text{on} \quad \mathcal{C}_0(\zeta) \quad , \\ &\psi_0 = \frac{\overline{\Omega_0}}{2\pi} \log |\boldsymbol{\xi}| + \mathrm{o}(1) \quad , \quad \text{when} \quad |\boldsymbol{\xi}| \to \infty \quad ; \end{aligned} \tag{13.249a}$$

$$\begin{aligned} &\frac{\partial^2 \varphi_0}{\partial \xi^2} + \frac{\partial^2 \varphi_0}{\partial \eta^2} = 0 \quad ; \quad \varphi_0 = \mathrm{o}(1) \quad , \quad \text{when} \quad |\boldsymbol{\xi}| \to \infty \quad ; \\ &\boldsymbol{D}\varphi_0 \cdot \boldsymbol{D}h + \boldsymbol{v}_\infty \cdot \boldsymbol{D}h = 0 \quad , \quad \text{on} \quad \mathcal{C}_0(\zeta) \quad . \end{aligned} \tag{13.249b}$$

We may write:

$$\psi_0 = \frac{\overline{\Omega_0}}{2\pi} \log |\boldsymbol{\xi}| - \frac{\boldsymbol{\mu}_0 \cdot \boldsymbol{\xi}}{2\pi |\boldsymbol{\xi}|^2} + \frac{\boldsymbol{\xi} \cdot \boldsymbol{N}_0 \cdot \boldsymbol{\xi}}{2\pi |\boldsymbol{\xi}|^4} + \mathrm{O}\left(\frac{1}{|\boldsymbol{\xi}|^3}\right) \quad ; \tag{13.250a}$$

$$\varphi_0 = -\frac{\boldsymbol{A}_0 \cdot \boldsymbol{\xi}}{2\pi |\boldsymbol{\xi}|^2} + \frac{\boldsymbol{\xi} \cdot \hat{\boldsymbol{B}}_0 \cdot \boldsymbol{\xi}}{2\pi |\boldsymbol{\xi}|^4} + \mathrm{O}\left(\frac{1}{|\boldsymbol{\xi}|^3}\right) \quad , \tag{13.250b}$$

but instead of (12.78), only

$$\overline{\Omega_0} = \iint\limits_{\text{ext}\, \mathcal{C}_0(\zeta)} \Omega_0(\tau, \xi, \eta) d\xi \, d\eta \tag{13.251}$$

holds for the isobaric surfaces which intersect the relief. In a similar way, for the isobaric surfaces, only the following is valid:

$$\frac{\partial \overline{\Omega_0}}{\partial \tau} = 0 \quad .$$

Near infinity, (12.79) must be replaced by:

$$v_0 = v^{\infty}(\zeta) + \frac{\overline{\Omega_0}}{2\pi}\frac{k\times\xi}{|\xi|^2} - \frac{1}{2\pi|\xi|^2}\left\{ k\times\mu_0 \right.$$

$$\left. -\,2\frac{(\mu_0\cdot\xi)(k\times\xi)}{|\xi|^2} + A_0 - \frac{2(A\cdot\xi)\xi}{|\xi|^2}\right\} + \mathrm{O}\left(\frac{1}{|\xi|^3}\right) \quad . \tag{13.252}$$

However, (12.80) may be used provided that χ_0 is defined by:

$$k\times D\psi_0 + D\varphi_0 = D\chi_0 \quad ,$$

such that:

$$\chi_0 = \frac{\overline{\Omega_0}}{2\pi}\mathrm{arctg}\frac{\eta}{\xi} + \frac{(k\times\mu_0)\cdot\xi - A_0\cdot\xi}{2\pi|\xi|^2}$$

$$+\,\frac{\xi\cdot N_0\cdot(k\times\xi) + \xi\cdot\hat{B}_0\cdot\xi}{2\pi|\xi|^4} + \ldots \quad .$$

Thus, instead of (12.81), we find:

$$\mathcal{H}_2 = -\,\frac{\gamma}{\mathrm{Bo}}\,\frac{\left[k\times\left(\mathrm{S}\partial\mu_0/\partial\tau - v^{\infty}\overline{\Omega_0}\right) - \mathrm{S}\partial A_0/\partial\tau\right]\cdot\xi}{2\pi|\xi|^2}$$

$$-\,\frac{\gamma}{2\mathrm{Bo}}\left(\frac{\overline{\Omega_0}}{2\pi}\right)^2\frac{1}{|\xi|^2} - \frac{\gamma}{2\pi\mathrm{Bo}|\xi|^2}\left\{ v^{\infty}\cdot[(k\times\mu_0) - A_0]\right.$$

$$-\,2\frac{[(k\times\mu_0) - A_0]\cdot\xi(v^{\infty}\cdot\xi)}{|\xi|^2}$$

$$\left. +\,\frac{\xi\cdot\mathrm{S}(\partial N_0/\partial\tau)\cdot(k\times\xi) + \xi\cdot\mathrm{S}(\partial\hat{B}_0/\partial\tau)\cdot\xi}{|\xi|^2}\right\} + \mathrm{O}\left(\frac{1}{|\xi|^3}\right) \; ; \tag{13.253}$$

Of course, a number of terms in (13.253) disappear for the cross-over altitude (in this case we have: $\mathrm{S}(\partial\mu_0/\partial t) = \overline{\Omega_0}v^{\infty}$, $A_0 = 0$, and $\hat{B}_0 = 0$,).

The following point is now stressed: from (12.93, 95–97) which remain valid whether or not there is a relief, it can be seen that in order to compute v_2, and then $\mathcal{H}_4$, $\partial^2\mathcal{H}_2/\partial\zeta^2$ is necessary. Let us then consider the highest isobaric surface which intersects the relief at the cross-over altitude [the reader is reminded that there is a one-to-one correspondance between pressure and altitude according to $\zeta \to \mathcal{H}_0(\zeta)$]. From (13.253), we can see that $\mathcal{H}_2$ cannot be expected to have as many derivatives as needed at the cross-over altitude. This is an indication that *the low Mach number expansion cannot be uniformly valid in the vicinity of the cross-over altitude*. This phenomenon has been studied in a slightly different context by Brighton (1978) and Hunt and Snyder (1980). We will return to this matter later.

For the moment, let us consider the outer expansion while assuming that:

$$h(\xi,\eta) \quad \text{vanishes in the vicinity of} \quad |\xi| = \infty \quad .$$

Equations (12.82–85) may be used unchanged, but $\boldsymbol{v}_0$ must be changed to:

$$\boldsymbol{v}_0 = \boldsymbol{v}^{\infty}(\zeta) + \mathrm{M}_{\infty}\frac{\overline{\Omega_0}}{2\pi}\frac{\boldsymbol{k}\times\tilde{\xi}}{|\tilde{\xi}|^2} - \mathrm{M}_{\infty}^2\frac{1}{2\pi|\tilde{\xi}|^2}\left\{\boldsymbol{k}\times\boldsymbol{\mu}_0 + \boldsymbol{A}_0 - 2\frac{(\boldsymbol{\mu}_0\cdot\tilde{\xi})(\boldsymbol{k}\times\tilde{\xi}) + (\boldsymbol{A}_0\cdot\tilde{\xi})\tilde{\xi}}{|\tilde{\xi}|^2}\right\} + \mathrm{O}\left(\mathrm{M}_{\infty}^3\right) \quad , \qquad (13.254)$$

and:

$$\mathcal{H}_0 + \mathrm{M}_{\infty}^2\mathcal{H}_2 + \ldots = \mathcal{H}_0(\zeta) + \mathrm{M}_{\infty}^3\left\{-\frac{\gamma}{2\pi|\tilde{\xi}|^3}\left[\boldsymbol{k}\times\left(\mathrm{S}\frac{\partial\boldsymbol{\mu}_0}{\partial\tau} - \overline{\Omega_0}\boldsymbol{v}^{\infty}\right) - \mathrm{S}\frac{\partial\boldsymbol{A}_0}{\partial\tau}\right]\cdot\tilde{\xi}\right\} + \mathrm{M}_{\infty}^4\mathrm{O}\left(\frac{1}{|\tilde{\xi}|^2}\right) + \ldots \quad .$$

Again, $a = 1$ must be chosen in (12.83) and (12.88) holds along with (12.87, 89). The leading solution for the outer expansion is:

$$\tilde{\boldsymbol{v}}_0 = \frac{\overline{\Omega_0}}{2\pi}\frac{\boldsymbol{k}\times\tilde{\xi}}{|\tilde{\xi}|^2} \quad , \quad \tilde{w}_0 = \tilde{\mathcal{H}}_0 = \tilde{T}_0 = 0 \quad , \qquad (13.255)$$

exactly as in (12.89). The validity of (13.255) again relies on $\partial\overline{\Omega_0}/\partial\tau = 0$. Of course, when no vertical vorticity exists beforehand, $\overline{\Omega_0}$ is zero and this leading velocity field is zero. Again, (12.90) is still valid but (12.91) must be changed to:

$$\begin{aligned}
\tilde{\boldsymbol{v}}_1 &\cong -\frac{1}{2\pi|\tilde{\xi}|^2}\left\{\boldsymbol{k}\times\boldsymbol{\mu}_0 + \boldsymbol{A}_0 - 2\frac{1}{|\tilde{\xi}|^2}\left[(\boldsymbol{\mu}_0\cdot\tilde{\xi})(\boldsymbol{k}\times\tilde{\xi}) + (\boldsymbol{A}_0\cdot\tilde{\xi})\tilde{\xi}\right]\right\} + \ldots \;; \\
\tilde{\mathcal{H}}_1 &\cong -\gamma\frac{1}{2\pi|\tilde{\xi}|^2}\left[\boldsymbol{k}\times\left(\mathrm{S}\frac{\partial\boldsymbol{\mu}_0}{\partial\tau} - \overline{\Omega_0}\boldsymbol{v}^{\infty}\right) - \mathrm{S}\frac{\partial\boldsymbol{A}_0}{\partial\tau}\right]\cdot\tilde{\xi} + \ldots \;, \qquad (13.256)\\
&\text{when} \quad |\tilde{\xi}| \to 0 \quad .
\end{aligned}$$

It follows that solution (12.92) is no longer valid and a comment is in order. Here the position with respect to the cross-over altitude does not matter. Once there is a relief, the solution in the far field to the order just above the leading one is no longer the continuation of the solution at order one horizontal distance. This far field solution mixes up all the altitude levels and requires a full solution to the equation for $\tilde{\mathcal{H}}_1$ that is not a solution for which the last term in this equation vanishes identically. We will not attempt to write this solution here. Suffice it to say that the technique presented after (12.111) should be used (see also Subsect. 12.5.1). We will end this discussion here since further research is needed concerning higher order terms in the expansions.

2. Let us now turn our attention to the cross-over altitude. The first point to

investigate concerns the behavior of the leading solution near cross-over. We will assume that the relief may be expanded according to:

$$h(\xi,\eta) = h_{\mathrm{m}} - h^{(2)}(\xi,\eta) - h^{(3)}(\xi,\eta) + \ldots \quad , \tag{13.257}$$

where $h^{(i)}$ is a polynomial of degree i which is homogeneous with respect to $\xi - \xi_0$, $\eta - \eta_0$, and where ξ_0, η_0 is the point of maximum altitude of the relief. If ζ_{m} is the pressure level at the top of the relief, namely $\mathcal{H}_0(\zeta_{\mathrm{m}}) = h_{\mathrm{m}}$, then an ellipse is obtained for the approximate form of $\mathcal{C}_0(\zeta)$:

$$h^{(2)}(\xi,\eta) = \frac{1}{\mathrm{Bo}\zeta_{\mathrm{m}}} T_{0\mathrm{m}}(\zeta - \zeta_{\mathrm{m}}) \quad . \tag{13.258}$$

The potential φ_0 is first considered. It can be shown that when $\zeta \to \zeta_{\mathrm{m}}$, we have:

$$\varphi_0 = (\zeta - \zeta_{\mathrm{m}})^{1/2} \phi_0\left(\frac{\xi - \xi_{\mathrm{m}}}{\sqrt{\zeta - \zeta_{\mathrm{m}}}} \quad , \quad \frac{\eta - \eta_{\mathrm{m}}}{\sqrt{\zeta - \zeta_{\mathrm{m}}}}\right) \cdot \boldsymbol{v}_\infty(\zeta) \quad , \quad \zeta > \zeta_{\mathrm{m}} \quad ;$$

$$\varphi_0 \equiv 0 \quad , \quad \zeta < \zeta_{\mathrm{m}} \quad . \tag{13.259}$$

Now ψ_0 must be considered. As a first approximation, the following may be obtained:

$$\psi_0^{(\mathrm{a})}(\tau,\xi,\eta,\zeta) = \frac{1}{2\pi} \iint\limits_{E_0(\zeta)} \Omega_0(\tau,\xi,\eta,\zeta)\left(\log\left[(\xi - \xi')^2 + (\eta - \eta')^2\right]^{1/2}\right) d\xi'\, d\eta'$$

where $E_0(\zeta)$ stands for the whole plane when $\zeta < \zeta_{\mathrm{m}}$ and for the exterior of $\mathcal{C}_0(\zeta)$ when $\zeta > \zeta_{\mathrm{m}}$. It will be seen later, however, that the above is not fully satisfactory.

Starting from the above approximation, ψ_0 may be sought in the following form:

$$\psi_0 = \psi_0^{(\mathrm{a})} + \psi_0^{(\mathrm{h})} \quad ,$$

where:

$$\frac{\partial^2 \psi_0^{(\mathrm{h})}}{\partial \xi^2} + \frac{\partial^2 \psi_0^{(\mathrm{h})}}{\partial \eta^2} = 0 \quad ,$$

$$\psi_0^{(\mathrm{h})} + \psi_0^{(\mathrm{a})} = 0 \quad , \quad \text{on} \quad \mathcal{C}_0(\zeta) \quad .$$

It is again easily seen that to leading order:

$$\psi_0^{(\mathrm{h})} = (\zeta - \zeta_{\mathrm{m}})^{1/2} \boldsymbol{\Psi}_0\left(\frac{\xi - \xi_{\mathrm{m}}}{\sqrt{\zeta - \zeta_{\mathrm{m}}}} \quad , \quad \frac{\eta - \eta_{\mathrm{m}}}{\sqrt{\zeta - \zeta_{\mathrm{m}}}}\right) \cdot \boldsymbol{D}\psi_0^{(a)}(\tau, \xi_{\mathrm{m}}, \eta_{\mathrm{m}}, \zeta_{\mathrm{m}}) \quad ,$$

provided we ignore $\psi_0^{(\mathrm{a})}(t, \xi_{\mathrm{m}}, \eta_{\mathrm{m}}, \zeta_{\mathrm{m}})$ which does not contribute to the velocity. Finally, it is deduced that:

$$\begin{aligned} \boldsymbol{v}_0 = {} & v_1\left(\frac{\xi - \xi_{\mathrm{m}}}{\sqrt{\zeta - \zeta_{\mathrm{m}}}} \quad , \quad \frac{\eta - \eta_{\mathrm{m}}}{\sqrt{\zeta - \zeta_{\mathrm{m}}}}\right) \boldsymbol{v}_\infty(\zeta_{\mathrm{m}}) \\ & + \boldsymbol{v}_2\left(\frac{\xi - \xi_{\mathrm{m}}}{\sqrt{\zeta - \zeta_{\mathrm{m}}}} \quad , \quad \frac{\eta - \eta_{\mathrm{m}}}{\sqrt{\zeta - \zeta_{\mathrm{m}}}}\right) \cdot \boldsymbol{D}\psi_0^{(\mathrm{a})}(\tau, \xi_{\mathrm{m}}, \eta_{\mathrm{m}}, \zeta_{\mathrm{m}}) \\ & + \boldsymbol{v}_0^{(R)}(\tau,\xi,\eta,\zeta) \quad . \end{aligned} \tag{13.260}$$

As a consequence, we find that:

$$\mathcal{H}_2 = \mathcal{F}\left(\frac{\xi - \xi_m}{\sqrt{\zeta - \zeta_m}} \quad , \quad \frac{\eta - \eta_m}{\sqrt{\zeta - \zeta_m}} \quad , \quad \tau\right) + \mathcal{H}_2^{(R)} \quad , \quad (\zeta > \zeta_m) \quad ,$$

$$\mathcal{H}_2 = \mathcal{H}_2^{(R)} \quad , \quad \zeta < \zeta_m \quad , \tag{13.261}$$

and $\partial\mathcal{H}_2/\partial\zeta$ is seen to be singular when $\zeta \downarrow \zeta_m$.

We now wish to derive a limiting process capable of removing this singularity. We set:

$$\boldsymbol{\xi} = \boldsymbol{\xi}_m + M_\infty^\alpha \hat{\boldsymbol{\xi}} \quad , \quad \zeta = \zeta_m + M_\infty^\beta \hat{\zeta} \quad . \tag{13.262a}$$

The limiting process that must be considered here is:

$$M_\infty \to 0 \quad ; \quad \hat{\boldsymbol{\xi}} \quad \text{and} \quad \hat{\zeta} \quad \text{fixed} \quad . \tag{13.262b}$$

According to (13.257), the relief is:

$$h = h_m - M_\infty^{2\alpha} h^{(2)}(\hat{\boldsymbol{\xi}}) + O\left(M_\infty^{3\alpha}\right) \quad , \tag{13.263}$$

where $h^{(2)}$ is a homogeneous polynomial of degree 2. Considering the fact that $\mathcal{H}_0(\zeta_m + M_\infty^\beta \hat{\zeta}) = h_m + o(M_\infty^\beta)$ and comparing with (13.263), we suspect that $\beta = 2\alpha$. However, both α and β are left free for the time being. From (12.1c), it is immediately seen that the following must be set:

$$\boldsymbol{v} = \hat{\boldsymbol{v}} \quad , \quad w = M_\infty^{\beta-\alpha} \hat{w} \quad .$$

From (12.1a) where the Coriolis term has been neglected, the following is obtained:

$$M_\infty^{-\alpha}\left\{(\hat{\boldsymbol{v}} \cdot \hat{\boldsymbol{D}})\hat{\boldsymbol{v}} + \hat{w}\frac{\partial \hat{\boldsymbol{v}}}{\partial \hat{\zeta}}\right\} + \frac{\text{Bo}}{\gamma M_\infty^{2+\alpha}} \hat{\boldsymbol{D}}\mathcal{H} + \ldots = 0$$

which leads to:

$$\mathcal{H} = \mathcal{H}_0(\zeta_m) + M_\infty^2 \hat{\mathcal{H}} = h_m + M_\infty^2 \hat{\mathcal{H}} \quad . \tag{13.264}$$

When compared to $\mathcal{H}_0(\zeta_m + M_\infty^\beta \hat{\zeta}) = h_m + O(M_\infty^\beta)$, the above suggests that $\beta = 2$ and $\alpha = 1$. However, this is ignored and we set:

$$T = T_0(\zeta_m) + M_\infty^\delta \hat{T} = T_m + M_\infty^\delta \hat{T} \quad , \tag{13.265}$$

so that (12.1b) yields:

$$M_\infty^{\delta-\alpha}\left(\hat{\boldsymbol{v}} \cdot \hat{\boldsymbol{D}}\hat{T} + \hat{w}\frac{\partial \hat{T}}{\partial \hat{\zeta}}\right) - M_\infty^{\beta-\alpha}\frac{\gamma - 1}{\gamma}\hat{w}\frac{T_m}{\zeta_m} + \ldots = 0 \quad , \tag{13.266}$$

whereas $T = -\text{Bo}\zeta(\partial\mathcal{H}/\partial\zeta)$ leads to:

$$M_\infty^\delta \hat{T} + M_\infty^{2-\beta} \text{Bo}\zeta_m \frac{\partial \hat{\mathcal{H}}}{\partial \hat{\zeta}} + \ldots = 0 \quad . \tag{13.267}$$

Let us now consider the conditions on the relief which is assumed to be located at:

$$\hat{\zeta} = \hat{\zeta}_s(\tau, \hat{\boldsymbol{\xi}}) \Leftarrow \text{relief} \quad .$$

The first condition is that $h = \mathcal{H}$ on the relief and from (13.263, 264), this leads to:

$$\mathrm{M}_\infty^2 \hat{\mathcal{H}}\big(\tau, \hat{\boldsymbol{\xi}}, \hat{\zeta}_s(\tau, \hat{\boldsymbol{\xi}})\big) = -\mathrm{M}_\infty^{2\alpha} h^{(2)}(\hat{\boldsymbol{\xi}}) \Rightarrow \alpha = 1$$

and:

$$\hat{\mathcal{H}}\big(\tau, \hat{\boldsymbol{\xi}}, \hat{\zeta}_s(\tau, \hat{\boldsymbol{\xi}})\big) = -h^{(2)}(\hat{\boldsymbol{\xi}}) \quad . \tag{13.268}$$

The second condition on the relief comes from (12.7) which reads:

$$\mathrm{M}_\infty^\beta \mathrm{S} \frac{\partial \hat{\zeta}_s}{\partial \tau} + \mathrm{M}_\infty^{\beta-1} (\hat{\boldsymbol{v}} \cdot \hat{\boldsymbol{D}} \hat{\zeta}_s - \hat{w}) = 0 \quad .$$

If the above is matched with (13.261), we can conclude that $\beta = 2$ and this implies that $\delta = 0$. Summing up, we have:

$$\begin{aligned} &\boldsymbol{\xi} = \boldsymbol{\xi}_\mathrm{m} + \mathrm{M}_\infty \hat{\boldsymbol{\xi}} \quad ; \quad \zeta = \zeta_\mathrm{m} + \mathrm{M}_\infty^2 \hat{\zeta} \quad ; \quad h = h_\mathrm{m} - \mathrm{M}_\infty^2 h^{(2)}(\hat{\boldsymbol{\xi}}) \quad , \\ &\boldsymbol{v} \equiv \hat{\boldsymbol{v}} \quad , \quad w = \mathrm{M}_\infty \hat{w} \quad , \quad T \equiv \hat{T} \quad , \quad \mathcal{H} = h_\mathrm{m} + \mathrm{M}_\infty^2 \hat{\mathcal{H}} \quad , \end{aligned} \tag{13.269a}$$

and the following equations are thus obtained:

$$\begin{aligned} &(\hat{\boldsymbol{v}} \cdot \hat{\boldsymbol{D}}) \hat{\boldsymbol{v}} + \hat{w} \frac{\partial \hat{\boldsymbol{v}}}{\partial \hat{\zeta}} + \hat{\boldsymbol{D}} \left(\frac{\hat{\mathcal{H}}}{\gamma/\mathrm{Bo}} \right) = 0 \quad , \\ &\hat{\boldsymbol{D}} \cdot \hat{\boldsymbol{v}} + \frac{\partial \hat{w}}{\partial \hat{\zeta}} = 0 \quad ; \\ &\hat{\boldsymbol{v}} \cdot \hat{\boldsymbol{D}} \hat{T} + \hat{w} \frac{\partial \hat{T}}{\partial \hat{\zeta}} = 0 \quad , \\ &\hat{T} + \mathrm{Bo} \zeta_\mathrm{m} \frac{\partial \hat{\mathcal{H}}}{\partial \hat{\zeta}} = 0 \quad . \end{aligned} \tag{13.269b}$$

The boundary condition on the relief is:

$$\hat{\mathcal{H}}_s + h^{(2)} = 0 \quad , \quad \hat{\boldsymbol{v}}_s \cdot \hat{\boldsymbol{D}} \hat{\zeta}_s = \hat{w}_s \quad , \tag{13.269c}$$

where the following notation has been used:

$$\hat{f}\big(\tau, \hat{\boldsymbol{\xi}}, \hat{\zeta}_s(\tau, \hat{\boldsymbol{\xi}})\big) \equiv \hat{f}_s(\tau, \hat{\boldsymbol{\xi}}) \quad .$$

It is emphasized that (13.269c) is only valid for $\hat{\zeta}_s > 0$.

According to (13.261), matching with the solution away from the cross-over altitude requires that:

$$\begin{aligned} &\hat{\mathcal{H}} \cong \left(\frac{\partial \mathcal{H}_0}{\partial \zeta} \right)_\mathrm{m} \hat{\zeta} + \mathcal{H}_2^{(R)} + \mathcal{F} \left(\frac{\hat{\xi}}{\sqrt{\hat{\zeta}}} \ , \ \frac{\hat{\eta}}{\sqrt{\hat{\zeta}}} \ , \tau \right) \quad , \quad \hat{\zeta} \to +\infty \quad ; \\ &\hat{\mathcal{H}} \cong \left(\frac{\partial \mathcal{H}_0}{\partial \zeta} \right)_\mathrm{m} \hat{\zeta} + \mathcal{H}_2^{(R)} \quad , \quad \hat{\zeta} \to -\infty \quad , \end{aligned} \tag{13.270}$$

where $\hat{\zeta} \to +\infty$ means $\hat{\zeta} \to +\infty$ with $\hat{\boldsymbol{\xi}}/\sqrt{\hat{\zeta}}$ fixed. As a matter of fact, we cannot consider $\hat{\zeta} \to +\infty$ with $\hat{\boldsymbol{\xi}}$ fixed because we would be brought inside the relief. It is easily verified that when $\hat{\zeta} \to -\infty$,

$$\hat{\boldsymbol{v}} \to \boldsymbol{v}_{0\mathrm{m}} \quad , \quad \hat{\mathcal{H}} \to \left(\frac{\partial \mathcal{H}_0}{\partial \zeta}\right)_{\mathrm{m}} \hat{\zeta} + \mathcal{H}_{2\mathrm{m}} \quad , \quad \hat{T} \to T_{0\mathrm{m}} \quad , \quad \hat{w} \to 0 \quad ,$$

is consistent with (13.269b). Here $\boldsymbol{v}_{0\mathrm{m}}$ is a constant vector and $\mathcal{H}_{2\mathrm{m}}$ and $T_{0\mathrm{m}}$ are constant values which are expected to be the values of T_0 and $\mathcal{H}_2$ right at the top of the relief according to the solution of (13.247). This is consistent with (13.270) if we assume that $\mathcal{H}_2^{(R)} = \mathcal{H}_{2\mathrm{m}}$.

We must now examine the behavior of the solution to (13.269b,c) when $\hat{\zeta} \to +\infty$ with $\hat{\boldsymbol{\xi}}/\sqrt{\hat{\zeta}}$ fixed. It is assumed that:

$$\hat{w} \to 0 \quad , \quad \hat{\boldsymbol{v}} \to \hat{\boldsymbol{V}}(\hat{\boldsymbol{x}}, \tau) \quad , \quad \hat{\mathcal{H}} \cong \left(\frac{\partial \mathcal{H}_0}{\partial \zeta}\right)_{\mathrm{m}} \hat{\zeta} + \hat{H}(\hat{\boldsymbol{x}}, \tau) \quad ;$$
$$\hat{T} \cong T_{0m} + \mathrm{O}\left(\frac{1}{\hat{\zeta}}\right) \quad , \tag{13.271}$$

where the following notation has been used:

$$\hat{\boldsymbol{x}} = \frac{\hat{\boldsymbol{\xi}}}{\sqrt{\hat{\zeta}}} \quad , \quad \hat{\nabla} = \left(\frac{\partial}{\partial \hat{x}} \, , \, \frac{\partial}{\partial \hat{y}}\right) \quad , \quad \hat{\boldsymbol{x}} = (\hat{x}, \hat{y}) \quad ,$$

and the last equation in (13.269b) has been taken care of. In order to verify (13.271), the following must be checked:

$$(\hat{\boldsymbol{V}} \cdot \hat{\nabla}) \hat{\boldsymbol{V}} + \hat{\nabla}\left(\frac{\hat{H}}{\gamma/\mathrm{Bo}}\right) = 0 \quad , \quad \hat{\nabla} \cdot \hat{\boldsymbol{V}} = 0 \quad . \tag{13.272}$$

Before verifying (13.272), we will first consider the boundary condition along the relief (13.269c). Using (13.271), we obtain:

$$\hat{\zeta}_s \left\{ \left(\frac{\partial \mathcal{H}_0}{\partial \zeta}\right)_{\mathrm{m}} + h^{(2)}(\hat{\boldsymbol{x}}_s) \right\} + \hat{H}(\hat{\boldsymbol{x}}_s, \tau) = 0 \quad ,$$
$$\hat{\boldsymbol{V}}_s \cdot \hat{\nabla} \hat{\zeta}_s = 0 \quad ; \quad \hat{\boldsymbol{x}}_s \equiv \frac{\hat{\boldsymbol{\xi}}}{\sqrt{\hat{\zeta}_s}} \quad , \tag{13.273}$$

where $\hat{H}$ in the first equation of (13.273) is negligible when taking the limit $\hat{\zeta} \to +\infty$. The function $\hat{\zeta}_s(\tau, \hat{\boldsymbol{\xi}})$ is derived from

$$h^{(2)}\left(\frac{\hat{\boldsymbol{\xi}}}{\sqrt{\hat{\zeta}_s}}\right) = -\left(\frac{\partial \mathcal{H}_0}{\partial \zeta}\right)_{\mathrm{m}} \quad ,$$

and from the fact that $h^{(2)}$ is quadratic, we may extract a closed form formula for $\hat{\zeta}_s$, namely

$$\hat{\zeta}_s = -\left\{\left(\frac{\partial \mathcal{H}_0}{\partial \zeta}\right)_m\right\}^{-1} h^{(2)}(\hat{\xi}) \quad . \tag{13.274}$$

Two consequences result from the above: a) $\hat{\zeta}_s$ does not depend on τ; and b) $\hat{\zeta}_s$ is the same (for $\hat{\zeta} \to +\infty$) as what would result from (13.262a) applied to $\mathcal{H}_0(\zeta_{s0}(\xi,\eta)) = h(\xi,\eta)$. Now, in order for (13.273) to hold, it is only necessary that:

$$\hat{\boldsymbol{V}} \cdot \hat{\boldsymbol{\nabla}} h^{(2)} = 0 \quad , \quad \text{along} \quad h^{(2)} + \left(\frac{\partial \mathcal{H}_0}{\partial \zeta}\right)_{\mathrm{m}} = 0 \quad . \tag{13.275}$$

Consequently, $\hat{\boldsymbol{V}}$ and $\hat{H}$ are obtained by solving (13.272) with (13.275) and $\hat{\boldsymbol{V}} \to \boldsymbol{V}_{0\mathrm{m}}$ when $|\hat{\boldsymbol{x}}| \to \infty$. Here $\boldsymbol{V}_{0\mathrm{m}}$ means the velocity field value according to (13.247) right at the top of the relief. It is stressed that the solution for $\hat{\boldsymbol{V}}$ and $\hat{H}$ is not necessarily irrotational. More precisely:

$$\boldsymbol{k} \cdot (\hat{\boldsymbol{\nabla}} \times \hat{\boldsymbol{V}}) = \hat{\Omega} \Rightarrow \hat{\boldsymbol{V}} \cdot \hat{\boldsymbol{\nabla}} \hat{\Omega} = 0 \quad . \tag{13.276}$$

However, the true vorticity is $\hat{\Omega}/\sqrt{\hat{\zeta}}$ and this does not match with an $O(1)$ vorticity according to (13.247). As a matter of fact, our rough analysis leading from (13.259) to (13.261) fails to consider the singular vorticity field due to convection by a singular velocity field!

The analysis should thus be reconsidered starting from the expression for $\psi_0^{(a)}$. It is conjectured that the proper correction to (13.261) would be $\mathrm{O}[(\zeta - \zeta_{\mathrm{m}})^{1/2}]$ and that it would match with the higher order term $\mathrm{O}(M_\infty)$ for $\boldsymbol{v}$ in (13.269a). Whether or not the solution for $\hat{\boldsymbol{V}}$ and $\hat{H}$ is rotational must be considered as an open question. The same is true concerning the possible occurrence of an $\mathrm{O}(M_\infty)$ rather than an $\mathrm{O}(\mathrm{M}_\infty^2)$ correction to $\hat{\boldsymbol{v}}_0$ due to vorticity effects near the top of the relief.

This topic requires then further research. However, it should be fairly obvious that a fully consistent theory of low Mach number flows over a relief can be realized. Two points must be stressed: the first concerns separation which occurs almost inevitably on the lee side of the relief; the second deals with the possible occurrence of the local Boussinesq state (see, for instance, Zeytounian and Guiraud, 1984). We add that the non-uniformity at the top of the relief may also occur near the curve where it matches with the flat environment. More generally speaking, some kind of non-uniformity may occur whenever $h(\xi,\eta)$ is not smooth.

3. Our starting point here is the unsteady Euler equations without the Coriolis terms. The following boundary condition on the relief is added:

$$\sigma(\boldsymbol{v} \cdot \boldsymbol{D}h) = w \quad , \quad \text{on} \quad z = \sigma h(x,y) \quad , \tag{13.277a}$$

and the condition at infinity:

$$\boldsymbol{v} \to \boldsymbol{V}_\infty(z) \quad , \quad p = p_\infty(z) \quad , \quad \varrho \to \varrho_\infty(z) \quad , \quad \text{when} \quad (x^2+y^2)^{1/2} \to \infty \quad , \tag{13.277b}$$

with $(dp_\infty/dz)+\mathrm{Bo}\varrho_\infty = 0$. An upper boundary condition is lacking but one will not be specified here.

The low Mach number flow for the above configurations was studied by Drazin (1961). The analysis was worked out again by Brighton (1977) who included the Coriolis effect and also did some laboratory experiments to model the atmospheric situation. This last work is reported on in Brighton (1978) and Hunt and Snyder (1980) performed further experiments.

Our basic assumption here is $\mathrm{M}_\infty \to 0$:

$$\begin{pmatrix} \boldsymbol{v} \\ w \\ p \\ \varrho \end{pmatrix} = \begin{pmatrix} \boldsymbol{v}_0 \\ w_0 \\ p_0 \\ \varrho_0 \end{pmatrix} + \mathrm{M}_\infty^2 \begin{pmatrix} \boldsymbol{v}_2 \\ w_2 \\ p_2 \\ \varrho_2 \end{pmatrix} + \dots \quad . \tag{13.278}$$

For the leading approximation, we obtain:

$$\begin{aligned} & w_0 = 0 \quad , \quad p_0 = p_0(z) \quad , \quad \varrho_0 = \varrho_0(z) \quad , \quad \frac{dp_0}{dz} + \mathrm{Bo}\varrho_0 = 0 \quad ; \\ & \mathrm{S}\frac{\partial \boldsymbol{v}_0}{\partial t} + (\boldsymbol{v}_0 \cdot \boldsymbol{D})\boldsymbol{v}_0 + \frac{\mathrm{Bo}}{\gamma}\boldsymbol{D}p_2 = 0 \quad , \\ & \boldsymbol{D} \cdot \boldsymbol{v}_0 = 0 \quad ; \\ & \frac{\partial p_2}{\partial z} + \mathrm{Bo}\varrho_2 = 0 \quad , \end{aligned} \tag{13.279a}$$

with:

$$\begin{aligned} & \boldsymbol{v}_0 \cdot \boldsymbol{D}h = 0 \quad , \quad \text{on} \quad z = \sigma h(\boldsymbol{x}) \quad , \\ & |\boldsymbol{x}| \to \infty : \ \boldsymbol{v}_0 \to \boldsymbol{V}_\infty(z) \quad , \quad p_2 \to 0 \quad , \quad \varrho_2 \to 0 \quad . \end{aligned} \tag{13.279b}$$

We have a two-dimensional flow over each plane z = const. The flow passes round (without slip) the cross-section of the relief formed by the intersection of this same plane z = const. The solution is obviously steady. Drazin (1961) and Brighton (1977) developed a closed form for the irrotational flow. We can see that vorticity cannot be generated by the model and may only persist in an unsteady solution. The above authors also considered higher order approximations. Drazin (1961) recognized the necessity of using local expansions for $z - z_\mathrm{m} = \mathrm{O}(\delta)$ where δ is a small parameter (our M_∞) and z_m is the altitude at the top of the relief. Such local expansions were also recognized as necessary when $(x^2+y^2)^{1/2} = \mathrm{O}(\delta^{-1})$. The situation near the top is quite analogous to the one discussed after (13.261). The behavior, near infinity in the horizontal direction should bear some resemblance to that discussed in Sect. 12.3 and following (13.255).

Coming back to (13.279), it is pointed out that Riley, Liu and Geller (1976) dealt numerically with a separated flow model. The wonderful experimental work reported on by Brighton (1978) and Hunt and Snyder (1980) relating to Drazin's model is also mentioned. The main conclusion of the above is the rather strong evidence that in the limit $\delta \to 0$, the flow is constrained to stay in horizontal planes and to experience a kind of two-dimensional separation when going around a cross-section of the relief in one of the above-mentioned horizontal planes. There is also convincing evidence that the extent of the cross-over region is O(δ) around the relief's maximum altitude. There is no clear proof of local Boussinesq states. Whether or not such local Boussinesq states are relevant to low Mach number flows remains a question open to debate.

It is pointed out that the lectures by P.A. Bois (1984a) provide a means of studying the problem at hand, namely, the behavior when $\mathrm{M}_\infty \to 0$, and for the linear version, when $\sigma \ll 1$. There is no evidence to indicate that Boussinesq waves occur. As a matter of fact, the solution looks as follows:

$$\begin{pmatrix} v \\ w \\ p \\ \varrho \end{pmatrix} = \begin{pmatrix} V_\infty \\ 0 \\ p_\infty \\ \varrho_\infty \end{pmatrix} + \sigma \iint_{-\infty}^{+\infty} h(x', y') G\big(x - x', \, y - y', \, z, \, \mathrm{M}_\infty\big) dx' \, dy' \quad .$$

Moreover, Bois' analysis provides, via a Fourier analysis with respect to horizontal variables, a way of studying $G(x, y, z, \mathrm{M}_\infty)$ when $\mathrm{M}_\infty \to 0$. This subject, however, will be left to future research while emphasizing that it relies heavily on a joint, yet unpublished Zeytounian-Guiraud project.

13.4.4 Asymptotic Formulation of the Rayleigh-Bénard Problem via the Boussinesq Approximation for Expansible Liquids [19]

The archetypal problem (referred to as the Rayleigh-Bénard problem) that we are going to consider here is one of a natural internal convection between two flat surfaces having different temperatures. Let T_0 be an (absolute) constant reference temperature which is the temperature of the environment. We will assume that the lower surface is at the temperature $T_1 = T_0 + \Delta T_0 = \text{const}$, whereas the upper surface remains at $T_0 < T_1$. We will also suppose as given the temperature difference $\Delta T_0 > 0$ and the distance d_0 which separates the two surfaces. Our dilatable liquid is characterized from the physics point of view by the state relation:

$$\begin{aligned} &\varrho = \varrho(T) \Rightarrow e = e(T) \quad , \quad \frac{De}{DT} = c(T) \quad , \\ &\lambda = \lambda(T) \quad , \quad \mu = \mu(T) \quad , \quad k = k(T) \quad . \end{aligned} \tag{13.280}$$

In (13.280), ϱ is the density, e is the specific internal energy, $c(T)$ is the specific heat of the liquid and λ, μ, k are the viscosity and heat conduction coefficients.

Generally speaking, the convection velocity $\boldsymbol{u}$ (of components u_i in an orthonormal Cartesian reference frame $\boldsymbol{e}_1$, $\boldsymbol{e}_2$, $\boldsymbol{e}_3$), ϱ, p and T satisfy the equations:

$$\frac{D}{Dt}\log \varrho + \frac{\partial u_j}{\partial x_j} = 0 \quad ; \quad \frac{D}{Dt} \equiv \frac{\partial}{\partial t} + u_j \frac{\partial}{\partial x_j} \quad ; \tag{13.281a}$$

$$\varrho \frac{Du_i}{Dt} + \frac{\partial p}{\partial x_i} + \varrho f_i = \frac{\partial}{\partial x_i}\left(\lambda \frac{\partial u_j}{\partial x_j}\right) + \frac{\partial}{\partial x_j}\left[\mu\left(\frac{\partial u_i}{\partial x_j} + \frac{\partial u_j}{\partial x_i}\right)\right] \quad ; \tag{13.281b}$$

[19] According to Zeytounian (1983).

$$\varrho c(T)\frac{DT}{Dt} + p\frac{\partial u_j}{\partial x_j} = \frac{\partial}{\partial x_i}\left(k\frac{\partial T}{\partial x_i}\right) + \lambda\left(\frac{\partial u_j}{\partial x_j}\right)^2 + \frac{\mu}{2}\left(\frac{\partial u_i}{\partial x_j} + \frac{\partial u_j}{\partial x_i}\right) \quad , \tag{13.281c}$$

where $f_i \equiv g\delta_{i3}$ with g being the (constant) magnitude of the force of gravity, and δ_{ij}, the Kronecker symbol ($\delta_{ij} \equiv 0$, if $i \neq j$ and $\delta_{ii} \equiv 1$). The coordinates x_i ($i = 1, 2, 3$) are chosen in such a way that $x_3 = 0$ simulates the lower surface, and $x_3 = d_0$, the upper surface. The Cartesian coordinates x_1 and x_2 identify any point in the plane $x_3 = 0$. The above mathematical formulation is assumed to be "exact" and will serve as the starting point of the forthcoming analysis. The reader will remark that we have not written lateral conditions (in x_1 and x_2) or initial conditions (in $t = 0$):[20] to do so would be unrealistic since we suspect the asymptotic model obtained hereafter to be a principal model which should be completed by local models which are valid in the vicinity of $t = 0$, and also at infinity when $(x_1^2 + x_2^2)^{1/2} \approx \infty$.

Let us then designate by ϱ_0, λ_0, μ_0, k_0 and c_0 the values of $\varrho(T)$, $\lambda(T)$, $\mu(T)$, $k(T)$ and $c(T)$ for $T = T_0 = \text{const}$. We define:

$$\beta(T) = -\frac{1}{\varrho}\frac{d\varrho}{dT}, \quad \alpha(T) = -\frac{1}{\mu}\frac{d\mu}{dT}, \quad \gamma(T) = -\frac{1}{k}\frac{dk}{dT}, \quad \Gamma(T) = -\frac{1}{c}\frac{dc}{dT},$$

where $\beta(T)$ is the coefficient of volume expansion of the liquid, and $\alpha(T)$ and $\gamma(T)$ are the viscous expansion and conduction expansion coefficients respectively. The following asymptotic analysis takes place in four stages.

1. Let us introduce the temperature and density perturbations:

$$\vartheta = \frac{T - T_0}{\Delta T_0} \quad \text{and} \quad \omega = \frac{\varrho - \varrho_0}{\Delta\varrho_0} \quad , \tag{13.282}$$

related to the convection phenomenon. As a consequence of (13.280, 282), we have:

$$\omega = \vartheta + \tau_0\vartheta^2 + \ldots \quad , \tag{13.283}$$

once the following has been taken:

$$\Delta\varrho_0 \equiv -\varrho_0\beta_0\Delta T_0 \quad , \quad \beta_0 \equiv -\frac{1}{\varrho_0}\left(\frac{d\varrho}{dT}\right)_{T=T_0}$$

and also once the following dimensionless parameter has been introduced:

$$\tau_0 = \frac{\Delta T_0}{2}\left(\frac{d\log\varrho\beta}{dT}\right)_{T=T_0} \equiv \frac{\Delta T_0}{2}\left\{\frac{1}{\beta_0}\left(\frac{d\beta}{dT}\right)_{T=T_0} - \beta_0\right\} \quad . \tag{13.284}$$

In what is to come, the dimensionless parameter:

$$\varepsilon_0 \equiv \beta_0\Delta T_0 \quad , \tag{13.285}$$

is going to play a major role in the asymptotic analysis. It is emphasized that:

[20] On this topic, the reader might consult Zeytounian (1984).

$$\frac{\tau_0}{\varepsilon_0} \equiv \frac{1}{2}\left\{\frac{1}{\beta_0^2}\left(\frac{d\beta}{dT}\right)_{T=T_0} - 1\right\} \quad . \tag{13.286}$$

At this stage, it is remarked that the state relation of our liquid can be given approximately in the form:

$$\omega = \vartheta \quad , \quad \text{when} \quad \tau_0 \to 0 \quad , \tag{13.287}$$

on the condition that:

$$\left|\frac{1}{\beta_0^2}\left(\frac{d\beta}{dT}\right)_{T=T_0} - 1\right|$$

remains bounded when $\varepsilon_0 \to 0$. Hence, $\tau_0 \to 0$ and $\varepsilon_0 \to 0$ in such a way that the similarity relation (13.286) is satisfied.

Relation (13.287) is the basis of the Boussinesq approximation. In this case:

$$\varrho(T) \approx \varrho(T_0)\{1 - \beta_0(T - T_0)\} \quad . \tag{13.288}$$

2. It can now be seen by taking into account:

$$\log \varrho = \log \varrho_0 + \frac{\Delta \varrho_0}{\varrho_0}\omega + \dots$$

and from (13.283) that:

$$\frac{\partial u_j}{\partial x_j} = \varepsilon_0\left\{\frac{D\vartheta}{Dt} + \tau_0\frac{D\vartheta^2}{Dt} + \dots\right\} \quad .$$

When $\varepsilon_0 \to 0$, the following incompressibility equation is again found:

$$\frac{\partial u_j}{\partial x_j} = 0 \quad , \tag{13.289}$$

which is compatible with (13.287) within the framework of the Boussinesq approximation.

3. For the forthcoming asymptotic analysis, it is best to use *dimensionless* values in (13.281b, 281c). But first, let us represent the pressure p (*with* dimensions) in the form:

$$p = g\varrho_0 d_0\left(1 - \frac{x_3}{d_0}\right) + \Delta p_0 \pi \quad ,$$

where Δp_0 is the pressure fluctuation created at the time of the convection phenomenon and must be determined from internal coherence conditions. Finally, π is the non-dimensional pressure perturbation.

Let us now introduce the dimensionless values:

$$\overline{x}_i = \frac{x_i}{d_0} \quad , \quad \overline{u}_i = \frac{u_i}{\nu_0/d_0} \quad , \quad \overline{t} = \frac{t}{d_0^2/\nu_0} \quad , \quad \overline{p} = \frac{p}{g\varrho_0 d_0} \quad ,$$

$$\overline{\lambda} = \frac{\lambda}{\lambda_0} \quad , \quad \overline{\mu} = \frac{\mu}{\mu_0} \quad , \quad \overline{k} = \frac{k}{k_0} \quad , \quad \overline{c} = \frac{c}{c_0} \quad , \tag{13.290}$$

where $\nu_0 = \mu_0/\varrho_0$. We designate $\mathrm{Pr} \equiv \nu_0/\kappa_0$, with $\kappa_0 = k_0/c_0\varrho_0$, the Prandtl number, and by:

$$\mathrm{Ra} \equiv \mathrm{Pr}\,\mathrm{Gr} = \frac{\nu_0}{\kappa_0}\,\frac{g\beta_0 \Delta T_0 d_0^3}{\nu_0^2} \quad , \tag{13.291}$$

the Rayleigh number, where:

$$\mathrm{Gr} = \frac{g\beta_0 \Delta T_0 d_0^3}{\nu_0^2} \equiv \frac{\varepsilon_0}{\sigma_0} \quad , \tag{13.292}$$

is the Grashof number. The parameter $\sigma_0 = \nu_0^2/gd_0^3 = (\nu_0/d_0)^2/gd_0$ plays the role of the square of a Froude number based on the characteristic velocity ν_0/d_0 and the length d_0.

Let us now introduce the dimensionless parameter:

$$\mathcal{A}_0 = \frac{\Delta p_0/g\varrho_0 d_0}{\sigma_0} \equiv \frac{\delta_0}{\sigma_0} \quad , \quad \delta_0 = \frac{\Delta p_0}{g\varrho_0 d_0} \quad . \tag{13.293}$$

Instead of the momentum equation (13.281b), the following non-dimensional equation is obtained:

$$\left(1 - \varepsilon_0\vartheta + \ldots\right)\frac{D\overline{u}_k}{D\overline{t}} + \mathcal{A}_0\frac{\partial \pi}{\partial \overline{x}_k} = \left(\frac{\partial^2}{\partial \overline{x}_1^2} + \frac{\partial^2}{\partial \overline{x}_2^2} + \frac{\partial^2}{\partial \overline{x}_3^2}\right)\overline{u}_k$$
$$+\varepsilon_0\left\{\frac{\lambda_0/\varrho_0}{\nu_0}\frac{\partial}{\partial \overline{x}_k}\left(\frac{D\vartheta}{D\overline{t}}\right) - \frac{\alpha_0}{\beta_0}\frac{\partial}{\partial \overline{x}_j}\left[\vartheta\left(\frac{\partial \overline{u}_k}{\partial \overline{x}_j} + \frac{\partial \overline{u}_j}{\partial \overline{x}_k}\right)\right] + \ldots\right\} \quad , \tag{13.294}$$

where $\alpha_0 \equiv -(1/\mu_0)(d\mu/dT)_{T=T_0}$, $k = 1$ and 2, whereas $j = 1, 2$ and 3.

Next results for $\overline{u}_3$, the non-dimensional equation:

$$\left(1 - \varepsilon_0\vartheta + \ldots\right)\frac{D\overline{u}_3}{D\overline{t}} + \mathcal{A}_0\frac{\partial \pi}{\partial \overline{x}_3} - \frac{\mathrm{Ra}}{\mathrm{Pr}}\vartheta + \ldots$$
$$= \left(\frac{\partial^2}{\partial \overline{x}_1^2} + \frac{\partial^2}{\partial \overline{x}_2^2} + \frac{\partial^2}{\partial \overline{x}_3^2}\right)\overline{u}_3 + \varepsilon_0\left\{\frac{\lambda_0/\varrho_0}{\nu_0}\frac{\partial}{\partial \overline{x}_3}\left(\frac{D\vartheta}{D\overline{t}}\right)\right.$$
$$\left. - \frac{\alpha_0}{\beta_0}\frac{\partial}{\partial \overline{x}_j}\left[\vartheta\left(\frac{\partial \overline{u}_3}{\partial \overline{x}_j} + \frac{\partial \overline{u}_j}{\partial \overline{x}_3}\right)\right] + \ldots\right\} \quad , \quad j = 1, 2, 3 \quad . \tag{13.295}$$

Finally, from the energy equation (13.281c), we obtain the following non-dimensional equation (for ϑ):

$$\left(1 - \varepsilon_0\frac{\Gamma_0}{\beta_0}\vartheta\right)\left(1 - \varepsilon_0\vartheta + \ldots\right)\frac{D\vartheta}{D\overline{t}} + \varepsilon_0\left\{B_0(1 - \overline{x}_3) + \eta_0\mathcal{A}_0\pi\right\}\frac{D\vartheta}{D\overline{t}}$$
$$= \frac{1}{\mathrm{Pr}}\left(\frac{\partial^2}{\partial \overline{x}_1^2} + \frac{\partial^2}{\partial \overline{x}_2^2} + \frac{\partial^2}{\partial \overline{x}_3^2}\right)\vartheta - \varepsilon_0\frac{1}{\mathrm{Pr}}\frac{\gamma_0}{\beta_0}\frac{\partial}{\partial \overline{x}_i}\left(\vartheta\frac{\partial \vartheta}{\partial \overline{x}_i}\right) + \ldots$$
$$+\frac{1}{2}\eta_0\left(\frac{\partial \overline{u}_i}{\partial \overline{x}_j} + \frac{\partial \overline{u}_j}{\partial \overline{x}_i}\right)^2 + \ldots \quad ; \quad i, j = 1, 2, 3 \quad , \tag{13.296}$$

where:

$$B_0 = \frac{g d_0}{c_0 \Delta T_0} \equiv \frac{\eta_0}{\sigma_0} \quad , \quad \text{with} \quad \eta_0 = \frac{(\nu_0/d_0)^2}{c_0 \Delta T_0} \quad .$$

The parameter η plays the role of the square of a Mach number based on the temperature fluctuation ΔT_0. It will also be observed that:

$$\Gamma_0 \equiv -\frac{1}{c_0}\left(\frac{dc}{dT}\right)_{T=T_0} \quad \text{and} \quad \gamma_0 = -\frac{1}{k_0}\left(\frac{dk}{dT}\right)_{T=T_0} \quad .$$

In the non-dimensional equations (13.294–296), we have written only the dominant terms which are necessary for the derivation of Boussinesq limiting equations.

4. Let us now analyze the results obtained. First of all, it is obvious that:

$$\mathcal{A}_0 = \mathrm{O}(1) \quad \text{and} \quad \mathrm{Gr} = \frac{\mathrm{Ra}}{\mathrm{Pr}} = \mathrm{O}(1) \quad , \tag{13.297}$$

in (13.294–296). This is a trivial consequence of the choice which led to the "least possible degenerate" limiting equations (when $\varepsilon_0 \to 0$ with fixed $\overline{t}$, $\overline{x}_i$). However, when:

$$\varepsilon_0 \to 0 \quad \text{with fixed} \quad \overline{t}, \overline{x}_i \quad , \tag{13.298}$$

it follows that:

$$\sigma_0 \to 0 \quad , \quad \delta_0 \to 0 \quad \text{and} \quad \eta_0 \to 0 \quad , \tag{13.299}$$

since we have the following similarity relations:

$$\sigma_0 = \frac{\mathrm{Pr}}{\mathrm{Ra}}\varepsilon_0 \quad , \quad \delta_0 = \frac{\mathcal{A}_0}{\mathrm{Gr}}\varepsilon_0 \quad , \quad \eta_0 = \frac{B_0}{\mathrm{Gr}}\varepsilon_0 \quad , \tag{13.300}$$

B_0 also being assumed to be O(1). We will return to B_0 later on.

Therefore, when the following quantities are assumed to remain bounded when $\varepsilon_0 \to 0$:

$$\frac{\alpha_0}{\beta_0} \quad , \quad \frac{\Gamma_0}{\beta_0} \quad \text{and} \quad \frac{\gamma_0}{\beta_0} \quad ,$$

then through the limiting process (13.298) and with the similarity conditions (13.300), the following Boussinesq equations are obtained for the limiting functions:

$$\begin{pmatrix} \boldsymbol{v}_0 \\ w_0 \\ \vartheta_0 \\ \pi_0 \end{pmatrix} = \lim_{\varepsilon_0 \to 0}^{P} \begin{pmatrix} \overline{u}_1 \boldsymbol{i} + \overline{u}_2 \boldsymbol{j} \\ \overline{u}_3 \\ \vartheta \\ \pi \end{pmatrix} \quad , \tag{13.301}$$

where:

$$\lim_{\varepsilon_0 \to 0}^{P} \equiv \left(\varepsilon_0 \to 0 \quad \text{with} \quad \overline{t}, \overline{x}_i, \mathcal{A}_0, B_0 \text{ and Gr fixed}\right) \quad .$$

Once the bars have been omitted from the non-dimensional values, the Boussinesq equations can be written for $\boldsymbol{v}_0$, w_0, ϑ_0 and π_0 in the following form:

$$\boldsymbol{D}\cdot\boldsymbol{v}_0+\frac{\partial w_0}{\partial x_3}=0 \quad , \quad \boldsymbol{D}\equiv\frac{\partial}{\partial x_1}\boldsymbol{i}+\frac{\partial}{\partial x_2}\boldsymbol{j} \quad , \tag{13.302a}$$

$$\frac{\partial \boldsymbol{v}_0}{\partial t}+(\boldsymbol{v}_0\cdot\boldsymbol{D})\boldsymbol{v}_0+w_0\frac{\partial \boldsymbol{v}_0}{\partial x_3}+\boldsymbol{D}\pi_0=\left(\boldsymbol{D}^2+\frac{\partial^2}{\partial x_3^2}\right)\boldsymbol{v}_0 \quad ; \tag{13.302b}$$

$$\frac{\partial w_0}{\partial t}+\boldsymbol{v}_0\cdot\boldsymbol{D}w_0+w_0\frac{\partial w_0}{\partial x_3}+\frac{\partial \pi_0}{\partial x_3}-\mathrm{Gr}\,\vartheta_0=\left(\boldsymbol{D}^2+\frac{\partial^2}{\partial x_3^2}\right)w_0 \quad ; \tag{13.302c}$$

$$\frac{\partial \vartheta_0}{\partial t}+\boldsymbol{v}_0\cdot\boldsymbol{D}\vartheta_0+w_0\frac{\partial \vartheta_0}{\partial x_3}=\frac{1}{\mathrm{Pr}}\left(\boldsymbol{D}^2+\frac{\partial^2}{\partial x_3^2}\right)\vartheta_0 \quad , \tag{13.302d}$$

once it has been assumed that $\Delta p_0=\varrho_0(\nu_0/d_0)^2$, i.e., $\delta_0\equiv\sigma_0$. It will be seen that ΔT_0 must satisfy the following double inequality:

$$\frac{(\nu_0/d_0)^2}{c_0}\ll\Delta T_0\ll\frac{1}{\beta_0} \tag{13.303}$$

and hence, we also have:

$$\frac{\Delta p_0}{\Delta T_0}\ll\varrho_0 c_0 \quad . \tag{13.304}$$

Inequality (13.304) indicates that the characteristic *pressure fluctuation* Δp_0 is *always smaller than the characteristic temperature fluctuation* ΔT_0. This property justifies to a certain extent the state relation $\varrho=\varrho(T)$ with $e=e(T)$ which was adopted at the beginning. As long as $|(1/\beta_0^2)(d\beta/dT)_{T=T_0}-1|$ remains bounded when $\varepsilon_0\to 0$, this state relation leads to (13.287).

The following boundary conditions with respect to x_3 must be imposed on the Boussinesq equations (13.302):

$$\boldsymbol{v}_0=0 \quad , \quad w_0=0 \quad \text{on} \quad x_3=0 \quad \text{and} \quad x_3=1 \quad , \tag{13.305a}$$

$$\vartheta=1 \quad , \quad \text{on} \quad x_3=0 \quad \text{and} \quad \vartheta=0 \quad , \quad \text{on} \quad x_3=1 \quad . \tag{13.305b}$$

5. In studies which are relative to hydrodynamic instability, it is recommended to work with homogeneous conditions. We should, therefore, introduce the new temperature perturbation:

$$\mathrm{Pr}\,\Xi=\vartheta+x_3-1\Rightarrow\Xi=0 \quad , \quad \text{on} \quad x_3=0 \text{ and } x_3=1 \quad . \tag{13.306}$$

The following new pressure perturbation will appear in this case:

$$\Pi=\pi+\mathrm{Gr}\,x_3\left(\frac{x_3}{2}-1\right) \quad . \tag{13.307}$$

Thus the Rayleigh-Bénard convective instability problem consists in elucidating the stability of the following basic (dimensional) free convection:

$$\boldsymbol{v} \equiv 0 \quad , \quad w \equiv 0 \quad ;$$
$$T = T_0 + \Delta T_0\left(1 - \frac{x_3}{d_0}\right) \quad ; \tag{13.308}$$
$$p = g\varrho_0 d_0\left(1 - \frac{x_3}{d_0}\right) + \Delta p_0 \, \mathrm{Gr}\left(1 - \frac{x_3}{d_0}\right)\frac{x_3}{d_0} \quad ,$$

with respect to the perturbations $\boldsymbol{v}$, w, Ξ, Π which satisfy the following Boussinesq-type boundary problem (written in *non-dimensional form*):

$$\boldsymbol{D}\cdot\boldsymbol{v} + \frac{\partial w}{\partial x_3} = 0$$
$$\frac{\partial \boldsymbol{v}}{\partial t} + (\boldsymbol{v}\cdot\boldsymbol{D})\boldsymbol{v} + w\frac{\partial \boldsymbol{v}}{\partial x_3} + \boldsymbol{D}\Pi = \left(\boldsymbol{D}^2 + \frac{\partial^2}{\partial x_3^2}\right)\boldsymbol{v} \quad ;$$
$$\frac{\partial w}{\partial t} + \boldsymbol{v}\cdot\boldsymbol{D}w + w\frac{\partial w}{\partial x_3} + \frac{\partial \Pi}{\partial x_3} - \mathrm{Ra}\,\Xi = \left(\boldsymbol{D}^2 + \frac{\partial^2}{\partial x_3^2}\right)w \quad ; \tag{13.309}$$
$$\frac{\partial \Xi}{\partial t} + \boldsymbol{v}\cdot\boldsymbol{D}\Xi + w\frac{\partial \Xi}{\partial x_3} - \frac{1}{\mathrm{Pr}}w = \frac{1}{\mathrm{Pr}}\left(\boldsymbol{D}^2 + \frac{\partial^2}{\partial x_3^2}\right)\Xi \quad ,$$
$$\boldsymbol{v} = 0 \quad , \quad w = 0 \quad , \quad \Xi = 0 \quad , \quad \text{on } x_3 = 0 \text{ and } x_3 = 1 \quad .$$

Should it prove necessary, the following initial conditions can be assigned to (13.309):

$$t = 0: \quad \boldsymbol{v} = 0 \quad , \quad w = 0 \quad \text{and} \quad \Xi = 0 \quad . \tag{13.310}$$

There are no lateral conditions in x_1 and x_2 since the plane surfaces $x_3 = 0$ and $x_3 = 1$ are assumed to be infinite horizontal planes! Although this hypothesis is, in fact, very restrictive, it remains coherent most particularly when considering the exponential-type asymptotic stability within the framework of a linearized theory. In this case, the seeking of the *neutral stability curve* is synonymous with seeking a *non-zero* solution of the following eigenvalue problem:

$$\left[\left(\frac{d^2}{dx_3^2} - K_0^2\right)^3 + \mathrm{Ra}\,K_0^2\right]W(x_3) = 0 \quad ;$$
$$W = \frac{dW}{dx_3} = \left(\frac{d^2}{dx_3^2} - K_0^2\right)^2 W = 0, \quad \text{for } x_3 = 0 \text{ and } x_3 = 1\,, \tag{13.311}$$

where $K_0 = \text{const}$. The horizontal, dimensionless, wave number K_0 is related to a linear dimension of the convection cells in the plane $x_3 = 0$. In (13.311), we have adopted the hypothesis $\mathrm{Pr} \equiv 1$ which is only slightly restrictive.

6. An important point is related to the value of Ra which is very large compared to unity! It would thus be very interesting to elucidate the limiting form of problem (13.309, 310) when $\mathrm{Ra} \to +\infty$. As a first step, just such a study could be carried out on linear problem (13.311) which has a known solution. It is pointed out that according to (13.292), $\mathrm{Ra} \gg 1$ signifies (when $\mathrm{Pr} \equiv 1$) that $\varepsilon_0 \gg \sigma_0$. In other

words, the limiting form of the basic equations (13.294–296) must be clarified when:

$$\varepsilon_0 \text{ fixed} \quad ; \quad \sigma_0 \to 0 \quad , \quad \text{then} \quad \varepsilon_0 \to 0 \quad . \tag{13.312}$$

As a matter of fact, the above is somewhat more complicated given relations (13.300) since it is necessary to compare δ_0 and η_0 to σ_0. If we wish for $\mathcal{A}_0 = 1$, then $\delta_0 = \sigma_0$ and likewise if $B_0 = 1$, then $\eta_0 = \sigma_0$. Hence, the following limiting situation must be considered:

$$\begin{aligned} &\varepsilon_0 \text{ fixed} \quad ; \quad \delta_0 \to 0 \quad , \quad \eta_0 \to 0 \quad \text{and} \quad \sigma_0 \to 0 \quad , \\ &\text{in such a way that} \quad \mathcal{A}_0 \equiv 1 \quad \text{and} \quad B_0 \equiv 1, \quad \text{then} \quad \varepsilon_0 \to 0 . \end{aligned} \tag{13.313}$$

Unfortunately, both (13.312, 313) lead us to impose $\vartheta = 0$ at the limit. The explanation of this strong degeneracy when $\mathrm{Ra} \to \infty$ is related to a double-scale singular structure in x_3, at least in the archetypal problem (13.311).

7. Let us now turn back to the expression $B_0 = \eta_0/\sigma_0$. Since $\sigma_0 \ll 1$, the hypothesis that $\eta_0 \ll 1$ generally leads us to suppose that $B_0 = \mathrm{O}(1)$. The established Boussinesq-type asymptotic theory thus remains valid if:

$$d_0 \sim \frac{c_0 \Delta T_0}{g} \quad . \tag{13.314}$$

This estimate of d_0 is only meaningful if η_0 is of the order of $\sigma_0 \ll 1$. On the other hand, if η_0 is *of the order unity*, then:

$$\varepsilon_0 B_0 \equiv \eta_0 \frac{\varepsilon_0}{\sigma_0} = \eta_0 \, \mathrm{Gr} \quad .$$

In this case, (13.302d) for ϑ_0 must be replaced by the following more complete limiting equation:

$$\begin{aligned} &\{1 + \mathrm{Gr}\, \eta_0 (1 - x_3)\} \left\{ \frac{\partial \vartheta_0}{\partial t} + \boldsymbol{v}_0 \cdot \boldsymbol{D} \vartheta_0 + w_0 \frac{\partial \vartheta_0}{\partial x_3} \right\} \\ &= \frac{1}{\mathrm{Pr}} \left(\boldsymbol{D}^2 + \frac{\partial^2}{\partial x_3^2} \right) \vartheta_0 + \frac{1}{2} \eta_0 \left(\frac{\partial u_i}{\partial x_j} + \frac{\partial u_j}{\partial x_i} \right)_0^2 \end{aligned} \tag{13.315}$$

where $(u_1, u_2)_0 \equiv \boldsymbol{v}_0$ and $(u_3)_0 \equiv w_0$. In this case, the following estimate must be considered instead of (13.314):

$$d_0 \sim \frac{\nu_0}{(c_0 \Delta T_0)^{1/2}} \quad . \tag{13.316}$$

Appendix

The Hydrostatic Forecasting Equations for Large-Synoptic-Scale Atmospheric Processes

In this Appendix, we formulate concisely but consistently the system of so-called hydrostatic non-adiabatic, viscous model equations for large, non-tangent, synoptic-scale atmospheric processes. For this we use full, exact Navier-Stokes atmospheric equations in spherical coordinates and three small parameters, namely the Mach number, the inverse Reynolds number and the hydrostatic parameter (the quotient between the height scale H and the horizontal length scale L). The first two are related in a definite way to the last.

We note that in a quite realistic meteorological situation, the non-adiabatic and viscous effects are only important in a thin layer in the vicinity of the ground and we formulate corresponding large-scale, synoptic, boundary layer equations. A brief account is given concerning the initial and upper boundary conditions.

A.1 The Governing Equations

The hydrostatic large-synoptic-scale model is necessary if we want to produce a weather prediction for more than three to four days on a sufficiently large area of the earth's sphere.

At present, the hydrostatic equations in most common use unfortunately do not seem asymptotically consistent from the point of view adopted here, in that the limiting model equations derived are not the ones which can issue from the full, exact Navier-Stokes atmospheric equations [see in Chap. 2 the equations (2.112)] under an expansion with respect to $\varepsilon = H/L \ll 1$, where L is assumed to be of the same order as the earth's radius a_0 and $L/a_0 = \delta = O(1)$; H is the height of the troposphere, which is used as a characteristic vertical length. Concerning the "non-consistent" hydrostatic equations, see, for instance, the review article by Cullen (1983).

The complete, consistent derivation of the hydrostatic model equations by the matched asymptotic expansion technique uses, concurrently with the main small parameter ε, two others:

$$\mathrm{M}_\infty^2 = \frac{U_0^2}{\gamma R T_\infty(0)} \quad \text{and} \quad \alpha \equiv \mathrm{Re}^{-1} = \frac{\mu_0/\varrho_\infty(0)}{U_0 L} \quad .$$

Here μ_0 is the eddy dynamic viscosity on the ground and we assume that the eddy viscosity $\mu(p)$, thermal eddy conductivity $k(p)$ and heat source $Q(p)$ are functions only of the pressure p. Doing this, we consider only a mean, standard distribution for μ, k and Q and ignore variations therefrom for the perturbed atmosphere.

Our main similarity relations are the following:

$$\frac{\varepsilon}{\gamma \mathrm{M}_\infty^2} = \ni \quad \text{and} \quad \frac{\varepsilon^2}{\alpha} = \mathrm{Re}_\perp \quad , \tag{1}$$

where $\ni$ and $\mathrm{Re}_\perp$ are similarity parameters of order O(1), and our main limiting process is then

$$\varepsilon \to 0 \quad , \quad \text{with} \quad \ni \quad , \quad \mathrm{Re}_\perp \quad \text{and} \quad t, x, y, z \quad \text{fixed} \quad . \tag{2}$$

Here t is the non-dimensional time and

$$x = \frac{\cos \varphi_0}{\delta}\lambda \quad , \quad y = \frac{\varphi - \varphi_0}{\delta} \quad , \quad z = \frac{r-1}{\varepsilon\delta} \quad , \tag{3}$$

where λ is the positive longitude in the eastward direction, φ the latitude ($\varphi_0 = \mathrm{const}$ is a reference latitude) and r the distance to the center of the earth ($r \equiv 1$ on flat ground).

If we use dimensionless variables (3) and an advective time scale (L/U_0), we may write the full Navier-Stokes atmospheric equations in the following concise manner:

$$\frac{D\varrho}{Dt} + \varrho\left\{\frac{\partial w}{\partial z} + \frac{1}{1+\varepsilon\delta z}[\boldsymbol{D}\cdot\boldsymbol{v} - \delta \tan\varphi v + 2\varepsilon\delta w]\right\} = 0 \quad ; \tag{4}$$

$$\varrho\left\{\frac{D\boldsymbol{v}}{Dt} + \left(\frac{1}{\mathrm{Ro}}\frac{\sin\varphi}{\sin\varphi_0} + \frac{\delta\tan\varphi}{1+\varepsilon\delta z}u\right)(\boldsymbol{k}\times\boldsymbol{v})\right\}$$

$$+\frac{\ni}{1+\varepsilon\delta z}\frac{1}{\varepsilon}\boldsymbol{D}p = \frac{1}{\mathrm{Re}_\perp}\frac{\partial}{\partial z}\left[\mu\frac{\partial \boldsymbol{v}}{\partial z}\right] + \mathrm{O}(\varepsilon) \quad ; \tag{5}$$

$$\ni\left(\frac{\partial p}{\partial z} + \mathrm{Bo}\varrho\right) = \mathrm{O}(\varepsilon^2) \quad ; \tag{6}$$

$$\varrho\frac{DT}{Dt} - \frac{\gamma-1}{\gamma}\frac{Dp}{Dt} = \frac{1}{\mathrm{Re}_\perp}\frac{1}{\mathrm{Pr}}\left\{\frac{\partial}{\partial z}\left[k\frac{\partial T}{\partial z}\right] + \mathrm{Pr}\frac{\gamma-1}{\gamma}\frac{\varepsilon}{\ni}\mu\chi\right.$$

$$\left.+\mathrm{Bo}\,\sigma_{00}\frac{d\mathcal{R}}{dz}\right\} + \mathrm{O}(\varepsilon^2) \quad ; \tag{7}$$

$$p = \varrho T \quad , \tag{8}$$

where $\mathrm{O}(\varepsilon)$ stands for terms vanishing under the limiting process (2). In these equations (4–8)

$$\frac{D}{Dt} = \frac{\partial}{\partial t} + \frac{\boldsymbol{v}\cdot\boldsymbol{D}}{1+\varepsilon\delta z} + w\frac{\partial}{\partial z} \quad , \tag{9}$$

where $\boldsymbol{v}$ is the horizontal velocity vector with components u and v, while $\boldsymbol{D}$ is the horizontal gradient with components

$$\frac{\cos\varphi_0}{\cos\varphi}\frac{\partial}{\partial x} \quad \text{and} \quad \frac{\partial}{\partial y} \quad .$$

Notice that the vertical component velocity w has been non-dimensionalized with εU_0. The assumption made about H implies that the Boussinesq number $\mathrm{Bo} = gH/RT_\infty(0)$ is of order unity. Finally, the expression for the dissipation function is

$$\chi = \frac{1}{(1+\varepsilon\delta z)^2}\left|\frac{\partial}{\partial z}\left(\frac{\boldsymbol{v}}{1+\varepsilon\delta z}\right)\right|^2 + \mathrm{O}(\varepsilon^2) \quad , \tag{10}$$

and

$$\varphi = \varphi_0 + \delta y \quad , \quad \boldsymbol{D}\cdot\boldsymbol{k} = 0 \quad .$$

A.2 The Hydrostatic Model Equations

The general hydrostatic model equations are obtained through the limiting process (2) applied to the Eqs. (4–8), with (10), keeping the other parameters fixed during the process.

First, under the limiting process (2), it is found that the limiting form of the vertical component of the momentum equation (6) is

$$\frac{\partial p}{\partial z} + \mathrm{Bo}\,\varrho = 0 \Rightarrow \frac{\partial}{\partial z} = -\mathrm{Bo}\,\varrho\frac{\partial}{\partial p} \quad , \tag{11}$$

with an error of order $\mathrm{O}(\varepsilon^2)$. This makes it possible, in a standard way (see Subsect. 2.4.3) to change the independent variables t, x, y and z to τ, ξ, η, ζ, with $\tau \equiv t$, $\xi \equiv x$, $\eta \equiv y$ and $\zeta \equiv p$, and then

$$z = \mathcal{H}(\tau, \xi, \eta, \zeta)$$

has to be considered one of the unknown functions. We keep the notation $\boldsymbol{D}$ for the horizontal gradient on the isobaric $\zeta = \text{const}$ surface (with the components $(\cos\varphi_0/\cos\varphi)(\partial/\partial\xi)$ and $\partial/\partial\eta$) and we set

$$\frac{D}{D\tau} = \frac{\partial}{\partial\tau} + \boldsymbol{v}\cdot\boldsymbol{D} + \omega\frac{\partial}{\partial\zeta} \quad , \tag{12}$$

where $\omega = Dp/Dt$ is the vertical pseudo-velocity, that is, the rate of variation of pressure following the particles. Afterwards, we expand $\mathcal{H}$, T, ϱ, $\boldsymbol{v}$ and ω according to the following scheme:

$$\begin{aligned} \mathcal{H} &= \mathcal{H}_0(\zeta) + \varepsilon\mathcal{H}_1 + \ldots \quad ; \\ T &= T_0(\zeta) + \varepsilon T_1 + \ldots \quad ; \\ \varrho &= \varrho_0(\zeta) + \varepsilon\varrho_1 + \ldots \quad ; \\ \boldsymbol{v} &= \boldsymbol{v}_0 + \ldots \quad ; \\ \omega &= \omega_0 + \ldots \quad . \end{aligned} \tag{13}$$

Obviously $\mathcal{H}_0$, T_0 and ϱ_0 do not depend on the horizontal variables ξ, η, and so we can assume that they do not depend on τ either. Although this does not follow directly from the equations, it will be found to be consistent with the constancy of $\mathcal{R}(p)$, which has been assumed previously. Of course, we have

$$\mathrm{Bo}\zeta\frac{d\mathcal{H}_0}{d\zeta} + T_0 = 0 \quad \text{and} \quad \varrho_0 = \frac{\zeta}{T_0} \quad , \tag{14}$$

but we do not know yet how T_0 depends on ζ. On reflection, we see that as a consequence of the energy equation, the standard atmosphere with $(\mathcal{H}_0, T_0, \varrho_0)$ a function of ζ alone, is obtained from a heat balance (see, for example, Sect. 11.1):

$$\varrho_0 k\frac{dT_0}{d\zeta} = \sigma_{00}\mathcal{R}(\zeta) \quad . \tag{15}$$

However, if we want to derive the limiting energy equation in a coherent way, then it is necessary to define a new similarity parameter

$$\mathcal{N} \equiv \frac{\alpha_0^0}{\ni} = \mathrm{O}(1) \quad , \tag{16}$$

where α_0^0 is a characteristic stability parameter for the standard atmosphere:

$$-\frac{\zeta}{\mathrm{Bo}}\left[\frac{dT_0}{d\zeta} - \frac{\gamma-1}{\gamma}\frac{T_0}{\zeta}\right] \equiv \alpha_0^0 \Gamma_0(\zeta) \quad . \tag{17}$$

In this case, instead of (2), we must consider the following limiting process:

$$\varepsilon \to 0 \quad , \quad \text{with} \quad \ni, \mathrm{Re}_\perp, \mathcal{N} \quad \text{and} \quad \tau, \xi, \eta, \zeta \quad \text{fixed} \quad . \tag{18}$$

According to the scheme (13–18) we find, from the basic, exact Eqs. (4–8) (rewritten in the τ, ξ, η and ζ variables, for $\boldsymbol{v}, \mathcal{H}, T, \omega$ and ϱ), first that

$$\frac{D_0}{D\tau}\boldsymbol{v}_0 + \left(\frac{1}{\mathrm{Ro}}\frac{\sin\varphi}{\sin\varphi_0} + \delta\tan\varphi u_0\right)(\boldsymbol{k}\times\boldsymbol{v}_0) + \ni \boldsymbol{D}\mathcal{H}_1$$
$$= \frac{\mathrm{Bo}^2}{\mathrm{Re}_\perp}\frac{\partial}{\partial\zeta}\left(\varrho_0\mu\frac{\partial\boldsymbol{v}_0}{\partial\zeta}\right) \quad ; \tag{19a}$$

$$\boldsymbol{D}\cdot\boldsymbol{v}_0 + \frac{\partial\omega_0}{\partial\zeta} = \delta\tan\varphi v_0 \quad . \tag{19b}$$

Going to higher order, we then find:

$$T_1 + \mathrm{Bo}\zeta\frac{\partial\mathcal{H}_1}{\partial\zeta} = 0 \quad , \quad \varrho_1 = -\frac{\varrho_0}{T_0}T_1 \quad ; \tag{19c}$$

$$\frac{D_0 T_1}{D\tau} - \frac{\gamma-1}{\gamma}\frac{T_1}{\zeta}\omega_0 - \mathcal{N}\mathrm{Bo}\frac{\Gamma_0(\zeta)}{\zeta}\omega_0$$
$$= \frac{\mathrm{Bo}^2}{\mathrm{Re}_\perp}\frac{1}{\mathrm{Pr}}\left\{\frac{\partial}{\partial\zeta}\left[\varrho_0 k\left(\frac{\partial T_1}{\partial\zeta} - \frac{d\log T_0}{d\zeta}T_1\right)\right]\right.$$
$$\left. + \mathrm{Pr}\frac{\gamma-1}{\gamma}\frac{1}{\ni}\varrho_0\mu\left|\frac{\partial\boldsymbol{v}_0}{\partial\zeta}\right|^2\right\} \quad . \tag{19d}$$

The system of Eqs.(19) is the limiting system of hydrostatic model equations, where

$$\frac{D_0}{D\tau} = \frac{\partial}{\partial\tau} + \boldsymbol{v}_0 \cdot \boldsymbol{D} + \omega_0 \frac{\partial}{\partial\zeta} \quad .$$

If for the basic, exact equations (in the τ, ξ, η, ζ variables) we set, on flat ground,

$$\left.\begin{aligned} \boldsymbol{v} = 0 \quad , \quad \omega = \mathrm{Bo}\varrho \frac{\partial \mathcal{H}}{\partial \tau} \quad , \\ \varrho k \frac{\partial T}{\partial \zeta} = \sigma_{00} \mathcal{R}(\zeta) \end{aligned}\right\} \quad \text{on } \mathcal{H} = 0 \quad , \tag{20}$$

then we find for the limiting system (19) the following boundary conditions:

$$\boldsymbol{v}_0 = 0 \quad , \quad \omega_0 = 0 \quad , \quad \frac{\partial T_1}{\partial \zeta} = \frac{d \log T_0}{d\zeta} T_1 \quad , \quad \text{on } \zeta = 1 \quad , \tag{21}$$

if use is made of the condition $\mathcal{H}_0(1) = 0$, on the ground.

Considering the hydrostatic model equations (19), we must give only the initial values of $\boldsymbol{v}_0$ and $\mathcal{H}_1$, and they have nothing to do with the corresponding initial conditions for the basic, exact equations. Consequently, it is necessary to formulate for the hydrostatic model equations (19) an adjustment problem analogous to the one that was considered in Sect. 7.5 of Chap. 7 for the primitive equations. This problem has been examined concisely in the framework of Navier-Stokes equations by Zeytounian (1980) and also in Sect. 7.7 of Chap. 7.

We leave unspecified the conditions at high altitude and in the horizontal directions for the basic, exact equations in as much as this does not alter the derivation of the limiting hydrostatic model equations (19).

If in the hydrostatic model equations (19) we suppose that $\delta \to 0$, then it is necessary also to assume that $\mathrm{Ro} \to 0$. In this case we can derive the corresponding limiting equations for the quasi-geostrophic and ageostrophic models if we consider the following limiting process:

$$\begin{aligned} \delta \to 0 \quad \text{and} \quad \mathrm{Ro} \to 0 \quad , \text{ with } \quad \beta = \frac{\delta}{\tan \varphi_0 \mathrm{Ro}} \\ \text{and} \quad \tau, \, \xi, \, \eta, \, \zeta \quad \text{fixed} \quad . \end{aligned} \tag{22}$$

A.3 The Large-Scale, Synoptic, Boundary Layer Equations

In a quite realistic meteorological situation, the non-adiabatic and viscous effects are important only in a thin layer in the vicinity of the ground. Mathematically, this is the consequence of assuming $\mathrm{Re}_\perp \gg 1$ in the hydrostatic full model equations (19). When $\mathrm{Re}_\perp \gg 1$, it is necessary to consider two new limiting processes:

$$\mathrm{Re}_\perp \to \infty \quad , \quad \text{with} \quad \tau, \, \xi, \, \eta \quad \text{and} \quad \zeta \text{ fixed} \quad , \tag{23}$$

and

$$\mathrm{Re}_\perp \to \infty \quad , \quad \text{with} \quad \tau, \xi, \eta \quad \text{and} \quad \hat{\zeta} = \frac{\zeta - 1}{\mathrm{Re}_\perp^{-1/2}} \quad \text{fixed} \quad . \tag{24}$$

According to (23), we find, first, the generalized, non-tangent primitive equations:

$$\begin{aligned}
&\frac{D_0 \boldsymbol{v}_0}{D\tau} + \left(\frac{1}{\mathrm{Ro}} \frac{\sin \varphi}{\sin \varphi_0} + \delta \tan \varphi u_0\right)(\boldsymbol{k} \times \boldsymbol{v}_0) + \ni \boldsymbol{D}\mathcal{H}_1 = 0 \quad ; \\
&\boldsymbol{D} \cdot \boldsymbol{v}_0 + \frac{\partial \omega_0}{\partial \zeta} = \delta \tan \varphi v_0 \quad ; \\
&T_1 + \mathrm{Bo}\zeta \frac{\partial \mathcal{H}_1}{\partial \zeta} = 0 \quad ; \quad \varrho_1 = -\frac{\varrho_0}{T_0} T_1 \quad ; \\
&\frac{D_0 T_1}{D\tau} - \frac{\gamma - 1}{\gamma} \frac{T_1}{\zeta} \omega_0 = \mathcal{N} \mathrm{Bo} \frac{\Gamma_0(\zeta)}{\zeta} \omega_0 \quad .
\end{aligned} \tag{25}$$

For Eqs. (25) we have the possibility of assuming only:

$$\omega_0 = 0 \quad , \quad \text{on} \quad \zeta = 1 \quad . \tag{26}$$

In order to derive the proper boundary layer equations associated with Eqs. (25), we must consider (24) and define a new $\hat{\omega}_0 = \omega_0/\mathrm{Re}_\perp^{-1/2}$ as a function of τ, ξ, η and $\hat{\zeta}$. We set $\hat{\boldsymbol{v}}_0$, $\hat{\mathcal{H}}_1$, $\hat{T}_1$ and $\hat{\varrho}_1$ for $\boldsymbol{v}_0$, $\mathcal{H}_1$, T_1 and ϱ_1 considered as functions of $\tau, \xi, \eta, \hat{\zeta}$.

First, we find that

$$\frac{\partial \hat{\mathcal{H}}_1}{\partial \hat{\zeta}} = 0 \Rightarrow \hat{\mathcal{H}}_1 = \mathcal{H}_1(\tau, \xi, \eta, 1) \equiv \mathcal{H}_1^1(\tau, \xi, \eta) \quad , \tag{27}$$

and we obtain for $\hat{\boldsymbol{v}}_0$ the following boundary layer equation:

$$\begin{aligned}
&\frac{\hat{D}_0 \hat{\boldsymbol{v}}_0}{D\tau} + \left(\frac{1}{\mathrm{Ro}} \frac{\sin \varphi}{\sin \varphi_0} + \delta \tan \varphi \hat{u}_0\right)(\boldsymbol{k} \times \hat{\boldsymbol{v}}_0) + \ni \boldsymbol{D}\mathcal{H}_1^1 \\
&\quad = \mathrm{Bo}^2 \varrho_0(1)\mu(1) \frac{\partial^2 \hat{\boldsymbol{v}}_0}{\partial \hat{\zeta}^2} \quad .
\end{aligned} \tag{28}$$

In (28) we have used the fact that μ (which characterizes the standard atmosphere) has no boundary layer structure. In (28)

$$\frac{\hat{D}_0}{D\tau} = \frac{\partial}{\partial \tau} + \hat{\boldsymbol{v}}_0 \cdot \boldsymbol{D} + \hat{\omega}_0 \frac{\partial}{\partial \hat{\zeta}} \quad \text{and} \quad \hat{\boldsymbol{v}}_0 = \hat{u}_0 \boldsymbol{i} + \hat{v}_0 \boldsymbol{j} \quad .$$

For (28), we have the following two boundary conditions:

$$\hat{\boldsymbol{v}}_0 = 0 \quad , \quad \text{on} \quad \hat{\zeta} = 0 \quad ; \tag{29a}$$

$$\lim_{\hat{\zeta} \to +\infty} \hat{\boldsymbol{v}}_0 = \boldsymbol{v}_0(\tau, \xi, \eta, 1) \equiv \boldsymbol{v}_0^1(\tau, \xi, \eta) \quad . \tag{29b}$$

In the boundary layer, instead of (19b), we have

$$\boldsymbol{D} \cdot \hat{\boldsymbol{v}}_0 + \frac{\partial \hat{\omega}_0}{\partial \hat{\zeta}} = \delta \tan \varphi \hat{v}_0 \quad , \quad \text{with}$$

$$\hat{\omega}_0 = 0 \quad , \quad \text{on} \quad \hat{\zeta} = 0 \quad , \tag{30}$$

and for $\hat{\varrho}_1, \hat{T}_1$ we find

$$\hat{\varrho}_1 = -\frac{\varrho_0(1)}{T_0(1)}\hat{T}_1 \quad , \tag{31}$$

$$\frac{\hat{D}_0\hat{T}_1}{D\tau} = \frac{\mathrm{Bo}^2}{\mathrm{Pr}}\varrho_0(1)k(1)\left\{\frac{\partial^2\hat{T}_1}{\partial\hat{\zeta}^2} + \mathrm{Pr}\frac{\gamma-1}{\gamma}\frac{1}{\ni}\frac{\mu(1)}{k(1)}\left|\frac{\partial\hat{\boldsymbol{v}}_0}{\partial\hat{\zeta}}\right|^2\right\} \quad . \tag{32}$$

Equation (32) for $\hat{T}_1$ should be resolved with the following two boundary conditions:

$$\frac{\partial\hat{T}_1}{\partial\hat{\zeta}} = 0 \quad , \quad \text{on} \quad \hat{\zeta} = 0 \quad ;$$

$$\lim_{\hat{\zeta}\to+\infty} \hat{T}_1 = T_1(\tau,\xi,\eta,1) \equiv T_1^1(\tau,\xi,\eta) \quad . \tag{33}$$

Equations (28, 30–32), with (29) and (33) are the boundary layer model equations and conditions for the large-scale synoptic motions.

Consequently, whenever a numerical, adiabatic weather prediction by primitive equations (25), with (26), has been achieved, we have the opportunity to carry over a numerical "non-adiabatic", boundary layer weather prediction from the model equations and boundary conditions (28–33).

Concerning initial conditions for (25), see Sect. 7.5 of Chapt. 7. Following El Mabrouk and Zeytounian (1984), we note that the condition

$$\lim_{\zeta\to 0}\left\{\frac{\zeta^2}{\Gamma_0(\zeta)}\mathcal{H}_1\frac{\partial\mathcal{H}_1}{\partial\zeta}\right\} = 0 \quad . \tag{34}$$

gives a necessary condition for posing the initial boundary-value problem related to (25) correctly. The adjustment problem for the boundary layer equations (28, 30–32), is at present still incompletely formulated despite the result of Zeytounian (1980).

Finally, assuming that the generalized non-tangent, primitive model equations (25), with (26, 34), and initial conditions (which are derived from the adjustment to hydrostatic balance for the adiabatic atmosphere), have been solved, we know $\boldsymbol{v}_0$, $\mathcal{H}_1$, ϱ_1, ω_0, T_1 and we may compute $\boldsymbol{v}_0^1$, $\mathcal{H}_1^1$, T_1^1, as functions of τ, ξ and η. Then the boundary layer equations (28, 30–32) for $\hat{\boldsymbol{v}}_0$, $\hat{\omega}_0$ and $\hat{T}_1$ have to be solved with the boundary conditions (29) and (33) on $\hat{\zeta} = 0$ and $\hat{\zeta} \to +\infty$, and also the initial conditions at $\tau = 0$. These last initial conditions may be viewed as the result of the final phase of the adjustment process to boundary layer equations which are consistent only for $\tau = \mathrm{O}(1)$.

References

Ahmed, B.M., Eltayeb, I.A. (1980): Philos. Trans. R. Soc. London A **1435**, **298**, 45–85

Arakawa, A. (1962): Proc. Int. Symp. on Num. Weather Pred. in Tokyo, 1960. Met. Soc. of Japan, pp. 122–144 (Russian ed. Gidromet. Izd., Leningrad 1967)

Arnold, V.I. (1965): Doklay Akad. Nauk SSSR (in Russian) **162**(5)

Atkinson, B.W. (1981): *Meso-scale Atmospheric Circulations* (Academic Press, London)

Baer, F. (1977): Beitr. Phys. der Atmosphäre **50**(3), 350–366

Batchelor, G.K. (1953): Q. J. R. Meteorol. Soc. **79**, 224

Beer, T. (1974): *Atmospheric Waves* (Adam Hilger, Bristol)

Bengtsson, L., Ghil, M, Källen, E. (eds.) (1981): Dynamic Meteorology. Data Assimilation Methods. Applied Math. Sci., Vol. 36 (Springer, Berlin, Heidelberg)

Benney, D.J. (1964): J. Fluid Mech. **18**, 385–391

Berestov, A.L., Monin, A.S. (1980): Adv. Mechanics **3**(3), 3–34 (in Russian)

Birkoff, G. (1960): *Hydrodynamics* (a study in logic, fact and similitude) (Princeton Univ. Press, Princeton, NJ)

Bois, P.A. (1976): J. Mec. **15**, 781

Bois, P.A. (1979a): J. Mec. **18**, 395

Bois, P.A. (1979b): Techniques asymptotiques pour les problèmes de gaz pesants avec application aux ondes atmosphériques. Thèse, Université de Paris VI

Bois, P.A. (1982): J. Rech. Atmos. **16**(2), 97

Bois, P.A. (1984a): Asymptotic theory of Boussinesq waves in the atmosphere. Pub. IRMA, Univ. de Lille I, Vol. VI, Fasc. 4, No. 2

Bois, P.A. (1984b): Geophys. Astr. Fluid Dyn. **29**, 267–303

Bolin, B. (1953): Tellus **5**(3), 373–385

Boussinesq, J. (1903): Théorie analytique de la chaleur. Vol. II (Gauthier-Villars, Paris) (see Leçons 34 and 35)

Blumen, W. (1968): J. Atmos. Sci. **29**, 929

Blumen, W. (1972): Rev. Geophys. Space Phys. **10**(2), 485–528

Brighton, P.W.M. (1977): "Boundary Layer and Stratified Flow over Obstacles", Ph. D. Diss., University of Cambridge

Brighton, P.W.M. (1978): Q. J. R. Meteorol. Soc. **104**, 289–307

Burtsev, A.I. (1969): WMO Regional Training Seminar (1965). Lectures on Numerical Short-Range Weather Prediction (Hydrometeo. Izd., Leningrad) pp. 210–220

Bykov, V.V. (1962): Izv. Akad. Nauk SSSR, Ser. Geofiz. (3), 418–423

Cahn, A. (1945): J. Meteorol. **2**(2), 113–119

Chapman, S., Lindzen, R.S. (1970): *Atmospheric Tides* (Gordon and Breach, New York)

Charney, J.G. (1948): Geofys. Publ. **17**, (2), 17 (MR **14**, 428)

Charney, J.G. (1949): J. Meteorol. **6**(6), 371

Charney, J.G. (1962): Proc. Int. Symp. on Num. Weather Pred. in Tokyo, 1960, Met. Soc. of Japan, pp. 82–111 (Russian ed. Gidromet. Izd., Leningrad 1967)

Charney, J.G. (1971): J. Atmos. Sci. **28**, 1087–1095

Coddington, E.A., Levinson, N. (1955): *Theory of Ordinary Differential Equations* (McGraw-Hill, New York)

Cole, J.D. (1968): *Perturbation Methods in Applied Mathematics* (Blaisdell, Waltham, Mass.)

Courant, R., Hilbert, D. (1966): *Methods of Mathematical Physics*, Vol. II (Interscience, New York)

Crow, S.C. (1970): Stud. Appl. Math. **49**, 21–44

Cullen, M.J.P. (1983): J. Comput. Phys. **50**, 1

Darnaudet, F., Bois, P.A. (1984): C.R. Acad. Sci., series II **299**(18), 1233–1236

Darnaudet, F. (1985): "Calcul d'ondes de relief tridimensionelles par voie asymptotique et numérique", Thèse, Université de Paris VI
Darrozes, J.S. (1972): Fluid Dyn. Trans. **6**(II), 119
Dikij, L.A. (1961): Izv. Akad. Nauk SSSR, Ser. Geofiz. (5), 756
Dikij, L.A. (1969): *The Theory of Oscillations of the Earth's Atmosphere* (Gidrometeoizdat, Moscow) (in Russian)
Dikij, L.A. (1976): *Stabilité Hydrodynamique et Dynamique de l'Atmosphère* (Gidrometeoizdat, Leningrad) (in Russian)
Dodd, R.K., Eilbeck, J.C., Gibbon, J.D., Morris, H.C. (1982): *Solitons and Nonlinear Wave Equations* (Academic Press, London)
Dorodnitsyn, A.A. (1940): Works of GGO, No. 31 (in Russian)
Dorodnitsyn, A.A. (1950): Tr. Tsentr. Inst. Prognozov **21** (48), 3–25
Drazin, P.G. (1961): Tellus **13**, 239–251
Drazin, P.G., Reid, W.H. (1981): *Hydrodynamic Stability* (Cambridge Univ. Press, Cambridge)
Dubreil-Jacotin, M.L. (1935): Atti Accad. Lincei Rend. Sci. Fis. Mat. Nat. **21**(6), 344–346
Dutton, J.A., Fichtl, G.H. (1969): J. Atmos. Sci. **26**, 241
Dutton, J.A. (1974): J. Atmos. Sci. **31**(2), 242
Eady, E.T. (1949): Tellus **1**, 33–52
Eckart, C. (1960): *Hydrodynamics of Oceans and Atmospheres* (Pergamon, Oxford)
Eckaus, W. (1979): *Asymptotic Analysis of Singular Perturbations* (North-Holland, Amsterdam)
Eliassen, A. (1949): Geofys. Publ. **17**, 3
El Mabrouk, M., Zeytounian, R.Kh. (1984): Revue Roum. Math. Pures Appl. **29**(3), 235
Ertel, H. (1941): Meteorol. Z. **58**(3), 77
Ertel, H. (1942): Meteorol. Z. **59**, 277–281
Eskinazi, S. (1975): *Fluid Mechanics and Thermodynamics of Our Environment* (Academic Press, New York)
Favard. J. (1963): Cours d'analyse de l'Ecole Polytechnique. **3**(2), 128–131 and 153–155
Fjelstad, S.E. (1958): Geofys. Publ. **20** (7), 1
Fjortoff, R. (1950): Geofys. Publ. **17**(6), 1–52
Friedlander, S. (1980): *An Introduction to the Mathematical Theory of Geophysical Fluid Dynamics* (North-Holland, Amsterdam)
Fromm, J.E. (1968): J. Comput. Phys. **3**, 176–189
Fromm, J.E. (1969): Phys. Fluids **12**, Suppl. II, pp. II,3–II,12
Germain, P. (1977): *Fluid Dynamics* (Gordon and Breach, London) pp. 1–147
Giese, J.H. (1951): J. Math. Phys. **1**(30), 31–35
Gjevik, B., Marthinsen, T. (1977): Q. J. R. Meteorol. Soc. **104**, 947–957
Goots, St. (1985): Une modèlisation asymptotique des courants océaniques. Thèse, Université de Lille I
Gough, D.O. (1969): J. Atmos. Sci. **26** (3), 448–456
Greenspan, H.P. (1968): *The Theory of Rotating Fluids* (Cambridge Univ. Press, Cambridge)
Grimshaw, R. (1972): J. Fluid Mech. **54**, 193–207
Grimshaw, R. (1975): J. Fluid Mech. **70**, 287–304
Guiraud, J.P. (1979): C.R. Acad. Sci. A **288**, 435–437
Guiraud, J.P. (1983): Some examples of application of asymptotic techniques to the derivation of models for atmosphere flows. CISM, Von Karman Session, October 1983, Udine (Italy)
Guiraud, J.P., Zeytounian, R.Kh. (1979): Geophys. Astr. Fluid Dyn. **12**(1/2), 61–72
Guiraud, J.P., Zeytounian, R.Kh. (1980): Geophys. Astr. Fluid Dyn. **15**(3/4), 283–295
Guiraud, J.P., Zeytounian, R.Kh. (1982): Tellus **34**(1), 50–54
Gustafsson, B., Sundström, A. (1978): SIAM J. Appl. Math. **35**(2), 343–357
Gutman, L.N. (1972): Introduction to the nonlinear theory of mesoscale meteorological processes (Gidromet. Izd., Leningrad) Israel Program for Sci. Transl., Jerusalem
Hide, R. (1971): J. Fluid Mech. **49**(4), 745–751
Hoskins, B.J. (1975): J. Atmos. Sci. **32**, 233–242
Houghton, J.T. (1977): *The Physics of Atmospheres* (Cambridge Univ. Press, Cambridge)
Howarth, L. (1951): Q. J. Mech. Appl. Math. **4**(2), 157–169
Huppert, H.E., Miles, J.W. (1969): J. Fluid Mech. **35**, 481–496
Hunt, J.C.R. (1980): Environmental fluid mechanics. IUTAM Conference, pp 13–31
Hunt, J.C.R., Snyder, W.H. (1980): J. Fluid Mech. **96**(4), 671

Ionescu-Bujor, C. (1961): "Etude intrinsèque des écoulements permanents et rotationnels d'un fluide parfait", Thèse, Faculté de Sciences de l'Université de Paris
Kadechnikov, V.M. (1980): Izv. Akad. Nauk SSSR, Fiz Atmos. Okeana **16**(6), 563–571
Kaplun, S. (1967): *Fluid Mechanics and Singular Perturbations* (Academic Press, New York)
Karpman, V.I. (1975): *Non-linear Waves in Dispersive Media* (Pergamon, Oxford)
Kevorkian, J., Cole, J.D. (1981): *Perturbation Methods in Applied Mathematics* (Springer, New York)
Kibel, I.A. (1940): Izv. Akad. Nauk SSSR, Ser. Geogr. Geofiz. (5), 627–638
Kibel, I.A. (1955a): Dokl. Akad. Nauk SSSR **104**(1), 60–63
Kibel, I.A. (1955b): Dokl. Akad. Nauk SSSR **100**.(2), 247–250
Kibel, I.A. (1963): *An Introduction to the Hydrodynamical Methods of Short Period Weather Forecasting* (Translation) (Macmillan, London)
Kibel, I.A. (1970): In *Mekhanika v SSSR za 50 let*, Vol. 2 (Izd. Nauka, Moscow) pp. 561–583, (in Russian)
Kochin, N.E., Kibel, I.A., Roze, H.V. (1963): *Hydromécanique Théorique* (Phys. Math. ed., Moscow) (in Russian)
Kozhevnikov, V.N. (1963): Izv. Akad. Nauk SSSR, Ser. Geofiz. (7), 1108–1116
Kreiss, H.-O. (1970): Commun. Pure Appl. Math. **23**, 277–298
Kuo, H.L. (1949): J. Meteorol. **6**(2), 105–122
Kuo, H.L. (1973): Adv. Appl. Mech. **13**, 247–330
Kurgansky, M.V. (1980): Izv. Akad. Nauk SSSR, Fiz. Atmos. Okeana **16**(12), 1243–1249
Landau, L.D., Lifshitz, E.M. (1959): *Fluid Mechanics* (Pergamon, Oxford) (Engl. transl.)
Laplace, P.S. (1775): Recherches sur plusieurs points du système du monde. Mem. Acad. R. Sci., Paris (1778), oeuvres 9:9ff, 1775
Lattes, R. (1967): Quelques Méthodes de Résolution de Problèmes aux limites de la Physique Mathématique (Gordon and Breach, New York)
Leblond, P.H., Mysak, L.A. (1978): *Waves in the Ocean* (Elsevier, Amsterdam)
Leonov, A.I., Miropolsky, Yu.Z. (1975a): Izv. Akad. Nauk SSSR, Fiz. Atmos. Okeana **11**(5), 491–502
Leonov, A.I., Miropolsky, Yu.Z. (1975b): Izv. Akad. Nauk SSSR, Fiz. Atmos. Okeana **11**(11), 1169–1178
Lerat, A., Peyret, R. (1973): C.R. Acad. Sci. A **276**, 759–762; ibid. A **277**, 363–366
Lighthill, J. (ed.) (1967): A discussion on nonlinear theory of wave propagation in dispersive systems. Proc. R. Soc. London A 299, No. 1456
Lighthill, J. (1978): *Waves in Fluids* (Cambridge Univ. Press, Cambridge)
Long, R.R. (1953): Tellus **5**(1), 42–58
Long, R.R. (1954): J. Atmos. Sci. **21**, 197–200
Long, R.R. (1965): Tellus **17**, 46–52
Longuet-Higgins, M.S. (1968): Philos. Trans. R. Soc. London A **262**, 511–607
Longuet-Higgins, M.S., Pond, G.S. (1970): Philos. Trans. R. Soc. London, A **266**, 193–223
Luke, J.C. (1966): Proc. R. Soc. London A **292**, 403–412
Majda, A., Osher, S. (1975): Commun. Pure Appl. Math. **28**, 607–675
Martchuk, G.I. (1974): Numerical solution of problems of atmospheric and oceanic dynamics (Gidromet. Izd., Leningrad) (In Russian)
Mihaljan, J. (1962): Astrophys. J. **136**, 1126–1133
Miles, J.W. (1968): J. Fluid Mech. **33**(4), 803–814
Miles, J.W. (1969): Proc. Twelfth Int. Congress Applied Mechanics, Stanford (Springer, Berlin) pp. 50–76
Miles, J.W. (1974): J. Fluid Mech. **66**, 241–260
Miles, J.W. (1980): Annu. Rev. Fluid Mech. **12**, 11–43
Miller, M.J., Thorpe, A.J. (1981): Q. J. R. Meteorol. Soc. **107**(453), 615–628
Miranker, W.L. (1981): *Numerical Methods for Stiff Equations and Singular Perturbation Problems* (Reidel, Dordrecht, Holland)
Monin, A.S. (1958): Izv. Akad. Nauk SSSR, Ser. Geofiz. (4), 497
Monin, A.S. (1972): *Weather Forecasting as a Problem in Physics* (MIT Press, Cambridge, Mass.) (Engl. Transl.)
Morel, P. (ed.) (1973): *Dynamic Meteorology* (Reidel, Dordrecht, Holland)
Nayfeh, A.H. (1973): *Perturbation Methods* (Wiley, New York)
Neiland, V.Ya. (1969): Izv. Akad. Nauk SSSR, Mekh. Zhidk. Gaza (4), 53–57
Noe, J.M. (1981): Sur une théorie asymptotique de la convection naturelle. Thèse de 3ième cycle, Univ. de Lille I

Obukhov, A.M. (1949): Izv. Akad. Nauk SSSR, Ser. Geogr. Geofiz. **13**(4), 281–306
Ogura, Y., Phillips, N.A. (1962): J. Atmos. Sci. **19**, 173–179
Oliger, J., Sundström, A. (1978): SIAM J. Appl. Math. **35**(3), 419–446
Outrebon, P. (1981): Correction de Fromm pour les schémas S_β^α et applications au phénomène d'adaptation au quasi-statisme en Météorologie. Thèse de 3ième cycle, Université de Paris 6
Palm, E., Foldvik, A. (1959): Geofys. Publ. **21**(6), 1–30
Pedlosky, J. (1979): *Geophysical Fluid Dynamics* (Springer, New York)
Pekelis, E.M. (1976): Izv. Akad. Nauk SSSR, Fiz. Atmos. Okeana **12**(5), 470–477
Pekeris, C.L. (1975): Proc. R. Soc. London A **344**, 81–86
Phillips, N.A. (1960): Tellus **12**, 121–126
Phillips, N.A. (1963): Rev. Geophys. **1**(2), 123–176
Phillips, N.A. (1967): Thermal Convection. A Colloquium, N. CAR, TN. 24
Phillips, N.A. (1970): Models for weather prediction, Annu. Rev. Fluid Mech. **2**, 251–292
Phillips, O.M. (1977): *The Dynamics of the Upper Ocean*, 2nd ed. (Cambridge Univ. Press, Cambridge)
Queney, P. et al. (1960): The Airflow over Mountains. Geneva, WMO. Tech. Note, No. 34
Queney, P. (1977): La Météorologie **8**, 113–143; continuation in **9**, 111–163 (1978)
Redekopp, L.G. (1977): J. Fluid Mech. **82**, 725–746
Redekopp, L.G., Weidman, P.D. (1978): J. Atmos. Sci. **35**, 790–804
Reid, W.H. (ed.) (1971): Mathematical Problems in the Geophysical Sciences 1 – Geophysical Fluid Dynamics. Lectures in Appl. Maths. Vol. 13 (American Math. Soc., Providence, RI)
Richardson, L.F. (1922): *Weather Prediction by Numerical Processes* (Cambridge, reprinted by Dover, New York 1966)
Riley, N. (1965): Mathematika **12**, 161–175
Riley, N., Liu, H.T., Geller, E.W. (1976): U-S, EPA, Env. Monitoring Ser. Rep., No. EPA-600/4-76-021 Res. Tri. Park, NC
Roberts, P.H., Soward, A.M. (eds.) (1978): *Rotating Fluids in Geophysics* (Academic Press, London)
Rosenblat, S. (1959): J. Fluid Mech. **5**, 206–220
Rossby, C.G. (1938): J. Mar. Res. **1**, 239–263
Rossby, C.G. et al. (1939): J. Mar. Res. **2**, 38–55
Sadokov, V.P. (1969): WMO Regional Training Seminar (1965), Lectures on Numerical Short-Range Weather Prediction, (Hydrometeo. Izd., Leningrad) pp. 251–281
Sawyer, J.S. (1962): Q. J. R. Meteorol. Soc. **88**(378), 412
Schlichting, H. (1932): Phys. Z. **33**, 327–335
Scorer, R.S. (1949): Q. J. R. Meteorol. Soc. **76**, 41–56
Scorer, R.S. (1957): J. Fluid Mech. **2**, 583–594
Scorer, R.S. (1978): *Environmental Aerodynamics* (Wiley, New York)
Smith, F.T. (1973): J. Fluid Mech. **57**(4), 803–824
Smith, F.T., Sykes, R.I., Brighton, P.W.M. (1977): J. Fluid Mech. **83**(1), 163–176
Smith, F.T., Brighton, P.W.M., Jackson, P.S., Hunt, J.C.R. (1981): J. Fluid Mech. **113**, 123–152
Smith, R.B. (1979): The influence of mountains on the atmosphere, Adv. Geophys. **21**, 87–230
Spiegel, E.A., Veronis, G. (1960): Astrophys. J. **131**,442–447
Stewartson, K., Williams, P.G. (1969): Proc. R. Soc. London A **312**, 181–206
Stuart, J.T. (1966): J. Fluid Mech. **24**(4), 673–687
Sykes, R.I. (1978): Proc. R. Soc. London A **361**, 225–243
Tolstoy, I. (1973): *Wave Propagation* (McGraw-Hill, New York)
Triebel (1972): *Höhere Analysis* (VEB Deutscher Verlag der Wissenschaften, Berlin)
Truesdell, C. (1954): *The Kinematics of Vorticity* (Indiana Univ. Press, Bloomington)
Van Dyke, M. (1952): J. of Applied Math. and Phys. III **23**, 343–353
Van Dyke, M. (1962): J. Fluid Mech. **14**, 161–177
Van Dyke, M. (1964): *Perturbation Methods in Fluid Mechanics* (Academic Press, New York)
Veltichev, I. (1965): Trudy, World Meteorological Centre, Moscow, Vol. 8, p. 45 (in Russian)
Veronis, G. (1956): Deep-Sea Res. **3**(3), 157
Veronis, G. (1963): J. Mar. Res. **21**, 110–124 and 199–204
Viviand, H. (1970): J. Mec. **9**, 573–599
Voulfson, A.N. (1981): Izv. Akad. Nauk SSSR, Fiz. Atmos. Okeana **17**(8), 873–876
Wiin-Nielsen, A. (1978): Tech. Report No. 14 (ECMRWF, Reading, England)
White, A.A. (1977): Q. J. R. Meteorol. Soc. **103**(437), 383–396
Whitham, G.B. (1970): J. Fluid Mech. **44**, 373–395

Whitham, G.B. (1974): *Linear and Non-linear Waves* (Wiley, New York)
Wurtele, M.G. (1957): Beitr. Phys. der Atmosphäre **29**, 242–252
Yih, C.-S. (1957): La Houille Blanche (3), 439–444
Yih, C.-S. (1967): J. Fluid Mech. **29**, 539–544
Yih, C.-S. (1980): *Stratified Flows* (Academic Press, New York)
Yudin, M.I. (1963): *New Methods on Problems of Short-Range Weather Prediction* (Gidromet. Izd., Leningrad) Chap. 10 (in Russian)
Zeytounian, R.Kh. (1964): Trudy, World Meteorological Centre, Moscow, Vol. 3, pp. 19–74 (in Russian)
Zeytounain, R.Kh. (1965): Trudy, World Meteorological Centre, Moscow, Vol. 6, pp. 65–76 (in Russian)
Zeytounian, R.Kh. (1966): Izv. Akad. Nauk SSSR, Fiz. Atmos. Okeana **2**, 61–64
Zeytounian, R.Kh. (1968a): J. Mec. **7**(2), 231–247
Zeytounian, R.Kh. (1968b): Etude hydrodynamique des phénomènes mésométéorologiques, Lectures edited by la Direction de la Météorologie Nationale, Paris
Zeytounian, R.Kh. (1969a): Phys. Fluids, Suppl. II, **12**(12,II), 46–50
Zeytounian, R.Kh. (1969b): Study of Wave Phenomena in the Steady Flow of an Inviscid Stratified Fluid. R.A.E. Library Translation No. 1404
Zeytounian, R.Kh. (1970a): In AGARD Conf. Proc., Vol. 48 – The Aerodynamics of Atmospheric Shear Flow, p. 11
Zeytounian, R.Kh. (1970b): Quelques aspects de la théorie des couches limites laminaires compressible instationnaires. Note Technique ONERA, No. 162
Zeytounian, R.Kh. (1971): C.R. Acad. Sci. A **272**, 1670
Zeytounian, R.Kh. (1974a): Arch. Mech. Stosow. **26**(3), 499–509
Zeytounian, R.Kh. (1974b): *Notes sur les écoulements rotationnels de fluides parfaits*. Lecture Notes in Physics, Vol. 27 (Springer, Berlin, Heidelberg)
Zeytounian, R.Kh. (1976): La météorologie du point de vue du mécanicien des fluides, Fluid Dyn. Trans. **8**, 289–352
Zeytounian, R.Kh. (1977a): J. Eng. Math. **11**(3), 241–247
Zeytounian, R.Kh. (1977b): Lecture Notes in Mathematics, Vol. 594 (Springer, Berlin, Heidelberg) pp. 518–539
Zeytounian, R.Kh. (1979): Izv. Akad. Nauk SSSR, Fiz. Atmos. Okeana **15**(5), 498–507
Zeytounian, R.Kh. (1980): C.R. Acad. Sci. A **290**, 567–569
Zeytounian, R.Kh. (1981): Publications IRMA, Université de Lille I, **3**(1)
Zeytounian, R.Kh. (1982a): Izv. Akad. Nauk SSSR, Fiz. Atmos. Okeana **18**(6), 583–601
Zeytounian, R.Kh. (1982b): Asymptotic Phenomena in Meteorology. Proc. of the BAIL II Conference, ed. by J.J.H. Miller (Boole Press, Dublin) pp. 124–138
Zeytounain, R.Kh. (1983a): J. Appl. Mec. and Technical Phys. **2**, 53–61 (in Russian)
Zeytounain, R.Kh. (1983b): Modèles asymptotiques pour les écoulements de fluides pesants en rotation – Application à la Météorologie. Conférence générale invitée au 6e Congrès Francais de Mécanique. Lyon 5–9 Sept. 1983, pp.3.1–3.42
Zeytounian, R.Kh. (1983c): C.R. Acad. Sci. I, **297**, 271–274
Zeytounian, R.Kh. (1983d): Asymptotically consistent models for atmospheric flows. CISM, Von Karman session, October 1983, Udine (Italy)
Zeytounian, R.Kh. (1984): C.R. Acad. Sci. I, **299**(20), 1033–1036
Zeytounian, R.Kh. (1985): Recent Advances in asymptotic modelling of tangent atmospheric motions, Int. J. Engng. Sci. **23**(11), 1239–1288
Zeytounian, R.Kh. (1986): *Les Modèles Asymptotiques de la Mécanique des Fluides I*. Lecture Notes in Physics, Vol. 245 (Springer, Berlin, Heidelberg)
Zeytounian, R.Kh., Guiraud, J.P. (1980): C.R. Acad. Sci. B **290**, 75–77
Zeytounian, R.Kh., Guiraud, J.P. (1984): BAIL III Conference. Lecture Notes of an Int. Short Course, ed. by J.J.H. Miller (Boole Press, Dublin) pp. 95–100

Subject Index